HIGH-PERFORMANCE
SYSTEM DESIGN

Books of Related Interest from IEEE Press . . .

NONVOLATILE SEMICONDUCTION MEMORY TECHNOLOGY: A Comprehensive Guide to Understanding and Using NVSM Devices
Edited by William D. Brown and Joe E. Brewer
1998   Hardcover   616 pp   IEEE Order No. PC5644   ISBN 0-7803-1173-6

INTEGRATED CIRCUIT MANUFACTURABILITY: The Art of Process and Design Integration
Edited by José Pineda de Gyvez and Dhiraj K. Pradhan
1999   Hardcover   336 pp   IEEE Order No. PC4481   ISBN 0-7803-3447-7

LOW-VOLTAGE/LOW-POWER INTEGRATED CIRCUITS AND SYSTEMS: Low-Voltage Mixed-Signal Circuits
Edited by Edgar Sánchez-Sinencio and Andreas G. Andreou
1999   Hardcover   592 pp   IEEE Order No. PC4341   ISBN 0-7803-3446-9

THE ESSENCE OF LOGIC CIRCUITS, Second Edition
Stephen H. Unger
1997   Hardcover   360 pp   IEEE Order No. PC5590   ISBN 0-7803-1126-4

ELECTRONIC AND PHOTONIC CIRCUITS AND DEVICES
Edited by Ronald W. Waynant and John K. Lowell
1999   Softcover   232 pp   IEEE Order No. PP5748   ISBN 0-7803-3496-5

HIGH-PERFORMANCE SYSTEM DESIGN

Circuits and Logic

Edited by

Vojin G. Oklobdzija
University of California, Davis

IEEE Solid-State Circuits Society, *Sponsor*

IEEE PRESS

IEEE Press Series on Microelectronic Systems
Stuart K. Tewksbury, *Series Editor*

The Institute of Electrical and Electronics Engineers, Inc., New York

This book and other books may be purchased at a discount
from the publisher when ordered in bulk quantities. Contact:

IEEE Press Marketing
Attn: Special Sales
Piscataway, NJ 08855-1331
Fax: (732) 981-9334

For more information about IEEE PRESS products,
visit the IEEE Press Home Page: http://www.ieee.org/organizations/pubs/press

© 1999 by the Institute of Electrical and Electronics Engineers, Inc.
3 Park Avenue, 17th Floor, New York, NY 10016-5997

Printed in the United States of America

10   9   8   7   6   5   4   3   2   1

ISBN 0-7803-4716-1

IEEE Order Number: PC5765

Library of Congress Cataloging-in-Publication Data
Oklobdzija, Vojin G.
   High-performance system design : circuits and logic / Vojin G.
Oklobdzija.
     p. cm.
   "IEEE Solid-State Circuits Society, sponsor."
   Includes bibliographical references and index.
   ISBN 0-7803-4716-1
   1. Metal oxide semiconductors, Complementary. 2. Logic circuits.
3. Low voltage integrated circuits. 4. High performance processors—
Design and construction. I. IEEE Circuits and Systems Society.
II. IEEE Solid–State Circuits Society. III. Title.
TK7871.99.M44037 1999   98-32107
621.3815—dc21   CIP

To my colleagues and friends

whose work is presented in these pages.

Contents

CHAPTER 2   ADVANCES IN BICMOS AND BIPOLAR CIRCUITS   87

CHAPTER 3   DESIGN FOR LOWER POWER   169

CHAPTER 4   CLOCK SUBSYSTEM   261

CHAPTER 5   HIGH-PERFORMANCE ARITHMETIC UNITS   **405**

Preface

This selected reprint volume is intended for graduate students in electrical and computer engineering as well as practicing engineers, and provides the knowledge and techniques necessary for a good and successful design. This selection of articles deals with issues in logic and circuit design of complex and high-performance systems. These issues range from implementation technology and circuit techniques to pipelining and clocking, as well as system and architectural matters such as mapping an algorithm into a particular implementation technology.

The book is suitable for a one-semester course in Advanced Digital Design and, with some omissions, it is possible to cover the material in one quarter. I have taught such a course since 1992, and the material covered has changed and evolved over the years as technology changed very rapidly. However, a core set of papers remained in this collection regardless of publication date. This book is oriented around this material and supplemented with contemporary results; the core material is presented in its original form because it contains the fundamentals upon which later results were built. Therefore, the useful life of this edited volume should be much longer than a normal technology cycle.

This book resulted from teaching a course in Advanced Logic Design. However, the idea for this course, and consequently this book, came from many papers and design ideas that I collected over the years at the IBM T. J. Watson Research Center and other places in the industry. After I introduced those papers to the academic world, I realized that much of this material was never taught to students. It often left them struggling for important bits of knowledge as they acquired their experience in the industry. Therefore, I decided to create a course and teach this material. Over the years in teaching this course in an academic setting and in short industrial courses through the University of California Extension, the material evolved into a comprehensive set that is covered in this book.

At the suggestion of the IEEE Press editors John Griffin and Russ Hall, I wrote an introduction to each chapter that briefly explains the chapter's main points and ideas. From the introduction, the reader can expand on those ideas by reading the collection of papers that follow.

Chapter 1 contains a set of papers on advanced circuits that are used to implement high-speed logic. This chapter covers dynamic and differential CMOS, and the majority of papers are fundamentally necessary for understanding the new developments and circuit techniques used in today's high-speed processors. The latter part of this chapter focuses on new pass-transistor circuit techniques such as CPL, DPL, DVL, and SRPL which are showing promising results.

Chapter 2 includes papers on BiCMOS technology and advanced ECL circuits that are combining bipolar and MOS technology. Those circuit techniques were used in very high-performance systems and are still finding use in the applications that require a mix of bipolar and MOS technology. They are characterized by excellent driving abilities, while the static power consumption has been somewhat reduced.

Chapter 3 describes the techniques for achieving low-power and relates the recent developments in circuits and logic designed to satisfy those requirements. Attention is also given to a research area of energy-recovery logic and development of the appropriate logic families.

Chapter 4 deals with the clocking of systems, clock generation, and clock distribution techniques. It contains papers dealing with timing issues in high-performance systems, as well as design of the high-performance latches and flip-flops as one of the most important components.

Lastly, Chapter 5 contains papers on VLSI algorithms and computer arithmetic that show the relationship between implementation and choice of algorithm. The papers deal with the issues in realizing a fast and technology-optimal ALU, as well as a fast parallel multiplier. Issues involving design of a floating-point unit as well as square root and divide are also covered. Those papers emphasize the importance of choosing the right algorithm and its proper mapping into the selected technology.

Vojin G. Oklobdzija
University of California, Davis

ACKNOWLEDGMENTS

I would like to acknowledge the contributions of many of my students who took this course during the years of my teaching the Advanced Logic Design course at the University of California, as well as the contribution of the industrial participants in the University of California Extension courses.

I would also like to acknowledge many of my colleagues who pointed out interesting papers to be included in this book, in particular, Larry Heller of IBM, Takayasu Sakurai of Toshiba Corporation and the University of Tokyo, Tadahiro Kuroda, Parameswar Iya, and Matsaka Matsui of Toshiba Corporation. I would like to thank all the members of the Hitachi Central Research Laboratories for their inspiring work and discussions over the years. I would especially like to thank Kazuo Yano, Katsuro Sasaki, Norio Okhubo, and Makoto Suzuki and the members of the Fujitsu Research Laboratories, in particular, Gensuke Goto, Atsuki Inoue, and Tetsuo Nakamura.

I am grateful to Hamid Partovi of AMD for his input and enthusiastic discussions on latches and flip-flops and for sharing his knowledge with me. I am thankful to William Bowhil, Bruce Gieske, and Ronald Preston of Digital Equipment Corporation for inspiring discussions. I am also grateful to Ian Young of Intel Corporation.

Finally, I would like to express my deep gratitude to IBM and all the members of the IBM T. J. Watson Research Laboratories for all the discussions, hard work, and friendships that we shared during my tenure. IBM was the best place to learn about computer design. In particular, I would like to thank Gregory Grohowsky, Ravi Nair, Earl Barnes, Norman Raver, Richard Matick, Sharon Chuang, Monty Denneau, Don Weingarten, C. J. Tan, Steve Unger, Stan Schuster, and Bennett Robinson.

I thank Joan Shao and Richard Tsina of the University of California Berkeley Extension for providing me with many opportunities to teach this course to the industrial participants.

Finally, I thank my son, Stanisha, for correcting my English.

Vojin G. Oklobdzija
University of California, Davis

Advances in CMOS Circuits

ADVANCES IN CMOS CIRCUITS

This section contains selected papers describing the advances being made in CMOS circuits and logic. It starts with a review paper in which Akira Masaki argues in favor of deep sub-micron CMOS over the use of bipolar ECL logic. The paper presents the main arguments explaining why high-speed logic should be using room-temperature CMOS instead of ECL. The paper was written in 1992 and its assertion that CMOS would become a dominant logic technology has proven true today.

At the time CMOS was introduced, however, it was not considered a viable technology for computer logic, whereas bipolar and n-MOS (in a later stage) were dominating the processor and microprocessor development. First consideration as a logic technology was given to CMOS with the introduction of Belmac-32 from Bell Laboratories. Belmac-32 was also the first 32-bit microprocessor on a single chip. The CMOS circuits and technology used in this processor were quite different from the ordinary CMOS known earlier. Even after Belmac-32 was introduced, a debate raged as to what kind of CMOS would be a suitable replacement for n-MOS in computer logic. The evolution of CMOS is described in the next seven papers. It is interesting that some of those circuit techniques were dormant for a while only to find a use in modern high-performance processors today.

The second paper (by Krambeck, Lee and Law from Bell Laboratories) is on CMOS-Domino circuits. This paper was a first serious departure from the static CMOS known in its classical sense as a circuit consisting of a p-type switching network using a function f and an n-type implementation of a complementary function $\bar{f}$ shown in Fig.1.1.

As shown in Fig. 1.1, the CMOS circuit requires twice as many transistors as n-MOS in order to implement the same function. To make this comparison even worse for CMOS, the switching time of the p-transistor is twice as long as that of the n-transistor owing to the different carrier mobility of n- and p-type carriers in silicon. Therefore, the transition form

low-to-$high$ output voltage (in CMOS) takes twice as long as the corresponding $high$-to-low transition. In order to make the transition times symmetric, the p-transistor switching network is usually built from devices that are twice as large as the n-type transistors. In such a case, we are dealing with a circuit that either has non-symmetric transition times or is considerably larger than if the same function f was implemented in n-MOS logic.

CMOS Domino Logic

To alleviate the CMOS problems, researchers from AT&T Bell Laboratories suggested building CMOS logic that contains only n-type transistors implementing the switching function f. This logic is a dynamic type because two clock-phases are necessary for its proper operation. First, there is a *precharge phase* during which the clock is low, followed by the *evaluation phase* during which the clock is high, as shown in Fig. 1.2. During the precharge phase, all of the nodes N are precharged via the p-type transistor Q_1. The nodes N will stay charged if there is no discharge path in the switching function f during the evaluation phase when the transistor Q_2 is conducting. However, if the nodes N were taken directly as outputs, thus driving the inputs of the next logic blocks, all of the subsequent blocks would start to discharge immediately following the precharge phase, simply because all of the outputs are at the *logic one*. Therefore, the final state of the functional blocks would be undetermined. Domino Logic resolves this problem by passing all of the nodes N through the regular CMOS inverters. Only the output of an inverter can drive the next logic block. In Domino Logic, all of the outputs are at *logic zero* immediately following the precharge stage. Therefore, no discharge can exist in the logic blocks that are themselves driven by other Domino Logic blocks. The evaluation phase starts with the logic blocks being driven by the *primary inputs*. Some of the blocks will be selectively discharged, if a path to ground is established by the logic function represented by that particular block. This would change their outputs

1

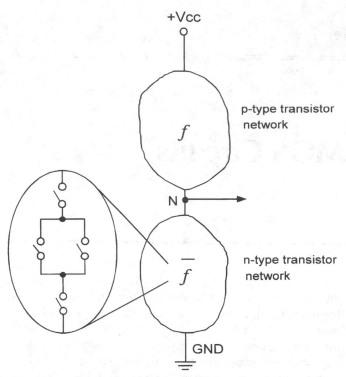

+Vcc

p-type transistor network

N

n-type transistor network

$\overline{f}$

GND

Fig. 1.1. Switch representation of static CMOS circuit implementing a function *f*.

old voltage of the inverter INV, and its output will assume the logic one. The logic one value of the output might in turn discharge the next block (though it was not supposed to do so) leading to the erroneous value. A techniques used to alleviate the *charge redistribution* employs a small feedback *p*-type transistor Q_f connected to the node *N* whose function is to replenish the charge lost in the redistribution and return of the node *N* to the *logic one* value. If the amount of redistributed charge was not large, this might be sufficient. However, depending on the amount of charge lost from the node N, a negative voltage *glitch* of a different magnitude will result on the node *N*. This glitch may be amplified in the inverter (INV), and it might be sufficient to discharge the next logic block..

The problem posed by the noninverting property of the CMOS Domino Logic is treated in the paper by Goncalves and DeMan in which they propose a new type of Domino Logic, termed NORA, consisting of *p*-type as well as *n*-type switching networks that are employed alternatively. The operation of NORA logic is illustrated in Fig. 1.4. Both *n*-type logic and *p*-type logic are in the precharge phase where the output node of the *n*-type switching function is precharged to logic one, while the output of the *p*-type switching function is discharged to logic *zero*. This is achieved by clocking the *n*-type and *p*-type switching functions with the two opposite clock phases. During the evaluation phase, *n*-type logic blocks will discharge their output to logic zero, if there exists a path in the *n*-transistor switching function connecting the output to ground. In turn, this may create a path to V_{cc} in a *p*-type transistor switching network of the next block bringing the output to logic one.

The change would propagate from the primary inputs to the primary outputs alternating between the *n*-type and *p*-type switching block. If a particular output is destined to be an input in the next logic block of the same type, the inverters make its operation very similar to Domino circuits which consist of *n*-type and *p*-type switching networks. A tristate buffer at the output of a logic section holds its logic value during the precharge phase, thus enabling pipelining of the logic section. NORA logic is also sensitive to charge redistribution, and the large *p*-type switching transistor blocks do not add to its performance. Though a very interesting concept, even today this type of logic has not found many applications.

The CMOS Domino family was further enhanced by Heller, Griffin, and Thoma of IBM, which resulted in CVSL logic. The IBM authors developed two types of CMOS logic which they called Cascode Voltage Switch: *static* and *dynamic*, (see Fig. 1.5). The first of these types of CMOS logic: static CVSL, consists of two *n*-type transistor switching functions *f* and $\overline{f}$, which are connected to V_{cc} via a cross-coupled *p*-type transistor combination. The advantages of this logic over regular CMOS are obvious. CVSL implements two functions *f* and $\overline{f}$, as CMOS does except that in CMOS *f* is implemented with a *p*-type transistor switching network. Given the inferior speed of *p*-type transistors, the *p*-type transistor switching function will usually end up being twice as large as the one implemented with the *n*-type transistors. In terms of the transistor numbers, CVSL contains two *p*-type transistors in the cross-coupled combination

from logic zero to *one* driving the inputs of subsequent logic blocks. Now if this change in turn creates a discharge path in the subsequent logic blocks, they will discharge, changing their outputs from *zero* to *one*. This change would further propagate through the logic from the primary inputs to the primary outputs like falling dominos. This is where the logic obtained its name: *Domino Logic*.

The benefits of CMOS Domino Logic were not obtained without a cost. First, part of the cycle used for precharge is essentially lost because no logic operation is possible during this time. The ability to rapidly discharge the output node *N* via the *n*-type MOS transistor switching network was supposed to offset this loss. This feature is dependent on implementation and is not always true. Furthermore, CMOS Domino is inherently *noninverting* logic which represents difficulties. (E.g., XOR gate implementation with Domino as described by Krambeck, Lee, and Law is not possible). This inability is to some extent compensated by using dual polarity latches at the inputs.

Yet another feature plagued CMOS Domino Logic: the *charge redistribution problem* described in the paper by Oklobdzija et al. The charge redistribution mechanism is illustrated in Fig.1.3. Charge redistribution is manifested by the loss of charge from the node *N*, owing to the creation of paths leading into the previously discharged parts of the transistor switching network *f*, but not to the ground node. The output of the Domino Logic block was to stay at logic zero level, because there is no path in the switching network *f* leading to ground. However, owing to the distribution of charge from the node *N* into the various nodes in the switching network, the voltage of the node *N* will drop. This voltage drop might exceed the threshold

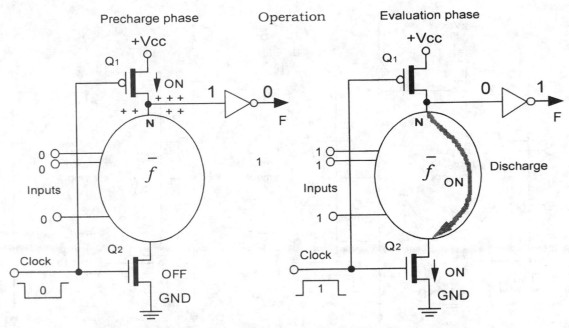

Fig. 1.2. CMOS Domino Logic [Krambeck, Lee, and Law].

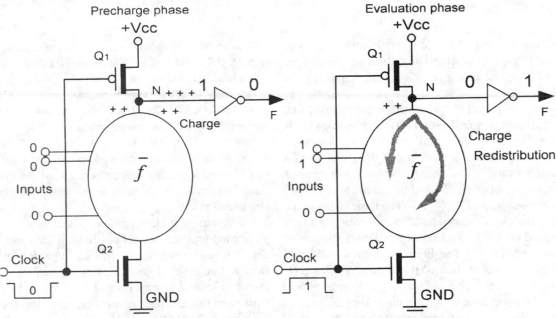

Fig. 1.3. Charge redistribution in CMOS Domino Logic [Oklobdzija et. al.].

that are in excess of the total number of transistors used by regular CMOS. In terms of size, CVSL is not larger than regular CMOS, although this is dependent on the complexity of the logic function implemented. Certainly, an operation involving only the n-type transistor switching function is faster. Furthermore, Heller, Griffin, Davis, and Thoma showed that implementation of f and $\bar{f}$ does not necessarily mean duplication. Therefore a number of transistors in the switching functions f and $\bar{f}$ can be shared, setting duplication only as an upper limit. Creation of CVSL circuits was supported by an automated synthesis tool (one of the first of that kind) that would create opti-

mal and shared n-type transistor switching functions f and $\bar{f}$ as shown in the example in Fig. 1.5a.

The second type of the new CMOS logic, *dynamic CVSL,* is a clocked version of static CVSL, and in essence it represents a dual-output CMOS Domino Logic. Transistor Qf is added to ensure more robust operation of the dynamic CVSL circuit. If charge redistribution occurs, resulting in the loss of charge at the inverter's input, the output voltage will drop toward zero. This will in turn activate the transistor Qf and the charge will be restored, pulling the output back to logic one. In order to discharge this node, the pass-transistor network

3

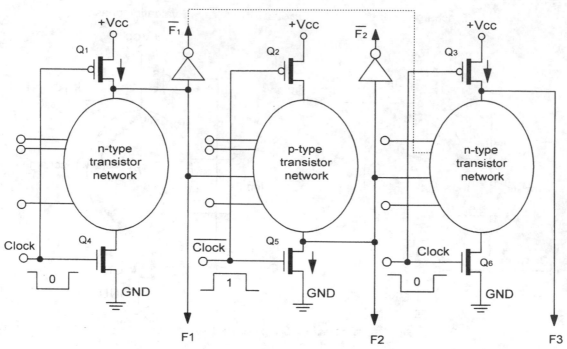

Fig. 1.4. Operation of NORA logic [Goncalves, DeMan].

implementing the function $\bar{f}$ has to overcome the transistor Qf. Qf is made to be a weak transistor; thus, its dimensions are smaller than the rest of the transistors implementing the switching function.

An interesting new concept is presented in the paper by Leo Pfennings on Differential Split-Level (DSL) CMOS Logic. This is essentially a very clever enhancement of CVSL, which allowed the use of one-micron devices at that time, and under a reduced voltage operation. The logic transistors were made of $L_{eff} = 1\mu$ devices, operating at 2.5 V, while the entire circuit operates on the commonly used $V_{DD} = 5$ V. The logic swing of the interconnection lines was reduced to 2.5 V, allowing for a sub-nanosecond speed (in 1985). The essence of DSL logic can be best understood from Fig.1.6. The difference between DSL and CVSL is in introduction of the two n-MOS transistors Q_3 and Q_4 between the load transistors Q_1 and Q_2 acting as source followers. Therefore, the impedance seen from the side of the logic is low, permitting a fast change of state, while their output resistance is high, allowing for the use of small devices Q_1 and Q_2 representing the *load* transistors. This enables a fast change of the logic state, which is further facilitated by the reduced voltage swing of $1/2\ V_{DD}$. The load transistors Q_1 and Q_2 are not completely turned off because their gates are being driven from the lower voltage potential. This makes an active loop capable of switching the state in a rapid manner. However, it also represents a power problem because DSL contains a static power component.

A comparative analysis of conventional CMOS versus Differential Cascode Voltage Switch Logic (which includes CVSL,

DSL, and NORA), both static and dynamic, is presented in the paper by Chu and Pulfrey. The analysis is somewhat limited because it involves a section of a full adder only. Their conclusion is that DCVS logic is faster than conventional CMOS, although this advantage is counterbalanced by a somewhat larger circuit and more active power consumption. The fastest logic technique seems to be DSL, but the problems with DSL involve static power dissipation and increased sensitivity on circuit parameters in order to make the operation reliable. In dynamic operation, differential logic seems to have a speed advantage over single-ended logic.

Comparative studies similar to that of Chu and Pulfrey were conducted extensively at IBM in the early 1980s in order to reach an important decision regarding which CMOS family should be adopted for future products. These studies involved several large pieces of the control logic besides the parts from the data path as test examples. The findings, which were not published, were in some agreement with those of Chu and Pulfrey. Although they showed that CVSL has an advantage over regular CMOS, this advantage was not sufficient to justify extensive changes in the computer-aided design tools and methodology, test generation in particular. Therefore, regular CMOS was chosen, but the real deciding factor was the status of the existing CAD tools, not the performance of the logic families compared. In spite of its advantages, CVSL failed to become a technology of the future. The choice is often influenced by other factors, such as CAD and testability as was the case. Those issues are covered in later chapters of this book.

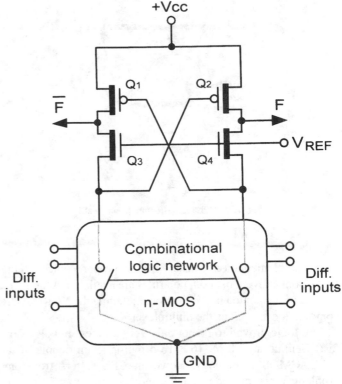

Fig. 1.6. DSL logic circuitm [Pfennings et al.].

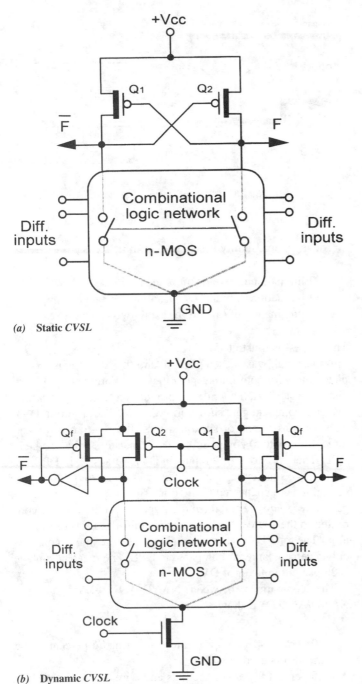

(a) Static *CVSL*

(b) Dynamic *CVSL*

Fig. 1.5. *CVSL* Circuit. [Heller et. al.].

Pass-Transistor Circuits

As the technology improved and started moving into the deep submicron region, use of regular CMOS reached its limits. The problems associated with ever-increasing requirements for performance and a higher level of integration required that other types of logic be examined. Extensive consideration was given to the use of pass-transistors' design style resurrected in several recent papers. Several experimental prototypes were built at the industrial research institutions that possessed the state-of-the-art technology.

This section begins with a relatively old paper by Whitaker, published in 1983, which shows how the logic functions can be efficiently and optimally built using pass-transistor logic. This paper contains some fundamental pass transistor building blocks that became very popular in VLSI design practices. In addition it established a methodology for synthesis of pass-transistor functions and presents a modified Karnaugh map that employs the pass variables, as well, aside from logic zeroes and ones.

In 1990, researchers from Hitachi Research Center in Japan published the structure known as Complementary Pass-Transistor Logic (CPL). The CPL structure was significant because it brought back the pass-transistor efficiency in implementation of logic circuits. The logic function built from the pass-transistors not only efficiently utilized the space on silicon, but also resulted in a very fast logic, which is also characterized by low-power consumption.

Yano and his co-workers went one step further than CVSL by decoupling the *p*-transistor latch combination and replacing it with two separate inverters. The logic is still dual-ended, which means that both the function *f* and its complement $\bar{f}$ are present at the outputs; in addition, the logic was built from the pass-transistor networks, as shown in Fig. 1.7. If we are to implement an AND gate, a NAND output will be readily available. There-

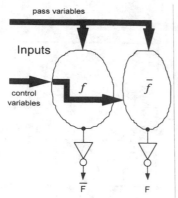

Fig. 1.7. CPL logic structure [Yano et al.].

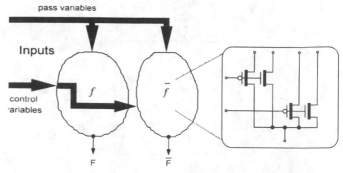

Fig. 1.8. DPL logic structure [Ohkubo].

fore, complementation consists of the proper choice of the signals only, given that both polarities are available. A family of gates is implemented in this fashion, including the XOR/XNOR combination as well as the multiplexer.

CPL logic proved to be not only very efficient but also very fast, yielding an 3.8 nS 16×16-b multiplier in double metal 0.5 μ CMOS technology. However, CPL suffers from one problem. When passed through a series of pass-transistors, the signal voltage is degraded by one V_T (*threshold voltage drop*). This brings both transistors in the output inverter to the conducting region, causing static current to flow from V_{cc} to GND and thus resulting in static power dissipation. To alleviate this problem, Hitachi researchers used two types of transistors: *logic transistors* (with $V_T = 0$ V) and *inverter transistors* (with $V_T = 0.4$ V and -0.4 V). Although this reduced static power dissipation and delay, it increased the process complexity and the sensitivity to noise. In the new version of CPL, the problem of the "threshold drop" was alleviated by using a special type of inverter which has the ability to restore the voltage level to its full potential.

Another pass-transistor logic, which evolved from the same group of researchers, is Double Pass-transistor Logic (DPL), Shown in Fig. 1.8. DPL was originally developed to overcome the threshold drop problems of CPL and to provide an alternative pass-transistor logic. In creating the switching network f, DPL uses both n-MOS and p-MOS transistors in parallel. This eliminates the problem of the threshold drop, and the use of inverters after each logic block is not necessary, thus enhancing the speed of DPL. Hitachi has shown two very fast implementations using DPL: one a 1.5 nS 32-b ALU and the other a 4.4 nS 54×54-b parallel multiplier.

Oklobdzija and Duchene have taken DPL a step further in developing the logic family, which they call DVL (Dual Value Logic). The new logic family was obtained from DPL by eliminating the redundant branches and rearrangement of signals. These simplifications still preserve full-swing operation of DPL and improve its speed. The speed improvement is a direct result of eliminating of one branch containing one transistor. This minimizes the capacitive load and the number of inputs applied to the previous gate.

The new logic family is achieved in three steps:

a. Elimination of redundant branches in DPL
b. Elimination of branches via signal rearrangement
c. Combination of (a) and (b) using two faster halves

The process is illustrated in Fig. 1.9 *a, b, c.*

A faster half was chosen from (a) and (b), resulting in a complete gate with both output polarities (c). Fortunately (a) produces a faster NAND, while (b) produces a faster AND, which makes a complete gate shown in Fig. 1.9c. The resulting DVL gate contains a total of six transistors (three p-transistors and three n-transistors) compared to four transistors of each type in DPL. There is a total of 9 inputs in DVL versus 12 in DPL, resulting in a smaller capacitive load of DVL gates. Of those inputs, three are connected to the transistor source and six to the gate (three to p-type and three to n-type). (In DPL four are connected to the source, four to p-type, and four to n-type transistors.) The total DVL gate area (taking resizing into account) is only 5 percent larger as compared to DPL. The speed advantage is 20 percent in favor of DVL.

The comparison between NAND/AND DPL gate and NAND/AND DVL shows

- 20 percent speed improvement, utilizing 75 percent of the transistors used in DPL.
- 25 percent fewer connections and wires as compared to a DPL gate.

A similar method is used to build the NOR/OR gates.

Further development of the differential CMOS family is presented in the paper by Lai and Hwang. They introduced pass-transistor logic in the CVS logic tree in order to eliminate the problem of current spikes. Lai and Hwang modified CVSL by using pass-variables instead of the ground connection at the opposite end of the pass-transistor block implementing the function f, while leaving the cross-coupled p-transistor latch at the opposite end.

The resulting logic (CVSL-PG) showed better performance than that of CVSL. This was demonstrated by implementation of a 2 nS 64-bit adder in 0.5 μ CMOS technology.

DPL

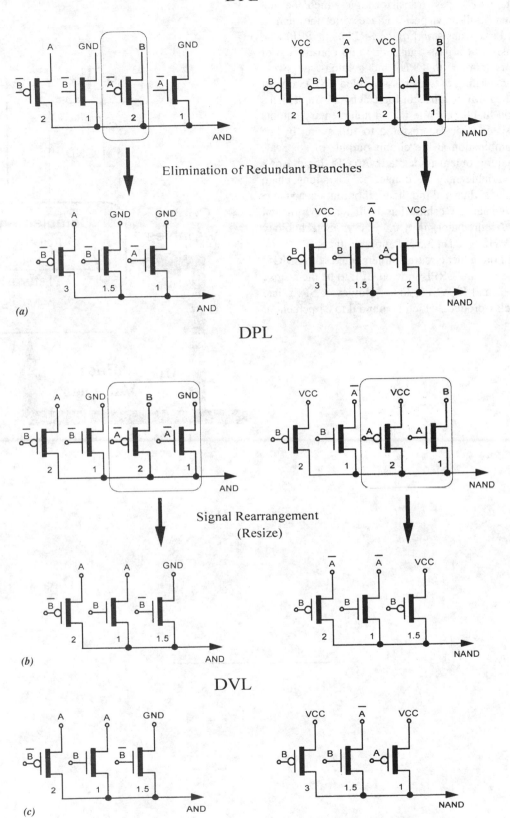

Elimination of Redundant Branches

(a)

DPL

Signal Rearrangement
(Resize)

(b)

DVL

(c)

Fig. 1.9. Transformation of DPL into DVL. *(a)* Elimination of redundant branches, *(b)* Signal Rearrangement, *(c)* Resulting DVL Gate [Oklobdzija, Duchene].

Researchers from Toshiba Corporation developed their version of differential CMOS pass-transistor logic which does not suffer from degraded pull-down performance which they named Swing Restored Pass-Transistor Logic (SRPL). In SRPL the generic gate consists of a pass-transistor logic constructed of n-MOS transistors (similar to CPL) and a latch-type swing restoring circuit consisting of two cross-coupled CMOS inverters (Fig. 1.10). The n-MOS transistor logic network implements any Boolean logic function, while the complementary outputs of the pass-transistor logic are restored to full swing by the cross-coupled combination at the circuit output. In this way, SRPL solves a major problem associated with CPL logic. However, the input variable may be connected to a long chain through several gates, thus making the total output capacitance of the circuit quite large. Toshiba has built an experimental MAC (Multiply Accumulator) in a $0.4\,\mu$ CMOS technology achieving a 150 MHz speed at 3.3 V supply voltage.

Comparisons of full adder circuits implemented with CMOS, CPL, DPL, DCVSPG, and SRPL showed CPL to be the fastest, followed by SRPL and DCVSPG logic. However, SRPL had the best power-delay product, which amounted to 21 percent of that of CMOS.

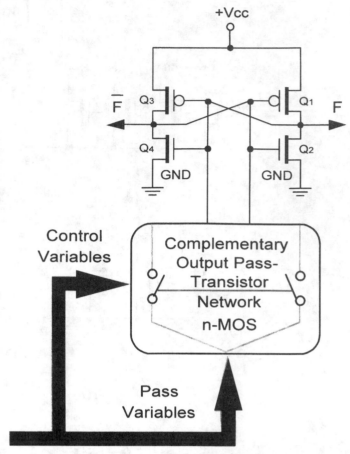

Fig. 1.10. Generic SRPL Gate [Parameswar et al.].

Deep-Submicron CMOS Warms up to High-Speed Logic*

AKIRA MASAKI

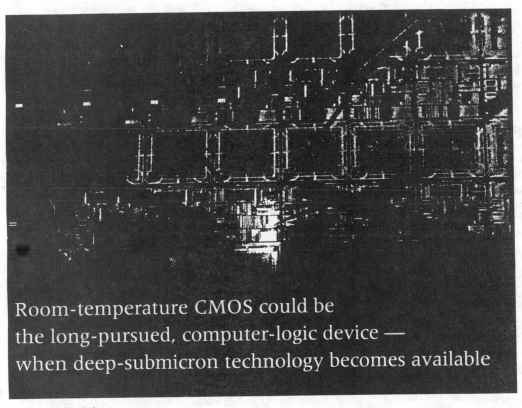

Room-temperature CMOS could be
the long-pursued, computer-logic device —
when deep-submicron technology becomes available

Silicon bipolar devices, particularly those in emitter-coupled logic (ECL), have been accepted as the most useful of the various high-speed technologies since ICs were first introduced into computers. But the possibilities for improving ECL will eventually reach their limits. In anticipation of that time, many alternative technologies have been proposed, including low-temperature CMOS, GaAs, HEMT, and even Josephson junction devices. It is difficult to compare these technologies and predict which technology will be ECL's successor.

Despite the predictive difficulties, it will be indicated in this article that room-temperature CMOS, which heretofore has not been widely recognized as producing very-high-speed devices, can be the

---

*This article is based on "Possibilities of CMOS Mainframe and its Impact on Technology R&D," an invited paper presented by the author at the 1991 VLSI Symposium on Technology.

Reprinted from *IEEE Circuits and Devices Magazine*, pp. 18-24, November 1992.

Table 1 Logic Function Capability (4-bit ALU)			
	CMOS (NAND)	ECL	Ratio
Total circuit count	98	57	1.7
Circuit stages in signal paths (avg.)	6.1	2.9	2.1

long-pursued post-ECL technology. If it becomes possible to implement high-speed computer logic with room-temperature CMOS, we will no longer need to worry about comparing technologies. Even if equivalent system-performance can be obtained by other technologies, CMOS will win industrial favor because the advanced semiconductor technology established through development of DRAM is directly applicable to CMOS. In addition, the technological compatibility with personal computers, workstations, and small computers benefits all aspects of R&D in high-speed CMOS computers. The ability to scale CMOS down will also allow circuit designs to be inherited through design generations. Last but not least, CMOS's low power dissipation cannot be matched by other technologies, even when applied to high-speed computer logic. As a consequence of the preceding factors, the integration scale achievable with CMOS is the highest of any of the candidate technologies.

Integrating High-Speed ICs
The scale of integration is often regarded as a non-critical parameter for high-speed computer logic because circuit speed is the primary concern. But integration scale has steadily increased by ten times every five years over the past quarter century (Fig. 1). This rate of increase is the same as that for DRAMs, which is usually stated as four times every three years. This clearly tells us that increases in integration scale have been the source of decreasing cost-performance ratio in computers.

It is not yet known whether this trend will, or should, continue. If it does continue, the gate count per chip will reach two million in the year 2000. Novel high-speed devices will not be able to cope with this complexity in a practical way, and it will not easy for conventional ECL either. Even if individual gate-circuit power is as little as 2 mW, a 2 megagate chip will consume 4 kW — an unmanageably large power.

Fortunately, silicon technology itself

should be capable of realizing the required scale of integration since there is no sign that the rate of increase of DRAM-chip integration is slowing. Therefore, in addition to meeting performance requirements, we should make a conscious effort to increase the integration of logic chips.

CMOS can realize the integration but its speed has been insufficient, at least at room temperature. Recently, a fairly fast circuit speed was obtained by decreasing the MOSFETs gate-length to approximately 0.2 μm [1,2]. However, it is not yet clear that very-high-speed computer logic can be made with such CMOS devices. By referring to extensive theoretical developments, we hope to shed some light on this question throughout the remainder of this article [3-9].

Estimating Logic Performance
Device performance is usually demonstrated with data from a very lightly loaded ring oscillator, but such data does not directly relate to system performance. While most experts agree when they estimate memory performance, estimating the performance of logic systems has always been much

more controversial. Various factors should be considered in such estimates, including the logic-function capabilities of basic circuits, the wire length of logic signal nets, and the power of CMOS circuits in system environments.

Logic Function Capability
In assessing the differences in logic-function capabilities among circuits [3,5,6], the essential questions are:

• How many circuits are required in all?
• In each technology, how many circuit stages are required for the critical signal path in a specific functional block?

The real potential of a technology can be grasped only after the second question is fully answered, because it is the power and delay of the functional block, not the unit circuit, that should be minimized. Unfortunately, it is difficult to obtain a general solution to this problem because the results of a design depend strongly upon the logical characteristics of the functional block, as well as upon the differences in the individual abilities of designers. Nonetheless, differences in the logic-func-

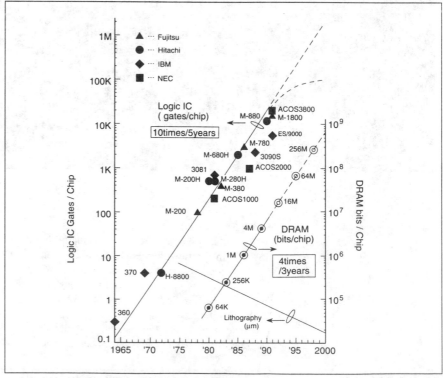

1. Although the integration scale of very-high-speed computer-logic chips is often regarded as non-critical, it has been increasing at the same rate as DRAMs for a quarter of a century.

tion capabilities of various circuits cannot be neglected.

We can demonstrate the point by comparing the total circuits and the average number of circuit stages in the signal paths of carefully designed 4-bit arithmetic and logical units built with ECL and simple CMOS NAND (Table 1). The circuit count and number of circuit stages required to implement a specific logic function are smaller for ECL. Results obtained in various case studies [5,6] lead us to conclude that simple CMOS NAND requires roughly twice as many total circuits and circuit stages as ECL.

Wire Length of Logic Signal Nets
Delays caused by wire capacitance significantly affect circuit performance. Estimating the wire length of logic signal nets is therefore indispensable when evaluating performance in system environments.

A simplified structure for a logic-circuit cell array, which could be an LSI chip, a module substrate, or a printed circuit board, provides a basis for wire-length estimates (Fig. 2). The wire length of the signal net, L_w, is $n_{pp} \times l_{pp} \times p$, where n_{pp} is the number of pin-to-pin (or terminal-to-terminal) wires per signal net, and l_{pp} is the terminal-to-terminal wire length expressed in terms of the average center-to-center spacing of the cells, p.

An equation for estimating l_{pp} [10], which is applicable to two-dimensional square arrays, was derived from the Rent's Rule equation [11]. By extending this work, we obtained equations for estimating various types of two- and three-dimensional packaging structures [7]. One of these is extremely useful:

$$l_{pp} = \frac{1 - 2^{2(r-1)}}{6\,(1 - 2^{2(r-1)L})} \cdot$$

$$\left\{ \frac{7\,(2^{(2r-1)L} - 1)}{2^{2r-1} - 1} - \frac{1 - 2^{(2r-3)L}}{1 - 2^{2r-3}} \right\}$$

where L is the base 2 logarithm of the number of circuit cells in the x and y directions, and r is the exponential coefficient in the Rent's Rule equation.

Power Consumption of CMOS Circuits
A CMOS circuit consumes energy only during switching. Estimating its power consumption in a realistic system environment therefore requires us to obtain the cir-

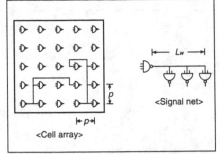

2. The signal net model for a logic-circuit cell array provides a convenient basis for wire-length estimates.

cuit's switching frequency in actual systems. To help us analyze the problem, let's introduce a quantity (k) that is defined as the average switching period (T) of the circuits in a system, divided by the circuit delay in the system (t_{sd}) [4,5]. The definition of k is similar to that of the CMOS switching factor defined in [12].

The value of k is independent of the hardware technology; it depends solely upon the logical structure of the system as long as the performance potential of the cir-

cuit is fully utilized. If sophisticated logic is used, k becomes smaller; that is, switching occurs more frequently.

There are several methods for obtaining the value of k [4,5]:

(1) Investigate the logical structure of the system.
(2) Measure the switching frequency of logic signals in system operation.
(3) Measure power dissipation in systems built with CMOS.
(4) Count the number of times the circuits switch by simulating the logical operation of the system.

Experimental values obtained with methods 2 and 3 are in good agreement with the estimated values from methods 1 and 4 (Table 2). We can therefore conclude that k is within the range of 20 to 200 for most computer systems. We will assume k to be 40.

Load-driving Capability and Wire Delay
Load-driving capability is one of the most important characteristics of computer-logic

Table 2: K Values (T/tsd)			
	Minicomputer		Large-scale
	A	B	computer
(1) Analysis of logic Structure	180	120	60 ~ 20
(2) Measurement of switching frequency	200		
(3) Measurement of power of CMOS computer		170 ~130	
(4) Logic Simulation			80 ~ 30

Table 3. Device Parameters for Very-High-Performance CMOS Logic		
	Intrablock 3NAND	Interblock buffer
Drain voltage V_{DD}	2V	
Gate length L_g	0.2 µm	
Gate oxide thickness t_{ox}	5 ~ 6 nm	
Gate width W_g	15 µm	75 µm
Drain current I_{DS} (n-ch)	7 mA	35 mA
(p-ch)	3.5 mA	18 mA
Input capacitance C_{in}	0.05 pF	0.25 pF
Wire capacitance	0.2 pF/mm	0.2 pF/mm
Wire Resistance	100Ω/mm	6Ω/mm
Switching energy (circuit)	0.5 pJ	2.5 pJ
(load)	3.0 pJ/pF	3.0 pJ/pF
Driving capability	400 ps/pF	60 ps/pF
Circuit delay (F0=1, C_w=0 pF)	80 ps	40 ps
(F0=3, C_w=0.2 pF)	200 ps	

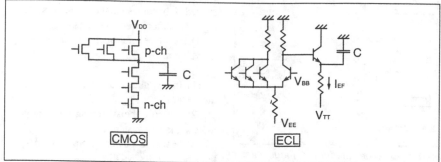

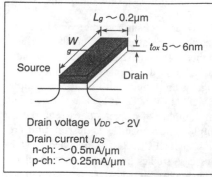

3. Driving capability is directly proportional to signal current divided by the signal voltage. CMOS's signal voltage is as large as 5 V and its driving current is small; ECL's signal voltage is as small as 0.5 V and the technology can drive a large current. The resulting load-driving capability is one reason for ECL's popularity.

4. In a deep-submicron CMOS device structure, drain current is increased by substantially decreasing gate length, L_g, and gate-oxide thickness, t_{ox}, thus increasing driving capability.

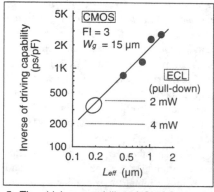

5. The driving capability of CMOS circuits has been improving steadily. The estimated driving capability of the 0.2-μm-gate CMOS is derived from the device characteristics for deep-submicron CMOS reported so far [1,2].

circuits and devices. ECL has been favored because of its large driving capability; conventional CMOS is weak in this area.

Driving capability is directly proportional to signal current divided by the signal voltage. CMOS's signal voltage is as large as 5 V and its driving current is small. On the other hand, ECL can drive a large current and its signal voltage is as small as 0.5 V (Fig. 3).

But things are changing. The drain current of a CMOS device can be increased by decreasing gate length L_g and gate-oxide thickness t_{ox} (Fig. 4). A fairly large drain current has been reported at a drain voltage as low as 2 V when gate length is decreased to approximately 0.2 μm [1,2]. Such "deep-submicron" devices represent the latest in a steady stream of improvements in the driving capability of CMOS circuits (Fig. 5).

The driving capability of ECL circuits, on the other hand, is mainly determined by the current flowing through the emitter-follower pull-down resistor when the output signal changes from high to low. Therefore, the circuit's driving capability is determined primarily by power dissipation and cannot benefit from device miniaturization. Since ECL circuit power cannot be increased in the future, the driving capability of CMOS will be equivalent to that of ECL when deep-submicron devices become available. The driving capability of ECL may still be improved by adopting active pull-down circuit techniques, but CMOS is attractive because its device structure and circuit configuration are very simple.

The delay caused by long signal nets on printed circuit boards and module substrates often determines system performance. Driving such long nets has been a strong

point of ECL. Usually, such signal nets are treated as controlled-impedance-terminated (CIT) nets, in which case signals reach the far end of the net with minimum delay. ECL is suitable for driving CIT nets.

In CMOS VLSI, long nets are implemented on the chip. Wire resistance is large, and resistance-capacitance (RC) nets are used. Delay per unit length of RC nets is larger than that of CIT nets, but wire length is shorter for CMOS VLSI so total delay can be competitive.

Case Study
Device Requirements and VLSI Chip Model

Based on our experience and the data reported so far, we expect certain basic characteristics for 0.2-μm gate-length, room-temperature CMOS (Table 3)[1,2]. To evaluate the possibilities of CMOS, we assume a particular VLSI chip model (Fig. 6). In actuality, almost half of the chip area will be occupied by memory circuits, but in this case study we assume that the chip consists of logic only. As a result, we are considering the worst-case wire length.

In the model, we assume a CMOS chip area of 20 mm x 20 mm, a reasonable assumption considering that a 15 mm x 15 mm ECL chip is used today [13]. The chip is hypothetically divided into 100 identical blocks. Actually, the block size will not be identical, especially when standard-cell design methodology is applied, but the assumption is valid for evaluating performance and chip area. Ten thousand CMOS circuits can be integrated in a 2-mm x 2-mm block. This leaves sufficient area for interblock buffers and chip input-output (I/O) circuits, because a 3-input CMOS NAND using 15-μm-gate-width FETs can

Table 4: Packaging Density of Future CMOS* VLSI Chips and Modern** ECL Modules			
		Packaging density	Wire Length
1st level	CMOS (block)	1,250/mm²	
	ECL (chip)	100/mm²	
	Ratio	12.5	1/3.5
2nd level	CMOS (chip)	125,000/cm²	
	ECL (module)	4,000/cm²	
	Ratio	30	1/5.5
*deep-submicron			
**early 1990's			

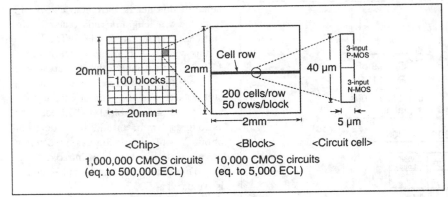

6. A chip model for future CMOS VLSI computer logic.

<Chip>
1,000,000 CMOS circuits
(eq. to 500,000 ECL)

<Block>
10,000 CMOS circuits
(eq. to 5,000 ECL)

be placed in a 40 μm x 5 μm area when 0.3-μm lithography is used. We assume the gate width of 15 μm, which would be unreasonably large for a microprocessor design, to obtain a large driving capability equivalent to ECL.

In our model, one million CMOS circuits with large driving capability are implemented on a chip. The chip is equivalent to 500,000 ECL circuits in terms of its logic-function capabilities. Implementing the same logic by the most modern ECL technology would result in a much larger module (Fig. 7). Let's compare the packaging density and wire length of the ECL module with the CMOS VLSI (Table 4). Since the integration scale of a modern ECL chip is 5,000 to 20,000 gates [14], 25 to 100 ECL chips would be necessary.

One CMOS block on the VLSI chip is equivalent to one ECL chip in terms of gate count. Since the density of the modern ECL is about 100 gate/mm² [15-17], the CMOS is 12.5 times denser. Therefore, the wire length of the CMOS is 1/3.5 that used in ECL. One CMOS chip is equivalent to one ECL module. Since the density of the modern ECL module is 3,500-4,000 gate/cm² [14], an 11 cm x 11 cm ceramic substrate is necessary. The CMOS is 30 times denser, and its wire length is 1/5.5 that used in ECL.

The estimated wire length of the CMOS VLSI chip is shown in Table 5 using the equations reported in [7]. A Rent's Rule coefficient of r = 2/3 is used. Two cases are shown for the intrablock nets. The first case (left column) assumes that CMOS circuits are uniformly distributed in the block, and applies to estimating total wire length. The second case (right column) applies when four CMOS circuits are used as a cluster to implement one ECL function, and applies to estimating critical path delay.

If the average number of terminal-to-terminal wires per signal net, n_{pp}, is 2, that is, FO = 2, wire length per signal net is about 240 μm for intrablock and 9 mm for interblock. Therefore, the total wire length in a block is about 2.4 m since each block has 10,000 signal nets. If wire channel pitch is 1.5 μm (that is, width = spacing = 0.75 μm), two signal layers are required for the intrablock wiring because approximately 50 percent of the total channel length is usable. As for the interblock wiring, the total wire length in a chip is about 100 m since the number of output terminals of a 5,000-ECL equivalent-circuit block is estimated to be 100 - 110. Since the interblock wire resistance causes a significant delay, a wire channel pitch as large as 6 μm is assumed for estimating performance. In this case, three signal layers are required for the interblock wiring. The required number of metal layers is at least six, including power and ground layers. This requirement is not easy to meet, but is not unrealistic for the future since 4-metal-layer chips are already used today [14].

Delay and Power Consumption of the CMOS VLSI Chip

Let's compare the delays of the modern ECL and the deep-submicron CMOS (Fig. 8). The lightly-loaded circuit delay of ECL is 70 ps [14]. The corresponding delay of CMOS is 40 ps for a 1-input inverter and 80 ps for a 3-input NAND gate, as shown in Table 3. The delay of the ECL with 3 fan-outs and 2-mm wire is 220 ps [17]. The corresponding wire length for CMOS is 0.6 mm (from Table 4). Using Table 3, the delay of

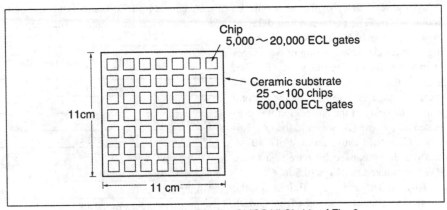

7. A hypothetical ECL module equivalent to the CMOS VLSI chip of Fig. 6.

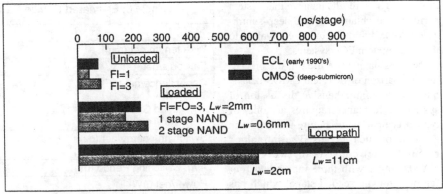

8. Projected delays in deep-submicron CMOS VLSI.

13

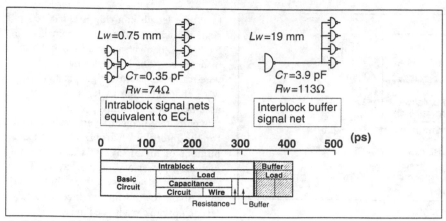

9. Critical-path delay of the CMOS VLSI.

the CMOS is estimated to be 170 ps for a 1-stage NAND. If a 2-stage circuit is needed to obtain logic-function capability equivalent to that of an ECL, the delay is 250 ps.

In the case of ECL, if a signal net is as long as one side of the module substrate, the transmission delay is about 800 ps because the dielectric constant of the substrate is 5.7 - 5.9 [14]. Assuming that the delay of a buffer circuit used for driving the wire is twice the lightly loaded circuit delay, the total delay of the net would be about 940 ps. The corresponding delay of the CMOS would be about 620 ps, assuming a buffer circuit consisting of two stages of inverters for driving the 2-cm wire. The gate widths of the transistors are 15 μm for the first stage and 75 μm for the second stage. The delay of the buffer circuit is estimated as 140 ps. The wire and load capacitance C_T would cause a delay of 240 ps, and the delay caused by wire resistance RW is estimated as 240 ps (0.5 $R_w C_T$).

Although a larger coefficient for the worst case should be assumed for CMOS than for ECL, delays of the CMOS are comparable to the ECL in the various paths in Fig. 8. These results indicate that system performance obtained by the deep-submicron CMOS will be roughly equivalent to the most modern ECL systems.

Various paths were used in the preceding comparison of ECL and CMOS delays, but a more comprehensive discussion of system performance requires a model of typical critical paths. Figure 9 shows one example of such models. The intrablock circuit model consists of two CMOS NAND stages with three inputs and four fan-outs for the second stage, which has associated wires. This is approximately equivalent to one ECL stage in critical paths. Total capacitance of this net is estimated to be 0.35 pF, using the wire length in Table 5 and the circuit characteristics in Table 3. The length of this wire is 0.75 mm and its resistance is 74 Ω. The net's total delay is about 290 ps.

Using interblock buffer circuits decreases the total delay in communicating with other blocks. The buffer circuit shown in Fig. 9 is a simple CMOS inverter with 75-μm-gate-width FETs. Fan-outs of four and associated wires are assumed. Total capacitance of this net is estimated as 3.9 pF, using the wire length in Table 5 and the circuit characteristics in Table 3. The length of this wire is 19 mm and its resistance is 113 Ω. This net's total delay is about 490 ps. Since the input capacitance of this circuit is large and the wire connecting the circuit with an intrablock circuit could be as long as 1 mm, an additional delay of about 200 ps should be assumed when driving this circuit with an intrablock circuit. Thus, total delay for the buffer circuit is estimated as 690 ps.

To obtain CMOS delays corresponding to ECL critical paths, a portion of the buffer delay has to be added to the intrablock delay. From the data reported in [18] the number of ECL equivalent circuit stages is estimated to be about six, considering that the block is equivalent to 5,000 ECL circuits. To obtain the "system delay," one sixth of the buffer-circuit delay should be added to the intrablock circuit delay.

The delay is about 400 picoseconds under typical process parameters and system environments. Assuming a coefficient of 1.5 for the worst case, the value for estimating the system performance is about 600 ps. Since the chip integration is very large, almost all portions of the critical

Table 5. Estimated Wire Lengths – CMOS VLSI Chip			
	Intrablock		Interblock
Circuits/cell	1	4	
Area (mm x mm)	2 x 2	2 x 2	20 x 20
No. of cells or blocks/area	200 x 50	50 x 50	10 x 10
Wire length lpp x p (μm)	121	189	4,660
Total wire length (mm)	2,415		97,860
Channel pitch (μm)	1.5		6.0
Number of layers required	2		3

Table 6. Estimated Device Parameters of the CMOS VLSI Chip		
	Intrablock	Interblock
Wire length/net (mm)	0.24	9.3
Fan-outs/net	2	2
Capacitance/net (pF)		
Wire	0.048	1.86
Load Circuits	0.10	0.10
Net Total	0.148	1.96
Switching energy/net (pJ)	0.94	8.4
/chip (μJ)	0.94	0.088
Avg. Switching Period (ns)	25	25
Power/net (μW)	38	340
/chip (W)	38	3.6

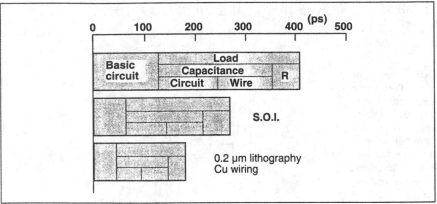

10. Applying advanced technology (such as silicon-on-insulator (SOI) and copper wiring) to deep-micron CMOS VLSI would substantially reduce delays.

paths can be implemented on one chip. The delay we've obtained can thus be directly compared to the system delay of ECL machines, which consists of on-chip circuit and loading delay, chip I/O delay, and wire delay on module substrates and/or printed circuit boards.

The switching period is estimated to be about 25 ns, using $k = 40$ and the system delay a little larger than 600 ps. The power of the CMOS chip can be estimated using the switching period, the device and basic circuit characteristics in Table 3, and the wire lengths in Table 5. As shown in Table 6, the total power of the intrablock circuits and interblock buffers are 38 W and 4 W, respectively. Total chip power will be about 50 W, including chip I/O circuits. If equivalent logic is implemented by using 2-mW ECL circuits, the total power would be more than 20 times larger.

Discussion

If the deep-submicron device in Table 3 is realized, we will be able to obtain system performance from CMOS computer-logic circuits equivalent to today's most modern ECL systems. In addition CMOS has other possibilities.

The effects of applying advanced process and device technology are shown in Fig. 10. The uppermost bar shows the delay breakdown of the CMOS VLSI described above. The second bar shows the effects of applying fully depleted silicon on insulator devices. Better device performance is obtainable by realizing fully depleted SOI MOSFETs [19, 20]. If basic circuit speed is doubled (at x 1.5 original drain current) using the same level of lithography, the system delay decreases by one third. If further miniaturization is realized and low-

resistance material such as copper is used for wiring, the delay will decrease to less than one half of the uppermost bar.

In the case of CMOS, system performance increases almost directly with improvements in device performance. This should provide strong motivation for developing advanced deep-submicron devices, including the technologies we've mentioned. The development of such devices should increase the possibility that room-temperature CMOS will become the long-pursued post-ECL high-speed technology. **CD**

Akira Masaki [SM] is with the Device Development Center, Hitachi, Ltd., Tokyo, Japan.

References

1. B. Davari et al., "A high performance 0.25 μm CMOS technology," *1988 International Electron Device Meeting Technical Digest*, pp. 56-59, December 1988.

2. M. Miyake, T. Kobayashi, and Y. Okazaki, "Subquarter-micrometer gate-length p-channel and n-channel MOSFET's with extremely shallow source-drain junctions," *IEEE Trans. on Electron Devices*, vol. 36, no .2, pp. 392-398, February 1989.

3. A. Masaki, Y. Harada, and T. Chiba, "200-gate ECL masterslice LSI," *1974 International Solid-State Circuits Conference Dig. Tech. Papers*, pp. 62-63, 230, February 1974.

4. A. Masaki *et al.*, "Comparison of MOS basic logic circuits," *IECE Trans.*, vol.58-D, no.7, pp. 397-404, July 1975.

5. A. Masaki and T. Chiba, "Design aspects of VLSI for computer logic," *IEEE Trans. Electron Devices*, vol. ED-29, no.4, pp. 751-756, April 1982.

6. A. Masaki et al., "Perspectives on hardware

technologies for very-high-performance computers," *Proc. IEEE International Conference on Computer Design*, pp. 561-564, October 1984.

7. A. Masaki and M. Yamada, "Equations for estimating wire length in various types of 2-D and 3-D system packaging structures," *IEEE Transactions on Components, Hybrids, and Manufacturing Tech.*, vol. CHMT-10, no. 2, pp. 190-198, June 1987.

8. A. Masaki, "Electrical resistance as a limiting factor for high performance computer packaging," *IEEE Circuits and Devices Magazine*, vol. 5, no. 3, pp. 22-26, May 1989.

9. A. Masaki, "Possibilities of CMOS mainframe and its impact on technology R&D," *1991 VLSI Symposium on Technology Dig. Tech. Papers*, pp. 1-4, May 1991.

10. W.E. Donath, "Placement and average interconnection length of computer logic," *IEEE Trans. Circuits Syst.*, vol. CAS-26, no. 4, pp. 272-277, April 1979.

11. B.S. Landman and R.L. Russo, "On a pin versus block relationship for partitions of logic graphs," *IEEE Trans. Comput.*, vol. C-20, no. 12, pp. 1469-1479, December 1971.

12. P.W. Cook, D.L. Critchlow, and L.M. Terman, "Comparison of MOSFET logic circuits," *IEEE J. Solid-State Circuits*, vol. SC-8, no. 5, pp. 348-355, October 1973.

13. H. Tokuda et al., "A 100k-gate ECL standard-cell LSI with layout system," *1990 International Solid-State Circuits Conference Dig. Tech. Papers*, pp. 94-95,272, February 1990.

14. Nikkei Microdevices, no. 65, pp. 145-151, November 1990.

15. Nikkei Microdevices, no. 64, p. 79, October 1990.

16. Nikkei Electronics, no. 509, pp. 111-113, September 17, 1990.

17. Nikkei Microdevices, no. 48, pp. 38-61, June 1989.

18. M. Nakagawa and M. Yamada, "A study on the relationship between integration scale and number of circuit stages in computer logic," 1981 IECE, paper 402, book 2, p.167.

19. D. Hisamoto et al., "A fully depleted lean-channel transistor (DELTA) — a novel vertical ultra thin SOI MOSFET —," *1989 International Electron Device Meeting Technical Digest*, pp. 833-836, December 1989.

20. G. Shahidi *et al.*, "Fabrication of CMOS on ultrathin SOI obtained by epitaxial lateral overgrowth and chemical-mechanical polishing," *1990 International Electron Device Meeting Technical Digest*, pp. 587-590, December 1990.

High-Speed Compact Circuits with CMOS

R. H. KRAMBECK, MEMBER, IEEE, CHARLES M. LEE, AND HUNG-FAI STEPHEN LAW, MEMBER. IEEE

Abstract—Characteristics of various CMOS and NMOS circuit techniques are described, along with the shortcomings of each. Then a new circuit type, the CMOS domino circuit, will be described. This involves the connection of dynamic CMOS gates in such a way that a single clock edge can be used to turn on all gates in the circuit at once. As a result, complex clocking schemes are not needed and the full inherent speed of the dynamic gate can be utilized. The circuit is most valuable where gates are complex and have high fan-out such as in arithmetic units. Examples are shown of the use of domino circuits in an 8-bit ALU, where simulations indicate a speed advantage of 1.5 to 2 over traditional circuits, and in a 32-bit ALU where a worst case add in 124 ns was projected and a time less than 100 ns was achieved.

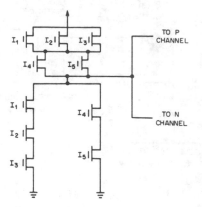

Fig. 1. Fully complementary MOS 32AOI gate. No static power but high-output capacitance and area.

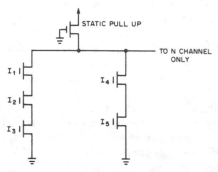

Fig. 2. Pseudo-NMOS 32AOI gate. Low output capacitance and area, but static pull-up current consumes power and slows pull-down.

I. INTRODUCTION

THIS paper will describe some new design techniques which can substantially reduce area and increase speed for circuits made with CMOS technology. These techniques combine, in a unique way, the speed and power advantages of dynamic circuits with the stability and ease of use of static circuits.

In a fully complementary CMOS circuit the logic function of each gate is implemented twice. For example, a combinational gate that does the AND/OR invert (AOI) function for one 3-input AND, and one 2-input AND (32 AOI), is shown in Fig. 1. The five n-channel transistors have all the information needed to implement the function and so do the five p-channel transistors. The advantage of having both arrays is that except for the very brief period when the output or the inputs are making transitions no current flows and no power is consumed.

The problem with this fully complementary approach is that for complex gates of the type shown in Fig. 1, substantial amounts of area can be wasted. For example, the same function could be made with six transistors in static NMOS or pseudo-NMOS as shown in Fig. 2. (Pseudo-NMOS refers to a design technique which gives circuits identical to NMOS circuits except for the use of a p-channel transistor as the load instead of an n-channel transistor.)

As a result of the extra area and extra transistors, the capacitive load on gates of a fully complementary circuit are considerably higher than the loads on a pseudo-NMOS or NMOS circuit. Each output goes to both a p-channel and an n-channel transistor in every gate it drives. P-channels are generally twice the size of n-channels to obtain more balanced rise and fall times [1]. As a result, the total gate load on each output will be three times higher. Parasitics do not increase that much but overall capacitance is at least a factor of two higher.

It would appear from this that pseudo-NMOS or NMOS would be much faster than CMOS but this is not the case. The problem is that pull-up current always flows in the pseudo-NMOS circuit even if the gate is pulling down. This slows the pull-down. Making the pull-up current very small does not solve this problem because then the pull-up would be very slow. In fact minimization of the sum of rise time and fall time occurs when pull-up current is one half the pull-down. Thus, at most only one-half as much current is available in a pseudo-NMOS circuit as there is in a CMOS circuit using the same size transistors. In actual circuits the sum of rise and fall time is somewhat worse than this for pseudo-NMOS because for noise immunity the pull-up is usually chosen somewhat smaller than half the pull-down.

As a result, the speed of CMOS and pseudo-NMOS are very close. The CMOS has twice the capacitance but also twice the available current. The tradeoff in choosing one or the other is between the low power of the CMOS and the low area of the pseudo-NMOS.

The remainder of this paper will show first how dynamic circuits have combined both low-capacitance and high-current

Manuscript received March 10, 1981; revised November 6, 1981.
The authors are with Bell Laboratories, Murray Hill, NJ 07974.

Reprinted from *IEEE Journal of Solid-State Circuits*, Vol. SC-17, No. 3, pp. 614-619, June 1982.

16

capability, but at a cost in circuit stability and operational complexity. Next, new techniques will be described which maintain the above advantage of dynamic circuits while still keeping the stability and simplicity of static circuits. Finally, some specific examples will be presented.

II. DYNAMIC CIRCUITS

Many dynamic circuit schemes have been described [2], but they all show some basic features in common. Basically, they involve precharging the output node to a particular level (usually high for NMOS), while the current path to the other level (ground for NMOS) is turned off. Changing of inputs to the gate must occur during this precharge phase. At the completion of precharge, the path to the high level is turned off by a clock and the path to ground is turned on. Then depending on the state of the inputs, the output will either float at the high level or will be pulled down. Fig. 3 illustrates how this is done for the 32 AOI gate described earlier. The advantage of a dynamic circuit is that the load capacitance is comparable to static pseudo-NMOS but the full pull-down current is available. Therefore, the gate should respond roughly twice as fast as either pseudo-NMOS or full CMOS. In addition, there is no static current path so power would be much closer to CMOS than to static pseudo-NMOS. (There is still some power penalty compared to CMOS because each gate must be precharged high every cycle even if its output is to continue low.)

However, there are serious problems involved in realizing these apparent speed advantages in real circuits. This happens because useful circuits generally have several logic gates in series and in the dynamic approach; no gate can be activated until its inputs have stabilized. There are many ways to clock the gates so that this occurs, and an example is shown in Fig. 4. A detailed description of the operation of this circuit is given in [2] and will not be repeated here. Basically, each gate goes through a precharge when transistors A and B are on, an evaluation when transistors B are on and A is off, and a hold period when transistors B are off. It is required that when a gate is in the evaluation mode, the gate driving it must be in the hold mode. There are four types of gates distinguished by the phase in which evaluation occurs. The one shown is type 3. This means that gate type 2 can drive either type 3 or type 4 but not type 1. Similar restrictions apply to each circuit type. This requires some additional care in design but is not a major problem. There are two reasons why the speed of this circuit will not be double that of a static circuit. First, each gate has two additional transistors in the pull-down path which reduces the available current considerably. For a 1- or 2-input gate this could easily be a factor of two. Second, the time allowed for a gate to stabilize must be chosen so that even the gate with the longest delay can settle down. This can cause substantial time waste on the faster gates because they must be allocated a full time slot. In addition, the difficulties of generating the four clocks and synchronizing them throughout the circuit to a small fraction of a gate delay are formidable. In practice considerably more than one gate delay would be needed between successive edges to assure a full gate delay in worst case. Overall then, in a circuit of reasonable complexity, the dynamic approach would not be any faster than

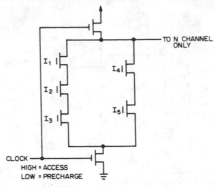

Fig. 3. Dynamic pseudo-NMOS gate. Low output capacitance and no static pull-up, but inputs must be valid before access begins.

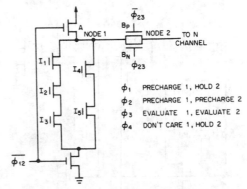

Fig. 4. Four phase dynamic pseudo-NMOS. The shortest clock phase must be long enough so that the slowest gate in the circuit can complete its evaluation. This results in considerable dead time.

static though it would have power advantages compared to pseudo-NMOS or NMOS.

III. CMOS DOMINO CIRCUIT

The CMOS domino circuit shares some characteristics with dynamic circuits. In particular, each output is precharged high while the path to ground is opened and the precharge is stopped while the path to ground is activated. The critical difference is that the transition from precharge to evaluation is accomplished by means of a single clock edge applied simultaneously to all gates in the circuit. This greatly simplifies clocking and permits utilization of the full inherent speed of the gates.

A single domino circuit gate is shown in Fig. 5. It consists of two parts. The first looks like a dynamic pseudo-NMOS gate and is clocked in the same way as such a gate, with a precharge phase followed by an evaluation phase. The second part is a static CMOS buffer. Only the output of the static buffer is fed to other gates of the circuit; the output of the dynamic gate goes only to the buffer. During precharge, the dynamic gate has a high output so the buffer output is low. This means that during precharge, all circuit nodes which connect the output of one domino gate to the input of another are low, and therefore the transistors they drive are off. In addition, during evaluation a domino gate can make only a single transition, namely from a low to high. Because of the nature of the dynamic gate which drives it, it is impossible for the buffer to go from high to low during evaluation. (Since the dynamic gate cannot go high, the buffer cannot go low.) As a result there

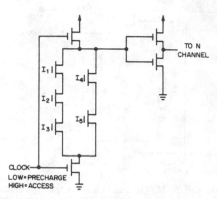

Fig. 5. Domino CMOS circuit. No static power, low area, with simple single edge clocking for all gates in the circuit.

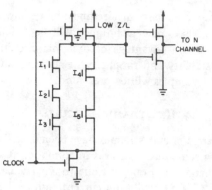

Fig. 7. Domino CMOS circuit with an additional pull-up device to permit static or low-frequency operation.

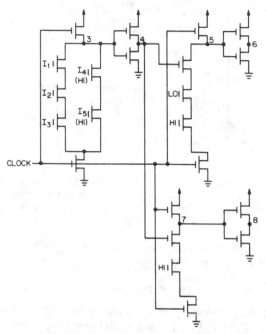

Fig. 6. An example of a domino CMOS circuit showing how a single clock activates all clocks simultaneously.

can be no glitches at any nodes in this circuit. All nodes can make at most only a single transition and then must stay there until the next precharge. This is reminiscent of the behavior of a row of dominos toppling into one another, and hence the proposed name.

Since there is no need to worry about glitches and since during precharge all domino outputs turn off the transistor they drive, all gates may be switched from precharge to evaluate with the same clock edge. An example of how this works is shown in Fig. 6. During precharge, nodes 3, 5, and 7 are all high so nodes 4, 6, and 8 are low. When precharge ends node 4 goes high which causes node 8 to go high. Node 6 remains low during evaluation.

As will be described in more detail in the next section many types of circuits when made with domino gates can be significantly faster than a corresponding circuit made with other techniques. The circuit has the low power of a dynamic circuit since there is never a dc path to ground. Also, the full pull-down current is available to drive the output nodes. At the same time the load capacitance is much smaller than for

CMOS because most of the p-channel transistors have been eliminated from the load. Meanwhile, the use of a single clock edge to activate the circuit provides simple operation and full utilization of the speed of each gate. (There is no dead time between output valid and operation of the next gate in the circuit.)

One limitation of this circuit technique is that all of the gates are noninverting. This may seem serious since an XOR is not possible, but actually very complex circuits can be implemented including an arithmetic logic unit (ALU) with two levels of carry look ahead (to be described later). This is feasible because the domino gate is fully compatible with standard CMOS gates and the needed CMOS XOR can be driven by the last domino circuit.

Another limitation is that each gate must be buffered. This has not been a problem in the circuits designed so far because buffers would have been needed anyway to achieve maximum speed. The need for buffers indicates that this circuit technique is most valuable in logic involving many gates with high fan-out.

IV. STATIC DOMINO CIRCUIT

In some applications it is desirable to have a static capability to allow lower frequency operation or to avoid the risk of storing data on floating nodes. This can be obtained in a domino circuit by the addition of a low current pull-up transistor as shown in Fig. 7. This functions as a means of removing charge which accumulates on the output node as a result of leakage or noise.

This transistor would be chosen small enough so there is no significant impact on pull-down current and so the power consumed during the evaluation phase is tolerable. A value of 10 μA is reasonable. This would require a p-channel transistor that is 20 μm long and 4 μm wide. For a chip with 2000 pull-up devices at 5 V, power consumption during evaluation would be 100 mW, if all gates are being pulled down. Average power would depend on the application but would be significantly less.

Another way to implement the static circuit is to include the static pull-up transistor shown in Fig. 7, but to have no clocked precharge transistor. This can be done if the time between evaluation phases is relatively long so precharge can be accomplished by the weak static pull-up transistor.

18

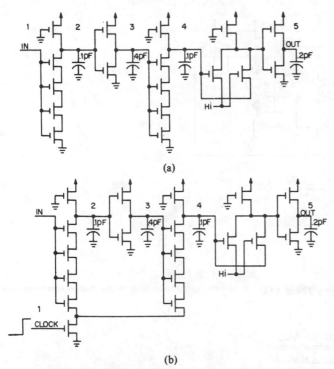

(a)

(b)

Fig. 8. (a) Part of an 8-bit ALU critical path using static pseudo-NMOS. (b) Same critical path with domino clocking.

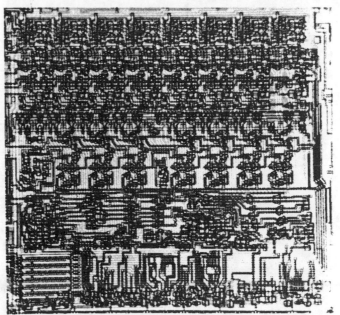

Fig. 9. Photograph of an 8-bit ALU. Ground switch is on lower left.

V. AN 8-BIT ALU

The first use of the domino circuit was on an 8-bit arithmetic logic unit (ALU) of an 8-bit microprocessor [3]. This happened because simulations indicated that adequate performance could not be obtained with a pseudo-NMOS circuit, while full CMOS was too area-consuming. The circuit of part of the critical path of ALU using pseudo-NMOS is shown in Fig. 8(a). The ALU in domino CMOS uses 690 transistors, and with a 15 μm pitch for metal and polysilicon, the area is 6000 mils^2. A similar transistor density in full CMOS would have required an additional 3000 mils^2 which was not available. A photograph of the ALU is shown in Fig. 9. The large structure on the lower left is the clocked ground switch which turns on the ALU when it is the evaluate phase. In a strip along the right side are the p-channel static load devices.

A SPICE [4] simulation of the simple pseudo-NMOS critical path predicted a worst case propagation delay of 450 ns which exceeded the chip requirement of 250 ns. A SPICE simulation of the domino circuit which is shown in Fig. 8(b) predicted 215 ns and so the design was made this way. Note that this circuit is like the one in Fig. 7 except that the clocked pull-up transistor has been eliminated and only the static one remains. This was done because the time between accesses of this ALU are so long that the low Z/L static transistor is sufficient to do the precharge. Table I shows propagation delays predicted by the simulation for both pseudo-NMOS and domino CMOS. The very slow pull-up time dominates the pseudo-NMOS CMOS delay. This happens because even though optimum speed is obtained with pull-up current equal to one-half pull-down current, noise margins forced a smaller ratio resulting in slow pull-ups. A histogram of measured propagation delay for 116 circuits that were fabricated is shown in Fig. 10. This

TABLE I
WORST CASE DELAYS IN 8-BIT ALU

| | Static Circuit Delay (nSec) | | Domino Circuit Delay (nSec) | |
Node	In Goes High	In Goes Low	In High	In Low
1	0	0	0	0
2	40	270	100	0
3	10	10	25	0
4	40	80	25	0
5	50	50	65	0
TOTAL	140	410	215	0

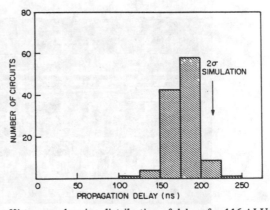

Fig. 10. Histogram showing distribution of delays for 116 ALU circuits.

histogram confirmed the high-speed predictions made by the simulation and verified the operation of the domino CMOS circuit.

VI. A 32-BIT ALU

For a more complex example, a critical path in a 32-bit ALU [5] will now be discussed. This circuit uses 3300 transistors and does a 32-bit add as well as other arithmetic and logic

19

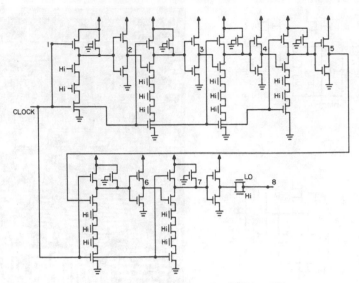

Fig. 11. Critical path through a 32-bit ALU.

TABLE II
WORST CASE DELAYS IN 32-BIT ALU

Node	Delay (nSec)
2	13
3	29
4	16
5	22
6	16
7	21
8	7
TOTAL	124

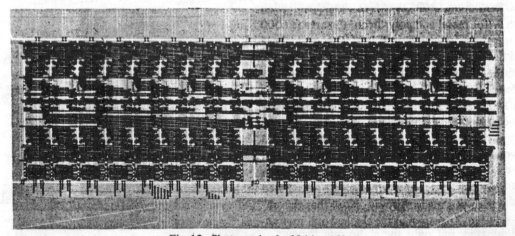

Fig. 12. Photograph of a 32-bit ALU.

function. The critical path in the domino CMOS path is shown in Fig. 11. Simulations have been made for this path and Table II gives propagation delays at various nodes on the critical path. The predicted worst case total propagation delay is 124 ns for $V_{DD} = 4.75$ V and a junction temperature of 105°C. This circuit was fabricated and a photograph of it is shown in Fig. 12. Process parameters of test transistors on the wafer were measured and using these a propagation delay of 104 ns was predicted. The actual delay was 97 ns.

VII. SUMMARY

A new compact, high-performance circuit design technique has been described for use with CMOS technology. This domino CMOS technique gives circuits with areas comparable to static NMOS or pseudo-NMOS, but gives a speed improvement of a factor of 1.5 to 2. This is achieved without resorting to any multiphase clocks and the static stability of the circuit can be maintained.

REFERENCES

[1] S. M. Kang, "A design of CMOS polycells for LSI circuits," *IEEE Trans. Circuits Syst.*, vol. CAS-28, pp. 838–843, Aug. 1981.

[2] W. M. Pensey and L. Lau, *MOS Integrated Circuits*. New York: Van Nostrand, 1972, pp. 260–282.

[3] J. A. Cooper, J. A. Copeland, R. H. Krambeck, D. C. Stanzione, and L. C. Thomas, "A CMOS microprocessor for telecommunications applications," in *Dig. ISSCC*, Feb. 1977.

[4] L. W. Nagel and D. O. Pederson, "Simulation program with integrated circuit emphasis," in *Proc. 16th Midwest Symp. Circuit Theory*," Waterloo, Ont., Canada, Apr. 1973.

[5] B. T. Murphy, R. Edwards, L. C. Thomas, and J. J. Molinelli, "A CMOS 32-bit single chip microprocessor," in *Dig. ISSCC*, Feb. 1981.

Design-Performance Trade-Offs in CMOS-Domino Logic

VOJIN G. OKLOBDZIJA AND ROBERT K. MONTOYE, MEMBER, IEEE

Abstract —This paper is a study of the charge-sharing problem and its effect on the performance of CMOS-Domino logic. Several solutions to the charge-sharing problem are examined, and the results are verified by simulation. Thus the charge-sharing problem in CMOS-Domino logic was identified and alternate approaches were evaluated.

I. INTRODUCTION

With increased interest in CMOS, Domino-type logic has been gaining favor due to its n-MOS-like performance (i.e., n-channel dominant delay) and CMOS-like power consumption [1], [2], and favorable testability relative to CMOS [3]. However, Domino logic presents a charge-redistribution problem that continues to impair its usability.

This logic family was developed during the course of implementation of BELMAC-32 microprocessor and the first paper on Domino logic was published by the authors from Bell Laboratories [2]. However, the logic originally published had several drawbacks. For example, inversion was not possible, making the implementation of EXCLUSIVE OR (XOR) function difficult. The original circuit implementation was very sensitive to the charge-redistribution problem, causing spurious results.

The authors from IBM developed a logic family, Cascode Voltage Switch (CVS), which further advanced the status of Domino logic [1]. This is a complete logic family because both polarities of each function output are available at every stage. In fact, this is a two-rail logic that offers a self-checking feature at no extra cost. Their logic family comes in two versions: static and dynamic. The dynamic version can be treated as part of the CMOS-Domino logic family. The static version of the CVS logic implements the p-MOS latches at the output nodes, which triggers a regenerative action to bring the nodes to their full logic one and logic zero values. An extension of this technique is Differential Split Level (DSL) logic, which claims a performance improvement of ten times over regular static CMOS, but consumes more power as reported [4].

In this paper, we will consider CMOS Domino and dynamic CVS version for the purpose of analysis and identification of the charge-redistribution problem.

II. OPERATION

Principles of operation of Domino-type logic are outlined by the circuit example shown in Fig. 1. This logic family evolved from the dynamic n-MOS (p-MOS) circuits and therefore retained two phases of operation: "precharge" and "evaluate" (designated PCHG and EVAL, respectively, in this paper). The basic logic function implemented with this type of logic consists of: clock circuitry (transistors Q_2, Q_{p1}), n-MOS transistor network SWF implementing given Boolean function f, and inverter. During the PCHG phase of the clock, the p-MOS transistor Q_{p1} is ON while the n-MOS transistor Q_2 is OFF. Node N4 is charged to V_{dd} and the output from the inverter is at the voltage level close to 0 v. This situation occurs at that time at every logic block including those whose outputs are connected to the inputs X_i of this particular block. Registers are designed in the same way so that all of their outputs are logic zero value during the

Manuscript received September 9, 1985; revised December 20, 1985.
The authors are with the IBM T. J. Watson Research Center, Yorktown Heights, NY 10598.
IEEE Log Number 8607667.

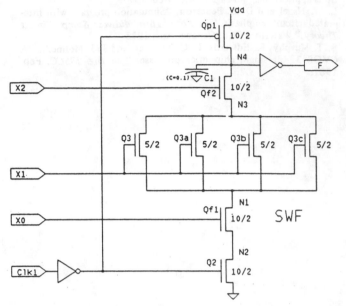

Fig. 1. Domino circuit. A p-channel transistor is indicated by a circle at the gate. W/L ratios are indicated next to the transistor. Waveforms on the internal nodes $N1$, $N3$, and $N4$ are shown in Fig. 2.

PCHG phase. As a consequence, all of the inputs X_i of the particular block, and all of the other blocks, are close to 0 v during the PCHG-phase. Therefore during the PCHG phase there is no electrical path from the "top" node (N_4) to the "bottom" node (N_1) and only the "top" node (N_4) is storing charge. When the clock turns to the EVAL phase, transistor Q_2 is ON creating the path from the node N4 through the switching network SWF to the ground. If the condition for the existence of an electrical path in the network SWF (between the nodes N4 and N2) is established by the signal values of the inputs to the SWF, node N4 is discharged to ground, which in turn makes the output of the inverter F the logic ONE. This value is the input to the subsequent logic block(s) and can cause the output of the block(s) to switch to ONE. This signal change is propagated in the "domino" fashion.

A. Charge-Redistribution Problem

From the operation of the Domino logic, it is clear that the charge is stored only at the "top" node (N4 in Fig. 1.) and the nodes N_1, N_2, N_3 are not charged during the PCHG phase. They might have been discharged during the previous cycle and thus have no charge. Therefore, during the evaluation phase, there may be an electrical path to several discharged nodes (causing charge redistribution) without an electrical path to ground. If there is sufficient charge redistribution (i.e., the ratio of the capacitance at the top node of the tree C_t) the uncharged capacitance internal to the tree C_i reduces the voltage below the inverter threshold I_{ih}

$$V_{dd} \times \frac{C_t}{C_i + C_t} \leq I_{th}.$$

This charge redistribution will cause the inverter at the output of the tree to falsely switch, thus placing the incorrect value on the line causing other groups to discharge falsely. One such example (shown in Fig. 2) is generated by simulation of the single Domino-logic stage using the Toggle circuit simulation package [5]. We are observing in this case the behavior of the logic block during the period of two full cycles: PCHG–EVAL, PCHG–EVAL. During the first cycle inputs $X0$, $X1$, and $X2$ are set to

Reprinted from *IEEE Journal of Solid-State Circuits*, Vol. SC-21, No. 2, pp. 304-306, April 1986.

22

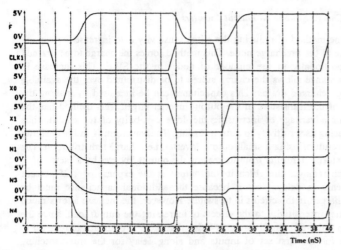

Fig. 2. Waveforms from the Domino circuit example in Fig. 1. The effect of charge redistribution is seen on the node $N4$.

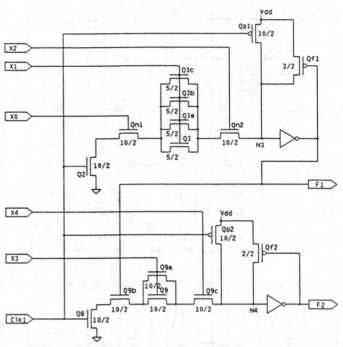

Fig. 3. An example of CVS logic which was used to simulate charge redistribution. p-channel transistors are distinguished by the circle associated with the gate symbol. Voltages on the nodes $N3$ and $N4$ are shown in Fig. 4

logic one causing the discharge of the entire network during the EVAL phase. Following the PCHG phase in the second cycle, the node $N4$ is precharged to the value of 5.000 V. In the second cycle, the inputs are set to $X0 = 0$, $X1 = 1$, and $X2 = 1$ creating an electrical path between the nodes $N4$, $N3$, and $N1$, but stopping short of the node $N2$ which is connected to ground during the EVAL phase. This situation causes redistribution of the charge between the nodes $N4$, $N3$, and $N1$. From the waveforms in Fig. 2, we can observe that the voltage on the node $N4$ falls to 1.0016 V. The voltage on node $N1$ has risen to 1.0005 V and the voltage on node $N3$ has risen to 1.0009 V. The input combination ($X_0 = 0$, $X_1 = 1$, and $X_2 = 1$) is supposed to produce the logic ZERO value at the output F_1. However, because of the redistribution of charge between the node N_4 and the nodes N_1, N_2, N_3 the voltage at node N_4 is only 1.0016 V. This produces an erroneous value of logic ONE at the output F_1.

III. PROBLEM ELIMINATION METHODS

In this section we examine the techniques used to alleviate the problem caused by charge redistribution and evaluate the trade-offs in reliability and circuit performance.

Two methods of reducing the charge-sharing problem are addressed. The first of these, incorporating feedback into the tree, reduces the charge-sharing problem by injecting charge into the tree during evaluation. The second method selectively increases the storage capacity of the precharge node in proportion to the number of nodes to which the charge can be redistributed.

A. Feedback Transistor (Dynamic CVS Logic ONE)

One method used by dynamic CVS logic to alleviate the charge redistribution problem is to place an additional p-MOS transistor Qf in parallel with the precharge transistor. The gate of this transistor is connected to the output of the inverter so that feedback from the output is obtained (Fig. 3). In this way the inverter-transistor combination acts as a "latch" that locks on the state where the output of the inverter F is at the ZERO logic value. Because the output $F = 0$ is "latched", it takes more current from the node $N4$ to pull the node to ground since the charge is being continuously replenished by the device Q_f. When charge redistribution occurs, transistor Q_f serves the purpose of replenishing the charge lost in the process of redistribution to the other nodes. We can distinguish two cases.

1) During the charge redistribution, the total sum of the currents to node N_4 is such that node N_4 will recover the charge lost by redistribution. The current from the node N_4 which is due to redistribution of charge, decreases exponentially in time. The current to the node N_4 through the transistor Q_f will also exponentially decrease in time but have a larger time constant.

As a result, the current calculated with reference to the node N_4 is negative at the beginning and is equal to the charge taken from the node and distributed to other nodes. Later, this current becomes positive, bringing the charge lost in redistribution back to the node N_4. It decreases exponentially to ZERO. This produces a voltage "spike" or "glitch" at the node N_4 which is of the amplitude that never exceeds the threshold for logic ONE at the inverter input.

2) In the second case, the amount of charge lost during the initial period is such that produces the voltage spike of sufficient amplitude to change the output value to logical ONE. This in turn cuts off the transistor Qf preventing it from replenishing the lost charge to the node $N4$. In this case, the fault is permanent and the output will stay at the erroneous value of logic ONE instead of ZERO.

Let us consider the first case and the consequences of the voltage "spike." The voltage "spike" at the node $N4$ is propagated through the inverter producing a positive voltage "spike" at the output F. The effect of this positive voltage "spike" can be twofold.

1) It can cause complete discharge in the next logic block creating the erroneous value to appear at its output. This value is propagated further in a "domino" fashion.

2) The voltage "spike" can cause a similar "spike" at the output of the next logic block. This spike is further propagated, in which case it can be:

a) of the smaller amplitude and therefore dissipated in the logic;
b) amplified through the consequent stages and therefore eventually resulting in a permanent error being propagated (much like the case 1);
c) fanned out in different directions, in which some will dissipate the spike and some will amplify it and distribute the erroneous reading.

These cases are illustrated in Fig. 4.

However, feedback transistor Q_f serves as a load transistor when the node $N4$ is forced to ground. Therefore the operation is now that of a ratioed circuit, and as such the transistor Q_f serves as a load and its L/W ratio has to be adjusted to the cumulative (L/W) eff ratio of the maximum length electrical path to ground of the switching block SWF. L/W ratio represents resistance of the feedback transistor in the circuit.

Let us define β to be the ratio of $(L/W)f/(L/W)$eff. This ratio represents the resistance of the feedback device compared to the resis-

23

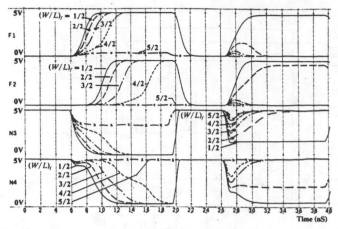

Fig. 4. Signals at the nodes $F1$, $F2$, $N3$, and $N4$ for the circuit in Fig. 3, as a function of time for various sizes $(W/L)_f$ of the feedback transistors Q_{f1} and Q_{f2}.

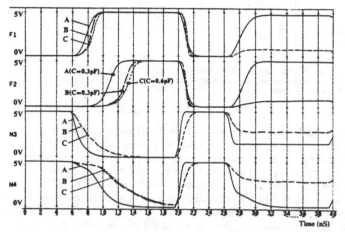

Fig. 5. Waveforms on the nodes $F1$, $F2$, $N3$, and $N4$ for the circuit in Fig. 3. without feedback transistors Q_{f1} and Q_{f2}. The size of the output inverters is varied: A—small inverter; B—output inverter size increased 7 times over case A; C—output inverter of the size in case B with the output load on the nodes $F1$ and $F2$ doubled.

tance of the switching network SWF. We can distinguish three cases:

1) β too large, in which case the feedback device is not very effective except for very small "glitches;"
2) β in the range on 0.9, which was determined to be the optimal value with respect to adequate "glitch" protection; and
3) β too small, in which case the output F acts as being stuck at zero, because the SWF block is too weak to pull the node $N4$ to ground.

However, the "safe" or "glitch-free" operation is dependent on the proper choice for β. The range of β values in which the circuit is effective in "glitch" suppression imposes the restriction on the maximal length of the possible electrical paths in the SWF network. Additionally, a restriction is placed on the number of nodes in the tree that can be at 0 V and cause redistribution.

Given the example shown, in which a SWF with three devices in a series and a total of six devices can provide glitch immunity only with careful feedback device tuning, it is clear that the range of effectiveness of the feedback device is very limited.

Another impact of the feedback device is on performance. This is visible in Fig. 4, as the delay increases markedly as a larger feedback device (smaller β) is used, until, as previously mentioned, the circuit fails to operate β below 0.6. Thus the reduction in glitch sensitivity is paid for by a degradation in performance. This difficulty becomes apparent as larger SWF's are used, since the path to ground involves more active devices, and there are more total devices in the circuit.

B. Charge Storage on Output Inverters

One method of reducing the glitch sensitivity is to increase the capacitance of the precharge node. This method forces the charge to be drained proportionally to the number of nodes in the SWF. This increase in capacitance allows the charge stored to be distributed over the available nonprecharged drains. This capacitance can be increased by making the size of the transistors in the output inverter larger. This method has the major advantage of increasing the drive capability of the circuit, thus reducing its sensitivity to output loading. The waveform A in Fig. 5 shows the result of using a small output inverter, i.e., charge redistribution. The waveform B results from increasing the output inverter by a factor of 7, which removes the glitch in the output at the expense of a 25-percent increase in delay. This is because the additional capacitance must be discharged if the circuit is to switch. However, the increase allows much greater insensitivity to the effects of increasing the output load, since the output driver is much larger. Notice that there is both charge redistribution for the first set of inputs and along delay for the true switching signal. The final waveform C shows the result of the circuit with the larger output inverter and double the expected loading, i.e., 0.6 pf. Note that the delay for this circuit is only slightly larger than for the circuit having the output inverter driving the smaller load.

The technique of increasing the output inverter size to reduce the effects of charge sharing is limited in range to acceptable output inverter sizes and delays. It has the potential to reduce charge sharing in cases where the problem is not too severe. However, the internal switching delay may grow significantly with larger trees. The additional side effect of increasing the switching speed of the output load may offset some of this weakness.

IV. Conclusion

The charge-sharing problem may be combatted using several methods. Two solutions to this problem were described and analyzed. It was concluded that feedback devices are helpful in a limited range and increasing output inverter sizes are additionally helpful to reduce the problems of charge redistribution. In practice, these two methods can be used to produce a large family of glitch-free trees. Further study is required to make Domino logic viable for a wider range of switching functions.

References

[1] L. G. Heller, W. R. Griffin, J. W. Davis, and N. G. Thoma, "Cascode voltage switch logic: A differential CMOS logic family," in *Proc. ISSCC* (San Francisco, CA), Feb. 22–24, 1984.
[2] R. H. Krambeck *et al.*, "High-speed compact circuits with CMOS," *IEEE J. Solid-State Circuits*, vol. SC-17, no. 3, June 1982.
[3] V. G. Oklobdzija and P. G. Kovijanic, "On testability of CMOS-Domino logic," in *Proc. 14th Int. Conf. Fault-Tolerant Computing* (Orlando, FL), June 20–22, 1984.
[4] L. C. M. G. Pfennings *et al.*, "Differential split-level logic for sub-nanosecond speeds," in *Proc. ISSCC 1985* (New York, NY), Feb. 14, 1985.
[5] J. F. Beetem *et al.*, "A large-scale MOSFET circuit analyzer based on waveform relaxation," in *Proc. ICCD'84* (Port Chester, NY), Oct. 8–11, 1984.

NORA: A Racefree Dynamic CMOS Technique for Pipelined Logic Structures

NELSON F. GONCALVES, STUDENT MEMBER, IEEE, AND HUGO J. DE MAN, SENIOR MEMBER, IEEE

Abstract—This paper describes a new dynamic CMOS technique which is fully racefree, yet has high logic flexibility. The circuits operate racefree from two clocks ϕ and $\bar{\phi}$ regardless of their overlap time. In contrast to the critical clock skew specification in the conventional CMOS pipelined circuits, the proposed technique imposes no restriction to the amount of clock skew. The main building blocks of the NORA technique are dynamic CMOS and C^2MOS logic functions. Static CMOS functions can also be employed. Logic composition rules to mix dynamic CMOS, C^2MOS, and conventional CMOS will be presented. Different from Domino technique, logic inversion is also provided. This means higher logic flexibility and less transistors for the same function. The effects of charge redistribution, noise margin, and leakage in the dynamic CMOS blocks are also analyzed. Experimental results show the feasibility of the principles discussed.

I. INTRODUCTION

IN the conventional CMOS technique there is an inherent redundancy of information. For each n-type device there is a corresponding p-type device. In fact, a complete logic function is built with the n devices and repeated with the p devices. As a consequence of this approach, substantial amounts of silicon are wasted, especially for complex logic. Also, power dissipation and speed are degraded by the extra area and extra transistors.

Another important problem of CMOS technique is clock races in pipelined circuits. To latch the information between two pipelined sections, transmission gates are usually employed. In

Manuscript received November 8, 1982; revised January 18, 1983. This work was supported in part by Fundaçao de Amparo a Pesquisa do Estado de São Paulo, Brazil.

The authors are with the Department Elektrotechniek-ESAT, Katholieke Universiteit Leuven, B 3030-Heverlee, Belgium.

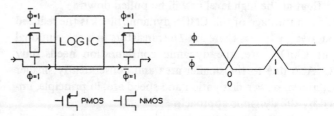

Fig. 1. Signal races in CMOS pipelined circuits.

CMOS logic, these transmission gates are generally implemented with p-n gates in parallel and controlled by clocks ϕ and $\bar{\phi}$, as shown in Fig. 1. The use of single gates (p- or n-type) is to be avoided in CMOS due to power dissipation and low noise margin as a result of clock feedthrough and bulk effect. CMOS p-n transmission gates, controlled by clocks ϕ and $\bar{\phi}$, suffer from signal races. As depicted in Fig. 1, this results from unavoidable overlap of the clock phases during the clock transitions. During the phase overlaps, all the transmission gates are switched on, which may cause illegal flow of information, depending on the ratio between the gate delay and the clock skew.

This race problem is usually bypassed by a careful synchronization of the two clock phases within a small fraction of the gate delay (a few nanoseconds). This skew clock control is extremely difficult, especially for high speed technologies, for unmatched clock loads or for distributed clock VLSI circuits [1]. This leads to highly critical and untestable designs. A possible solution to the clock race is the use of four clock phases which, however, requires too much silicon area.

To overcome the redundancy of information in the conventional CMOS, dynamic circuit schemes have been proposed in

Reprinted from *IEEE Journal of Solid-State Circuits*, Vol. SC-18, No. 3, pp. 261-266, June 1983.

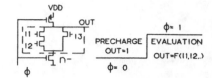

Fig. 2. An n-type dynamic CMOS logic block.

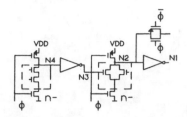

Fig. 4. Domino circuit.

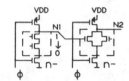

Fig. 3. Internal delay race problem.

the literature [2]–[4]. Fig. 2 shows the dynamic CMOS building block. The desired logic function is implemented using only n-type devices. The logic tree is connected to V_{DD} and groundthrough clocked transistors. There are two modes of operation. First, for phase $\phi = 0$ the output node is precharged to a high level while the current path to ground is turned off. Then, for phase $\phi = 1$, the path to the high level is turned off by the clock and the path to ground is turned on. Therefore, depending on the state of the inputs, the output node will either float at the high level or will be pulled down.

A clear advantage of this CMOS dynamic block is the reduced silicon area. Whereas there are $2n$ transistors in a conventional n-input CMOS gate, the dynamic configuration needs only $n + 2$. Also due to the smaller area and consequently smaller capacitances, power dissipation and speed are, in principle, improved by the dynamic approach.

A strong limitation of this dynamic structure is the impossibility of cascading the logic blocks for implementing complex logic. Consider, for instance, the circuit in Fig. 3. During the precharge phase, nodes $N1$ and $N2$ are set up to the high level "1." In the evaluation phase ($\phi = 1$), internal delay in block 1, associated with a "1" → "0" transition of node $N1$, can cause an incorrect discharge of node $N2$. This occurs because, during the evaluation phase and while node $N1$ is still "1," there is a direct path between node $N2$ and ground. When this path is eliminated by the effective transition of node $N1$ to "0," the precharge information of node $N2$ could already be gone. We define such a race as the "internal delay problem."

In the Domino technique, Krambeck *et al.* [4] have solved the internal race by placing a static inverter after every dynamic block, as indicated in Fig. 4. During the precharge phase, the outputs of all the static inverters are set up to a low level. Consequently, all the n-type transistors driven by these inputs are set up to an OFF condition. Now, during the evaluation phase, internal delays cannot incorrectly discharge the dynamic storage nodes since during the entire delay period the path to ground is turned off.

A limitation of the Domino technique is the lack of inverted signals. The combination of the dynamic block with the static inverter gives a noninverted signal. This decreases logic flexibility and, therefore, usually requires more transistors for a given logic function. Besides this inconvenience, no provisions are made to overcome the clock race problem.

In the next section, a new technique called NORA is presented which overcomes the above deficiencies. In Section III, the properties of the NORA CMOS technique are analyzed and proved. Logic composition rules to mix dynamic, static, and C^2MOS [5] logic functions are also derived. The dynamic CMOS limitations are described in Section IV. Experimental results and major conclusions are presented in Sections V and VI, respectively.

II. NORA CMOS TECHNIQUE

The main building blocks of NORA technique are shown in Fig. 5. The logic functions are implemented using n-type and p-type dynamic CMOS and C^2MOS blocks. Conventional (static) CMOS function blocks can also be eventually employed. Logic composition rules to combine these functions, preserving the racefree properties, will be presented in Section III. As it will further be shown, to guarantee a fully racefree operation in pipelined circuits, the storage of information must always be performed by a C^2MOS function block (C^2MOS latch stage). In a previous paper [6], the NORA (*NO RAce*) technique was called n-p-CMOS, due to the possible employment of n- and p-dynamic blocks. We decided to change the name because the p-dynamic block is not essential to the racefree principle; it is only used to increase the logic flexibility.

The pipelined circuit in Fig. 5 is defined as a ϕ-section. For phase $\phi = 0$ $\bar{\phi} = 1$, the ϕ-section is in the precharge phase. The outputs of all the n- and p-dynamic blocks are precharged to "1" and "0," respectively. Also during this phase, the ϕ-section inputs are in a sampling mode, i.e, these inputs are set up.

For phase $\phi = 1$ $\bar{\phi} = 0$, the ϕ-section is in the evaluation phase. The ϕ-section inputs are held constant, and the outputs of all the dynamic blocks are evaluated as a function of the ϕ-section inputs and of the internal inputs[1]. From these output results, those which must be transferred to the next pipelined section are stored in C^2MOS latch stages.

In the circuit of Fig. 5, notice the following characteristics (see Section III).

1) Inverted and noninverted signals are provided. When direct coupling between dynamic blocks is desired, the logic function is implemented by alternating p- and n-logic blocks. If the inverter is required, a Domino like connection is employed, i.e, sequences of the same block type are used (n-

---

[1] For convenience, the inputs of a dynamic block have been separated into section inputs and internal inputs. The section inputs are set up during the precharge phase. The internal inputs are set up during the evaluation phase. For instance, in Fig. 5:

IN—section inputs
$N2, N3$—internal inputs.

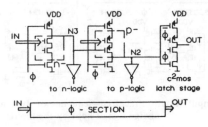

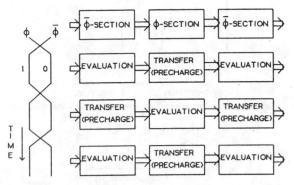

Fig. 5. NORA-CMOS pipelined circuit—ϕ-section.

Fig. 6. Pipelined system.

inverter-n or p-inverter-p). Compared with the Domino technique, this means higher logic flexibility and less transistors for the same function.

2) n-p as well as p-n sequences are possible and the sequences can be of arbitrary logical depth. Therefore, many logic levels can be operated in only half a clock period.

By interchanging ϕ and $\bar{\phi}$ in the circuit of Fig. 5, a $\bar{\phi}$-section is obtained. A sequence of ϕ- and $\bar{\phi}$-sections makes a pipelined system, as shown in Fig. 6. For phase $\phi = 0$ $\bar{\phi} = 1$, the ϕ-sections are precharged while the $\bar{\phi}$-sections are in the evaluation phase. The ϕ-section outputs are held constant by the C^2MOS latch stages. Then, for phase $\phi = 1$ $\bar{\phi} = 0$, the ϕ-sections are in the evaluation phase and the $\bar{\phi}$-sections are precharged. Now, the $\bar{\phi}$-section outputs, evaluated in the previous phase, are held constant in such a way that the ϕ-sections can use this information to compute the corresponding results. In this way, there is a complete flow of information; with the information travelling from one $\bar{\phi}$-section to the next ϕ-section, from this to the next $\bar{\phi}$-section, and so on.

III. NORA RACEFREE PROPERTIES AND LOGIC COMPOSITION RULES

In this section the racefree properties of the NORA technique will be carefully analyzed. Logic composition rules to combine dynamic, conventional, and C^2MOS function blocks will also be derived.

A. Internal Delay Racefree Property

The internal delay racefree property is defined as the capability of the dynamic block to keep its precharge signal during the delay time of the previous blocks to set up the internal inputs. It is easy to prove that a dynamic block will have the internal delay racefree property if the following conditions occur:

1) During the precharge phase, the internal inputs are set up in such a way they cut off their corresponding transistors.

2) During the evaluation phase, the internal inputs are glitch-free, i.e, these inputs can make only one transition.

From the above conditions the following results can be derived:

a) When the number of "static" inversions between two dynamic blocks is even, complementary type of logic blocks must be used for these two blocks (n-p or p-n). For instance in Fig. 5, this corresponds to alternate p- and n-logic blocks when the direct coupling between dynamic blocks is desired.

b) The same type of dynamic blocks (n-n or p-p) must be used when the number of "static" inversions is odd. In Fig. 5, this corresponds to Domino-like connections: n-inverter-n or p-inverter-p.

c) Normally, after mixing dynamic blocks with static CMOS, the circuit should be kept static up to the C^2MOS latch stage. Static functions can also be used after the C^2MOS stage. This should be done because in general "static" functions driven by dynamic blocks are not glitch-free. (Exceptions are the inverter and in some cases the NAND, NOR ....)

These logic composition rules can easily be implemented in a CAD system like Dialog [7] for automatic checking of the logic design consistency.

B. Clock Racefree Properties

As indicated in Fig. 6, to have a working pipelined system the results generated during the evaluation phase must be held constant until the end of the transfer phase. The latched information should not be altered by the precharge signal or by input variations. It will now be proven that after the evaluation phase a NORA pipelined section keeps its output results in spite of high–high or low–low clock overlaps (clock skew).

For simplicity, let us initially consider that all the circuits in the pipelined section are built only with dynamic blocks; the two exceptions being the C^2MOS latch stage and the static inverter for connecting complementary dynamic blocks. For this circuit, two possible cases should be analyzed.

Case I—Precharge Racefree

During the evaluation phase, the dynamic block which precedes the C^2MOS latch stage has its precharge signal modified by the inputs. Such a situation is indicated in Fig. 7 for an n-type and a p-type dynamic block.

As indicated in Fig. 7, the alteration of the output information is controlled by only one of the phases ϕ or $\bar{\phi}$. Therefore, these outputs are not influenced by the other phase. The outputs are, for instance, completely immune to the overlap of the phases. This kind of output latch control by only one phase (ϕ or $\bar{\phi}$) is completely different from the conventional case with transmission gates, where the output latch is controlled simultaneously by the two phases ϕ and $\bar{\phi}$. In contrast to the critical clock skew specification of the conventional transmission gates (few nanoseconds), the NORA technique imposes no restriction to the amount of clock skew.

Note: Although the NORA circuit is immune to the overlap

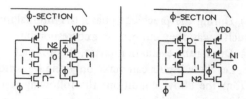

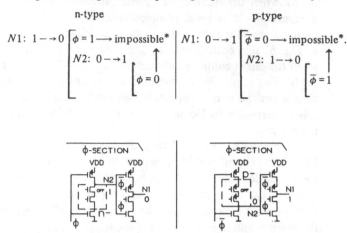

Fig. 7. Precharge racefree—precharge signal altered by the inputs:

$$
\begin{array}{cc}
\text{n-type} & \text{p-type} \\
\end{array}
$$

$N1:\ 1 \to 0 \begin{bmatrix} \phi = 1 \longrightarrow \text{impossible*} \\ N2:\ 0 \to 1 \qquad \uparrow \\ \qquad\qquad \phi = 0 \end{bmatrix}$ $N1:\ 0 \to 1 \begin{bmatrix} \overline{\phi} = 0 \longrightarrow \text{impossible*}. \\ N2:\ 1 \to 0 \qquad \uparrow \\ \qquad\qquad \overline{\phi} = 1 \end{bmatrix}$

Fig. 8. Input variation racefree—precharge signal kept by the inputs.

of the clock phases, there could still be signal races for clock signals with very slow rise and fall times (10 to 20 times the gate delay). In contrast to the clock overlap race this kind of race is easily eliminated since it does not require control of the clock skew and, also, the available time margin is much larger.

Case II—Input Variation Racefree

The other possible case, i.e, when the dynamic block keeps the precharge signal, is illustrated in Fig. 8.

If the dynamic block keeps the precharge signal, at least one of the logic transistor should be driven off. If this transistor is controlled by an internal input, the dynamic block which generates this input has also kept its precharge signal. This occurs because the internal inputs are precharged in such a way that the corresponding driven transistors are off. Therefore, there must be at least one sequence of dynamic blocks with precharge signals preserved. Fig. 9 depicts this sequence. Again, as shown in Fig. 9, the alteration of the output information is controlled by only one of the phases ϕ or $\overline{\phi}$. Therefore, they are not influenced by the overlap of the phases.

For the case being analyzed, the racefree property has been derived from the interelation between a dynamic CMOS block and a C^2MOS latch stage. Let us now show that the input variation racefree property can also be derived by the action of two C^2MOS stages: "A NORA pipelined circuit is input variation racefree if the total number of inversions (static and dynamic) between two C^2MOS latch stages is even." The proof is indicated in Fig. 10. This racefree property can also be used to solve the clock race condition of some conventional CMOS circuits. An important circuit which can be built using the above property is the shift register.

Combining the racefree properties derived from two C^2MOS functions and from C^2MOS with dynamic block, the following result can be proven.

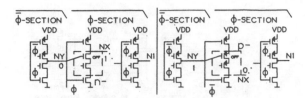

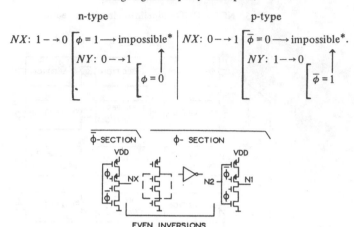

Fig. 9. Input variation racefree—sequence of dynamic blocks with precharge signals kept by the inputs:

$$
\begin{array}{cc}
\text{n-type} & \text{p-type} \\
\end{array}
$$

$NX:\ 1 \to 0 \begin{bmatrix} \phi = 1 \longrightarrow \text{impossible*} \\ NY:\ 0 \to 1 \qquad \uparrow \\ \qquad\qquad \phi = 0 \end{bmatrix}$ $NX:\ 0 \to 1 \begin{bmatrix} \overline{\phi} = 0 \longrightarrow \text{impossible*}. \\ NY:\ 1 \to 0 \qquad \uparrow \\ \qquad\qquad \overline{\phi} = 1 \end{bmatrix}$

EVEN INVERSIONS

Fig. 10. Input variation racefree—even inversions between two C^2MOS latch stages:

$$
\begin{array}{cc}
\text{"1" modification} & \text{"0" modification} \\
\end{array}
$$

$N1:\ 1 \to 0 \begin{bmatrix} \phi = 1 \longrightarrow \text{impossible*} \\ N2, NX:\ 0 \to 1 \qquad \uparrow \\ \qquad\qquad \phi = 0 \end{bmatrix}$ $N1:\ 0 \to 1 \begin{bmatrix} \overline{\phi} = 0 \longrightarrow \text{impossible*}. \\ N2, NX:\ 1 \to 0 \qquad \uparrow \\ \qquad\qquad \overline{\phi} = 1 \end{bmatrix}$

Consider a NORA pipelined section, built with dynamic, conventional, and C^2MOS function blocks. Consider all the chains of function blocks of this pipelined section, starting in a C^2MOS input stage (C^2MOS latch stage of the previous pipelined section) and ending in a C^2MOS output latch stage. The NORA pipelined section is clock racefree if, for every chain, the following conditions are satisfied:

1) Precharge racefree:

a) There is an even number of inversions between the C^2MOS output stage and the last dynamic block (see Fig. 11).

2) Input variation racefree:

b1) There is a dynamic block in such a way that there is an even number of inversions between this dynamic block and the C^2MOS input stage (see Fig. 12); or

b2) the total number of inversions between the two (input, output) C^2MOS stages is even (see Fig. 10).

If the pipelined section does not satisfy the clock race conditions, generally, circuit modifications can be easily included. By way of example, consider the nonracefree pipelined section indicated in Fig. 13(a). For this example, the following circuit modifications would eliminate the race condition:

1) conversion of one static function to dynamic function [see Fig. 13(b)];

2) conversion of one static function to C^2MOS function [see Fig. 13(c)];

3) placement of one static function after the C^2MOS latch

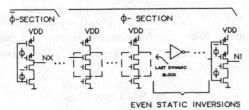

Fig. 11. Precharge racefree—even inversions between the C²MOS output stage and the last dynamic block.

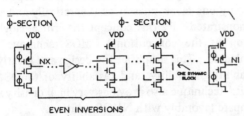

Fig. 12. Input variation racefree—even inversions between the C²MOS input stage and one dynamic block.

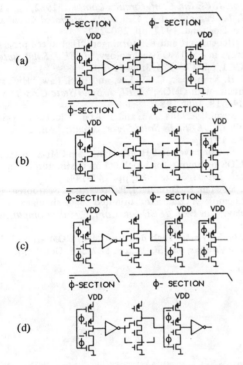

Fig. 13. Elimination of signal races. (a) Circuit with race of signals. (b) Conversion to dynamic CMOS function. (c) Conversion to C²MOS function. (d) Placement after the C²MOS output stage.

stage, provided that the racefree property of the next pipelined section would not be destroyed [see Fig. 13(d)].

IV. DYNAMIC CMOS LIMITATIONS

In this section the limitations of the NORA technique will be presented. These limitations are directly related to the dynamic storage of information and, therefore, they are common to all the dynamic techniques.

A. Charge Redistribution

The output signal of the dynamic blocks relies on storage nodes. As indicated in Fig. 14, by commutation of an OFF transistor to an ON state, a charge redistribution effect may ap-

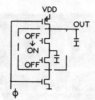

Fig. 14. Charge redistribution in dynamic blocks.

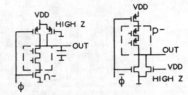

Fig. 15. Dynamic CMOS for low operating frequency.

pear between the output capacitance and the parasitic logic tree capacitances.

Normally, there will be no charge redistribution between the precharged node and the logic tree nodes controlled by section inputs. This occurs because these inputs are set up during the precharge phase and, therefore, the logic tree nodes will also be precharged. Yet, some charge redistribution effect will exist, if the precharge period after input set up is too small. This extra period of precharge generally does not result in speed limitation for the pipelined system due to the small capacitances of the logic trees.

For the internal inputs, such attenuation of the charge redistribution does not exist, since these inputs are set up only after the precharge period. In this case, the charge redistribution must be minimized by layout and by proper logic tree arrangement. The transistors driven by internal inputs must be placed as far as possible from the output storage node.

B. Leakage and Noise Margin

Another limitation of the dynamic CMOS techniques is the leakage of the storage nodes. Due to clock feedthrough, power supply variation, noise, etc., the inputs of the dynamic block can be altered from the ideal zero and V_{DD} values. Consequently, the logic transistors are driven to weak inversion. This leakage effect imposes a limit to the lowest operating frequency and to the noise margin of the circuit.

For lower frequency applications, a possible solution [8] is the addition of a high impedance transistor, as shown in Fig. 15.

V. RESULTS

In Fig. 16, a microphotograph of the chip designed to characterize the NORA technique is shown. It contains serial full adders, subtractors, shift registers, a 4-bit serial–parallel multiplier and some special structures to analyze charge redistribution, leakage, and clock feedthrough. Fig. 17 shows a NORA serial full adder containing only 32 transistors. The pipelined output sum is generated using only 20 transistors, compared with 28 if conventional CMOS is employed. The circuit area is $130 \times 318 \ \mu m^2$, giving a density of 770 transistors/mm² in a 5 μm technology, which compare favorably with an NMOS solution. The threshold levels of the n-type and p-type devices

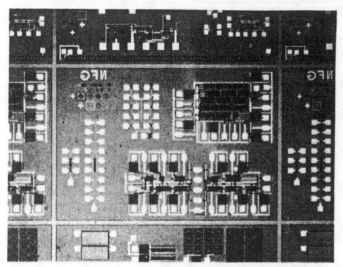

Fig. 16. Microphotograph of the chip to characterize the NORA-CMOS technique.

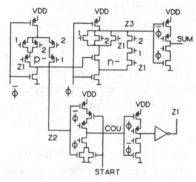

Fig. 17. NORA-CMOS serial full-adder—circuit diagram.

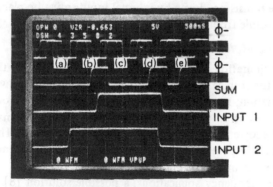

Fig. 18. NORA-CMOS serial full-adder—experimental results for very large clock skew. (a) $0 + 0 + 0$. (b) $1 + 0 + 0$. (c) $1 + 1 + 0$. (d) $1 + 1 + 1$. (e) $0 + 0 + 1$.

and the devices have been tested on the wafer probe up to 10 MHz. More careful measurements about speed and noise margin are under investigation and will be presented in a later publication.

VI. CONCLUSION

A new dynamic CMOS technique has been presented. The NORA technique provides high logic flexibility, high speed circuits, and compact chip areas. A new concept of latch control by only one clock phase was theoretically and experimentally demonstrated. By this concept the critical clock skew specification of the conventional CMOS technique is completely eliminated. This simplifies the design and greatly increases the reliability, feasibility, and testability of CMOS circuits, The NORA technique also provides very high density layouts, which compare favorably with NMOS solutions.

REFERENCES

[1] M. Shoji, "Electrical design of BELLMAC-32A microprocessor," in *Proc. IEEE Int. Conf. Circuits Comput.*, 1982, pp. 112–115.
[2] W. M. Pensey and L. Lau, *MOS Integrated Circuits.* New York: Van Nostrand, 1972, pp. 260–282.
[3] E. Hebenstreit and K. Horninger, "High-speed programmable logic arrays in ESFI SOS technology," *IEEE J. Solid-State Circuits*, vol. SC-11, pp. 370–374, June 1976.
[4] R. H. Krambeck, C. M. Lee, and H. S. Law, "High-speed compact circuits with CMOS," *IEEE J. Solid-State Circuits*, vol. SC-17, pp. 614–619, June 1982.
[5] Y. Suzuki, K. Odagawa, and T. Abe, "Clocked CMOS calculator circuitry," *IEEE J. Solid-State Circuits*, vol. SC-8, pp. 462–469, Dec. 1973.
[6] N. F. Goncalves and H. De Man, "n-p-CMOS: A racefree dynamic CMOS technique for pipelined logic structures," in *ESSCIRC Dig. Tech. Papers*, Sept. 1982, pp. 141–144.
[7] H. De Man, D. Dumlugol, P. Stevens, G. Schrooten, and I. Bolsens, "Logmos: A transistor oriented logic simulator with assignable delays," in *Proc. IEEE Int. Conf. Circuits Comput.*, 1982, pp. 42–45.
[8] R. G. Stewart, "High density CMOS ROM arrays," *IEEE J. Solid-State Circuits*, vol. SC-12, pp. 502–506, Oct. 1977.

are $+1$ V and -1 V, respectively. Fig. 18 shows experimental results for very large clock skew = 150 ns at 1 MHz clock frequency, without disturbing the circuit operation. Notice that during the evaluation phase $\phi = 1$ $\bar{\phi} = 0$, the results are obtained and then hold constant until the end of the transfer phase $\phi = 0$ $\bar{\phi} = 1$. Also from experimental results, the minimum working frequency was less than 1 kHz at room temperature, indicating that the current leakage due to weak inversion is not a critical limitation. The measured power disssipation of the serial full-adder was 17 μW/MHz at a supply voltage of 5 V. The circuits were designed for a maximum operating frequency of 14 MHz,

Cascode Voltage Switch Logic: A Differential CMOS Logic Family

LAWRENCE G. HELLER, WILLIAM R. GRIFFIN

IBM GENERAL TECHNOLOGY DIVISION, ESSEX JUNCTION, VT

JAMES W. DAVIS, NANDOR G. THOMA

IBM SYSTEM PRODUCTS DIVISION, BOCA RATON, FL

IMPORTANT CRITERIA for choosing a suitable VLSI logic family include power, delay, logic circuit density, device/process complexity, and compatibility with design automation tools. This paper will describe a differential CMOS logic family — Cascode Voltage Switch Logic (CVSL).

Logic design leverage is achieved in CVSL by cascoding differential pairs of MOS devices into powerful combinational logic tree networks capable of processing complex Boolean logic functions within a single circuit delay. Logic trees with N-high cascoding of differential pairs of NMOS devices are capable of processing Boolean functions with up to (2^N-1) input variables. CVSL has been found to offer a performance advantage of up to 4X compared to CMOS/NMOS primitive NAND/NOR logic families, while maintaining the expected low power characteristics of CMOS circuitry. Potentially, CVSL is twice as dense as primitive NAND/NOR logic, and is compatible with existing design automation tools. Combinational logic trees can be designed in cascoded high-peformance NMOS devices with unstacked PMOS devices used sparingly as pull-up devices in load and buffer circuitry. Optimization of the PMOS devices and the criticality of the PMOS to NMOS spacing can therefore be relaxed, relieving the device/process complexity burden for CVSL designs.

The CVSL circuit concept, in its differential form, is illustrated in Figure 1. Depending on the differential inputs, either node N1 or node N2 is pulled down by the NMOS combinational logic tree network. Regenerative action sets the PMOS latch to static outputs $Q, \overline{Q}$ of full differential V_H and ground logic levels. The logic trees are free of direct current after the latch sets. Since the inputs drive only the NMOS tree devices, input gate capacitance loading is typically a factor of 3X smaller than CMOS circuits that require complementary N-channel and P-channel devices to be driven.

The logic trees networks can be designed automatically using existing logic minimization algorithms[1]. An example of the efficiency of a differential CVSL circuit, requiring 12 devices, is shown in Figure 2. Device redundancy is naturally reduced by the functional power of the differential logic trees. Implementation of the same Boolean function in primitive CMOS NAND gates, as indicated in Figure 3, required 5 NAND gates and 28 devices, not counting the additional inverters to provide the complementary inputs. It will be noted that performance leverage in CVSL is enhanced by a reduction in the number of circuit delays compared to primitive logic. The Boolean function Q can also be implemented with 16 devices in a cascoded fully CMOS circuit.

The differential version of Figure 2, however, requires 6 fewer large P-channel devices and has considerably less input capacitance.

An experimental masterslice chip, personalized at the metal and contact levels, has been designed. The chip contains 10,880 *brickwalled* NMOS differential pairs, forming 1088 CVSL trees. The image design is compatible with existing automatic placement[2] and wiring algorithms. A photomicrograph of the chip, implemented in a $2\mu m$ CMOS technology, is shown in Figure 4. The macro in the upper right-hand corner contains 150 CVSL logic trees, which were designed top-down and automatically placed and wired from a high level language description. Other experiments include a 4b carry look-ahead ALU and several ring oscillators. Delays in the ring oscillators loaded with 0.3pF of wiring capacitance were measured in the 1 to 2ns range.

Performance of the circuit of Figure 1 is limited by the set time of the PMOS latch. A high-performance clocked CVSL circuit is illustrated in Figure 5. The outputs $Q, \overline{Q}$ are precharged low when the clock phase PC is low and data are propagated in a *domino* mode[3] when PC goes high. Feedback devices T1, T2 hold the internal nodes N1, N2 statically high prior to switching within the logic tree. The feedback devices reduce charge sharing noise within the tree and improve the noise margin, with only a small sacrifice in performance. During switching either N1 or N2 is pulled down and either device T1 or T2 is shut off. No direct current flows after switching. The logic invert function is implicit in this clocked differential CVSL circuit, a clear advantage over other incomplete domino type logic families. All logic functions can be implemented, including, for example, the XOR. A 4-way clocked XOR is illustrated in Figure 6.

Acknowledgments

The authors would like to acknowledge the work of R.S. Chelemer and J.J. Preli for their automatic placement and wiring support, the logic synthesis effort of R.D. Kilmoyer, and many others for helpful discussions.

[1] Brayton, R.K. and McMullen, C., "Decomposition and Factorization of Boolean Expressions," *Proc. IEEE ISCA*, Rome, Italy; May, 1982.

[2] Kirkpatrick, S., Gelatt, Jr., C.D. and Vecchi, N.P., "Optimization by Simulated Annealing", *Science*; May 13, 1983.

[3] Krambeck, R.H., Lee, C.M. and Law, H.S., "High-Speed Compact Circuits with CMOS," *IEEE J. Solid State Circuits*, Vol. SC-17, No. 3; June, 1982.

Reprinted from *Digest of Technical Papers, IEEE International Solid-State Circuits Conference*, pp. 16-17, February 1984.

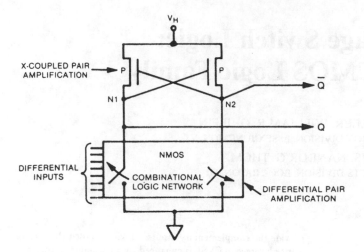

FIGURE 1—Basic CVSL circuit.

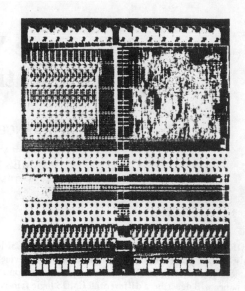

FIGURE 4—Photomicrograph of masterslice chip.

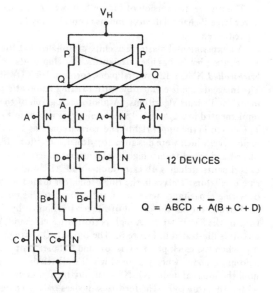

12 DEVICES

$$Q = A\overline{B}\,\overline{C}\,\overline{D} + \overline{A}(B + C + D)$$

FIGURE 2—CVSL implementation of Q.

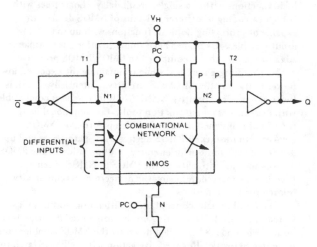

FIGURE 5—Clock CVSL.

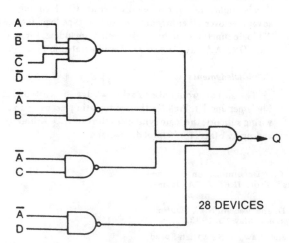

28 DEVICES

FIGURE 3—CMOS NAND implementation of Q.

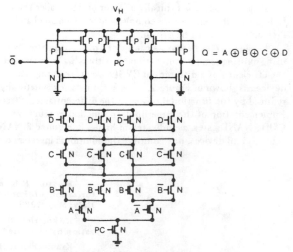

$$Q = A \oplus B \oplus C \oplus D$$

FIGURE 6—Clocked CVSL 4-way XOR.

Differential Split-Level CMOS Logic for Subnanosecond Speeds

LEO C. M. G. PFENNINGS, WIM G. J. MOL, JOSEPH J. J. BASTIAENS, AND JAN M. F. VAN DIJK

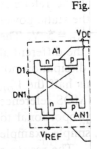

Abstract —Subnanosecond gate delays (0.8 n) have been measured on complex logic gates (e.g., sum functions of a full adder) designed in the differential split-level (DSL) CMOS circuit technique. This high speed has been achieved by reducing the logic swing (2.4 V) on interconnect lines between logic gates, by using current controlled cascoded cross-coupled NMOS–PMOS loads, by using combined open NMOS drains as outputs, and by employing shorter channel lengths ($L_{eff} = 1 \mu$m) for the NMOS devices in the logic trees with reduced maximum drain–source voltages to avoid reliability problems. Extra ion implantation protects these transistors from punchthrough.

I. INTRODUCTION

THE GENERAL rule: "Enhancement Depletion NMOS is faster than CMOS" was a challenge to explore speed improvements in CMOS circuit techniques. Differential split-level (DSL) [1] CMOS logic makes a compromise between the static power dissipation of E/D NMOS and the dynamic power dissipation of CMOS.

This paper will describe the switching behavior and the speed improvements owing to the DSL circuit technique. This circuit technique shows similarity with differential cascode voltage switch logic (CVSL) [2] but the electrical behavior of the DSL circuit technique is essentially different.

The DSL circuit technique was implemented in a double-metal 2.5-μm CMOS n-well process with conventional 2.5-μm projection lithography, except for the polysilicon. In this mask only the gate length of the n-channel transistors in the logic trees is decreased to 1.5 μm ($L_{eff} = 1$ μm), while the other polysilicon details and pitches remain constant. Therefore a stepper exposure is used for the polysilicon definition. An extra deep boron implant protects the short-channel NMOS transistors from punchthrough.

DSL incorporates a reduced maximum drain–source voltage $V_{DS \, max}$ in the logic trees resulting in less hot-electron-induced degradation of these devices. Decreasing $V_{DS \, max}$ reduces the lateral electric field near the drain edge of these NMOS transistors [3], [4]. This allows the use of shorter channel lengths without lifetime reduction.

Manuscript received March 28, 1985; revised May 28, 1985.
The authors are with Philips Research Laboratories, Prof. Holstlaan, P.O. Box 80.000, 5600 JA Eindhoven, The Netherlands.

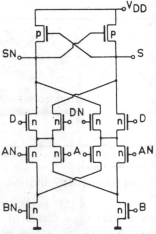

Fig. 1. Sum part of a differential CVSL full adder.

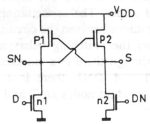

Fig. 2. Differential CVSL basic circuit.

II. SHORT SUMMARY OF CVSL

Fig. 1 shows the circuit diagram of the sum function of a full adder designed in differential cascode voltage switch logic. The cross-coupled PMOS loads and the differential logic NMOS trees are typical for this circuit technique. To explain the switching behavior of the CVSL circuit technique we replace the differential logic NMOS trees by two NMOS transistors as shown in Fig. 2.

Now suppose we switch input D from a low to a high level, starting with input D low and input DN high. Then node SN is at a high level of V_{DD} and node S at a low level of 0 V so PMOS $P1$ is on and $P2$ is off. If we now switch the inputs D and DN then NMOS $N1$ turns on and $N2$ turns off. This is ratioed logic because transistor $N1$ has to discharge node SN, while $P1$ is still on. $P1$ switches off, after $P2$ has switched on and node S has reached a high level. So during switching both $N1$ and $P1$ (or $N2$ and $P2$ depending on the input transition) conduct, causing relatively large current spikes and additional delay.

Reprinted from *IEEE Journal of Solid-State Circuits*, Vol. SC-20, No. 5, pp. 1050-1055, October 1985.

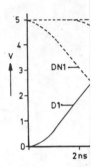

Fig. 3

```
5
4
V
  3    DN1
  2    D1
  1
  0
       2ns
```

Fig. 4. Swi

Fig. 3 show
circuits. The
segments repr
work, i.e., ful
parasitic capac
Fig. 4 sho
shows that al
the inputs D
complementar
D3, respective

III. D

The DSL p
transistors N
part and the
by a referenc
plus the thre
guarantee opt
of the PMOS
F and FN in
switch input
input D low
of V_{DD} and
determines th
F has a low
weakly on. T
transistor N1
$V_{DD.}$ for nod
for node SN
NMOS N1 t
The half
discharged a

ACKNOWLEDGMENT

The authors would like to thank A. T. Van Zanten and H. J. M. Veendrick for their contributions to this subject.

REFERENCES

[1] L. C. M. G. Pfennings, W. G. J. Mol, J. J. J. Bastiaens, and J. M. F. Van Dijk, "Differential split-level CMOS for sub-nanosecond speeds," in *ISSCC Dig. Tech. Pap.*, vol. XXVIII, 1985, pp. 212–213, 351.
[2] L. G. Heller and J. W. Davis, "Cascode voltage switch logic," in *ISSCC Dig. Tech. Pap.*, vol. XXVII, 1984, pp. 16, 17.
[3] C. Hu, "Hot-electron effects in MOSFETs," in *Tech. Dig. IEDM*, 1983, pp. 176–181.
[4] C. Hu, S. C. Tam, F.-C. Hsu, P.-K KD, T.-Y. Chan, and K. W. Terrill, "Hot electron-induced MOSFET degradation-model, monitor, and improvement," *IEEE Trans. Electron Devices*, vol. ED-32, p. 375, 1985.
[5] F. M. Klaassen, "Design and performance of micro-size devices," *Solid-State Electron.* vol. 21, pp. 565–571, 1978.
[6] B. Eitan and D. Frohman-Bentchkowsky, "Surface conduction in short channel MOS devices as a limitation to VLSI scaling," *IEEE Trans. Electron Devices*, vol. ED-29, p. 154, 1982.
[7] ——, "Hot electron injection into the oxide in n-channel MOS devices," *IEEE Trans. Electron Devices*, vol. ED-28, p. 328, 1981.

A Comparison of CMOS Circuit Techniques: Differential Cascode Voltage Switch Logic Versus Conventional Logic

KAN M. CHU AND DAVID L. PULFREY, MEMBER, IEEE

Abstract —Differential cascode voltage switch (DCVS) logic is a CMOS circuit technique which has potential advantages over conventional NAND/NOR logic in terms of circuit delay, layout density, power dissipation, and logic flexibility. In this paper a detailed comparison of DCVS logic and conventional logic is carried out by simulation, using SPICE, of the performance of full adders designed using the different circuit techniques. Specifically, comparisons are made between a static full CMOS design and two different implementations of static DCVS circuits, and, in the dynamic case, between two conventional NORA implementations and DCVS forms of both NORA and DOMINO logic. The parameters compared are: input gate capacitance, number of transistors required, propagation delay time, and average power dissipation. In the static case, DCVS appears to be superior to full CMOS in regards to input capacitance and device count but inferior in regards to power dissipation. The speeds of the two technologies are similar. In the dynamic case, DCVS can be faster than more conventional CMOS dynamic logic, but only at the expense of increased device count and power dissipation.

I. INTRODUCTION

DIFFERENTIAL cascode voltage switch (DCVS) logic is a recently proposed CMOS circuit technique which is claimed to have advantages over traditional NAND/NOR circuit techniques in terms of circuit delay, power dissipation, layout area, and logic flexibility [1]. DCVS also has an inherent self-testing property which can provide coverage of both stuck-at and dynamic faults [2]. A further attraction of DCVS circuits is the fact that they can be readily designed using straightforward procedures based on Karnaugh maps (K-maps) and tabular methods [3].

All these worthwhile features would appear to make DCVS logic a very promising CMOS circuit technique. To investigate this possibility, we have compared DCVS logic and more conventional CMOS logic forms using the full-adder circuit as a test vehicle. The full adder is suited to this purpose as it is a common, yet reasonably complex, building block in digital circuits. The comparison reported here uses SPICE simulations to assess the performance parameters of area, input loading, speed, and power dissipation. Area is represented by the number of transistors needed to implement the adder and loading is quantified

Manuscript received August 29, 1986; revised January 26, 1987. This work was supported by the Natural Sciences and Engineering Research Council of Canada.

The authors are with the Electrical Engineering Department, University of British Columbia, Vancouver, B.C. V6T 1W5, Canada.

IEEE Log Number 8714872.

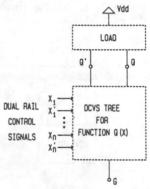

Fig. 1. Block diagram of a DCVS circuit. The load circuitry is connected to nodes Q and Q'.

in terms of the input gate capacitance. Speed is assessed by simulating the worst-case propagation time. Power dissipation is computed at the maximum frequency of operation of each circuit.

II. CIRCUIT TECHNIQUES FOR DCVS LOGIC

The basic DCVS circuit comprises two parts: a binary decision tree and a load (see Fig. 1). The tree is specified such that:

1) when the input vector $x = (x_1, \cdots, x_n)$ is the true vector of the switching function $Q(x)$, then the output Q is disconnected from node G and the node Q' is connected to G; and
2) when $x = (x_1, \cdots, x_n)$ is the false vector of $Q(x)$, then the reverse holds.

There are two trees required to implement a full adder, one to perform the sum and one to perform the carry function (see Fig. 2). These circuits, which were designed using the K-map procedure described in [3], are used as the tree circuits for all the DCVS circuit forms examined in this paper. The various DCVS forms differ in their load circuitry, as is now described.

The load for a static DCVS circuit is the simple latch shown in Fig. 3. Depending on the differential inputs, either node Q or Q' is pulled down by the DCVS tree network. Regenerative action sets the PMOS latch to static outputs Q and Q' of V_{DD} and ground or vice versa. The

Reprinted from *IEEE Journal of Solid-State Circuits*, Vol. SC-22, No. 4, pp. 528-532, August 1987.

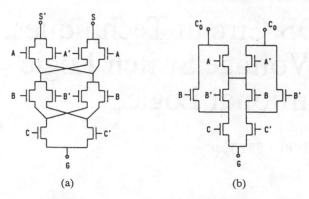

(a)   (b)

Fig. 2.   The DCVS trees for a full adder. (a) The circuit providing the sum, $S(A, B, C) = A + B + C$. (b) The circuit yielding the carry, $C_o(A, B, C) = AB + BC + CA$.

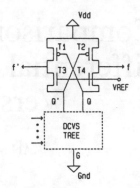

Fig. 4.   The load for a static DSL circuit.

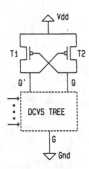

Fig. 3.   The load for a static DCVS circuit.

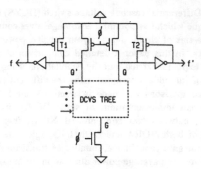

Fig. 5.   The load and circuit arrangement for a DCVS DOMINO circuit.

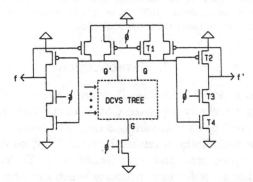

Fig. 6.   The load and circuit arrangement for a DCVS NORA pipelined section.

logic trees do not pass any direct current after the latch sets.

A variation of this static DCVS circuit is the differential split-level (DSL) logic circuit [4] shown in Fig. 4. Two n-transistors $T3$ and $T4$ with their gates connected to a reference voltage V_{REF} are added to reduce the logic swing at nodes Q and Q'. If V_{REF} is set to $V_{DD}/2 + V_{th}$, where V_{th} is the threshold voltage of the n device, then the nodes Q and Q' are clamped at $V_{DD}/2$. Suppose node Q is pulled down from 2.5 V (i.e., assume $V_{DD} = 5$ V) to a low level. $T1$ switches from its low-current state to its high-current drive state very quickly, because $T4$ is initially OFF. The voltage on node f' goes up to 5 V because $T1$ is fully ON. Node Q' is raised up to 2.5 V until $T3$ is in the cutoff mode. DSL circuits would be expected to be about two times faster than standard DCVS circuits on account of the need for logic swings of only half the rail-to-rail voltage difference. This should result in a reduction by two times of the charges needed to be manipulated in the circuit.

Turning now to dynamic operation of DCVS circuits, consider first the DOMINO [5] configuration of Fig. 5 [1]. Nodes Q and Q' are precharged to high during the precharge phase ($\phi = 0$) and either node Q (node f) or Q' (f') discharges to low during the evaluation phase ($\phi = 1$). Transistor $T1$ (or $T2$) is a high impedance p transistor which serves as the feedback device to maintain the high logic level at node Q' (or Q), where charges may be lost due to charge sharing [6].

For dynamic operation of pipelined architectures, NORA (NO RACE) techniques [7] are suitable for imple-

menting logical functions. In its original form the NORA structure consists of n- and p-logic gates to enhance logic flexibility. The p-logic gates usually cause long delay times and consume large areas. Using DCVS logic in the NORA technique will eliminate p-logic gates because of the inherent availability of complementary signals. The general structure of a DCVS NORA pipelined section consisting of only one dynamic gate is shown in Fig. 6. This type of circuit technique is suitable for use in a heavily pipelined logic design, as in the case, for example, of a newly developed 8×8 pipelined multiplier [8].

As Fig. 6 indicates, the load circuitry is symmetrical, and thus, for analysis purposes, only one side of it need be considered. During the evaluation phase ($\phi = 1$), node Q is either floating or discharged depending on the inputs. The output register acts as a clocked inverter, and the output can be either high or low. During the precharge phase ($\phi = 0$), the ground path of the register is blocked. If the output resulting from the previous evaluation is high, then

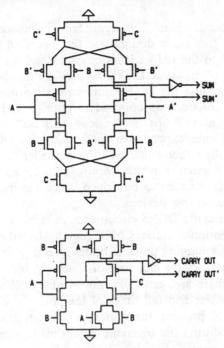

Fig. 7. Circuits for a static CMOS full adder.

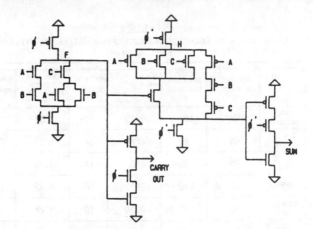

Fig. 8. Circuit for a conventional NORA full adder.

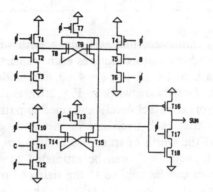

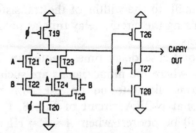

Fig. 9. Circuit for a modified NORA full adder (from [9]).

the output continues to be high regardless of the voltage of Q. If the output is low (i.e., node Q has never been discharged) and transistor $T1$ is ON, then the output continues to be low because no charges can be added through $T2$. Thus for a ϕ section of a pipeline, the output changes freely when ϕ is high and is latched at the falling edge of ϕ.

III. CONVENTIONAL CMOS CIRCUIT TECHNIQUES

To provide a basis for comparison of the DCVS circuits described in Section II, conventional CMOS designs operating under static and dynamic conditions need to be considered.

The circuit used here for a static CMOS full adder is shown in Fig. 7. Two subcircuits are identified, one to generate the sum signal and one to generate the carry out signal. The three-way EXCLUSIVE-OR gate in the sum circuit has the highest stack level and largest parasitic capacitance, and thus determines the worst-case delay time of the adder. This circuit is relatively fast compared to other possible static full CMOS implementations because the complemented outputs are obtained through only one gate delay from the complementary inputs.

Two versions of conventional approaches to dynamic CMOS full-adder design were studied. One, a conventional NORA adder with serial n- and p-logic blocks, is shown in Fig. 8. The other circuit, a modified NORA adder [9], is shown in Fig. 9. It contains a special three-way XOR gate to generate the sum signal.

IV. COMPARISON OF THE FULL ADDERS

To compare the performance of the various forms of full adders, each of the circuits described in Sections II and III was simulated using SPICE. The conventional CMOS cir-

cuits simulated were those shown in Figs. 7–9. Four different DCVS circuits, two static and two dynamic, were generated by connecting the full-adder tree of Fig. 2 to the load circuits shown in Figs. 3–6. The DCVS circuits were laid out on a Metheus $\lambda700$ workstation in accordance with design rules for the single-metal 3-μm CMOS process of Northern Telecom, Ottawa, Canada [10]. The areas occupied by the DCVS circuits were about 2.2×10^{-4} and 3.5×10^{-4} cm^2 for the static and dynamic versions, respectively. The results of SPICE simulations from the schematics of all the circuits are summarized in Table I.

The input gate capacitance gives a measure of the input loading of the circuit. This parameter is, for the case of transistors of fixed length (3 μm in this case), determined by the number of transistors and their widths. A general guideline used in the first iteration of a design was to size the transistors in a tree network such that the equivalent conductance of any single discharging path was the same as the conductance of a minimum-size ($W = 3$ μm in our

41

TABLE I

TABLE I
Comparison of Simulation Results for Different Types of
Full Adders

PROPERTY / CIRCUIT TECHNIQUE	INPUT GATE CAPACITANCE (fF)	OUTPUT LOAD CAPACITANCE (fF)	# OF P-DEVICES / # OF N-DEVICES	WORST CASE DELAY TIME (ns)	AVERAGE POWER DISSIPATION AT MAX. FREQ. (mW)	NORMALIZED POWER-DELAY PRODUCT
STATIC FULL CMOS	155	155	15/15	20	0.58	1.00
STATIC DCVS	85	85	4/18	22	1.11	2.11
STATIC DSL	85	85	4/22	14	1.35	1.63
NORA	110	220	12/10	18	0.83	1.29
MODIFIED NORA	45	90	8/20	10	1.24	1.06
DCVS NORA	85	170	12/28	10	1.55	1.34
DCVS DOMINO	85	170	12/24	9	1.75	1.36

case) n transistor. For example, a path with four serially connected transistors requires each transistor contained in that path to be 12 μm ($= 4 \times 3$ μm) wide. Consider the right half of the network in Fig. 2(b); that the number of transistors (or stack level) contained in path $A'BC'$ is three implies that each transistor in this path should be 9 μm wide. If the width of transistor C' is 9 μm, then the width of B' in path $B'C'$ can be estimated as 4.5 μm. Similar principles can be applied to the sizing of transistors in the charging or discharging paths in other circuits. The final form of a design was arrived at by making adjustments, usually small, to the widths of the transistors on the basis of minimizing the circuit delay time as predicted by SPICE simulations.

The worst-case delay times quoted in Table I refer to situations where the input signals are such that the circuit operation is likely to be slowest. For example, in the conventional NORA circuit of Fig. 8, the speed performance will be poorest when $A = B = $ HI and $C = $ LO. In this case, during the evaluation phase ($\phi = 1$), node F needs to be pulled down, in order to turn on the p-channel transistor through which node H, via transistor C, is connected to the output stage to render the sum signal LO.

Power dissipation was computed using the procedure described by Kang [11]. The figures quoted in Table I refer to average power dissipation at the maximum frequency of operation of each circuit, i.e., as determined by the worst-case delay times. The power-delay product, normalized to the static full CMOS case, is also shown in Table I.

The output load capacitances used in the simulations are meant to represent typical load conditions. A fan-out of two was used for the dynamic designs as these circuits are buffered and would be expected to be able to drive larger loads than the static gates.

V. Discussion

Considering, first, the static designs, it appears that the DSL technique yields a significantly faster circuit than do the other two techniques. This is to be expected due to the need for logic swings which are only one-half of the rail-to-rail value. The significant differences between the other two static designs are the increased power dissipation and the reduced device count and input gate capacitance of the static DCVS circuit. The number of devices is less because the DCVS implementation uses only p-channel transistors, as opposed to both p- and n-channel devices, as pull-ups in the load and buffer circuitry. The input gate capacitance loading in the DCVS circuit is typically a factor of 2 or 3 times smaller than conventional CMOS circuits which require complementary n- and p-channel devices to be driven, since the inputs drive only n-channel tree devices.

The static DCVS circuit consumes more power than the conventional static CMOS circuit because the charging and discharging times of nodes Q and Q' in Fig. 3 depend on the turn-on and turn-off paths within the DCVS tree and these are, generally, not symmetrical. An asymmetry in the rise and fall times of the potential at nodes Q and Q' will prolong the period of current flow through the latch during the transient state, thus increasing the power dissipation.

The apparent attractiveness of the static DSL circuit in regards to speed is negated somewhat by three possible problems which may arise when using this technique. For example, with reference to Fig. 4, if node Q' is at 2.5 V, then $T2$ is partially ON and it is possible to destroy the low logic level that would otherwise have appeared on node f. Although reducing the size of the p device alleviates this problem, it decreases the output drive capability and results in longer delay. Thus a trade-off should be considered when the sizes of $T1$ and $T2$ are chosen. Another problem is due to the body effect existing in $T3$ and $T4$. Although the threshold voltage V_{th} is equal to 0.8 V in the Northern Telecom 3-μm CMOS process [10], SPICE simulations show that it is necessary to set V_{REF} equal to 4.2 V in order to clamp either of the nodes Q or Q' to 2.5 V. Also the clamped logic swing is sensitive to the stack level of the DCVS tree for a fixed V_{REF}. The third problem is that this circuit exhibits static power dissipation. There is a direct current path to ground through transistors $T1$ and $T3$ when Q' is low or through $T2$ and $T4$ when Q is low.

Turning now to the dynamic circuits, all the designs have a similar power-delay product. The DCVS circuits appear to have a speed advantage, but this is achieved at the expense of an increased device count. The conventional NORA circuit, Fig. 8, is characterized by a large input gate capacitance due to the wide transistors in the p-logic block, and a slow speed due to the use of two levels of gate delay and because half of the logic is performed by p transistors. Considerable improvement in these two areas is achieved by the modified NORA adder of Fig. 9. This circuit has two times smaller input gate capacitance and is nearly twice as fast as the serial NORA adder. The disadvantage of this circuit is that accidental discharge due to races is possible under certain conditions. For example, if $A = 0$, $B = 1$, and $C = 1$, the gate of $T14$ (or source of $T15$) and the gate of $T15$ (or source of $T14$) are pulled down. If the drain nodes of $T7$ and $T10$ do not pull down at similar rates so that a voltage difference of more than one threshold is developed across the gate nodes of $T14$ and $T15$, the drain node of $T13$ discharges accidentally. To avoid this requires careful sizing of the transistors along the discharging paths so that the conductance to ground and the capacitive load associated with each of the pull-down paths is equal. Tight process control and detailed simulation through circuit extraction are needed if this circuit is to be successfully implemented.

The DCVS NORA (Fig. 6) adder has smaller input gate capacitance and delay time than the conventional NORA adder, although the area consumed is larger. The large area stems from the symmetrical buffer circuits used to provide complementary outputs. The DCVS NORA circuit is as fast as the modified NORA adder, but only at the expense of a higher device count and increased input gate capacitance. However, the DCVS version of NORA is superior to the modified conventional version, in terms of circuit flexibility, due to its provision of complementary outputs, and reliability, due to the fact that accidental discharge cannot occur. The DCVS DOMINO (Fig. 5) adder is similar to the DCVS NORA adder in all the parameters evaluated in this comparison. It is the only kind of full-adder circuit which can be included in a DOMINO chain without causing race problems.

VI. Conclusions

The main conclusion to be drawn from this work is that DCVS logic offers opportunities for realizing faster circuits than are possible with conventional forms of CMOS logic, but this speed advantage is often gained at the expense of circuit area and active power consumption.

The fastest static logic technique investigated was the differential split-level (DSL) version of DCVS logic. The worst-case delay time for this implementation was 14 ns, while that of a conventional CMOS circuit was 20 ns. However, DSL may have some problems in terms of static power dissipation, security of charge storage, and sensitivity of the logic swing to the number of input signals.

In dynamic operation, DCVS versions of NORA and DOMINO circuits appear to be a few nanoseconds faster (9–10 versus 10–18) than their conventional counterparts. Further, DCVS logic may overcome the problem of accidental discharge, which appears to be a concern with one of the conventional NORA techniques evaluated in this study.

References

[1] L. G. Heller, W. R. Griffin, J. W. Davis, and N. G. Thoma, "Cascode voltage switch logic: A differential CMOS logic family," in *ISSCC Dig. Tech. Papers*, 1984, pp. 16–17.

[2] R. K. Montoye, "Testing scheme for differential cascode voltage switch circuits," *IBM Tech. Disc. Bull.*, vol. 27, pp. 6148–6152, 1985.

[3] K. M. Chu and D. L. Pulfrey, "Design procedures for differential cascode voltage switch circuits," *IEEE J. Solid-State Circuits*, vol. SC-21, pp. 1082–1087, Dec. 1986.

[4] L. C. Pfennings, W. G. J. Mol, J. J. J. Bastiaens, and J. M. F. Van Dijk, "Differential split-level CMOS logic for subnanosecond speeds," in *ISSCC Dig. Tech. Papers*, 1985, pp. 212–213; also *IEEE J. Solid-State Circuits*, vol. SC-20, pp. 1050–1055, Oct. 1985.

[5] R. H. Krambeck, C. M. Lee, and H. Law, "High-speed compact circuits with CMOS," *IEEE J. Solid-State Circuits*, vol. SC-17, pp. 614–619, June 1982.

[6] L. G. Heller, "Stabilizing cascode voltage switch logic," *IBM Tech. Disc. Bull.*, vol. 27, p. 6015, 1985.

[7] N. F. Goncalves and H. J. De Man, "NORA: A racefree dynamic CMOS technique for pipelined logic structure," *IEEE J. Solid-State Circuits*, vol. SC-18, pp. 261–266, June 1983.

[8] K. M. Chu, "Cascode voltage switch logic circuits," M.A.Sc. thesis, Univ. of British Columbia, Vancouver, Canada, 1986.

[9] A. H. C. Park, "CMOS LSI design of a high-throughput digital filter," M.Sc. thesis, Mass. Inst. of Technol., Cambridge, ch. 4, 1984.

[10] G. Puukila, "Canadian Microelectronics Corporation guide for designers using the Northern Telecom CMOS3 Process," Canadian Microelectronics Corp., Kingston, Ont., Rep. IC 85-6, 1985.

[11] S. M. Kang, "Accurate simulation of power dissipation in VLSI circuits," *IEEE J. Solid-State Circuits*, vol. SC-21, pp. 889–891, Oct. 1986.

Pass-transistor networks optimize n-MOS logic

Formal methods for transfer-gate logic design achieve minimum area, delay, and power for complex circuits

by Sterling Whitaker, *American Microsystems Inc., Santa Clara, Calif.*

☐ Conventional methods of designing logic chips employ blocks of discrete and small-scale integrated circuits—an approach that is now crumbling under the impact of very large-scale integration. Designers who set out to create high-performance cost-effective VLSI chips must minimize the power, delay, and area of MOS ICs. But traditional logic design, with its black-box representation of Boolean functions, does not shrink them.

However, these three parameters can be minimized by experimenting with the many combinatorial logic circuits, transfer gates, and MOS pass transistors that affect an IC's power consumption, speed, and size. Systematically designed pass-transistor networks can reduce complex functions to highly regular structures that operate more quickly than conventional n-channel MOS logic, fill only one third as much space, and consume only one eighth as much power.

Designing pass-transistor networks has been more a craft than a science. Early masters—Carver Mead and Lynn Conway among them—popularized pass transistors in latches, flip-flops, multiplexers, and some combinatorial logic structures [*Electronics*, Oct. 20, 1981, p. 102]. The more formal design techniques presented here extend the benefits of pass transistors to networks too complex to be realized intuitively, and they also permit informally designed pass networks to be verified in a systematic way.

The basics

Figure 1 shows the pass gate and its logical function. When control inputs are high the transistor conducts, passing the logic level at its input to its output. When the control input is low the transistor is off, leaving the output in a high-impedance (or undefined) state. Designers form pass-gate networks (see Fig. 1c) by joining together the outputs of pass transistors to feed the inputs of succeeding transistors.

A high signal level at a pass transistor's input is reduced by a threshold voltage at the output. This reduced level is passed, without further degradation, through additional pass transistors. As a result of this reduced signal, conventional buffers, not pass-gate outputs, drive control inputs.

Consider an example of formal pass-network design: an exclusive-NOR gate (Fig. 2). Added to an ordinary truth table is a column that indicates, for each input state, which input variables (or complements) can be passed to the output to get a desired function. When

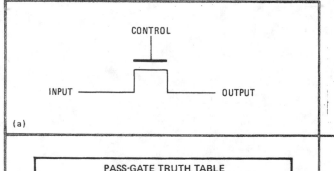

(a)

PASS-GATE TRUTH TABLE

INPUT	CONTROL	OUTPUT
0	0	UNDEFINED
1	0	UNDEFINED
0	1	0
1	1	1

(b)

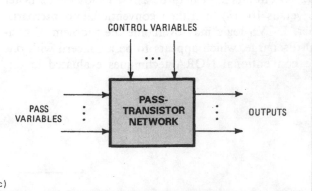

(c)

1. Pass gate. When an MOS transistor is operated as a pass gate, as shown in (a), the device passes the signal at its drain to its source. As the truth table given in (b) demonstrates, the output is in a high-impedance, or undefined, state when the transistor is shut off. Networks of pass gates make pass and control variables flow at right angles (c).

Reprinted with permission from *Electronics* - S. Whitaker, "Pass-transistor networks optimize n-MOS logic," Vol. 56, No. 19, pp. 144-148, September 1983. © 1983 Penton Publishing.

EXCLUSIVE NOR TRUTH TABLE			
A	B	OUTPUT	PASS FUNCTION
0	0	1	$\overline{A} + \overline{B}$
0	1	0	$A + \overline{B}$
1	0	0	$\overline{A} + B$
1	1	1	$A + B$

(a)

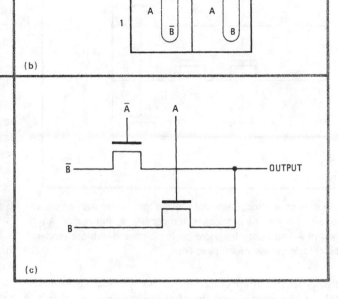

(b)

(c)

2. Straightforward logic. With the aid of a truth table (a) that includes pass functions, it is possible to modify conventional logic design for pass transistors. The pass functions are entered in a Karnaugh map (b) with pass variables looped together. The resulting exclusive-NOR gate needs just two devices (c).

both A and B equal 0 the output should be 1. Either $\overline{A}$ or $\overline{B}$ can be passed to the output. $\overline{A} + \overline{B}$ is called the pass function.

For each state of the input variables, pass functions are then entered into corresponding cells of a conventional Karnaugh map (Fig. 2b). To steer the pass variables to the output, they are grouped to produce the control functions that drive the pass-gate control inputs. The left loop in Fig. 2b groups $\overline{B}$ under the control of $\overline{A}$; the other one groups B under the control of A.

Figure 2c shows the resulting pass-transistor network, which has one pass-transistor delay and comprises two devices. The network's steady-state power dissipation—consisting only of leakage currents between the substrate and the transistors' source and drain regions—is negligible. Figure 3 shows the other basic logic functions—AND, NAND, OR, NOR, and exclusive-OR—that can be derived in the manner of exclusive-NOR.

Contrast this simplicity with the traditional n-MOS version, which comprises five transistors, is plagued by pull-down or pull-up delays, and has one node that dissipates steady-state power. Complementary-MOS gates do not dissipate steady-state power, but they do comprise eight transistors and have a pull-up or -down delay.

Synthesizing pass networks

Logic functions that are more complex can be approached as the simple gates are: with a truth table that includes the pass functions and with a modified Karnaugh map to derive control functions. The control-function groupings in the Karnaugh map for a pass network can be reduced with techniques that resemble minterm reduction in classical logic design. But the reduction rules differ in three ways from those of conventional Karnaugh maps.

3. Fundamentals. Like the exclusive-NOR gate of Fig. 2, the five other basic logic functions can be implemented with the assistance of just two transistors. The X and Y inputs in (a) take the values shown in the table (b) to produce the logic functions.

For one thing, a pass transistor's output is not defined when its control gate is not asserted, so a variable must be passed in every map state for which the output has been defined. (In a conventional map, only those states with true outputs must be looped.) Then, too, more than one variable may be passed in one state because the pass

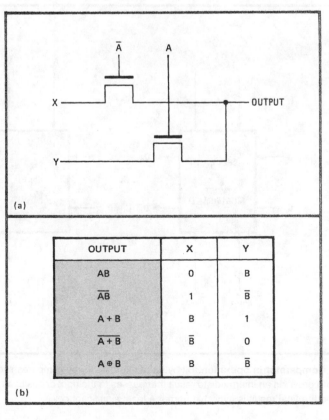

(a)

OUTPUT	X	Y
AB	0	B
$\overline{AB}$	1	$\overline{B}$
A + B	B	1
$\overline{A + B}$	$\overline{B}$	0
A ⊕ B	B	$\overline{B}$

(b)

45

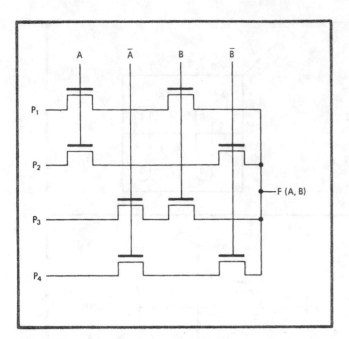

4. Function generator. Each state of the input word AB connects just one of the four pass lines, P_1 through P_4, to the output. Any function of A and B can therefore be implemented with the proper choice of logic values for the pass lines.

functions in the map ensure that the passed variables are all at the same logic level. Finally, once a "don't care" state has been included in a loop, it acquires a pass function determined by the variable being looped.

The input variables must be divided between two sets—pass variables and control variables—for modified Karnaugh maps to be used in pass-network design. Maps do not always make it clear whether or not such a division

is worth making, especially with more than a few variables. One thing can help: understanding the connection between pass networks and the canonical sum-of-products form of Boolean equations.

All possible functions of variables can be synthesized in a single pass-transistor network. Mead and Conway discussed such a function generator for an arithmetic and logic unit. Consider Fig. 4, for example. In each of four possible states of control variables A and B, one of pass lines P_1 through P_4 is connected to the output. To implement the function "A exclusive-OR B," for example, input 0110 is applied to the pass lines.

Input determines function

This circuit is useful because different functions can be implemented just by placing the proper input on the pass lines. In fact, the scheme can be extended to implement all the functions of N variables, which require 2^N pass lines and $N2^N$ transistors. Many random-logic designs do not need this flexibility. But they do need to implement the function with the least possible area, power, delay—and effort.

In the network shown in Fig. 4, properly assigning a third variable—C—or its complement, or the constants 1 or 0 to the pass lines achieves any particular function of three variables. (This claim can readily be verified with the canonical sum-of-products form for Boolean logic functions.) Of the eight minterms that can be formed from three variables, four at most enter any particular function. A fifth minterm can always be reduced in combination with one of the others. Each pass line of the network implements a three-term product. (The first is ABP_1.) Since every possible state of inputs

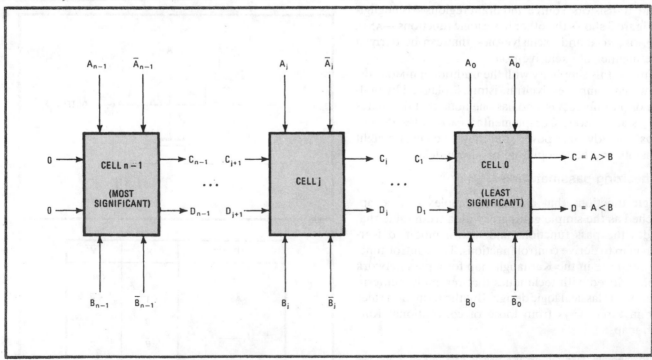

5. Comparator in operation. Bit by bit, an iterative array compares two digital words A and B. The comparison of the most significant bits (left) provides an intermediate result that passes to the right and is used in comparing the next bits. Each cell performs the same operations, and the final result is available from the least significant cell.

46

6. Map. For cell j of the comparator shown in Fig. 5, output C_j is mapped as a function of the corresponding bits of A and B and of outputs C_{j+1} and D_{j+1} from the previous cell. The loops (color) show that $\overline{A}_j + B_j$ passes C_{j+1}, while $A_j\overline{B}_j$ passes $\overline{D}_{j+1}$.

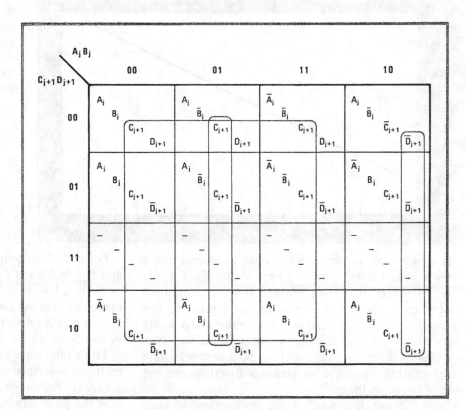

A and B leads to the selection of some pass line, the output is always defined.

Moreover, a particular function of N variables can always be implemented with $2(N-1)$ control lines, 2^{N-1} pass lines, and $N2^{N-1}$ transistors. This, however, is a worst case. Fewer than the maximum number of minterms may be present, and some can be reduced in combination.

When written in what might be called the "pass canonical form," Boolean equations can generally be translated straight into pass-transistor networks. Pass networks, as their structures show, directly implement a Boolean equation of the form:

$$F = P_1 F_1(C_1, \ldots, C_m)$$
$$+ \ldots + P_n F_n(C_1, \ldots, C_m)$$

where the F_i are the control functions formed with series and parallel combinations of pass transistors, the C_i are the control variables that drive the gates of those transistors, and the P_i are the pass variables.

The F_i are a complete and disjoint set: their logical sum is 1, and the logical product of any two is 0. These conditions guarantee that every state of the control variables selects one and only one pass variable. They are easy to meet, for on a Karnaugh map of the C_i the F_i are nonintersecting loops that cover the whole map. Of course, when a high-impedance output state is desired, the corresponding control function can deliberately be left out of the network.

As the above equation shows, no pass variable enters any control function, so an equation does not of necessity allow more than one pass variable. Several pass variables, however, usually produce a denser and faster network. A simple algorithm can be used to calculate the maximum number of pass variables for a Boolean equation written in reduced minterm form.

Simple logic

For simple logic functions, pass networks do significantly improve on the traditional implementation, but performance requirements limit their size and usefulness. A signal passing to a network's output travels through several device channels, each with on-resistance and capacitances to gate and substrate. Delay through the network therefore increases with the square of the

number of pass transistors. Conventional logic gates merely add delays in turn. The resistance-capacitor time constant associated with the channel of a single pass transistor is typically about 0.2 nanosecond. The delay of a conventional logic stage is about 2 ns.

Conventional NAND or NOR gates can implement any function in three levels of logic. Pass networks, however, may require conventional buffer stages at the output to restore logic levels and drive further stages. For a fair comparison, one logic-gate delay should therefore be added to the pass-network delay.

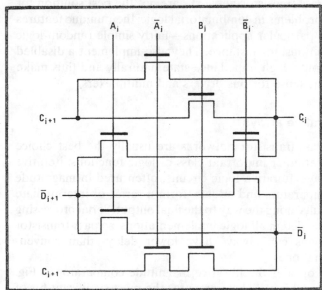

7. Iteration. One cell of the comparator takes just eight pass transistors. The layout occupies an area of 3,024 square micrometers, only a third of the area of a conventional n-channel MOS implementation, and the cell dissipates almost no standby power.

47

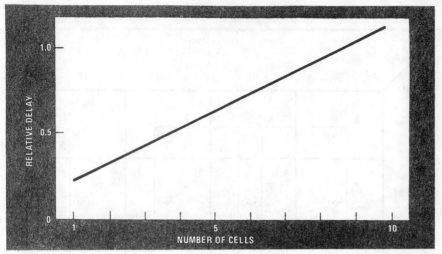

8. Delay comparison. The propagation delay of a pass-transistor implementation of the magnitude comparator increases with the number of stages. For more than eight stages, it turns out that conventional buffers are required with pass gates.

A signal can pass through five pass transistors and a conventional buffer stage in a time about equal to the delay involved in passing through three conventional logic stages. Those five transistors correspond to five control variables, which can pass at least one and possibly several pass variables. Functions of six or fewer variables are thus generated more quickly in pass networks than in conventional logic; more complex functions are not always suited to them.

Like conventional NAND logic, programmable logic arrays incur only three logic-stage delays, however many variables may team up in the function that is implemented. Unlike pass networks, which basically are wired-OR functions, both PLAS and conventional NAND logic independently form each minterm of a function, permitting a single minterm to be used in more than one output function. Several are often derived from one set of input variables, so cutting the number of transistors in PLAS and conventional NAND logic often offsets the additional area both need for ground connections and load devices and the larger number of transistors per function.

Pass-transistor networks are not the best solution for all problems in combinatorial logic; their unique features suit particular applications—fairly simple random-logic functions, for instance. They also implement a disabled output's high-impedance state naturally and thus make great sense for bus drivers and multiplexers.

Iterative arrays

Pass-transistor networks are usually the best choice for another important class of logic functions: iterative arrays. Iterative logic circuits, often used in magnitude comparators and adders, form a series of intermediate results along the way to the final output. Serial processing is needed in all logic implementations, so pass-transistor designs can always have lower delays than conventional ones.

Consider the iterative magnitude comparator in Fig. 5. A 1-bit cell in it compares the most significant bit of A and the most significant bit of B. Intermediate results C and D are then passed to the next-most significant cell, where they are used with the next bits of A and B to complete the comparison. The process continues until the final result is available from the least significant cell.

For cell j, the output C_j is 1 if A_j is greater than B_j and D_{j+1} is 0, or if C_{j+1} is 1. Output D_j is 1 if D_{j+1} is 1, or if A_j is less than B_j and C_{j+1} is 0. These conditions make the final output, C, high if A is greater than B. D is high if B is greater than A. If both C and D are low, A and B are equal.

Truth tables are constructed for C_j and D_j (with the methods described earlier) and the pass functions are entered in Karnaugh maps. The map for C_j (Fig. 6) shows how the pass variable C_{j+1} is looped under the control function $\overline{A}_j + B_{jj}$, while $\overline{D}_{j+1}$ is looped under the function $A_j\overline{B}_j$. The map for D_j is completed in a similar manner.

The resulting pass-network implementation of one cell of the magnitude comparator (Fig. 7) occupies 3,024 square micrometers—only 34% of the 8,840 mm² conventional logic designs need. Simulations of the two versions using parameters extracted from the layout are employed to compare circuit delays.

As mentioned previously, pass-transistor networks form RC delay lines, and signals on them obey a diffusion equation. The delay takes the form $An^2 + B$, where n is the number of cells in the comparator and A and B are constants. Delays through the conventional logic network take the form Cn. Constant C is larger than A, in part because pass networks cut capacitance by cutting the number of transistors needed to implement a function. This reduction increases the circuit's regularity, cuts the amount of wiring and gate capacitance, and completely eliminates depletion-load devices and their gate capacitance.

Figure 8 compares the delay of a pass network with that of a conventional circuit by plotting them as functions of the number of bits in the comparator. For 8 or fewer bits, pass networks are faster than conventional ones. For more bits, they can still be faster if conventional buffer stages are inserted when the delay from an additional pass-gate stage would exceed the delay through a buffer followed by the pass stage.

With 8 or fewer bits, the pass-transistor network's power consumption is negligible. For more than 8 bits, the conventional buffers added to the circuit contain two power-dissipating nodes. The conventional logic design has two power-consuming nodes in each cell, so its total power consumption is eight times that of the pass network. □

A 3.8-ns CMOS 16×16-b Multiplier Using Complementary Pass-Transistor Logic

KAZUO YANO, MEMBER, IEEE, TOSHIAKI YAMANAKA, MEMBER, IEEE, TAKASHI NISHIDA,
MASAYOSHI SAITO, MEMBER, IEEE, KATSUHIRO SHIMOHIGASHI, MEMBER, IEEE,
AND AKIHIRO SHIMIZU

Abstract — A 3.8-ns 257-mW CMOS 16×16-b multiplier with a supply voltage of 4 V is described. A complementary pass-transistor logic (CPL) is proposed and applied to almost the entire critical path. The CPL consists of complementary inputs/outputs, an nMOS pass-transistor logic network, and CMOS output inverters. The CPL is twice as fast as conventional CMOS due to lower input capacitance and higher logic functionality. Its multiplication time is the fastest ever reported, including bipolar and GaAs IC's, and it can be enhanced further to 2.6 ns with 60 mW at 77 K.

I. INTRODUCTION

THE SPEED of CMOS devices, which were used mainly in low-power high-density LSI's, has increased drastically with the rapid progress in miniaturization. CMOS speed is getting close to that of Si bipolar technology; for example, with submicrometer CMOS technology, a 9-ns 1-Mb SRAM [1] and a 7.4-ns 16×16-b multiplier [2] have been reported. However, the recent progress in fast engineering workstations and real-time digital-signal processing requires faster CMOS speed.

A multiplier is an essential element in any digital-signal processing circuit and constitutes the critical path in DSP and FPU LSI's. Recently, versatile microprocessor chips also have begun to contain multipliers, which is made possible by the rapid progress in integration technology. Therefore, the demand for improved multiplier performance is increasing. In addition, multipliers are designed and fabricated as benchmarks for demonstrating various high-speed technologies, e.g., Si bipolar [3], GaAs [4], and Josephson junction devices.

This paper describes a fast 3.8-ns 0.5-μm CMOS 16× 16-b multiplier that is implemented as a test vehicle for investigating a new circuit technique for high-speed CMOS-based logic circuits. A new family of advanced differential CMOS logic, called complementary pass-transistor logic (CPL), is proposed and fully utilized on almost the entire critical path to achieve very high speeds. The

Manuscript received September 1, 1989; revised December 4, 1989.
K. Yano, T. Yamanaka, T. Nishida, M. Saito, and K. Shimohigashi are with the Central Research Laboratory, Hitachi Ltd., Kokubunji, Tokyo 185, Japan.
A. Shimizu is with the Hitachi VLSI Engineering Corporation, Tokyo, Japan.
IEEE Log Number 8934072.

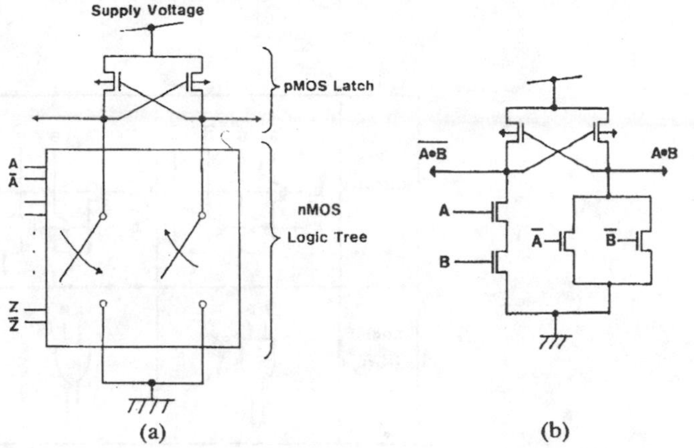

Fig. 1. Basic CVSL circuit. (a) Schematic structure. (b) AND/NAND circuit.

circuit techniques and the technology used in this multiplier have the potential for 100-MHz operation of 32- to 64-b floating-point multiplication, thus enabling very fast DSP's, FPU's and ASIC's. First, the concept of CPL is introduced in Section II, and then CPL implementation of the multiplier is described in Section III. Device fabrication is described in Section IV, and finally the performance of the fabricated multiplier is shown in Section V.

II. CPL: CONCEPT AND EXAMPLES

Several differential CMOS logic families such as cascode voltage switch logic (CVSL) [5] (Fig. 1) and differential split-level logic (DSL) [6] have been proposed for CMOS circuit speed improvement. These have the common features of complementary data inputs/outputs, an nMOS logic tree, and a pMOS cross-coupled load, which together can reduce input capacitance, increase logic functionality, and sometimes eliminate inverter circuits. Therefore, these logic families can increase speed. However, the actual advantage of CVSL circuits is less than that anticipated in the original paper, as clarified in [7]. This is because the pMOS cross-coupled latch cannot easily be inverted due to the regenerative property of the latch. High-speed inversion of the pMOS latch is possible only when the gate width of the pMOS is sufficiently small. However, a small

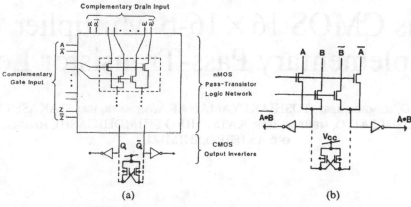

Fig. 2. Basic CPL circuit. (a) Schematic structure. (b) AND/NAND circuit.

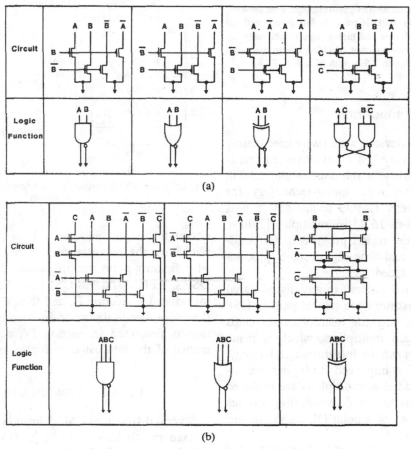

Fig. 3. CPL circuit modules. (a) Two-way logic. (b) Three-way logic.

gate width severely degrades the pull-up transit time. DSL is faster than conventional CMOS, however at the expense of static power consumption [7].

The main concept behind CPL is the use of an nMOS pass-transistor network for logic organization, and elimination of the pMOS latch, as shown in Fig. 2. CPL consists of complementary inputs/outputs, an nMOS pass transistor logic network, and CMOS output inverters. The pass transistors function as pull-down and pull-up devices. Thus the pMOS latch can be eliminated, allowing the advantage of the differential circuits to be fully utilized. Because the high level of the pass-transistor outputs (nodes Q and $\bar{Q}$) is lower than the supply voltage level by the threshold voltage of the pass transistors, the signals have to be amplified by the output inverters. At the same time, the CMOS output inverters shift the logic threshold voltage and drive the capacitive load. The logic threshold shift is necessary because the logic threshold voltage of the output inverter is lower than half the supply voltage, due to the lowering of the high signal level.

A pMOS latch can also be added to CPL, as shown in Fig. 2, to decrease static power consumption, as opposed to the conventional pull-up function. In this case, the pMOS gate width can be designed to be minimum, as long

50

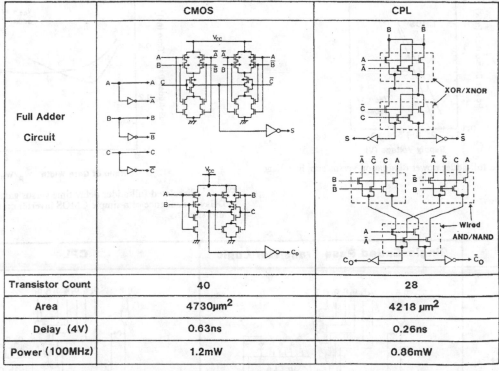

	CMOS	CPL
Full Adder Circuit		
Transistor Count	40	28
Area	4730μm²	4218 μm²
Delay (4V)	0.63ns	0.26ns
Power (100MHz)	1.2mW	0.86mW

Fig. 4. Comparison of CMOS full adder with CPL full-adder circuit.

as the pull-up function is completed in the given cycle time.

Arbitrary Boolean functions can be constructed from the pass-transistor network by combining four basic circuit modules: an AND/NAND module, an OR/NOR module, an XOR/XNOR module, and a wired-AND/NAND module. These are shown in Fig. 3(a) in which the XOR/XNOR module is used once [8]. One attractive feature of CPL is that the complementary outputs are produced by the simple four-transistor circuits. Because inverters are unnecessary in CPL circuits, the number of critical-path gate stages can be reduced. Note that these various functions are produced by an indentical circuit configuration with only a change of input configuration. This property of CPL is apparently suitable for masterslice design. Logic functions for three and greater inputs can also be easily constructed similarly to two-way logic. Examples of three-way input logics are shown in Fig. 3(b).

As an example of more complex logic circuits, a CPL full adder is shown in Fig. 4. A conventional CMOS full adder [9] is also shown in Fig. 4 for comparison. In the CPL full adder, both the sum logic and carry logic are structured on the CPL concept by combining the basic modules in Fig. 3. The sum logic comprises two XOR/XNOR modules, whereas the carry logic comprises three wired-AND/NAND modules. The output inverters are "overhead," in the sense that they are needed whether the circuit has one, two, or many inputs. Therefore designing with complex logic functions in a gate is adopted to minimize the overall device count and delay time.

Since pMOS can be eliminated in logic construction in the CPL, the input capacitance is about half that of the conventional CMOS configuration, thus achieving higher speed and lower power dissipation. Moreover, the powerful logic functionality of CPL due to the multilevel pass-transistor network realizes complex Boolean functions efficiently with a small number of MOS transistors, thus further reducing area and delay time. In fact, the transistor count in the CPL full adder is 28, whereas in the CMOS it is 40. The areas required for these full adders based on the half-micrometer design rule are 4218 μm² (37×114 μm²) for CPL and 4730 μm² (55×86 μm²) for conventional CMOS. The actual area reduction rate is smaller than the transistor-count reduction rate, because the interconnection area is not proportional to the transistor count. Further, the complementary input/output function eliminates the internal inverter to provide XOR input in the full adder, thus reducing the number of critical-path gate stages.

To compare the full-adder performance between CPL and CMOS, circuit simulations are performed using the half-micrometer device parameters at a supply voltage of 4 V. The 4-V supply was chosen because it is the maximum voltage at which half-micrometer CMOS devices are immune to hot-carrier degradation. The simulated worst-case delay time (which refers to situations where the input signals are such that circuit operation is slowest) of the CPL full adder is as short as 0.26 ns, which is 2.5 times faster than the conventional CMOS.

The simulated power dissipation as a function of supply voltage is shown in Fig. 5. The CPL consumes 30% less power than CMOS with a 4-V supply mainly due to smaller input capacitance. The effect of power reduction in CPL is more significant at lower supply voltages. This is because the logic swing of the pass-transistor outputs

51

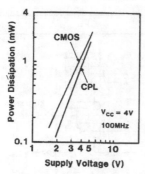

Fig. 5. Simulated full-adder power dissipation versus supply voltage.

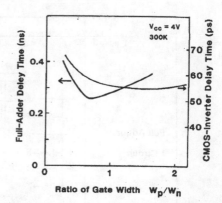

Fig. 6. Simulated full-adder delay time versus gate width ratio W_P/W_N. Delay dependence of a simple CMOS inverter on the gate width ratio is also shown.

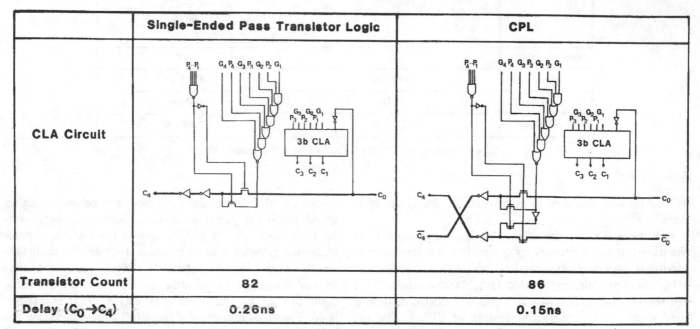

	Single-Ended Pass Transistor Logic	CPL
CLA Circuit		
Transistor Count	82	86
Delay ($C_0 \rightarrow C_4$)	0.26ns	0.15ns

Fig. 7. Comparison of 4-b carry-lookahead circuits between a CPL circuit and its single-ended counterpart. C_j is carry signal, G_j is carry-generate signal, and P_j is carry-propagate signal.

(nodes Q and $\overline{Q}$ in Fig. 2) is smaller than the supply voltage level. The dynamic power resulting from charging and discharging the capacitances in CPL circuits is given as

$$P = C_{OUT} \cdot V_{CC}^2 \cdot f + C_{INT} \cdot V_{CC}(V_{CC} - V_{TN}) \cdot f \quad (1)$$

where P is the dynamic power in the CPL circuit, C_{OUT} is the output capacitance driven by the output inverters, C_{INT} is the internal capacitance (at nodes Q and $\overline{Q}$ in Fig. 2), V_{CC} is the supply voltage, V_{TN} is the threshold voltage of the nMOS pass transistors, and f is the operating frequency. On the other hand, as is well known, the power dependence on supply voltage in CMOS circuits is described by squared dependence:

$$P' = C_{OUT}' \cdot V_{CC}^2 \cdot f \quad (2)$$

where P' is the dynamic power in the CMOS circuit, and C_{OUT}' is the total capacitance in the CMOS circuit. In the full-adder circuits in Fig. 4, the sum of C_{OUT} and C_{INT} is 20% smaller than C_{OUT}'. In addition, at low supply voltages the threshold voltage effect of the second term in (1) further reduces the dynamic power dissipation.

The following are key points in CPL circuit design:

1) controlling the threshold voltage of the nMOS pass transistors to a lower value than the threshold voltage of pMOS; and
2) designing the logic threshold voltage of the output CMOS inverter to a lower value than $V_{CC}/2$.

These points improve the speed, static power dissipation, and noise margins of CPL circuits by offsetting the lowering of the high signal level at the pass-transistor output nodes. The full-adder delay time as a function of gate width ratio between pMOS and nMOS in output inverters W_P/W_N has a minimum as shown in Fig. 6. The optimum gate width ratio exists at the ratio of about 0.75, which is much less than that of the ordinary CMOS inverters (1.5–2).

The CPL circuit can also be applied to carry-lookahead logic, which is indispensable for fast ALU's and multipliers. The CPL 4-b carry-propagate circuit and its single-ended counterpart are shown in Fig. 7. The logic functions

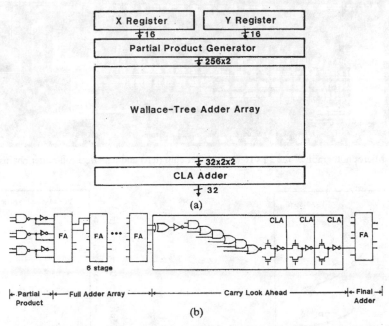

Fig. 8. (a) Block diagram of the 16×16-b multiplier. (b) Critical path of the 16×16-b multiplier.

of these carry-propagate circuits are expressed by

$$C_4 = G_4 + P_4 \cdot [G_3 + P_3 \cdot (G_2 + P_2 \cdot G_1)] + P_4 \cdot P_3 \cdot P_2 \cdot P_1 \cdot C_0 \tag{3}$$

$$C_4 \equiv A + B \cdot C_0 \tag{4}$$

where C_0 is the carry input from the lower carry-lookahead unit, C_4 is the carry output to the higher carry-lookahead unit, P_j's ($j = 1$–4) are the carry-propagate signals, and G_j's are the carry-generate signals. The OR logic in (4) can be expressed by the simple pass-transistor circuits in Fig. 7 by considering the following relationship:

$$A \cdot B = 0. \tag{5}$$

Within a single inverter delay time, the CPL carry-lookahead unit can quickly transfer the carry output to the upper unit (C_4 and $\overline{C_4}$) after receiving the carry input from the lower unit (C_0 and $\overline{C_0}$). Complementary carry data in the CPL unit can be inverted simply by twisting the carry lines. The simulated delay time is only 0.15 ns/4 b. By contrast, the single-ended counterpart requires two-stage inverter delay. Because in actual adders and multipliers these units are connected in series (for a 64-b adder, 16 units have to be connected) and constitute the critical path, the delay reduction of a factor of 2 significantly reduces the total adder delay time.

III. MULTIPLIER ARCHITECTURE

The 16×16-b multiplier was designed using a parallel multiplication architecture, as shown in Fig. 8(a). A Wallace-tree adder array and a CLA adder were used to minimize the critical-path gate stages. There are a total of 8500 transistors in an active area of 1.3×3.1 mm², whereas the area including bonding pads is 1.6×4.5 mm². The

transistor count is less than that of a full CMOS counterpart [2], mainly because the transistor count in the full adder is less than that of the CMOS full adder. The critical path consists of a partial-product generator, six full adders, lookahead carry logic, three carry-propagate circuits, and a final full adder as shown in Fig. 8(b).

IV. DEVICE FABRICATION

The multiplier and the ten-stage full-adder chains are fabricated with double-level-metal 0.5-μm CMOS technology. The minimum feature size is 0.5 μm, and the gate oxide thickness is 12.5 nm. In this fabricated 0.5-μm CMOS device optimized for CPL, the nMOS pass transistors are designed to have a threshold voltage of 0 V, whereas the other nMOS and pMOS have a threshold voltage of 0.4 and −0.4 V, respectively. This threshold control reduces the static power dissipation and delay time. The drain saturation current per 10-μm width is 4.6 mA for nMOS and −2.6 mA for pMOS. The interconnection metal consists of first-level W and second level Al. The W was adopted for its high immunity to electromigration.

V. PERFORMANCE RESULTS

A microphotograph of the multiplier and the full-adder chains is shown in Fig. 9. Ten-stage full-adder chains using CPL full adders and CMOS full adders were designed and fabricated to compare the actual performance. The measured worst delay time as a function of supply voltage is shown in Fig. 10.

The delay dependences are similar for CMOS and CPL. Thus CPL can be used at supply voltages at least as low as those for CMOS. The results agree well with the simula-

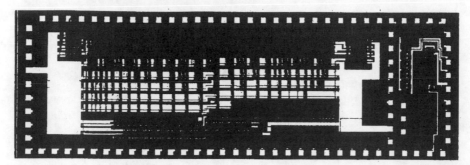

Fig. 9. Microphotograph of the 16×16-b multiplier chip (left) and ten-stage full-adder chains (right).

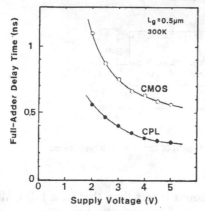

Fig. 10. Measured full-adder delay time versus supply voltage.

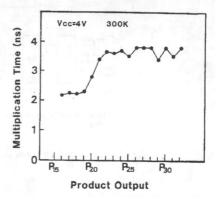

Fig. 12. Measured multiplication time versus product output.

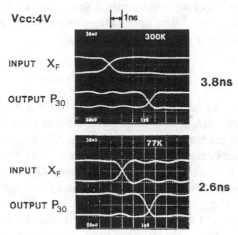

Fig. 11. Measured waveforms of the multiplier.

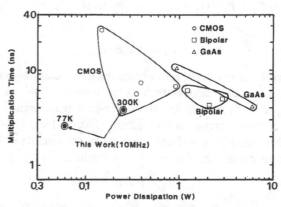

Fig. 13. Comparison of delay and power dissipation of high-speed 16×16-b multipliers.

tion results, experimentally verifying the advantage of CPL circuits.

The completed multiplier chips were probe tested at wafer level. The fully functional chips were mounted on the 68-pin pin-grid-array ceramic package, followed by waveform observation of the clock inputs and the product outputs through source-follower circuits on the chip. The test was performed up to the maximum frequency limit (250 MHz) of the pulse generator. Speed performance was measured using the worst-case pattern, $FFFF \times 8001 - 7FFF \times 8001$. Circuit simulation was also performed to confirm that this pattern consists of a series of worst-case operations of the full adder and the CLA. The multiplication time was measured at both room and liquid-nitrogen temperatures. The latter was considered for high-end ap-

plications, such as supercomputers [10], [11]. The maximum multiplication times were 3.8 and 2.6 ns at room and liquid-nitrogen temperatures, respectively. The measured waveforms are shown in Fig. 11. The observed signal amplitude is larger at lower temperatures because the on-chip source-follower gain is larger. The multiplication time versus product output is shown in Fig. 12. P_{30} gives the longest delay time as expected from our simulations. There is very little variation on the product output from P_{22} to P_{32}, which means the carry propagates very quickly in the CPL carry-lookahead circuit. Power dissipation was 257 and 60 mW at 300 and 77 K, respectively, for 10-MHz operation with a pattern of $FFFF \times FFFF - 0000 \times FFFF$.

The multiplication times at both room and liquid-nitrogen temperatures are compared with the published 16×16-b multipliers in Fig. 13. The minimum multiplication time before this work was 4.1 ns with a power dissipation of 6.2 W and was realized by GaAs high electron mobility

TABLE I
FEATURES OF THE 16×16-b MULTIPLIER

Architecture	Wallace Tree + CLA
Technology	0.5-μ m CMOS
Gate Length	0.5 μ m
Gate Oxide Thickness	12.5 nm
Metal Line/Space W	0.8/0.8
Al	1.0/1.0
Active Area	1.3×3.1 mm^2
Transistor Count	8500
Multiplication Time(4 V)	3.8 ns (300 K), 2.6 ns (77 K)
Power Dissipation(10 MHz)	257 mW (300 K), 60 mW (77 K)

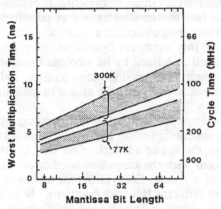

Fig. 14. Calculated floating-point multiplication time versus mantissa bit length.

transistors (HEMT's). Previously, CMOS multipliers dissipated power one order of magnitude lower than the others. However, the multiplication time was 2–3 times larger than those of fast multipliers. By contrast, the present CPL's multiplication times at both room and liquid-nitrogen temperatures are faster than those of any other devices. In addition, the power dissipation is much lower than those of the other devices. The features of this multiplier are summarized in Table I.

The estimated performance of a multiplier using CPL and the half-micrometer CMOS devices is shown in Fig. 14, considering that the bit length is enlarged and floating-point multiplication functions are included. The architecture is assumed to be a combination of Booth's algorithm and Wallace-tree adder array. The bit length and the depth of the carry-lookahead unit are assumed to be optimized for the mantissa bit length. The variations of process, supply voltage, and temperature are considered simply by multiplying the empirical factor by the typical delay time. The multiplication time increases with increasing mantissa bit length. However, the 64-b floating-point multiplication time is estimated to be 10 ns at room temperature and 6 ns at liquid-nitrogen temperature. Thus, operation at over 100 MHz is possible even at room temperature if the multiplier architecture and the carry-lookahead circuit configurations are carefully optimized for high-speed operation.

VI. CONCLUSIONS

This paper described a fast 16×16-b multiplier using a new differential CMOS logic family, CPL. In CPL, differential logic is constructed without pMOS latching load, enabling a speed more than twice as fast as conventional CMOS. The power dissipation is also smaller due to smaller input capacitance. The multiplier is the fastest ever reported at both 300 K and 77 K, proving that the half-micrometer CMOS technology fully utilizing CPL has a speed which is at least competitive with those of other fast devices with a much smaller power dissipation at room temperature, and is faster at liquid-nitrogen temperature. These results also demonstrate that half-micrometer CMOS devices fully utilizing CPL have a performance potential of a 100-MHz repetition rate for floating-point multiplication by carefully optimizing the multiplier architecture for high-speed operation. Therefore, very high-speed MPU's, DSP's, FPU's, and ASIC's are possible.

ACKNOWLEDGMENT

The authors wish to thank Y. Sakai and O. Minato from Hitachi Semiconductor Development Center and T. Masuhara, T. Nakagawa, T. Baji, K. Kaneko, T. Sawase, and K. Ishibashi from Hitachi Central Research Laboratory for their useful suggestions and discussions. The authors are also greatly indebted to K. Yagi and the device processing staff members from Hitachi Central Research Laboratory for their support throughout the sample fabrication, and to A. Kawamata from Hitachi VLSI Engineering Corporation and K. Ueda from Hitachi Central Research Laboratory for their layout design.

REFERENCES

[1] K. Sasaki et al., "A 9ns 1Mb CMOS SRAM," in ISSCC Dig. Tech. Papers, 1989, pp. 34–35.
[2] Y. Oowaki et al., "A 7.4ns CMOS 16×16 multiplier," in ISSCC Dig. Tech. Papers, 1987, pp. 52–53.
[3] M. Suzuki, M. Hirata, and S. Konaka, "43ps/5GHz bipolar macrocell array LSIs," in ISSCC Dig. Tech. Papers, 1988, pp. 70–71.
[4] K. Kajii et al., "A 40 ps high electron mobility transistor 4.1K gate array," in IEEE 1987 Custom Integrated Circuit Conf., pp. 199–202.
[5] L. G. Heller, W. R. Griffin, J. W. Davis, and N. G. Thoma, "Cascode voltage switch logic: A differential CMOS logic family," in ISSCC Dig. Tech. Papers, 1984, pp. 16–17.
[6] L. C. M. G. Pfennings, W. G. J. Mol, J. J. J. Bastiaens, and J. M. F. van Dijk, "Differential split-level CMOS logic for sub-nanosecond speed," in ISSCC Dig. Tech. Papers, 1985, pp. 212–213.
[7] K. M. Chu and D. L. Pulfrey, "A comparison of CMOS circuit techniques: Differential cascode voltage switch logic versus conventional logic," IEEE J. Solid-State Circuits, vol. SC-22, pp. 528–532, 1987.
[8] T. Kengaku, Y. Shimazu, T. Tokuda, and O. Tomisawa, IECE Japan, 2-83, 1987.
[9] M. Uya, K. Kaneko, and J. Yasui, "A CMOS floating point multiplier," in ISSCC Dig. Tech. Papers, 1984, pp. 90–91.
[10] S. Hanamura et al., "Operation of bulk CMOS devices at very low temperatures," in 1983 Symp. VLSI Tech., pp. 46–47.
[11] T. Vacca et al., "A cryogenically cooled CMOS VLSI supercomputer," VLSI Syst. Design, vol. VIII, no. 7, pp. 80–88, 1987.

Lean Integration: Achieving a Quantum Leap in Performance and Cost of Logic LSIs

KAZUO YANO, YASUHIKO SASAKI, KUNIHITO RIKINO* AND KOICHI SEKI

ULSI Research Center, Hitachi Central Research Laboratory, Kokubunji, Tokyo 185, Japan
*Hitachi Device Engineering, Kokubunji, Tokyo 185, Japan

ABSTRACT

Lean integration aims at a fundamental change in top-down design by following the path from CISC to RISC. The central idea is a lean cell, which has a tree-shaped nMOS network with input ports placed at the end of an every branch of the tree. A lean cell has flexibility of transistor-level circuit design and full compatibility with conventional cell-based design. An extremely simple lean-cell library with only 7 cells and a synthesis tool called "Circuit Inventor," which uses the lean cells, are developed and they are compared with the conventional "complex" CMOS library that has over 60 cells. The results show that the area, the delay, and the power dissipation are improved by lean integration and performance-cost ratio is improved by a factor of three.

INTRODUCTION

Due to recent progress in top-down LSI design using logic synthesis and HDL, the cell-library design is becoming a key factor in achieving high performance ASICs and MPUs. This is quite natural if we recall that the cell library corresponds to the "instruction sets" if we compare the LSI design to the CPU design. Recently even the evaluation method of cell-library quality has been seriously considered [1]. A conventional CMOS cell library usually has over 60 cells even if we limit the cells to the combinational logic. We postulate that this complex library is, compared to a "CISC," causing unnecessary silicon and engineering costs, and a much leaner LSI design method just like RISC should be conceived. In fact, creating and maintaining the library is becoming such a serious burden in LSI design that it is creating opportunities for companies to provide library-design services.

In this work, we propose lean integration, a completely new cell-library architecture which achieves a quantum leap in performance and cost for logic LSIs. We investigate the impact of this method by comparing overall figures with those of CMOS.

LEAN CELL

The proposed lean-cell library is compared with a conventional CMOS library in Table 1 and Fig. 1. The lean-cell library has only 7 cells, which is far smaller than a conventional library. The number of essential logic cells is also less and it is only 3, Y1, Y2, Y3 (Fig. 1). The other 4 cells are simple inverters.

Another feature of lean cells is that they are defined by the transistor network topology, a binary tree-shaped nMOS network, rather than the Boolean function. The name "lean" came from the fact that the situation is just like transistors are directly connected to the cell ports of the cell. Although the cell works as a multiplexer, the essential advantage of the lean cell is that transistor-level circuit engineering is possible by using this cell. In fact the transistors act as pass-transistors, or source-grounded configuration, or source-follower configuration depending on the input configuration. By contrast, the conventional cell is defined by its Boolean function, which is often chosen based on previous case studies, and the inner circuit configuration is simply the means to meet the function requirement.

The advantage of a lean cell is that it changes the function by changing the configuration of the cell input ports (Fig. 3). The drain port, the end of a branch of the tree, has the freedom to be connected with the output of another cell, or a power supply line, or a ground line. Different input configurations correspond to different Boolean functions. Note that a very complex logic function is achieved by a single cell. This functionality came from that the transistor-level circuit engineering already described.

Despite this transistor-level flexibility of the lean cell, it is fully compatible with the framework of cell-based design. As a result the delay of the cell can be defined as a function of the load capacitance. This is made possible by the output inverter, which separates the inputs from the output. The feedback inverter and the pull-up pMOS, both consisting of minimum-size MOSFETs, are added to avoid DC leakage current in the CMOS inverter.

Because preparing and updating the lean-cell library requires only small engineering cost, it is much easier to adopt the state-of-the-art process technology even in a tight schedule constraint. Therefore, the lean cell encourages concurrent interaction between logic designers and process engineers. By contrast, major revision of the conventional library, which includes cell-layout data, logic-synthesizer data, and automatic place and router data for more than 60 cells, requires much more engineering effort and is sometimes unrealistic.

The area of a logic block is reduced by using lean cells. The logic area is given by the following equation.

$$\text{Logic area} = \text{Net count} \times \frac{1}{\text{Nets per cell}} \times \text{Cell area} \quad (1)$$

where net count is mainly determined by the logic synthesis algorithm, nets per cell (NPC) is mainly determined by the cell architecture, and cell area is mainly determined by the

Reprinted from *Proceedings of the IEEE Custom Integrated Circuits Conference*, pp. 26.5.2 - 26.5.4, May 1994.

technology level. This relation corresponds to the well-known relation used in CPU design:

CPU time =

$$\text{Instruction count} \times \frac{1}{\text{Instructions per cycle}} \times \text{Cycle time} \quad (2)$$

In Eq.(1) a logic function is considered to be a box which reduces the number of nets, or nodes. The lean cell, which has a large NPC without increasing the cell area, has a high capability of reducing nets. Therefore the number of cells required to build a logic block is smaller than that of the CMOS, resulting in smaller block area. A similar argument holds for power consumption, which leads to lower power consumption.

The delay of the logic block is also reduced by the lean cells. Very complex logic functions, which are not included in conventional CMOS libraries, can be achieved by using only a single lean cell. Therefore, the number of critical-path cells is reduced. In addition, complex CMOS gates with large parasitic capacitance are slow and have low current drive capability. By contrast, parasitic capacitance of lean cells is small and high current drive capability is possible due to the output inverter.

The transition from a CMOS library to the lean-cell library is just like the transition from CISC to RISC (Table. 2). A lean-cell library corresponds to the small instruction set of RISCs. The instructions of the RISC were not convenient for designers who were accustomed to the conventional orthogonal instructions of CISCs. However, this has been overcome by using an optimized compiler. A lean cell has a similar characteristic. Because it is somewhat like directly controlling the transistor behavior from outside of the cell, logic designers, who are not familiar with the details of individual circuit may become reluctant to deal with them. However, logic synthesis based on lean cells solves this problem. The high performance of RISCs is explained by the large number of instructions per cycle. The high performance-cost ratio of lean integration is explained by its larger nets per cell. The simple instruction set of RISCs and the especially good expectation of a relation between the performance and the instruction sequence help the compiler to provide highly efficient instructions. The lean cells also help the synthesizer to provide area- and delay-effective net lists.

We also developed a logic synthesis tool called "Circuit Inventor", which fully utilizes the lean cell characteristics. Circuit Inventor accepts an HDL description, creates net lists based on lean cells and gives those data to the layout tool. It expresses the required logic function in a compact form by using a reduced BDD (Binary Decision Diagram) [2] and conducts various optimizations. BDD has the same network topology as lean cells and efficient mapping to cells is possible. Automatic insertion of optimized inverters into heavily-loaded nodes is possible. The details of the internal algorithm of Circuit Inventor will be described elsewhere.

EXPERIMENTS AND DISCUSSION

The performance of lean cells is compared with that of CMOS cells. Two types of benchmark logic are chosen. One is a 4-b adder/subtracter, which represents arithmetic logic, and the other is 7-input 4-output random logic, which is created by assigning random numbers to the output of the truth table. CMOS logic is synthesized by using a popular commercial logic synthesis tool. 0.5-μm process with 3-level metal is assumed and poly-cell-type layout style is used. Metal 1 is assigned to the intra-cell wiring, metal 2 is assigned to Y-direction inter-cell wiring, and metal 3 is assigned to X-direction inter-cell wiring. The cell input/output ports are formed as through-holes between metal 1 and metal 2. The critical path and the wiring load was extracted from the layout data (Fig. 4) and the delay was obtained by using circuit simulation. The power consumption was determined by using circuit simulation of the total circuit.

The results are dramatic (Table. 3). The lean cells show higher figures in all respects including area, delay, and power consumption for either benchmark. If we define the performance-cost ratio of a cell architecture by using the product of the area, delay, and power, the lean cells has 3-4 times learger ratio.

The average NPC is actually boosted in the lean integration from 2.8 to 4.7 (Table 4), which is the major contributor of the area reduction. The delay per cell is smaller in the lean cells, which contributes the delay reduction.

The dependence of the delay on supply-voltage is another important aspect of the technology choice. The lean cells become slower than CMOS cells at supply voltages below the critical value, because of the influence of threshold voltage of the nMOS. However, under practical conditions, where the extrapolated threshold voltage is smaller than $V_{CC}/2.7$, the lean cells are always faster than the CMOS cells as shown in Fig. 5.

CONCLUSIONS

Lean integration, which provides a quantum leap in performance and cost of ASICs and MPUs is proposed. This new design method goes far beyond marginal improvements and reaches 3-4 times improvement in performance-cost ratio and gives much higher competitiveness in "lean LSIs" and in systems that use LSI chips.

ACKNOWLEDGEMENTS
The authors would like to thank E. Takeda, K. Uchiyama, S. Narita, T. Noguchi, N. Kageyama, M. Tonomura of Hitachi Central Research Laboratory for their valuable discussions.

[1] H. Harvey-Horn, "User-defined benchmarks help evaluate IC physical libraries," Electronics Design, Oct., 80 (1993)
[2] R. E. Bryant, "Graph-based algorithm for Boolean function manipulation," IEEE Computers, Vol-C35 (8), 677(1986)

Table 1 Cell lists of covential and proposed lean-cell library

Conventional CMOS Library (61 cells)		Lean-Cell Library (7 cells)
INVERTER	4AND	2OR/2AND	Y1
INVERTER_P2	4AND_P2	2OR/2AND_P2	Y2
INVERTER_P4	2OR	3OR/2AND	Y3
INVERTER_P8	2OR_P2	3OR/2AND_P2	INVERTER
2NAND	3OR	2ANDx2/2OR	INVERTER_P2
3NAND	3OR_P2	2ANDx2/2OR_P2	INVERTER_P4
2NOR	4OR	3ANDx2/2OR	INVERTER_P8
2NOR_P2	4OR_P2	3ANDx2/2OR_P2	
3NOR	2AND/2NOR	2ANDx3/3OR	
3NOR_P2	2ANDx2/2NOR	2ANDx3/3OR_P2	
4NOR	3AND/3NOR	2ANDx2/3OR	
4NOR_P2	2OR/2NAND	2ANDx2/3OR_P2	
2XOR	2OR/3NAND	2ORx2/2AND	
2XOR_P2	3OR/2NAND	2ORx2/2AND_P2	
2XNOR	2AND/2OR	2ANDx4/4OR	
2XNOR_P2	2AND/2OR_P2	2ANDx4/4OR_P2	
2AND	3AND/2OR	8NAND	
2AND_P2	3AND/2OR_P2	8NAND_P2	
3AND	2AND/3OR	8AND	
3AND_P2	2AND/3OR_P2	8AND_P2	
	8OR		

(_P2, _P4, _P8 represent x2, x4, x8 powered cells)

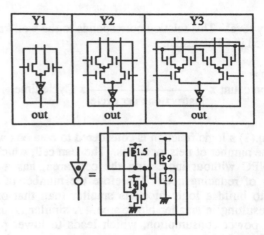

Fig. 1 Circuit diagram of lean cells and the output inverters

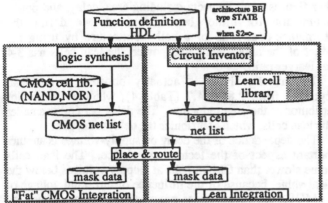

Fig. 2 Conventional "fat" CMOS integration
vs. lean integration

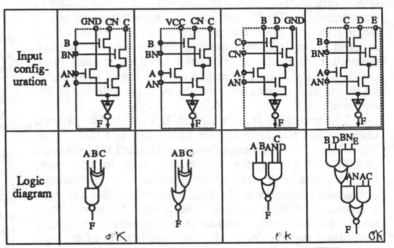

Fig. 3 Various logic fuctions of the lean cell "Y2"

Table 2 Transition from CMOS "fat" library to lean-cell library is compared to "CISC->RISC" transition

	CISC	RISC		"fat" CMOS	lean
No. instructions	many instructions ⟶	fewer instructions	No. cells	many cells ⟶	fewer cells
instruction fuction	orthogonal self-contained instructions ⟶	Load/Store architecture	Cell logic function	self-contained logic(NAND,NOR) ⟶	Select/Amplify
Instructions/cycle	low(~1/3) ⟶	high(~1)	Nets/cell	low(~2.8) ⟶	high(~4.7)
programming	assembler ⟶	high-level language	design	schematic ⟶	HDL

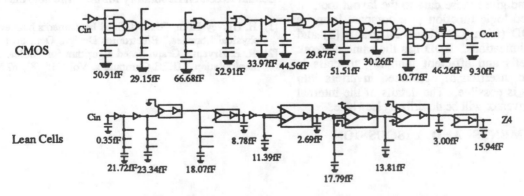

Fig. 4 synthesized critical path circuit of the 4-b adder/subtractor

Table 3 Summary of benchmark design

	CMOS	Lean
4-bit ADD-SUB		
Layout		
AREA	84288.6μm² (1.0)	48589.2μm² (0.55)
Delay Time	3.620ns (1.0)	2.691ns (0.74)
Tr.Count	828 (1.0)	545 (0.66)
Gate Width	7583μm (1.0)	3091μm (0.41)
Net Count	386 (1.0)	400 (1.04)
Power	6.08mW/MHz (1.0)	3.84mW/MHz (0.63)
Cell Count	133 (1.0)	85 (0.64)
Critical Path	12 (1.0)	10 (0.83)
7in Random Logic		
Layout		
AREA	86786.04μm² (1.0)	60819.44μm² (0.70)
Delay Time	2.284ns (1.0)	1.590ns (0.70)
Tr.Count	800 (1.0)	644 (0.81)
Gate Width	7530μm (1.0)	3741μm (0.50)
Net Count	385 (1.0)	473 (1.23)
Power	5.87mW/MHz (1.0)	3.58mW/MHz (0.61)
Cell Count	136 (1.0)	98 (0.72)
Critical Path	10 (1.0)	10 (1.0)

Table 4 Comparison of figures per cell

	4-bit ADD-SUB		7-in Random Logic	
	CMOS	Lean	CMOS	Lean
Tr. Count /Cell	6.22	6.41	5.88	6.57
Gate Width / Cell	57μm	36μm	55μm	38.2μm
Net Count / Cell	2.9	4.7	2.8	4.8
Area / Cell	633μm²	571μm²	638μm²	620μm²
Delay Time / Cell (Average wire length)	0.302ns (351μm)	0.269ns (133μm)	0.286ns (334μm)	0.197ns (206.5μm)
Power / Cell	45.7μW/MHz	45.2μW/MHz	43.2μW/MHz	36.5μW/MHz

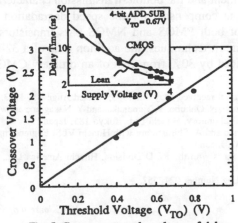

Fig. 5 Crossover supply voltage of delay
between CMOS and lean cells

A 1.5-ns 32-b CMOS ALU in Double Pass-Transistor Logic

Makoto Suzuki, *Member, IEEE*, Norio Ohkubo, Toshinobu Shinbo, Toshiaki Yamanaka, *Member, IEEE*,
Akihiro Shimizu, Katsuro Sasaki, *Member, IEEE*, and Yoshinobu Nakagome, *Member, IEEE*

Abstract—This paper describes circuit techniques for fabricating a high-speed adder using pass-transistor logic. Double pass-transistor logic (DPL) is shown to improve circuit performance at reduced supply voltage. Its symmetrical arrangement and double-transmission characteristics improve the gate speed without increasing the input capacitance. A carry propagation circuit technique called conditional carry selection (CCS) is shown to resolve the problem of series-connected pass transistors in the carry propagation path. By combining these techniques, the addition time of a 32-b ALU can be reduced by 30% from that of an ordinary CMOS ALU. A 32-b ALU test chip is fabricated in 0.25-μm CMOS technology using these circuit techniques and is capable of an addition time of 1.5 ns at a supply voltage of 2.5 V.

I. INTRODUCTION

ENHANCING the performance of macros is essential to the construction of high-performance microprocessors where macrointensive design is used to achieve a high MIPS performance. Of the many data path macros, ALU's, or adders, are the key components in processor chips for ALU's in execution units, floating-point adders, and final carry propagation adders in floating-point multipliers and digital signal processing units. A number of fast adder architectures have been proposed in the long history of computer arithmetic [1]–[7], some of which use pass-transistor logic for carry propagation [6], [7]. Pass-transistor logics gain their speed advantage over CMOS due to their high logic functionality. However, a problem with this architecture is the series connection of the pass transistors in the carry propagation path.

This paper describes circuit techniques for realizing a faster adder using pass-transistor logic. A carry propagation technique called conditional carry selection (CCS) has been developed to solve the series connection problem, and double pass-transistor logic (DPL) has been developed to improve circuit performance at reduced supply voltage. A symmetrical arrangement and the double-transmission characteristics of the DPL gate compensate for the speed degradation due to the usage of both PMOS and NMOS pass transistors. Applying these circuit techniques, the addition time of a 32-b ALU can be reduced by 30% from that of an ordinary CMOS ALU. A

Manuscript received May 18, 1993; revised August 5, 1993.
M. Suzuki, N. Ohkubo, T. Yamanaka, and Y. Nakagome are with the Central Research Laboratory, Hitachi Ltd., Tokyo 185, Japan.
T. Shinbo and A. Shimizu are with Hitachi VLSI Engineering Corporation, Tokyo 187, Japan.
K. Sasaki is with the R&D Division, Hitachi America Ltd., Brisbane, CA 94005-1819.
IEEE Log Number 9212552.

1.5-ns 32-b ALU has been developed using 0.25-μm CMOS technology and these circuit techniques [8].

Double pass-transistor logic and its characteristics are discussed in Section II. The conditional carry-selection circuit is described in Section III. Section IV describes the architecture and simulated results of a fabricated 32-b ALU test chip. Some experimental results are shown in Section V, and the conclusions are summarized in Section VI.

II. DOUBLE PASS-TRANSISTOR LOGIC

Several pass-transistor logic families for macrocell design have been proposed for improving the performance of CMOS circuits. Complementary pass-transistor logic (CPL) [9] is one example; it has been applied to the full adders in multiplier circuits and has been shown to result in high speed due to its low input capacitance and high logic functionality. However, when implementing CPL, particularly in reduced supply voltage designs, it is important to take into account the problems of noise margins and speed degradation. These are caused by mismatches between the input signal levels and the logic threshold voltage of the CMOS inverters, which fluctuates with process variations. DPL is a modified version of CPL that meets the requirement of reduced supply voltage designs.

A. DPL Gate

A basic circuit diagram of a DPL gate is shown in Fig. 1. By simply exchanging the input nodes, two-input AND/NAND, OR/NOR, XOR/XNOR gates and multiplexers can be constructed. The DPL gate consists of complementary inputs/outputs and is thus a dual rail logic like CPL. Dual rail logic has been widely used for other logic families, such as clocked CVSL [10], and for self-timed logic [11] and bipolar DCS logic [12]. DPL gates consist of both NMOS and PMOS pass transistors, in contrast to CPL gates, where only NMOS pass transistors are used.

Fig. 2 compares the construction of XOR gates in CPL, CMOS, and DPL pass-transistor logics. The CPL gate consists only of NMOS transistors, resulting in low input capacitance and high-speed operation. However, the above-mentioned problems are caused by the high output signal level being lower than the supply voltage V_{CC} by the NMOS threshold voltage V_{th}. The usual way to avoid this is to use CMOS pass-transistor logic. Full-swing operation is attained by simply adding PMOS transistors in parallel with the NMOS transistors. However, this addition results in increased input capacitance.

Reprinted from *IEEE Journal of Solid-State Circuits*, Vol. 28, No. 11, pp. 1145-1150, November 1993.

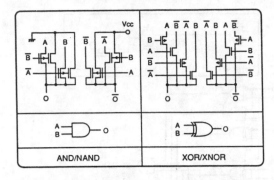

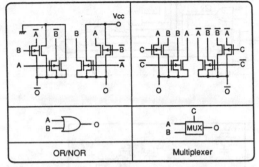

Fig. 1. Double pass-transistor logic (DPL) gates.

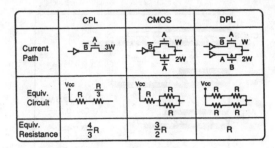

Fig. 3. Comparison of equivalent resistance for CPL, CMOS, and DPL pass-transistor logics.

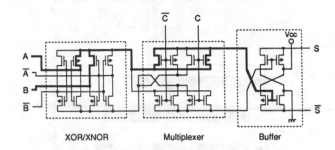

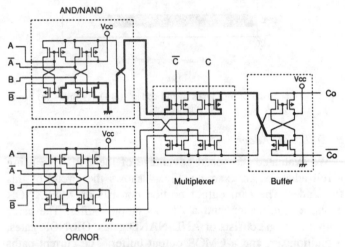

Fig. 4. DPL full adder.

Fig. 2. Comparison of CPL, CMOS, and DPL pass-transistor logics for XOR gates.

In the DPL gate, the inputs to the gates of the PMOS transistors are changed from A to B. This arrangement compensates for the speed degradation of CMOS pass-transistors in two ways. First, it is a symmetrical arrangement whereby any input is connected to the gate of one MOSFET and the source of another. In the case of the XOR/XNOR, as can be seen in Fig. 1, it is perfectly symmetrical. Any of the inputs A, $\overline{A}$, B, and $\overline{B}$ is connected to the gates of the NMOS and PMOS and to the sources of the NMOS and PMOS. This results in a balanced input capacitance and reduces the dependence of the delay time on data.

Secondly, it has double-transmission characteristics. The truth tables in Fig. 2 show how the pass transistors operate for the XOR function. In this table, the column labeled Pass shows which signals are passed and performs the XOR function. For example, in the DPL gate, both A and B are passed when A and B are low. In both the CPL and CMOS implementations, the gate input A or $\overline{A}$ controls the pass transistors. When A is

low, B is passed, and $\overline{B}$ is passed when A is high. In the DPL gate, on the other hand, there are two types of pass transistors: one is controlled by A and the other by B. The A-controlled pass transistors operate in the same way as CPL and CMOS. For the B-controlled pass transistors, when B is low, A is passed, and $\overline{A}$ is passed when B is high. As a result, there are always two current paths driving the buffer stage.

Fig. 3 compares the equivalent resistance of the pass transistors. In order to compare the driving source impedance, the equivalent resistance includes that of the CMOS buffer with the same input capacitance. This comparison is rather simplified, but it qualitatively illustrates the double-transmission property. In the DPL design, the widths of the NMOS and PMOS pass transistors are one-third and two-thirds, respectively, of the NMOS pass transistor in the CPL gate, so the input capacitance and the gate area are nearly the same for all these architectures. As shown in Fig. 3, the resistance, including that of the CMOS buffer of the previous stage, is smallest for the DPL gate due to its double-transmission property.

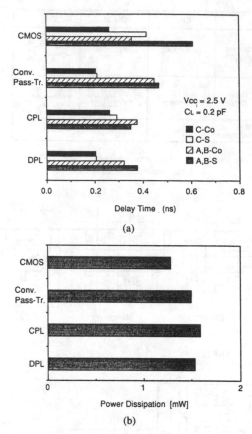

(a)

(b)

Fig. 5. Comparison of four types of full adders. (a) Delay times. (b) Power dissipation.

B. DPL Full Adder

We evaluated the speed advantage of the DPL gate using a full adder as an example. Fig. 4 shows the circuit of this full adder. The sum output portion consists of XOR/XNOR gates, a multiplexer, and a CMOS output buffer. The carry output portion consists of AND/NAND gates, OR/NOR gates, a multiplexer, and a CMOS output buffer. The current paths for the $\overline{S}$ and $\overline{Co}$ outputs when A, B, and C are all low, for example, are shown by the bold lines. These current paths include two pass transistors, and there are two current paths for each output, as discussed above.

Fig. 5 compares the simulated delay times and power dissipation of four kinds of full adders: CMOS [13], conventional CMOS pass-transistor logic [7] arranged in a dual rail structure, CPL [9], and DPL, with a load capacitance of 0.2 pF. For the slowest path that determines the speed of, for example, a multiplier, the DPL full adder is as fast as CPL, 18% faster than the conventional pass-transistor logic, and 37% faster than CMOS. As for the carry output delays ($C - Co$ and $A - Co$) that determine the ALU speed, the DPL full adder is the fastest of all. The power dissipation is simulated for 250-MHz operation with 0.2 pF loaded on each output regardless of whether it is single or dual rail logic. Under these conditions, the pass-transistor architectures show slightly higher power dissipation than CMOS because they have dual rail structure and double the load capacitance. The load capacitance determines which architecture dissipates the least power, and at lower load capacitance the dual rail pass-transistor architectures dissipate less power than CMOS.

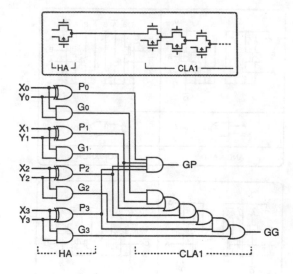

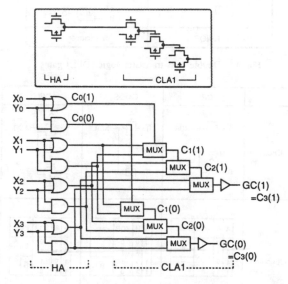

Fig. 6. 4-b carry look-ahead circuits. (a) Conventional AND–OR carry look-ahead circuit. (b) Conditional carry select (CCS) circuit.

III. CONDITIONAL CARRY SELECTION CIRCUIT

The most important component of high-speed ALU's is a look-ahead carry circuit. We have developed a new look-ahead carry scheme, called conditional carry selection (CCS). Fig. 6 compares a 4-b implementation of this scheme with a conventional AND-OR carry look-ahead circuit. In the conventional circuit [Fig. 6(a)], generated carry signals (G_j) are propagated (P_j) through an AND-OR circuit chain to form a group-generate (GG) signal, which is expressed as

$$C_3 = G_3 + P_3 \cdot [G_2 + P_2 \cdot (G_1 + P_1 \cdot G_0)]$$
$$+ P_3 \cdot P_2 \cdot P_1 \cdot P_0 \cdot C_{-1}$$
$$= GG + GP \cdot C_{-1} \qquad (1)$$
$$GG = G_3 + P_3 \cdot [G_2 + P_2 \cdot (G_1 + P_1 \cdot G_0)] \qquad (2)$$
$$GP = P_3 \cdot P_2 \cdot P_1 \cdot P_0. \qquad (3)$$

Thus, the 4-b carry look-ahead circuit involves three AND-OR circuits in its critical path. Furthermore, using pass-transistor

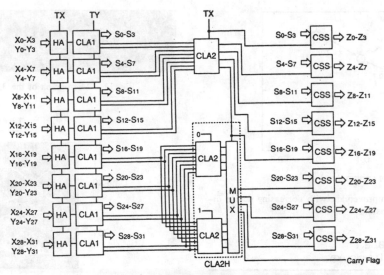

Fig. 7. Block diagram of the 32-b ALU.

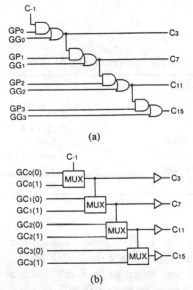

(a)

(b)

Fig. 8. Four-block carry look-ahead circuit CLA2. (a) Conventional AND-OR carry look-ahead circuit. (b) Conditional carry select (CCS) circuit.

logic, there are seven pass transistors connected in series, as shown in the inset figure above.

On the other hand, in the CCS architecture conditional carry signals for each bit $C_j(0)$ (assuming an incoming group carry of 0) or $C_j(1)$ (assuming an incoming group carry of 1) are selected by the multiplexers depending on the conditional carry signals of the previous bit, $C_{j-1}(0)$ or $C_{j-1}(1)$, as expressed by

$$C_j(k) = G_j + P_j \cdot C_{j-1}(k)$$
$$= G_j = X_j \cdot Y_j \quad \text{(if } C_{j-1}(k) = 0) \quad (4)$$
$$= G_j + P_j = X_j + Y_j \quad \text{(if } C_{j-1}(k) = 1) \quad (5)$$
$$k = 0 \text{ or } 1.$$

This conditional carry selection procedure finally forms the conditional group carries $GC(0)$ and $GC(1)$. In this way, the critical carry propagation path can be constructed by three multiplexers instead of three AND-OR gates. As shown in Fig. 6(b), the CCS architecture also avoids the series connection of

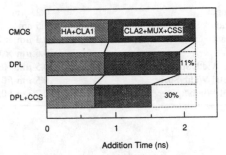

Fig. 9. Simulated comparison of 32-b ALU addition times.

pass transistors in the multiplexers, as used in a transmission-gate conditional-sum adder [7], Manchester carry chain [14], or the AND-OR carry look-ahead circuit of Fig. 6(a).

IV. 32-b ALU ARCHITECTURE

Fig. 7 shows a block diagram of the 32-b ALU based on carry select architecture [3]. DPL gates are used for all the circuits from the half adders (HA) to the final conditional-sum selection (CSS) circuits. The CCS architecture is applied not only to the 4-b carry look-ahead circuit CLA1 but also to the block carry look-ahead circuit CLA2, where four AND-OR circuits can be replaced with four multiplexers, as shown in Fig. 8. The CSS circuit consists of a multiplexer that selects the conditional sums, $S_j(0)$ or $S_j(1)$, according to the incoming block carry signal. In the carry look-ahead circuit, the upper 16 bits are processed by a conditional carry selection method whereby block carry signals are generated by CLA2, assuming the carry of the lower 16 bits C_{15} to be 0 or 1, and are then selected by the multiplexer according to the incoming true carry. This architecture enhances parallelism and results in fast operation. This is because the carry signals of the upper 16 bits are calculated in parallel with those of the lower 16 bits, and the carry signals of the upper 16 bits are generated after the delay time of a single multiplexer. The TX and TY signals select the function of the ALU.

Fig. 9 compares the simulated addition times of 32-b ALU's. An ordinary CMOS ALU uses the carry look-ahead circuits

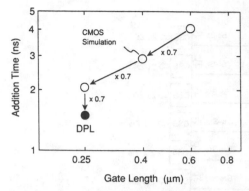

Fig. 10. Simulated addition time improvement with decreasing device dimensions.

TABLE I
PROCESS TECHNOLOGY

Technology	0.25-μm CMOS Triple Metal
MOSFET	
Gate Length	0.25 μm
Gate Oxide	6.5 nm
Contact/Via 1	0.3 μm $\times$ 0.3 μm
Via 2	0.6 μm $\times$ 0.6 μm
First Metal Width/Space	0.5 μm / 0.4 μm
Second Metal Width/Space	0.5 μm / 0.4 μm
Third Metal Width/Space	0.7 μm / 0.6 μm

of Fig. 6(a) and Fig. 8(a) and the architecture of Fig. 7, with the CMOS combinational gates like a four-input AND-OR-NOT gate. The combination of DPL with a conventional AND-OR carry look-ahead circuit reduces the addition time by 11% from that of an ordinary CMOS ALU, and the combination of DPL with a CCS carry look-ahead circuit reduces the addition time by 30%. Fig. 10 shows the simulated reduction of addition time with decreasing device dimensions. The CMOS simulation was done for an ordinary CMOS ALU with decreasing supply voltages for each generation—5 V, 3.3 V, and 2.5 V, respectively. The addition time is reduced by 30% for each generation. Therefore, the 30% improvement of the DPL and CCS architecture corresponds to a one-generation advance in process technology.

V. EXPERIMENTAL RESULTS

The 32-b ALU test chip described above was fabricated using 0.25-μm triple-metal CMOS technology. The major process parameters are summarized in Table I. Actually, this test chip was fabricated on the same wafer as an SRAM test chip [15]. An i-line stepper was used for all layers. The first metal is tungsten, and the second and third metals are aluminum. A micrograph of the test chip is shown in Fig. 11. It measures 1.58 mm $\times$ 0.38 mm (0.6 mm^2). We also designed an ordinary CMOS ALU as discussed in Section IV that measures 2.1 mm $\times$ 0.26 mm (0.55 mm^2). The area penalty of the DPL ALU, compared with the CMOS ALU, is thus 10%. This device is capable of performing 32-b additions in 1.5 ns at a supply voltage of 2.5 V, as shown in the

Fig. 11. Micrograph of the 32-b ALU test chip.

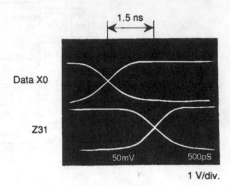

Fig. 12. Measured waveforms of the 32-b ALU test chip (1.58 mm $\times$ 0.38 mm).

TABLE II
CHARACTERISTICS OF THE 32-b ALU TEST CHIP

Organization	32-b ALU
Architecture	Carry Select Addition
Circuit	DPL
	CCS CLA
Addition Time	1.5 ns
Power Dissipation	8 mW (at 50 MHz)
Supply Voltage	2.5 V
Chip Size	0.60 mm^2

waveforms of Fig. 12. The power dissipation was measured at the maximum frequency limit (50 MHz) of the pattern generator and was found to be 8 mW with about 12% gate activity. This extrapolates to 107 mW at a 1.5-ns cycle time. The characteristics of this 32-b ALU test chip are summarized in Table II. Fig. 13 shows how the addition time depends on the supply voltage. The solid lines show the simulated results for CMOS and DPL ALU's, and the circles show the results for the DPL ALU measured at room temperature. The DPL ALU has an excellent low-voltage performance and agrees well with the results of circuit simulation. DPL AND/NAND and OR/NOR ring oscillators have also been fabricated, showing speed improvements of 15% and 30% compared with CMOS NAND and NOR ring oscillators, respectively.

VI. CONCLUSION

Double pass-transistor logic (DPL) has been developed to improve circuit performance at reduced supply voltage. Its symmetrical arrangement and double-transmission characteristics compensate for the speed degradation arising from the use of PMOS and NMOS pass transistors. A carry propagation circuit technique called conditional carry selection (CCS) has been developed to solve the problem of series-connected pass transistors in the carry propagation path. By combining these

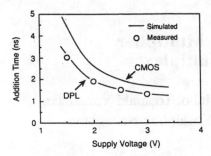

Fig. 13.   Variation of the addition time of 32-b ALU's with supply voltage.

circuit techniques, the addition time of the 32-b ALU can be reduced by a substantial 30% from that of an ordinary CMOS ALU. A 1.5-ns 32-bit ALU has been developed using 0.25-μm CMOS technology and these circuit techniques. It should also be possible to apply the proposed DPL gates and CCS adder architecture to other data path macros, such as floating-point units, resulting in processing units with very high performance.

ACKNOWLEDGMENT

The authors wish to thank Dr. K. Shimohigashi, Dr. T. Nishimukai, Dr. T. Nagano, and Dr. M. Hiraki for their useful discussions, and K. Ueda and K. Takasugi for their assistance and support. The authors are also greatly indebted to T. Nishida, N. Hashimoto, N. Ohki, and H. Ishida for their assistance with the process technology and device fabrication.

REFERENCES

[1] J. Sklansky, "An evaluation of several two-summand binary adders," *IRE Trans. Electron. Comput.*, vol. EC-9, pp. 213–226, June 1960.
[2] J. Sklansky, "Conditional-sum addition logic," *IRE Trans. Electron. Comput.*, vol. EC-9, pp. 226–231, June 1960.
[3] O. J. Bedrij, "Carry-select adder," *IRE Trans. Electron. Comput.*, vol. EC-11, pp. 340–346, June 1962.
[4] C. L. Chen, "2.5-V bipolar/CMOS circuits for 0.25-μm BiCMOS technology," *IEEE J. Solid-State Circuits*, vol. 27, pp. 485–491, Apr. 1992.
[5] K. Yano *et al.*, "3.3-V BiCMOS circuit techniques for 250-MHz RISC arithmetic modules," *IEEE J. Solid-State Circuits*, vol. 27, pp. 373–381, Mar. 1992.
[6] H. Hara *et al.*, "0.5-μm 3.3-V BiCMOS standard cells with 32-kilobyte cache and ten-port register file," *IEEE J. Solid-State Circuits*, vol. 27, pp. 1579–1584, Nov. 1992.
[7] A. Rothermel *et al.*, "Realization of transmission-gate conditional-sum (TGCS) adders with low latency time," *IEEE J. Solid-State Circuits*, vol. 24, pp. 558–561, June 1989.
[8] M. Suzuki *et al.*, "A 1.5 ns 32 b CMOS ALU in double pass-transistor logic," *ISSCC Dig. Tech. Papers*, pp. 90–91, Feb. 1993.
[9] K. Yano *et al.*, "A 3.8-ns CMOS 16 × 16-b multiplier using complementary pass-transistor logic," *IEEE J. Solid-State Circuits*, vol. 25, pp. 388–395, Apr. 1990.
[10] L. G. Heller *et al.*, "Cascode voltage switch logic: A differential CMOS logic family," *ISSCC Dig. Tech. Papers*, pp. 16–17, Feb. 1984.
[11] J. Yetter *et al.*, "A 100 MHz superscalar PA-RISC CPU/coprocessor chip," in *1992 Symp. VLSI Circuits Dig. Tech. Papers*, pp. 12–13, June 1992.
[12] E. B. Eichelberger *et al.*, "Differential current switch—high performance at low power," *IBM J. Res. Develop.*, vol. 35, no. 3, pp. 313–320, May 1991.
[13] M. Uya *et al.*, "A CMOS floating point multiplier," *IEEE J. Solid-State Circuits*, vol. SC-19, pp. 697–702, Oct. 1984.
[14] N. Weste and K. Eshragian, *Principles of CMOS VLSI Design: A Systems Perspective*. Reading, MA: Addison-Wesley, 1988.
[15] T. Yamanaka *et al.*, "A 2.3 μm^2, single-bit-line SRAM cell with high soft-error-immune structure," presented at 1993 Symp. VLSI Technology, May 1993.

A 4.4-ns CMOS 54X54-b Multiplier Using Pass-transistor Multiplexer

Norio Ohkubo, Makoto Suzuki, *Toshinobu Shinbo, Toshiaki Yamanaka,
*Akihiro Shimizu, **Katsuro Sasaki, and Yoshinobu Nakagome

Central Research Laboratory, Hitachi Ltd.,
1-280 Higashi-Koigakubo, Kokubunji, Tokyo 185, Japan.

*Hitachi VLSI Engineering Corporation, Kodaira, Tokyo 187, Japan.
**R&D Division, Hitachi America Ltd., Brisbane, CA 94005-1819

Abstract

A 54 X 54-b multiplier using pass-transistor multiplexer has been fabricated by 0.25-μm CMOS technology. To enhance the speed performance, a new 4-2 compressor and a carry look-ahead adder (CLA) both featuring the use of pass-transistor multiplexers have been developed. The new circuits have a speed advantage over conventional CMOS circuits because the number of critical-path gate stages is minimized due to the high logic functionality of pass-transistor multiplexers. The active size of the 54 X 54-b multiplier is 3.77 mm X 3.41 mm. The multiplication time is 4.4 ns at 2.5 V power supply.

Introduction

Enhancing the performance of floating point operation is indispensable for current high-performance microprocessors. In particular, high speed multiplication operation is becoming one of the keys in RISCs, DSPs, graphics accelerators and so on, because of increasing demand from multimedia applications. Recent high-end microprocessors call for an operation frequency of 200 MHz or over, and a multiplier will be required to operate in one clock cycle. However, no 54 X 54-b multiplier with a delay time less than 5 ns has yet been reported [1][2]

This paper describes a 54 X 54-b multiplier macro developed for the mantissa multiplication of two double-precision numbers as outlined in the IEEE standard. To reduce the multiplication time, a new 4-2 compressor and a carry look-ahead adder (CLA) featuring pass-transistor multiplexers have been developed. The new circuits gain a speed advantage over conventional CMOS circuits because the number of critical-path gate stages is minimized due to

the high logic functionality of pass-transistor multiplexers. The 54 X 54-b multiplier was fabricated by triple-metal 0.25-μm CMOS technology.

Architecture

The block diagram of the 54 X 54-b multiplier is shown in Fig. 1. We used Booth's algorithm, Wallace's tree and a conditional carry-selection (CCS) adder [3]. The number of partial products is halved by Booth's algorithm. Without propagation of the carry, partial products are summed by Wallace's tree. The summed results are added by CCS adder with high-speed carry propagation.

Wallace's tree used the 4-2 compressor, which has five inputs and three outputs. Carry-out (Co) is connected to the next 4-2 compressor's carry-in (Ci), as shown in Fig. 1. Without propagating the carry to the higher bit, the 4-2 compressor can add four partial products because the carry-out (Co) does not depend on the carry-in (Ci). By using the 4-2 compressor, only four addition stages are needed for Wallace's tree as shown in Fig. 1.

Circuit and Layout Design

The 4-2 compressor circuits using pass-transistor multiplexers are shown in Fig. 2. Since the pass-transistor multiplexer circuit, as shown in Fig. 3, has high logic functionality, a full adder circuit is constructed by three pass-transistor multiplexers. The 4-2 compressor is constructed of two full adders, and there are four critical-path gate stages, as shown in Fig. 2(a). This circuit is faster than the conventional CMOS circuit. For further speed improvement, we used an improved 4-2 compressor. The number of critical-path gate stages for this circuit becomes

Reprinted from *Proceedings of the IEEE Custom Integrated Circuits Conference*, pp. 26.4.1-26.4.4, May 1994.

three by exploiting parallelism, as shown in Fig. 2(b). The simulated delay comparison for these 4-2 compressor circuits is shown in Fig. 4. The proposed circuit reduces the propagation delay time by 18% from that of full-adder-based circuit. The construction of Wallace's tree is shown in Fig. 5. By using the 4-2 compressor, the construction of Wallace's tree can be simplified.

The carry look-ahead adder (CLA) in the final adder also uses pass-transistor multiplexers, as shown in Fig. 6. We have already reported a new look-ahead carry scheme called conditional carry-selection (CCS) [3]. The 4-bit CLA is constructed by three multiplexers and is faster than the conventional pass-transistor-based design by avoiding series-connected pass-transistors in the carry propagation path. To apply this scheme to the final 108-bit adder, the 4-bit CLA is modified to an 8-bit CLA, as shown in Fig. 6. The new 8-bit CLA achieves four critical-path gate stages by exploiting the parallelism. It reduces the 108-bit addition time to 1.52 ns.

Fabrication

The chip was fabricated by triple-metal 0.25-μm CMOS technology. Table 1 shows the process technology. The 1st metal is tungsten, and the 2nd and 3rd metals are aluminum. It operates from supply voltage of 2.5 V. Figure 7 shows a micrograph of the chip. 100,200 transistors are integrated in the active area of 3.77 mm X 3.41mm.

Evaluation

The simulated multiplication time of the 54 X 54-b multiplier is shown in Fig. 8. The multiplication time is 4.4 ns with a 2.5 V power supply. It shows excellent characteristics at such a low voltage because of the pass-transistor-multiplexer-based design where both NMOS and PMOS are turned on. The Characteristics of this 54 X 54-b multiplier test chip are summarized in Table 2. The measured waveforms of Wallace's tree are shown in Fig. 9. The measured delay was almost the same as that of the simulated value.

Figure 10 shows the multiplication time plotted against the device dimensions. The multiplication time with the full-adder-based circuit is estimated to be 5.1 ns. Therefore, this multiplier achieves 14% improvement in multiplication time due to the use of the new circuits with pass-transistor multiplexers.

Conclusions

A new 4-2 compressor and CLA using pass-transistor multiplexers have been developed to shorten multiplication time. The multiplication time of the 54 X 54-b multiplier is reduced by 14% due to the reduction of the critical-path gate stages by using pass-transistor multiplexers. A 4.4-ns multiplication time was achieved with 0.25-μm CMOS technology.

Acknowledgments

The authors wish to thank Dr. K. Shimohigashi, Dr. T. Nishimukai, Dr. E. Takeda, and Dr. T. Nagano for their useful discussions, and K. Ueda and K. Takasugi for their assistance and support. The authors are also greatly indebted to T. Nishida, A. Fukami, N. Ohki, and H. Ishida for their assistance with the process technology and device fabrication.

References

[1] G. Goto et al., "A 54 X 54-b regularly structured tree multiplier," IEEE J. Solid-State Circuits, vol. 27, pp. 1229-1236, September 1992.

[2] J. Mori et al., "A 10-ns 54 X 54-b parallel structured full array multiplier with 0.5-μm CMOS technology," IEEE J. Solid-State Circuits, vol. 26, pp. 600-606, April 1991.

[3] M. Suzuki et al., "A 1.5ns 32b CMOS ALU in double pass-transistor logic," in 1993 ISSCC Dig. Tech. Papers, pp. 90-91, February 1993.

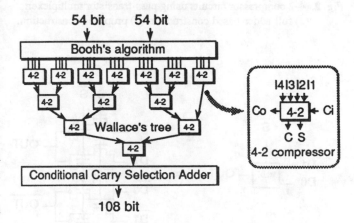

Fig. 1. Block diagram of the 54 X 54-b multiplier using pass-transistor multiplexer.

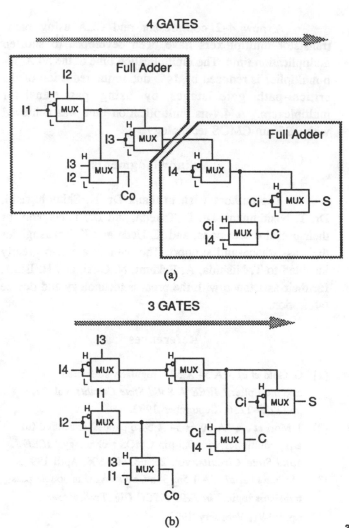

(a)

(b)

Fig. 2. 4-2 compressor circuits using pass-transistor multiplexer:
(a) full-adder-based construction (b) proposed construction.

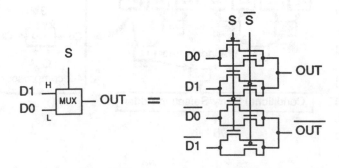

Fig. 3. Pass-transistor multiplexer circuit.

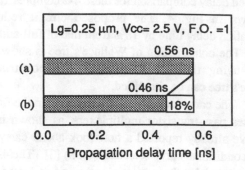

Fig. 4. Simulated comparison of 4-2 compressor circuits:
(a) full-adder-based construction (b) proposed construction.

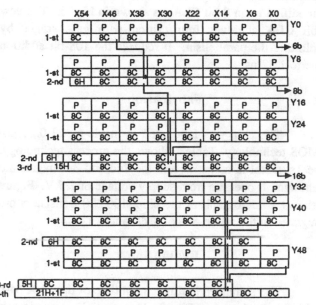

P : 8X4 Partial product generators
C : 4-2 Compressor
H : Half adder
F : Full adder

Fig. 5. Construction of Wallace's tree.

Table 1. Process technology

Technology	0.25-μm CMOS Triple metal
Gate length	0.25 μm
Gate oxide	6.5 nm
1st Metal Width/Space	0.5 μm / 0.4 μm
2nd Metal Width/Space	0.5 μm / 0.4 μm
3rd Metal Width/Space	0.7 μm / 0.6 μm

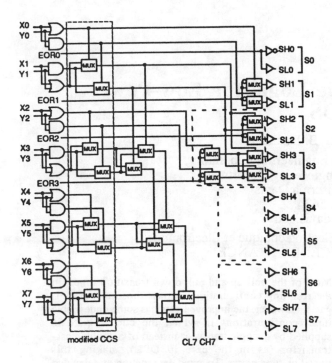

Fig. 6. 8-bit CLA using pass-transistor multiplexer.

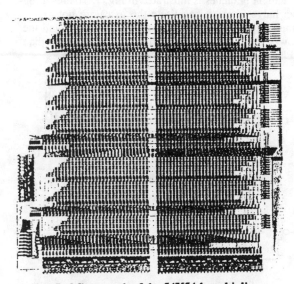

Fig. 7. Micrograph of the 54X54-b multiplier.

Table 2. Characteristics of the 54X54-b multiplier

Organization	54X54-b multiplier
Multiplication time	4.4 ns
Active Area	3.77 X 3.41 mm
Transistors	100200

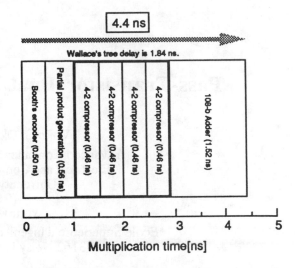

Fig. 8. Multiplication time of 54X54-b multiplier.

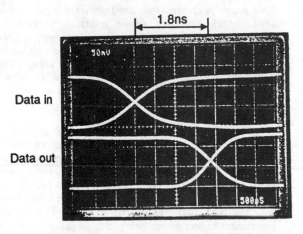

Fig. 9. Measured waveforms of the Wallace's tree.

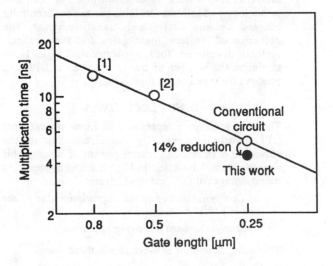

Fig. 10. 54X54-b multiplication time versus device dimension.

69

Pass-Transistor Dual Value Logic for Low-Power CMOS

Vojin G. Oklobdzija, B. Duchêne*

Advanced Computer System Engineering Laboratory,
Electrical and Computer Engineering Department
University of California Davis
(916) 752-5634
vojin@ece.ucdavis.edu

*Ecole Superieure d'Ingenieurs en Electrotechnique et Electronique
93162 Noisy le Grand CEDEX FRANCE

ABSTRACT

This paper presents new pass-transistor logic termed DVL which contains fewer transistors than its counterpart DPL yet maintaing comparable performance. A method for synthesis of such networks is also developed and demonstrated in this paper. The new logic is characterized by good speed and low power. The simulations and tests were performed using 1-μm CMOS.

I. INTRODUCTION

New logic CMOS families using pass-transistor circuit techniques have recently been proposed with the objective of improving speed and power consumption [1-6]. This logic (in most cases) passes the charge between the nodes rather than charging the nodes from V_{CC} and then discharging them to GND. This feature contributes to less power being used as compared to the regular CMOS. The Double Pass-Transistor Logic (DPL), developed by Hitachi demonstrated an 1.5nS 32-bit ALU and 4.4nS 54-bit multiplier in 0.25 μm technology [4,5]. However, DPL has not yet been fully adopted because of its high transistor count. The objective of the new logic gates and the synthesis method developed for pass-transistor logic is to minimize the number of transistors used in DPL and preserve the speed of the logic.

II. NEW LOGIC GATES

The new logic gate represents an improvement over DPL family achieved by the elimination of the redundant branches and rearrangement of signals. This simplification, illustrated in Fig. 1, 2 and 3, preserves the advantages of DPL gates which are:

  a) Compensation of speed degradation due to the use of pMOS transistors.

  b) Straightforward full swing operation.

This simplification is achieved by in three steps:

A. *Elimination of the redundant branches*

This simplification is achieved by eliminating the redunant branches (shown in shaded area) from DPL.

Most of the pull up and pull down transition times, in the resulting configuration, surpass those of the DPL gates. However, the improved gate has some undesirable input configurations in which the current path is supplied by a single transistor instead of a double pass-transistor (as in the case in DPL), making this transition time worse. To avoid degradation of delay due to the use of just one pMOS transistor, the particular transistor width is increased. The elimination of redundant branches is illustrated in Fig. 1. The resulting two halves (which constitute the gate) are not of the same speed. The faster half is NAND (60pS) and the slower is AND (70pS), which is still being faster than DPL (75pS).

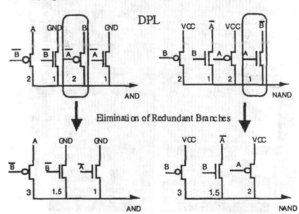

Fig. 1. Elimination of redundant branches

B. *SiMrrangement*

The use of two parallel pMOS transistors is avoided by simple signal re-arrangement because the two pMOS transistors contribute more to the delay than one pMOS transistor in parallel with one nMOS transistor. This is especially true in a pull up operation. The current is always provided by one nMOS transistor alone or by one nMOS and one pMOS in parallel. The AND/NOR DPL gate (in Fig. 2.) is obtained from NOR/OR DPL configuration whose inputs are simply inverted. Signal rearrangement applied to AND/NOR DPL gate results

Reprinted from *Proceedings of the 1995 International Symposium on VLSI Technology*, pp. 341-344, May/June 1995.

in an AND gate configuration which is faster than DPL (60pS vs. 75pS), where AND is a faster half.

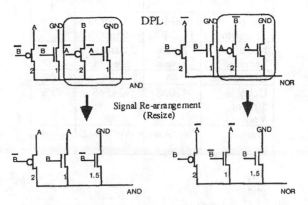

Fig.2. Signal Re-arrangement

C. Selection of the faster halves

Finally we take a faster half from Fig.1. and form Fig.2. The resulting AND/NAND complementary logic gate (shown in Fig.3.) is obtained by elimination of the redundant branches for the NAND and rearrangement of signals for the AND gates respectively. We named this logic: DVL (Dual Value Logic). The resulting AND/NAND DVL gate contains a total of 6 transistors as compared to DPL consisting of 4 transistors of each type. There is a total of 9 inputs in DVL versus 12 in DPL resulting in a smaller capacitive load of DVL gate. In DVL 3 inputs are connected to the transistor source and 6 to the gate (3 to p-type and 3 to n-type). In DPL 4 inputs are connected to the source and 8 to the gate (4 to p-type and 4 to n-type transistors).

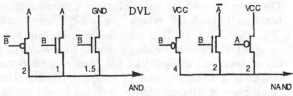

Fig. 3. Resulting DVL Gate

The comparison between NAND/AND DPL gate and DVL shows:

- *20% speed improvement, using 75% of transistors used in DPL gate.*

- *25% less connections and wires than in DPL gate. The 4% area penalty comparison to DPL is quite negligible.*

Similar arguments can be used to build the NOR/OR gates from DPL gates.

III. SYNTHESIS METHOD

The synthesis method for DVL is based on the method used to create the logic gates described before. At present, there is no known algorithm to find minimal multi-stage logic circuits. The new proposed method for synthesis of DVL, is based on transistors instead of logic gates. In place of conjointly assembling several basic gates, functions are synthesized at the transistor level. In addition, the programming of this method has been developed to prove the efficiency of the theory presented.

The key point of DVL synthesis consists of employing Karnaugh-Map at the transistor level. Thus, we are not cascading several logic gate levels (NAND/AND or NOR/OR) but building functions by directly using several transistor levels in series. However, the choice of *pseudo Karnaugh-Map* in the programming for less than 8 inputs was done because its explanation is simple, but a *pseudo Quine McCluskey* technique could have been adopted instead.

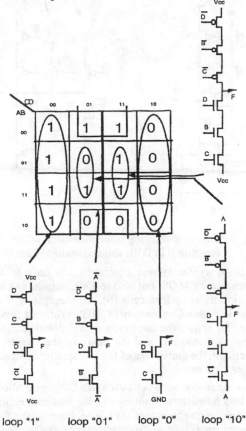

Fig. 4. Loops allowed for 4 inputs

Usually, the general Karnaugh-Map is covered by loops of "0" or "1", in such a way that a minimized Sum of Products form is obtained. In our case, four classes of loops are allowed (as illustrated in Fig. 4) to directly synthesize a part of the final circuit. Accumulating all loops necessary to cover the Karnaugh-Map yields the resulting circuit.

IV. RESULTS

The best way to compare efficiency of the presented algorithm is via synthesizing and simulating circuits obtained using our automated algorithm, and comparing it to circuits produced by CPL and DPL using the concept of logic gates.

$$\overline{F} = \overline{B}C + AB\overline{C}$$

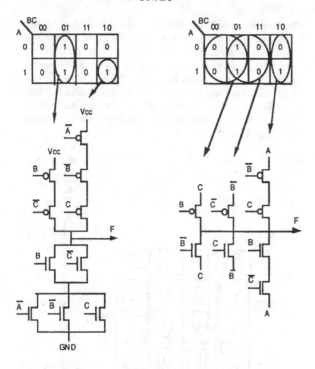

Fig. 5. Example showing implementation of the function F in DVL and conventional CMOS

The DVL synthesis was compared not only to the Conventional CMOS, but also to DPL circuits [4] and CPL circuits using lean cells [6]. An example of DVL synthesis versus Conventional CMOS synthesis (given in Fig. 5.) shows the improvement in global size, the number of transistors and the delay of the circuit. In each circuit, the global size of DVL is smaller compared to other circuits.

The comparison with AND/NAND DPL gate shows 25% less transistors resulting in 25% less connections and wires in an equivalent DVL gate, keeping the total transistor area constant. Similar methods can be used to build the NOR/OR gates.

A. Comparison with CMOS

A comparison between DVL and conventional CMOS is given in Table I, while the simulation results (for the given function $F2 = \overline{\overline{B}C + AB\overline{C}}$) are shown in Fig. 6.

TABLE I.
COMPARISON BETWEEN DVL AND CONVENTIONAL CMOS

Function F2	CMOS	DVL	Savings
No. of Transistors	10nMOS 10pMOS	8pMOS 8nMOS	20%
No. of Levels	3 gate levels	2 transistor levels	
Global size	44	36	18%
Delay @ 50% of Vout	430pS	245pS	43%
Transitor ratio	Wp/Wn=2	Wp/Wn=2	

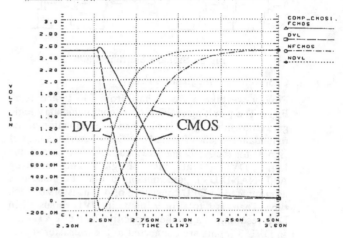

Fig. 6. Delay Comparison between DVL and Conventional CMOS for a 3 inputs function F2

B. Comparison with DPL

The function used for comparison is a three variable function $F_2(A,B,C)$, where $F2 = \overline{\overline{B}C + AB\overline{C}}$ and $\overline{F2} = \overline{B}C + AB\overline{C}$. This function was implemented using 4 DPL gates in two logic levels. Afterwards this circuit was built using DVL. The load applied to the output is a standard load of two gate inputs. The comparison results are shown in Table II and timing simulation for the function F_2 are shown in Fig. 7.

TABLE II.
COMPARISON BETWEEN DPL AND DVL FOR A 3 INPUTS FUNCTION F2

Function F2	DPL	DVL	Savings
No. of Transistors	16pMOS 16nMOS	8pMOS 8nMOS	50%
No. of Levels	2 gate levels	2 transistor levels	
Global size	48	30	37.5%
Delay @50%	290pS	120pS	58.6%
@80%	350pS	240pS	31.4%
Transistor ratio	Wp/Wn=2	Wp/Wn=2 2 nMOS=1.5	

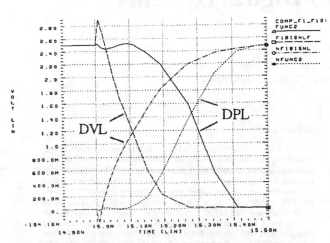

Fig. 7. Simulated delays for the 3 inputs DPL function F2 and DVL implementation of F2

C. Comparison with CPL

The function F published by Yano in [6] was synthesized for DVL and compared to CPL which uses lean cells and special inverters [6]. Delays were measured for 1 cell, 2 cells and 3 cells cascaded. DVL circuit is made with conventional inverters. The comparisons were made using the output load of 15 FF in both cases.

TABLE III.
COMPARISON BETWEEN CPL AND DVL FOR A 4 INPUTS
FUNCTION.

	1 cell	2 cells	3 cells	Cell size
	375pS (100%)	760pS (100%)	1,150p S (100%)	105µ (100%)
Circuit F in DVL	380pS (101%)	660pS (86%)	950pS (82%)	108µ (103%)

The comparison between CPL and DVL is shown in Table III and the delays of the two cascaded CPL and DVL cells are compared in Fig. 8.

CONCLUSION

DVL logic family has been developed which has advantages over standard CMOS as well as new pass-transistor families such as DPL and CPL. However, the exact speed improvement is dependent on each particular circuit. The power consumption is also reduced for 30-50% over conventional CMOS. Generation of DVL is supported by an automated synthesis tool based on the algorithm developed in the course of this work.

ACKNOWLEDGMENT

The authors acknowledge the contributions of P. Lee, W. Wong and H. Yu.

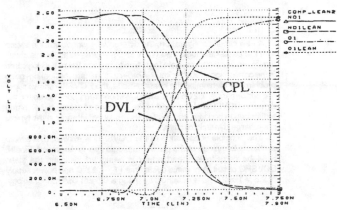

Fig. 8. Delays comparison between 2 cascaded DVL and CPL cell.

REFERENCES

[1] F.S. Lai and W. Hang, "Differential Cascode Voltage Switch with Pass Gate Logic Tree for High Performance CMOS Digital Systems", *1993 International Symposium on VLSI Technology, Systems and Applications*, pp358-362, May 1993.

[2] Yano, K, et al, "A 3.8 ns CMOS 16X16-b Multiplier Using Complementary Pass-Transistor Logic", *IEEE J. Solid State Circuits*, vol 25, p388-395, April 1990.

[3] Akilesh Parameswar, et al, "A Swing Restored Pass-Transistor Logic Based Multiply and Accumulate Circuit for Multimedia Applications", *Proceedings of the IEEE 1994 Custom Integrated Circuit Conference, San Diego, California, May 1-4, 1994*.

[4] Makoto Suzuki, et al, "A 1.5 ns 32 b CMOS ALU in Double Pass-Transistor Logic", *1993 ISSCC Dig. Tech. Papers*, pp90-91, February 1993.

[5] Ohkubo, N., et al, "A 4.4nS CMOS 54x54-b Multiplier Using Pass-Transistor Multiplexer",*Proceedings of the IEEE 1994 Custom Integrated Circuit Conference, San Diego, California, May 1-4, 1994*.

[6] Yano, K, et al, "Lean Integration : Achieving a Quantum Leap in Performance and Cost of Logic LSIs", *Proceedings of the IEEE 1994 Custom Integrated Circuit Conference, San Diego, California, May 1-4, 1994*.

Differential Cascode Voltage Switch with the Pass-Gate (DCVSPG) Logic Tree for High Performance CMOS Digital Systems

F.S. LAI AND W. HWANG

IBM T.J. WATSON RESEARCH CENTER

P.O. BOX 218, YORKTOWN HEIGHTS, NEW YORK 10598

Abstract

A new circuit configuration, the differential cascode voltage switch with the pass–gate logic tree (DCVSPG), is presented. In this circuit family, we use the pass–gate logic tree to replace the nMOS logic tree in the conventional DCVS circuit in order to eliminate the floating–node problem. By eliminating the floating–node, the DCVSPG shows superior performance, silicon area and power consumption. Moreover, the dynamic DCVSPG also provides the leverage of relieving the charge redistribution concern and reinforces the signal integrity in the typical pre–charge dynamic circuits.

The principle of operation of the DCVSPG is explained. A simple synthesis technique of the pass–gate logic tree is discussed. Finally, a 64–bit carry look–ahead adder is designed by using the static DCVSPG circuit. A nominal cycle time (T_a = 22°C and power supply of 2.5 V) of 2.0 ns is obtained by using a 0.5 μm CMOS technology.

I. Introduction

The conventional cascode voltage switch (DCVS) is claimed to have advantages over the traditional static CMOS NAND/NOR design in terms of circuit delay, layout area, logic flexibility and power dissipation [1,2]. For the dynamic implementation, DCVS also shows the superior logic implementation flexibility over the standard domino logic which is suffered from the lacking of inverting gates [3]. However, the conventional DCVS can cause floating–node in both legs of their nMOS logic tree. This floating–node causes the static DCVS to become ratioed logic which in turn creates current spikes and additional delay [4]. Although the ratioed logic problem can be solved in dynamic DCVS by using the pre–charging scheme. Unfortunately, the floating–node problem still exist in the dynamic DCVS, and it will trigger another problem such as the charge redistribution. The result of charge redistribution might develop a false logic evaluation. This makes the dynamic circuit very un–reliable. Complementary pass–transistor logic (CPL) [5] was developed to solve this floating–node problem. The loading in the CPL was chosen using static inverters instead of a cross–coupled pMOS latch in conventional DCVS to restore the signal. This results in a mismatch problem between the input signal level and the logic threshold voltage of the static CMOS inverter. It also caused the CPL to have poorer noise margin and speed degradation. Recently, the double pass–transistor logic (DPL) [6] was developed to solve the CPL problem at the expense of double transistor counts and silicon area. In this paper, a new DCVSPG circuit family is developed [1] to overcome the above mentioned drawbacks in DCVS and CPL. The regeneration problem in DCVS caused by the floating–node is solved by the pass–gate network.

In the following sections, the operation principles of static and dynamic DCVSPG circuits will be presented first. Then the comparison of sum circuits among DCVS, DCVSPG and static CMOS are discussed. Finally, the implementation of a 64–bit carry look–ahead adder by using a 0.5 μm CMOS technology will be described. Conclusions will be given in the last section.

II. DCVSPG Circuit Operation Principle

In this section, the basic operation of static and dynamic DCVSPG will be presented. The synthesis of pass–gate logic tree will also be described.

(i) Static Circuit

Fig. 1 shows the typical static DCVSPG to evaluate the AND function of $q = ab$. The an and bn are the complementary signals of input variables a and b respectively. Initially, we assume both of a and b signals are low, the N2 and N4 transistors all turn OFF. However, the an and bn signals are all high which in turn switch the N1 and N3 transistors ON. It shows that node q is low and node qn is high. This leads to the cross–coupled pMOS transistor P1 is ON and P2 is OFF. When both of a and b signals swing from low to high, the node q is instantly charged up to high through the N4 transistor. This makes the P1 transistor turn OFF while the node qn is discharging through N2 transistor. This is great contrast to the conventional DCVS circuit shown in Fig. 2. In the transition period, the node q kept low momentarily which let the P1 transistor ON while the node qn is discharging in the conventional DCVS circuit. This floating–node problem causes the conventional DCVS to have larger power consumption and propagation delay. Besides that, for the DCVSPG logic, the pMOS is only used as a load to bring up the full–swing signal, it is not the critical device in the pull–up operation. Therefore, the device size can be small. In the conventional DCVS, however, pMOS has to be twice as large as the nMOS device in order to get a comparable pull–up operation. This leads to a larger silicon area consumption.

Figs. 3 and 4 show the ASTAP simulation results of logic function AND of $q = ab$ for the conventional DCVS and DCVSPG with the function of the pMOS device width. In those simulations, we set the nMOS device width constant (W_n = 20 μm). It is very interesting to note that the rise time is a very strong function of the pMOS width for the DCVS circuit. It is obvious that the rise time decreases when pMOS device width increases. However, the rise time increases again when the pMOS device width increases beyond 10–20 μm. This indicates that the conventional DCVS has the ratioed circuit problem. An optimum pMOS device has to be carefully chosen. On the other hand, the rise time of DCVSPG is almost constant due to the pull–up behavior is mainly done by the nMOS instead of the pMOS. The overall performance of the DCVSPG is much superior to that of the DCVS at any pMOS width and loading.

(ii) Dynamic Circuit

Fig. 5 shows the AND logic function implementation of $q=ab$ for conventional dynamic DCVS and Fig. 6 for the dynamic DCVSPG. When clock signal ϕ is low, both of q and qn nodes are charged up to V_{dd} high level signal for both of these circuits. Both of circuits start to evaluate the logic function when ϕ is high. For the dynamic DCVS, the node qn is starting to discharge when signal a and b swing from low level 0 to high level V_{dd}. The stored charge in node qn will flow through the transistors N3, N4 and N5 to ground. Node q is presumably staying in V_{dd} voltage level due to the transistors N1 and N2 turn OFF. However, for the dynamic DCVSPG, the node qn is discharged through transistors N2 and N6 to node bn which is in the ground state. The node q is charged up through transistors N1 and N4 by node b which is in the V_{dd} state.

There are several distinct characteristics in DCVSPG in comparison with the conventional DCVS circuit. The advantages are no floating–node generation in both of logic tree legs, symmetrical logic topol-

Reprinted from *Proceedings of the 1993 International Symposium on VLSI Technology*, pp. 358-362, June 1995.

ogy and shorter logic stack height. Elimination of the floating–node prevents the static circuit from suffering the ratioed logic problem and the dynamic circuit from the charge redistribution problem. The symmetrical logic topology in the logic tree and the shorter logic stack height improve the circuit performance and power consumption [3].

(iii) Synthesis of Pass–Gate Logic Tree

The synthesis of conventional n–channel logic tree for DCVS had been discussed by using the modified Karnaugh map or the modified Quine–McCluskey tabular method [4]. The synthesis of pass–gate logic had also been explained [7]. However, the synthesis algorithm they showed is either not quite clear or too simple. Let us show a synthesis procedure by using a recursive minimization with the Karnaugh map.

In order to synthesis the logic function $F = \bar{a}\,\bar{b}\,\bar{c}\,\bar{d} + a(b + c + d)$, where $\bar{a}\,\bar{b}\,\bar{c}\,\bar{d}$ are an, bn, cn, dn in our figures, into the pass–gate logic, the Karnaugh map result is shown in Fig. 7. The ab is assumed to be the control variables and cd is the input function variables. By grouping the same output function pattern together, the pass–gate logic can be minimized as $F = \bar{a}\,\bar{b}\,(x_1) + b(x_2) + a(x_2)$ as shown in Fig. 7. The x_1 pattern is [1010] and x_2 pattern is [0011]. These two patterns can be continuously minimized as $x_1 = c(d) + \bar{c}(\bar{d})$ and $x_2 = c(1) + \bar{c}(0)$, where 1 is the V_{dd} state and 0 is the ground state. The final pass–gate logic function is then $F = \bar{a}\,\bar{b}\,[c(d) + \bar{c}(\bar{d})] + b[c(1) + \bar{c}(0)] + a[c(1) + \bar{c}(0)]$. The DCVSPG circuit is shown in Fig. 7(b). This circuit is much simpler and faster than the pure static CMOS implementation.

III. Comparison of The sum Circuits

In order to compare the circuit performance of various design techniques, the sum circuit of the full adder is being simulated using ASTAP. Fig. 8 shows the static and dynamic versions of the conventional DCVS design technique. The static and dynamic implementation of the DCVSPG is shown in Fig. 9. The static CMOS design of the sum circuit is also shown in Fig. 10.

The device width is designed following a basic rule that the conductance of all the discharging path are assumed to be the same as a conductance of a minimum size ($W_n = 3\,\mu m$) nMOS transistor. For example, in the conventional DCVS static circuit shown in Fig. 8(a), three transistors are seriesly connected along the discharge path. The transistor width is then chosen as $3\,\mu m \times 3 = 9\,\mu m$. For 8(b), however, the device size is then increased up to $3\,\mu m \times 4 = 12\,\mu m$ in the dynamic DCVS configuration. The pMOS device size is chosen as twice larger than that of the nMOS device.

The overall simulation results are shown in Table I. The load output capacitance is assumed that the circuit drives a chain of the similar circuits. With a fan–out of one for the static circuit, it is obvious that the output capacitance is the same with the input capacitance. The dynamic circuits, however, are buffered by the C²MOS latch and would be expected to be able to drive larger loads than that of the static gate. A fan–out of two was then used for the dynamic designs.

Considering, first, the static design, it appears that the static DCVSPG yields the best power–delay product. The DCVSPG has the lowest logic tree stack height such that its transistor size and input capacitance are the smallest. And yet its best performance is solely due to the pull–up and pull–down are all done by the high–performance nMOS transistor. The lowest power consumption of the static DCVSPG is also due to the very symmetrical charging and discharging times of nodes q and qn in Fig. 9(a). The asymmetrical charging and discharging periods, however, of the conventional static DCVS causes a prolong transient time when the transistor switches and thus dissipates more power. The DCVS and DCVSPG have the area advantages over the conventional static CMOS circuit due to the redundant pMOS transistors are reduced dramatically in the DCVS configuration.

For the dynamic circuits, it is interesting to note that the power–delay product of the dynamic DCVSPG is the same with the static counterpart in the sum circuit. Of course, its speed is almost twice faster than that of the static DCVSPG at the expense of the larger device count and silicon area.

IV. Static 64–bit Adder Architecture

The whole adder core is shown in Fig. 11. This architecture is implemented by the binary carry look–ahead algorithm [8,9]. There are total 12 rows shown here. The first row is the PG circuit to generate the p (propagation) and g (generate) signals. The last row is the sum circuit. There are total 10 rows to generate the carry signal. Inside the 10 rows of carry chain, the white rectangular and triangular are the buffer circuit and driver circuit respectively. The black rectangular is the merge circuit. Some of white rectangular cell can also route the signal from top to its left to feed into the black merge circuits. The sum bit 0 comes from the right hand side.

From the carry look–ahead theory, the sum bit can be written as
$$s_i = a_i \oplus b_i \oplus c_{i-1} \qquad (1)$$

The generation and propagation bits are defined as
$$g_i = a_i b_i \qquad (2)$$
$$p_i = a_i \oplus b_i \qquad (3)$$

The merge bit can be explained as
$$G_i = g_i + G_{i-1} P_i \qquad (4)$$
$$P_i = p_i P_{i-1} \qquad (5)$$

According to Fig. 11, the merge circuit takes the signals from the top cell and the right–hand signal and outputs signals into the bottom cell. so the g_i and p_i are the signals come from the top cell. However, the G_{i-1} and P_{i-1} are the signals from the right–hand white rectangular cell. By assuming $G_{-1} = c_{-1}$ and $P_{-1} = 0$, we can easily demonstrate that the G_i signal is actually the carry signal c_i with the following definition
$$c_i = g_i + p_i c_{i-1} \qquad (6)$$
With the definition of equations (2) and (3), the PG circuit is shown in Fig. 12. The merge circuit of equations (4) and (5) is shown in Fig. 13. At this circuit, $G_{i-1} n$ is the complementary signal of G_{i-1}. The sum circuit of equation (1) is shown in Fig. 14.

The ASTAP simulation results are shown in Fig. 15 by using a 0.5 μm CMOS technology. All the results are simulated at the nominal condition with $T_a = 22°C$ and power supply of 2.5 V. The propagation delay is around 2 ns. The total rows of Fig. 11 in actual ASTAP simulation is 15 stages. This includes 1 driver stage to drive a 0.3 pF capacitive load. This driver stage costs roughly 150 ps delay. Fig. 16 shows the 64–bit adder circuit performance in the function of power supply voltage.

V. Conclusions

The DCVSPG circuit family has been developed. It is shown to have superior performance to that of a conventional DCVS approach. By using the pass–gate logic tree instead of the conventional nMOS logic tree, the floating–node problem is eliminated. This leads to no ratioed logic problem, symmetrical logic topology and shorter logic stack height. A 2 ns 64–bit CMOS adder is achieved by using the static DCVSPG circuit family.

VI. References

[1] F. S. Lai and W. Hwang, to be published
[2] L. G. Heller et. al., Dig. of ISSCC, pp. 16–17, 1984.
[3] L. M. Chu and D. I. Pulfrey, IEEE JSSC, pp. 528–532, 1987.
[4] L. C. Pfennings et. al., IEEE JSSC, pp. 1050–1055, 1985.
[5] K. Yano et. al., IEEE JSSC, pp. 388–395, 1990.
[6] M. Suzuki et. al., Dig. of ISSCC , Paper 5.4, 1993.
[7] D. Radhakrishnan et. al., IEEE JSSC, pp. 531–536, 1985.
[8] R. P. Brent and H. T, Kung, IEEE Comp., pp. 260–264, 1982.
[9] B.W. Wei et. al., Rep. UCB 86/252, UC Berkeley, 1985.

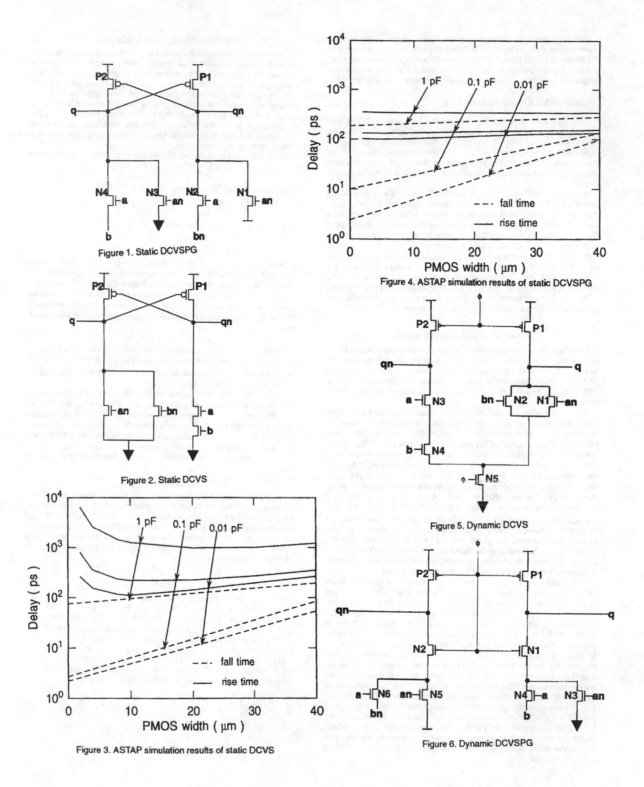

Figure 1. Static DCVSPG

Figure 2. Static DCVS

Figure 3. ASTAP simulation results of static DCVS

Figure 4. ASTAP simulation results of static DCVSPG

Figure 5. Dynamic DCVS

Figure 6. Dynamic DCVSPG

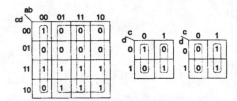

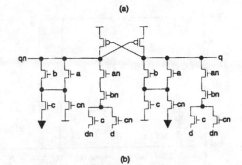

(a)

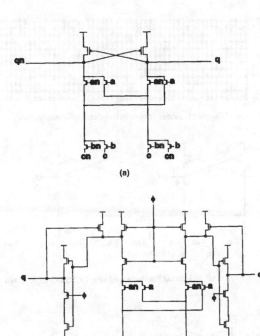

(a)

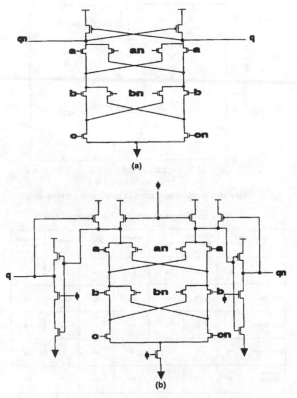

(b)

Figure 7. Synthesis of pass–gate logic

(a)

(b)

Figure 8. (a) Static DCVS (b) Dynamic DCVS

(b)

Figure 9. (a) Static DCVSPG (b) Dynamic DCVSPG

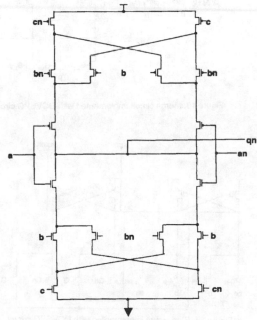

Figure 10. Static CMOS circuit

77

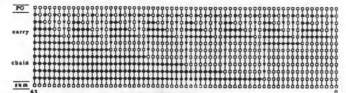

Figure 11. Adder core with the detailed construction

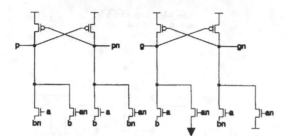

Figure 12. PG circuit implemented with DCVSPG logic family

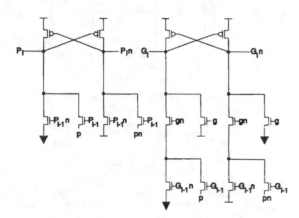

Figure 13. Merge circuit implemented with DCVSPG circuit

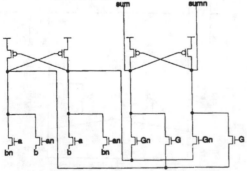

Figure 14. Sum circuit implemented with DCVSPG circuit

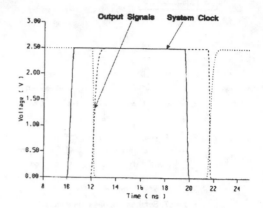

Figure 15. ASTAP simulation results

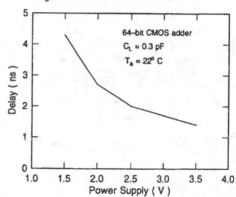

64-bit CMOS adder
C_L = 0.3 pF
T_a = 22° C

Figure 16. Circuit performance in function of power supply

Table I. Comparison of the sum circuit

	Gate Input Capacitance (fF)	Load Output Capacitance (fF)	P / N	Normalized Area	Delay (ps)	Power (µW)	Normalized Power-Delay Product
Static CMOS	108	108	8/8	1.00	327	117	1.00
Static DCVS	36	36	2/10	0.82	336	189	1.86
Static DCVSPG	24	24	2/8	0.51	210	48	0.28
Dynamic DCVS	48	96	6/15	1.37	145	119	0.44
Dynamic DCVSPG	36	72	6/14	0.86	122	60	0.26

A High Speed, Low Power, Swing Restored Pass-Transistor Logic Based Multiply and Accumulate Circuit for Multimedia Applications

AKILESH PARAMESWAR, HIROYUKI HARA, TAKAYASU SAKURAI

TOSHIBA CORPORATION

1, KOMUKAI TOSHIBA-CHO, SAIWAI-KU, KAWASAKI, JAPAN 210

Abstract

Swing Restored Pass-transistor Logic (SRPL), a high speed, low power logic circuit technique for VLSI applications is described. By the use of a pass-transistor network to perform logic evaluation, and a latch type swing restoring circuit to drive gate outputs, this technique renders highly competitive circuit performance. An SRPL based Multiply and Accumulate Circuit for multimedia applications is implemented in double metal 0.4μm CMOS technology.

Introduction

To date, the most widely used VLSI circuit design technique has been full CMOS. It has been attractive because it makes it easy to implement reliable circuits that have excellent noise margins. However, the continuing push for higher performance systems has, in recent years, brought the disadvantages of full CMOS to the fore, and a number of researchers have proposed alternative logic techniques (1-3). The majority of these have been static techniques because dynamic logic styles still suffer from charge sharing and noise margin problems, and difficulties in design and design for testability.

Complimentary Pass-transistor Logic (CPL) (1), uses a complimentary output pass-transistor logic network to perform logic evaluation, and CMOS inverters for driving of the outputs. This arrangement however suffers from leakage current through the inverter. Double Pass-transistor Logic (DPL) (2), uses both pMOS and nMOS devices in the pass-transistor network to avoid non-full swing problems, but it has high area and high power drawbacks. As the name suggests, Differential Cascode Voltage Switch with Pass Gate (DCVSPG) (3) is the same as the cascode voltage switch logic proposed in (4), but uses a pass-transistor network for logic evaluation. This logic style suffers from degraded pull down performance when used in a long chain.

In this paper, we propose a high speed, low power logic circuit technique that attempts to overcome these problems.

Swing Restored Pass-transistor Logic

Basic Circuit

The generic Swing Restored Pass-transistor Logic (SRPL) gate consists of two main parts as shown in Fig. 1. A complimentary output pass-transistor logic network that is constructed of n channel devices, and a latch type swing restoring circuit consisting of two cross coupled CMOS inverters. The gate inputs are of two types, pass variables that are connected to the drains of the logic network transistors, and control variables that are connected to the gates of the transistors. The logic network has the ability to implement any random Boolean logic function. Fig.2, for instance, shows the implementation of an SRPL full adder. The complimentary outputs of the pass-transistor logic network are restored to full swing by the swing restoration circuit.

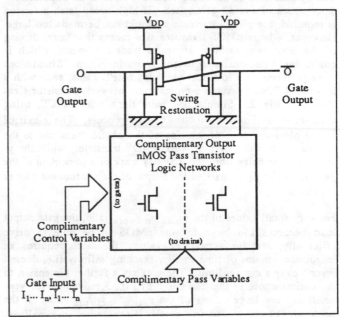

Figure 1 : Generic SRPL Gate

Gate Optimization

We have found that in the interests of speed, the nMOS transistors of the logic network farther away from the output should have larger drivability (i.e. size) than those closer to the output. This is because the transistors closer to the output pass smaller swing high signals due to the voltage drop across the transistors farther away from the output. The precise values for a given circuit depend on layout and other circuit considerations, so they must be determined by case by case simulation. Typical values are indicated in Fig. 2.

Reprinted from *Proceedings of the IEEE Custom Integrated Circuits Conference*, pp. 12.5.1-12.5.4, May 1994.

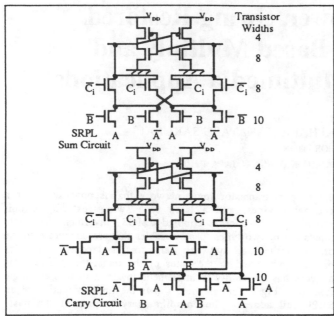

Figure 2 : Full Adder Circuit in SRPL

The optimization of the swing restoring latch is an important determinant of overall gate speed. If high speed latch inversion is required, the pMOS transistors should not be made too large. However, a large pMOS transistor size means that faster driving of the load is possible. Hence a trade-off exists, which is qualitatively demonstrated by the graph in Fig. 3. Simulations were performed on identical cascaded SRPL gates, each with a fan-out of 2, and assuming typical pass network transistor sizes shown in Fig 2. Simulations were done with SPICE, using parameters of Toshiba's 0.4μm CMOS process. The x-axis of Fig. 3 plots the ratio of the size of the pMOS transistor to the size of the topmost pass network nMOS transistor, while the y-axis plots the delay from the $0.5V_{DD}$ mark of a pass input of the gate to the $0.5V_{DD}$ mark of the output of the *subsequent* gate in the cascade.

For very small values of the $p_{latch}/n_{network}$ ratio, the gate output load becomes too large for the pMOS to be able to drive efficiently, and for very large values, the latch requires an inordinate amount of time to flip, reaching infinity (i.e. doesn't invert) over a certain limit. There exists a further dimension to the optimization, in that the curve of Fig. 3 moves up for very small or very large values of the $n_{latch}/n_{network}$ ratio. If the latch nMOS device is too small, discharging is penalized, whereas if it is too large, it introduces undue capacitive loading.

On the whole, the graph shows that though there is a trade off in determining the size of the pMOS transistor in the latch, there exists substantial design margin, making it easy to design circuits in SRPL. This design margin also means that SRPL circuits are quite robust against process variations, which might cause the threshold voltages of the transistors to fluctuate.

Performance Comparison with Competing Techniques

Full adders in CMOS, CPL, DPL, DCVSPG and SRPL were constructed, and simulated in the cascaded conditions shown in Fig. 4. Again, 0.4μm CMOS process parameters were used to perform SPICE simulations. The worst case waveforms for each

of the full adders are shown in Fig. 5. Other performance values are recorded in Tab. 1.

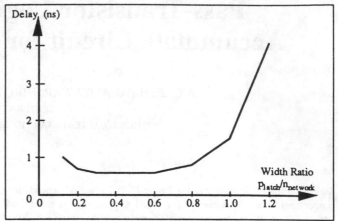

Figure 3 : Dependence of Delay on Transistor Widths

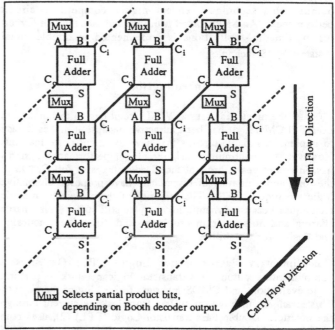

Figure 4 : Carry Save Addition of Partial Products

As Fig. 5 shows, CMOS has the slowest speed. Moreover, power consumption is quite high. The main reason for these poor performance figures is that the inefficient pMOS network of CMOS leads to a higher transistor count, larger gate area, and larger input capacitances due to the poor drivability of the pMOS transistor. DPL proves to be about 30% faster than CMOS, but this is at the expense of a higher transistor count, and more power consumed. DCVSPG is much faster than CMOS, but suffers from the problem that it cannot be used in the array structure of Fig. 4. The reason for this is that there is no pull down mechanism other than that through the pass-transistor networks. Thus for long chains of cascaded gates, the pull down becomes severely degraded as shown by the dotted line of Fig. 5.

CPL, as Fig. 5 clearly shows is the fastest of the five techniques. However, this is achieved at the expense of high power consumption. Furthermore, CPL suffers from the major drawback that it is a non-full swing technique. The non-full swing signals at the inputs of the inverters mean poor noise margins, particularly as the inverter threshold is susceptible to process variations. Moreover, CPL circuits consume static power because of the leakage current that is always flowing through one of the inverters of a gate. The inverter output never quite reaches V_{SS} as the curve of Fig 5. shows. Because a V_{DD} of 3.3V is high relative to a channel width of 0.4μm, the speed degrading effects of the leakage are not prominent. However, when V_{th} is a significant fraction of V_{DD}, as it will certainly be in the future, the fall time of the output lengthens, and CPL becomes slower than SRPL.

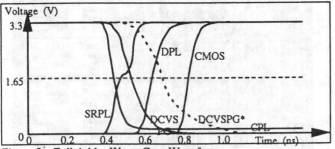

Figure 5 : Full Adder Worst Case Waveforms

SRPL has good speed performance. In the simulated conditions of Fig. 4, each SRPL circuit within the full adder fans out to only two other similarly sized circuits (carry and sum). This implies relatively light loading conditions, much less than the usual CMOS stage ratio of 3.5 or 4. It is important to note that this condition is not restricted to the simulated case. Low fanout is a very common occurrence in the design of VLSI circuits, particularly in data paths. In such conditions, it makes sense to connect the pass-transistor network output to the gate output, and to restore the swing with the cross coupled pair of inverters. The initial rise in voltage caused by the pass network output takes the gate output voltage a good margin above the $V_{th,n}$ of the transistors of the following gate, speedily setting up the correct logical path. Also because of the relatively light loading conditions, the inversion of the latch is faster, and so the $P_{latch}/_n$ network ratio can be made slightly larger. Thus a good pull up time through the *a priori* set up logical path of the following gate is achieved.

Table 1 : Comparison of Full Adder Circuits

	CMOS	CPL	DPL	DCVS PG	SRPL
Speed (ns)	0.82	0.44	0.63	0.53	0.48
Power at 100Mz (mW)	0.52	0.42	0.58	0.3	0.19
Power-Delay Product (normalized)	1.0	0.43	0.86	0.37	0.21
Transistor Count	40	28	48	24	28

As Tab. 1 shows, SRPL has the lowest power consumption and the lowest power-delay product of the different techniques. The main reasons for the low power are the low transistor count and the low input capacitance. Also, the fast inversion action of the latch quickly cuts off any d.c. path through the pass network.

In summary, SRPL shows itself to be a very competitive low power, high speed circuit technology. SRPL circuits will also occupy less area because of the lower transistor count. Particularly because the number of p channel devices is small, less area will be wasted on well boundary separation. The pMOS transistors are also smaller than, for instance, the CPL case, leading to slightly better area performance. This promising logic technique was used to construct a Multiply and Accumulate Circuit (MAC) for multimedia applications.

Multiply and Accumulate Circuit

The multiply and accumulate operation is crucial to a wide range of signal processing applications. With the increasing level of integration of processors dedicated to multimedia, it has become essential that high speed MAC macrocells be provided on chip. However, high speed is not the sole imperative. System portability is also a key issue, and hence low power is also very important. The MAC presented in this paper was designed with these requirements foremost in mind.

MAC Architecture

The overall circuit is shown in Fig. 6. The multiplier and multiplicand are 16-bit wide, whereas the accumulated result has a bit width of 32. A pipelined scheme was not implemented because the frequency of operation was expected to be more than sufficient to cover even the most advanced multimedia applications. Furthermore, pipelining introduces problems of complicated control and timing, and extra area and power required by the pipeline registers.

A Booth decoding scheme was used to obtain 8 partial products, which are added in a carry save manner as shown in Fig. 4. Each full adder row receives a running sum and carry from the row above. The very top adder of each column of the summation receives one of its inputs from the accumulated total of the previous cycle, which is fed back as shown in Fig. 6. A Wallace tree architecture for partial product addition was not used because the such an architecture would lead to larger power consumption due to the larger area and wiring requirements. Each of the full adders in the partial product summation array is constructed using the SRPL technique described above.

The final CLA adder cum register to which the partial product summation array outputs its carry saved result uses the same design as that of (5), where a dynamic sense amplifying scheme is used to perform both carry propagation, and latching of the final result. This design is ideally suited to the MAC design because of the high speed addition followed by the instantaneous latching. The complimentary outputs of the SRPL based summation array perfectly match the complimentary input requirements of the sense amplifying technique used by the final adder. It should be noted though that the dynamic sense amplifying technique used in (5) is completely different from the static swing restoring technique proposed in this paper.

Performance

The MAC was fabricated using a double metal 0.4μm process as summarized in Tab. 2. The chip photomicrograph is shown in Fig. 7. As Tab. 2. shows, the MAC operates at a maximum frequency of 150MHz, which more than sufficient for multimedia applications. Moreover, the power consumed is only 34mW at

81

this frequency, satisfying the other important multimedia requirement. The 150MHz operating frequency translates to a one cycle delay time of 6.7ns. For comparison, the MAC was simulated with a CPL partial product addition array. The simulated delay time was 6.3ns.

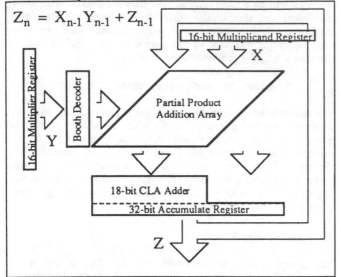

$$Z_n = X_{n-1} Y_{n-1} + Z_{n-1}$$

16-bit Multiplicand Register

X

16-bit Multiplier Register

Y

Booth Decoder

Partial Product Addition Array

18-bit CLA Adder

32-bit Accumulate Register

Z

Figure 6 : Multiply and Accumulate Circuit

Though the SRPL MAC is 0.4ns slower than the CPL version, it should be remembered that CPL is the fastest technique ever reported, being nearly twice as fast as CMOS. Moreover, the power consumed by the CPL version was estimated to be more than twice that consumed by the SRPL MAC. In addition, as has been mentioned, CPL suffers from margin problems that will be exacerbated by the future reduction of the supply voltage, and this reduction in V_{DD} will also lead to speed degradation.

Conclusion

A new high speed, low power logic circuit technology was proposed, and used to implement a multiply and accumulate circuit in double metal 0.4µm CMOS. The MAC achieves a frequency of 150MHz, and 34mW and shows much promise for multimedia applications.

Table 2 : MAC Characteristics

Technology	CMOS process
n Channel Length	0.4µm (Eff. 0.39µm)
p Channel Length	0.5µm (Eff. 0.47µm)
Gate Oxide Thickness	9 nm
No. of Metal Layers	2
Power Supply Voltage	3.3 Volts
Operating Frequency	150MHz
Latency	0 cycles
Power Consumed at 150MHz	34 mW
Active Area	0.98 mm²

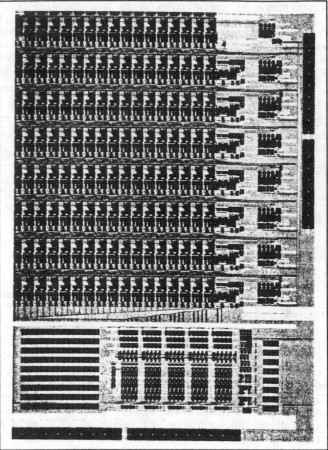

Figure 7 : MAC Photomicrograph

Acknowledgements

The authors would like to gratefully thank the assistance and encouragement of Fumihiko Sano, Yoshinori Watanabe, Masataka Matsui, Hidetoshi Koike, Fumitomo Matsuoka, Masakazu Kakumu and Kenji Maeguchi

References

(1) K. Yano et al , "A 3.8ns CMOS 16x16 Multiplier Using Complimentary Pass-transistor Logic," vol. 25, no. 2, pp.388-395, April 1990.

(2) M. Suzuki et al , "A 1.5ns 32bit CMOS ALU in Double Pass-transistor Logic," 1993 IEEE International Solid-State Circuits Conference, pp. 90-91.

(3) F.S. Lai and W. Hwang, "Differential Cascode Voltage Switch with Pass Gate Logic Tree for High Performance CMOS Digital Systems," 1993 International Symposium on VLSI Technology, Systems and Applications, pp. 358-362.

(4) L.G. Heller, W.R. Griffin, J.W. Davis and N.G. Thoma, "Cascode Voltage Switch Logic : A Differential CMOS Logic Family," 1984 IEEE International Solid-State Circuits Conference, pp. 16-17

(5) M. Matsui et al , "Sense-amplifying pipeline flip-flop scheme for 200MHz video de/compression macrocells," in press (1994 IEEE International Solid-State Circuits Conference).

0.5V SOI CMOS Pass-Gate Logic

TSUNEAKI FUSE, YUKIHITO OOWAKI, MAMORU TERAUCHI,SHIGEYOSHI WATANABE,
MAKOTO YOSHIMI, KAZUNORI OHUCHI, AND JUN'ICHI MATSUNAGA

ULSI RESEARCH LABORATORIES

TOSHIBA CORPORATION, KAWASAKI, JAPAN

Demand for low-power ULSIs for mobile electronic equipment is increasing rapidly. To reduce power consumption, lower operating voltage and minimized device size (or count) is essential. To lower the actual threshold voltage and lower the operation voltage, SOI MOSFET with gate-body connection is proposed [1]. However, the circuit architecture that affords the maximum advantage of the body controlled SOI MOSFET is not reported. This SOI CMOS pass-gate logic offers the lowest operation voltage and reduced transistor dimensions.

Figure 1 shows conventional and proposed pass-gate logic. In the conventional complementary pass-gate logic (CPL, Figure 1a), the high-level signal of the pass-gate network is less than the supply voltage, Vcc [2]. This is because the pass-gate turns off when the source voltage reaches Vcc-Vt, where Vt is the threshold voltage of pass-gate which is increased by the body-effect. The drive capability of the network is degraded due to the channel resistance of pass-gates, so the output signal from the pass-gate network is amplified by using the buffer. In SOI CMOS pass-gate logic (Figure 1b, c), the body of SOI pass-gate is connected to the input signal given to the gate. Low threshold voltage for the on-state pass-gate and high threshold voltage for the off-state pass-gate is realized, and the increase in the threshold voltage due to the body-effect is suppressed. Two types of buffer suitable for the SOI pass gate logic are examined. The buffer used in the Type A logic is composed of two CMOS inverters and a pMOS latch circuit, as shown in Figure 1b. The body of the MOSFET is connected to the gate (gate-body connection, GBC scheme). For the buffer used in the Type B logic, pull-up pMOSFETs are cross-coupled [3]. The body of the cross-coupled pull-up pMOSFET is connected to the buffer input, (input-body connection, IBC scheme), as shown in Figure 1c. Figure 2 shows the full-adder delay versus supply voltage. For SOI pMOS / nMOSFETs, the absolute value of the threshold voltage is 0.4V at 0V body-bias, and is 0.17V at 0.5V body-bias. Due to the low threshold voltage for the on-state MOSFET in the GBC scheme, the Type A full-adder reduced the delay to 1/3 of that of the conventional SOI CPL at 0.5V. Lowest operation voltage, V_{CCmin}, is improved by 0.17V by the GBC scheme, where V_{CCmin} is defined as the supply voltage which gives 2ns delay.

Transistor dimension are optimized for Type A (GBC) and Type B (IBC) pass-gate logic. The major difference between Type A and Type B logics is that the pass-gate network drives only two nMOSFETs in the Type B logic, while the pass-gate network drives two nMOSFETs and two pMOSFETs in the Type A logic. Figure 3 shows the full-adder delay versus the gate-width of the pass-gate network. By use of optimized buffer dimensions (Wp/Wn=0.6, Wu/Wn=0.4 for the Type B logic, and Wp/Wn=2.0, Wu/Wn=1.1 for the Type A logic), the optimum pass-gate width of the Type B logic is 0.6Wn, while that for the Type A logic is 1.3Wn. As a result, the total transistor dimension of the Type B logic is less than half that of the Type A logic.

The buffer using cross-coupled pull-up pMOSFETs reduces total transistor dimensions. There are two design options to control the body-bias. Figure 4 shows two types of buffer chain using the cross-coupled pull-up pMOSFET. One type uses the GBC scheme, and the other uses the IBC scheme (Figure 4). In the buffer using the GBC scheme, the on-state pull-up pMOSFET keeps high threshold voltage until the output node responds. This is because the body of the pull-up pMOSFET is connected to the output node. In the buffer using the IBC scheme, on the other hand, the threshold voltage of the on-state pull-up pMOSFET decreases before the output node responds. As a result, the buffer using the IBC scheme operates with high-speed and small short-circuit current, compared with the buffer using the GBC scheme.

A 40-stage buffer chain is used to measure the speed advantage of the buffer using the IBC scheme, in the Type B logic. A micrograph of the test chip is shown in Figure 5. Figure 6 shows the ratio of the buffer delay with the IBC scheme to that with the GBC scheme versus supply voltage. Measured threshold voltage of the SOI MOSFET is 0.58V at the body-bias of 0V, and 0.35V at the body-bias of 0.5V, respectively. The IBC scheme is 36% faster than the GBC scheme at 0.5V, and the V_{CCmin} is improved by 0.08V. Type B logic using the IBC scheme is 10 times faster than the CPL, and the minimum operation voltage is improved by 0.25V.

Multiplication is useful for estimating the logic performance. In the pass-gate full-adder using the buffer with the IBC scheme, the dissipation is reduced by 0.5V operation and reduced transistor dimensions. For a 16x16b multiplier using a Wallace-tree adder and CLA adder, the simulated multiplication time is 18ns at 0.5V. And the power-delay product is 70pJ including 50pF I/O, was more than an order of magnitude improvement for the CPL (Figure 7).

Acknowledgments:

The authors are grateful to H. Tago, T. Mizuno, and Y. Ushiku for helpful discussions and thank T. Arikado, A. Hojo, and H. Hara for encouragement.

References:

[1] Assaderaghi, F., et al., "A Dynamic Threshold Voltage MOSFET (DTMOS) for Ultra-Low Voltage Operation," IEDM Techical Digest, pp. 809-812, Dec., 1994.

[2] Yano, K., et al., "A 3.8-ns CMOS 16μ 16-b Multiplier Using Complementary Pass-Transistor Logic," IEEE J. Solid-State Circuits, vol. 25, pp. 388-395, April, 1990.

[3] Heller, L. G., et al., "Cascade Voltage Logic: A Differential CMOS Logic Family," ISSCC Digest of Techical Papers, pp. 16-17, Feb., 1984.

Reprinted from *Digest of Technical Papers, IEEE Journal of Solid-State Circuits Conference*, pp. 88-90, February 1996.

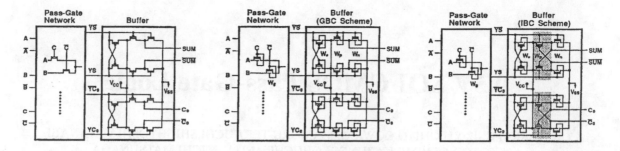

Figure 1: Conventional and proposed pass-gate logic.
(a) CPL, (b) Type A, (c) Type B.

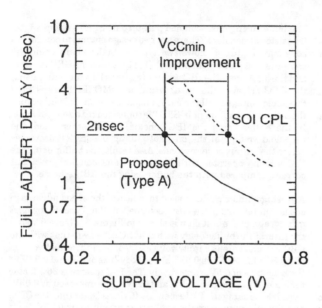

Figure 2: Simulated full-adder delay vs. supply voltage.

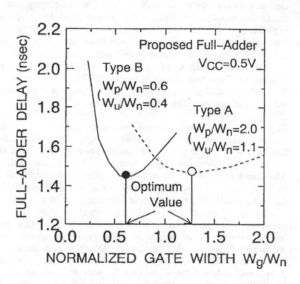

Figure 3: Gate-width optimization for pass-gate logic.
Figures 4 and 5: See page 424.

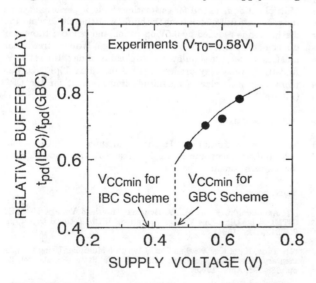

Figure 6: Measured buffer delay vs. supply voltage.

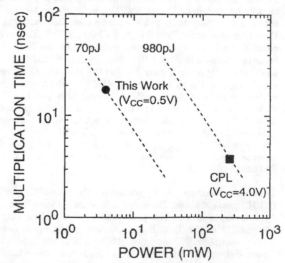

Figure 7: 16x16b multiplier power-delay product.

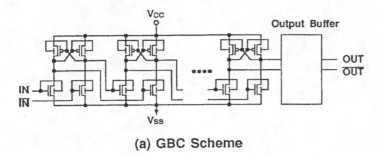

(a) GBC Scheme

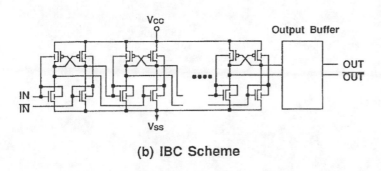

(b) IBC Scheme

Figure 4: Buffer chain schematics: (a) GBC (b) IBC .

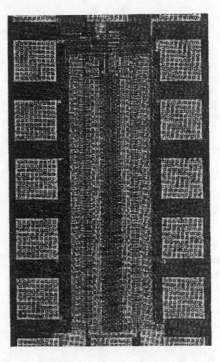

Figure 5: Chip micrograph.

Chapter 2

Advances in BiCMOS
and Bipolar Circuits

ADVANCES IN BiCMOS AND BIPOLAR CIRCUITS

BiCMOS Circuits

Requirements for higher-performance that have been driving the development of microprocessors and high-speed digital electronics led to the integration of bipolar transistors into CMOS logic. Incorporation of the bipolar transistor for the purpose of enhancing performance was first applied to the fast static memory (cache) before it proliferated into digital logic. To fabricate a bipolar device requires only a few extra process steps added to the CMOS process. The BiCMOS fabrication process consists of the steps taken from the bipolar process and fused with those required for the CMOS process. Those extra steps do come at cost. For each processing step being added to the fabrication process, the yield decreases, which in turn means a higher cost of a product. BiCMOS technology is to be found only in such applications where the benefits of BiCMOS can offset the cost added to the product: high-performance processors or a mixed analog and digital process where bipolar process steps are already in place. Even when used in logic, BiCMOS gates are used sparsely and only in the speed-critical areas.

Cost is one of the limiting factors which explains why BiCMOS technology is not widely used. The projections show that BiCMOS is just a transition stage in CMOS development and that with further scaling down of CMOS device features BiCMOS will gradually lose its place. This position is further supported by problems associated with scaling the supply voltage down. That is, as the supply voltage is lowered, following the scaling of the device features, BiCMOS starts losing its speed advantages over CMOS. The last paper in this section treats the BiCMOS operation at the low-supply voltage.

Figure 2.1 shows a cross section of a BiCMOS chip, which illustrates how a bipolar NPN transistor is integrated into a CMOS structure. The figure shows that the bipolar transistor does not take much space as compared to MOS transistor. However, integration of a bipolar transistor requires additional process steps.

The process steps needed for a BiCMOS process are listed in Fig. 2.2 next to the steps needed for a bipolar and a CMOS process. The number of steps is increased from 13 to 17 with respect to a CMOS process and from 15 to 17 with respect to the bipolar process. Those extra steps increase BiCMOS fabrication costs by approximately 30 to 50 percent, depending on process yield.

The logic diagram of a basic BiCMOS NAND gate is shown in Fig. 2.3. Bipolar transistors Q_1 and Q_2 are used to speed up the pull-up and pull-down transistors, respectively. When both inputs A and B are at the logic "one" level, the base of the transistor Q_1 is at ground level, keeping Q_1 *OFF*. At the same time, the path through transistors MN_3 and MN_4 turns the transistor Q_2 *ON*. When one of the inputs (or both) is at logic "zero," the transistor Q_1 is turned *ON* because the current is supplied to its base by either of the transistors MP_1 or MP_2. Although the output of the gate is at the logic one the transistor MN_5 is ON, preventing the transistor Q_2 from being ON as well as discharging any charge that might have been left in the base of Q_2.

A major drawback of this circuit configuration is that the signal swing of the output is reduced by 2 V_{be}, that is, the maximal value of the output $V_{out_{max}} = V_{cc} - V_{be}$ and the minimal value of the output $V_{out_{min}} = V_{be}$. Often, a resistor is inserted between the base of the bipolar transistor Q_1 and its emitter as shown in Fig. 2.4. Though in this way a full-output signal swing is assured, this technique has a major disadvantage: The base potential of the upper transistor (Q_1) falls with the output level owing to large base to emitter resistance (R_1), which cannot discharge its base and turn the upper transistor Q_1 OFF during a substantial part of the transition period. This in turn creates current spikes caused by the fact that both Q_1 and Q_2 are conducting during part of the output signal transition from "high-to-low."

Thus, both BiCMOS configurations suffer from serious drawbacks, the *N*-type (Fig. 2.3) from not having the full signal swing and the *R*-type (Fig. 2.4) from having the current spikes and increased power dissipation.

The first paper in this section analyzes the full-swing techniques for BiCMOS logic, where two different resistor configurations are considered. One has a base-emitter shunt resistor (as shown in Fig. 2.4), whereas in the other configuration the shunt resistor is connected between the collector and emitter of the bipolar transistor.

BiCMOS Schematic Cross-section

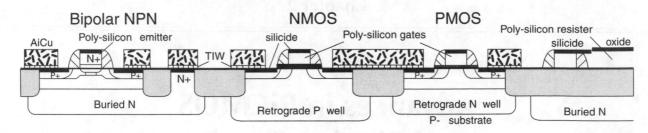

Fig. 2.1. BiCMOS cross section.

Another way of overcoming the problem associated with the lack of a full-output signal swing and current spikes during the signal transition is analyzed in the third paper by Nishio et al. The authors were able to effectively combine the two techniques, *R*-type and *N*-type, into a configuration that utilizes a feedback from the output to the input of the gate. This feedback, consisting of the two inverters, is used to insert a sufficient delay from the output signal, which is controlling the connection from the base to the ground of the bipolar output transistor. In their configuration, when the input changes from *high* to *low,* the path to ground is disconnected so that the output transistor receives its full base current. When the input changes from low to high, the path to ground is connected, discharging the base of the bipolar transistor and preventing the penetration current in the output stage. In this way, a sufficiently fast transition is achieved by preventing the leakage current from the base of the output transistor and eliminating the penetration current, thus achieving the low-power operation.

The application of BiCMOS logic is described in the paper by Nakatsuka et al. presenting Hitachi's 32-b microprocessor, which is one of the first to use BiCMOS technology. The speed-up achieved in the various parts of the microprocessor by the selective use of BiCMOS gates in the critical path is shown and compared to the case where only CMOS is used. The largest speed-up was in the register file (from 6.2 nS to 2.5 nS), followed by the ALU carry path (11.5 nS to 5.0 nS). The BiCMOS circuits are used selectively to drive the nodes with large capacitive load since the advantage of BiCMOS over CMOS diminishes when the capacitive load is very small. The BiCMOS cells are also used to accelerate address decoding of the ROM and RAM whose word lines have very large capacitance. In the de-

BiCMOS Process Steps

CMOS	BiCMOS	Bipolar
	Buried layer	Buried layer
	Epi	Epi
Well	Well	
Isolation	Isolation	Isolation
Field adjust	Field adjust	Field adjust
	Deep N+	Deep N+
Gate oxide	Gate oxide	
Poly	Poly	
LDD	LDD	Resistor
	Int. base	Int. base
P+ S/D	P+ S/D	Ext. Base
N+ S/D	N+ S/D	Emitter
Cnt	Cnt	Cnt
Mtl 1	Mtl 1	Mtl 1
Via	Via	Via
Mtl 2	Mtl 2	Mtl 2
Pass	Pass	Pass

Fig. 2.2. BiCMOS process steps.

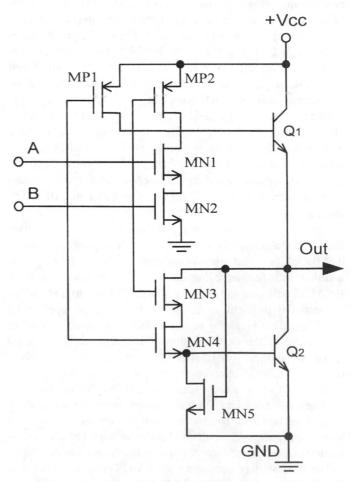

Fig. 2.3. BiCMOS NAND gate.

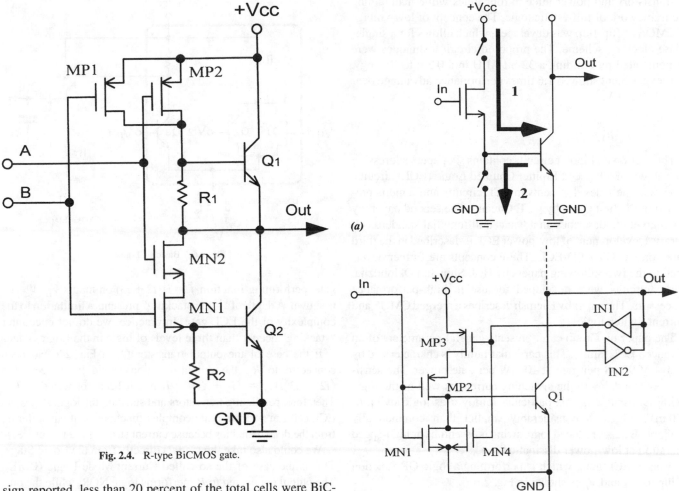

Fig. 2.4. R-type BiCMOS gate.

Fig. 2.5. (a) Transient saturation (principle), (b) TS-FS-BiCMOS configuration (pull-down section only). Adapted from Hiraki et al., "A 1.5-V Full-Swing BiCMOS Logic Circuit," *IEEE Journal of Solid-State Circuits,* Vol. 27, pp. 1569, (November 1991).

sign reported, less than 20 percent of the total cells were BiCMOS, yet the resulting performance increase was 1.5 to 2 times.

The sixth paper in this section on BiCMOS, by Hiraki et al., deals with the problem of scaling down the supply voltage below 2 V. It has been known that the conventional BiCMOS will start losing its performance advantage over CMOS when the supply voltage drops below 3 V. Other advanced BiCMOS families such as C-BiCMOS, QC-BiCMOS, and BiCMOS will start losing their performance advantage over CMOS when the supply voltage drops below 2 V. The BiCMOS logic, developed by Hiraki et al., TS-FS-BiCMOS (Transiently Saturated Full Swing BiCMOS), maintains its performance advantage over CMOS when the power supply voltage drops even further and at 1.5 V supply voltage operates twice as fast as CMOS. The transient saturation of bipolar transistors is achieved with a sophisticated feedback loop from the output to the base of the bipolar transistors. The principle of transient saturation is illustrated in Fig. 2.5a and b. The bipolar transistor is driven into saturation by the current injected by the p-MOS transistor connected to its base. As soon as the output changes, the minority carriers are discharged from the base via the path to ground designated as 2 in Fig. 2.5a.

The operation of TS-FSBiCMOS circuit is shown in Fig. 2.5. It is a noninverting logic and consequently a zero at the input produces a zero at the output and vice versa. When the input changes from one to zero, the transistor MP_2 turns *ON* (MP_3 is

already ON) and the charge is being injected into the base of the transistor Q_1. This causes the output to change rapidly from one to zero. This change is reflected in the transistors MN_4 and MP_3 via the inverter IN_1 as one, turning the transistor MP_3 OFF and MN_4 ON. Thus, the transistor MN_4 will cause discharge of the minority carriers from the base of Q_1 to the ground turning the transistor Q_1 OFF. The logic zero at the output is now being held by a cross-coupled pair of inverters IN_1 and IN_2. The pull-up part of these circuits is symmetric, and it operates in the same manner. Thus, in TS-FS-BiCMOS, bipolar transistors saturate only transiently, and static power dissipation is avoided because DC current is cut off by the transistor MN_3. This circuit achieves full-swing operation and operates twice as fast as CMOS at 1.5 V supply voltage.

The final paper in this set is the paper by Yano et al., which describes circuit techniques for 250 MHz RISC arithmetic modules. In order to take advantage of BiCMOS circuits, the design style favors large and complex functional blocks before the BiCMOS stage. This design style is referred to as Feedback Massive-input Logic (FML). Use of FML reduces the number of transistors and power three to four times while maintaining the framework of fully static logic. The concept of low-voltage BiCMOS D-flip-flop was developed, which allows for a single-phase clocking scheme. The proposed circuit techniques were demonstrated by building a 32-bit ALU in a 0.3 u technology. This experiment showed 1.6 times performance advantage over CMOS at 3.3 V.

Bipolar Circuits

The section on bipolar circuits contains six papers addressing the subject of advanced Emitter Coupled Logic (ECL) circuits. Review of new developments in ECL circuits and logic is presented in the first paper by C. T. Chuang. The second paper, by J. Greub et al., describes an advanced differential standard cell library. Development of low-power ECL is described in the third paper on AC-CS-APD-ECL. These concepts are further developed in the two following papers by H. J. Shin and Oklobdzija. Those circuits were developed for use in high-performance processors. The paper by Hiemsh describes a merged CMOS and Current Switch logic.

The paper by Greub et al. presents general parameters of an advanced ECL family. This particular family is characterized by delays of 90 pS per gate at 10 mW per gate power. The sensitivity of the delay to the switching current as well as the logic swing is discussed. This particular family uses logic swing of 250 mV, which is considerably smaller than commercially available ECL. A reduced logic swing is essential for high-speed operation at low-power dissipation.

A basic ECL gate, which is performing a logic OR function of inputs V_{i1} and V_{i2} is shown in Fig. 2.6.

The circuit consists of two stages: the logic stage and the output stage. The logic function in the logic stage is implemented by "current steering" where the current I_s from the current source is "steered" between two branches (containing R_1 and R_2): the current in the first branch (containing R_1) is controlled by the input signals, while another one (containing R_2) is controlled by the reference voltage. The Logic Stage shown in Fig. 2.6 performs logical OR, which is achieved by connecting transistors Q_1 and Q_2 in parallel in the first branch. Increasing the logic fan-in (resulting in a three- or four-input NOR logic gate) is simple and consists of paralleling the transistors in the R_1 branch.

The Output Stage is an Emitter Follower, which serves to amplify the signals and provide a sufficient drive at the output. The connection of the Output Stage to the Logic Stage also determines the logic function at the output node.

In our example, the base of the output transistor Q_4 is connected to R_2, therefore performing OR function. Had the base of the output transistor Q_4 been connected to R_1, the corresponding output function would have been NOR. Therefore, this gate could perform the OR or NOR function, or both (if we connect two output stages to R_1 and R_2); thus a differential output stage is easy to realize.

Although OR/NOR is relatively simple to implement, implementation of the AND/NAND function requires "stacking" the logic levels. This is illustrated in Fig. 2.7, where a basic ECL

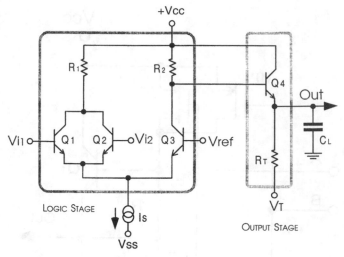

Fig. 2.6. Basic ECL gate.

gate, performing function $\bar{f} = i1(i2 + i3)$ on inputs V_{i1}, V_{i2}, V_{i3} is shown. A depth of "logic stacking" presents a limitation to the complexity of the ECL gate. In practice, we do not encounter "stacking" deeper than three levels of logic in the Logic Stage.

If the base of the output transistor Q_1 in Fig. 2.7 had been connected to R_2, the output function would have been $\bar{f} = (i2 + i3)\overline{i1}, f = (i_2 + i_3) = i_1$ had it been connected to R_3). Therefore, paralleling transistors and stacking the logic stages, an ECL cell can perform quite complex functions, without suffering from the delay penalties because current steering is a fast process.

We could also take the signals directly from the Logic Stage. This is the case of the so-called Current Mode Logic (CML). The output connected to R_1 performs logic NOR, while the output connected to R_2 performs logic OR function (Fig. 2.6).

CML is saving on transistors that would otherwise be used for the Output Stage but is paying for that with inferior driving capabilities. CML can therefore be used only where the signals are local and not driving larger loads. The paper by Greub et al. advocates precisely that point. These authors have shown that with clever use of a combination of CML and ECL, remarkable speed and reasonable power can be achieved. The cells they used are classified into low power = 6 mW/gate, medium = 10 mW/gate and high power = 14 mW/per gate. In their design, most of the power is dissipated in the output stage by the cell that is driving a larger capacitive load. While this cannot be avoided (unavoidable energy is needed to load the capacitance of the interconnections), a standard ECL gate uses more power for its output stage than is really necessary.

If we analyze the circuit shown in Fig. 2.6, we may observe that the transistor Q_4 in the output stage is always conducting. The voltage swing at the output varies approximately between $V_{cc} - V_b$ (when Q_4 is in saturation) and $V_{cc} - V_{be} - I_sR_2$ (when Q_4 is slightly ON). The voltage swing on the output node is adjusted by the voltage drop on the resistor R_2, and this drop is reflected to the output voltage V_o. Therefore, as $\Delta V_o = (V_{cc} - V_{be}) - (V_{cc} - V_{be} - I_sR_2) = I_sR_2$. Given that the current in this branch varies between I_s and 0 (because of current steering), the output voltage swing $\Delta V_o = R_2 I_s$ and is adjustable by the

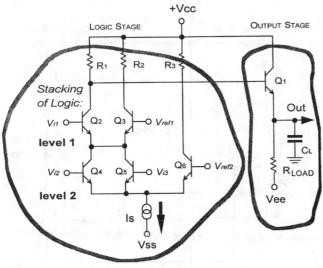

Fig. 2.7. Complex ECL gate.

value of the resistor R_2 as well as the value of the constant current of the current source I_s.

The speed of the ECL gate depends on the ability of the output stage to rapidly charge and discharge the output load capacitance C_L. The charging of C_L is dependent on the ability of transistor Q_4 to provide sufficiently large current to C_L, which in turn depends on how well the transistor Q_4 is "driven," that is, how much charge is delivered to the base of Q_4. Driving of the transistor Q_4 is dependent on the value of the resistor R_2. As a rule of thumb, the current in the resistor R_2 should be about one-fifth of the desired output current in the transistor Q_4. The reason is that in saturation the current amplification factor β of the output transistor Q_4 is degraded and can be as small as five.

On the other hand, the ability to discharge the capacitive load C_L ("high-to-low" signal transition) is entirely dependent on the

value of the resistor R_T placed between the emitter of Q_4 and terminal voltage V_T. If this resistor is too large, this transition will be too slow, therefore increasing the cumulative delay of the gate. However, if resistor R_T is too small, a substantial current will exist in R_T, which is to be provided by the transistor Q_4. In addition, the transistor Q_4 needs to provide sufficient current to charge C_L. This requires a substantial drive to be provided by the transistor Q_4, and the power consumed in the output stage can be quite large.

As the frequency of operation that is required of today's processors and digital systems increases, the relative comparison of the power consumption between ECL and CMOS becomes interesting. This is illustrated in the charts provided by T. Kuroda which are shown in Fig. 2.8. The power consumption of CMOS (which is considered a "low-power" technology) increases linearly with the frequency of operation and the square of the signal voltage swing:

$$P_{CMOS} = K\,C\,\Delta V^2 f$$

The power consumption of ECL circuits (considered "power consuming") is relatively independent of the frequency, and it depends only on the product of the signal voltage swing and the current in the logic and output stage:

$$P_{ECL} = K(I_S + I_o)\Delta V$$

At the frequency of operation somewhere between 100 and 200 MHz, the power consumption of CMOS surpasses that of ECL even when both are utilizing advanced technology and advanced circuits (Fig. 2.8). Given that the speed of the commercial processors today is in this range, use of ECL is not to be so easily ruled out. Whether we should still characterize CMOS as low-power technology is an open question, as is its use in high-performance processors.

Advanced ECL circuits that are given consideration for VLSI should be characterized with sufficiently low power so that the

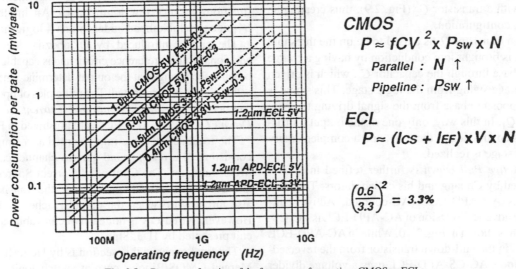

Fig. 2.8. Power as a function of the frequency of operation: CMOS vs ECL. Based on data provided from "Workshop on the Future of BiCMOS Logic Circuit." Presented at *1993 Symposium on VLSI Circuits*, Kyoto, Japan, May 19–21.

91

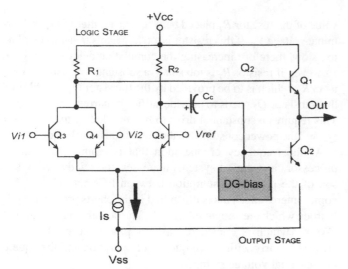

Fig. 2.9. APD-ECL from IBM. Adapted from C. T. Chuang, et. al., "High-Speed Low Power ECL Circuit with AC-Coupled Self-Biased Dynamic Current Source and Active-Pull-Down Emitter-Follower Stage," *IEEE Journal of Solid-State Circuits,* Vol. 27, pp. 1207, (August 1992).

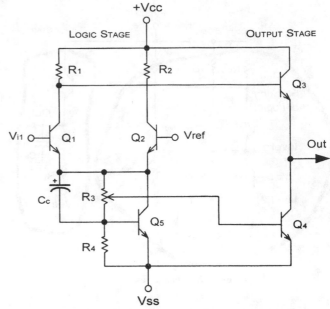

Fig. 2.10. AC-APD-ECL from IBM. Adapted from C. T. Chuang, et. al., "High-Speed Low Power ECL Circuit with AC-Coupled Self-Biased Dynamic Current Source and Active-Pull-Down Emitter-Follower Stage," *IEEE Journal of Solid-State Circuits,* Vol. 27, pp. 1207, (August 1992).

VLSI chip can be cooled with conventional cooling techniques, preferably air. Therefore, the signal swing ΔV should be sufficiently small. The signal swing is a factor that influences not only the speed, but also the power consumption, giving ECL an edge over CMOS. Let us assume, for example, that we are comparing a 3 V CMOS circuit with an ECL circuit whose signal swing is 300 mV. Because of the ΔV^2 factor influencing power consumption in CMOS (versus the ΔV factor in ECL), ECL has an order of magnitude advantage in terms of power, which is achieved by reducing the signal swing to a value as low as 200 mV.

A review of advanced ECL logic is given in the paper by C. T. Chuang. In this paper, the author describes several configurations to be considered for use in advanced ECL logic which are applying an Active Pull-Down (APD) configuration in the output stage (Fig. 2.9). One characteristic of those circuits is that the problem with the resistor R_T (Fig. 2.6 and Fig. 2.7) is solved by replacing it with a transistor Q_2 (Fig. 2.9), thus creating an active pull-down configuration.

The Pull-Down transistor Q_2 is biased to be on the threshold of conducting. It is brought into conduction by having a charge injected into its base through the capacitor C_c, which is driven by the signal voltage swing from the logic stage. This signal is arriving in the opposite phase from the signal driving the base of the transistor Q_1. In this way, only one of the output transistors Q_1 and Q_2 is conducting at the time, and a complementary push-pull output stage is realized.

The idea of *Active Pull-Down* is further refined in several schemes presented by Chuang and his collaborators. The most promising one is AC-APD-ECL (AC-coupled, Active Pull-Down ECL). The advanced version of AC-APD-ECL is AC-CS-APD-ECL, which is shown in Fig. 2.10. While in AC-APD-ECL charge is injected in the pull-down transistor from the reversed-biased storage diode, AC-CS-APD-ECL uses a voltage divider in the emitter branch of the current switch to bias the Active Pull-Down transistor. AC-CS-APD-ECL shows better performance

than AC-APD-ECL, and it achieves the speed of 66 pS at 3.0 mW per gate with the load of 0.3 pF as reported in the paper by Chuang et al. This represents an improvement of 1.62 times over ECL and 1.43 times over AC-APD-ECL using 0.8μ design rules of an advanced IBM bipolar process.

The next two papers by H. J. Shin and Oklobdzija present novel advanced ECL circuits. Each one has advantages over others as well as disadvantages. The circuit presented by H. J. Shin permits "Emitter-Dotting," a significant feature, which allows for implementation of the "Wired-OR" function on the outputs. The same "Wired-OR" function can be accomplished with regular ECL logic shown in Figs 2.6 and 2.7. ECL derives its power from its ability to implement quite powerful functions in very few logic stages. An implementation of a 64-bit carry-lookahead adder in just 4 logic stages is possible.

The circuit introduced by Oklobdzija does not require an additional voltage reference circuit and is capable of self-adjusting because the changes at the output determine when the pull-down action should be terminated. The principles of operation of the circuit termed *Feedback Controlled Current Source-Active Pull Down ECL* (FCCS-APD-ECL) are illustrated in Fig. 2.11.

The feedback delay is implemented as a simple RC constant consisting of a resistor and gate to channel capacitance of an MOS transistor. The Controlled Current Source is turned OFF and ON in the opposite phase form, the transistor Q_1, and it is also turned OFF when the output reaches its logic low level. However, Wired-OR is not easily achievable, as it is in the circuit presented by H. J. Shin.

The final paper in this section is by Heimsh et al. presenting Merged CMOS/Bipolar current switch logic. The concept of merging MOS and bipolar transistors in an ECL circuit structure to achieve high-speed while reducing the power by an or-

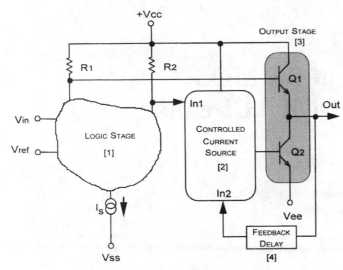

Fig. 2.11. FCCS-APD-ECL circuit operation [Oklobdzija].

der of magnitude is presented. The paper shows how to use bipolar transistors in the time-critical parts of the logic while employing CMOS in the noncritical sections. The comparisons are done on a 16-bit adder where a bipolar differential pair is used to propagate the carry signal, which is the time-critical section of the adder. Use of MOS transistors in the lower levels of the differential ECL structure allowed for the reduction of the power supply voltage without driving the bipolar transistors into saturation. The circuit presented in this paper shows another way of looking at BiCMOS circuits. It represents an interesting concept that can be used in the environment where the bipolar transistor already exists, such as read-channel disk drive electronics or high-speed communication, where the front end of the receiver is already using bipolar transistors.

An Overview of BiCMOS
State-of-the-Art Digital Circuits

HYUN J. SHIN

IBM THOMAS J. WATSON RESEARCH CENTER
YORKTOWN HEIGHTS, NY

Abstract—State-of-the-art BiCMOS digital circuits for logic applications are reviewed in this paper. Applications have diversified into many areas such as gate arrays, ASICs, and data and signal processors. Leading-edge BiCMOS circuits that have been demonstrated include gate arrays with more than 100K gates operating at or above 100 MHz; standard-cell macros for ASICs and processors in a 0.5μm 3.3-V BiCMOS technology; and a 1000-MIPS 32-bit microprocessor and a video signal processor, both having a clock frequency of 250 MHz. BiCMOS progress is being driven by the use of fast high-gain bipolar transistors to improve circuit speed. Most BiCMOS digital circuitry is CMOS-dominant with add-on of specific BiCMOS or bipolar circuitry in appropriate places.

1. INTRODUCTION

Since its introduction in the late 1970s [1], BiCMOS technology has been progressing quite remarkably [2]. Behind this enormous growth is the idea that CMOS circuits, when complemented by bipolar devices, would provide near-ideal, highly integrated, low-power high-speed digital or mixed-analog/digital VLSI circuits that can cover a wide range of applications; BiCMOS would be the one technology to exploit the advantages of CMOS, i.e., high integration-level and low power, with the strengths of bipolar transistors, such as fast switching speed, good drive capability, and large transconductance (g_m), or gain.

Naturally, in the early stage of development, BiCMOS technologies were simple enhancements to CMOS with minimally added process complexity and relatively low-performance npn bipolar transistors [1,3,4]. Therefore, the first applications of BiCMOS were mostly limited to selective use of bipolar devices in CMOS-dominant circuits for improving drive capability of an output buffer, accuracy of a reference generator, or voltage gain of a sense amplifier [5–7]. To go beyond these limited applications in the early BiCMOS technologies, starting from the mid-1980s, substantial efforts have been directed to the development of advanced BiCMOS technologies that offer optimized, small-size, high-performance bipolar transistors

[2]. Despite some penalty in increased processing steps and cost, this new breed of sophisticated CMOS-based BiCMOS with uncompromised bipolar devices has led to the proliferation of high-speed digital VLSI and analog applications, including gate arrays, memories, application-specific integrated-circuits (ASICs), and general-purpose microprocessors and signal processors [8–10].

There also have been recent efforts to develop bipolar-based BiCMOS technologies for high-performance bipolar-intensive logic applications to reduce power and increase integration density, especially to increase on-chip memory capacity [9, 11]. Since maintaining high-frequency characteristics (e.g., cut-off frequency f_T) of the bipolar transistors is essential in these applications, the bipolar devices should not be compromised by the addition of the CMOS devices to be used in high-density memory cells. However, the difference between the bipolar-based and the CMOS-based BiCMOS is getting less obvious [2] as new technologies strive to optimize both types of devices.

For BiCMOS technology assessment, a frequently used figure of merit is the relative performance advantage of a BiCMOS logic gate over the gate in pure CMOS; it is typically represented by the gate delay as a function of load capacitance (as in Fig. 3). Until recently (to the 0.8-μm-technology generation), the most widely used BiCMOS gate circuits were the totem-pole type configured with an emitter-follower pull-up and gated-diode pull-down circuitry, as shown in Fig. 1. Figure 2 shows that typical gate delays of these conventional BiCMOS circuits are about 2 times better than those of CMOS in each technology generation marked by the corresponding pattern size, especially for 2.0- to 0.8-μm 5-V technologies [8]. Interestingly, converting CMOS into BiCMOS of equal technology generation for this 2× performance improvement takes only about a year or so, while developing the next CMOS generation for a similar speed enhancement needs approximately three to four years. The CMOS delays improve by

Reprinted from *BiCMOS Integrated Circuit Design: With Analog, Digital, and Smart Power Applications,* edited by M. I. Elmasry, pp. 185-192, IEEE Press. © 1994 IEEE.

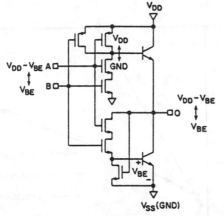

Fig. 1 Conventional BiCMOS logic circuit with emitter-follower pull-up and gated-diode pull-down circuitry ([14]).

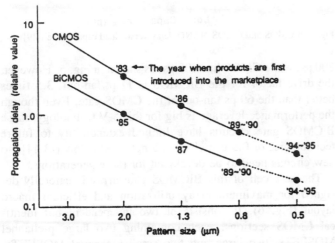

Fig. 2 CMOS/BiCMOS propagation delay versus pattern size ([8]).

a factor of 2 in every generation, roughly following the constant-voltage scaling rule.

Beyond the 0.8-μm into deep sub-micron generations, however, the performance leverage of conventional BiCMOS circuits over CMOS is questionable as shorter channel-length, scaled technologies require correspondingly lower power-supply voltages to ensure reliability of MOS and bipolar devices. Table 1 [12] shows a steady and quite rapid trend in power-supply voltage reduction into the future in accordance with the continued introduction of next-generation technologies: once in approximately every four years. With reduced

TABLE 1 Low-voltage CMOS technology projections ([12]).

	Year				
Technology	1992	1994	1996	1998	2000
Channel length, μm	0.5	0.4	0.3	0.2	0.1
Dynamic RAM Bits, Mb	16	64	64	256	256
Supply voltage, V	3.3/5.0	3.3	2.5/3.3	2.5	2.5
Logic Gates × 1000	200	400	600	1000	2000
Supply voltage, V	3.3	3.3	2.5	2.5	1.5
Processor speed, MHz	50	100	150	200	500

supply voltages, the conventional BiCMOS gate circuit of Fig. 1 that features gated-diode pull-down circuitry quickly loses its speed advantage because it needs at least $2 V_{BE} + V_{Tn}$ of supply voltage before turning on the pull-down path [13]. Here, V_{BE} is the base-emitter turn-on voltage (typically 0.8 V) and V_{Tn} is the threshold voltage of the gating nMOSFET. For example, as plotted in Fig. 3, BiCMOS is comparable in delay with CMOS for a 3.6-V 0.5-μm technology, but becomes much worse than CMOS with a 2.5-V 0.25-μm process. In essence, the fast-paced progress in technology has cast a sudden shadow on the future of BiCMOS in digital circuits. The conventional BiCMOS gates will no longer be usable, and viable BiCMOS circuit techniques that still have leverage with this unprecedented reduction of supply voltage are needed. To solve this critical problem, a number of BiCMOS gate circuits that may be extendible into the far sub-micron range with continued speed leverage over CMOS have been proposed [14–18]. These gates primarily utilize pull-down circuits, which are better than the gated diode, as well as full-swing techniques to maximize voltage drive. However, these new circuits require either a more sophisticated complementary bipolar technology [14,18] or use many devices in a large area [16–18]. Some of them are not sufficiently fast or provide only noninverting logic functions [15,18]. So, the question remains:Is it worth investing in BiCMOS rather than pure CMOS, or will the return be marginal with all the increased complexity and cost of processes and decreased integration density?

The answer to this question can be "yes" if BiCMOS or bipolar circuit techniques other than the conventional BiCMOS gates are cleverly applied to specific places in digital circuits. Because each bipolar device requires one V_{BE} for operation (and this value tends to rise as the feature size is reduced), techniques to minimize stacking of bipolar transistors become important. Furthermore, exploiting the main advantages of bipolar devices, such as high speed, high gain, and excellent

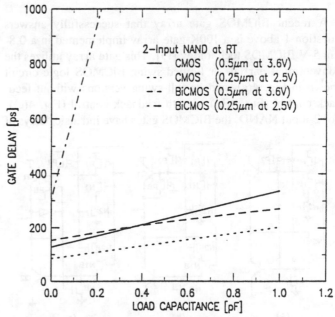

Fig. 3 Extendibility of conventional BiCMOS gate versus CMOS.

drivability, for reducing signal swing and noise and improving sensitivity is beneficial. In fact, these ideas have spurred the development of BiCMOS digital designs where bipolar usage is highly specific and selective. This trend is likely to continue the realization of BiCMOS digital circuits with performance beyond what pure CMOS can offer but with minimal cost penalty.

This paper is an overview of the highlights of the most recent advances in BiCMOS digital circuits. In Section 2, state-of-the-art gate arrays are reviewed. BiCMOS ASICs are looked at in Section 3, and more general purpose data and signal processors are discussed in Section 4.

2. Gate Arrays

Although the main driving force for BiCMOS development has been improved performance of BiCMOS over CMOS, demonstration of gate arrays that employ fast BiCMOS gates for high-speed logic applications has been rather rare. This is because there are many difficult questions to be answered before efficient and cost-effective BiCMOS gate arrays are designed:

1. Which BiCMOS gate circuits offer maximum performance with reasonable processing cost, circuit complexity, and gate density?
2. Are these BiCMOS circuits extendible to future technology generations that will require appropriately lower power-supply voltages, so that no major redesign effort is needed?
3. What are optimum basic cell images for maximum array utilization; versatile macro design such as latches, register files, and memories; and CMOS/BiCMOS gate inter-mixing?
4. How easily can existing CMOS gate-array macros be converted to BiCMOS macros?

A recent BiCMOS gate array that successfully answers question 1 above is a 100K-gate array implemented in a 0.8-μm 5-V BiCMOS technology [19]. This gate array utilizes the conventional partial- or limited-swing BiCMOS logic circuit shown in Fig. 4(a), and its full-swing versions without feedback control (Fig. 4(b)) or with feedback control (Fig. 4(c)). For 2-input NAND, the BiCMOS gates have intrinsic delays of

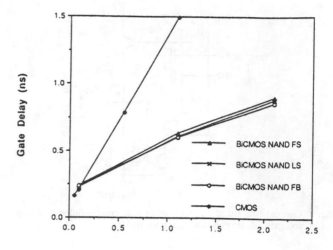

Load Capacitance (pF)

Fig. 5 BiCMOS and CMOS NAND delay versus load capacitance ([19]).

200 ps, while CMOS had 100 ps, as shown in Fig. 5. However, the drive factor for the BiCMOS is 17 ps/fan-out, 3.5 times better than the 60 ps/fan-out of the CMOS gate. Even though the performance leverage is big for this 5-V technology, these BiCMOS gate circuits have limited extendibility to future technologies of 0.5-μm channel length and below [13,14], so new circuits need to be developed for these generations.

The base cell of this BiCMOS gate array is carefully designed for maximum array utilization and efficient macro layout (Fig. 6). It consists of two independent but identical CMOS sections, each containing two large p-channel MOSFETs, two large and two small n-channel MOSFETs, and a middle bipolar section having two npn transistors as well as four small (two n- and two p-channel) MOSFETs. The bipolar section is equally accessible by each CMOS section. So, in one base cell, two gates can be implemented, one of which may be BiCMOS. Also, due to the small MOSFETs, efficient BiCMOS latch and register-file macros can be designed. The base-cell size is 46.8×32 ($= 1500$) μm² and, with this cell, 106K gates can be populated within a 9×9-mm² array core (Fig. 7). With triple metal layers and local interconnect, this gate array has achieved a remarkably high array utilization of 92% in a test application to a 74K-gate digital filter. The input/output interface is ECL 100K compatible. These design points

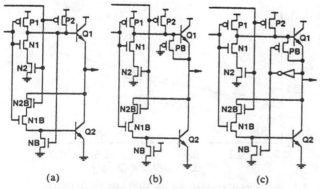

Fig. 4 BiCMOS 2-input NAND circuits: (a) limited swing (LS); (b) full swing (FS); and (c) feedback control (FB) ([19]).

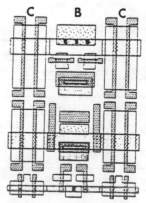

Fig. 6 A BiCMOS base-cell layout ([19]).

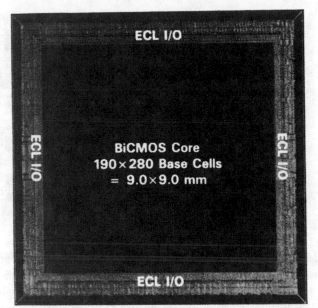

Fig. 7 A 106K-BiCMOS gate array photograph (chip size = 11.4 mm^2) ([19]).

are similarly addressed in the 200K-gate array implemented with a comparable BiCMOS circuit and technology [20].

The 100K-gate array has been applied to a 100-MHz 64-tap FIR filter [21], which dissipates maximum power of 6 W. The filter utilizes 187K MOSFETs and 22K bipolar transistors in an area of 49 mm^2, equivalent to 55K gates or approximately two-thirds of the array core. BiCMOS buffer circuits are selectively used to speed up the critical paths, including the clock and data-bus trees and the carry-save multiply-adder circuit, as shown in Fig. 8. The fast BiCMOS buffer also shows less speed degradation than CMOS at high chip temperature. The performance of this gate-array BiCMOS filter is comparable to a contemporary CMOS filter custom-designed with a module generator, which suggests that the advantage of gate-array design—quick turnaround time—is maintained without sacrificing the performance using BiCMOS technology.

Another state-of-the-art BiCMOS gate array worth noting because it considers questions 1–3 above is a 237K-gate array realized in a more advanced 0.5-μm, 13-GHz f_τ, 3.3-V tech-

nology [22]. The gate circuit called BiPNMOS (shown in Fig. 9), which is a feedback-controlled full-swing version of the BiNMOS gate, is used. The BiNMOS circuit is one of the earliest BiCMOS circuits [6] and uses a bipolar transistor only for pull-up drive and n-channel MOSFETs for pull-down [15]. As shown in Fig. 10, this gate has a delay of 230 ps (for 2-input NAND with a fan-out of 7 at 3.3 V), 35% faster than pure CMOS, and is suitable for supply voltages down to 2.5 V.

This BiCMOS gate array features an optimized 1400-μm^2 basic cell containing eight small MOSFETs (two p-channel, six n-channel). With the small MOSFETs, memory macros, including a high-speed static RAM, a high-density RAM, a ROM, and a CAM, can be effectively designed, as demonstrated by a 16K-bit 2.7-ns RAM and a CAM using one basic cell for one memory cell, a 4.0-ns high-density RAM with two memory cells per basic cell, and a 3.2-ns ROM integrated on a test chip. A single-ended bit-line sensing scheme with double bit lines and ECL-like BiCMOS sense amplifiers is used to improve the speed.

A gate array that seeks minimal effort for converting existing CMOS gate macros into BiCMOS (specifically in response to question 4 above) has also been described [23]. Although the gate array requires a Schottky diode technology and the gate delay is marginally better than that of CMOS, an existing library of CMOS gates can be used to form BiCMOS macros without modification.

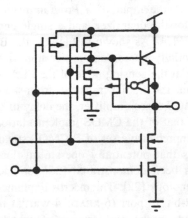

Fig. 9 A BiPNMOS gate ([22]).

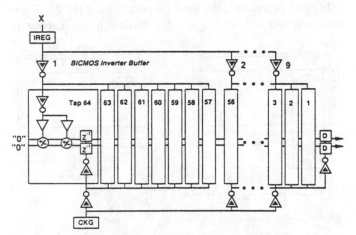

Fig. 8 A 64-tap BiCMOS FIR filter data/clock path ([21]).

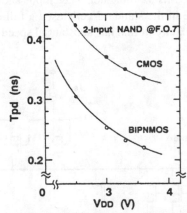

Fig. 10 Supply-voltage dependence of BiPNMOS and CMOS gate delays ([22]).

3. APPLICATION-SPECIFIC INTEGRATED-CIRCUITS

For high-speed ASICs, standard-cell macros custom-designed in BiCMOS have been emerging steadily [24, 25]. This is because the use of BiCMOS technology in upgrading performance of functional macros through custom tuning with BiCMOS circuits is believed to be more economical than the use of the next-generation CMOS. Also, unlike basic BiCMOS gate circuits whose extendibility to advanced technologies with lower supply voltages is often questionable, the BiCMOS standard-cell macro circuits are likely to be retained in future technologies and maintain the performance leverage over the CMOS because they are individually tailored to specific needs, such as improving sensing gain and speed, reducing voltage swing, and increasing load drivability. One thing to notice in BiCMOS ASICs is that, for the same reason, the fraction of bipolar devices used relative to the total devices is quite small compared to BiCMOS gate arrays.

A good example is the 54-MHz Differential Pulse Code Modulation (DPCM) coders for HDTV systems implemented in a 2.5-μm BiCMOS technology that is about 30% more expensive than CMOS of the same lithography generation [24]. Even though the minimum feature size is relatively large, the high-speed operation is possible due to architectural modification of the coders as well as some BiCMOS circuit techniques employed. The BiCMOS circuits utilized are a conventional totem-pole-type buffer for driving a significant capacitive load at the input of a Programmable Logic Array (PLA), used for the quantizer, and a single-ended sense amplifier for the PLA. As shown in Fig. 11, the BiCMOS sense circuit is essentially a common-emitter amplifier, which limits voltage swing at the sensing node of the PLA to about 0.8 V and yet prevents the bipolar transistor from saturating. By using this BiCMOS sense amplifier, the delay of the PLA is cut to nearly half that of the CMOS implementation.

A more comprehensive set of BiCMOS standard-cell macros for ASICs that potentially operate at clock rates up to 200 MHz has been demonstrated [25] for a 0.5-μm 3.3-V BiCMOS technology [22]. The macros implemented include a 3.0-ns 72×32-bit 10-port (6-READ, 4-WRITE) register file, a 5.0-ns 32K-byte cache memory, a 2.5-ns table look-aside buffer, and a 3.0-ns 32-bit adder. The chip is shown in Fig. 12. The basic gate in the standard cells used a full-swing version of the BiNMOS circuit that has substantial speed leverage over

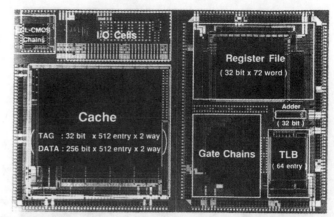

Fig. 12 A 0.5-μm BiCMOS standard-cell chip micrograph ([25]).

CMOS at 3.3 V. Interestingly, the register macro does not utilize BiCMOS circuits except for a fast-recovering bit-line load. A CMOS self-aligned threshold inverter is used as a single-ended bit-line sense amplifier. In the cache macro, a circuit technique of intermixing ECL and CMOS is adopted to improve speed, as depicted in Fig. 13. As an extension of this technique to basic BiCMOS standard cells for future 2-V technology generations, a direct-coupled ECL/CMOS circuit has also been explored here.

4. DATA AND SIGNAL PROCESSORS

Compared with the ASICs discussed in Section 3, general-purpose BiCMOS processors for data and digital signal processing have shown impressive progress in recent years [27–32]. Since the first demonstration of a 70-MHz 32-bit microprocessor in a 1.0-μm BiCMOS technology in 1989 [26], a 1000-MIPS microprocessor operating at 250 MHz and a 250-MHz video signal processor have been reported [29,32]. These BiCMOS processors have been implemented with custom circuit macros that utilize bipolar or BiCMOS circuits only selectively in appropriate places; in fact, the usage of bipolar transistors in the chips ranges merely from 0.4% to 2.7%. This use of a very small proportion of bipolar devices, however, maximizes the performance without degrading the integration level per unit chip area.

In these processors, the most frequently used BiCMOS circuits are buffered logic gates for driving heavy capacitive

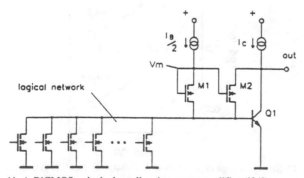

Fig. 11 A BiCMOS unlocked small-swing sense amplifier ([24]).

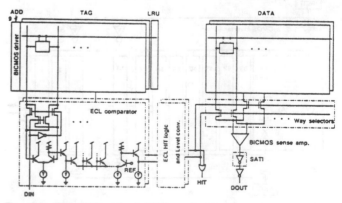

Fig. 13 A Cache macro circuit ([25]).

loads, which differ depending on the technology generation due to the reasons discussed in Section 1. Another BiCMOS circuit utilized is the bipolar-based sense amplifier and bit-line clamp circuit for memory macros that limits the bit-line swing to about 100 mV on READ to minimize noise while maintaining access time and speeding recovery from WRITE. This BiCMOS sense circuit has been adopted for the register file of the 32-bit RISC superscalar microprocessor [27]. With a 0.8-μm 5-V BiCMOS process, the register file can be accessed within 3.0 ns and the microprocessor operates at 40 MHz. In the companion 50-MHz cache controller for the microprocessor [28], BiCMOS techniques are found in the transistor-transitor-logic (TTL) output driver, phase-locked loops (PLLs) for clocks, and cache tag comparator. Basically, they apply the conventional totem-pole BiCMOS gate circuit. Fast BiCMOS gates for a phase/frequency detector in the PLL minimize false glitches and reset periods, resulting in locking up to 90 MHz. The BiCMOS 17-bit tag comparator in Fig. 14 uses a wired-OR emitter-follower for pull-up and an npn pull-down transistor controlled by the NMOS NOR circuitry. This circuit, limiting the highly capacitive NOR output to a swing of only one V_{BE}, makes comparisons in 1.5 ns.

In a 250-MHz 1000-MIPS superscalar microprocessor implemented in a more advanced 0.3-μm 3.3-V BiCMOS technology (Fig. 15) [29], the quasi-complementary BiCMOS (QC-BiCMOS) circuit shown in Fig. 16 [16] is used. The QC-BiCMOS gate uses a quasi-pnp device emulated as a combination of a p-channel MOSFET and an npn transistor for pull-down. (This is also used in the merged BiCMOS gate [17].) Although the circuit is complicated, requiring many devices, it offers more than twice the speed of CMOS at 3.3 V and is extendible to 2.5-V technologies. As illustrated in Fig. 17, this QC-BiCMOS gate is applied to latches and decoders in critical paths of the processor for speed and load drivability. An ECL sense amplifier is used selectively in the cache memory and table look-aside buffer (TLB). Also, QC-BiCMOS-feedbacked massive-input logic [30] is utilized in the arithmetic logic unit (ALU), to achieve addition in 1.2 ns, which is 60% faster than

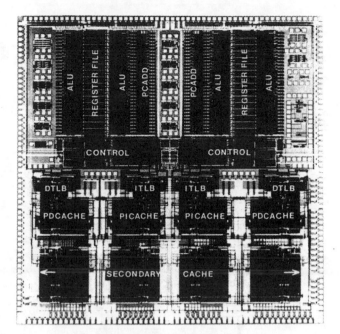

Fig. 15 A 1000-MIPS superscalar BiCMOS microprocessor ([29]).

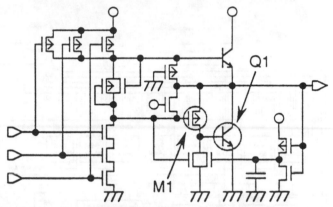

Fig. 16 A quasi-complementary BiCMOS logic circuit ([29]).

CMOS. The QC-BiCMOS-feedbacked massive-input logic gate provides high-speed complex logic with very large fan-in, at the expense of high circuit complexity.

Applying buffered BiCMOS logic gates for driving large capacitive loads is effective in realizing high-performance vector and signal processors. For example, the 64-bit vector-pipelined processor (Fig. 18) implemented in a 0.8-μm 5-V BiCMOS technology [31] operates at 100 MHz to achieve 200-MFLOPS, simply by utilizing BiNMOS-type gates for the Booth decoder in the multiplier/divider, the decoder in the barrel shifters, the word-line decoders in the register file, the clock drivers, etc. With the same technology, a much higher speed (250 MHz) was obtained for the 16-bit video signal processor shown in Fig. 19 [32]. A pipelined ALU with a carry-look-ahead adder, featuring BiNMOS-type logic gates, as well as a pipelined convolver/multiplier with a Booth decoder, utilizing a faster version of the conventional BiCMOS gate, are key contributors for the speed-up. Among the 1.13 million transistors in this video processor chip, only 5,000, or 0.4%, are bipolar devices.

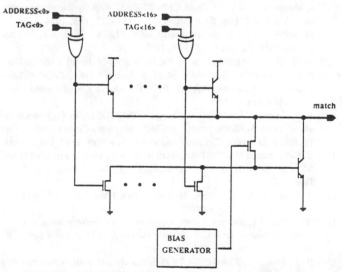

Fig. 14 A 17-bit tag comparator for 50-MHz BiCMOS cache controller ([28]).

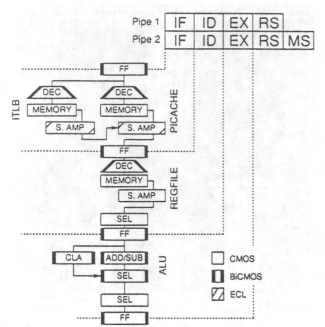

Fig. 17 Pipeline stages and critical paths of 1000-MIPS microprocessor ([29]).

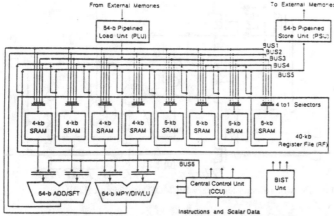

Fig. 18 A block diagram of 200-MFLOPS BiCMOS vector-pipelined processor ([31]).

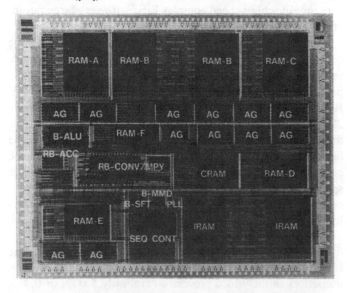

Fig. 19 A photograph of 250-MHz BiCMOS video signal processor ([32]).

5. Conclusions

There has been explosive progress in recent years in both BiCMOS digital circuits and BiCMOS technologies. Versatile BiCMOS circuit techniques for maximizing performance have been exploited in a variety of applications in high-speed high-density gate arrays, ASICs, and data/signal processors. Innovative BiCMOS logic gates and circuit techniques, such as full-swing techniques for gate circuits and small-swing techniques for sensing, extendible to the next few generations of sub-micron BiCMOS technologies that require supply-voltage scaling, are actively being developed to prolong the life and widen the spectrum of BiCMOS in digital applications.

Acknowledgments

The author would like to thank Prof. M. I. Elmasry at the University of Waterloo, C.-T. Chuang, and C. J. Anderson for their encouragement. Special thanks are due to L. M. Terman and P. J. Lim for their valuable advice and editorial help.

References

[1] G. Zimmer et al., "A fully implanted NMOS, CMOS, bipolar technology for VLSI of analog-digital systems," *IEEE Trans. Electron Devices*, vol. ED-26, pp. 390–396, April 1979.

[2] R. H. Havemann and R. J. Eklund, "Overview of BiCMOS device and process integration," in M. I. Elmasry (ed.), *BiCMOS Integrated Circuit Design*, Part I: Analysis and Design, New York: IEEE Press, 1993.

[3] P. A. Sullivan et al., "High performance bipolar transistors in a CMOS process," *IEEE Trans. Electron Devices*, vol. ED-29, pp. 1679–1680, Oct. 1982.

[4] C. S. Yue et al., "Improved bipolar transistor performance in a VLSI CMOS process," *IEEE Electron Device Lett.*, vol. EDL-4, pp. 294–296, Aug. 1983.

[5] E. L. Hudson and S. L. Smith, "An ECL compatible 4K CMOS RAM," in *ISSCC Dig. Tech. Papers*, Feb. 1982, pp. 248–249.

[6] O. Minato et al., "A Hi-CMOSII 8K × 8b Static Ram," in *ISSCC Dig. Tech. Papers*, Feb. 1982, pp. 256–257.

[7] J. Miyamoto et al., "A 28 ns CMOS SRAM with bipolar sense amplifiers," in *ISSCC Dig. Tech. Papers*, pp. 224–225, Feb. 1984.

[8] M. Kubo et al., "Perspective on BiCMOS VLSI's," *IEEE J. Solid-State Circuits*, vol. 23, pp. 5–11, Feb. 1988.

[9] A. G. Eldin, "An overview of BiCMOS state-of-the-art static and dynamic memory applications," in M. I. Elmasry (ed.), *BiCMOS Integrated Circuit Design*, Part III: BiCMOS Memory Applications, New York: IEEE Press, 1993.

[10] B. H. Leung, "Analog circuit design in BiCMOS technology: An overview," in M. I. Elmasry (ed.), *BiCMOS Integrated Circuit Design*, Part IV: BiCMOS Analog Circuit Applications, New York: IEEE Press, 1993.

[11] Y. Kobayashi et al., "SST-BiCMOS technology with 130-ps CMOS and 50-ps ECL," in *Symp. VLSI Technology Dig. Tech. Papers*, pp. 85–86, June 1990.

[12] B. Prince and R. H. Salters, "ICs going on a 3-V diet," *IEEE Spectrum*, vol. 29, pp. 22–25, May 1992.

[13] H. J. Shin, "Performance comparison of driver configurations and full-swing techniques for BiCMOS logic circuits," *IEEE J. Solid-State Circuits*, vol. 25, pp. 863–865, June 1990.

[14] H. J. Shin, "Full-swing BiCMOS logic circuits with complementary emitter-follower driver configuration," *IEEE J. Solid-State Circuits*, vol. 26, pp. 578–584, April 1991.

[15] A. E. Gamal et al., "BiNMOS: A basic cell for BiCMOS sea-of-gates," in *Proc. CICC*, pp. 8.3.1–8.3.4, May 1989.

[16] K. Yano et al., "Quasi-complementary BiCMOS for sub-3-V digital circuits," *IEEE J. Solid-State Circuits*, vol. 26, pp. 1708–1719, Nov. 1991.

[17] R. B. Ritts et al. "Merged BiCMOS logic to extend the CMOS/BiCMOS performance crossover below 2.5-V supply," *IEEE J. Solid-State Circuits*, vol. 26, pp. 1606–1614, Nov. 1991.

[18] M. Hiraki et al., "A 1.5-V full-swing BiCMOS logic circuit," in *ISSCC Dig. Tech. Papers*, pp. 48–49, Feb. 1992.

[19] J. D. Gallia et al., "High-performance BiCMOS 100K-gate array," *IEEE J. Solid-State Circuits*, vol. 25, pp. 142–149, Feb. 1990.

[20] Y. Enomoto et al., "A 200K gate 0.8μm mixed CMOS/BiCMOS sea-of-gates," in *ISSCC Dig. Tech. Papers*, pp. 92–93, Feb. 1990.

[21] T. Yoshino et al., "A 100-MHz 64-tap FIR digital filter in 0.8-μm BiCMOS gate array," *IEEE J. Solid-State Circuits*, vol. 25, pp. 1494–1501, Dec. 1990.

[22] H. Hara et al., "0.5-μm 2M-transistor BiPNMOS channelless gate array," *IEEE J. Solid-State Circuits*, vol. 26, pp. 1615–1620, Nov. 1991.

[23] T. Horiuchi et al., "A new BiCMOS gate array cell with diode connected bipolar driver," in *Proc. IEEE 1990 Bipolar Circuits and Technology Meeting*, pp. 124–127, Sept. 1990.

[24] A. Rothermel et al., "BiCMOS circuits for DPCM coders in HDTV systems," *IEEE J. Solid-State Circuits*, vol. 25, pp. 1470–1475, Dec. 1990.

[25] H. Hara et al., "0.5-μm BiCMOS standard-cell macros including 0.5W 3ns register file and 0.6W 5ns 32KB cache," in *ISSCC Dig. Tech. Papers*, pp. 46–47, Feb. 1992.

[26] T. Hotta et al., "A 70MHz 32b microprocessor with 1.0μm BiCMOS macrocell library," in *ISSCC Dig. Tech. Papers*, pp. 124–125, Feb. 1989.

[27] F. Abu-Nofal et al., A three-million-transistor microprocessor, in *ISSCC Dig. Tech. Papers*, pp. 108–109, Feb. 1992.

[28] B. Joshi et al., "A BiCMOS 50MHz cache controller for a superscalar microprocessor," in *ISSCC Dig. Tech. Papers*, pp. 110–111, Feb. 1992.

[29] O. Nishii et al., "A 1,000MIPS BiCMOS microprocessor with superscalar architecture," in *ISSCC Dig. Tech. Papers*, pp. 114–115, Feb. 1992.

[30] K. Yano et al., "3.3-V BiCMOS circuit techniques for 250-MHz RISC arithmetic modules," *IEEE J. Solid-State Circuits*, vol. SC-27, pp. 373–381, March 1992.

[31] F. Okamoto et al., "A 200-MFLOPS 100-MHz 64-b BiCMOS vector-pipelined processor (VPP) ULSI," *IEEE J. Solid-State Circuits*, vol. 26, pp. 1885–1893, Dec. 1991.

[32] J. Goto et al., "250-MHz BiCMOS super-high-speed video signal processor (S-VSP) ULSI," *IEEE J. Solid-State Circuits*, vol. 26, pp. 1876–1884, Dec. 1991.

Performance Comparison of Driver Configurations and Full-Swing Techniques for BiCMOS Logic Circuits

HYUN J. SHIN, MEMBER IEEE

Abstract —Driver configurations and full-swing techniques for several types of BiCMOS logic circuits have been compared among each other to examine their performance in scaled technologies. Among three driver configurations (common emitter, gated diode, and emitter follower) analyzed, the emitter-follower type is most advantageous for scaled power-supply voltage circuits. Also, full-swing techniques boost the circuit performance and base–emitter shunting is more favorable than collector–emitter shunting.

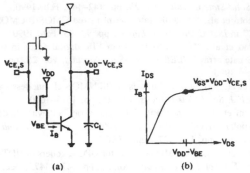

Fig. 1. (a) Common-emitter (CE) BiCMOS driver configuration and (b) its base current during pull-down.

I. Introduction

Rapidly growing attention is being given to BiCMOS technologies and circuits for integrated circuits with high speed and high integration level. The potential of BiCMOS technologies has been demonstrated in various BiCMOS circuits including static RAM's [1], [2], microprocessors [3], and gate arrays [4], [5]. However, in gate array applications, there are concerns about the leverage of BiCMOS logic circuits over pure CMOS as the power-supply voltage is reduced in scaled technologies.

Since the leverage and scalability of BiCMOS circuits depend on circuit details, a comparison of possible circuit techniques is important. In this paper, generic driver configurations and full-swing techniques for BiCMOS logic circuits are compared to assess the limitations of conventional BiCMOS circuits and provide directions in scaled technologies.

II. BiCMOS Driver Configurations

In BiCMOS logic circuits, three generic types of drivers can be identified as shown in Figs. 1, 2, and 3. All of them use switching MOSFET's to supply the base current and BJT's to drive output nodes. Operation of the pull-down circuitry (drawn with thick lines) is described here, but the complementary pull-up circuitry (in thin lines) is included for comprehensiveness. Each circuit in Figs. 1, 2, and 3 is assumed to be driven by the same circuit.

Fig. 1(a) shows a common-emitter (CE) configuration where the MOSFET operates in the common-source mode and the BJT in the common-emitter mode. Initially the MOSFET is assumed to be OFF and the base node of the BJT is discharged to GND by a device (not shown). When the MOSFET becomes ON, because the drain–source voltage of the MOSFET is V_{DD}, a large drain current flows and the base node will be charged up rapidly. As soon as the base potential rises above the BJT turn-on voltage ($\equiv V_{BE}$), the BJT becomes active and sinks a large current from the output. The base (or drain) current at this moment is still large because the MOSFET is in the

saturation region with the drain–source bias of $V_{DD} - V_{BE}$ as shown in Fig. 1(b). When the output is fully discharged, the BJT will be saturated with the collector potential $V_{CE,S}$ and the base potential $V_{BE,S}$.

This circuit has a logic swing close to the full power-supply voltage (i.e., $V_{DD} - V_{CE,S}$), so the gate–source drive for the MOSFET is large. Because this driver type has a large near-constant base current during switching, it can be very fast. However, it has a substantial dc power consumption due to the drain current that flows during the nonswitching period. Also the speed will be degraded because of the BJT saturation unless the base–collector junction is clamped using a low-drop diode such as a Schottky. Another disadvantage of this circuit is that its logic function is noninverting, and inversion must be obtained with an additional circuit.

Fig. 2 shows a gated-diode (GD) driver and its base current. The MOSFET acts as a switch between the base and collector. If the input level is high (V_{IH}), the MOSFET turns on and the base node is charged up by a current from the output until the BJT becomes active. Because the output is discharged and the base potential increases, the gate–source drive as well as the drain–source voltage decreases as the output falls, and the MOSFET current is reduced rapidly, as indicated in Fig. 2(b). When the BJT becomes active, the base potential becomes roughly fixed and the gate–source bias stays at $V_{IH} - V_{BE}$. The drain–source voltage decreases as a large collector current quickly discharges the load capacitance. Finally, the drain potential reaches V_{BE} and the base current stops flowing. The BJT is diode-connected and the low level of the output is V_{BE}. Similarly, the logic high level is $V_{DD} - V_{BE}$.

This driver does not have a BJT saturation problem but has a partial logic swing (from V_{BE} to $V_{DD} - V_{BE}$). So $V_{IH} = V_{DD} - V_{BE}$ and the base current is roughly proportional to $V_{DD} - 2V_{BE} - V_T$, where V_T is the threshold voltage of the MOSFET with a substrate bias of V_{BE}. As a result, the speed of this circuit degrades quite rapidly as the power-supply voltage is reduced. This circuit in a complementary form also suffers from a poor balance between the rise and fall times, because the pull-down

Manuscript received June 13, 1989; revised November 20, 1989.
The author is with the IBM Thomas J. Watson Research Center, Yorktown Heights, NY 10598.
IEEE Log Number 9035669.

Reprinted from *IEEE Journal of Solid-State Circuits,* Vol. 25, No. 3, pp. 863-865, June 1990.

102

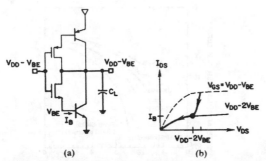

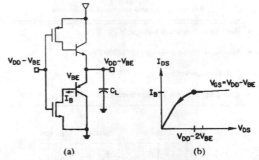

Fig. 2. (a) Gated-diode (GD) driver configuration and (b) its base current trajectory during pull-down.

Fig. 3. (a) Emitter-follower (EF) driver configuration and (b) its base current trajectory during pull-down.

circuitry consists of an n-channel MOSFET and an n-p-n BJT, both fast, and the pull-up circuitry is composed of a p-channel MOSFET and a p-n-p BJT, both comparatively slow. In addition, this driver is sensitive to the body effect and parasitic resistances at the base and emitter. The voltage drop across the resistors increases the source potential above V_{BE}, which reduces the gate–source drive and increases the threshold voltage.

An emitter-follower (EF) type driver is shown in Fig. 3(a). The MOSFET operates as a common-source inverter which drives the emitter follower. Because of the emitter-follower output stage, this circuit has a partial logic swing like the gated-diode driver. When the input is high, initial MOSFET current is determined by the gate bias $V_{DD} - V_{BE}$ and drain potential $V_{DD} - V_{BE}$, assuming that the base–emitter voltage was zero through a clamp device (not shown). A large current discharges the base node and the base–emitter drop increases until the BJT turns on. Then the output starts to follow the voltage transient at the base node. The drain current decreases as the drain–source voltage decreases.

Compared to the gated-diode driver, this circuit has a better base drive because of a larger gate–source drive voltage. Therefore, this driver is better for scaled power supplies. Also, this circuit is insensitive to the body effect and parasitic resistances. Like the common-emitter driver in Fig. 1, this has a good rise/fall balance because the pull-down circuitry consists of an n-channel MOSFET and a p-n-p BJT and the pull-up path has a p-channel MOSFET and an n-p-n BJT. Furthermore, in this configuration, it is possible to physically merge the MOSFET and BJT into a compact structure for less area and parasitics; e.g., the base of a p-n-p BJT and the drain of an n-channel MOSFET may share a common n-type diffusion area. The BJT driving the output does not saturate.

III. Full-Swing Techniques

As analyzed in the previous section, the gated-diode and emitter-follower drivers have small gate–source drive voltages due to the partial swing. If logic swings are increased to the full power-supply voltage, the speed of these circuits will improve. Additionally, the circuits with full swing allow the use of a lower MOSFET threshold voltage for a faster speed. They also have a better noise margin and interface well with pure CMOS circuits.

One technique to achieve the full swing is to use a resistive shunt network between the collector and emitter of the BJT, as illustrated in Fig. 4(a) for the emitter-follower and gated-diode drivers. The shunt network can be a simple resistor, a MOS-

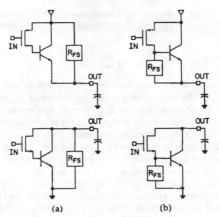

Fig. 4. Full-swing techniques: (a) collector–emitter shunt and (b) base–emitter shunt.

FET, or a CMOS positive feedback circuit. This network pulls the output all the way to the power-supply level after the BJT stops providing current, and holds it during the nonswitching period.

Because the shunt element adds current to the output, the final stage of the transient will be faster than a driver without the shunt element. However, a main drawback of this technique is a crossover current flowing through the shunt element when, e.g., the output begins to fall from the high level. This current not only increases power dissipation but also slows down the transient. Although the crossover current can be made smaller using a shunt element with higher resistance, longer time is needed for the full swing.

Another full-swing technique is shown in Fig. 4(b). A resistive network shunts the base and emitter of the BJT after the BJT is cut off and brings the output to the power-supply level through the MOSFET that is already ON. As before, effective resistance of the network remains low during the nonswitching period.

This technique does not have a crossover current flowing through the shunting element. Also the shunt element clamps the base–emitter junction of the BJT when the MOSFET is OFF for the opposite output transition. However, if a passive resistor is used for shunting, it bypasses a part of the MOSFET current flowing into the base during the main transient period, and the speed will be degraded. This problem is minimized by increasing the resistance, but the output will reach the full supply level more slowly.

- Zero-Bias Threshold Voltage: $V_{T0} \cong 0.2V_{DD}$.

- Minimum Turn-off Margin for MOSFETs: 0.2 V.

- BJT Turn-on Voltage: $V_{BE} = 0.7$ V.

- Clamped Collector Voltage before Saturation: $V_{CE,m} = 0.3$ V.

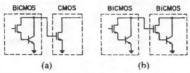

(a)   (b)

Fig. 5. (a) BiCMOS-to-CMOS interface when BiCMOS circuits are intermixed with pure CMOS. (b) BiCMOS-to-BiCMOS interface without intermixing.

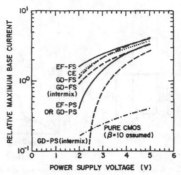

Fig. 6. Base currents versus power-supply voltage for the BiCMOS drivers.

IV. COMPARISON

For theoretical scaling, the power-supply voltage also scales as line width decreases. To predict performances of the above drivers in future technologies, maximum active base currents versus supply voltage are compared on a relative scale. This is approximately equivalent to comparing drive currents because the output current is proportional to the base current through the effective current gain of the BJT. In this comparison, the MOSFET transconductances are set to be equal for the different circuits. Also, one set of MOSFET threshold voltages is used for a circuit and the substrate bias effect on the MOSFET threshold voltage is ignored for simplicity. Identical BJT's with zero parasitics are assumed for the different circuits. The parameters used are listed in Table I.

For the gated-diode and emitter-follower circuits, both partial-swing (PS) and full-swing (FS) cases are considered. For full swing, current through the shunt element is ignored. Also, for the gated-diode driver, as illustrated in Fig. 5, a case of intermixing the BiCMOS circuit with pure CMOS is compared with a case without intermixing. Without intermixing, the MOSFET threshold voltages can be minimized with minimum noise margin at the interface between the BiCMOS circuits. If intermixed, however, the interface between the BiCMOS output and CMOS input requires higher threshold voltages to turn off the MOSFET's in the CMOS stage with enough margin. The other driver configurations are not affected by the intermixing.

The relative maximum base currents are plotted in Fig. 6. Without full-swing techniques, the common-emitter (CE) type provides more base current than the emitter-follower (EF) type or the gated-diode (GD) configuration for all supply voltages, and the ratio becomes larger as the supply voltage is reduced. Note here that the gated-diode circuit without intermixing has the same base current as the emitter-follower circuit.

However, the gated-diode type with intermixing shows much less current than the one without intermixing. The current decreases quite rapidly as the supply voltage decreases and the circuit even fails to operate below 2.3 V. Furthermore, if this BiCMOS driver is compared to a pure CMOS driver with an assumption that the effective current gain of the BJT ($\equiv \beta$) is 10, it eventually becomes inferior to CMOS in drive capability. The effective current gain is much smaller than a peak dc gain because it represents large-signal transient characteristics including high-level injection and, even, saturation. Also, it accounts for the fact that the channel width of a CMOS driver is adjusted wider than that of a BiCMOS driver, to make the circuit area equal for a fair comparison.

Now, with full-swing techniques, base currents of the emitter-follower and gated-diode circuits are boosted substantially and become larger than or comparable to the current of the common-emitter driver. The gated-diode circuits provide less currents than the emitter-follower type. Note that the gated-diode circuit intermixed with pure CMOS is still worse than the one without intermixing.

Although details of the curves will change in real circuits due to parasitic components, the body effect, and complicated real-device characteristics, the relative trends will be maintained.

V. CONCLUSION

For future BiCMOS technologies with reduced power-supply voltages, the emitter-follower drivers are advantageous in terms of drive capability and power consumption. The gated-diode configuration is inferior to others for driving purposes. Also, conventional BiCMOS logic circuits utilizing the partial-swing gated-diode drivers will quickly lose their performance leverage over CMOS circuits, as the supply voltage is lowered. Although the common-emitter type has good drive, it may not be a good choice for VLSI circuits because of the saturation and power dissipation problems. Full-swing techniques are effective in enhancing the drive currents of BiCMOS circuits, and base–emitter shunting is more favorable than collector–emitter shunting due to less crossover current.

ACKNOWLEDGMENT

The author would like to thank C. T. Chuang, C. L. Chen, L. M. Terman, and D. Gaffney for their encouragement and advice during this work.

REFERENCES

[1] H. Tran et al., "An 8ns BiCMOS 1Mb ECL SRAM with a configurable memory array size," in ISSCC Dig. Tech. Papers, Feb. 1989, pp. 36–37.
[2] M. Matsui et al., "An 8 ns 1Mb ECL BiCMOS SRAM," in ISSCC Dig. Tech. Papers, Feb. 1989, pp. 38–39.
[3] T. Hotta et al., "A 70MHz 32b microprocessor with 1.0 μm BiCMOS macrocell library," in ISSCC Dig. Tech. Papers, Feb. 1989, pp. 124–125.
[4] Y. Nishio et al., "A BiCMOS logic gate with positive feedback," in ISSCC Dig. Tech. Papers, Feb. 1989, pp. 116–117.
[5] J. Gallia et al., "A 100K gate sub-micron BiCMOS gate array," in Proc. IEEE CICC, May 1989.

A Feedback-Type BiCMOS logic Gate

YOJI NISHIO, FUMIO MURABAYASHI, SHOICHI KOTOKU, ATSUO WATANABE,
SHOJI SHUKURI, AND KATSUHIRO SHIMOHIGASHI, MEMBER, IEEE

Abstract —This paper will report on the development of a feedback-type
BiCMOS logic gate using a 0.5-μm BiCMOS technology. The propagation delay time of a three-input NAND gate with a 0.93-pF load is 245 and
290 ps at a supply voltage of 4.5 and 4 V, respectively. These values are
about 1.4–1.2 times better than the 0.8-μm BiCMOS gate operating at
5 V. A power dissipation of 0.4 mW was obtained with a 0.93-pF load, 4-V
supply voltage, and 14-MHz operation. The power dissipation is comparable to that of a CMOS gate.

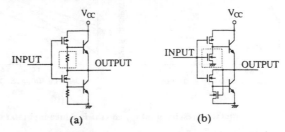

Fig. 1. Conventional BiCMOS logic gates: (a) R type and (b) N type.

I. INTRODUCTION

BiCMOS technology is being actively applied to memories and
logic VLSI's [1]–[6] because of their high speed, density, and
function, and their low power dissipation characteristics. As
process technology is refined, the supply voltage must be reduced
due to the lower endurance of the devices to high voltages and
the larger power dissipation of the LSI chips. Therefore, as the
MOS drain current decreases because of the lower supply voltage, it is important that the base current from the MOS in a
totem-pole type BiCMOS logic gate is high enough for high-speed
switching, while maintaining low power dissipation characteristics. Also, as the threshold voltage of the MOS is reduced, a full
logic swing function is necessary, even for BiCMOS gates, to
ensure that a dc current does not flow in the next gate and that
the speed of the multi-input gate does not decrease.

In order to solve those problems, some BiCMOS logic gates
have been developed.

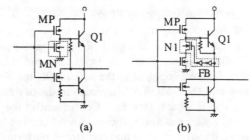

Fig. 2. Proposed BiCMOS logic gates: (a) $R + N$ type and (b) FB type.

II. CONVENTIONAL BiCMOS LOGIC GATES

Fig. 1 shows the conventional BiCMOS inverters [1]–[3]. The
main difference lies in the method of discharging the base of the
bipolar transistor. Namely, in the R type it is done by the
resistances, and in the N type it is done by NMOS transistors.
Both variations have been applied to gate arrays and memories.
They have many advantages over conventional gates such as
CMOS or ECL gates, but they also have several weak points.

In the R type, high-value resistances are used for speed, but
when the input level rises and the output level falls from a high
level, the base potential of the upper bipolar transistor falls with
the output level due to the large resistance of the bypass circuit.
So, it is difficult for the upper bipolar transistor to turn off and
the penetration current becomes large. Therefore, the power
dissipation is approximately twice as large as for the CMOS gate.

In the N type, the output high level V_{OH} is $V_{CC} - V_{BE}$, and the
low level V_{OL} is $GND + V_{BE}$. Since the gate voltage V_{GS} of the

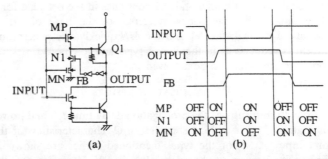

Fig. 3. Circuit operation of FB type: (a) FB-type inverter and (b) timing
chart.

OFF-state MOS is V_{BE}, a dc current will flow if the MOS
threshold voltage is low. Also, the gate voltage of the ON-state
MOS of the quiescent input terminal in the next stage is smaller
by the amount V_{BE}. As the ON-resistance value becomes larger,
the speed of the multi-input gate becomes slower.

III. PROPOSED BiCMOS LOGIC GATES

Fig. 2 shows the proposed BiCMOS logic gates. These show
inverters. Multi-input logic gates can be constructed easily by
changing the connection of PMOS and NMOS transistors.

The $R + N$ type gate adds a discharging NMOS MN to the R
type. So, this gate offers improved power dissipation characteristics over the R type. Also, as this gate has a full logic swing due
to the resistance between the base and the emitter, the above-
mentioned drawbacks of the N type are eliminated. But when the
input falls and the base current is supplied to the base of $Q1$, the
driving PMOS current flows partly into the discharging NMOS
in the same way as the N type. In the FB type, the leakage

Manuscript received March 29, 1989; revised June 19, 1989.
Y. Nishio, F. Murabayashi, S. Kotoku, and A. Watanabe are with Hitachi
Research Laboratory, Hitachi Ltd., 4026 Kuji-cho, Hitachi-shi, Ibaraki-ken
319-12, Japan.
S. Shukuri and K. Shimohigashi are with the Central Research Laboratory,
Hitachi Ltd., Kokubunji, Tokyo 185, Japan.
IEEE Log Number 8930188.

Reprinted from *IEEE Journal of Solid-State Circuits*, Vol. 24, No. 5, pp. 1360-1362, October 1989.

105

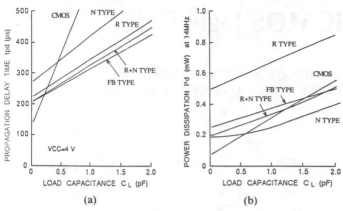

Fig. 4. Simulation results of (a) t_{pd} and (b) P_d (three-input NAND).

current is stopped by NMOS $N1$ and the feedback CMOS inverters.

Fig. 3 gives circuit operation of the feedback-type inverter. Operation of the multi-input gate is similar to an inverter.

The FB signal is a delayed signal of the output due to the feedback CMOS inverters. When the input falls and the driving PMOS current is being supplied to the base of $Q1$, the FB signal is initially low and the NMOS $N1$ continues in the OFF state for a while due to the feedback inverters. Consequently, the NMOS leakage current is stopped and sufficient driving current from the MP is supplied into the base, so that high-speed performance can be achieved.

On the other hand, when the input rises and $Q1$ turns OFF, the FB signal is initially high and $N1$ continues in the ON state for a while due to the feedback inverters. So, the base of $Q1$ is discharged through the $N1$ and the MN quickly, the penetration current is reduced as in the $R + N$ type, and low power dissipation characteristics can be achieved too.

IV. SIMULATION RESULTS

Simulation results of the propagation delay time t_{pd} and power dissipation P_d versus load capacitance C_L characteristics for the three-input NAND of the types mentioned earlier are shown in Fig. 4. The power supply voltage V_{CC} is 4 V. For reference, the characteristics of the CMOS which has a cell size about the same as the BiCMOS are shown too. The features just described are well demonstrated. Namely, the delay time of the N type is large. In this case, the potential of the two quiescent input terminals of the N type are fixed at 3.3 V. The power dissipation of the R type is about twice as much as the other types, because of its large penetration current. BiCMOS gates are slower than CMOS at very low load capacitance, since the effect of the bipolar transistor does not appear at very low load capacitance.

The FB type is the fastest for $C_L \geq 0.25$ pF, and it has low power dissipation characteristics like the other types.

V. EXPERIMENTAL RESULTS

The proposed BiCMOS logic gates were fabricated by using a 0.5-μm BiCMOS device.

The performance of the three-input NAND was evaluated by measuring various ring-oscillator frequencies and currents.

A part of the pattern of the 31-stage ring oscillator in the FB type is shown in Fig. 5. The feedback CMOS inverters which are constructed by using tungsten wire are placed between the driving PMOS and NMOS transistors. The feedback inverters use the minimum-size MOS transistors, since the inverters are only used

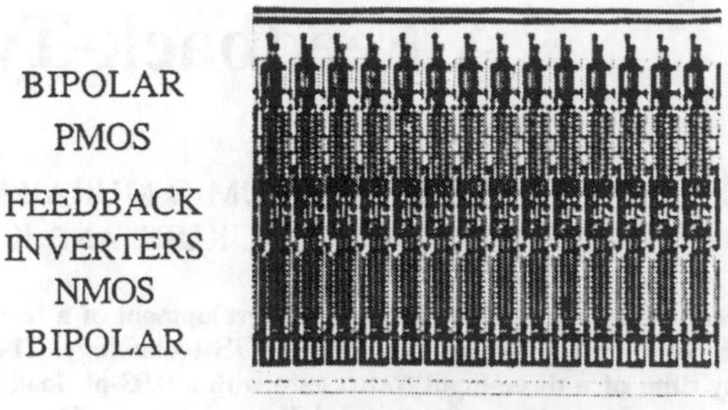

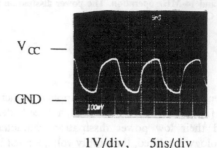

Fig. 5. Pattern of ring oscillator.

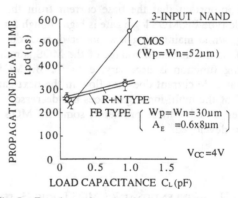

1V/div. 5ns/div

Fig. 6. Waveform of ring oscillator.

Fig. 7. Experimental results of propagation delay time.

for making a delayed signal of the output. Bipolar transistors are arranged on both sides of the MOS transistors. The cell height of the FB type is larger than that of the $R + N$ type by about 8 μm, or about 6 percent.

Fig. 6 shows a waveform of the ring oscillator. As the oscillator frequency is very high and the value of resistance between the base and emitter is large, the output level does not reach the V_{CC} or GND level. But under ordinary use, the level reaches the V_{CC} or GND level due to the resistance between the base and emitter.

Fig. 7 shows the experimental results of the propagation delay time versus load capacitance characteristics for the three-input NAND at 4 V. The MOS channel width of the driving MOS is 30 μm, and the emitter size is 0.6×8 μm. For reference, the characteristics of the CMOS which has a cell size almost the same as the BiCMOS are shown too. The MOS channel width is 52 μm.

As expected, the FB type is the fastest for $C_L \geq 0.25$ pF. Under the present device conditions, the t_{pd} is about 290 ps at 0.6 pF. Even if the cell size of the $R + N$ type is equal to the FB type, the speed performance of the FB-type gates is superior to that of the $R + N$ type gates in simulation. Also, even in fine technology such as 0.5 μm, the speed performance of the BiCMOS gate is superior to that of the CMOS for $C_L \geq 0.25$ pF.

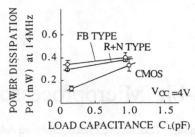

Fig. 8. Experimental results of power dissipation.

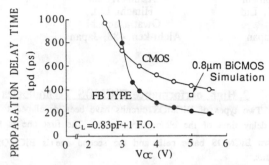

Fig. 9. Voltage dependence of propagation delay time.

Fig. 8 shows the experimental results for the power dissipation P_d characteristics.

The P_d of the BiCMOS is comparable to the CMOS at 1 pF. Operating frequency is 14 MHz. This frequency value corresponds to the average frequency of the internal gates, when the clock frequency is about 100 MHz.

As process technology is refined, the supply voltage needs to be reduced due to the lower hot-carrier endurance voltage of the MOS transistor. So, the supply voltage dependence of the propagation delay time was studied.

Fig. 9 shows the measured voltage dependence of t_{pd} for the FB type and CMOS three-input NAND. The propagation delay time in the BiCMOS increases rapidly as the supply voltage becomes less than 3.3 V, and at 3 V, the speed performance is almost equal to that of the CMOS. By changing the supply voltage from 4 to 4.5 V, the t_{pd} becomes 245 ps from about 290 ps.

The speed performance of the 4-V low-power version 0.5-μm BiCMOS is superior to that of the 5-V version 0.8-μm BiCMOS. Therefore, the endurance voltage of the device must be guaranteed for over 4-V operation.

VI. CONCLUSION

Some drawbacks of conventional BiCMOS logic gates are shown, and in order to solve their problems, a new feedback-type BiCMOS logic gate was proposed. By using a 0.5-μm BiCMOS, the t_{pd} was 290 and 245 ps at 4 and 4.5 V, respectively. Performance of a 4-V low-power version 0.5-μm BiCMOS was superior to that of a 5-V version 0.8-μm BiCMOS.

ACKNOWLEDGMENT

The authors wish to thank S. Asai, T. Masuhara, H. Sunami, K. Miyata, I. Masuda, T. Bandoh, K. Yagi, T. Nishida, T. Nagano, and H. Maejima for their encouragement and support.

REFERENCES

[1] S. C. Lee et al., "Bi-CMOS circuits for high performance VLSI," in Proc. Symp. VLSI Technol., Sept. 1984, pp. 46–47.
[2] Y. Nishio et al., "A subnanosecond low power advanced bipolar-CMOS gate array," in Proc. ICCD, Oct. 1984, pp. 428–433.
[3] I. Masuda et al., "High-speed logic circuits combining bipolar and CMOS technology," Trans. Inst. Electron. Commun. Eng. Japan, vol. J67-C, no. 12, pp. 999–1005, Dec. 1984.
[4] K. Ogiue et al., "A 15 ns/250 mW 64K static RAM," in Proc. ICCD, Oct. 1985, pp. 17–20.
[5] H. Tran et al., "An 8 ns BiCMOS 1 Mb ECL SRAM with a configurable memory array size," in ISSCC Dig. Tech. Papers, Feb. 1989, pp. 36–37.
[6] T. Hotta et al., "A 70 MHz 32b microprocessor with 1.0 μm BiCMOS macrocell library," in ISSCC Dig. Tech. Papers, Feb. 1989, pp. 124–125.

A High Performance BiCMOS 32-bit Microprocessor

Yasuhiro Nakatsuka, Takashi Hotta, Ryuichi Satomura, Syuichi Nakagami, Takashi Moriyama, Shigemi Adachi,
Shigeya Tanaka, Tadaaki Bandoh   Tetsuo Nakano, Atsuo Hotta   Shoji Iwamoto

Hitachi Research
Laboratory, Hitachi, Ltd.
4026 Kuji-cho, Hitachi-shi
Ibaraki-ken 319-12, Japan

Device Development
Center, Hitachi, Ltd.
Ome-shi
Tokyo 198, Japan

Asahi Works
Hitachi, Ltd.
Owariasahi-shi
Aichi-ken 488, Japan

ABSTRACT

We have developed the world's first BiCMOS 32-bit single chip microprocessor. It integrates 529 K transistors into a 12.98 mm square chip and typically realizes a 70 MHz frequency. The frequency is 1.5 to 2 times faster than that of today's CMOS microprocessors.

The microprocessor is designed with two design philosophies: (1) to reduce the number of inter-chip communication signals in the critical paths; and (2) to use basic cells optimally so as to allow fabrication into a single chip. The microprogram is divided into two parts and often-used microinstructions are stored in the ROM on the chip to reduce inter-chip communications. The TLB is also integrated in the microprocessor to reduce the inter-chip communication signals for memory access. Because of chip size and logic complexity constraints, the percentage of BiCMOS basic cells is limited; less than 20% of the basic cells can be BiCMOS ones.

1. Introduction

The BiCMOS (Bipolar-CMOS) technology combines CMOS logic circuits and bipolar transistors into BiCMOS cells. This technology realizes VLSIs which have characteristics of CMOS logic circuits (high density and low power consumption) and bipolar transistors (high driving capability and high sensitivity). The BiCMOS elementary circuits have been shown to double the performance of CMOS elementary circuits without increasing total area or power consumption.[1] - [3] Although the elementary technologies have been established, a single chip BiCMOS microprocessor is unknown.

Recently the performance of super-minis or super-workstations is growing rapidly. Using only CMOS technology, however, it would be difficult to obtain microprocessors which operate at more than 33 MHz. On the other hand, ECL logic circuits can realize very high performance. But, as they need a very large cooling system, they are not suitable to workstations. Only the BiCMOS microprocessor can satisfy both high performance and low power consumption needs.

Accordingly, our goal was to realize a BiCMOS microprocessor operating 1.5 to 2 times faster than a CMOS one. To cope with the growing performance of super-minis or super-workstations, we set the operating frequency at 70 MHz.

In **section 2**, the BiCMOS elementary technologies are discussed. In **section 3**, the design philosophies are shown, according to the analysis of the characteristics of BiCMOS circuits. In **sections 4** and **5**, the acceleration technologies of the microprocessor considering the characteristics of BiCMOS circuits are described.

2. High performance BiCMOS Technologies

Two types of BiCMOS circuits have been developed to shorten the delay time of the elementary circuits. The first one is the well known BiCMOS basic cells and the second one is BiCMOS sense circuits.

A BiCMOS basic cell consists of a CMOS basic cell and a bipolar booster, as shown in **Fig. 1**. As a bipolar transistor has high driving capability, it is efficiently used at the output stage which has a large load capacitance. On the other hand, using a CMOS logic circuit at the input stage realizes the low power consumption. Moreover, if only a small number of these basic cells are used among CMOS basic cells which have high integration density, the total integration density remains high.[1]

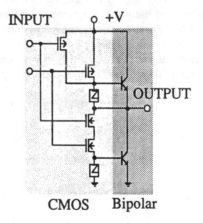

Fig. 1 BiCMOS Basic Cell

Figure 2 shows the delay time of 2-NAND basic cells versus their load capacitance. In the figure, the Double and Quad CMOS basic cells have output MOS transistors whose sizes are, respectively, two and four times larger than that of the CMOS basic cell. If the load capacitance is more than 1.0 pF, a BiCMOS basic cell will operates 1.5 times faster than any CMOS basic cells.

The BiCMOS basic cells are normally used in the random logic part and the circuits having a bipolar booster are also used to accelerate address decoding of the ROM and RAM whose word lines have very large load capacitance.

Reprinted from *Proceedings of the IEEE International Conference on Computer Design: VLSI in Computers and Processors*, pp. 358-360, 1989.

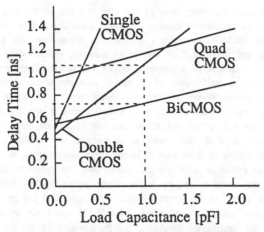

Fig. 2 Delay Time of Basic Cells

A BiCMOS sense circuit consists of an NMOS logic circuit and one bipolar transistor as **Fig. 3** shows. The logic circuit operates as a current switch which controls the base current of the bipolar transistor. As its output voltage is kept to within 0.8 V by the bipolar transistor, the NMOS logic circuit can operate very fast. As the cut off frequency of the bipolar transistor is very high, it can amplify the low swing, high speed signal to a full swing signal with a small delay time. In this structure, the BiCMOS circuit can operate twice as fast as a CMOS one. The BiCMOS sense circuit can be applied to the various parts of a microprocessor, by designing an appropriate NMOS logic circuit. For example, it can be used as the sensing amplifier of ROMs, RAMs, and register files. The NMOS logic circuit consists of switches which are controlled by the data of the ROM, SRAM and register file. It can also be used for the carry propagation circuit in the ALU.[2] - [5]

Analysis showed that the CMOS ROM access time of 18.0 ns was reduced to 9.2 ns, CMOS ALU carry generation time of 11.5 ns was reduced to 5.0 ns, and CMOS register file access time of 6.2 ns was reduced to 2.5 ns as shown in **Fig. 4**.

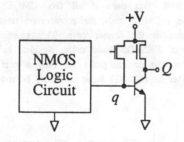

Fig. 3 BiCMOS Sense Circuit

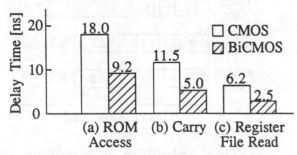

Fig. 4 Delay Time of Sense Circuits

3. Design Philosophies

To realize a high performance microprocessor, the delay time of the critical paths has to be shortened. There are three typical critical paths in the CPU. They are the path for microinstruction fetching and the next microinstruction address calculation, the path for microinstruction decoding and execution of a register-register instruction, and the path for cache memory access for execution of a register-memory instruction. These critical paths generally consist of three parts, macrocell part, I/O driver part, and random logic part. The macrocell is a large scale cell which has regular logic such as bit sliced logic. It is used as memory in the first and third critical paths, register-file and ALU in the second critical path. The I/O driver drives the inter-chip communication signal lines such as the address/data buses in the first and third critical paths. The random logic part consists of basic cells, such as an address controller for the microprogram memory in the first critical path, or the data path controller in the second and third critical paths.

The effects of BiCMOS circuits on these three parts were analyzed. The macrocells used the BiCMOS address decoder and BiCMOS sense circuits. They operated twice as fast as CMOSs, but their cell area was almost the same, because the NMOS logic circuits in the BiCMOS sense circuits were much larger than one bipolar transistor. The ratios of the bipolar transistors used were only 0.8 % for the ROM and 1.7 % for the data structure macrocell which contained ALUs, register file, and dynamic data buses. Therefore, no problems should be encountered using BiCMOS technologies in the macrocell part.

The I/O driver had almost the same speed as a CMOS one, when it had a TTL signal level, which is often used by devices interfacing the microprocessor. It took more than 10 ns for an inter-chip communication. The BiCMOS basic cell was faster than CMOS one, when it had a large load capacitance (cf. **Fig. 2**). But it depended on the logic design whether the basic cell had a large load capacitance or not. On the other hand, the size of the BiCMOS basic cell was much larger than the CMOS one. Therefore it would not be easy to use BiCMOS basic cells.

Based on this analysis, we adopted two design philosophies. The first one was the reduction of the number of inter-chip communication signals and the second one was the optimal usage of BiCMOS basic cells.

The former was concerned with the system structure and the latter was concerned with the size of the chip. They are discussed in **sections 4** and **5**, respectively.

4. System structure

Since the inter-chip communication time of BiCMOS VLSIs is as slow as that of the CMOS, the microprocessor needs to be fabricated into a single chip. Although the integration density of the VLSI is increasing, the whole system cannot be integrated in a chip. The choice of which elements to integrate is very important for reducing the number of inter-chip communication signals and determining the operating frequency of the microprocessor. Our choice was whether to include the control storage (CS), cache memory, and translation lookaside buffer (TLB) or not.

The CS stores the microinstructions. In ordinary microprocessors, it is implemented as a ROM on the chip. But this approach is difficult since our microprocessor has a large volume of microinstructions. However, if the CS is outside the chip, it takes more than 35 ns, including the inter-chip communication time of about 10 ns, to access the microinstruction even if we use 15 ns high speed BiCMOS static RAMs for the CS. As the access time of the CS determines the operating frequency, we divided the microinstructions into two parts. The first part is a set of microinstructions which are used frequently and located in the ROM on the chip. The second part is a set of microinstructions which are located in the RAMs outside the chip and it takes more than 5 cycles to get a microinstruction from them. In this way, high frequency and large capacity of the CS are realized at the same time.

Since the gap in speeds between the high performance microprocessor and the interfacing systems such as the main memory is enlarged, the system performance will drop greatly if this gap cannot be filled by an architectural counterplan. The main technology we adopted was to use a large capacity cache memory which had a low miss-hit ratio. As the speed of the microprocessor was 1.5 to 2 times faster than the ordinary CMOS microprocessors which have 16 or 32 K bytes cache memories, their miss-hit ratio must be less than half the ordinary one. Therefore the size of the cache memory was set as 64 K bytes which cannot be integrated into the chip due to area constraint.

The TLB is often integrated in the memory management unit (MMU) for an ordinary microprocessor. The MMU also integrates the control logic for the cache memory. But with this system structure, there are four inter-chip communications on the cache memory access as shown in Fig. 5. Therefore, we did not use the MMU chip, but integrated the TLB and control logic for the cache memory into the microprocessor to reduce the number of inter chip communications (Fig. 6).

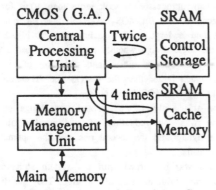

Fig. 5 Structure of Previous System

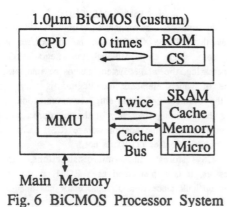

Fig. 6 BiCMOS Processor System

Although the delay time of the critical paths in the chip was shortened by reducing inter-chip communication signal lines and by using the BiCMOS technology, it still took about two and a half cycles to get the data from the cache memory, which would slow down the execution of a memory-register instruction. Therefore, we overlapped the half cycle of a cache memory access with the following access which allowed the cache memory access of data every two cycles. And to shorten execution cycles, a memory access request was put at the end of the address calculation stage. As it takes two cycles to fetch such an instruction, the number of the execution cycles is then four. Without these technologies, it would take three cycles to fetch the instruction and five cycles to execute it, i.e. one memory request cycle, three access cycles, and one execution cycle. As the memory request and execution are processed concurrently with the instruction fetching, it takes six cycles to execute the instruction. Therefore, the performance of the memory-register instruction was accelerated 1.5 times by the introduced technologies.

To shorten the access time of the static RAMs for the cache memory, they were directly connected to the microprocessor, independent of the system bus, by a high speed private bus which connected the main memory to the microprocessor. Both the private bus and the system bus have 32-bit data width. But the system bus is an address/data multiplexed bus because of the pin constraint.

5. Chip Size Constraint

The microprocessor consists of macrocell and random logic parts. As the size of the macrocells is determined, the total chip area depends on the size of the random logic part. The area increase using BiCMOS basic cells is not negligible, because the size of basic cells are much larger than that of the CMOS basic cells. Therefore, the number of usable BiCMOS basic cells is limited by the chip size constraint. **Figure 7** shows the delay time of a sample critical path of the random logic part consisting of five basic cells. If BiCMOS basic cells are adopted instead of the two CMOS cells which have large load capacitance, the delay time will be shortened to 84%. But even if all five CMOS basic cells are replaced by the BiCMOS cells, the acceleration ratio of the path is almost the same as the former case. This example indicates the optimal usage of BiCMOS basic cells. Besides, as the basic cells which belong to the critical paths are only a part of the random logic, the number of BiCMOS basic cells can be reduced.

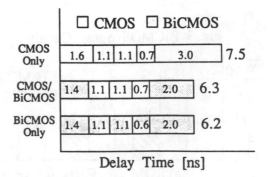

Fig. 7 Delay Time of Basic Cells

From our experience, a processor of this class requires at least 6 K cells to realize such control logic. The chip size and logic cells constraints provide a guideline as to how many BiCMOS gates can be used. Our analysis showed the number to be less than 20 % of the total cells (**Fig. 8**). According to this constraint, the number of BiCMOS cells actually used for the microprocessor was about 18 % of the total number of cells in the control logic part. This result indicated that BiCMOS basic cells must be used to drive a large load capacitance such as inter-block signal lines or lines having many fan-outs. If the number of usable BiCMOS basic cells is exceeded, the signal lines in the critical paths have a priority to be driven by them.

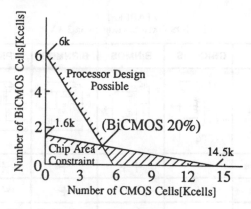

Fig. 8 Usage of BiCMOS Basic Cells

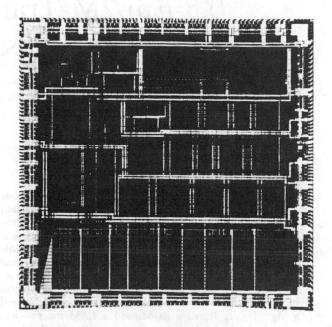

Fig. 9 Photograph of the BiCMOS Microprocessor

6. Chip Specifications

Figure 9 shows a photograph of the microprocessor chip and **Table 1** lists its specifications. The chip die size is 12.98 mm square and 529 K transistors are integrated on it. The typical operating frequency is 70 MHz. The consuming power is about 2.1 W at 40 MHz. Since the ratio of bipolar transistors used is only 1.5 %, high density, high speed, and low power consumption are realized. It integrates the ROM and TLB, as mentioned before, and the 32-bit data structure which includes the register file, and arithmetic and logic units. The microinstructions are read from the ROM or RAMs outside the chip. The microprocessor has a five-stage pipeline structure, and can execute a resistor-resistor instruction in a cycle. As it takes two cycles to fetch an instruction or an operand, the memory-resistor instructions are executed in four cycles.

Table 1 Chip Specifications

Item	Value
Operating Frequency	70MHz
Peak Performance	70MIPS
Power Consumption	2.1W(40MHz)
LSI Technology	1.0μmHi-BiCMOS
Die Size	12.98mmx12.98mm
Number of Transistors	529K Tr (Bipolar 8K Tr)
Number of Basic Cells	5.99K Cells (BiCMOS 18%)

Conclusions

We have developed a 70 MHz 32-bit single chip custom microprocessor using 1.0μm BiCMOS technology. It is the world's first BiCMOS single chip microprocessor and its frequency is the fastest among 32-bit microprocessors.

To reduce the number of inter-chip communication signals, a new system structure was introduced. The microprocessor integrates microprogram ROM and TLB for MMU. Therefore no inter-chip communications for microinstruction access are needed and only two times inter-chip communications for cache memory access. It has a cache memory bus separated from the system bus, and the cache memory consists of standard SRAMs only.

The optimal number of BiCMOS basic cells was also discussed. Less than 20% of cells could be used due to chip size and logic complexity constraints.

References

[1] Nishio, Y., et al, "0.45 ns 7k Hi-BiCMOS Gate Array with Configurable 3-port 4.6k SRAM", IEEE Custom Integrated Circuits Conference, pp203-204, May 1987.

[2] Hotta, T., et al., "CMOS/bipolar Circuits for 60 MHz Digital Processing", IEEE J. Solid-State Circuits, vol. SC-21, no.5, pp808-813, Oct.1986; also in ISSCC Dig.Tech. Papers, pp190-191, Feb. 1986.

[3] Hotta, T., et al., "1.3μm CMOS/bipolar Macrocell Library for the VLSI Computer", IEEE J. Solid-State Circuits, vol. SC-23, no.2, pp500-506, Apr. 1988.

[4] Hotta, T., et al. ,"A 70 MHz 32b Microprocessor with 1.0μm BiCMOS Macrocell Library", IEEE International Solid State Circuits Conference, THAM9.7, pp124-125, Feb. 1989.

[5] Tanaka, S., et. al., "A BiCMOS 32-bit Execution Unit for 70 MHz VLSI Computer", IEEE Custom Integrated Circuits Conference, pp10.8.1-10.8.4, May 1989.

0.5-μm 2M-Transistor BiPNMOS Channelless Gate Array

Hiroyuki Hara, Takayasu Sakurai, *Member, IEEE*, Makoto Noda, Tetsu Nagamatsu,
Katsuhiro Seta, Hiroshi Momose, *Member, IEEE*, Youichirou Niitsu,
Hiroyuki Miyakawa, and Yoshinori Watanabe

Abstract —A channelless gate array has been realized using 0.5-μm BiCMOS technology integrating more than 2-million transistors on a 14×14.4-mm^2 chip. A small-size PMOS transistor and a small-size inverter are added to the conventional BiNMOS gate to form the BiPNMOS gate. The gate is suitable for 3.3-V supply and achieves 230-ps gate delay for a two-input NAND with full-swing output. Added small-size MOS transistors in the BiPNMOS basic cell can also be used for memory macros effectively. A test chip with four memory macros—a high-speed RAM, a high-density RAM, a ROM, and a CAM macro—was fabricated. The high-speed memory macros utilize bipolar transistors in bipolar middle buffers and in sense amplifiers. The high-speed RAM macro achieves an access time of 2.7 ns at 16-kb capacity. On the other hand, the high-density RAM macro is rather slow but the memory cell occupies only a half of the BiPNMOS basic cell using a proposed single-port memory cell.

TABLE I
BiCMOS GATE SUITABLE FOR 3.3 V

	CBiCMOS	BiNMOS	BiRNMOS	BiPNMOS
Circuit				
Process	−	++	+	++
Speed	+	+	+	++
Full-swing	+	−	+	++

I. INTRODUCTION

IN ORDER TO ensure high reliability of 0.5-μm gate-length MOSFET's, a power supply voltage of 3.3 V is adopted. Reducing the supply voltage is also necessary for decreasing the power dissipation of highly integrated system LSI's. The low supply voltage, however, makes the conventional totem-pole BiCMOS gate lose its speed advantage over CMOS. In order to maintain the speed advantage over CMOS even at the low supply voltage, a BiNMOS gate [1] and a complementary-BiCMOS gate [2] have been proposed.

Table I shows candidates for the low-supply-voltage BiCMOS gate. In all these gates, improvement is done on the pull-down circuit of an original BiCMOS gate, since the pull-down part is the cause of the low speed in the original BiCMOS gate at low supply voltage. The C-BiCMOS gate requires a p-n-p transistor with a high cutoff frequency, as well as an n-p-n transistor, and hence suffers from an increase in process cost. On the other hand, the BiNMOS gate can be realized with only an

n-p-n transistor, but the output does not reach V_{DD}; instead it stays at the $V_{DD} - V_{BE}$ level, where V_{BE} is a base–emitter turn-on voltage. This increases the delay time and leakage current of the next gate, and hence decreases noise margins. A BiRNMOS gate, which has a resistor inserted between the emitter and base of the pull-up bipolar transistor, achieves a full-swing output. However, if the resistance is set small to realize high-speed pull-up, it bypasses the base current and hence degrades the speed. Here, we propose a BiPNMOS gate which is suitable for a low supply voltage. The BiPNMOS gate achieves high speed, low process cost, and full-swing capability. The BiPNMOS gate can maintain a speed advantage over CMOS at 3.3-V supply voltage.

Another problem concerns memory macros in a BiCMOS gate array. High-speed and high-density memory macros are necessary for building a high-performance system on a chip. However, not many kinds of memory macros have been introduced in a BiCMOS gate array, which prevents the user from making highly value-added VLSI's easily. Besides, recent gate arrays tend to use six-transistor CMOS memory cells for a RAM macro [4], [5] but bipolar transistors arranged in BiCMOS basic cells are not utilized effectively. This inefficiency is solved by new circuit technologies, a double bit-line structure [7] with bipolar middle buffer and a BiCMOS sense amplifier with dotted-emitter structure [6], [8]. As for variety of memory macros, several kinds of memory macros are

Manuscript received April 19, 1991; revised July 22, 1991.

H. Hara, T. Sakurai, M. Noda, T. Nagamatsu, K. Seta, H. Momose, Y. Niitsu, and H. Miyakawa are with the Semiconductor Device Engineering Laboratory, Toshiba Corporation, 1, Komukai-Toshiba-cho, Saiwai-ku, Kawasaki 210, Japan.

Y. Watanabe is with Toshiba Microcomputer Engineering Corporation, 1, Komukai-Toshiba-cho, Saiwai-ku, Kawasaki 210, Japan.

IEEE Log Number 9102947.

Reprinted from *IEEE Journal of Solid-State Circuits*, Vol. 26, No. 11, pp. 1615–1619, November 1991.

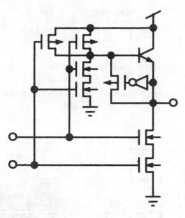

Fig. 1. BiPNMOS gate.

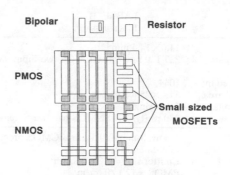

Fig. 3. Basic cell layout of BiPNMOS gate array.

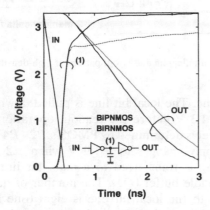

Fig. 2. Output waveforms of BiPNMOS and BiRNMOS gates.

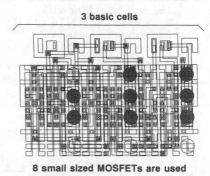

Fig. 4. D-type flip-flop cell layout pattern.

realized including RAM's, a ROM, and a CAM. All these macros use developed BiCMOS circuits.

II. BiPNMOS Gate

Fig. 1 shows a two-input NAND BiPNMOS gate. A small-size PMOS transistor and an inverter are added to the conventional BiNMOS gate. The added PMOS transistor realizes the full swing of the output, while not decreasing the base current of the n-p-n transistor and hence not degrading the speed. When an output terminal is low, the added PMOS is OFF and all current flowing through the PMOS becomes the base current efficiently. When the voltage of the output terminal reaches the threshold voltage of the feedback inverter, the added PMOS turns on and starts pulling up the output to V_{DD}. Even after the bipolar transistor turns off at $V_{DD} - V_{BE}$, the added PMOS keeps ON and realizes a full-swing output. Fig. 2 shows a comparison of the output waveforms of the BiPNMOS gate and the BiRNMOS gate. In this case, the value of the resistor of the BiRNMOS is set to 50 kΩ. As seen from the figure, the full swing of BiRNMOS gate is achieved very slowly. On the other hand, the BiPNMOS gate offers high-speed pull-up to V_{DD}. This reduces the delay and the leakage current of the next gate and increases noise margins.

Fig. 3 shows a layout pattern of a BiPNMOS basic cell. A bipolar transistor and small-size MOS transistors are added to the pure CMOS basic cell. These transistors are not only used to construct the basic BiPNMOS gates but also can be used in flip-flops, memory macros, and other cells effectively. Fig. 4 shows an example layout pattern of a D-type flip-flop cell. This D-type flip-flop cell with the BiPNMOS drivers is constructed on three basic cells, and utilizes eight out of 12 small-size MOS transistors.

III. Memory Macros

Recent gate arrays tend to include small MOS transistors in a basic cell to implement a high-density six-transistor CMOS RAM cell. In the present BiPNMOS gate array, small MOS transistors laid out in the BiPNMOS basic cells can be utilized for six-transistor RAM cells. Moreover, the bipolar transistors are utilized for word-line drivers and sense amplifiers efficiently. Four memory macros, namely high-speed RAM, high-density RAM, ROM, and CAM, are designed. A summary of these macros is listed in Table II. One memory cell of the high-speed RAM macro occupies one basic cell. The high-density RAM macro is rather slow but the memory cell occupies half of the basic cell. As for the ROM macro, eight memory cells can be realized in one basic cell and the CAM macro is realized employing one basic cell per bit.

Fig. 5 shows a circuit diagram of the high-speed RAM macro. A RAM cell has a separate write port and a read port and can be used as a two-port RAM cell enabling WRITE and READ operation at the same time. WRITE operation is performed through the DATA line connected to NMOS $N3$, and the dimensions of $N3$ must be selected bigger than $N1$ for stability. High-speed opera-

113

TABLE II
FEATURES

Chip size	14.0×14.4 mm^2
Internal gate count	237 120 raw gates (CMOS two-input NAND gate)
I/O cell count	1044
Basic cell area	54.4×25.6 μm^2
Supply voltage	3.3 V single
Gate delay	230 ps (typ., BiPNMOS two NAND, fan-out = 7)
Process	1-poly, 3-metal, 0.5-μm BiCMOS process
Bipolar	Poly-Si emitter
	Emitter size = 0.8×2.8 μm^2
MOS	PMOS = 12.1/0.6 μm
	NMOS = 12.1/0.5 μm
High-speed RAM	1 memory cell/1 basic cell
	T_{acc} = 2.7 ns (typ. at 512 word $\times$ 32 b)
High-density RAM	2 memory cells/1 basic cell
	T_{acc} = 4.0 ns (typ. at 256 word $\times$ 32 b)
ROM	8 memory cells/1 basic cell
	T_{acc} = 3.2 ns (typ. at 256 word $\times$ 32 b)
CAM	1 memory cell/1 basic cell
	Data to match = 2.0 ns (typ. at 64 entry $\times$ 32 b)

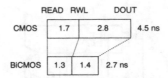

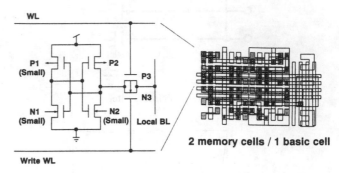

Fig. 6. Simulated access time of BiCMOS RAM macro compared to CMOS RAM.

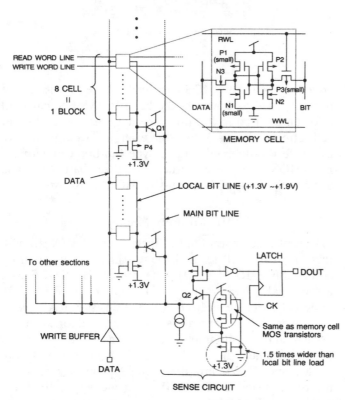

Fig. 7. Circuit diagram and layout pattern of high-density RAM macro.

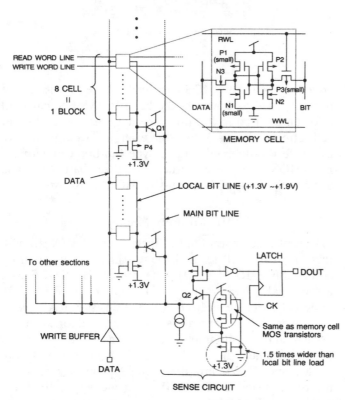

Fig. 5. Circuit diagram of fast RAM macro.

tion in the READ mode is achieved by the following three techniques:

1) double bit lines with bipolar middle buffers,
2) 0.6-V bit-line swing,
3) BiCMOS sense amplifiers with emitter-dotted shared transistors.

Readout operation from a memory cell is carried out through a PMOS ($P3$) connected to the BIT line. A small transistor is used for $P3$ in order to limit the swing of the local bit line. The local bit line is pulled down to the low level of +1.3 V before the readout operation. The local bit-line level is determined by PMOS's $P2$–$P4$ and NMOS $N2$, and only changes to +1.9 V when $P2$ is ON. This 0.6-V swing is transferred to a main bit line through a bipolar middle buffer ($Q1$). The number of memory cells connected to the local bit line is eight (one block). The bipolar middle buffer effectively drives the highly capacitive main bit line. The bipolar middle buffer ($Q1$) is also used as a component of a differential sense amplifier using an emitter dotting technique. The other side of the amplifier is made with the bipolar transistor ($Q2$) whose base is controlled by a self-tracking reference voltage generator. The reference generator uses MOS transistors, which are the same as a memory cell, and 1.5 times wider MOS compared with a local bit-line load to generate a proper reference voltage. The reference generator tracks process, voltage, and temperature variation in a self-tracking way. Fig. 6 shows simulated delay component distributions of the BiCMOS RAM macro and a pure CMOS RAM macro. The delay from READ signal to read word line (RWL) is reduced from 1.7 to 1.3 ns by using BiCMOS drivers. More drastic delay reduction is achieved in the delay from the bit line to the output of the sense amplifier. The current consumption of a sense circuit is 1.3 mA per bit.

In some applications, density is more important than speed. For these applications, a high-density RAM macro, which implements two memory cells in one basic cell, is more suitable. Fig. 7 shows the circuit diagram and the cell layout pattern. The RAM cell realizes high density to adopt a single-port cell and to use the small-size MOS transistors. In READ mode only a PMOS transfer gate turns on to ensure cells stability. In WRITE mode both PMOS and NMOS transfer gates turn on to ensure stable WRITE operation. The sensing scheme is the same as that of the high-speed RAM macro.

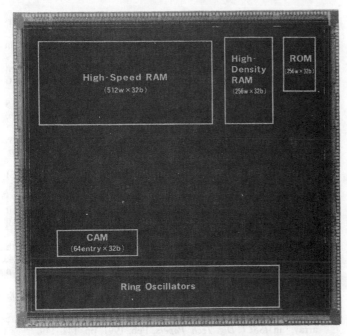

Fig. 8. Microphotograph of evaluation chip.

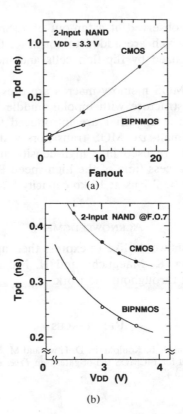

(a)

(b)

Fig. 9. (a) Propagation delay time versus fan-out. (b) Supply voltage dependence of delay time.

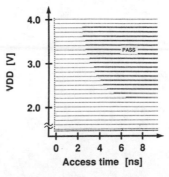

Fig. 10. Schmoo plot of high-speed RAM macro.

The ROM macro adopts the double bit line and the same sensing scheme as the RAM's. Four PMOS ROM cells and four NMOS ROM cells can be built using one basic cell. The local bit line for the PMOS ROM cell and that for the NMOS ROM cell are separate, because the bit-line levels are different. Content addressable memory (CAM) is another important macro. Small-size MOS transistors are used for the six-transistor memory cells, so a single CAM cell can be constructed with only one basic cell. The CAM macro employs a double match-line architecture and a BiCMOS pull-up circuit to drive the second match line [13]. The key features of the CAM macro are listed in Table II together with a summary of the present chip.

IV. EXPERIMENTAL RESULT

Fig. 8 shows a microphotograph of the fabricated 14.0×14.4-mm^2 chip. The test chip contains ring oscillators to evaluate the speed for the BiPNMOS gate and the CMOS gate, and memory macros. The test chip is fabricated using a 0.5-μm triple-metal BiCMOS process. Although the gate length for the MOSFET's is 0.5 μm, the basic design rule of 0.8 μm is adopted to achieve high speed and high yield at the same time. A poly-Si emitter bipolar transistor with a cutoff frequency of 13 GHz is adopted, and the rapid thermal annealing (RTA) technique is applied to realize a stable poly-single interface. The features of the present chip are shown in Table II, together with the key process parameters of the 0.5-μm BiCMOS technology. Fig. 9(a) shows a speed dependence of the BiPNMOS gate on fan-out. A BiPNMOS two-input NAND gate realizes a propagation delay time of 230 ps with a fan-out of 7 at room temperature. The BiPNMOS gate shows 35% smaller delay than the pure CMOS gate

even at 3.3 V. Fig. 9(b) shows the speed dependence of the BiPNMOS gate and the pure CMOS gate on V_{DD}. The speed advantage over the CMOS gate is observed down to 2.5 V. Fig. 10 shows a schmoo plot of the high-speed RAM macro and the typical access time of 2.7 ns is seen at 512 word $\times$ 32 b. The speeds of the high-density RAM and ROM macros are 4.0 and 3.2 ns, respectively. These memory macros achieve a high-speed operation using BiCMOS circuit technologies, even in a high-memory-capacity range.

V. CONCLUSIONS

A BiPNMOS gate is proposed that assures full-swing and high-speed operation at 3.3 V. The BiPNMOS gate reduces the delay and the leakage current of the next gate and increases noise margins. A two-input NAND gate delay

115

of 230 ps is observed at a fan-out of 7. Small-size MOS transistors, which are added to construct the BiPNMOS gate, are utilized by flip-flop cells and memory macros effectively.

Four BiCMOS memory macros are presented. A double bit-line structure with bipolar middle buffers and a BiCMOS sense amplifier are introduced for high-speed operation. Small-size MOS transistors and a single-port memory cell are used in a high-density memory macro. The typical access time of the high-speed RAM macro is measured to be 2.7 ns at 16-kb capacity.

ACKNOWLEDGMENT

The authors would like to express their appreciation to Dr. T. Iizuka, K. Maeguchi, and H. Nakatsuka for encouragement throughout the work.

REFERENCES

[1] A. E. Gamal, J. L. Kouloheris, D. How, and M. Morf, "BiNMOS: A basic cell for BiCMOS sea-of-gates," in *Proc. CICC*, May 1989, pp. 8.3.1–8.3.4.

[2] H. J. Shin *et al.*, "Full-swing complementary BiCMOS logic circuits," in *Proc. BCTM*, 1989, pp. 229–232.

[3] Y. Nishio *et al.*, "A BiCMOS logic gate with positive feedback," in *ISSCC Dig. Tech. Papers*, Feb. 1989, pp. 116–117.

[4] H. Veendrick, D. Elshout, D. Harberts, and T. Brand, "An efficient and flexible architecture for high-density gate arrays," in *ISSCC Dig. Tech. Papers*, Feb. 1990, pp. 86–87.

[5] Y. Enomoto, T. Sasaki, S. Tsutsumi, and S. Tone, "A 200k gate 0.8 μm mixed CMOS/BiCMOS sea-of-gates," in *ISSCC Dig. Tech. Papers*, Feb. 1990, pp. 92–93.

[6] L. Tamurat, T. S. Yang, D. Wingard, M. Horowitz, and B. Wooley, "A 4ns BiCMOS translation-lookaside buffer," in *ISSCC Dig. Tech. Papers*, Feb. 1990, pp. 66–67.

[7] T. Sakurai *et al.*, "Double word line and bit line structure for VLSI RAMs," in *Proc. 15th Conf. Solid State Devices & Materials*, 1983, pp. 269–272 (A-7-6).

[8] T. S. Yang, M. A. Horowitz, and B. A. Wooley, "A 4-ns 4K×1-bit two-port BiCMOS SRAM," *IEEE J. Solid-State Circuits*, vol. 23, no. 5, pp. 1030–1040, Oct. 1988.

[9] H. Hara *et al.*, "A 350 ps 50K 0.8 μm BiCMOS gate array with shared bipolar structure," in *Proc. CICC*, May 1989, pp. 8.5.1–8.5.4.

[10] J. Gallia *et al.*, "A 100K gate sub-micron BiCMOS gate array," in *Proc. CICC*, May 1989, pp. 8.6.1–8.6.4.

[11] C. L. Chen, "Level-shifted and voltage-reduced 0.5 μm BiCMOS circuits," in *ISSCC Dig. Tech. Papers*, Feb. 1990, pp. 236–237.

[12] H. Hara *et al.*, "0.5 μm 2M-transistor BiPNMOS channelless gate array," in *ISSCC Dig. Tech. Papers*, Feb. 1991, pp. 148–149.

[13] T. Nagamatsu *et al.*, "A 1.9ns BiCMOS CAM macro with double match line architecture," in *Proc. CICC*, May 1991, pp. 14.3.1–14.3.4.

A 1.5-V Full-Swing BiCMOS Logic Circuit

Mitsuru Hiraki, *Member, IEEE*, Kazuo Yano, *Associate Member, IEEE*, Masataka Minami, Kazushige Sato,
Nozomu Matsuzaki, Atsuo Watanabe, Takashi Nishida, Katsuro Sasaki, *Member, IEEE*, and
Koichi Seki, *Member, IEEE*

*Abstract—*A BiCMOS logic circuit applicable to sub-2-V digital circuits has been developed for the first time. A transiently saturated full-swing BiCMOS (TS-FS-BiCMOS) logic circuit operates twice as fast as CMOS at 1.5-V supply. A newly developed transient-saturation technique, with which bipolar transistors saturate only during switching periods, is the key to sub-2-V operation because a high-speed full-swing operation is achieved to remove the voltage loss due to the base–emitter turn-on voltage. Furthermore, both small load dependence and small fan-in dependence of gate delay time are also attained with this technique. A two-input gate fabricated with 0.3-μm BiCMOS technology verifies the performance advantage of TS-FS-BiCMOS over other BiCMOS circuits and CMOS at sub-2-V supply.

I. Introduction

THE switching speed of BiCMOS circuits is seriously deteriorated as the supply voltage is reduced. This difficulty limits the performance of low-voltage BiCMOS ULSI's. In conventional BiCMOS circuits (Fig. 1(a)), which are widely used at 5-V supply, the gate–source voltage of $M1$ is reduced by $2V_{BE}$. This voltage loss seriously degrades the switching speed of the circuit when the supply voltage is reduced to about 3 V [1]. This voltage loss is caused by two factors. One factor is that the fast output voltage transition obtained by the bipolar transistor $Q1$ is V_{BE} smaller than the supply voltage. The other factor is that the source node voltage of $M1$ rises up to V_{BE}.

To improve the low-voltage performance of BiCMOS circuits, several types of BiCMOS circuits have been proposed, such as complementary BiCMOS (C-BiCMOS) [2], quasi-complementary BiCMOS (QC-BiCMOS) [3]–[5], merged BiCMOS (MBiCMOS) [6], and BiNMOS [7], and the performance advantage of these BiCMOS circuits has been demonstrated at 3-V supply. The idea behind these BiCMOS circuits is to ground the source node of $M1$ to remove one V_{BE} voltage loss. As an example, Fig. 1(b) shows the configuration of C-BiCMOS. Although the voltage rising of the source node is prevented, these circuits also suffer from the one V_{BE} voltage loss in the out-

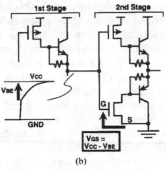

(a)

(b)

Fig. 1. Voltage loss in reported BiCMOS circuits: (a) a conventional BiCMOS circuit, and (b) C-BiCMOS.

put voltage transition. This is because these circuits have the same pull-up driver configuration as the conventional BiCMOS circuit. Because of this voltage loss, the switching speed is seriously degraded even with these circuits when the supply voltage is further reduced to about 2 V. For sub-2-V BiCMOS ULSI's, therefore, a high-speed full-swing circuit technique is essential.

This paper describes a novel full-swing BiCMOS logic circuit that breaks through the limit for sub-2-V operation [8]. This full-swing BiCMOS logic circuit operates twice as fast as CMOS at 1.5-V supply. The newly developed transient-saturation technique enables high-speed full-swing operation. First, the concept of the full-swing BiCMOS circuit using this transient saturation technique is presented in Section II. Next, some simulated performance of the circuit are described in Section III, and design issues of the circuit are discussed in Section IV. Then, some experimental results are shown in Section V. Finally, Section VI summarizes this work.

Manuscript received April 6, 1992; revised June 18, 1992.
M. Hiraki, K. Yano, M. Minami, N. Matsuzaki, T. Nishida, K. Sasaki, and K. Seki are with the Central Research Laboratory, Hitachi Ltd., 1-280 Higashi-koigakubo, Kokubunji-shi, Tokyo 185, Japan.
K. Sato and A. Watanabe are with the Device Development Center, Hitachi Ltd., Tokyo 198, Japan.
IEEE Log Number 9202809.

Reprinted from *IEEE Journal of Solid-State Circuits*, Vol. 27, No. 11, pp. 1568-1573, November 1991.

II. Circuit Concept

To achieve a high-speed full-swing operation in Bi-CMOS circuits, full-swing operation of bipolar transistors must be realized. The saturation type circuit shown in Fig. 2(a) achieves full-swing operation of a bipolar transistor. But the switching speed of a saturated bipolar transistor is slow because once the excess minority carriers are charged into the base, they remain there until the next switching period [9], [10]. The full-swing BiCMOS circuit that we propose is based on the concept of saturating the bipolar transistor transiently. The idea of this transient-saturation technique is illustrated in Fig. 2(b). The transient-saturation technique enables the bipolar transistor to achieve a high-speed full-swing operation because the excess minority carriers are first charged into the base and then discharged out of the base immediately after the output voltage transition.

The configuration of the full-swing BiCMOS circuit using the transient-saturation technique is shown in Fig. 3(a). Simulated waveforms are also provided in Fig. 3(b) to explain the transient operation of this circuit. Since this is a noninverting circuit, the output is high when the input is high. When the input falls, the pull-down transition occurs as follows. $MP2$ turns on, and the base of $Q1$ is charged through the current path from V_{CC} to $B1$. $Q1$ continues driving the load until the output voltage nearly reaches zero. Thus, $Q1$ achieves full-swing operation. Although $Q1$ saturates, this does not slow the next pull-up transition because the excess minority carriers of $Q1$ are discharged immediately after the pull-down transition through $MN4$, which turns on as the feedback signal FB is pulled high. Note that the node voltage of $B1$ falls nearly to zero immediately after the pull-down transition. The pull-up transition is easily understood because of the symmetrical circuit configuration. Thus, bipolar transistors saturate transiently, only in the switching period. Therefore, the proposed BiCMOS circuit is called transiently-saturated full-swing BiCMOS (TS-FS-BiCMOS).

Static power dissipation is avoided because $MN3$ cuts off the dc current path after the pull-up transition and $MP3$ cuts it off after the pull-down transition. An arbitrary non-inverting logic function is realized by replacing $MN1$ and $MN2$ with nMOS logic, and $MP1$ and $MP2$ with pMOS logic. To realize an inverting logic function in TS-FS-BiCMOS circuits, a CMOS gate should be placed as the stage before a TS-FS-BiCMOS buffer.

The static output voltage is held by the small-size CMOS latch composed of $INV1$ and $INV2$. Although both $Q1$ and $Q2$ are off in steady state, this circuit suffers little from a switching noise at the output node. The reason for this is as follows. If the switching noise is large enough to flip the CMOS latch, the bipolar transistor ($Q1$ or $Q2$) begins to drive the load again. As a result, the output voltage quickly goes back to the static level. If the switching noise is small enough not to flip the CMOS latch, it does not flip the next stage gate, either.

To verify the advantage of the transient-saturation technique, the circuit operation was simulated for the config-

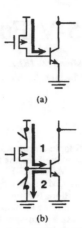

Fig. 2. Full-swing circuit technique: (a) saturation type (conventional), and (b) transient-saturation type (proposed).

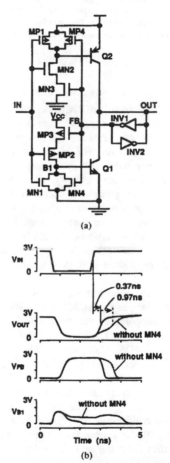

Fig. 3. (a) Configuration of TS-FS-BiCMOS buffer, and (b) simulated voltage waveforms.

uration without $MN4$. With this configuration, $Q1$ continues to saturate even after the pull-down transition. The simulated waveforms are shown with hatched lines in Fig. 3(b). These show that the configuration without the transient-saturation technique results in about 2.5 times longer delay time owing to the excess minority carriers.

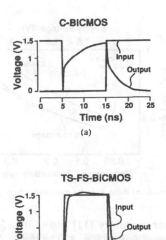

C-BiCMOS

(a)

TS-FS-BiCMOS

(b)

Fig. 4. Simulated input and output voltage waveforms of (a) C-BiCMOS, and (b) TS-FS-BiCMOS.

Fig. 4 shows the simulated output voltage waveforms of C-BiCMOS and TS-FS-BiCMOS operating at 1.5-V supply for the step function input. In C-BiCMOS, fast output voltage transitions obtained by the bipolar transistors are V_{BE} smaller than the supply voltage and the following transitions by additional elements are slow (Fig. 4(a)). This slow transition not only degrades the driving capability but also causes large leakage current in the next stage gate, especially at sub-2-V supply voltages. In contrast, high-speed full-swing operation is achieved in TS-FS-BiCMOS even at 1.5-V supply (Fig. 4(b)).

III. CIRCUIT PERFORMANCE

SPICE simulation was performed to compare the performance of TS-FS-BiCMOS with those of CMOS, conventional BiCMOS, C-BiCMOS, QC-BiCMOS, and BiNMOS. Configurations of the simulated circuits are shown in Fig. 5. Here, all the circuits have an input capacitance of 60 fF. QC-BiCMOS (Fig. 5(e)) is a circuit that uses a quasi-p-n-p connection instead of a p-n-p transistor of C-BiCMOS. BiNMOS (Fig. 5(f)) is a circuit whose pull-down driver is composed of nMOS logic. The device parameters used in the simulation are listed in Table I. In this simulation, 0.3-μm technology was assumed.

Fig. 6 shows the simulated dependence of the gate delay time on supply voltage. TS-FS-BiCMOS is faster than any other BiCMOS circuit below 2-V supply. Moreover, TS-FS-BiCMOS is twice as fast as CMOS at 1.5-V supply. This has been impossible in any other BiCMOS circuit. For example, the conventional BiCMOS circuit loses its performance advantage over CMOS at about 3 V because of $2V_{BE}$ voltage loss. C-BiCMOS, QC-BiCMOS, and BiNMOS lose their performance advantage over CMOS at about 2 V owing to one V_{BE} voltage loss. Fig.

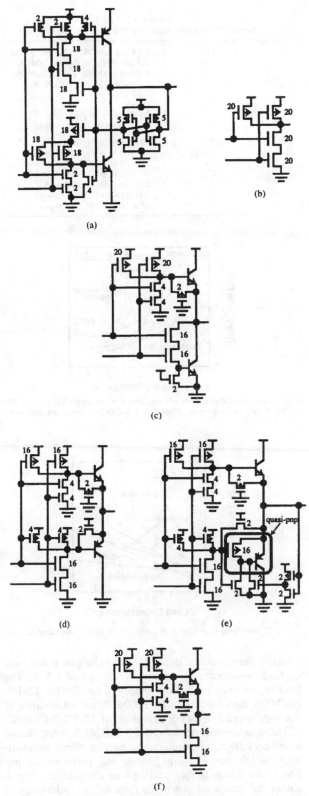

Fig. 5. Compared two-input gates. (a) TS-FS-BiCMOS, (b) CMOS, (c) conventional BiCMOS, (d) C-BiCMOS, (e) QC-BiCMOS, and (f) BiNMOS. The numbers indicate gate widths of MOSFET's.

TABLE I
DEVICE PARAMETERS USED FOR SPICE SIMULATION

MOS	L_g	0.35 μm (nMOS), 0.35 μm (pMOS)
	L_{eff}	0.3 μm (nMOS), 0.3 μm (pMOS)
	I_{DS}	4.8 mA (nMOS), 2.4 mA (pMOS)
		($V_{DS} = V_{GS} = 3.3$ V, $W = $ 10 μm)
	T_{OX}	10 nm
	C_j	20 fF ($W = 10$ μm)
	V_{th}	0.3 V (nMOS), −0.3 V (pMOS)
Bipolar transistor	H_{FE}	89 (n-p-n), 89 (p-n-p)
	F_{TMAX}	15 GHz (n-p-n), 6 GHz (p-n-p)
	C_{JE}	12.4 fF (n-p-n), 23.8 fF (p-n-p)
	C_{JC}	10.6 fF (n-p-n), 16.8 fF (p-n-p)
	R_E	25 Ω (n-p-n), 25 Ω (p-n-p)
	$R_{bb'}$	300 Ω (n-p-n), 250 Ω (p-n-p)
	R_c	42 Ω (n-p-n), 42 Ω (p-n-p)

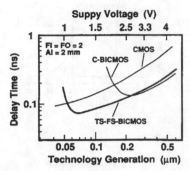

Fig. 8. Simulated delay time for each technology generation.

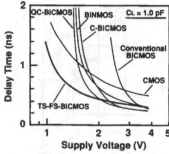

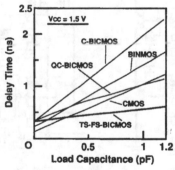

Fig. 6. Simulated dependence of gate delay time on supply voltage. TS-FS-BiCMOS is a two-AND gate. CMOS and BiCMOS circuits are two-NAND gates.

Fig. 7. Simulated dependence of gate delay time on load capacitance.

7 shows the simulated dependence of the gate delay time on load capacitance at the supply voltage of 1.5 V. The load dependence at this low supply is smaller for TS-FS-BiCMOS than for any other BiCMOS circuit because of the high-speed full-swing operation of TS-FS-BiCMOS.

These performances of TS-FS-BiCMOS were shown based on 0.3-μm technology. To predict circuit performance in the post-0.3-μm generation, performance improvement due to device scaling was also studied. Fig. 8 shows the simulated gate delay time for the technology of each generation. The device parameters were determined by assuming that dimensions in the devices were scaled proportionally. The supply voltage was scaled consider-

ing hot-carrier immunity [11] and is indicated along the upper horizontal axis of Fig. 8. The threshold voltages of the MOSFET's were also scaled. The simulation results shows that the performance of TS-FS-BiCMOS improves down to less than 0.1 μm, whereas the performance of C-BiCMOS cannot be improved at less than 0.2 μm because of its base–emitter voltage loss.

IV. DESIGN ISSUES

Since TS-FS-BiCMOS uses the feedback signal to discharge the excess minority carriers, the feedback CMOS inverter must be carefully designed. Fig. 9 shows the simulated gate delay time T_d as a function of the input transition interval T_p. Here, W_n denotes the gate width of nMOS in the feedback CMOS inverter. When the driving capability of the feedback CMOS inverter is too small (when $W_n = 1$ μm, for example), a fatal increase in gate delay time is observed for short cycle time operations. This is because the base-carrier discharging is not yet completed when the next transition starts. To suppress this delay-time increase, the CMOS inverter must be designed so that it has enough driving capability. For example, the delay-time increase can be reduced down to only 10% by increasing W_n to 5 μm, which corresponds to one-fourth the gate width of $MP1$ or $MP2$ driving the bipolar transistor.

In Table II, the device count and gate area of TS-FS-BiCMOS are compared with those of C-BiCMOS and CMOS for two-input gates. The gate area of TS-FS-BiCMOS is 1.4 times larger than that of C-BiCMOS and 3.5 times larger than that of CMOS, owing to the larger number of devices of TS-FS-BiCMOS. But this disadvantage of TS-FS-BiCMOS can be considerably relieved by the feature shown in Fig. 10. Fig. 10 shows the simulated dependence of gate delay time on fan-in for TS-FS-BiCMOS and CMOS. Fan-in dependence of TS-FS-BiCMOS is smaller than that of CMOS. It should be noted that this feature of TS-FS-BiCMOS allows a very complex logic function to be achieved with a single-stage gate in TS-FS-BiCMOS whereas several-stage gates are needed in CMOS for the same logic function. Therefore, the overall area of TS-FS-BiCMOS can be drastically reduced in spite of its gate area disadvantage. Fig. 11 is provided to explain why the fan-in dependence of TS-FS-BiCMOS

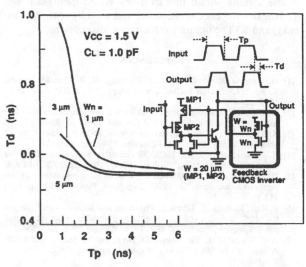

Fig. 9. Simulated delay time as a function of input transition interval.

TABLE II
DEVICE COUNT AND GATE AREA OF TWO-INPUT GATES

| | Device Count | | |
	MOS	Bipolar	Relative Area
TS-FS-BiCMOS	16	2	3.5
C-BiCMOS	10	2	2.5
CMOS	4	—	1

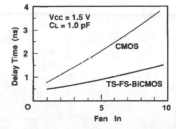

Fig. 10. Simulated dependence of gate delay time on fan-in.

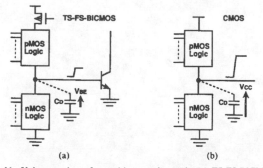

Fig. 11. Voltage swing of parasitic capacitance in (a) TS-FS-BiCMOS, and (b) CMOS. C_D denotes parasitic capacitance.

delay is smaller than that of CMOS delay. The difference of fan-in dependence comes from the difference of the voltage swing of the parasitic capacitances. As fan-in increases, the parasitic capacitance increases both in TS-FS-

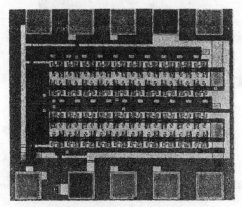

Fig. 12. Photomicrograph of a test chip.

BiCMOS and CMOS, and this causes the increase of the gate delay time. But the voltage swing of the parasitic capacitance is only one V_{BE} in TS-FS-BiCMOS instead of as large as V_{CC} in CMOS. This results in smaller fan-in dependence for FS-FS-BiCMOS than for CMOS.

V. EXPERIMENTAL RESULTS

TS-FS-BiCMOS, conventional BiCMOS, C-BiCMOS, QC-BiCMOS, BiNCMOS, and CMOS were fabricated using 0.3-μm BiCMOS technology with triple-poly-Si and double-metal process. The gate lengths of both nMOS and pMOS were 0.4 μm and the gate oxide thickness was 9 nm. The threshold voltages of nMOS and pMOS were 0.5 and −0.1 V, respectively. Bipolar transistors of both n-p-n type and p-n-p type were fabricated in vertical structure with poly-Si emitters for all the BiCMOS circuits except p-n-p transistors of TS-FS-BiCMOS. P-n-p transistors of TS-FS-BiCMOS were fabricated in a lateral structure to simplify the fabrication process. This is because the collector of the p-n-p transistor needs to be isolated from the p-type substrate in TS-FS-BiCMOS. Cutoff frequencies of the vertical n-p-n and p-n-p transistors were 8 and 6 GHz, respectively. The cutoff frequency of the lateral p-n-p transistor was 500 MHz.

Fig. 12 shows a photomicrograph of a test chip used for gate delay time measurement. The measurement was performed using a 17-stage gate chain. Fig. 13 shows the measured waveforms of the input and the output of a loaded TS-FS-BiCMOS gate chain operating at 1.5-V supply. Fig. 14 shows the measured dependence of the gate delay time on supply voltage for TS-FS-BiCMOS, conventional BiCMOS, C-BiCMOS, QC-BiCMOS, and BiNMOS. TS-FS-BiCMOS is faster than any other BiCMOS circuit below 2-V supply. These measured results demonstrate the performance advantage of TS-FS-BiCMOS over other BiCMOS circuits at sub-2-V supply. Here, the delay time of TS-FS-BiCMOS shown in Fig. 14 is pull-down delay time. The pull-up delay time of TS-FS-BiCMOS was about 2.5 times longer than the pull-down delay time because of the low switching speed of the lateral p-n-p transistor. In a practical chip, therefore, vertical p-n-p transistors should be used in TS-FS-BiCMOS

121

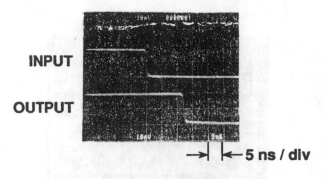

INPUT

OUTPUT

→ |← 5 ns / div

Vcc = 1.5 V

Fig. 13. Measured voltage waveforms of a TS-FS-BiCMOS gate chain.

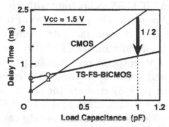

Fig. 14. Measured dependence of gate delay time on supply voltage.

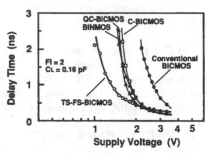

Fig. 15. Measured dependence of gate delay time on load capacitance.

for high-speed pull-up operation. Fig. 15 shows the measured dependence of the gate delay time on load capacitance for TS-FS-BiCMOS and CMOS operating at 1.5-V supply. As predicted in the circuit simulation, the fabricated TS-FS-BiCMOS is twice as fast as CMOS even at 1.5-V supply.

VI. SUMMARY

A BiCMOS circuit applicable to 1.5-V digital circuits has been developed for the first time. The transient-saturation technique is the key to sub-2-V operation because high-speed full-swing operation is achieved with this technique. Both small load dependence and small fan-in dependence of gate delay time are also attained with this technique. The simulated and measured results show that TS-FS-BiCMOS is faster than any other BiCMOS circuit below 2-V supply and is twice as fast as CMOS even at 1.5-V supply. Therefore, TS-FS-BiCMOS is the most promising circuit for BiCMOS ULSI's in sub-2-V regime.

ACKNOWLEDGMENT

The authors would like to thank K. Shimohigashi and T. Nagano for encouragement, and H. Higuchi, M. Suzuki, and S. Tachibana for technical discussions.

REFERENCES

[1] H. Momose, K. M. Cham, C. I. Drowley, H. R. Grinolds, and H. S. Fu, "0.5 micron BiCMOS technology," in IEDM Tech. Dig., 1987, pp. 838–840.
[2] H. J. Shin, "Full-swing logic circuits in a complementary BiCMOS technology," in Dig. Symp. VLSI Circuits, 1990, pp. 89–90.
[3] K. Yano et al., "Quasi-complementary BiCMOS for sub-3-V digital circuits," in Dig. Symp. VLSI Circuits, 1991, pp. 123–124.
[4] K. Yano et al., "3.3-V BiCMOS circuit techniques for 250-MHz RISC arithmetic modules," in Proc. CICC, 1991, pp. 15.8.1–15.8.4.
[5] O. Nishii et al., "A 1,000MIPS BiCMOS microprocessor with superscalar architecture," in ISSCC Dig. Tech. Papers, 1992, pp. 114–115.
[6] P. Raje, R. Ritts, K. Cham, J. Plummer, and K. Saraswat, "Merged BiCMOS: A device and circuit technique scalable to the sub-micron sub-2-V regime," in ISSCC Dig. Tech. Papers, 1991, pp. 150–151.
[7] A. E. Gamal, J. L. Kouloheris, D. How, and M. Morf, "BiNMOS: A basic cell for BiCMOS sea-of-gates," in Proc. CICC, 1989, pp. 8.3.1–8.3.4.
[8] M. Hiraki et al., "A 1.5-V full-swing BiCMOS logic circuit," in ISSCC Dig. Tech. Papers, 1992, pp. 48–49.
[9] H. J. Shin, "Performance comparison of driver and full-swing techniques for BiCMOS logic circuits," IEEE J. Solid-State Circuits, vol. 25, no. 3, pp. 863–865, 1990.
[10] S. H. Embabi, A. Bellaouar, M. I. Elmasry, and R. A. Hadaway, "New full-voltage-swing BiCMOS buffers," IEEE J. Solid-State Circuits, vol. 26, no. 2, pp. 150–153, 1991.
[11] E. Takeda, Y. Ohji, and H. Kume, "High field effects in MOSFETs," in IEDM Tech. Dig., 1985, pp. 60–63.

3.3-V BiCMOS Circuit Techniques for 250-MHz RISC Arithmetic Modules

Kazuo Yano, *Member, IEEE*, Mitsuru Hiraki, *Member, IEEE*, Shohji Shukuri, Makoto Hanawa, *Member, IEEE*, Makoto Suzuki, *Member, IEEE*, Satoru Morita, Atsushi Kawamata, Nagatoshi Ohki, Takashi Nishida, and Koichi Seki, *Member, IEEE*

Abstract—A quasi-complementary BiCMOS gate for low-voltage supply is applied to a 3.3-V RISC data path. For a parallel RISC processor, the major issues are the construction of arithmetic modules in a small number of transistors and the shortening of the cycle time as well as the delay time. The feedbacked massive-input logic (FML) concept is proposed to meet these requirements. It reduces the number of transistors and the power within the framework of fully static logic 3–4 times. A low-voltage BiCMOS D-flip-flop is also conceived to allow the single-phase clocking scheme, which is favorable for high-frequency operation of RISC's. To demonstrate these circuit techniques, a 32-b ALU is designed and fabricated using 0.3-μm BiCMOS to demonstrate 1.6 times performance leverage over CMOS at 3.3 V.

I. INTRODUCTION

BiCMOS technology has been employed in a wide spectrum of ULSI's because of its fast speed (twice as fast speed as CMOS with almost the same integration level), its capability to interface with the ECL level, and its excellent stability and sensitivity to small input signals in sense amplifiers. Although BiCMOS, to date, has been effectively applied to ECL-interfaced SRAM's [1] and DRAM's [2], its application to fast microprocessors is recent [3], [4]. One strong motivation for this is that the RISC concept is becoming a major factor in fast engineering workstations. Fast device technologies, such as GaAs, bipolar ECL, and BiCMOS, are expected to be applied in microprocessors [5], [6]. BiCMOS is particularly advantageous for parallel-processing architecture, which will play an important role in next-generation processors because of its low-power and high-density characteristics.

However, how much the degraded performance of

Manuscript received August 8, 1991; revised November 4, 1991.

K. Yano is with the Center for Solid-State Electronics Research, Arizona State University, Tempe, AZ 85287-6206 on leave from the Central Research Laboratory, Hitachi Ltd., Kokubunji, Tokyo 185, Japan.

M. Hiraki, S. Shukuri, M. Hanawa, M. Suzuki, T. Nishida, and K. Seki are with the Central Research Laboratory, Hitachi Ltd., Kokubunji, Tokyo 185, Japan.

S. Morita, A. Kawamata, and N. Ohki are with Hitachi VLSI Engineering Corporation, Kodaira, Tokyo 187, Japan.

IEEE Log Number 9105614.

BiCMOS at low supply voltages [7], [8] affects the value of deep-submicrometer BiCMOS is still unclear. Although new circuits, such as the BiCMOS [9] and C-BiCMOS [9], [10], have been proposed and studied to break conventional limitations, the BiNMOS has a small performance leverage over the CMOS and the advantages of the C-BiCMOS are considerably offset by the additional process complexity and cost for the p-n-p bipolar fabrication. Therefore, a new circuit is still strongly desired.

Another "new" approach to solve this situation is to use quasi-complementary BiCMOS (QC-BiCMOS) [11]–[14]. Although the basic circuit configuration of QC-BiCMOS itself has been examined by many designers since the early stage of BiCMOS development [11], it has not played an important role in actual integrated circuits mainly because of its limited performance leverage over CMOS. However, as we demonstrated in a previous paper [13], [14], QC-BiCMOS gates with two additional circuit techniques (one is a separation between the base of the pull-up bipolar and the base of the quasi-p-n-p, the other is a carefully designed base discharging circuit and full-swing input/output) can have a sufficient performance leverage over CMOS with an excellent low-voltage performance. We demonstrated this by fabricating 0.3-μm three-input NAND gates. However, the feasibility of the QC-BiCMOS in actual larger circuits is unknown. In addition, issues for applying QC-BiCMOS to fast microprocessors are not clarified. Particularly, anticipated high-machine-cycle operation (approaching nominally 250 MHz for 0.3-μm region) will surely pose new issues on circuit design.

This paper discusses the main issues which one inevitably encounters in designing fast low-voltage BiCMOS RISC arithmetic modules. Using this as a vehicle, new circuit techniques to solve these issues are explored. The advantages of the circuit techniques are demonstrated in an experimental 32-b ALU. The issues and low-voltage circuit are discussed in Section II. The design and the fabrication of an experimental ALU are discussed in Section III, followed by the conclusions.

Reprinted from *IEEE Journal of Solid-State Circuits*, Vol. 27, No. 3, pp. 373–380, March 1992.

II. Low-Voltage Circuit Techniques for 3.3-V RISC

For realizing a high-performance RISC, the conventional precharged circuits, which are widely used in the fast arithmetic modules [15], [16], are unfavorable, because a short cycle time as well as a short gate delay is essential in fast RISC's. In CISC's, precharging can be done while the other circuit is operating because of the longer logic path and the longer machine cycle. However, a RISC, which owes its performance to a shorter machine cycle, cannot afford the additional precharging time. In addition, it is difficult to maintain sufficient noise immunity in a dynamic circuit at a high frequency around 250 MHz, which is our nominal target machine cycle. However, it is also difficult to get enough performance only by the simple static logic in many situations. This makes us conclude that a new circuit which reduces both the delay time and cycle time is expected for high-machine cycle RISC's.

Another important issue for a high-performance RISC is to reduce area and power. Indeed, small area and small power have been pursued in all integrated circuits. However, in the deep-submicrometer RISC's they are directly related with the performance. This is because in a 0.3-μm region (or less), implementing parallel processing elements on a chip is a very attractive (and will probably be a very common) option to achieve high performance. Below eight processors, the number of processors is expected to linearly enhance the total performance of a computer system. In this case the area and power of each processor, which determine the number of processors on a chip, are unprecedentedly important in the deep-submicrometer region. When one sees the BiCMOS gate in this light, its larger area than the CMOS gate may be a considerable disadvantage. Therefore reducing area and power dissipation without degrading the performance is the next challenge for 0.3-μm-level BiCMOS circuit designers.

Based on this consideration, we looked for a way to reduce area and power within the framework of fully static logic. Our approach to solve these issues is to use a circuit named feedbacked massive input logic (FML) shown in Fig. 1. The main features of this circuit are: 1) the separation between the nMOS logic circuit and the pMOS logic circuit by the MOS transistors ($MN1$, $MP1$ in Fig. 1(a)) controlled by the feedback signal from the output; and 2) large input logic is formed in a single stage. The basic operation is explained using the five-input NAND gate as an example (Fig. 1(b)). We start from the case when inputs A, B, C, and D are high level, and input E changes from low to high. Before the transient, $MN6$ and $MN11$ are off. At this moment, nodes $N1$, $N4$ and outputs $N7$, $N5$ are high. When E changes to high, $MN11$ turns on, and therefore node $N4$ is discharged through the path of $MN7$–$MN11$. Because $MP11$ is off at this moment, the voltage of $N4$ goes down to low level without the charge stored in the parasitic capacitance $C7$ (which originates from drain capacitance of $MP6$–$MP10$) being discharged. This is fast because the parasitic capacitance $C7$ is not discharged. The equivalent resistance $Z2$ is designed to be high so that the charge stored in $C7$ would not be discharged at this time. Then $MP13$ turns on, and node $N6$ is charged to turn $Q2$ on. This is followed by the discharging of output $N7$. Simultaneously node $N1$ is discharged through the path of $MN1$–$MN6$ to the ground level. $Q1$ stays off. After the delay of the inverters $I2$ and $I1$, $N5$ becomes low. Then the charge stored in $C7$ is discharged. Simultaneously $MN1$ turns off, however, $N1$ was already discharged, therefore node $N1$ stays low. The operation when input E changes from high to low is similar to this procedure.

In this operation, because the discharging of $N4$ is almost free from the large parasitic capacitances $C7$, the delay is small even if the number of inputs is large. The advantage of this feedback circuit is shown in the simulated result shown in Fig. 2. In this simulation the 0.3-μm BiCMOS parameters shown in Table I are used. These parameters are based on our 0.3-μm process having a double-polysilicon self-aligned structure. The figure shows that the effect of the feedback is larger when the number of inputs is large. In the seven-input logic, the delay with feedback is 1.4 times smaller than that without feedback.

Discharging paths $Z1$ and $Z2$ requires careful design. $Z1$ is necessary to avoid $N1$ being a floating node when A–E are all high and the output is low. If the impedance of $Z1$ is too large, the noise coupled into $N1$ could raise the base of $Q1$. On the other hand, if the impedance of $Z1$ is comparable to the impedance of $MN1$, the effect of the feedback circuit is lost. The impedance of $Z1$ is determined considering this trade-off relation.

Good low-voltage performance down to 2.5 V is possible because of the quasi-complementary-type BiCMOS buffer [12]–[14], which avoids the gate–source voltage loss, which has the been the origin of the degraded low-voltage performance in conventional BiCMOS circuits. In the QC buffer a voltage loss occurs only in the drain-source voltage in the pMOS transistor ($MP2$ in Fig. 1 (a)); the resultant drain-current degradation is small. Large current drive is attained even at low supply voltage. The simulated delay dependences show that the FML circuit is faster than the CMOS circuit down to 2 V (Fig. 3), which has been impossible in conventional BiCMOS circuits (including C-BiCMOS). Note that this excellent low-voltage performance is achieved without a p-n-p transistor.

Because this FML can incorporate very complex logic functions in a single gate, the number of transistors, area, and power are drastically reduced. The simple example shown in Fig. 4 illustrates how the large input logic works to reduce the number of transistors. The six-input AND (or six-input NAND in negative logic) can be constructed with two stages of three-input NAND and two-input NOR. However, constructing the same logic in a single six-input gate reduces the number of MOS transistors to almost half and the number of bipolar transistors to about one-third.

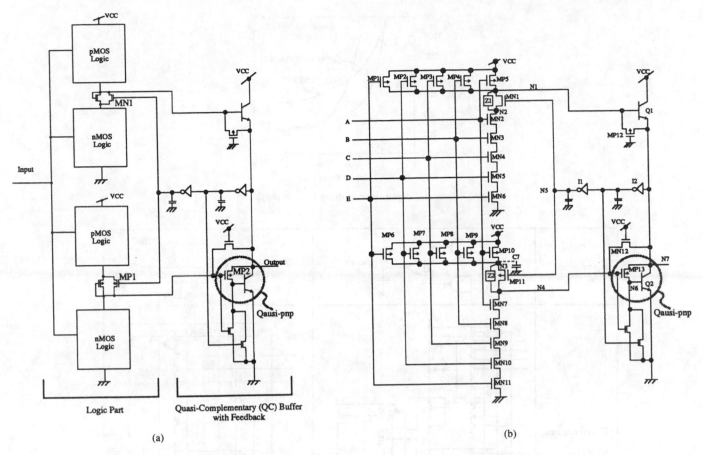

Fig. 1. Circuit diagram of (a) feedbacked massive input logic (FML) and (b) FML five-input NAND logic.

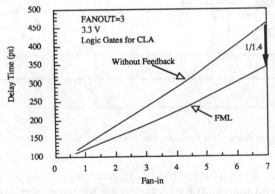

Fig. 2. Simulated dependence of delay time on fan-in.

TABLE I
0.3-μm BiCMOS DEVICE PARAMETERS USED IN CIRCUIT SIMULATIONS

MOSFET	gate oxide thickness	8 nm
	drain current	2.8 mA (nMOS) 2.0 mA (pMOS)
	drain-substrate capacitance	10 fF
Bipolar Transistor	cutoff frequency	24 GHz
	knee current	5 mA
	junction capacitance	
	C_{TE}	6.5 fF
	C_{TC}	5.0 fF
	base resistance	250 Ω

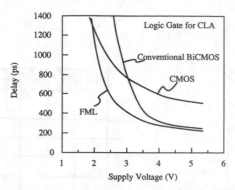

Fig. 3. Simulated dependence of delay time on supply voltage.

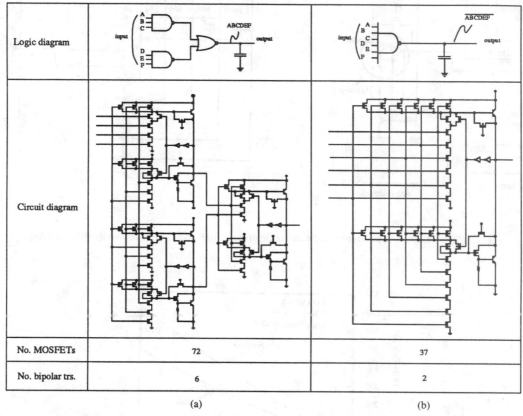

Logic diagram	(a)	(b)
Circuit diagram		
No. MOSFETs	72	37
No. bipolar trs.	6	2

(a)   (b)

Fig. 4. Two different circuit designs for six-input NAND (AND): (a) two-stage form, and (b) single-stage form.

The effect of reducing the number of transistors by a large fan-in logic is observed in general logics. The effect is evaluated quantitatively using a 16-b carry lookahead unit as a benchmark. The performance of a carry lookahead unit strongly affects the performance of critical paths in ALU's and multipliers. Various 16-b carry lookahead units having different average fan-ins are designed as shown in Fig. 5. The total gate width, transistor count, number of gate stages on the critical path, and total delay are evaluated as shown in Figs. 6 and 7. With increase in average fan-in, the numbers of bipolar transistors and MOSFET's decrease rapidly. For example, if we compare the transistor count in a 7-fan-in design with that of a 3-fan-in design, which is the typical design point for CMOS, the number of bipolar transistors is $1/2.5$ and the number of MOSFET's is $1/2$. The total gate width is also reduced to $1/1.5$. Because the total gate width is a good figure for the power dissipation, the power in the 7-fan-in design is about $1/1.5$ times smaller than that of the three-input logic. Although the feedback circuits in FML increase the area, compared to the simple QC-BiCMOS, reduced area due to the large fan-in design overwhelms the increase. It should be noted that the delay increase of FML is very small even when the average fan-in is large (Fig. 7). By contrast, the delay increases very rapidly in CMOS with an increase in fan-in. This is because with an increase in fan-in, the average load capacitance increases, which severely degrades CMOS speed. The dependence of speed advantage due to the feedback technique in Fig. 7 seems different from that of Fig. 2. This is because the fan-out is fixed in the comparison in Fig. 2, whereas the fan-out is dependent on the average fan-in in Fig. 7. In

126

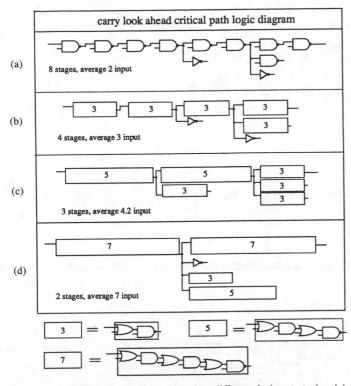

Fig. 5. Critical path logic diagrams for four different design strategies: (a) eight-stage form with average 2 fan-in, (b) four-stage form with average 3 fan-in, (c) three-stage form with average 4.2 fan-in, and (d) two-stage form with average 7 input. The supply voltage is 3.3 V.

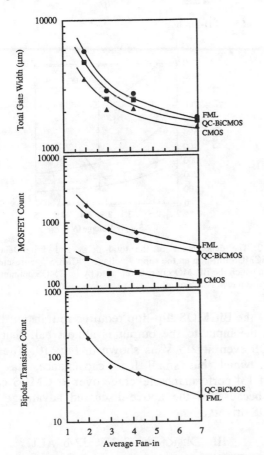

Fig. 6. The dependence of total MOS gate width, MOSFET count, and bipolar transistor count on average fan-in, which corresponds to the various logic configurations in Fig. 5.

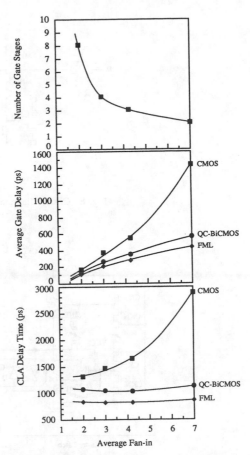

Fig. 7. The dependence of the number of gate stages, average gate delay, and total carry lookahead delay on average fan-in, which corresponds to the various logic configurations in Fig. 5.

general, the advantage of the feedback technique is naturally larger when the fan-out is smaller.

Because the FML circuit operation is static, it is possible to design the cycle time to be the same as the delay time of the critical path. This is beneficial for the high-machine-cycle design of RISC's. The smaller number of transistors due to the FML circuit enables implementing a larger number of processor units in a chip, which again enhances the effective performance of the processor.

Another important issue in high-machine-cycle RISC circuit design is to obtain a fast flip-flop that can be operated within the single-phase clocking environment. In RISC's the pipeline pitch is relatively shorter than those of CISC's. Therefore, the delay of a flip-flop easily contributes to large overhead. This tendency is emphasized when the superpipeline technique or vector execution units [17] are introduced in a microprocessor. When the machine cycle is as short as 4 ns (250 MHz), the conventional two-phase nonoverlapping clocking scheme is difficult because of the additional margin needed to avoid overlapping between different phases. A single-phase clocking scheme is desirable in this respect.

A conventional BiCMOS flip-flop [18], which improves the current drive capability of a CMOS flip-flop by adding a BiCMOS output buffer, is shown in Fig. 8. However, in this conventional circuit, the master latch is pure

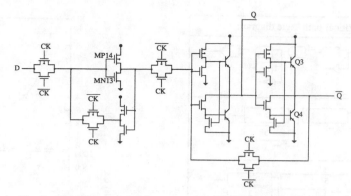

Fig. 8. Circuit diagram of the conventional BiCMOS D-flip-flop [18].

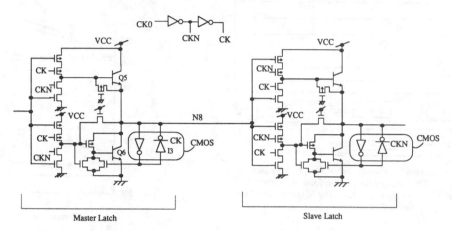

Fig. 9. Circuit diagrams of the new QC-BiCMOS D-flip-flop.

CMOS, which has no performance improvement due to BiCMOS. In addition, when there is a clock skew between CK and $\overline{CK}$, the output driven by the bipolar transistors $Q3$ and $Q4$ is to be inverted by the MOSFET's $MN13$ and $MN14$. When this conflict occurs the operating speed is very low.

Our new D-type flip-flop, which can be used in a single clocking scheme, is shown in Fig. 9. This is based on the concept that the data are retained by the CMOS latch, whereas the critical path in this flip-flop is exclusively constructed from two-stage quasi-complementary Bi-CMOS. This enables excellent low-voltage operation. Even if there is a clock skew between CK and CKN, the output node is driven quickly by the QC buffer without being affected by the reluctant CMOS latch. The CMOS clocked inverter $I3$ retains the voltage level of $N8$. The superior current drive capability of $Q5$ (or $Q6$) is insensitive to the retaining current drive of $I3$. Input data are transferred quickly without interference from the retaining current. Conventionally the BiCMOS circuit is believed to have a performance advantage over the CMOS only when the load capacitance is large. Although the load capacitance of node $N8$ is small (about 30 fF) in this flip-flop, the performance advantage of BiCMOS over CMOS is relatively large.

The performance of this D-flip-flop is evaluated by circuit simulations as shown in Fig. 10. The total delay (0.31

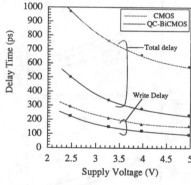

Fig. 10. The dependence of the total delay and write delay of the QC-BiCMOS flip-flop on the supply voltage. "CMOS" represents the circuit, in which the BiCMOS part is replaced by its CMOS counterpart. The load capacitance is 5 fan-out.

ns) of the BiCMOS flip-flop required to transmit the data from the input to the output is almost half that of the CMOS even at 2.5 V as shown in Fig. 10. The master latch, which has small load capacitance, has a factor-of-1.4 performance leverage over its CMOS counterpart because of the above-discussed advantage of Bi-CMOS drivers.

III. Design of 3.3-V 32-b ALU

A fully static 32-b ALU employing the above circuits is designed using 0.3-μm BiCMOS rule. The conditional

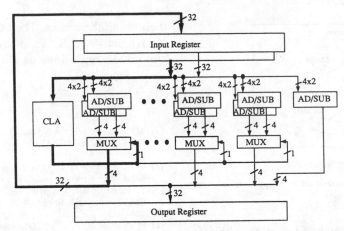

Fig. 11. Block diagram of the designed 32-b ALU. CLA: carry lookahead; AD/SUB: adder/subtractor; and MUX: multiplexer.

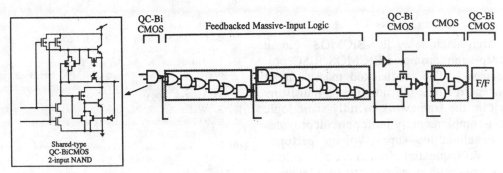

Fig. 12. The critical path of the 32-b ALU.

sum architecture, which selects the two 4-b adder results according to the carry signal generated by the carry lookahead unit, is used as shown in Fig. 11. The critical path consists of only six-state gates, which include QC-BiCMOS gates, FML gates (maximum nine-input gate is used), CMOS multiplexers, and QC-BiCMOS D-type flip-flops as shown in Fig. 12. Because 64 two-input NAND circuits are required in first stage of the carry lookahead unit, the shared QC-BiCMOS circuit discussed in [13] and [14] is used to reduce the area. In the following two-stage large-input carry lookahead logics, the FML technique is fully employed to reduce delay, area, and power. The features of the designed ALU are shown in Table II. Circuit simulation based on our 0.3-μm Bi-CMOS parameters shows that the adder delay of 1.2 ns is obtained at 3.3-V supply. The power consumption is as small as 39 mW at a frequency of 250 MHz (20% gate activity is assumed). These results are sufficient for 250-MHz RISC operation.

The ALU is fabricated using the self-aligned double-polysilicon 0.3-μm BiCMOS technology. The gate length is 0.3 μm, the gate-oxide thickness is 9 nm, the threshold voltage is 0.3 V, and the cutoff frequency of the n-p-n bipolar transistor is 20 GHz. The fine patterns, such as gate, contact, and first metal, are defined by electron-beam lithography, and the other patterns are defined by the i-line optical lithography. The interconnection consists of first-level tungsten, second-level aluminum, and third-level

TABLE II
32-b ALU DESIGN SUMMARY

Architecture	32 b, Conditional Sum
Technology	0.3-μm Double-Poly Si BiCMOS, T_{ox} = 9 nm
Interconnection	Three-Layer Metal (W/Al/Al)
Area	0.6 × 2.8 mm^2
Delay Time	1.2 ns (ADD), 0.31 ns (Register) at 3.3 V
Power	39 mW (250 MHz, at 3.3 V)

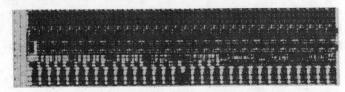

Fig. 13. Microphotograph of the fabricated ALU.

aluminum. More details of our device/process technology will be described in an another paper. A microphotograph of the fabricated ALU is shown in Fig. 13. The total number of transistors is 13 800, which includes 1000 bipolar transistors. Therefore the percentage of the number of bipolar transistors is only 7%, although almost the entire critical path consists of BiCMOS gates. The delays from the input signal (it is actually a regenerated signal by an on-chip inverter circuit) to the output signal are measured on the wafer level using the on-chip source-follower circuits. Excellent low-voltage performance with a good agreement with simulation is demonstrated as shown in

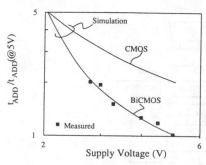

Fig. 14. The dependence of the normalized adder delay time on supply voltage.

Fig. 14. The 1.6-times performance leverage in this actual ALU over CMOS is confirmed, which opens the door to 0.3-μm BiCMOS RISC's and related ASIC's.

IV. Conclusions

Low-voltage high-machine-cycle BiCMOS circuit techniques for deep-submicrometer BiCMOS RISC processors were discussed. The feedbacked massive-input logic (FML) concept reduces the number of transistors and the power within the framework of fully static logic 2–3 times. A quasi-complementary buffer circuit provides unprecedentedly excellent low-supply-voltage performance even below 3 V. Sequential circuits are also able to operate at low voltages with good performance leverage over CMOS due to the novel QC-BiCMOS flip-flop. These demonstrate that the BiCMOS continues to have superior performance over CMOS in the sub-0.3-μm sub-3-V region, which promises 250-MHz-level parallel processing RISC's and ASIC's.

Acknowledgments

The authors wish to thank Dr. K. Shimohigashi, Dr. T. Masuhara, Dr. H. Sunami, Dr. M. Hirao, Dr. E. Takeda, and Dr. T. Nishimukai for their useful discussions and encouragement. The authors wish to express appreciation to T. Hayashi, T. Doi, and O. Nishii for their useful suggestions and comments for VLSI processor design. The authors are greatly indebted to Dr. Y. Kawamoto, Dr. T. Kure, Dr. F. Murai, and the processing staff members for their chip fabrication. The authors are also grateful to Dr. H. Higuchi and S. Tachibana for their valuable discussions and suggestions on BiCMOS circuit design.

References

[1] M. Suzuki et al., "A 3.5-ns, 500-mW, 16-kbit BiCMOS ECL RAM," IEEE J. Solid-State Circuits, vol. 24, no. 5, pp. 1233–1237, Oct. 1989.

[2] G. Kitsukawa et al., "A 23-ns 1-Mb BiCMOS DRAM," IEEE J. Solid-State Circuits, vol. 25, no. 5, pp. 1102–1111, Oct. 1990.

[3] T. Hotta et al., "A 70MHz 32b microprocessor with 1.0 μm BiCMOS macrocell library," in ISSCC Dig. Tech. Papers, 1989, pp. 124–125.

[4] M. Yamashita et al., "200-MHz 16-bit BiCMOS signal processor," in ISSCC Dig. Tech. Papers, 1989, 12.8.

[5] E. W. Brown et al., "Implementing SPARC in ECL," IEEE Micro, vol. 10, pp. 10–22, Feb. 1990.

[6] D. A. Whitmire, V. Garcia, and S. Evens, "A 32b GaAs RISC microprocessor," in ISSCC Dig. Tech. Papers, 1988, p. 34.

[7] H. Momose, K. M. Cham, C. I. Drowley, H. R. Grinolds, and H. S. Fu, "0.5 micron BiCMOS Technology," in IEDM Tech. Dig., 1987, pp. 838–840.

[8] M. Fujishima, K. Asada, and T. Sugano, "Evaluation of delay-time degradation of low-voltage BiCMOS based on a novel analytical delay-time modeling," IEEE J. Solid-State Circuits, vol. 26, no. 1, pp. 25–31, Jan. 1991.

[9] A. Watanabe, T. Nagano, S. Shukuri, and T. Ikeda, "Future BiCMOS technology for scaled supply voltage," in IEDM Tech. Dig., 1989, pp. 429–433.

[10] H. J. Shin et al., "Full-swing complementary BiCMOS logic circuits," in Proc. 1989 Bipolar Circuit Technology Meeting, 1989, pp. 229–233.

[11] G. Kitsukawa et al., "Low power, high-speed BiCMOS memory circuits," in Extended Abst. 1984 Int. Conf. Solid State Devices Mater. (Kobe), 1984, pp. 233–236.

[12] P. Raje, R. Ritts, K. Cham, J. Plummer, and K. Saraswat, "Merged BiCMOS: A device and circuit technique scalable to the sub-micron sub-2V regime," in ISSCC Dig. Tech. Papers, 1989, TPM 9.2.

[13] K. Yano et al., "Quasi-complementary BiCMOS for sub-3-V digital circuits," in Tech. Dig. 1991 VLSI Symp. Technology.

[14] K. Yano et al., Quasi-complementary BiCMOS for sub-3-V digital circuits," IEEE J. Solid-State Circuits, vol. 26, pp. 1708–1719, Nov. 1991.

[15] K. Makino, "High speed circuit technology for mainframe VLSI," in Proc. 1987 VLSI Circuit Symp., 1987, pp. 93–94.

[16] T. Hayashi et al., "The SDC cell—A novel design methodology for high-speed arithmetic modules using CMOS/BiCMOS precharged circuit," IEEE J. Solid-State Circuit, vol. 25, no. 2, p. 403, Apr. 1990.

[17] "INTEL 386TM32 bit architecture," Intel Japan, Apr. 11, 1989.

[18] S. C. Lee, D. W. Schucker, and P. T. Hickman, "BiCMOS technology for high-performance VLSI circuits," VLSI Design, pp. 98–100, Aug. 1984.

Advanced Bipolar Circuits

Designers now understand the underlying principles of high-speed,
low-power bipolar circuits— and performance is taking off

by Ching-Te Chuang

High-speed bipolar circuits have long been main stream for implementing high-performance mainframe computer systems. In recent years, high-end logic has been shifting from transistor-transistor logic (TTL) to emitter-coupled logic (ECL), current-mode logic (CML) [1], and non-threshold logic (NTL), with ECL being the predominant circuit technology. NTL is gradually finding its place in high-end masterslice designs because of its low power-delay product [2-4].

Power dissipation has long been known to limit VLSI applications of both ECL (Fig. 1a) and NTL (Fig. 1b) circuits. The power/speed limitation of both circuits derives primarily from the passive resistors in the delay path: the collector load resistor R_C limits the pull-up delay and the emitter-follower resistor R_{EF} limits the pull-down delay.

This article reviews the recent advances in high-speed, low-power bipolar circuits aimed at achieving superior power-delay performance and load-driving capability over conventional ECL and NTL circuits. We will examine the basic principles underlying power/speed improvement, among which are charge-buffering, dc/ac-coupled active pull-down schemes, and complementary push-pull approaches. We will then describe the utilization and combination of these basic principles to form various high-speed, low-power circuits in both npn-only and complementary circuit configurations, and we will discuss the design trade-offs of these circuits

Principles and Approaches

The essence of high-speed, low-power operation is the ability to achieve both a low standby current and a large dynamic current during the switching transient. This combination is easily achieved in CMOS circuits but not in current-controlled devices such as bipolar transistors. The current and logical voltage levels in bipolar circuits are typically set by passive resistors so large switching currents inevitably imply high dc power consumption. The unique property, or problem, of saturation in bipolar transistors imposes additional severe constraints on circuit configurations for high-speed, low-power operation.

One way of overcoming these constraints is by using the charge-buffering principle (Fig. 2a), which was first used in a complementary circuit configuration [5-7]. The principle involves buffering electrical charges in a charge-storage diode (CSD). This permits the generation of a large dynamic current for switching but requires only a very small dc current in standby. The scheme uses the large diffusion capacitance of the storage diode as a variable dynamic capacitor to provide large dynamic current, as shown by the current spikes in Fig. 2a, when the input changes. For optimum performance, the time constant of the charge-storage diode must be designed with regard for the loading and dynamic capacitance of the driven device. Dynamic-to-static current ratios of over 100 can be obtained with proper device and circuit designs.

A second approach is to replace the passive resistors with active devices. When applied to the emitter-follower resistor, this approach results in various active-pull-down schemes. The active-pull-down element acts like a switch in parallel with a small standby current source (Fig. 2b). The switch must be controlled by the logic stage to provide a large transient pull-down current consistent with the logic inputs. Dc coupling of the switch to the logic stage is preferred, but the incompatibility of voltage levels limits the configurations that can be used. Special device elements may be needed, which we will discuss later. Ac coupling, on the other hand, has the advantage of completely blocking the dc signal, which provides an easy interface with the logic stage. It does so, however, at the expense of introducing capacitors into the circuits, along with associated process, qualification, and reliability issues.

Complementary push-pull approaches typically offer the most significant improvement on power/speed and load-driving capabilities, although at the expense of complicated process technology. For the most familiar pnp (pull-up), npn (pull-down) configuration (Fig. 2c), both driving devices

Reprinted from *IEEE Circuits and Devices Magazine*, pp. 32-36, November 1992.

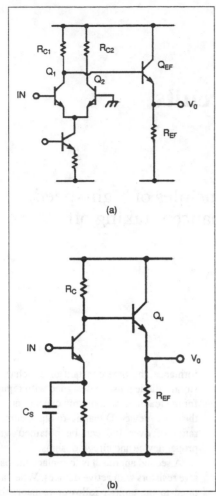

1. *Schematics of conventional ECL circuit (a), and conventional NTL circuit (b). In both circuits, pull-up delay is limited by collector load resistor R_C and pull-down delay is limited by emitter-follower resistor R_{EF}.*

will saturate if they are also used to set the logical voltage levels and must switch all the way from cut-off to saturation or vice versa. Performance is thus limited by both the saturation and device parasitics (since all the junction and parasitic capacitances have to be charged and discharged). But it's possible to use this configuration solely for delivering the large transient current required during switching and to set the logic level by other means, thus avoiding the saturation problem and improving the performance. For the npn (pull-up), pnp (pull-down) complementary emitter-follower configuration (Fig. 2d), the output voltage in the steady state always sits half-way between the voltages at the bases of the two transistors. In this case, both driving devices are biased at cut-in (nearly-on) condition during standby and one of them switches into the active

region to provide the switching current. This configuration thereby avoids saturation and the necessity to charge/discharge the junction/parasitic capacitance, and the switching speed is limited only by the base transit time (or, more properly, the diffusion capacitance) of the driving transistors. The two base voltages swing in the same direction with equal magnitude, except during the switching transient when one of the driving devices experiences base-emitter overdrive. Since the output voltage is in phase with the two base voltages, the inverting function has to be provided by the front-end logic stage, and this configuration is used solely for improving the load-driving capability.

Npn-Only, Active-Pull-Down Circuits
One example of a dc-coupled active-pull-down circuit is the JFET pull-down ECL circuit (Fig. 3a) [8]. This circuit utilizes a "free" p-channel JFET, which is readily available in any bipolar technology by modifying the base pinch-resistor structure. Thus modified, the JFET replaces the emitter-follower resistor R_{EF} as the pull-down device. The JFET acts as a variable resistor whose value is modulated by the gate voltage, which is in turn controlled by the logic stage. The channel resistance is reduced during the output high-to-low transition to provide a large transient pull-down current (Fig. 3b, Contour A). During the output low-to-high transition, the channel resistance is increased to reduce the bypassing current to V_T, thus improving the pull-up delay. A circuit of this type based on a 0.8-μm double-poly self-aligned bipolar technology [9, 10] at 1.0 mW/gate provides a loaded pull-down delay that is 24-percent less than that of a conventional ECL circuit. To maximize the modulation factor, the doping profile for the JFET channel region may have to be optimized separately from the base region for the npn transistor.

This circuit scheme is straightforward and does not require extra biasing devices or circuits for the pull-down element. However, this scheme's leverage tends to decrease with reduced voltage swing because the modulation factor becomes smaller (Fig. 3b, Contour B.) The leverage also decreases as the termination voltage V_T is raised to reduce the power consumption, again because of a smaller modulation factor. In this case, the current-voltage contour shifts to the left as V_{GD} becomes smaller (Fig. 3b, Contour C).

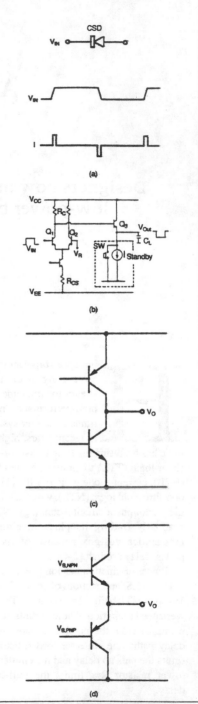

2. *Basic high-speed low-power principles and approaches. Charge-buffering (a): charge buffered (diffusion capacitance) in a storage diode is used to generate large dynamic current when input changes. Dc/ac-coupled active-pull-down (b): active-pull-down device acts as a switch in parallel with a small standby current source. Pnp (pull-up), npn (pull-down) complementary push-pull driver (c). Npn (pull-up), pnp (pull-down) complementary emitter-follower driver (d).*

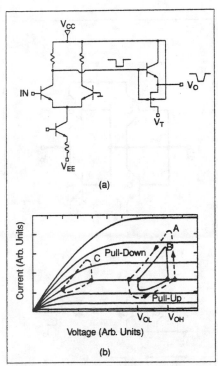

3. JFET pull-down ECL circuit (a). The JFET acts as a variable resistor whose channel resistance is modulated by the gate voltage controlled by the logic stage. See Ref. [8]. Current-voltage contour during transitions (b): optimized contour with large modulation factor (A); reduced voltage swing results in smaller modulation factor (B); raised V_T causes contour to shift to the left with smaller modulation factor (C).

The ac-coupled active-pull-down ECL (AC-APD-ECL) circuit [11, 12] (Fig. 4a) utilizes a capacitor to couple a transient voltage pulse to the base of an npn pull-down transistor. This reduces the dc power consumption in the emitter-follower stage, and the large transient pull-down current improves the pull-down delay. The capacitor completely blocks dc signals and alleviates the level-compatibility problem in interfacing with the logic stage, although extra biasing devices are needed to establish the standby current in the output stage. This circuit, which has also been known as the Turbo ECL circuit, was used in a 13,000-gate, gate array [13]. In this collector-node ac-coupled scheme, the coupling capacitor presents a load to the logic stage, and one has to wait until the logic stage switches for the transient signal to be coupled to the pull-down transistor. Thus, the current switch still requires substantial power to achieve fast switching.

The coupling capacitor can be moved

from the collector node to the common-emitter node of the switching transistors (Fig. 4a, Node A). The coupling then occurs before the switching of the logic stage, and the transient current through the capacitor has a speed-up effect on the logic stage (the transient current adds to the switching current when the input goes "high" and helps to turn off the switching transistor when the input goes "low"). Implementation of this scheme in the NTL circuit results in the so-called super push-pull logic (SPL) circuit (Fig. 4b) [14]. This circuit does not require a separate speed-up capacitor at the common-emitter node of the switching transistors because speed-up is already provided by the coupling capacitor.

Another approach is to replace capacitor coupling with charge-buffered coupling, which results in the charge-buffered active-pull-down ECL circuit (CB-APD-ECL) (Fig. 5) [15]. This charge-buffered coupling scheme provides a much larger dynamic current than can reasonably be achieved through capacitor coupling. An equivalent dynamic capacitance as large as several hundred femtofarads, which is impractical to implement as a "real" capacitor in integrated-circuit technology, can easily be provided by the diffusion capacitance of the charge-buffering diode. In 0.8-μm double-poly self-aligned bipolar technology at 1.0 mW/gate, the loaded-delay and load-driving capability of the ac-coupled active-pull-down ECL circuit is about 20 percent better than conventional ECL circuits. The charge-buffered active-pull-down ECL circuit is about 40 percent better.

While the aforementioned active-pull-down schemes improve pull-down delay by replacing the emitter-follower resistor R_{EF} with an active device, the pull-up delays of these circuits are still limited by the collector load resistor R_C. The pull-up delay, in general, can be improved by complementary push-pull schemes described below.

Complementary Push-Pull Circuits
The combination of charge-buffering and a complementary push-pull driver results in the charge-buffered-logic (CBL) circuit (Fig. 6) [5-7]. The front end of this circuit resembles diode-transistor-logic (DTL) except for the use of charge-storage diodes. Since the only dc standby current required for the circuit is the sustaining base currents for the pnp (pull-up) and npn (pull-down) transistors, the circuit can operate with extremely low dc current — as low as 10 μA.

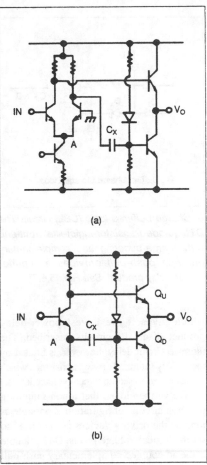

4. Collector-node ac-coupled active-pull-down ECL circuit, also known as Turbo ECL circuit (a). Capacitor is used to couple a transient voltage pulse to the pull-down transistor. See Refs. [11,12,13]. Emitter-node ac-coupled active-pull-down NTL circuit, also known as super push-pull logic circuit (b). Emitter-node coupling provides early coupling and speed-up effect. See Ref. [14].

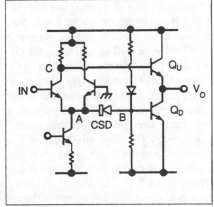

5. Charge-buffered active-pull-down ECL circuit. Charge-buffered coupling is used to provide a large equivalent dynamic capacitance. See Ref. [15].

133

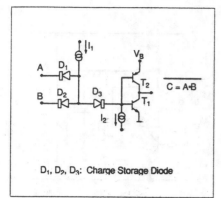

D₁, D₂, D₃: Charge Storage Diode

6. Charge buffered logic (CBL) circuit. The DTL (diode-transistor-logic)-like front-end with charge-buffering and complementary pnp-npn push-pull provides a superior power-delay product. See Ref. [5,6,7].

which gives it the lowest power-delay product of all the circuits described. The ultimate circuit delay, however, is limited by the heavily saturated driving transistors, which are also used to set the logical voltage levels.

We've mentioned that a non-saturating pnp-npn circuit configuration is possible as long as the driving devices are not used to set the logical voltage levels. One example is the ac-coupled complementary push-pull ECL (AC-PP-ECL) circuit, which utilizes two capacitors to couple a transient voltage pulse from the common-emitter node of the switching transistors to the bases of a pair of complementary pnp-npn push-pull transistors (Fig. 7) [16]. In this circuit scheme, the push and pull transistors are biased at cut-in condition. One of them is turned on heavily during the switching transient and then back to the cut-in condition, thus providing a high-speed non-saturating push-pull driver. During the output low-to-high transition, most of the pull-up current is supplied by the pull-up pnp transistor Q_U; only a very small fraction is supplied by the emitter-follower transistor Q_{EF}. The pull-up delay, measured from the time the output voltage crosses the reference voltage, is completely determined by the transient current through Q_U. The delay path of this circuit is from the input to the common-emitter node of the switching transistors (Node A in Fig. 7), through coupling capacitors C_P and C_N to the push-pull transistors Q_U and Q_D. Thus, the collector load resistor R_C is decoupled from the delay path, and a large resistor with a small switching current I_{CS} can be used without degrading the switching speed. The collector load

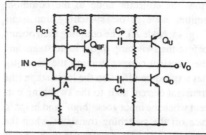

7. Ac-coupled complementary push-pull ECL circuit. Two capacitors are used to couple the transient voltage pulse to complementary push-pull transistors. The collector load resistor R_C is completely decoupled from the delay path. See Ref. [16].

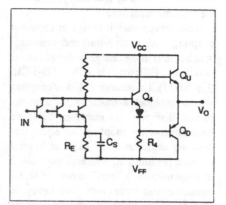

8. NTL with complementary emitter-follower driver (NTL-CEF). A double-diode voltage shifter (Q_4 and the series diode) is used to improve switching speed. See Ref. [17].

resistor is used merely for setting the final output voltage level through the emitter-follower transistor Q_{EF}. In a 0.8-μm complementary bipolar process at 0.5 mW/gate, the circuit offers more than twice the loaded-gate speed and load-driving capability of a conventional ECL circuit.

When the complementary emitter-follower driver configuration of Fig. 2d is implemented with an NTL front-end (Fig. 8), we obtain a circuit that is able to drive very large loads [17]. An additional emitter-follower transistor Q_4 is used to drive the base of the pull-down pnp transistor. The transistor Q_4 and the series diode provide a double-diode voltage shifter between the base nodes of Q_U and Q_D to ensure there is no current through the branch containing Q_4 when the input is "high." Implemented in a 0.8-μm complementary bipolar process at 1.0 mW/gate, the circuit offers a 2.4-times improvement in the pull-down delay of a loaded gate and 4.0-times improvement in load-driving capability over the conventional NTL circuit. The complementary emitter-

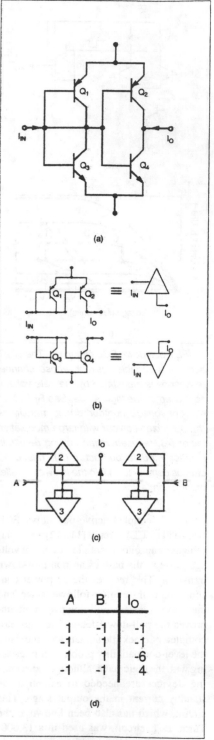

A	B	I_O
1	-1	-1
-1	1	-1
1	1	-6
-1	-1	4

(d)

9. A complementary current-mirror logic (CCML) circuit (a). Current is used as the input/output variable, and various logic functions can be implemented simply by changing the current-mirroring ratio. The logic symbol for the CCML (b). The I_{OUT}/I_{IN} ratio is shown within the triangular symbol. NOR gate, unnormalized output (c). Truth table of the NOR gate (d). See Ref. [18].

follower scheme can also be implemented in an ECL circuit, but the improvement will be less significant because of the front-end configuration [17].

All the circuits discussed up to now have voltage levels as their input and output variables. It is also possible to use currents as the input and output variable, as is the case in the complementary current-mirror logic (CCML) circuit (Fig. 9) [18]. This circuit operates primarily by current-mirroring between the input and output stage. The current summation property and the variable output current levels that can be obtained by changing the current-mirroring ratio provide additional degrees of freedom in design. Various logic functions can be implemented simply by changing the current-mirroring ratio. The NOR function, for example, can be implemented with four current mirrors and two different current-mirroring ratios (Fig. 9c). This NOR gate has unnormalized output current levels, as shown by its truth table (Fig. 9d). For instance, inputs of A = +1 and B = -1 yield a NOR = -1, while A = B = 1 yields a NOR = -6 output. In simple circuits some of the power advantage of CCML over ECL is lost because of the large number of gates needed to implement the function, but the additional arithmetic power of CCML allows very interesting and compact circuit design in more complex logic applications.

What Does it Mean?

All of the circuit schemes we've discussed offer significant improvements over standard bipolar logic-circuit approaches in power delay and load-driving capability, but it is important to understand the associated design trade-offs. The primary considerations are:

(1) Logic capability of the circuit. Emitter-dotting (dotted-OR), for example, is not available in most active-pull-down or complementary push-pull circuits.

(2) Special device elements (such as JFETs and capacitors) and complex processes (such as complementary bipolar).

(3) Design methodology (such as the use of current as the input/output variable in complementary current-mirror logic). One must also maintain power-supply and I/O compatibility within each circuit family (such as ECL and ECL-based active-pull-down and complementary push-pull circuits) so circuits can be readily mixed to offer additional freedom in design optimization.

The large variety of high-speed, low-power bipolar circuits that have emerged in recent years may seem confusing, but the development of these circuits is guided by some basic principles: charge-buffering, dc/ac-coupled active pull-down schemes, and complementary push-pull approaches. The various high-speed, low-power npn-only and complementary bipolar circuit configurations are formed by utilizing and combining these principles. With a good understanding of these principles and their associated design trade-off, designer will continue to improve bipolar circuits and move them into a performance range unattainable with CMOS circuits. **CD**

Dr. C. T. Chuang [SM] is manager of the High-Performance Circuits Group at the IBM T. J. Watson Research Center, Yorktown Heights, New York.

References

1. R. L. Treadway, "DC analysis of current mode logic," *IEEE Circuits and Devices Magazine*, 5, 2, pp. 21-35, 1989.

2. M. Suzuki, S. Horiguchi, and T. Sudo, "A 5-K gate bipolar masterslice LSI with 500 ps loaded gate delay," *IEEE J. Solid-State Circuits*, SC-18, 5, pp. 585-92, 1983.

3. M. Suzuki, H. Okamoto, and S. Horiguchi, "Advanced 5-K gate bipolar gate array with a 267 ps basic gate delay," *IEEE J. Solid-State Circuits*, SC-19, 6, pp. 1038-40, 1984.

4. H. Ichino, *et al.*, "A 50-ps 7K-gate masterslice
using mixed cells consisting of an NTL gate and an LCML macrocell," *IEEE J. Solid-State Circuits*, SC-22, 2, pp. 202-07, 1987.

5. S. K. Wiedmann, "Charge-buffered logic (CBL) - a new complementary bipolar circuit concept," *Dig. Tech. Papers*, pp. 38-39, 1985 Symp. VLSI Tech.

6. S. K. Wiedmann, "Potential of bipolar complementary device/circuit technology," *Tech. Digest*, IEDM, p. 96-99, 1987.

7. S. K. Wiedmann *et al.*, "Sub-300 ps CBL circuits," *IEEE Electron Device Letters*, 10, 11, p. 484-86, 1989.

8. H. J. Shin, P. F. Lu, and C. T. Chuang, "High-speed low-power JFET pull-down ECL circuit," *Proc. 1990 IEEE Bipolar Circuits and Technology Meeting*, pp. 136-39.

9. T. C. Chen, *et al.*, "A submicron high performance bipolar technology," *Dig. Tech. Papers*, pp. 87-88, 1989 Symp. VLSI Tech.

10. T. C. Chen, *et al.*, "A submicrometer high-performance bipolar technology," *IEEE Electron Device Letters*, EDL-10, 8, pp. 364-66, 1989.

11. K. Y. Toh, *et al.*, "A 23 ps/2.1 mW ECL gate," *Dig. Tech. Papers*, p. 224-25, 1989 ISSCC.

12. K. Y. Toh, *et al.*, "A 23-ps/2.1-mW ECL gate with an ac-coupled active-pull-down emitter-follower stage," *IEEE J. Solid-State Circuits*, SC-24, 5, pp. 1301-05, 1989.

13. B. Coy, A. Mai, and R. Yuen, "A 13,000 gate 3 layer metal bipolar gate array," *Proc. 1988 CICC*, pp. 20.1.1.-20.1.3.

14. M. Usami and N. Shiozawa, "SPL (super push-pull logic) - a bipolar novel low-power high-speed logic circuit," *Dig. Tech. Papers*, pp. 11-12, 1989 Symp. VLSI Circuits.

15. C. T. Chuang and K. Chin, "High-speed low-power charge-buffered active-pull-down ECL circuit," *Proc. 1990 IEEE Bipolar Circuits and Technology Meeting*, pp. 132-35.

16. C. T. Chuang and D. D. Tang, "High-speed low-power ac-coupled complementary push-pull ECL circuit," *Dig. Tech. Papers*, pp. 117-18, 1991 Symp. VLSI Circuits.

17. C. T. Chuang, "NTL with complementary emitter-follower driver: a high-speed low-power push-pull logic circuit," *Dig. Tech. Papers*, pp. 93-94, 1990 Symp. VLSI Circuits.

18. C. M. Horwitz and M. D. Silver, "Complementary current-mirror logic," *IEEE J. Solid-State Circuits*, 23, 1, pp. 91-97, 1988.

High-Performance Standard Cell Library and Modeling Technique for Differential Advanced Bipolar Current Tree Logic

Hans J. Greub, *Member, IEEE*, John F. McDonald, *Member, IEEE*, Ted Creedon, and Tadanori Yamaguchi

Abstract —A high-performance standard cell library for the Tektronix advanced bipolar process GST1 has been developed. The library is targeted for the 250-MIPS Fast Reduced Instruction Set Computer (FRISC) project. The GST1 devices have a minimal emitter size of 0.6 μm $\times$ 2.4 μm and a maximum f_t of 15.5 GHz. By combining advanced bipolar technology and high-speed differential logic, gate propagation delays of 90 ps can be achieved at a power dissipation of 10 mW. The fastest buffers/inverters have a propagation delay of only 68 ps. A 32-b ALU partitioned into four slices can perform an addition in 3 ns using differential standard cells with improved emitter-follower outputs and fast differential I/O drivers. A modeling technique for high-speed differential current tree logic is introduced. The technique gives accurate timing information and models the transient behavior of current trees.

I. Introduction

THIS PAPER describes an experimental standard cell library for the advanced bipolar process GST1 under development at Tektronix [1]. The cell library was designed for the Fast Reduced Instruction Set Computer (FRISC) project [2], [3]. The 32-b processor is partitioned into circuits with a maximum complexity of 1000 current tree gates since the process yield is too low for a single-chip implementation. The standard cell library was optimized for speed to achieve a processor cycle time of 4 ns. Differential ECL logic is used to lower propagation, interconnect delays, and switching noise.

Because of the low targeted complexity, a higher power dissipation could be accepted in the speed versus power trade-off than in previously reported advanced bipolar libraries [4], [5]. The current trees are built out of the smallest GST1 devices with ECL output drivers to lower interconnect loading delays. The resulting standard cells are characterized by a high drive capability combined with a low fan-in load. Each cell is made available with three different output drivers. The drive capability of the ECL output driver circuits was improved to reduce the need for high-power gates.

The transient behavior of the high-speed, high-power cells is not dominated by interconnect capacitance as current starved ECL. Thus transients and glitches intrinsic to the structure of current trees are visible and cannot be simulated with a simple behavioral logic model. A structural modeling technique for high-speed current tree logic has been developed to improve the delay accuracy and to capture transients that could lead to circuit failure.

To reduce I/O delays, high-speed differential drivers and receivers with a low logic swing of $\pm$250 mV are provided besides standard single-ended I/O circuits. The single-ended drivers are ECL 10K compatible and have a voltage swing of 865 mV. Low I/O delays are crucial for the carry propagation in the FRISC data path, which had to be partitioned into four 8-b slices. In particular, the 32-b ALU is on the most critical delay path of the processor and is, therefore, examined in detail.

II. Advanced Bipolar Circuit Technology

A. Advanced Bipolar Process

The GST1 advanced bipolar n-p-n transistor devices are built with a self-aligned polysilicon emitter–base (E–B) process with a coupling base implant. This results in shallow emitter and base junction depths [6]. Fig. 1 shows the structure of n-p-n devices and polysilicon resistors and Fig. 2 shows a SEM device cross section. The 1-μm trench isolation reduces the collector-to-substrate (C–S) capacitance and increases device density. The smallest devices have an emitter stripe of 0.6 μm $\times$ 2.4 μm and can be placed on a dense 8-μm $\times$ 12-μm grid. A self-aligned titanium-silicide layer on top of the polysilicon layer for emitter and collector contacts reduces the sheet resistance to 1 Ω/$\square$ and thereby provides an additional layer for short interconnect. The same polysilicon layer without the silicide is used for resistors. Two gold metal layers with a 4-μm pitch are available for interconnect. The advanced bipolar n-p-n devices have a maximum f_t of 15.5 GHz [1]. Ring-oscillator delays of 55 ps per stage have been measured. Further, dual 4-b analog-to-digital converters with a performance of 1.5

Manuscript received September 20, 1990; revised January 15, 1991. This work was supported in part by DARPA under Contract DAAL03-90-G-0187.

H. J. Greub and J. F. McDonald are with the Center for Integrated Electronics, Rensselaer Polytechnic Institute, Troy, NY 12181.

T. Creedon was with the Electronic Systems Laboratory, Tektronix, Inc., Beaverton, OR 97007. He is now with Kestrel Technologies, Lake Oswego, OR 97035.

T. Yamaguchi is with Integrated Circuit Operation, Tektronix, Inc., Beaverton, OR 97007.

IEEE Log Number 9143201.

Reprinted from *IEEE Journal of Solid-State Circuits*, Vol. 26, No. 5, pp. 749–762, May 1991.

136

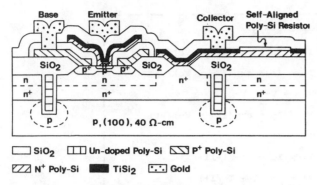

Fig. 1. Device structure.

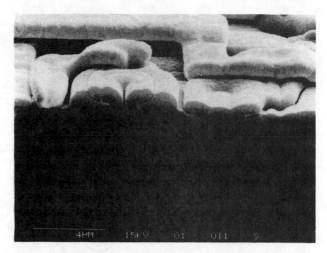

Fig. 2. SEM cross section of n-p-n device.

Gs/s have been demonstrated [7]. The dimensions and key parameters of the smallest GST1 device are summarized in Table I.

B. Current Switch

Fig. 3 shows the basic building block of current tree logic, the current switch (CSW). The input current into the common-emitter node is switched left or right depending upon the two base voltages. Using a simplified Ebers–Moll model for the bipolar transistor, the dc characteristics of a current switch buffer can be expressed in a closed form [8]. However, the effect of the parasitic emitter and base resistances should be included to obtain a good match. Unfortunately, the analysis does not yield a closed-form solution even if only the emitter resistance R_e is included.

Fig. 4 shows the delay of current switch buffers with and without 500 μm of interconnect capacitance as a function of switching current. The switching current was fixed at 400 μA since increasing the switching current any further would mainly increase power dissipation. The nominal switching current should be set below the optimal current to avoid operation in the region where delays increase rapidly with higher current and power. With a logic swing of ± 250 mV and a fan-out of 1, a current-mode logic (CML) buffer has a delay of 64 ps and an emitter-coupled logic (ECL) buffer with an emitter-follower current I_{ef} of 800 μA has a delay of 66 ps. The CML buffer dissipates only 2 mW but has a propagation delay sensitivity R_s of 400 Ω. The delay sensitivity R_s multiplied by the load capacitance C_l gives the incremental gate delay due to interconnect loading. A linear delay dependence is a good approximation for ECL or CML circuits [9]:

$$T_d = T_0 + \Delta t_d = T_0 + R_s \cdot C_l.$$

The ECL buffer has a power dissipation of 10 mW with an R_s of only 119 Ω. To save power, CML is used within standard cells where the interconnect length is short. The nominal voltage swing was fixed at ± 250 mV. This drives the current switch well beyond the points with maximum noise margin (gain = 1) and results in a voltage gain of 2.6 at 360 K. The voltage swing is determined by a trade-off between the delay sensitivity to capacitive loading and the desired voltage gain and noise margin of the logic. The

TABLE I
GST1 MINIMAL N-P-N DEVICE PARAMETERS

Size	8 μm $\times$ 12 μm
Emitter Size	0.6 μm $\times$ 2.4 μm
Current Gain h_{FE}	100
Emitter Resistance R_e	60 Ω
E–B Capacitance	6.7 fF
B–C Capacitance	7.5 fF
C–S Capacitance	9.0 fF
CutOff Frequency f_t, $V_{CB} = 0.85$ V, $T = 300$ K	12 GHz

current must be fully switched left or right at nominal input voltage levels, otherwise logic level degradation will occur if current switches are cascaded or stacked to build current trees. Fig. 5 shows the buffer delays as a function of logic swing V_l. The delays of CML buffers increase rapidly at high logic swings because the devices start to saturate.

The voltage swing is an important characteristic of a logic family since propagation and interconnect delays as well as switching noise increase with V_l. The ECL buffer delays with 500 μm of interconnect increase by 17% if the logic swing is changed from differential (± 250 mV) to single-ended levels (500 mV). If the interconnect capacitance C_l is large, the incremental gate delay Δt_d as a function of the logic swing V_l can be approximated by

$$\Delta t_{d_l} = K_l \cdot \frac{V_l \cdot C_l}{I_s} = K_l \cdot R_l \cdot C_l.$$

The constant K_i can be derived from a sensitivity analysis and depends upon the circuit configuration (ECL = 0.31 for $I_s = I_{ef}$, CML = 0.65) and the device technology [9]. To lower interconnect loading delays either the logic swing V_l must be lowered or the switching current I_s must be increased. Higher switching current implies, however, higher power dissipation. Hence, a low logic swing is the key to high-speed logic with low power dissipation! The switches must exhibit high gain and generate little switching noise to support low logic swings. Bipolar logic with a logic swing of only 250 mV has a big advantage over CMOS with a logic swing of 3–5 V in this respect.

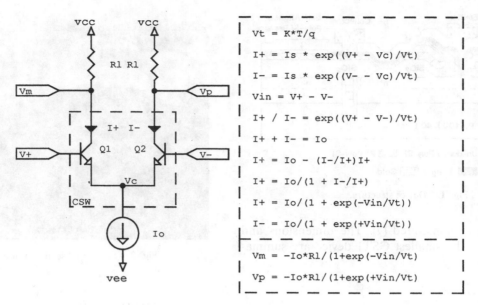

```
Vt = K*T/q

I+ = Is * exp((V+ - Vc)/Vt)

I- = Is * exp((V- - Vc)/Vt)

Vin = V+ - V-

I+ / I- = exp((V+ - V-)/Vt)

I+ + I- = Io

I+ = Io - (I-/I+)I+

I+ = Io/(1 + I-/I+)

I+ = Io/(1 + exp(-Vin/Vt))

I- = Io/(1 + exp(+Vin/Vt))

Vm = -Io*Rl/(1+exp(-Vin/Vt)

Vp = -Io*Rl/(1+exp(+Vin/Vt)
```

Fig. 3.   Bipolar current switch.

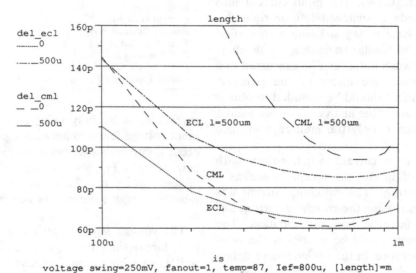

voltage swing=250mV, fanout=1, temp=87, Ief=800u, [length]=m

Fig. 4.   CML and ECL buffer delays versus switching current.

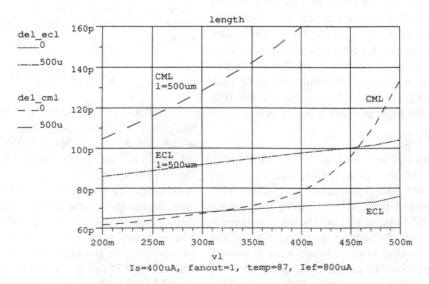

Is=400uA, fanout=1, temp=87, Ief=800uA

Fig. 5.   ECL and CML buffer delays versus logic swing.

138

Fig. 6 shows the switching noise (power) of differential ($V_l = \pm 250$ mV) and single-ended ECL buffers ($V_l = 500$ mV) for a positive and negative signal transition. The single-ended buffer generates considerable switching noise because of its higher logic swing and unbalanced load. Differential ECL logic produces only small switching transients and hence evades the delta-I noise problem common to high-speed logic.

C. Differential Current Tree Logic

The high speed and low switching noise of differential logic make it very attractive for bipolar [10], [11] or GaAs logic [12]. Differential GaAs logic is called source-coupled FET logic (SCFL). The high performance and efficient logic implementation of cascaded differential logic trees has led to the development of a similar CMOS logic family at IBM [13], called cascode voltage switch logic (CVSL).

Fig. 7 shows a differential AND/OR gate with three levels of series gating. An equivalent single-ended OR gate needs twice the voltage swing to obtain the same noise margin. Twice the voltage swing is sufficient, despite the fact that the generation of the reference voltages is sensitive to supply voltage drops on power rails, because doubling the voltage swing also doubles the maximum gain of the current switch. To obtain twice the voltage swing either the load resistance R_l or the switching current I_s must be increased by a factor of 2:

$$\text{gain}_{\max} = g_m \cdot R_l = \frac{I_s}{2 \cdot V_T} \cdot R_l = \frac{V_l}{2 \cdot V_T}.$$

The number of switches that can be stacked with standard ECL supply voltages is limited to three for ECL and to four for CML. The input signals for current switches at different levels must be offset by at least one base–emitter junction voltage $V_{BE0} \approx 0.85$ V to avoid saturating the bipolar devices. The nominal logic swing at each level is ± 250 mV.

Since a full current tree with three levels of current switches forms a 3-to-8 decoder, any Boolean function of three variables can be implemented in a single current tree by using collector dotting at the top level. An efficient logic implementation is obtained by eliminating current switches with both collectors connected together and by using collector dotting at level two for intermediate decoding states. A four-input multiplexer gate can also be implemented with a single current tree as shown in Fig. 8. By using feedback from the outputs of the current tree, data latches with any two-input gate at the input can be implemented as shown in Fig. 9. The feedback signals are taken from the top of the tree rather than from the output because of layout considerations.

Differential signals can be inverted with zero delay and power by exchanging the true and inverted signal pair connections at any input or output port. This reduces the number of cells in the standard cell library since dual gates like AND/OR are physically identical. Dual gates get mapped into the same cell during netlist generation.

Emitter followers are used to increase the drive capability of the gates and to shift output levels. A standard cell can drive only one output level because emitter followers tend to ring if they have to drive outputs at multiple levels. Since most of the power is dissipated in the emitter followers, each logic gate is available with three different strength drivers ($I_{ef} = 400$, 800, and 1200 μA).

The propagation delays of differential logic depend upon the path the current takes through the tree. The delay from inputs at a given level to the top of the tree can, therefore, depend upon input signals at higher levels. For example, in the differential AND gate shown in Fig. 7, the delay from the lowest level input depends upon whether the current flows through current switch $S2$ to q or through $S2$ and $S3$ to q or qb. The maximum propagation delays for a medium-power AND gate with a level-one output are 90 ps from level 1, 135 ps from level 2, and 180 ps from level 3. An equivalent three-input single-ended OR gate has a propagation delay of 95 ps for the OR output. The delay sensitivity R_s of a single-ended OR gate is 131 Ω for the rising edge and 257 Ω for the falling edge at a power dissipation of 10.5 mW. The medium-power differential AND/OR gate has a power dissipation of 10 mW and a delay sensitivity R_s of only 116 Ω. The differential gate has no decisive speed advantage over the single-ended gate at low loads, but the interconnect delay sensitivity of the differential gate is considerably lower and does not depend upon the signal transition.

While differential logic is faster than single-ended logic due to its low logic swing and can be efficiently implemented with current trees, there are also disadvantages. Twice as many signal interconnections must be routed. This increases the average interconnect length since the width of routing channels and feedthroughs doubles. Further, two emitter followers are needed for every gate, which increases power dissipation. However, differential logic requires no power for inverters or reference voltage generators and its sensitivity towards voltage drops on power rails is low.

Existing CAD tools can easily be modified to support three different signal offset levels, differential signal inversion, and checking for input-level violations that cause saturation in standard cells. However, the designer has to assign signals levels avoiding level violations and keeping the propagation delays on critical paths minimal. The standard cell router should support differential wiring. All differential wires should be routed right next to each other to obtain equal loading on differential nets. Parallel routing of differential signals further reduces crosstalk since crosstalk signals will couple almost equally to both wires and thereby produce mainly common-mode noise, which is largely rejected by current switches.

D. Emitter Followers and Buffers

Emitter followers have a tendency to ring, which leads to long settling times. Propagation delays are quite diffi-

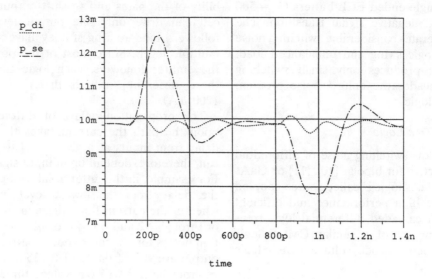

differential ECL (p_di): V1=250mV Is=400uA Ief=800uA Vcc-Vee=5V l=500um
single ended ECL (p_se): V1=500mV Is=400uA Ief=800uA Vcc-Vee=5V l=500um

Fig. 6.   Switching noise of single-ended and differential ECL buffers.

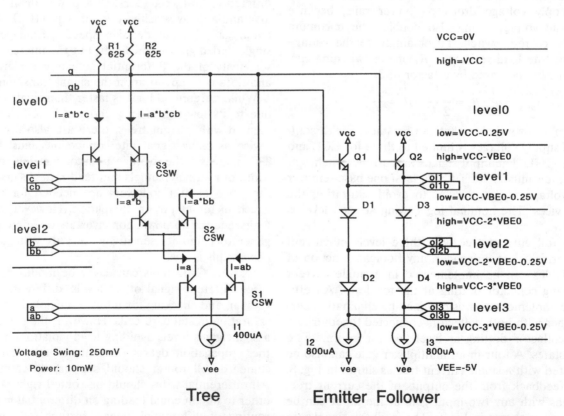

Tree   **Emitter Follower**

Fig. 7.   Differential three-input AND gate.

cult to model for input signals that arrive while the outputs have not yet settled. Therefore, emitter-follower and level-shifter configurations were developed to obtain faster settling times and lower interconnect delays.

The improved emitter followers have a damping resistor between the differential outputs to reduce ringing, as shown in Fig. 10. For level-2 and -3 emitter followers an f_t doubler circuit is used to reduce ringing and increase driving capability. The damping resistors cause a maximum loss of 20 mV in voltage swing since the current flowing through the emitter–base junction is higher for the transistor with a logic-high output signal. The buffers

140

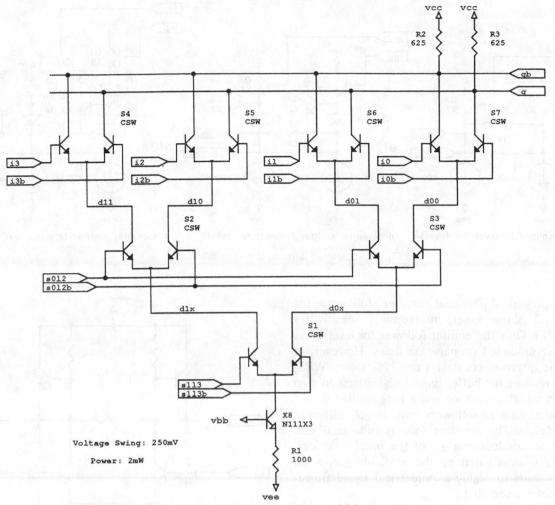

Fig. 8. Differential four-input multiplexor.

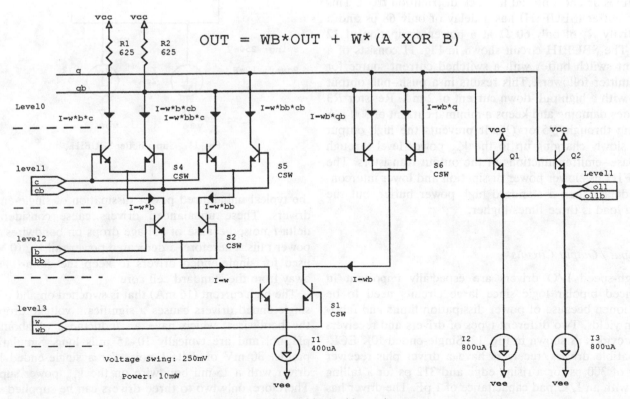

$$OUT = WB*OUT + W*(A\ XOR\ B)$$

Fig. 9. Differential latch with XOR inputs.

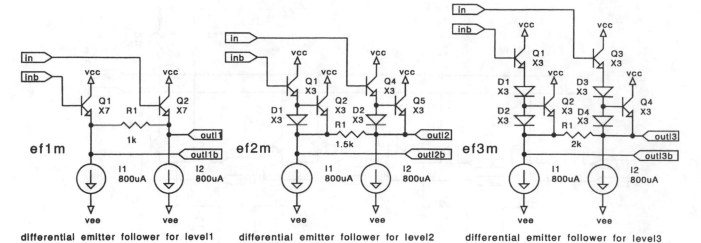

ef1m
differential emitter follower for level1

ef2m
differential emitter follower for level2

ef3m
differential emitter follower for level3

Fig. 10. Improved differential emitter followers.

with the improved differential emitter followers for level 2 and 3 show lower interconnect sensitivities (-21%, -23%). Only the emitter follower for level 1 has an 8 ps higher unloaded propagation delay. However, at high loads the interconnect delays are 17% lower. Without damping resistor the buffer has an underdamped step response with a high overshoot and a long settling time.

Highly loaded emitter followers have largely different rise and fall delays. The rise-time delay is quite small due to the high transconductance g_m of the bipolar devices. The fall time is dominated by the available pull-down current. This leads to highly asymmetrical signal transitions in current-starved ECL.

A special buffer is available for driving long interconnect lines as encountered in clock distribution trees. This super buffer (SBUF1H) has a delay of only 68 ps and a sensitivity R_s of only 60 Ω at a power dissipation of 12 mW. The SBUF1H circuit shown in Fig. 11 consists of a current switch buffer with a switched current source for the emitter followers. This results in a push–pull output stage with a high pull-down current of 2 mA. Resistor $R3$ provides damping and keeps a minimal current of 800 μA flowing through $Q5$ or $Q6$. It prevents the high output from slowly charging up to the V_{cc} power level through the base–emitter junctions of the output transistors. The SBUF1H has lower power dissipation and lower interconnect delays than a standard high power buffer, but the fan-in load is three times higher.

E. Input / Output Circuits

High-speed I/O drivers are especially important in advanced bipolar logic since large circuits need to be partitioned because of power dissipation limits and fabrication yields. Two different types of drivers and receivers are provided as shown in Fig. 12. Single-ended 10K ECL-compatible drivers/receivers have a driver plus receiver delay of 300 ps for a rising edge and 312 ps for a falling edge with an I/O pad capacitance of 1 pF. The driver has

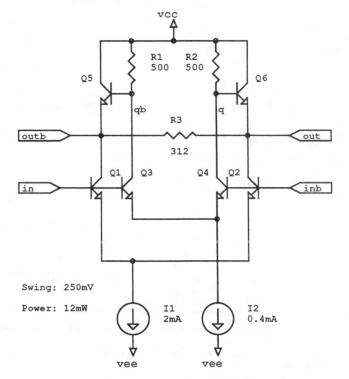

Swing: 250mV

Power: 12mW

Fig. 11. Super buffer SBUF1H.

the typical unbalanced power dissipation of single-ended drivers. These unbalanced drivers cause considerable delta-I noise because of voltage drops on bondwires and power rails. Therefore, a dedicated power rail V_{pp} (0 V) is used for single-ended drivers to keep the delta-I noise away from the standard cell core.

The high current (16 mA) that is switched on and off by single-ended drivers causes a significant voltage drop on the bondwires, which have an inductance of about 20 pH/mil and are typically 10–15 mils long. Simulations predict 30 mV of delta-I noise for a single-ended I/O driver with a 15-mil bondwire on the V_{pp} power supply. Therefore, only two to three drivers can be supplied with

142

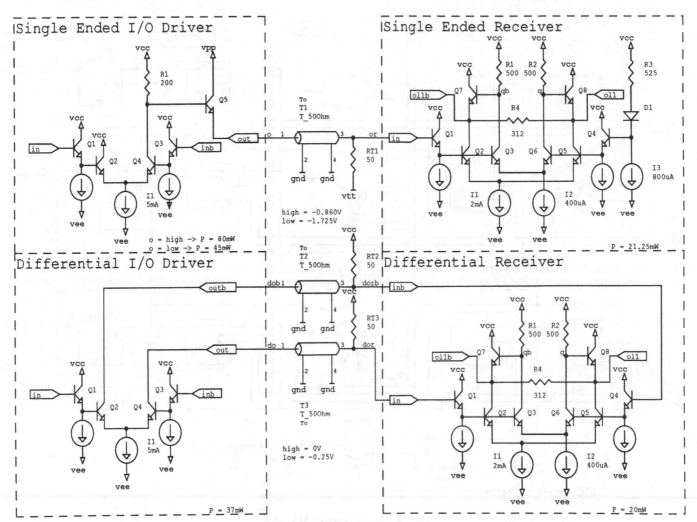

Fig. 12. I/O drivers and receivers.

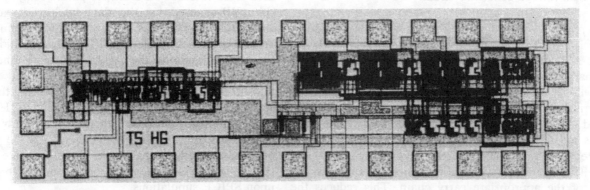

Fig. 13. Standard cell test circuit.

one V_{pp} power pad else the voltage drop on the bondwire and power rails can cause saturation of the output devices. By using tab bonding or a flip-chip die mount, the power supply inductance could be substantially reduced.

The second driver is a differential open-collector driver with a voltage swing of only ± 250 mV. The two transmission lines are terminated with 50-Ω resistors to V_{cc}. The differential driver plus receiver delay is only 220 ps with a pad capacitance of 1 pF. Differential drivers have the disadvantage of using up two I/O pads, however, since they have lower and balanced power dissipation fewer power pads per driver are required. The receivers use the same circuit configuration as the super buffer to drive the typically long interconnect from the chip periphery to the core. Fig. 13 shows a standard cell test chip with single-ended and differential I/O cells, a toggle flip-flop, and a

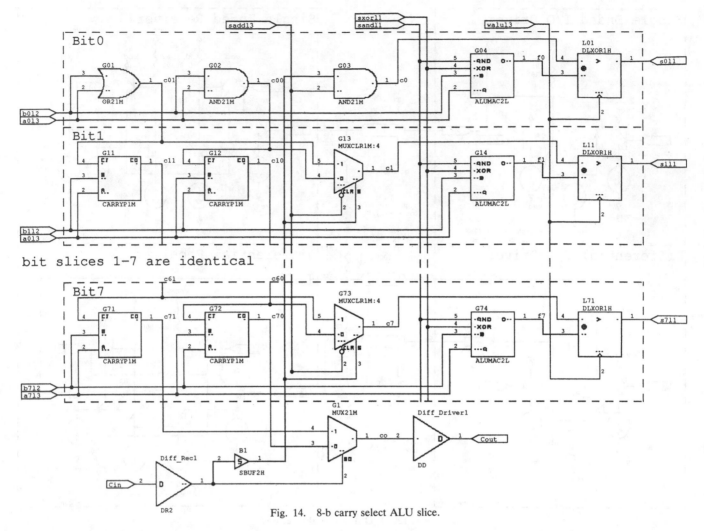

Fig. 14. 8-b carry select ALU slice.

structure to measure interconnect delays of ECL buffers, and a bias voltage generator circuit.

F. ALU Circuit

The 32-b ALU is on a critical path of FRISC since the data path had to be partitioned into four 8-b slices. The ALU has a 3-ns time slot to produce a 32-b result from the arrival of the level-2 operand. The carry select scheme is used to speed up carry propagation. The carry for each slice is calculated in two parallel carry chains, one for an assumed carry-in of one and the other for a carry-in of zero. The actual carry-in of the slice selects only the result of the appropriate carry chain. This reduces the fall-through time for the carry to a receiver, multiplexer, and driver delay if the carry chains have had time to settle. Further, the carry-in signal of the first slice must only be available on chip when the carry chains have settled. The carry select ALU can be implemented with only five current trees per bit as shown in Fig. 14.

The carry propagate gate CARRP1M and the multiplexer with clear MUXCLR1M are medium-power (10 mW) gates since they are on the critical path but drive only short interconnect. The programmable function gate ALUMAC2L generates the Boolean XOR, OR, or AND

function of the two operands. A low-power gate (6 mW) is used since it is not on a critical path. A high-power gate (14 mW) is used for the data latch with XOR inputs DLXOR1H since it has to drive long interconnect and is on a critical path. Differential I/O drivers and receivers are used to minimize the carry fall-through time.

The ALU can perform ADD, AND, OR, and XOR functions. A subtraction is performed by inverting the carry-in and operand *B*. The output latch DLXOR1H not only latches the result but also generates the sum by performing an XOR of the carry and the XOR of the two input operands generated by the ALUMAC2L gate. Table II shows worst-case propagation delays for a 32-b add based upon SPICE simulations.

The simulation results include an average on-chip interconnect length of 600 μm between the clusters of cells that form a bit slice. The carry-in receiver and carry-out driver are placed right next to each other to avoid routing the carry-in signal all the way across the chip. The four data-path slices are mounted right next to each other on a multichip module. The off-chip interconnect between slices is at most 8 mm long. The microtransmission lines on the multichip module have a polyimide dielectric with an ϵ_r of 3.2 resulting in a low interconnect delay of 6 ps/mm. The 32-b ADD delay is the silicon delay plus three chip-to-chip interconnect delays (3×48 ps) resulting in a worst-case delay of 2.79 ns. Assuming the clock skew can be controlled within ± 100 ps the ALU can perform a worst-case 32-b ADD within the allocated 3-ns time slot. By using carry select over a group of 3 and then 5 b the delay of the first slice could be reduced to 850 ps, resulting in a worst-case delay of only 2446 ps.

G. Standard Cell Library

The following list shows the differential standard cells used for the FRISC project. Many cells map into dual logic gates like AND and OR. Dual cells are available in the schematic library but are mapped onto the same cell during netlist expansion. Further, every input and output port of a differential cell can be inverted at no cost. Most cells are available with three different power levels ($\langle p \rangle$: low power = 6 mW, medium power = 10 mW, high power = 14 mW) and with three different output levels ($\langle l \rangle$: level 1, level 2, level 3). Master/slave latches dissipate an additional 2 mW. The library also includes a 32×8-b single-port memory cell for the register file of the processor [3], [14].

Combinational Cells

AND2$\langle p,l \rangle$	dual-input AND gate
XOR2$\langle p,l \rangle$	dual-input XOR gate
AND3$\langle p,l \rangle$	three-input AND gate
XOR3$\langle p,l \rangle$	three-input XOR/full adder
COMP$\langle p,l \rangle$	comparator with enable
ANDOR$\langle p,l \rangle$	ANDOR gate
ALUMAC$\langle pl \rangle$	programmable AND/XOR/OR gate
CARRYP$\langle p,l \rangle$	carry propagate gate

Multiplexer Cells

MUX2$\langle p,l \rangle$	dual-input multiplexer
MUXCLR$\langle p,l \rangle$	dual-input multiplexer with clear
MUX4$\langle p,l \rangle$	four-input multiplexer

Buffers and Level Shifters

BUF$\langle p,l \rangle$	buffer
SBUFH$\langle l \rangle$	super buffer
LS$\langle l \rangle$	level shifter

Storage Cells

SRF$\langle p \rangle \langle l \rangle$	set–reset flip-flop
DL$\langle p \rangle \langle l \rangle$	simple data latch
DLC$\langle p \rangle \langle l \rangle$	data latch with synchronous clear
DLAND$\langle p,l \rangle$	data latch with AND gate inputs
DLXOR$\langle p,l \rangle$	data latch with XOR gate inputs
DLMUX$\langle p,l \rangle$	data latch with MUX gate inputs
MSL$\langle p,l \rangle$	master/slave latch
MSAND$\langle p,l \rangle$	master/slave latch with AND gate inputs
MSMUX$\langle p,l \rangle$	master/slave latch with MUX gate inputs

I/O Cells

SEDS	single-ended driver ECL 10K
SER	single-ended receiver ECL 10K
DD	differential driver
DR	differential receiver

Special Cells

RF32$\times$8	32×8-b memory cell
SYNC	four-phase clock generator

TABLE II
WORST-CASE SILICON DELAYS FOR 32-b ADD

Chip/Circuit	Path	Delay
SLICE 1	A Op $\rightarrow$ C$_{out}$_sl1	1.196 ns
SLICE 2	C$_{out}$_sl1 $\rightarrow$ C$_{out}$_sl2	0.451 ns
SLICE 3	C$_{out}$_sl2 $\rightarrow$ C$_{out}$_sl3	0.451 ns
SLICE 4	C$_{out}$_sl3 $\rightarrow$ Sum_32	0.550 ns
32-b ALU	A Op $\rightarrow$ Sum_32	2.648 ns

TABLE III
TYPICAL LOGIC DELAYS

Current Switch Delay	45 ps
Level-1 Output	40 ps
Level-2 Output	45 ps
Level-3 Output	55 ps
Fan-out Penalty per Current Switch	5 ps

A simple delay model is given to the designer which allows quick evaluation of different circuit configurations. Table III gives approximate delay figures for the current switches and the emitter followers.

The fan-out penalty for a medium-power gate is only 5 ps. However, gates like the four-input multiplexer shown in Fig. 8 can have two current switches connected to the same cell input port. Only one of the current switches can, however, be active. A detailed delay model will be described in the following section. Table IV shows typical interconnect delays.

III. MODELING OF DIFFERENTIAL CURRENT TREE LOGIC

The design of high-speed digital circuits relies heavily on accurate circuit simulation to detect problems and predict performance before fabrication. For simulation at the circuit level, SPICE provides excellent results, however, its simulation speed is prohibitively slow for large digital circuits. Digital simulators use simple digital models and event-driven timing control [15], which allows simulation of very large circuits. However, most simulators are geared towards CMOS because of its dominance in the market place. As described in [16], single-ended bipolar transistor subcircuits can be mapped into equivalent logic gates that can be simulated on a conventional digital simulator. Another modeling technique transforms the transistor-level circuits into labeled weighed graphs [17] requiring a highly specialized simulation tool.

The model presented here uses a current switch, two or more transistors connected at a common-emitter node, as a model primitive, and allows the simulation of either differential or single-ended circuits. Only the mapping of transistors with a common-emitter node to a current switch is required to generate a simulation model from a device-level (SPICE) description. The structure of the tree models is the same as the physical structure. The current switch primitive can easily be added to digital simulators that support user extensions.

TABLE IV
INTERCONNECT DELAYS

Offset	Low-Power Gate 6 mW, $I_{ef} = 400\ \mu A$	Medium-Power Gate 10 mW, $I_{ef} = 800\ \mu A$	High-Power Gate 14 mW, $I_{ef} = 1.2$ mA
level 1	48 ps/mm	29 ps/mm	24 ps/mm
level 2	62 ps/mm	35 ps/mm	26 ps/mm
level 3	74 ps/mm	42 ps/mm	31 ps/mm

High-speed current tree logic has several properties that needed to be modeled. The signal path and therefore the delay from an input to the output can depend upon input signals at higher levels in the tree. Thus the propagation delays from a certain level input to the output can depend upon the state of other input signals. A simple behavioral model cannot capture these delay dependencies since the delays are calculated in most simulators before the simulation starts. However, they can easily be captured by a structural model based on current switches since the actual signal path through the tree is simulated. The current switch primitive can be described with a simple behavioral model that is easy to implement on most digital simulators. The output of a current tree can be independent of a signal at a lower level. For example, if the lowest input signal of an AND current tree is undefined, the output should still be low if any of the other input signals is low. This is very important because most digital simulators set all nodes initially to the undefined state, "x." Further, the treatment of glitches is important for latches. Clock signals generated by a gate with an unsymmetrical tree have short glitches at each differential signal transition. The two signals of a differential pair are both low or both high during the glitch. Latches must be able to capture valid data if the necessary setup and hold times have been observed, even if such glitches occur on clock lines.

A. Digital Current Switch Model

Asymmetrical current trees have nonsimultaneous output signal transitions and transient glitches. The output signals of a tree can be equal during transients even though no output change should occur according to the truth table. If such glitches occur on clock lines latched data can be disturbed. The current switch model must, therefore, handle differential and nondifferential input signal conditions as shown in Fig. 15.

The simulation of differential logic on the current switch level increases the number of nodes and elements in the netlist and will slow down simulations. However, the slower simulation time must be traded off against increased accuracy and the ability to capture transients which might affect circuit performance.

Simulation efficiency could be improved by representing each differential signal pair with a single digital node. The two differential current tree outputs (q, qb) can be converted into a single-ended signal with a differential-to-single-ended converter. This converter marks nondifferential outputs of the current tree with an unknown

Extended Truth Table

Inputs			Outputs	
COM	IN	INB	Q	QB
1	?	?	1	1
X	?	?	X	X
0	1	0	0	1
0	0	1	1	0
0	1	1	0	0
0	0	0	0	0
0	X	X	0	0
0	1	T	0	1
0	0	T	1	0
0	T	T	0	0
0	X	T	0	0

X=undefined, T=threshold, ?={0,1,X,T}

COM,Q,QB represent current levels (low=on)

Signal Strength: low > high

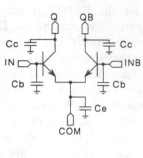

Fig. 15. Digital current switch model.

TABLE V
PARAMETERS OF CURRENT SWITCH MODEL

C_C	10 fF
C_B	40 fF
C_E	0 fF
T_d	40 ps + 400 $\Omega \cdot C_{load}$

logic signal. The current switch can be reduced for such a single-ended simulation of differential circuits to a four-terminal device. The single-ended modeling of differential signals reduces the number of nodes, but it requires invertor primitives for differential signal inversion. In the unlikely case that each gate output signal needs to be inverted, the total number of nodes will be larger due to the additional converters. The single-ended modeling of differential signals makes probing and saving of simulation results more efficient and allows the use of standard fault simulation and test-pattern generation software.

Negative logic is used to represent a current flowing in or out of the common-emitter node or the q and qb output nodes. Both outputs are active if the two inputs are equal and current is flowing into the common-emitter node. Latches would lose data just copied if the current switch connected to the clock signal would output no current for nondifferential inputs. However, if it outputs current on both sides no data is lost as long as the input current switch and the feedback current switch outputs agree. This will be the case as long as the data input is stable. The model will therefore correctly indicate a longer hold time and not loss of data. Sending current on both outputs for a nondifferential input condition reduces

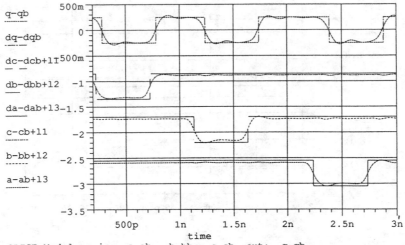

Fig. 16.   SPICE and digital XOR3T model.

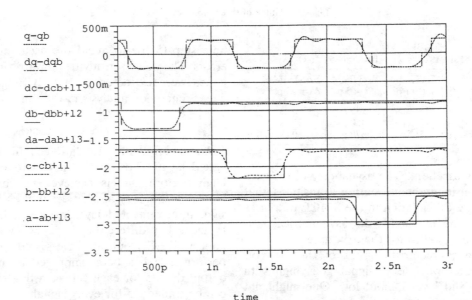

Fig. 17.   SPICE and digital AND3T model.

modeling pessimism in general since the tree output might not depend upon which way the current flows for a given set of input signal states.

The current switch model uses only a simple inertial delay model with capacitive load delays. If at least four signal strength values are available, the signal strength can be used to mark signal levels, which allows the detection of level violations causing saturation. The model includes capacitors to model input and output loading. It assumes differential input signals since the base capacitors are physically between base and emitter and can only be modeled as shown for differential input signals. However, most digital simulators support only capacitors connected to ground. The model parameters (Table V) depend on the operating conditions of the current switch like the switching current I_s, the voltage swing at the output, and V_{CB0} of the transistors. The capacitor C_B is 20% larger in an active current switch. This represents a

dynamic load change that is hard to simulate. However, it will be shown that a simple current switch model for all three levels can give excellent results since the dependencies are intrinsically small. In order to see the small differences the digital simulator would have to be run with a time step Δt below 5 ps. The match to a current switch at level 1 is most important since long and therefore critical signals are routed preferably on the topmost level to reduce propagation delays. The biggest simulation error is introduced by using a fixed C_B. Table V shows the parameters for the current switch with $I_s = 400\ \mu A$, $V_l = 250$ mV, and $V_{CB0} = 0.85$ V.

B. Modeling of Current Trees

Figs. 16 and 17 show a comparison of SPICE simulation results with a digital simulation of a three-input XOR tree and an AND/OR tree. The match for the symmetrical XOR

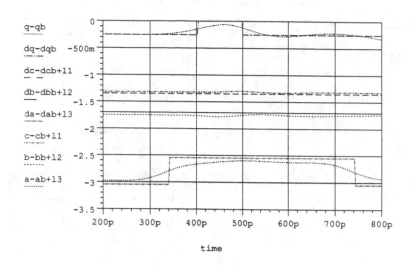

q-qb

dq-dqb

dc-dcb+l1

db-dbb+l2

da-dab+l3

c-cb+l1

b-bb+l2

a-ab+l3

SPICE Model: in: a,ab b,bb c,cb out: q,qb
DIGITAL Model: in: da,dab db,dbb dc,dcb out: dq,dqb

Fig. 18. Glitch of three-input AND tree.

TABLE VI
SPICE AND FASTSIM RESULTS WITH INTERCONNECT DELAYS

PATH	SPICE	FASTSIM $\Delta t = 1$ ps	FASTSIM $\Delta t = 5$ ps	FASTSIM $\Delta t = 25$ ps
A Op $\rightarrow C_{out}$	1196 ps	1185 ps	1185 ps	1300 ps
$C_{in} \rightarrow C_{out}$	451 ps	448 ps	450 ps	475 ps
$C_{in} \rightarrow$ Sum	550 ps	567 ps	570 ps	550 ps

(full adder) gate is excellent. The asymmetry of the AND tree results in nonsimultaneous output transitions that show up as rise-time degradation in the SPICE output as shown in Fig. 17. The match for the asymmetrical AND gate is clearly not as good as for the XOR gate.

Fig. 18 shows a characteristic glitch of the AND/OR current tree if the lowest level signal goes high with level-2 input high and level-1 input low. One might not expect an output transient for an AND gate with one input kept at a static low. However, the current has to propagate through two current switches after the level-3 input transition before the q output is pulled low again. SPICE shows only a signal level degradation, which can, however, lead to erroneous switching in a noisy environment. The digital model marks the glitch with nondifferential outputs allowing the detection of circuits that are sensitive to these transients. All current trees with unequal path delays from a particular current switch to the output show similar glitches. The three-input AND gate represents the worst case and should be used with caution on clock lines.

C. Modeling Accuracy Compared to SPICE

For verification of the accuracy of the standard cell models, the ALU slice shown in Fig. 14 was modeled with SPICE and FASTSIM, a digital simulator from Tektronix. Current switch and level-shifter primitives were added through its C-language interface. Table VI shows that excellent agreement (4% deviation) is possible. However, the digital simulator must be run with a sufficiently small time step (5 ps) to avoid the accumulation of rounding errors. The delay sensitivities towards interconnect capacitance were extracted from SPICE data by a six point linear regression analysis in the range 0–500 fF.

IV. CONCLUSION

An experimental standard cell library with a typical gate delay of 90 ps for a 10-mW gate has been developed. High performance is achieved by combining advanced bipolar technology and differential current tree logic design. Interconnect delay sensitivities have been reduced by using low differential logic swings of ± 250 mV and improved ECL output drivers and buffers. Power and performance has been improved by providing different output drivers for each cell such that speed versus power can be traded off for every signal.

I/O delays can be significantly reduced by using high-speed differential drivers with low logic swing of ± 250 mV and multichip packaging. Differential I/O circuits consume further less power and are balanced, thereby avoiding delta-I noise problems.

Modeling differential logic at the current switch level gives excellent delay accuracy and allows the designer to capture transients and glitches that could cause circuit failure. Further, the modeling approach can be implemented on conventional digital simulators.

ACKNOWLEDGMENT

The authors would like to thank Tektronix for their support of the FRISC project and the fabrication of the advanced bipolar circuits.

REFERENCES

[1] T. Yamaguchi et al., "Process and device performance of a high-speed double poly-Si bipolar technology using borosenic-poly process with coupling-base implant," IEEE Trans. Electron Devices, vol. 35, no. 8, pp. 1247–1255, Aug. 1988.

[2] H. J. Greub, J. F. McDonald, and T. Creedon, "Architecture of a 32bit fast reduced instruction set computer (FRISC) for implementation with advanced bipolar differential logic and wafer scale hybrid packaging," in *VLSI 87*. Amsterdam: North Holland, 1988, pp. 275–287.

[3] H. J. Greub, "FRISC—A fast reduced instruction set computer for implementation with advanced bipolar and hybrid wafer scale technology," Ph.D. dissertation, Rensselaer Polytech. Inst., Troy, NY, Dec. 1990.

[4] M. Franz *et al.*, "SH100E 10,000 gate ECL/15,00 gate array family with ECL/TTL I/O compatibility," in *Proc. IEEE CICC*, 1988.

[5] M. P. Depey *et al.*, "A 10K-gate 950MHz CML demonstrator circuit made with a 1-μm trench isolated bipolar silicon technology," *IEEE J. Solid-State Circuits*, vol. 24, no. 3, pp. 552–557, June 1989.

[6] H. K. Park *et al.*, "High-speed polysilicon emitter base bipolar transistor," *IEEE Electron Device Lett.*, vol. EDL-7, no. 12, pp. 658–660, Dec. 1986.

[7] V. E. Garuts, E. O. Traa, Y-C. S. Yu, and T. Yamaguchi, "A dual 4-bit, 1.5Gs/s analog to digital converter," presented at the Bipolar Circuit and Technology Meeting, Sept. 1988.

[8] R. L. Treadway, "DC analysis of current mode logic," *IEEE Circuits and Devices*, pp. 21–35, Mar. 1989.

[9] W. Fang, "Accurate analytical delay expressions for ECL and CML circuits and their applications to optimizing high-speed bipolar circuits," *IEEE J. Solid-State Circuits*, vol. 25, no. 2, pp. 572–583, Apr. 1990.

[10] H. Ichino *et al.*, "Super self-aligned technology (SST) and its applications," presented at the Bipolar Circuit and Technology Meeting, 1988.

[11] M. Suzuki, M. Hirata, and S. Konaka, "43-ps 5.2-GHz macrocell array LSI's," *IEEE J. Solid-State Circuits*, vol. 23, no. 5, pp. 1182–187, Oct. 1988.

[12] S. Shimizu *et al.*, "An ECL-compatible GaAs SCFL design method," *IEEE J. Solid-State Circuits*, vol. 25, no. 2, pp. 539–545, Apr. 1990.

[13] L. G. Heller and R. Griffin, "Cascode voltage switch logic: A differential CMOS logic family," in *ISSCC Dig. Tech. Papers*, Feb. 1984, pp. 16–17.

[14] H. J. Greub, J. F. McDonald, and T. Creedon, "Key components of the fast reduced instruction set computer (FRISC) employing advanced bipolar differential logic and wafer scale multichip packaging," presented at the Bipolar Circuit and Technology Meeting, 1988.

[15] E. G. Ulrich, "Exclusive simulation of activity in digital networks," *Commun. ACM*, vol. 12, no. 2, pp. 102–110, Feb. 1969.

[16] P. Kozak, A. K. Bose, and A. Gupta, "Design aids for the simulation of bipolar gate arrays," in *Proc. 20th DAC*, June 1983, pp. 286–292.

[17] I. N. Hajj and D. Saab, "Switch-level logic simulation of digital bipolar circuits," *IEEE Trans. Computer-Aided Design*, vol. CAD-6, no. 2, Mar. 1987.

High-Speed Low-Power ECL Circuit with AC-Coupled Self-Biased Dynamic Current Source and Active-Pull-Down Emitter-Follower Stage

C. T. Chuang, K. Chin, H. J. Shin, and P. F. Lu

Abstract—This paper presents an ECL circuit with ac-coupled self-biased dynamic current source and active-pull-down emitter-follower stage for low-power high-speed gate array applications. The circuit features an ac-coupled dynamic current source to improve the power–delay of the logic stage (current switch). A novel self-biasing scheme for the dynamic current source and the active-pull-down transistor with no additional devices and power in the biasing circuit is described. Based on a 0.8-μm double-poly self-aligned bipolar technology at a power consumption of 1.0 mW/gate, the circuit offers $1.62\times$ $(1.90\times)$ improvement in the speed (load driving capability) of a loaded gate compared with the conventional ECL circuit. The design considerations of the circuit are discussed.

I. INTRODUCTION

THE power dissipation of high-speed bipolar ECL circuits (Fig. 1(a)) has long been known to limit their VLSI applications. Recently, various active-pull-down schemes [1]–[4] have been actively pursued to reduce the power consumption and enhance the speed of ECL circuits. One example is the ac-coupled active-pull-down ECL (AC-APD-ECL) circuit (Fig. 1(b)) [1], [2] which utilizes a capacitor to couple a transient voltage pulse to the base of the pull-down n-p-n transistor. Implemented in a 0.8-μm 26-GHz double-poly self-aligned bipolar process [5], [6], this circuit has achieved unloaded gate delay of 23 ps at 2.1 mW/gate. With a slightly different biasing scheme, the circuit has also been known as the Turbo-ECL circuit and used in a 13 000-gate gate array [7]. While this scheme reduces the dc power consumption in the emitter-follower stage and improves the pull-down delay, there are still some constraints limiting the use of this circuit for low-power high-speed gate array applications. For example: 1) substantial power consumption is still needed in the current switch to achieve fast switching; 2) additional devices (R_3, R_4, and D_1 in Fig. 1(b); the values for R_3 and R_4 are typically in the range of several tens of kilohms to keep the biasing current at around several tens of microamperes so as to minimize the biasing power) are needed to implement the biasing circuit for the active-pull-down transistor, thus directly impacting the cell size; and 3) the power consumption for the biasing circuit is wasted.

This paper presents a new ECL circuit configuration featuring an ac-coupled dynamic current source and active-pull-down emitter-follower stage (AC-CS-APD-

Manuscript received November 26, 1991; revised March 16, 1992.
The authors are with IBM Research Division, Thomas J. Watson Research Center, Yorktown Heights, NY 10598.
IEEE Log Number 9201117.

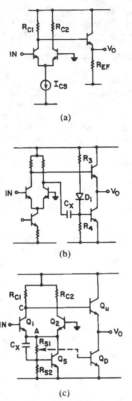

Fig. 1. Schematics of (a) conventional ECL circuit (the current source I_{CS} can be implemented either by a resistor or by a transistor current source), (b) AC-APD-ECL circuit, and (c) ECL circuit with ac-coupled self-biased dynamic current source and active-pull-down emitter-follower stage (AC-CS-APD-ECL).

ECL) for low-power high-speed gate array application. The dynamic current source provides a large dynamic current during the switching transient to improve the power–delay for the logic stage (current switch). In addition, the circuit utilizes a novel self-biasing scheme for the dynamic current source and the active-pull-down transistor and requires no additional devices and power in the biasing circuit, thus further improving the power–delay performance compared with other active-pull-down schemes.

II. CIRCUIT CONFIGURATION AND OPERATION

In the present scheme (Fig. 1(c)), resistors R_{S1} and R_{S2} form the current source as in a regular ECL circuit. An additional transistor Q_S and a capacitor C_X are added to

Reprinted from *IEEE Journal of Solid-State Circuits*, Vol. 27, No. 8, pp. 1207-1210, August 1992.

form a dynamic current source which provides a large dynamic current during the switching transient. Resistors R_{S1} and R_{S2} not only act as the steady-state current source but also form the biasing circuit for Q_S. Transistor Q_S is biased at near cut-in condition with essentially no dc current when the input is "low" and at lightly on condition when the input is "high." When the input rises to "high," the voltage at node A follows immediately once the input crosses the reference voltage. This transient signal is ac-coupled through C_X to turn on Q_S heavily, resulting in large dynamic current and hence fast pull-down of the collector node C. This scheme, therefore, improves the switching speed of the logic stage (current switch) substantially with no (or very little) additional power consumption. Notice that the difference in the base voltage of Q_S between the input "high" and input "low" states is slightly less than $(\Delta V \times R_{S2})/(2 \times (R_{S1} + R_{S2}))$ where ΔV is the logic swing of the circuit, since the voltage swing at node A is slightly less than half the logic swing due to the reason to be explained later. When the input falls from "high" to "low," the capacitor C_X helps to turn off Q_1, thus improving the switching speed as well.

The resistors R_{S1} and R_{S2} are also used to bias the active-pull-down transistor Q_D as shown in Fig. 1(c). The desired biasing level (and hence steady-state current) of Q_D is achieved by tapping the voltage at different points of R_{S1} as shown by the dashed arrow in Fig. 1(c). Notice that the whole biasing circuit (as well as its power consumption) for the active-pull-down transistor in Fig. 1(b) is completely eliminated. Thus, resistors R_{S1} and R_{S2} serve threefold functions: 1) as the steady-state current source for the current switch, 2) as the biasing circuit for the ac-coupled dynamic current source, and 3) as the biasing circuit for the ac-coupled active-pull-down transistor. It is worthwhile to point out that while a speedup capacitor can be connected between node A and V_{EE} to enhance the switching speed of the current switch, the present scheme requires a much smaller capacitor (hence easier to realize in integrated circuit technology) and is much more effective in generating a large dynamic switching current (rather than depending on the dynamic current through the speedup capacitor itself for fast switching, the capacitor C_X in the present scheme only has to provide the transient base overdrive to Q_S) as well as providing an active-pull-down emitter-follower stage.

Fig. 2 compares the net pull-up and pull-down currents of the present circuit with those for the conventional ECL circuit and the ac-coupled active-pull-down ECL (AC-APD-ECL) circuit during the switching transient at a power consumption of 2.0 mW/gate with FI/FO = 3 and C_L = 0.3 pF. These waveforms are based on a 0.8-μm double-poly self-aligned bipolar technology [5], [6] with pertinent measured device parameters listed in Table I. The device parameters and models have also been calibrated against the ECL circuit, the AC-APD-ECL circuit, as well as the static frequency divider performance. Clearly, the present circuit not only offers a much larger, sharper pull-down current but also improves the pull-up

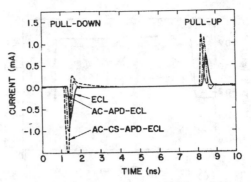

Fig. 2. Net pull-up and pull-down currents during the switching transient for the conventional ECL circuit, the AC-APD-ECL circuit, and the AC-CS-APD-ECL circuit (0.8-μm design rule, FI/FO = 3, C_L = 0.3 pF, 2.0 mW/gate).

TABLE I
TYPICAL TRANSISTOR PARAMETERS OF THE n-p-n TRANSISTOR AT 0.8- AND 0.5-μm DESIGN RULES

Design Rule	0.80 μm	0.50 μm
A_E (Wafer)	$0.4 \times 4.0~\mu m^2$	$0.25 \times 2.0~\mu m^2$
H_{FE}	100	120
Base Transit Time	6.0 ps	3.0 ps
C_{EB}	7.54 fF	5.96 fF
C_{CB}	3.80 fF	2.37 fF
C_{CS}	6.52 fF	4.82 fF
R_E	17.5 Ω	56 Ω
R_{BX}	164 Ω	200 Ω

current by quickly shutting off Q_1 and Q_D during the switching transient (the speed in this case is 127 ps for the conventional ECL circuit, 90 ps for the AC-APD-ECL circuit, and 79 ps for the present circuit). The pertinent waveforms during the switching transient are shown in Fig. 3. Note that in Fig. 3(a), the voltage scale for V_A (voltage at the common-emitter node of the switching transistors) is intentionally shifted down by one V_{BE} to show that it follows the input voltage (once the input crosses the reference voltage) with a swing less than half the logic voltage swing. This is because the V_{BE} drop of the input transistor (when the input is "high") is larger than the V_{BE} drop of the reference transistor (when the input is "low") due to the extra current from the dynamic current source Q_S as explained below. As can be seen in Fig. 3(b), the dynamic current source conducts only a very small current when the input is "low." As the input rises, the current through the dynamic current source increases with a current spike coming from the transient base overdrive through the capacitor coupling. The current through the dynamic current source, I_{QS}, can be seen to add to the steady-state switching current I_{RS1}, thus improving the switching speed of the logic stage. Finally, with the input at "high" and steady state reached, a large portion of the steady-state current for the current switch (which is substantially higher than the steady-state current with the input at "low") now flows through the dynamic current source Q_S.

151

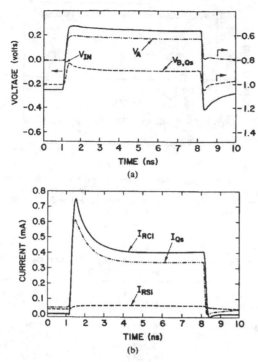

(a)

(b)

Fig. 3. Pertinent waveforms during the switching transient for the AC-CS-APD-ECL circuit. (a) Voltage waveforms at the input, the common-emitter node of the switching transistors (node A in Fig. 1(c)), and the base node of Q_S (see Fig. 1(c)). (b) The current flowing through the resistor current source I_{RS1}, the current flowing through the dynamic current source I_{Q_s}, and the current flowing through the load resistor I_{RC1}. (0.8-μm design rule, FI/FO = 3, $C_L = 0.3$ pF, 2.0 mW/gate.)

III. Circuit Performance

The power–delay characteristics for the conventional ECL circuit, the AC-APD-ECL, and the present circuit (AC-CS-APD-ECL) at 0.8-μm design rule (see Table I) are shown in Fig. 4. For the unloaded case (FI/FO = 1, Fig. 4(a)), the speed improvement at 1.0 mW/gate is about 30.2% with respect to ECL and about 13.5% with respect to AC-APD-ECL (46.9 ps for ECL, 37.8 ps for AC-APD-ECL, and 32.7 ps for the present circuit). For the loaded case (FI/FO = 3, $C_L = 0.3$ pF, Fig. 4(b)), speed improvements of 1.62× (129 versus 208 ps) at 1.0 mW/gate and 1.58× (66 versus 104 ps) at 3.0 mW/gate are obtained compared with the conventional ECL circuit. Also shown in Fig. 4(b) is the power–delay characteristic of the AC-APD-ECL circuit, which falls between the conventional ECL circuit and the present circuit (as expected!) with a speed improvement of about 1.43× at 1.0 mW/gate (145 ps for the AC-APD-ECL circuit versus 208 ps for the conventional ECL circuit). The scaled performances of the conventional ECL circuit and the present circuit at 0.5-μm design rules are shown in Fig. 5. For the unloaded case (FI/FO = 1, Fig. 5(a)), the present circuit achieves delays of 12.9 ps at 3.0 mW/gate and 23 ps at 1.0 mW/gate. For the loaded case (FI/FO = 3, $C_L = 0.3$ pF, Fig. 5(b)), the delays are 57 ps at 3.0 mW/gate and 107 ps at 1.0 mW/gate.

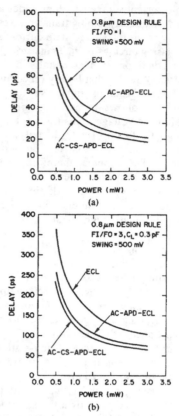

(a)

(b)

Fig. 4. (a) Unloaded (FI/FO = 1) power–delay characteristics, and (b) loaded (FI/FO = 3, $C_L = 0.3$ pF) power–delay characteristics for the conventional ECL circuit, the AC-APD-ECL circuit, and the AC-CS-APD-ECL circuit at 0.8-μm design rule.

The superior load driving capability of the circuit is illustrated in Fig. 6. At 0.8-μm design rule with FI/FO = 3 and 1.0 mW/gate, the circuit achieves a driving capability of 185 ps/pF, a 1.9× improvement over the 352 ps/pF for the conventional ECL circuit and a 1.18× improvement over the 218 ps/pF for the AC-APD-ECL circuit. The improvement over the AC-APD-ECL circuit results mainly from the fact that the power saved from eliminating the biasing circuit and from the use of the dynamic current source in the current switch can now be allocated to the output stage to drive the load.

IV. Discussion

As mentioned previously, the dynamic current source transistor Q_S is biased at near cut-in (lightly on) condition when the input is "low" ("high") and the difference in its base voltage between the input "high" and input "low" states is slightly less than $(\Delta V \times R_{S2})/(2 \times (R_{S1} + R_{S2}))$, where ΔV is the logic swing of the circuit. Since at very low power the current through Q_S for the input "high" state has to be kept low, Q_S may become completely cut off when the input is "low," thus limiting the dynamic overdrive current through the C_X coupling and impacting the switching speed. This design constraint can

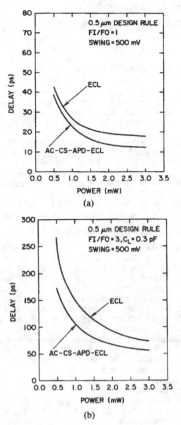

(a)

(b)

Fig. 5. (a) Unloaded (FI/FO = 1) power–delay characteristics, and (b) loaded (FI/FO = 3, C_L = 0.3 pF) power–delay characteristics for the conventional ECL circuit and the AC-CS-APD-ECL circuit at 0.5-μm design rule.

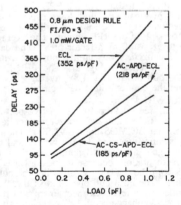

Fig. 6. Delays versus capacitive loading for the conventional ECL circuit, the AC-APD-ECL circuit, and the AC-CS-APD-ECL circuit (0.8-μm design rule, FI/FO = 3, 1.0 mW/gate).

be alleviated by allowing a larger base–emitter voltage across Q_S (via a larger $(R_{S2})/(R_{S1} + R_{S2})$ ratio) and adding a small series resistor in the emitter of Q_S to limit the current. This approach offers a smaller current ratio between the input "high" and "low" states for Q_S while still providing enough dynamic overdrive current through C_X coupling.

It is worthwhile to point out that the dynamic current source must be carefully designed to maintain sufficient margin for the output voltage low level. Since with the input at "high," a large portion of the steady-state current in the current switch flows through the dynamic current source Q_S, the output low level will be sensitive to the V_{BE} biasing and current gain of Q_S as well as the variation in supply voltage. This voltage level sensitivity can be alleviated by the addition of a small series resistor in the emitter of Q_S (as discussed in the previous paragraph) so that the steady-state current in Q_S is set by the resistor and therefore is much less sensitive to the variations in V_{BE} and current gain of Q_S.

V. Conclusion

In summary, we have described a new high-speed low-power ECL circuit with ac-coupled dynamic current source and active-pull-down emitter-follower stage. The ac-coupled dynamic current source improves the power–delay of the logic stage (current switch). A novel self-biasing scheme for the dynamic current source and the active-pull-down transistor with no additional devices/power in the biasing circuit was described. The superior power–delay performance and load driving capability of the circuit were illustrated and key aspects of the circuit discussed.

References

[1] K. Y. Toh et al., "A 23 ps/2.1 mW ECL gate," in ISSCC Dig. Tech. Papers, 1989, pp. 224–225.
[2] K. Y. Toh, C. T. Chuang, T. C. Chen, and J. D. Warnock, "A 23-ps/2.1-mW ECL gate with an ac-coupled active-pull-down emitter-follower stage," IEEE J. Solid-State Circuits, vol. 24, no. 5, pp. 1301–1306, Oct. 1989.
[3] H. Itoh, T. Saitoh, T. Yamada, M. Yamamoto, and A. Masaki, "Advanced ECL with new active pull-down emitter-followers," in Proc. 1988 IEEE Bipolar Circuits and Technology Meeting, pp. 23–25.
[4] C. T. Chuang and K. Chin, "A high-speed low-power charge-buffered active-pull-down ECL circuit," in Proc. 1990 IEEE Bipolar Circuits and Technology Meeting, pp. 132–135.
[5] T. C. Chen et al., "A submicron high performance bipolar technology," in Dig. Tech. Papers, Symp. VLSI Tech., 1989, pp. 87–88.
[6] T. C. Chen et al., "A submicrometer high-performance bipolar technology," IEEE Electron Device Lett., vol. 10, no. 8, pp. 364–366, Aug. 1989.
[7] B. Coy, A. Mai, and R. Yuen, "A 13,000 gate 3 layer metal bipolar gate array," in Proc. 1988 CICC, pp. 20.1.1–20.1.3.

Self-Biased Feedback-Controlled Pull-Down Emitter Follower for High-Speed Low-Power Bipolar Logic Circuits

Hyun J. Shin

IBM Thomas J. Watson Research Center, Yorktown Heights, NY 10598

I. INTRODUCTION

In high-performance bipolar/BiCMOS digital VLSI chips, basic circuits such as Emitter-Coupled Logic (ECL), Non-Threshold Logic (NTL), and memories need to be fast while consuming little power. The conventional passive-pull-down emitter follower (Fig. 1) usually used to drive the output in these circuits becomes less appropriate because its pull-down delay and dc power have a reciprocal relationship, although it is simple and allows versatile logic implementations through the collector-dotting and emitter-dotting. To solve the problem of the passive-pull-down emitter follower, various active-pull-down (APD) circuits have been proposed for low-power high-speed applications [1-3]. The popular technique used in these circuits (as shown in Fig. 2) is to set the steady-state pull-down current low through extra bias circuitry and utilize an out-of-phase signal from the logic stage through the ac (or capacitive) coupling to modulate the current in a push-pull manner during the input transition periods. This generally provides a large momentary sink current for pull-down that greatly enhances the speed and drive capability. These conventional APD circuits, however, suffer from increased complexity and gate area and lack of logic versatility. Because the technique requires a pair of in-phase and out-of-phase signals (X and $\bar{X}$) for the push-pull control, collector-dotting (even cascoding in some cases) is not allowed and delay in the preceeding logic stage increases from added loading on the out-of-phase signal node. Besides, emitter-dotting to implement wired-OR is not possible in these circuits. Furthermore, because the push-pull action is controlled by two inputs potentially skewed in time, the transient pull-down current may remain unnecessarily high after the output transition is finished or may go back to steady-state value prematurely before the output completes its transition. Although a circuit using only the in-phase input for the push-pull has been proposed to eliminate most of these drawbacks [4], its performance improvement is not substantial.

In this paper, a novel, self-biased, feedback-controlled, active-pull-down emitter follower that is simple and efficient, needs only one signal, allows the collector-dotting and emitter-dotting, and yet demonstrates a far superior performance is presented.

II. FEEDBACK-CONTROLLED PULL-DOWN CIRCUIT

The self-biased, feedback-controlled pull-down emitter follower (FPD-EF) for high-speed low-power bipolar logic is shown in Fig. 3. It consists of the emitter-follower transistor Q1, pull-down transistor Q2, current setting resistors R1 and R2, level-shifting/coupling diode D1, and optional clamping diode D2. The circuit is very effective because the biasing, inverting, level-shifting, and coupling functions are merged into a minimum number of devices connected in a simple topology. Compared to the conventional APD circuit in Fig. 2, this FPD-EF uses only the signal X from the preceeding logic stage and does not need capacitors (dc-coupled).

The steady-state bias current of FPD-EF is simply self-determined by the resistor values and supply voltages. As long as the circuit in the steady states ensures that Q1 and Q2 are not saturated for a given termination potential V_T, V_{CC}, and a signal swing, Q1 can be considered as a short circuit because it is ON passing all the Q2 current (Fig. 4). The voltage across R1 is, then, $V_{CC} - V_T - V_{BE2} - V_{D1}$ and the current in R1 ($\equiv I_{R1}$), which is sum of the bias current for the emitter follower ($\equiv I_{EF}$) and the current in D1-R2 ($\equiv I_{R2}$), becomes $(V_{CC} - V_T - V_{BE2} - V_{D1})/R1$. Because $I_{R2} = V_{BE2}/R2$, I_{EF} is easily obtained as $(V_{CC} - V_T - V_{BE2} - V_{D1})/R1 - V_{BE2}/R2$.

During the transient periods, the FPD-EF circuit operates in a push-pull mode that is efficiently controlled through the negative feedback, as depicted in Fig. 5. First, when X rises from low to high, the output OUT follows with a delay causing a momentary increase of v_{BE1} and corresponding surge of the current in Q1 ($\equiv i_{Q1}$). This current surge makes a large dip of Y that, in turn, is transferred to the base of Q2 through the shifting/coupling diode D1. (In case the dip is too large, the diode D2 may clamp Y and prevent Q1 from being saturated.) As a result, the current in Q2 ($\equiv i_{Q2}$) drops and the net output pull-up current i_{PU} increases. As the output rises to the high state, v_{BE1} and i_{Q1} return to the steady-state values and this push-pull action gradually comes to an end. Second, when X falls from high to low, because OUT follows with a delay, v_{BE1} is reduced for the transition interval. This successively turns Q1 off and raises Y. The surge of Y is coupled to the base of Q2, increasing i_{Q2} and the net output pull-down current i_{PD} substantially. Once the output falls to the low state, v_{BE1} and i_{Q1} return to the steady-state values and the push-pull operation stops. Therefore, the output transitions of

FPD-EF are much faster than those of the resistor-pull-down (RPD) circuit in Fig. 1 and well controlled by the input/output states in contrast to the conventional APD circuits. Due to a finite delay in the feedback loop, the FPD-EF output may ring with decay before it settles.

Because the FPD circuit needs one input X only, the collector-dotting and cascoding in the logic stage is well permitted. The emitter-dotting for FPD-EF is also allowed as illustrated in Fig. 6 by connecting the collectors of the emitter-follower transistors as well with an additional wire. This emitter-dotted circuit operates like the basic FPD circuit in Fig. 3 with (X1 + X2 + ... + Xn) replacing X. The performance of this emitter-dotted FPD-EF depends on the locality of the dots, because the parasitic capacitance of the collector node increases for wide-spread dots.

III. SIMULATION RESULTS

To demonstrate the advantages of the FPD-EF circuit, the FPD-ECL gates have been simulated and compared to the RPD-ECL gates based on a 0.5μm advanced npn Si-bipolar technology [5] with $V_{CC}/V_{EE}/V_T/V_R = 3.6/0/1.5/2.2$ V, at 65 °C. For the unloaded basic gates (fan-in (FI) = fan-out (FO) = 1 and C_L = 0 pF) with a 500mV signal swing, the gate delays are minimized when approximately 30% of the total gate power is allocated in the emitter-follower stages. The FPD-ECL delay is further optimized if 10% of the emitter-follower stage current (I_{R1}) is assigned to I_{R2}. As shown in Fig. 7, the FPD and RPD circuits have comparable delay-power characteristics; the average delay 30 ps at 1 mW/gate. The pull-down delay of FPD-ECL is a little slower than that of RPD-ECL because initial pull-down current from the high output state is smaller.

The performance leverage of the FPD circuit is remarkably demonstrated in Fig. 8 for a typical loaded case (FI = FO = 3 and C_L = 0.25 pF). In this case, the optimum power allocation for the emitter follower is about 50%. Note that the pull-down delay of FPD-ECL is essentially equal to the pull-up delay and, for 1 mW/gate, it is improved drastically from 247 ps (of RPD-ECL) to 92 ps by a factor of about 2.7. From the power savings point of view, the FPD-ECL circuit consumes nearly 2.8× less for the same delay. In terms of the average delay and power, the improvement is 2 times. Fig. 9 compares the output waveforms and net output currents of the FPD-ECL and RPD-ECL circuits. The FPD-ECL output rings a little and has a large and sharp pull-down current. The superior load drive capability of the FPD-EF circuit is illustrated in Fig. 10. At 1 mW/gate, the pull-down drive capability of FPD-EF is almost identical to the pull-up capability and enhanced by about 10× from 916 ps/pF of the RPD circuit. The average drive capability is improved to 98 ps/pF by a factor of 5. The delay-power characteristics of a localized, 3-way emitter-dot gate with a 600mV swing are similar to the ones in Fig. 8. The FPD-ECL gate maintains roughly 2.4× pull-down and 1.8× average speed leverage at 1 mW/gate over the emitter-dotted RPD circuit.

IV. CONCLUSION

The self-biased, feedback-controlled, active-pull-down emitter follower is a very efficient and superior circuit applicable to high-speed low-power bipolar/BiCMOS digital VLSIs. The circuit is effective because the biasing, inverting, level-shifting, and coupling functions are simply merged into a small number of devices and is versatile for logic implementations because it does not need any extra out-of-phase signal from the logic stage, allowing the collector-dotting and emitter-dotting. The push-pull operation of this novel circuit is precisely controlled by a feedback mechanism and results in a remarkable performance enhancement: 2.7× typical simulated pull-down speed and 10× drive capability over the conventional RPD-ECL.

ACKNOWLEDGMENT

The author would like to thank S. Dhong, C. Chuang, K. Toh, and M. Arienzo for their encouragement.

REFERENCES

[1] H. Itoh, et al., "Advanced ECL with new active pull-down emitter-followers," in *Proc. IEEE 1988 Bipolar Circuits & Tech. Meet.*, pp.23-25.
[2] K.-Y. Toh, et al., "A 23ps/2.1mW ECL gate," in *ISSCC Dig. Tech. Papers*, 1989, pp.224-225.
[3] C. T. Chuang, et al., "High-speed low-power darlington ECL circuit," in *Dig. Tech. Papers, Symp. VLSI Circuits*, 1992, pp.80-81.
[4] H. J. Shin, et al., "A high-speed low-power JFET pull-down ECL circuit," in *Proc. IEEE 1990 Bipolar Circuits & Tech. Meet.*, pp.136-139.
[5] G. Patton, et al., "63-75GHz f_T SiGe-base heterojunction bipolar technology," in *Dig. Tech. Papers, Symp. VLSI Tech.*, 1990, pp.49-50.

Reprinted from *Digest of Technical Papers, IEEE Symposium on VLSI Circuits*, pp. 27-28, May 1993.

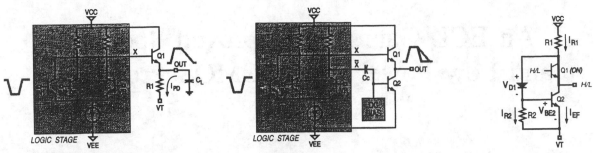

Fig. 1. Conventional Resistor-Pull-Down (RPD) Emitter Follower (EF) for ECL/NTL.

Fig. 2. Conventional Active-Pull-Down (APD) Emitter Follower.

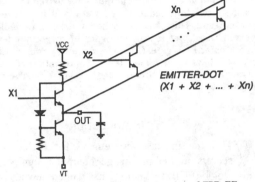

Fig. 4. Self-Biasing in FPD Emitter Follower.

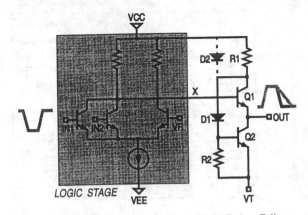

Fig. 3. Feedback-Controlled Pull-Down (FPD) Emitter Follower.

Fig. 6. Emitter-Dotting of FPD-EF.

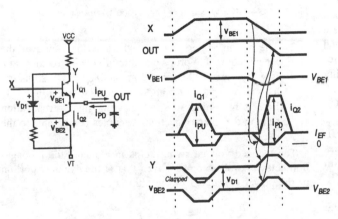

Fig. 5. Feedback-Controlled Push-Pull Operation of FPD-EF.

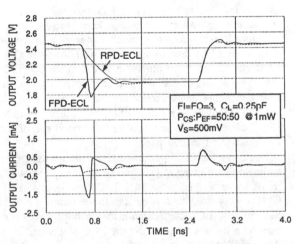

Fig. 9. Output Waveforms and Currents for Loaded ECL Gates.

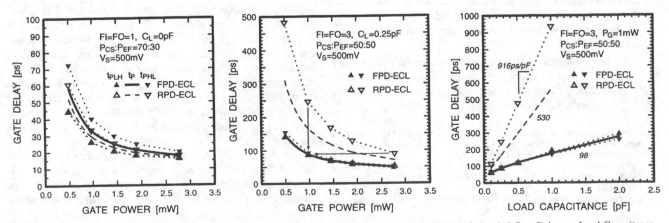

Fig. 7. Unloaded Basic ECL Gate Delay vs. Power Consumption per Gate.

Fig. 8. Loaded ECL (FI = FO = 3, C_L = 0.25 pF) Gate Delay vs. Power Consumption.

Fig. 10. Loaded Gate Delay vs. Load Capacitance.

An ECL Gate with Improved Speed and Low Power in a BiCMOS Process

Vojin G. Oklobdzija, *Fellow, IEEE*

Abstract—An emitter-coupled logic (ECL) gate exhibiting an improved speed-power product over the circuits presented in the past [1], [7]–[11] is described. The improvement is due to a combination of a push–pull output stage driven by a controlled current source, thus reducing the static and increasing the dynamic current. This circuit has better driving capabilities and improved speed, yet it uses an order of magnitude less power than a regular ECL gate. Due to its reduced power consumption, this gate allows for a higher level of integration of ECL logic. The realization of this circuit using a regular bipolar process is also possible.

I. Introduction

WITH the increase in demand for high-performance servers in the mid-range computer family, emitter-coupled logic (ECL) technology has gained new attention [3]. In part, this is due to its ability to achieve higher levels of integration, due to technological improvements, as well improvements in cooling techniques. Development of new ECL circuits characterized with much lower power consumption than regular ECL had its impact as well. Another area where ECL circuits are very attractive is the clock generation and distribution part of the chip. Recent advances in design for low power have demonstrated the use of a half-swing clocking scheme in reducing the power of the clock distribution network [5]. Such a network could make very good use of the ECL circuits treated in this paper.

Some useful ECL parameters and a discussion of ECL technology are given in [1], while a survey of recent advances in ECL circuits is given in [7]. Comparison of ECL with CMOS technology with respect to future trends is given in [2]. Several new ECL configurations have been developed since then with the aim of reducing the ECL power and increasing the speed [11], [12]. In all of the cases, it is essential to reduce the power consumed by the ECL circuit and yet maintain the driving capabilities and switching speed.

There are two inherent advantages of using ECL over CMOS:

1) Switching current in a bipolar differential pair is much faster than changing the voltage at the MOS transistor terminals. This advantage is applicable to the part that performs the logic operation.

Manuscript received February 16, 1995; revised May 25, 1995. This work was supported by Sun Microsystem Laboratories and in part by the Office of Research at the University of California, Davis.

The author is with Integration, Berkeley, CA, and the Department of Electrical and Computer Engineering, University of California, Davis, CA 95616 USA.

Publisher Item Identifier S 0018-9200(96)00110-2.

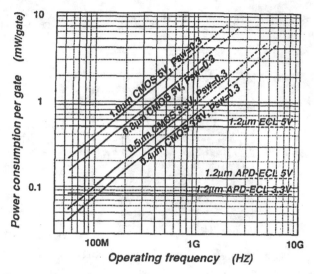

Fig. 1.   ECL-CMOS speed-power trade-off (as taken from [6]).

2) The output voltage swing, which is the swing of the signal required to propagate through the interconnection, is much smaller than in a comparable CMOS structure. This is of a particular importance today when the cycle time is becoming so short that the signal propagation time in the interconnections becomes comparable to the cycle.

The reduced voltage swing helps in reducing the propagation delay as well as power given that at the dynamic portion of the power consumed in the chip is

$$P_0 = f_0 \times C_L \times V_S^2.$$

At very high frequencies of operation (above 100–200 MHz), the power consumption of CMOS can be quite substantial. A dynamic portion of power (which increases linearly with frequency) starts to dominate. Some recently introduced CMOS processors are consuming amounts of power that are in the same order of ECL power [4]. Therefore it is a common misunderstanding to think of CMOS as low power technology in the high-performance domain. This observation is illustrated very well in a chart provided by Kuroda [6] and shown in Fig. 1.

The ECL power depends on the frequency of operation as well. However, this dependency is not as strong as it is in the case of CMOS. The reason for it is that the voltage swing in ECL is an order of magnitude smaller, thus making

Reprinted from *IEEE Journal of Solid-State Circuits*, Vol. 31, No. 1, pp. 77–83, January 1996.

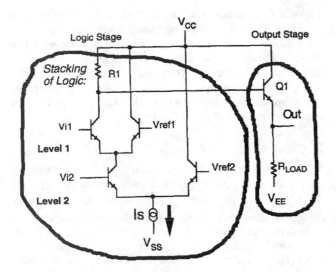

Fig. 2. Structure of a regular ECL gate.

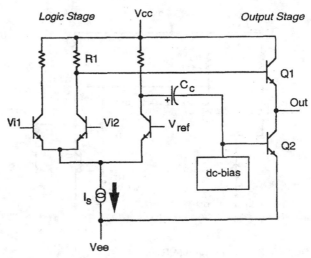

Fig. 3. APD-ECL from IBM [8].

this dependency factor ($C_L \times V_S^2$) two orders of magnitude smaller.

A very important part of an ECL gate is its driving stage because in the driving stage most of the power is used to provide sufficient signal driving capability. The speed of the gate depends on how much dynamic current the output drive stage provides.

In a regular ECL gate (Fig. 2), the switching speed is very much dependent on the value of the resistor R_{LOAD} in the path to V_{EE} [1]. This creates a direct relation between the static and dynamic current in the output stage. This resistor should be small, preferably, in order to rapidly discharge the load capacitance. However, when the output voltage is at the *high* level the current thorough R_{LOAD} can be substantial, thus amounting to substantial static power of the gate. Any increase in the dynamic current results in increased static power. To achieve rapid changes of the output voltage the output transistor has to be driven with the substantial current from the logic state (about five times due to degradation of β in that region). Consequently, any desired increase in speed is paid for by the substantial static power consumed by the gate.

Authors from IBM developed several ECL structures using active-pull down (APD) bipolar combination in the output APD-ECL [7]–[10]. The operation of IBM's APD-ECL [8] is illustrated in Fig. 3. A review of those techniques is given in [7] describes various structures in which a connection exists from the logic tree of the ECL structure to the "*pull-down*" output transistor which causes an extra amount of charge to be injected into the base of the "*pull-down*" transistor Q_2, thus speeding up the transition. Their circuit results in faster operation at lower power, compared to regular ECL.

Another refinement of the APD-ECL family developed by the same authors are the AC-APD-ECL [9] and the AC-CS-APD-ECL circuits [10], the latter being slightly faster [9]. Despite its remarkable performance, the AC-CS-APD-ECL has some shortcomings. The operating point of the output stage in AC-APD-ECL depends on the operating point of its logic stage. This dependency makes adjustments in the logic stage

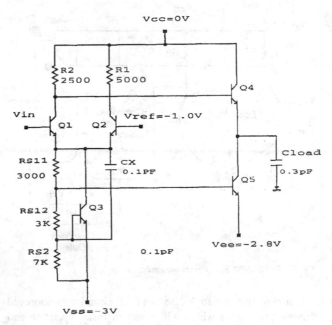

Fig. 4. AC-APD-ECL from IBM [10].

difficult, especially if one is trying to reduce power. Any adjustment of the operating point related to power involves both logic and output stages, which makes this process rather difficult. AC-CS-APD-ECL circuit is shown in Fig. 4. It is also difficult to cascode the logic higher than one level (cascoding is illustrated in Fig. 2). The emitter-dotting operation is not permitted either.

Recently, several other very effective ECL circuits were reported [11], [12], [15], most notably the circuit developed by H. J. Shin of IBM named FPD-EF-ECL [11]. This circuit has a remarkable simplicity (as shown in Fig. 5), while yielding several benefits such as the ability to permit emitter dotting. However, those schemes were not able to significantly reduce the static current in the output stage, which also affected the driving capability and in turn the speed of the gate.

157

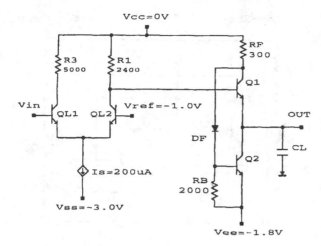

Fig. 5. FDP-EF-ECL from IBM [11].

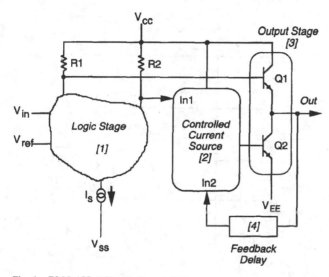

Fig. 6. FCCS-APD-ECL circuit operation.

Our objective was to develop an ECL circuit with improved driving capabilities which will result in faster operation and lower sensitivity to loading (increased fan-out capabilities). Yet, our objective was to design a gate with lower power compared to the previous work [7]–[12].

II. CIRCUIT OPERATION

We have developed a new circuit which utilizes feedback-controlled-current-source in an active-pull-down ECL configuration (FCCS-APD-ECL) [13], [14]. Conceptual operation of this circuit is summarized in Fig. 6, which illustrates the major parts of this circuit. The circuits consists of an ECL logic tree [1], controlled current source [2], output driver stage [3], (consisting of the transistors Q_1 and Q_2) and feedback stage [4]. The current in the current source is controlled by the state of the logic stage as well as the output state (via the feedback stage).

The controlled current source (CCS) injects current into the pull-down transistor in the output stage during transition

periods. In a steady-state, the current in Q_2 is reduced to its minimum value. Resulting circuit realization has a better power-delay product than ones previously reported [1], [7]–[11] because its output stage is operating in a truly push-pull mode of operation. The power consumed in the output stage is an order of magnitude lower than in its counterparts yet maintains comparable speed. The ability to drive large capacitive loads is substantially improved.

A. Operation

The circuit described in this paper uses a separate power supply in the output stage, electrically separating the logic stage from the output stage. There are several benefits resulting from this configuration

  a) The logic stage is decoupled from the noise which is usually generated in the output stage.

  b) The operating point of the output stage is not dependent on the logic stage, therefore the logic can be cascoded in more than one level, thus allowing for more complex logic operations within one ECL tree. This is not possible in AC-CS-APD-ECL [10].

  c) Reduced power supply voltage in the output stage results in an overall power reduction, given that the output stage is a main contributor to the power budget.

The current is injected into the base of Q_2 only when: the output is *high* and the voltage at the opposite end of the logic stage R_2 becomes *high*. It is obvious that this can happen only at the beginning of the transition from *high* to *low* when the current I_s in the logic stage is switched from R_2 to R_1 branch. Once the output reaches *low*, the current injected into the base of Q_2 will be cut-off (reduced).

During the logic *low* at the output, there is no current in Q_2 and the circuit is ready for the transition from *low* to *high*, which will occur when the current I_s is switched from R_1 to R_2 branch. This transition enables Q_1 to drive the output *high* while the voltage drop across R_2 will prevent the CCS from supplying the current to the base of Q_2. The behavior of the CCS can be summarized as:

  i) The current source is producing its maximal available current during the *high*-to-*low* transition of its output stage.

  ii) During the *low* to *high* transition, as well as in the steady state (low or high) the current source is in the reduced current mode providing just enough current to keep the transistor in slightly conducting mode.

High current injected by the CCS (2) into the output transistor Q_2 ensures sufficient current drive for the output transistor Q_2 to rapidly discharge the load capacitance C_L and drive the output node from *high* to *low* state. Output transistor Q_1 is driven directly from the logic stage. The current in the CCS is controlled by the logic stage (1) and it is in the opposite phase of the current driving the output transistor Q_1. This assures that the transistors Q_1 and Q_2 are driven in opposite phases thus eliminating direct current path from V_{cc} to V_{EE}.

In summary, *low* to *high* transition of the output stage is produced by driving Q_1. The output node will assume *high* value and after a reasonable delay through the feedback path

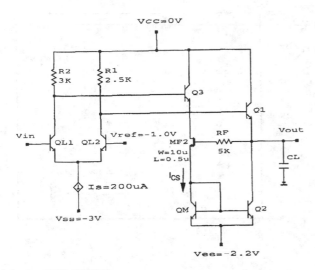

Fig. 7. FCCS-APD-ECL.

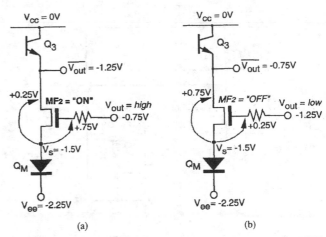

Fig. 8. Voltage on the terminals of the transistor MF$_2$ in ON (a) and OFF (b) state.

(4) one of the controls for the current path in CCS (2) will be enabled. This, however, will not result in the current in Q_2 because the signal from the logic stage (1) will keep CCS disabled.

During a *High* to *low* transition, the signal from the logic stage enables the current path in CCS (2). Given that the CCS is also enabled (by the output being *high*) the transistor Q_2 will be driven resulting in a fast transition from *high* to *low* at the output. After this transition the output value *low* will be passed with the same delay in the feedback stage (4) to the CCS (2). The CCS will be disabled, reducing the current driving Q_2. This sets the stage for a fast *low* to *high* transition because the output transistor Q_1 does not have to supply excess current to compensate for the current sunk by Q_2. The delay introduced by the feedback stage (4) is necessary to allow sufficient driving current during the *high* to *low* transition as well as to keep Q_2 off during *low* to *high* transition. This delay T_d is set to the maximal tolerable value of the signal edges, $T_d = \max(t_r, t_f)$.

The proposed circuit achieves full push-pull operation while being capable of maintaining reduced voltage swing (of 500 mV in our case). This is achieved by always keeping both transistors Q_1 and Q_2 slightly ON, thus the value of the resistor R_1 can adjust the ouput voltage swing.

III. IMPLEMENTATION

Adhering to the operation described in Section II. this gate can be realized in two ways, one being a purely bipolar realization. A BiCMOS implementation of FCCS-APD-ECL is shown in Fig. 7. This circuit uses a simple RC constant to achieve delay in (4). Feedback delay is achieved by a simple RC network consisting of a resistor R_F and a gate capacitance C_g of transistor MF$_2$.

The function of the transistor MF$_2$ is crucial to the operation of this circuit. The gate of MF$_2$ is connected to the output of the circuit and is driven by a full logic swing, therefore $V_g = [-0.75 \text{ V}, -1.25 \text{ V}]$. The drain of MF$_2$ is connected to the transistor Q_2 and it is driven by the exactly same

voltages, but in the opposite phase $V_D = [-1.25 \text{ V}, -0.75 \text{ V}]$. The source of the transistor MF2 is at the constant potential $V_S = -1.5$ V, which is one diode drop above the V_{EE} ($V_{EE} = -2.25$ V). Relative to the source, the voltage at the MF$_2$ terminals is shown in Fig. 8 (a) and (b).

The parameters of the transistor MF$_2$ are: $V_T = 0.4$ V and $L = 0.5\mu$ and $W = 10\mu$ which is achievable in a submicron technology. When the output is *high* the CCS is set for the *high-to-low* transition. When the output is *low* CCS is disabled and Q_2 is OFF enabling the true push-pull *low-to-high* transition of the output. A delay introduced by the feedback (4) consisting of R_F and C_g is necessary to assure a full transition from *high-to-low* before turning MF$_2$ and Q_2 OFF. This delay is adjusted to be equal to the worse case t_f time and it is not critical in the circuit operation.

The CCS (Fig. 7.) contains a "current mirror" consisting of transistors Q_M and Q_2. The current I_{cs} in the branch consisting of Q_3, MF$_2$ and Q_M, is "*mirrored*" into the "*pull-down*" output transistor Q_2. The maximal value of I_{cs} is limited by the maximal current transistors Q_3 and MF$_2$ can provide, which is determined by R_2 and β_{Q_3} as well as the channel resistance of MF$_2$, (RON). The maximal current in Q_2 is set by the ratio of the emitter areas of Q_2 and Q_M (mirror current). Though, I_{cs} is sensitive to the variations in V_{EE}, those variations are compensated due to the existence of a "negative feedback" from the output to the gate of MF$_2$. Any increase in I_{cs} (due to the variation of V_{EE}) is thus "reflected" in Q_2, thus lowering V_{out} and reducing I_{cs} (via R_F, MF$_2$ feedback). The value of V_{EE} is set to three "*diode drops*" ($V_D = 0.75$ V) and therefore it should not be difficult to keep relatively stable.

A bipolar realization of FCCS-APD-ECL is shown in Fig. 9. This allows using a simple bipolar process rather than BiC-MOS. However, this version has slightly inferior speed compared to the BiCMOS realization because the voltage swing in the logic stage (1) needed to drive the transistor Q_F to its cut off is larger, therefore degrading the speed of the logic part. Nevertheless, the difference is not considerable.

The output voltage levels across the transistor Q_F are determined by the voltage drop across R_1. The supply voltage

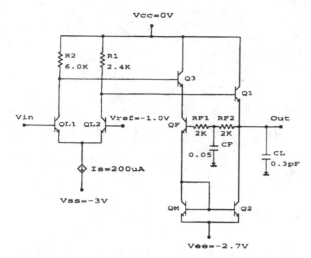

Fig. 9. Bipolar FCCS-APD-ECL.

TABLE I
TYPICAL TRANSISTOR PARAMETERS USED FOR SIMULATION

hfe	100
Tf	6.0pS
Cje	7.54fF
Cjc	3.8fF
Cjs	6.52fF
Re	17.5 ohm
Rb	164 ohm
fT	20GHz

V_{EE} is carefully set to be $V_{EE} = -2.7$ V. This brings the voltage at the emitter of Q_F to be $V_{Fe} = -1.95$ V. The voltage swing across R_1 is 1.2 V, making V_c of Q_F to be $V_c = -1.95$ V enough to turn Q_F off by driving its V_{ce} to zero.

IV. RESULTS AND COMPARISON

The performance of FCCS-APD-ECL circuit is assessed by simulation using transistor models based on a sub-micron process. For comparison purposes, the simulation parameters shown in Table I and published in [10] were used. The sub-micron transistor models used for simulation in [9]–[11] were not available. Therefore those circuits were re-simulated with our models in order to perform a relative comparison. The relative difference between our circuit and the circuits presented in [9]–[11] should be preserved, though the results are not an accurate representation of the real speed of those circuits.

FCCS-APD-ECL operates faster than the ones reported [7]–[10]. The power-delay product for FCCS-APD-ECL under no-load conditions is shown in Fig. 10(a) and under 0.3 pF load in Fig. 10(b).

FCCS-APD-ECL gate has also better driving capabilities as shown in Fig. 10(b). At 2 mW per gate and at an output load of 0.3 pF, the delay of the FCCS-APD-ECL is 101 pS versus 148 pS of FPD-EF-ECL, 153 pS of AC-CS-APD-ECL, and 166

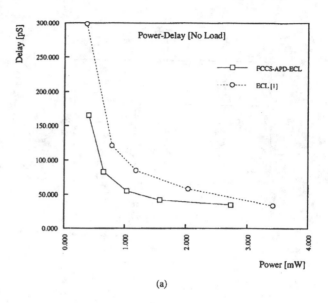

(a)

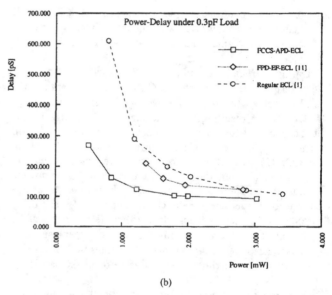

(b)

Fig. 10. Comparative power-delay characteristics: FCCS-APD-ECL, FPD-EF-ECL [11], regular ECL (a) Power-delay characteristic: 0.0 pF Load and (b) power-delay characteristic: 0.3 pF load.

pS of regular ECL as used in [1]. Comparison with AC-CS-APD-ECL (of IBM) was difficult because this circuit requires extensive tuning. However our circuit compares favorably for all of the simulated points, as shown in Fig. 10.

All the delay measurements reported are the $T_{delay} = \max[t_{rise}, t_{fall}]$. This is different from a more universal definition of $T_{delay} = \text{Average}[t_{rise}, t_{fall}]$. The reason for the new definition is that the t_{rise} and t_{fall} times in a regular ECL circuit are asymmetric with the delay dominated by $t_{fall} > t_{rise}$. The critical improvement is really in the t_{fall} time and we have chosen to emphasize this improvement by taking the worse delay, rather than diminish its effect by taking an average value.

Power consumption of the new gate is lower as compared to previously reported ones. The difference in power for the same delay can be observed in Figs. 10(a) and (b). For example, a

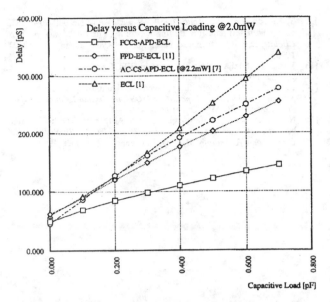

Fig. 11. Delay versus capacitive loading: ECL [1], AC-CS-APD-ECL [10], FPD-EF-ECL [11], FCCS-APD-ECL.

TABLE II
DRIVING CAPABILITY AT 2.0 mW PER GATE

FCCS-APD-ECL	133pS/pF
FPD-EF-ECL [8]	276pS/pF
AC-CS-APD-ECL [7]	328pS/pF
ECL [1]	403pS/pF

200 pS speed for 0.3 pF load is achieved with 0.7 mW of power versus 1.42 mW of FPD-EF-ECL and 1.7 mW for regular ECL. It is also significant to note that the power consumption of the new gate comes for the most part from its logic stage. The output stage has order of magnitude lower power consumption due to its complementary nature.

The sensitivity to capacitive load exhibited by the FCCS-APD-ECL gate as compared to [1], [10], and [11] is shown in Fig. 11.

The superior load driving capability of FCCS-APD-ECL is visible from Fig. 11, and the specific data is shown in Table II for the 2.0 mW per gate power.

The advantage of the new circuit is especially visible from the output current wave forms. Typical simulated voltage, current, and output power wave forms for our circuit (a), compared to regular ECL [1] (b), are shown in Fig. 12. During the steady state intervals the current in FCCS-APD-ECL circuit is very small. During the transitions this circuit exhibit large current peaks which are responsible for its excellent driving ability. The power generated in the output stages is also shown. When averaged over the signal period the power consumption is small which explains its CMOS-like power behavior.

V. CONCLUSION

The new ECL circuit (FCCS-APD-ECL) shows power advantage over the ones previously reported [7]–[11], yet main-

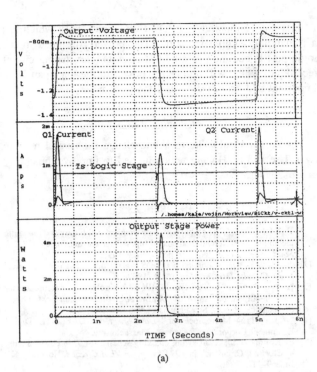

(a)

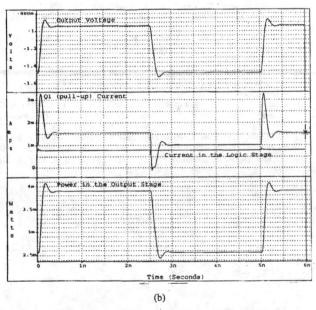

(b)

Fig. 12. Simulated output voltage, current and power response for the square pulse at the input, for FCCS-APD-ECL versus ECL [1] using 0.3 pF load (a) FCCS-APD-ECL (93 pS delay @ 3.0 mW per gate) and (b) ECL [1] (93 pS delay @5.6 mW per gate).

tains its speed. This is achieved through a combination of controlled current sources and careful tuning of signal levels and their timing relationships, resulting in increased dynamic and reduced static current in the output stage. Therefore, the major part of the power budget is consumed in the logic stage rather than the output stage, which sets this circuit apart from those previously developed. The ability of this circuit to handle higher capacitive loads is particularly important and offsets its inability for *wired-OR*. This is the main disadvantage

161

of this gate compared to FPD-EF-ECL gate [11], which has a remarkably simple structure. The power consumed in the logic part is common to all ECL circuits. Developing circuit techniques which will reduce the power in the logic part is an important and promising area for future work.

ACKNOWLEDGMENT

The author would like to acknowledge the careful reading by J. Lee and H. Hwang which helped to improve the clarity of this paper.

REFERENCES

[1] H. J. Greub et al., "High-performance standard cell library and modeling technique for differential advanced bipolar current tree logic," *IEEE J. Solid State Circuits*, vol. 26, no. 5, May 1991.

[2] A. Masaki, "Deep-submicron CMOS warms up to High-speed logic," *IEEE Circuits Devices Mag.*, Nov. 1992.

[3] N. Jouppi et al., "A 300 MHz 115 W 32 b bipolar ECL microprocessor with on-chip cache," in *IEEE 40th Int. Solid-State Circuit Conf.*, Digest of Technical Papers, San Francisco, Feb. 24–26, 1993.

[4] "Alpha 21164 Microprocessor," *Data Sheet*, EC-QAEPA-TE, Digital Equipment Corporation, Maynard, Massachusetts.

[5] H. Kojima et al., "Half-swing clocking scheme for 75% Power Saving in Clocking Circuitry," in *1994 Symp. on VLSI Circuits*, Honolulu, HI, June 9–11, 1994.

[6] T. Kuroda, "BiCMOS-Where is the Beef," Panel Discussion, in *1993 Symposium on VLSI Circuits*, Kyoto, Japan, May 19–21, 1993.

[7] C. T. Chuang, "Advanced bipolar circuits," *IEEE Circuits Devices Mag.*, Nov. 1992.

[8] K. Toh et al., "A 23-pS/2.1-mW ECL gate with an AC-coupled active pull-down emitter-follower stage," *IEEE J. Solid-State Circuits*, vol. 24, no. 5, Oct. 1989.

[9] C. T. Chuang et al., "High-speed low-power AC-coupled complementary push-pull ECL circuit," *IEEE J. Solid-State Circuits*, vol. 27, no. 4, Apr. 1992.

[10] ——, "High-speed low-power ECL circuit with AC-coupled self-biased dynamic current source and active-pull-down emitter-follower Stage," *IEEE J. Solid-State Circ.*, vol. 27, no. 8, Aug. 1992.

[11] H. J. Shin, "Self-biased feedback-controlled pull-down emitter follower for high-speed low-power bipolar logic circuits," in *1993 Symp. on VLSI Circuits*, Digest of Technical Papers, p. 27, Kyoto, Japan, May 19–21, 1993.

[12] T. Kuroda et al., "Capacitor-free level-sensitive active pull-down ECL circuit with self-adjusting driving capability," in *1993 Symp. on VLSI Circuits*, Digest of Technical Papers, p. 29, Kyoto, Japan, May 19–21, 1993.

[13] V. G. Oklobdzija, "New ECL gate in BiFET process," *Electron. Lett.*, vol. 29, no. 23, Nov. 1993.

[14] ——, "An ECL gate with improved speed and low power in a BiFET process," in *1994 IEEE Bipolar/BiCMOS Circuits and Technology Meeting*, Minneapolis, MN, Oct. 1994.

[15] N. Jouppi et al., "A fully-compensated APD circuit with 10:1 ratio between active and inactive current," in *1994 IEEE Bipolar/BiCMOS Circuits and Technology Meeting*, Minneapolis, MN, Oct. 1994.

Merged CMOS/Bipolar Current Switch Logic (MCSL)

WOLFGANG HEIMSCH, ASSOCIATE MEMBER, IEEE, BIRGIT HOFFMANN, ROLAND KREBS, ERNST G. MÜLLNER, BRUNO PFÄFFEL, AND KLAUS ZIEMANN

Abstract—A merged CMOS/bipolar current switch logic (MCSL) is presented. CMOS/ECL level conversion and logical operation are realized simultaneously. This circuit technique allows a supply voltage reduction to 3.3 V. A carry delay time of 150 ps/bit for a 4-bit BiCMOS full adder was measured. This is about five times faster than an optimized CMOS adder.

I. INTRODUCTION

IN RECENT YEARS, combined bipolar/CMOS technologies were applied to shorten the delay times through critical paths of complex circuits, since BiCMOS combines the high switching speed of bipolar ECL circuits with the low area, low power consumption, and ease of design of CMOS. One remarkable application of BiCMOS technology was a carry propagation circuit in a 32-bit arithmetic unit of a digital processing circuit [1], in which the system speed was increased by more than a factor of 2. In that case bipolar transistors were used as sense and buffer devices. To obtain higher speed improvement factors (bipolar switching time t_{bip}/MOS switching time t_{MOS}) for the whole circuit, the voltage swing ΔV of the logic circuits has to be minimized as shown in (1):

$$t = \frac{C \cdot \Delta U}{I}$$

$$\frac{t_{bip}}{t_{MOS}} = \frac{(C_{wire} + C_{bip}) \Delta V_{bip}}{(C_{wire} + C_{MOS}) \Delta V_{MOS}}. \qquad (1)$$

It can be seen that for a given current, t_{bip}/t_{MOS} depends directly on the ratio $(\Delta V_{bip}/\Delta V_{MOS})$ of the voltage swings at any node capacitance C. C is composed of the wiring capacitance C_{wire} and the input capacitance of any transistor at this node, C_{bip} or C_{MOS}. Therefore, in complex structures, the time-critical path which determines the overall switching speed has to be implemented in a logic with a small voltage swing, e.g. ECL, using bipolar transistors (Fig. 1) while the remaining larger part can be realized

Manuscript received April 3, 1989; revised June 7, 1989.
The authors are with Corporate Research and Development, Siemens AG, 8000 Munich 83, West Germany.
IEEE Log Number 8929852.

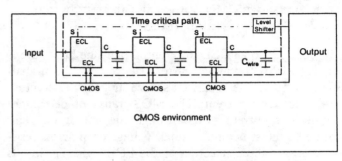

Fig. 1. Concept for using MCSL.

in CMOS logic. But then the problem of level conversion between the CMOS environment and the ECL parts arises. To avoid time, area, and power consumption at the interface between CMOS and ECL, a logic which performs level conversion and logical operation at the same time is desired. This logic must operate with a large input voltage swing to communicate with a CMOS environment and a small output voltage swing for the ECL logic in the time-critical path. The new merged CMOS/bipolar current switch logic (MCSL) closes this gap between CMOS logic with its large input and large output voltage swing and ECL logic with its small input and small output voltage swing.

MCSL is introduced in Section II of this paper and is applied to a BiCMOS ripple adder in Section III. The adder exhibits bipolar performance without any additional circuits for level conversion at the input. In contrast to a pure bipolar solution, area and power consumption are reduced by 50 percent for each bit. The advantage in area consumption results from the smaller number of transistors and the smaller spacing of the MOS part. Only 28 transistors in comparison to 48 transistors (considering emitter followers and level shifters) are necessary for each bit. The advantage in power consumption results from the smaller number of current paths. Only two gate and four emitter-follower currents instead of four gate and eight emitter-follower currents are necessary. In comparison to a pure CMOS adder cell, the MCSL adder exhibits a speed improvement by a factor of 5 while the area consumption is only three times larger.

Reprinted from *IEEE Journal of Solid-State Circuits*, Vol. 24, No. 5, pp. 1307-1311, October 1989.

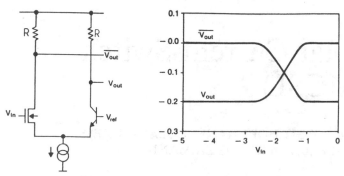

Fig. 2. Merged CMOS/bipolar current switch with transfer characteristic.

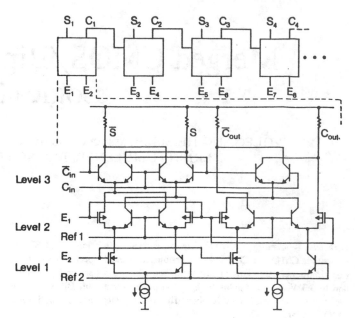

Fig. 3. MCSL ripple adder with full-adder cell.

II. MERGED CMOS/BIPOLAR CURRENT SWITCH LOGIC (MCSL)

In contrast to an ECL differential pair, the basic gate is a merged CMOS/bipolar current switch (Fig. 2). In analogy to an ECL gate, a current source supplies the structure with a constant current. The MOS transistor determines whether the current flows through the left or the right branch of the structure. A small voltage drop across resistor R allows the connection to the ECL logic. The inputs are controlled directly by a CMOS voltage swing. Therefore the structure is able to operate in a full CMOS input environment. Hence a level conversion and a logical operation are performed at the same time, thus reducing propagation delay time, area consumption, and power consumption. Compared to ECL the reference voltage is not critical because it must be centered with respect to the large input signal. The reference voltage is chosen such that the supply voltage is distributed equally over the gates and the current source. Supply voltage fluctuations are not critical because the effects of these variations are compensated for by an appropriate bias driver. The transfer characteristic of the MCSL gate can be seen in Fig. 2. The output resistance R is 200 Ω. The current source provides a current of 1 mA to the MCSL gate. The MOS transistor has a W/L of 10/1.5 and the bipolar h_{fe} was 100. To demonstrate the advantages of this MCSL logic, a BiCMOS ripple adder is fabricated.

III. COMPARISON OF MCSL, CMOS, AND ECL RIPPLE ADDERS

A. MCSL BiCMOS Ripple Adder

The ripple adder is a typical example of a circuit with a time-critical path. This latter is the carry path, because in each full-adder cell the two input bits and the carry of the previous stage are combined to form the sum and the carry for the next stage. Therefore the sum of the last bit cannot be performed until the carries of the previous stages are determined. The n-bit addition time is

$$t_{add} = (n-1)t_{carry} + t_{sum}. \qquad (2)$$

A BiCMOS adder cell using the new MCSL logic is shown

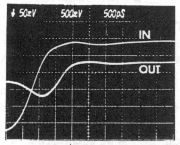

Fig. 4. BiCMOS adder: sum output first adder cell. Voltage scale: input 1 V/div; output 50 mV/div. Time scale: 500 ps/div.

in Fig. 3. The circuit for carry and sum is realized using three-level series gating. In the lower two levels the two input bits to be added (CMOS voltage swing) control the current path. The n-channel transistors used in the first level and the p-channel transistors used in the second level minimize the circuit area. In principle, it is possible to use either n-channel or p-channel transistors in both levels, but with the smaller gate–source voltage and body effect, this would lead to very large p-channel transistors in level 1 or n-channel transistors in level 2, thus increasing adding time. In the third level, which is controlled by the carry bit of the previous stage, only bipolar transistors are used as they provide a fast propagation of the carry bit to the next adder cell. The ECL sum output communicates directly with fast bipolar logic. An optional ECL/CMOS level converter [2] offers the possibility to work within a CMOS environment. The output voltage swing is adjusted by the output resistors and the corresponding current source.

The measurements of a 4-bit MCSL adder show a carry delay time of 150 ps/bit. The input and output signals of the MCSL and CMOS 4-bit adders are shown in Figs. 4 and 5. The power consumption per bit was 16 mW and is determined by two gate currents (Fig. 3) and four emitter-follower currents for sum s, $\overline{sum}$ $\bar{s}$, carry C_{out}, and $\overline{carry}$ $\overline{C_{out}}$. The area consumption per bit was 18 000

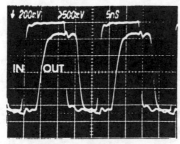

Fig. 5. CMOS adder: carry output fourth adder cell. Voltage scale: input 1 V/div; output 200 mV/div. Time scale: 2 ns/div.

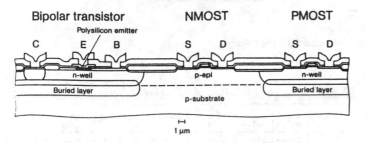

Bipolar transistor

Emitter size	$1.6 \times 10\ \mu m^2$
hfe	100
Transit time	18 ps
Re	11 Ω
Rc	40 Ω
Rb	325 Ω
C_{je}	60 fF
C_{jc}	78 fF
C_{js}	70 fF

MOS transistor

Gate oxide thickness	25 nm
N-channel L_{eff}	1.1 μm
V_{th}	0.87 V
P-channel L_{eff}	1.1 μm
V_{th}	-0.9 V

Fig. 6. BiCMOS process features.

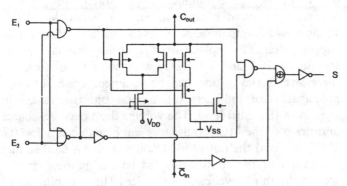

Fig. 7. Pure CMOS ripple adder cell.

μm^2. The MCSL adder results are based on a 1.5-μm BiCMOS process (Fig. 6) using double metal and triple polysilicon. The gate length of the MOS transistors is 1.5 μm. The bipolar transistors have a polysilicon emitter and a transit time of 18 ps.

To further evaluate the results of the MCSL adder, a pure CMOS and a pure ECL adder were fabricated.

B. CMOS Ripple Adder

In the carry path of the pure CMOS ripple adder (Fig. 7), an AND/NOR mixed gate is used. This results in a short delay time for the carry path at the expense of the sum path consisting of five gate delays. A carry delay time of

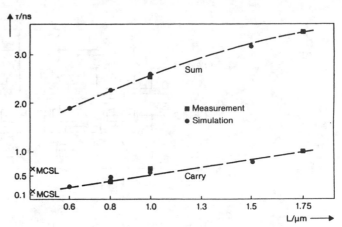

Fig. 8. Sum and carry delay times of different CMOS generations.

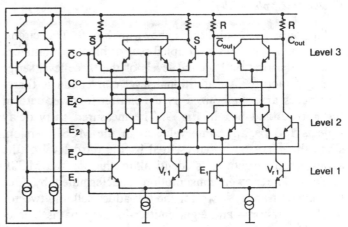

Fig. 9. Pure bipolar ripple adder cell.

TABLE I
COMPARISON OF MCSL, CMOS, AND ECL ADDERS

	ECL	MCSL	CMOS
Carry path	150 ps	150 ps	800 ps
Sum path	250 ps	700 ps	3.2 ns
16 bit addition	2.5 ns	2.95 ns	15.2 ns
Bipolar transistors per bit	48	28	—
Area per bit	32 000 μ^2	18 000 μ^2	5 900 μ^2
Gate currents	4	2	—
Emitter follower	8	4	—
Power per bit	32 mW	16 mW	0.2 mW at 10 MHz

800 ps, a factor of 5 slower than the MCSL version, was determined. In Fig. 8, the carry and sum delay times are plotted as a function of gate length. Fig. 8 shows that a 0.6-μm pure CMOS technology is necessary to obtain approximately the same result (200 ps) as for the conservative 1.5-μm BiCMOS MCSL adder. This clearly demonstrates the superiority of the MCSL BiCMOS adder over the pure CMOS version.

165

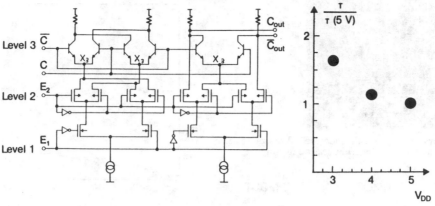

Fig. 10. MCSL ripple adder cell for reduced supply voltages.

C. ECL Ripple Adder

The pure ECL solution is shown in Fig. 9. This circuit also contains an emitter-coupled pair of bipolar transistors used in the first and second levels of the three-level series gating structure. An additional level conversion is therefore necessary, symbolized by the emitter-follower path in the surrounded box in Fig. 9. The measured carry delay time is 150 ps, the same value as in the MCSL version, since the carry path, represented by the third level of the series-gated structure, is identical to that of the MCSL version. The power consumption is determined by four gate current paths—two for the full-adder cell and two for level conversion—and eight emitter-follower paths. Therefore the power consumption increases by a factor of 2 compared with the MCSL version. The area consumption increases from 18 000 to 32 000 μm^2/bit. Metal 1 pitch was 3.5 μm.

D. Comparison of Adder Performances

The different adders are compared in Table I. The MCSL version shows the same carry delay times as the pure ECL version and is faster than the pure CMOS version by a factor of 5. The 16-bit addition also reflects the speed advantage in the carry path. The ECL/CMOS level conversion is not included in the results of Table I, because the ECL sum output should communicate directly with fast bipolar logic. An optional ECL/CMOS level converter offers the possibility of working directly within the CMOS environment. Then, for a load of 300 fF, 700 ps would have to be added to the addition time. MCSL shows a good compromise between the large area and power dissipation of the ECL version and the smaller values for CMOS. This clearly demonstrates the superiority of the MCSL version over the pure ECL or pure CMOS solutions.

IV. MCSL FOR REDUCED SUPPLY VOLTAGES

For future technologies it is necessary to reduce the supply voltages to lessen the high electric field effects arising from the small dimensions. This leads to a problem

TABLE II
SIMULATED RESULTS FOR THE DIFFERENT ADDER VERSIONS

Version	sum delay	carry delay	16 bit add.	power	circuit	Lgeo	supply voltage
V1	700ps	150ps	2.95ns	16 mW/bit	Fig.3	1.5um	5V
V2	500ps	150ps	2.75ns	2.7mW/bit	Fig.3	1.0um	5V
V3	550ps	185ps	3.32ns	1.5mW/bit	Fig.10	1.0um	5V
V4	780ps	210ps	3.93ns	1.0mW/bit	Fig.10	1.0um	3.3V

for three-level series gating using bipolar transistors because of saturation effects and the consequent dramatically increasing delay time. Three-level series gating is essential to obtaining high switching speed with a reasonable power consumption. With a voltage drop of 1 V at the current source (Fig. 9), 200-mV voltage swing at the resistor, and 0.8 V for $V_{be,on}$, a minimal supply voltage of 3.6 V is necessary to avoid saturation. Fig. 10 shows the MCSL version which guarantees an even further reduction in supply voltage. By replacing the remaining bipolar transistors in levels 1 and 2 (Fig. 3) with MOS transistors, the three-level series gating can be retained. To ensure fast propagation of the carry bit, bipolar transistors have to remain in the third level. The voltage drop between source and drain of the MOS transistors can be reduced below 0.8 V ($V_{be,on}$) and the potential at nodes X_3 is lowered so that saturation of the emitter-coupled pair of bipolar transistors in the third level can be avoided. The switching speed of the critical carry path in the upper level is unaffected by this measure and a short carry delay time is thus guaranteed. In Table II the simulated delay time for the carry signal and the sum signal as well as the power distribution of the MCSL adder cell are shown. The results of $V2$, $V3$, and $V4$ are based on a bipolar f_t of 16 GHz. The adder cell with MOS and bipolar transistors in levels 1 and 2 is the fastest ($V2$). Using the adder cell in Fig. 10 at a supply voltage of 5 V with MOS transistors only on levels 1 and 2, the delay time increases. But it should be noted that the power consumption decreases because the emitter followers are gone ($V3$). At a reduced supply voltage of 3.3 V the delay time further increases and the power consumption decreases ($V4$). The increase of the adding time is due to the fact that the widths of the MOS transistors have to be adapted to the lower supply voltages. Therefore the capacitive loads are higher on internal nodes and at the coupling between MOS gates and drain and source increases. The current to discharge the internal nodes is the same for the different supply voltages. This leads to an increase of the signal propagation time from level 1 to level 3 and also to an increase of the adding time. Despite this, the carry delay time remains short.

V. Conclusion

The merged CMOS/bipolar current switch logic (MCSL) was presented. CMOS/ECL level conversion and logical operation are realized simultaneously with merged bipolar/CMOS gates. A supply voltage reduction down to 3.3 V is possible. As an example of this circuit technique, an adder with a carry delay time of 150 ps was designed and measured. This is about five times faster than an optimized CMOS design at comparable feature size.

Acknowledgment

The authors would like to thank M. Stegherr, M. Böhner, and B. Zehner for helpful discussions, M. Wurm for circuit design and verification, and H. Klose for assembly technology.

References

[1] T. Hotta et al., "CMOS/bipolar circuits for 60-MHz digital processing," in ISSCC Dig. Tech. Papers, Feb. 1986, pp. 190–191.
[2] K. Ogiue et al., "13-ns, 500-mW, 64-kbit ECL RAM using HI-BICMOS technology," IEEE J. Solid-State Circuits, vol. SC-21, no. 5, pp. 681–684, Oct. 1986.
[3] W. Heimsch et al., "Merged CMOS/bipolar current switch logic (MCSL)," in ISSCC Dig. Tech. Papers, Feb. 89, pp. 122–123.

Chapter 3

Design for Low Power

Design for Low Power

This chapter covers the design of low power logic and systems that are operating with minimal use of energy for computation. Demand for reducing power in digital systems has not only been present in the systems that are designed or operating under conditions where low-power consumption is required but also in high-performance systems. The growth of high-performance microprocessors has also been limited by the power of the package using the inexpensive air-cooling techniques. The limit to that power has been set at about 50 watts. However, the increasing demand for performance (which has been roughly doubling every two years) is leaving an imbalance in the power dissipation increase, which is growing at approximately 10 watts per year. This growth is threatening to limit the performance increase of future microprocessors. As Kuroda and Sakurai state, the "CMOS ULSIs are facing a power dissipation crisis." The increase in power consumption for three generations of Digital Equipment Corporation, "Alpha" architecture high-performance processors, is given in Fig. 3.1.

The goal of achieving low-power operation can be approached on several levels.

Technology and Circuits

Most of the improvement on power savings is gained through *technology*. Scaling the device and process features, together with lowering the threshold and supply voltage, result in an order of magnitude savings in power. Indeed, this resulting power reduction has been a salient achievement during the entire course of processor *and* digital systems development. Had this not been the case, the increase in power from one generation to another would have been much larger, limiting the performance growth of microprocessors much earlier. The technology accounts for approximately 30 percent of the improvement in gate delay per generation. The resulting switching energy CV^2 has been improving at the rate of 0.5 times per generation. Given that the frequency of operation has been doubling for each new generation, the power factor $P = CV^2f$ remained constant

$(0.5 \times 2 = 1.0)$. It is the increased complexity of the VLSI circuits that goes largely uncompensated as far as power is concerned. It is estimated that the number of transistors has been tripling for every generation.

The power consumption can also be influenced by careful sizing of the transistors, variation in the transistor threshold voltage, as well as layout, placement, and routing. These issues are discussed in the paper by Tan.

Logic Design Style

Logic function and logic design style also influences the power. The important influencing factors are the number of inputs and internal nodes as well as the number of transitions. It is also important to design the logic so that the number of signal "glitches" (unnecessary transitions on the nodes before reaching the final value) will be minimized or eliminated.

Architectural Tradeoffs

Power optimization at the *architecture level* is achieved by techniques such as:

- The tradeoff between the use of parallelism or pipelining, and corresponding scaling of the supply voltage, which is made possible by the reduced speed of operation.
- Minimization of switching activity by appropriate choices of number representation.
- Ordering of input signals in order to minimize unnecessary switching.
- Choice of the appropriate computational structure.
- Resource sharing, such as busses and execution units.

Proper Choice of the Computational Algorithm

Proper choice of the computational algorithm, various transformations of the existing algorithm, and quantization in order to minimize the number of required operations can all be effective techniques in reducing the total energy used to accomplish a specified task.

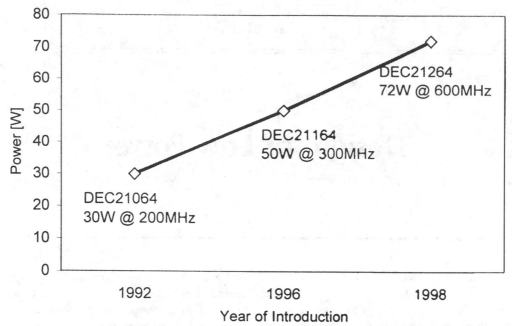

Fig. 3.1. Power consumption for "Alpha" architecture high-performance processors.

A general overview of those techniques is presented in the paper by Chandrakasan and Brodersen. The paper describes various approaches for minimizing power consumption of digital systems.

An overview of low-power ULSI circuit techniques is given in the second paper, by Kuroda and Sakurai. The authors discuss the effect of the scaling theory on the future growth of the power of ULSI chips. A "constant field scaling theory" (formulated by Dennard et al.) assumes that device voltages as well as device dimensions are scaled down by a scaling factor x. This theory promises limitless progress because it predicts that the power density will remain constant. In practice, however, neither the supply voltage nor a threshold voltage of the transistor had long been scaled. The constant field scaling theory is therefore not realistic.

The constant voltage theory assumes that the power supply remains constant, leading to an increase in the power density by a factor of x^3. This is certainly not acceptable, and in practice the power supply voltage has been scaling down with the device features. As a result, a theory has been formulated which assumes scaling of the power supply by a factor of $x^{0.5}$. This assumption seems to satisfy both the device reliability and circuit performance. With this scaling approach, the increase in power density grows by a factor of $x^{0.5}$. It is predicted that even with this scaling factor the power dissipation of CMOS would exceed that of ECL by the year 2000. A new philosophy of circuit design needed to alleviate this power crisis is presented in this paper. The authors give an extensive analysis to the CMOS pass-transistor logic. New logic families such as Swing Restored Pass-transistor Logic (SRPL) and Sense-Amplifying Pass-transistor Logic (SAPL) are presented, along with the methodology for synthesis of logic functions utilizing this type of circuit.

An extensive study of MPL (an advanced version of CPL as applied to multilevel pass-transistor logic) applied to 27 randomly selected circuits is presented in the paper by Sasaki et al. Their results show on average power improvement of 23 percent over conventional CMOS, while the area was reduced by 15 percent and speed by 12 percent. The total improvement over CMOS was reported to be 42 percent.

A performance study of several CMOS master-slave latches is reported in the paper by Ko et al. Several popular latches are compared in terms of speed, power, and area savings. The choice of an appropriate latch is important given that the control logic can consume 20 percent of the microprocessor power. This number is likely to increase in the future owing to super-scalar and super-pipelined implementation of modern processors. It was found that the use of low-area latch results in up to 122 percent more power. The latch proposed in this paper improves speed for 56 percent, resulting in a power increase of 6 percent, and a better power-delay product.

The study of power dissipation in the clock system, especially as it applies to highly pipelined systems was done by De Man et al. They compared C²MOS registers and true-single-phase registers (known as TSPC) that have been used in high-speed systems. Their results show that the application of TSP registers yields no significant saving in power compared to C²MOS registers. Given that any power savings owing to simpler clock networks is compensated by increased power dissipation in register circuits, they were the first to suggest the use of reduced-swing clock signals. In this paper they present a special driver for generation of the reduced-swing clock signals. Use of a reduced-swing clock resulted in power savings of up to 60 percent without reducing the circuit speed as reported by Kojima (*see* Chapter 4).

A comparison of the power-delay characteristics of CMOS adders was done by Nagendra et al. The best power-delay product is achieved by using simple speed-optimized structures such as the variable block adder VBA. In addition they have shown that the speed can be influenced by selective sizing of the transistors without large increases in the power dissipation.

The paper by Sakuta et al. treats the problem of designing a low-power multiplier for a DSP core. The authors have shown that balancing delays in a parallel multiplier reduces the power with minimal penalty in performance. This is due to the simultaneous arrival of the signal to the final adder and elimination of unnecessary transitions.

Horowitz et al. introduce the energy-delay product as a metric for evaluating the power efficiency of a design. An appropriate scaling of the supply voltage results in a lower power, however, at the expense of the speed of the circuit. The energy-delay curve shows an optimal operation point in terms of a design's energy efficiency. This point is reached by various techniques all of which discussed in this paper. The authors reached the interesting conclusion that the fabrication technology seems more important for the energy-delay than the machine's architectural features. This finding is consistent with the fact that the processors' performance has been increasing four times per generation. Although we would expect a sixfold increase in performance the frequency has been doubling per generation and the number of transistors has been tripling. This shows that the transistors have not been used efficiently and that the architectural features that are consuming this transistor increase have not been producing the desired effect in terms of the processors' energy efficiency.

A detail power analysis of a programmable DSP processor was shown in the paper by Kojima et al. The authors have shown a module break down of the power used in different DSP blocks. Contrary to many opinions it was found that the bus power is significantly smaller as compared to the data path. It was also shown in this paper how a simple switch of the multiplier inputs can reduce multiplier power by 4–8 times.

Reduction of the logic swing and of the clock signal excursions is another way of reducing the power, especially given that the power is directly proportional to the V_s^2 (where V_s is the signal swing). The power reduction owing to the reduced clock signal swing can be substantial, as shown by Kojima (see Chapter 4). In addition, this paper shows how to develop a "half-swing clocking" scheme.

Adiabatic Logic

The principles of *adiabatic logic* are based on the studies by Bennett and Landauer of IBM in which they show that computation without the use of energy is theoretically possible. They treat the case of *"reversible computing"* where a change of state can be achieved without dissipation of energy. A node is brought to a different state in a way that is similar to the adiabatic transitions in thermodynamics, which gave the name *adiabatic logic* to this type of computing.

A review of specific principles of adiabatic circuits is presented in the paper by Denker, and an approach to design adiabatic logic is described by Kramer et al. Though in its infancy, adiabatic logic shows an interesting approach to computing as an unconventional way of building the logic and systems. The paper by Athas et al. presents an application of the energy recovery techniques to a design of a 16-bit microprocessor powered by its clock. A resonant clock-driver used in this microprocessor recovers and reuses the energy that would otherwise be dissipated as heat. The microprocessor was fabricated, and the prototype demonstrated low-power dissipation running at close to 60 MHz. The last paper in this set presents another development of adiabatic logic with integrated single-phase power-clock supply. This logic reveals that only a single-phase power-clock is needed for a proper operation of the logic, thus greatly simplifying power-clock generation. Another feature of this logic is that it can work under two different modes of operation. It can operate as a regular clocked CMOS, or it can switch into the adiabatic mode. This logic represents an order of magnitude improvement in power dissipation over the regular CMOS logic at a relatively low frequency of operation.

Minimizing Power Consumption in Digital CMOS Circuits

ANANTHA P. CHANDRAKASAN AND ROBERT W. BRODERSEN, FELLOW, IEEE

An approach is presented for minimizing power consumption for digital systems implemented in CMOS which involves optimization at all levels of the design. This optimization includes the technology used to implement the digital circuits, the circuit style and topology, the architecture for implementing the circuits and at the highest level the algorithms that are being implemented. The most important technology consideration is the threshold voltage and its control which allows the reduction of supply voltage without significant impact on logic speed. Even further supply reductions can be made by the use of an architecture-based voltage scaling strategy, which uses parallelism and pipelining, to tradeoff silicon area and power reduction. Since energy is only consumed when capacitance is being switched, power can be reduced by minimizing this capacitance through operation reduction, choice of number representation, exploitation of signal correlations, resynchronization to minimize glitching, logic design, circuit design, and physical design. The low-power techniques that are presented have been applied to the design of a chipset for a portable multimedia terminal that supports pen input, speech I/O and full-motion video. The entire chipset that performs protocol conversion, synchronization, error correction, packetization, buffering, video decompression and D/A conversion operates from a 1.1 V supply and consumes less than 5 mW.

I. INTRODUCTION

In recent years, the desirability of portable operation of all types of electronic systems has become clear and a major factor in the weight and size of portable devices is the amount of batteries which is directly impacted by the power dissipated by the electronic circuits. In addition, the cost of providing power (and associated cooling) has resulted in significant interest in power reduction even in nonportable applications which have access to a power source. In spite of these concerns, until recently, there has not been a major focus on a design methodology of digital circuits which directly addresses power reduction, with the focus rather on ever faster clock rates and logic speeds. The approach which will be presented here, takes another viewpoint, in which all possible aspects of a system design are investigated with the goal of reducing the power consumption. These considerations range from the technology being used for the implementation, the circuit and logic topologies, the digital architectures and even the algorithms being implemented. What is assumed is that the application, which is desired to be implemented with low power is known, and tradeoffs can be made as long as the functionality required of this application is met within a given time constraint.

Maintaining a given level of computation or throughput is a common concept in signal processing and other dedicated applications, in which there is no advantage in performing the computation faster than some given rate, since the processor will simply have to wait until further processing is required. This is in contrast to general purpose computing, where the goal is often to provide the fastest possible computation without bound. One of the most important ramifications of only maintaining throughput is that it enables an architecture driven voltage scaling strategy, in which aggressive voltage reduction is used to reduce power, and the resulting reduction in logic speed is compensated through parallel architectures to maintain throughput. However, the techniques presented are also applicable to the general purpose environment, if the figure of merit is the amount of processing per unit of power dissipation (e.g., MIPS/W). Since in this case the efficiency in implementing the computation is considered and voltage scaling decreases the energy expended per evaluation.

The optimization to minimize area at all costs, has only been secondary to the fixation on increasing circuit speed and again our position is that this should be examined with respect to its effect on power consumption. Some of the techniques that will be presented will come at the expense of increased silicon area and thus the cost of the implementation will be increased. The desirability of this tradeoff can only be determined with respect to a given market situation, but in many cases a moderate increase in area can have substantial impact on the power requirements. It is clear that if power reduction is more important than increasing circuit clock rate, then the area consumed by large clock buffers, power distribution busses and predictive circuit architectures would be better spent to reduce the power dissipation.

Manuscript received May 1, 1994; revised July 24, 1994. This work was supported by ARPA. A. Chandrakasan's work was supported by an IBM fellowship.

A. Chandrakasan is with the Department of Electrical Engineering, Massachusetts Institute of Technology, Cambridge, MA 02139 USA.

R. W. Brodersen is with the Department of EECS, University of California, Berkeley, CA 94720 USA.

IEEE Log Number 9408188.

Reprinted from *IEEE Proceedings*, Vol. 83, No. 4, pp. 498–523, 1995.

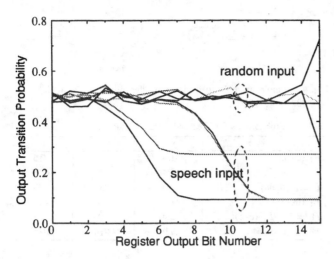

Fig. 1. Dependence of activity on statistics: Correlated versus random input.

Since CMOS circuits do not dissipate power if they are not switching, a major focus of low power design is to reduce the switching activity to the minimal level required to perform the computation. This can range from simply powering down the complete circuit or portions of it, to more sophisticated schemes in which the clocks are gated or optimized circuit architectures are used which minimize the number of transitions. An important attribute which can be used in circuit and architectural optimization is the correlation which can exist between values of a temporal sequence of data, since switching should decrease if the data is slowly changing (highly correlated). An example of the difference in the number of transitions which can be obtained for a highly correlated data stream (human speech) versus random data is shown in Fig. 1—the transition activity for a few registers in an FIR filter design. For an architecture which does not destroy the data correlation, the speech data switches 80% less capacitance than the random input. In addition, the sequencing of operations can result in large variations of the switching activity due to these temporal correlations.

After a brief summary of the sources of power dissipation in CMOS circuits, techniques will be presented to reduce power dissipation at the four levels of CMOS system design: technology, circuits, architectures, and algorithms. This discussion will be followed by an application of these techniques to a demanding multimedia application.

II. FACTORS INFLUENCING CMOS POWER CONSUMPTION

There are three major sources of power dissipation in digital CMOS circuits which are summarized in the following equation [1]:

$$P_{\text{avg}} = P_{\text{switching}} + P_{\text{short-circuit}} + P_{\text{leakage}}$$
$$= \alpha_{0 \to 1} C_L \cdot V_{dd}^2 \cdot f_{clk} + I_{sc} \cdot V_{dd} + I_{\text{leakage}} \cdot V_{dd}. \quad (1)$$

The first term represents the switching component of power, where C_L is the load capacitance, f_{clk} is the clock frequency and $\alpha_{0 \to 1}$ is the node transition activity factor (the average number of times the node makes a power consuming transition in one clock period). The second term is due to the direct-path short circuit current, I_{sc}, which arises when both the NMOS and PMOS transistors are simultaneously active, conducting current directly from supply to ground. Finally, leakage current, I_{leakage}, which can arise from substrate injection and subthreshold effects, is primarily determined by fabrication technology considerations.

A. Capacitive Voltage Transitions

The switching or dynamic component of power consumption arises when the capacitive load, C_L, of a CMOS circuit is charged through PMOS transistors to make a voltage transition from 0 to the high voltage level, which is usually the supply, V_{dd}. For an inverter circuit, the energy drawn from the power supply for this positive going transition is $C_L V_{dd}^2$, half of which is stored in the output capacitor and half is dissipated in the PMOS device. On the V_{dd} to 0 transition at the output, no charge is drawn from the supply, however the energy stored in the capacitor ($\frac{1}{2} C_L V_{dd}^2$) is dissipated in the pull-down NMOS device. If these transitions occur at a clock rate, f_{clk}, the power drawn from the supply is $C_L V_{dd}^2 f_{clk}$. However, in general, the switching will not occur at the clock rate (except for clock buffers), but rather at some reduced rate which is best described probabilistically. $\alpha_{0 \to 1}$ is defined as the average number of times in each clock cycle that a node with capacitance, C_L, will make a power consuming transition (0 to 1), resulting in an average switching component of power for a CMOS gate to be,

$$P_{\text{switching}} = \alpha_{0 \to 1} C_L V_{dd}^2 f_{clk}. \quad (2)$$

Since the energy expended for each switching event in CMOS circuits is $C_L V_{dd}^2$, it has the extremely important characteristic that it becomes quadratically more efficient as the high transition voltage level is reduced. An effective capacitance, $C_{\text{effective}}$, can be defined which includes the transition activity occurring and is equal to $\alpha_{0 \to 1} C_L$, yielding an average transition energy of $C_{\text{effective}} V_{dd}^2$. Clearly, after the high transition level is reduced by the maximum extent possible, the next step is to minimize $C_{\text{effective}}$, through the choice of logic function, logic style, circuit topology, data statistics, and the sequencing of operations. In addition, an optimization can be made both statically, in which the probabilities of transition are made with respect to the Boolean operation and dynamically which includes the effect of spurious "glitching" transitions which can occur before the logic settles out to its static value.

1) Logic Function: The type of logic function, NAND, NOR or a more complex gate, will all yield different static transition probabilities. Let us first consider a static 2-input NOR gate and assume that only one input transition is possible during a clock cycle and also assume that the inputs to the NOR gate have a uniform input distribution of high and low levels. This means that the four possible states for inputs A and B (00, 01, 10, 11) are equally likely. For a

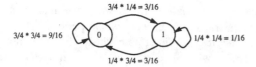

Fig. 2. State transition diagram for a 2 input NOR gate assuming random inputs.

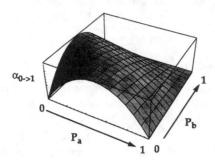

Fig. 3. Transition probability for a two input NOR gate as a function of input statistics.

NOR gate, the probability that the output is **0** is $\frac{3}{4}$ and that it will be **1** is $\frac{1}{4}$. The 0 to 1 transition probability of a CMOS gate is given by the probability that the output is in the **0** ($=\frac{3}{4}$) state multiplied by the probability the next state is **1** ($=\frac{1}{4}$). For a NOR gate this translates to

$$\alpha_{0\to 1} = p(0)p(1) = p(0)(1-p(0)) = \frac{3}{4}(1\frac{3}{4}) = \frac{3}{16}. \quad (3)$$

The state transition diagram annotated with transition probabilities is shown in Fig. 2. Note that the output probabilities are no longer uniform.

For a 2-input static XOR gate on the other hand, the $0 \to 1$ transition probability once again assuming uniformly distributed inputs is given by $p(0)(1-p(0)) = \frac{1}{2}(1-\frac{1}{2}) = \frac{1}{4}$.

2) Logic Style: One basic choice of logic style is between static and dynamic CMOS. For a dynamic implementation of the 2 input NOR gate which is implemented as an NMOS-tree with PMOS precharge, power is consumed during the precharge operation for the times when the output capacitor was discharged the previous cycle. Since all inputs are equi-probable, there is then a 75% probability that the output node will discharge immediately after the precharge phase, yielding a transition probability of 0.75. The corresponding probability is smaller for a static implementation (as computed in the previous subsection). Note that for the dynamic case, the activity depends only on the signal probability, while for the static case the transition probability depends on previous state. If the inputs to a static CMOS gate do not change from the previous sample period, then the gate does not switch. This is not true in the case of dynamic logic in which gates can switch. For a dynamic NAND gate, the activity is $\frac{1}{4}$ (since there is a 25% chance that the output will be discharged) while it is $\frac{3}{16}$ for a static implementation. Other considerations with respect to logic style is the amount of capacitance on the switching node and the sensitivity to supply voltage reduction.

3) Signal Statistics: In the previous sections, the NOR gate was analyzed assuming random inputs. However, signals in a circuit are typically not always random and

Fig. 4. Simple example to demonstrate the influence of circuit topology on activity.

Table 1 Probabilities for Tree and Chain Toplologies

	O1	O2	F
P_1 (chain)	1/4	1/8	1/16
$P_0 = 1 - P_1$ (chain)	3/4	7/8	16/16
$P_{0\to 1}$ (chain)	3/16	**7/64**	15/256
P_1 (tree)	1/4	1/4	1/16
$P_0 = 1 - P_1$ (tree)	3/4	3/4	15/16
$P_{0\to 1}$ (tree)	3/16	**3/16**	15/256

it is necessary to consider the effect of signal statistics on power. To illustrate the influence of signal statistics on power consumption, consider once again a 2 input NOR gate, and let P_a and P_b be the probabilities that the inputs A and B are **1**. In this case the probability that the output node is a **1** is given by:

$$P_1 = (1 - P_a)(1 - P_b). \quad (4)$$

Therefore, the probability of transitioning from 0 to 1 is

$$\alpha_{0\to 1} = P_0 P_1 = (1 - (1 - P_a)(1 - P_b))(1 - P_a)(1 - P_b). \quad (5)$$

Fig. 3 shows the transition probability as a function of P_a and P_b. From this plot, it is clear that understanding the signal statistics and their impact on switching events can be used to significantly impact the power dissipation.

4) Circuit Topology: The manner in which logic gates are interconnected can have a strong influence on the overall switching activity. There are two components to switching activity: a static component (which does not take into account the timing behavior and is strictly a function of the topology and the signal statistics) and a dynamic component (which takes into account the timing behavior of the circuit). To illustrate this point consider two alternate implementations of $F = A \cdot B \cdot C \cdot D$ as shown in Fig. 4. First consider the static behavior assuming that all primary inputs (A, B, C, D) are uncorrelated and random (i.e., $P_{1(a,b,c,d)} = 0.5$). For an AND gate, the probability that the output is in the **1** state is given by

$$P_1 = P_a P_b. \quad (6)$$

Therefore the probability that the output will make a 0 to 1 transition is given by

$$\alpha_{0\to 1} = P_0 P_1 = (1 - P_1)P_1 = (1 - P_a P_b)P_a P_b. \quad (7)$$

Given this, the signal and transition probabilities can be computed for the two topologies shown in Fig. 4 which is summarized in Table 1.

The results indicate that the chain implementation will have an overall lower switching activity than the tree implementation for random inputs. However, looking strictly at the static behavior of the circuit is not adequate,

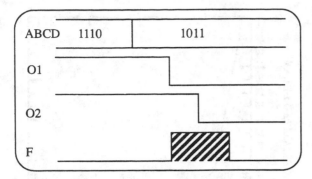

Fig. 5. Circuit imbalances result in spurious transitions.

and it is also important to consider the timing behavior to accurately make power tradeoffs of various topologies. Timing skew between signals can cause spurious transitions (also called glitches) resulting in extra power. To illustrate this tradeoff, consider once again the two topologies in Fig. 4. For the chain implementation with an input transition of $1110 \rightarrow 1011$ for $ABCD$, assuming a unit delay for each gate, ignoring dynamic switching effects, the output node will not make a transition. However, due to timing skew through the logic, the output will make an "extra" transition; i.e., O2 will only be valid 2 units after the inputs arrive, causing the output AND gate to evaluate with the new input D and the previous value of O2 (Fig. 5). The tree implementation, on the other hand, is balanced and is glitch free. This example demonstrates that a circuit topology can have a smaller static component of activity while having a higher dynamic component.

The actual waveforms illustrating the glitching behavior in static CMOS circuits are shown in Fig. 6, which is the SPICE simulation of a static 16-b adder, with all bits of IN0 and the CIN of the LSB going from "zero" to "one," and with all the bits of IN1 set to "zero." For all bits, the resultant sum should be zero; however, the propagation of the carry signal causes a "one" to appear briefly at most of the outputs. These spurious transitions dissipate extra power over that strictly required to perform the computation. Note that some of the bits only have partial glitching. The number of these extra transitions is a function of input patterns, internal state assignment in the logic design, delay skew, and logic depth. Though it is possible with careful logic design to eliminate these transitions (for example using balanced paths as described in Section V-D), dynamic logic intrinsically does not have this problem, since any node can undergo at most one power-consuming transition per clock cycle.

B. Short-Circuit Component of Power

The previous section analyzed the switching component of power consumption which corresponds to the amount of energy required to charge parasitic capacitors. The switching component of power is independent of the rise and fall times at the input of logic gates. Finite rise and fall times of the input waveforms however result in a direct current path between V_{dd} and GND which exist for a short period of time during switching. Specifically, when

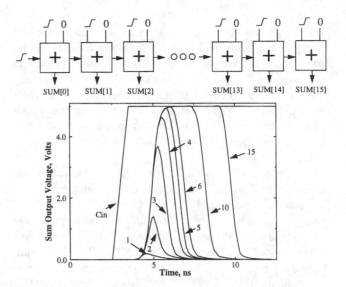

Fig. 6. Waveforms for a 16-b adder demonstrating glitching behavior.

the condition $V_{tn} < V_{in} < V_{dd} - |V_{tp}|$ holds for the input voltage, where V_{tn} and V_{tp} are NMOS and PMOS threshold voltages, there will be a conductive path open between V_{dd} and GND because both the NMOS and PMOS devices will be simultaneously on. Short circuit currents are significant when the rise/fall time at the input of a gate is much larger than the output rise/fall time. This is because the short-circuit path will be active for a longer period of time. To minimize the total average short-circuit current, it is desirable to have equal input and output edge times [2]. In this case, the power consumed by the short-circuit currents is typically less than 10% of the total dynamic power. An important point to note is that if the supply is lowered to be *below* the sum of the thresholds of the transistors, $V_{dd} < V_{Tn} + |V_{Tp}|$, the short-circuit currents can be eliminated because both devices will not be on at the same time for any value of input voltage. As will be shown later, the architecture driven voltage scaling that is proposed allows supply voltages which satisfy this criteria.

C. Leakage Component of Power

There are two types of leakage currents: reverse-bias diode leakage on the transistor drains and subthreshold leakage through the channel of an off device.

The diode leakage occurs when a transistor is turned off and another active transistor charges up/down the drain with respect to the former's bulk potential. For example, consider an inverter with a high input voltage, in which the NMOS transistor is turned on and the output voltage is driven low. The PMOS transistor will be turned off, but its drain-to-bulk voltage will be equal to $-V_{dd}$ since the output voltage is at 0 V and the bulk for the PMOS is at V_{dd}. The resulting current will be approximately $I_L = A_D J_S$, where A_D is the area of the drain diffusion, and J_S is the leakage current density, set by the technology and weakly dependent on the supply voltage. For a typical CMOS process, J_S is approximately 1–5 pA/μm^2 (25° C), and the minimum A_D is 7.2 μm^2

for a 1.2 μm minimum feature size. J_S doubles with every 9° increase in temperature. For a 1 million transistor chip, assuming an average drain area of 10 μm^2, the total leakage current is on the order of 25 μA. While this is typically a small fraction of the total power consumption in most chips, it could be significant for a system application which spends much of its time in standby operation, since this power always being dissipated even when no switching is occurring.

The second component of the leakage power is the subthreshold leakage which occurs due to carrier diffusion between the source and the drain when the gate-source voltage, V_{gs}, has exceeded the weak inversion point, but is still below the threshold voltage V_t, where carrier drift is dominant [3]. In this regime, the MOSFET behaves similarly as a bipolar transistor, and the subthreshold current is exponentially dependent on the gate-source voltage V_{gs}. The current in the subthreshold region is given by

$$I_{ds} = \kappa e^{(V_{gs} - V_t)/(nV_T)} \left(1 - e^{V_{ds}/V_T}\right) \qquad (8)$$

where κ is a function of the technology, V_T is the thermal voltage (KT/q) and V_t is the threshold voltage. For $V_{ds} \gg V_T$, $(1 - e^{-Vds/VT}) \approx 1$; that is, the drain to source leakage current is independent of the drain-source voltage V_{ds}, for V_{ds} approximately larger than 0.1 V.

Associated with this is the subthreshold slope S_{th}, which is the amount of voltage required to drop the subthreshold current by one decade. The subthreshold slope can be determined by taking the ratio of two points in this region (from (8)):

$$\frac{I_1}{I_2} = e^{(V_1 - V_2)/(nV_T)} \qquad (9)$$

which results in $S_{th} = nV_T \ln (10)$. At room temperature, typical values for S_{th} lie between 60–90 mV/(decade current), with 60 mV/dec being the lower limit. Clearly, the lower S_{th} is, the better, since it is desirable to have the device "turnoff" as close to V_t as possible and this is one important advantage of silicon-on-insulator technologies which have an S_{th} near 60 mV [4]. The choice of optimal V_t for low voltage applications that trades-off subthreshold leakage and switching currents is presented in Section III.

III. Technology Optimization

As noted in (1), the energy per transition is proportional to V_{dd}^2. Therefore, it is only necessary to reduce the supply voltage for a quadratic improvement in the power-delay product of CMOS logic. Unfortunately, we pay a speed penalty for a V_{dd} reduction (as seen from Fig. 7), with the delays drastically increasing as V_{dd} approaches the threshold voltages of the devices. Even though the exact analysis of the delay is quite complex if the nonlinear characteristic of a CMOS gate are taken into account, it is found that a simple first-order derivation adequately predicts the experimentally determined dependence and is given by

$$T_d = \frac{C_L \times V_{dd}}{I} = \frac{C_L \times V_{dd}}{\frac{\mu C_{ox}}{2}(W/L)(V_{dd} - V_t)^2}. \qquad (10)$$

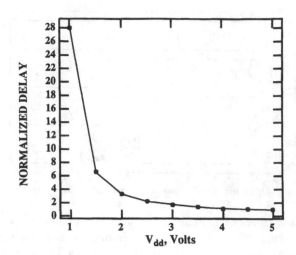

Fig. 7. Normalized delay versus V_{dd} for a typical gate in a standard CMOS process.

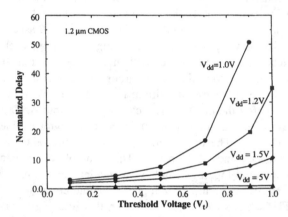

Fig. 8. Effect of threshold reduction on the delay for various supply voltages.

Since the objective is to reduce power consumption while keeping the throughput of overall system fixed, compensation for these increased delays at low voltages is required. In Section V-A, an architecture driven voltage scaling strategy is presented in which parallel and pipelined architectures are used to compensate for the increased gate delays at reduced supply voltages and meet throughput constraints. Another approach to reduce the supply voltage without loss in throughput is to modify the threshold voltage of the devices. Low threshold voltage devices have been used to implement inverter circuits at supply voltages as low as 200 mV [5]. Reducing the threshold voltage allows the supply voltage to be scaled down (and therefore lower switching power) without loss in speed. For example, a circuit running at a supply voltage of 1.5 V with $V_t = 1$ V will have approximately the same performance as the circuit running at a supply voltage of 0.9 V and a $V_t = 0.5$ V using the simple first order theory of (10). Fig. 8 shows a plot of normalized delay versus threshold voltage for various supply voltages.

Since a significant power improvement can be gained through the use of low-threshold MOS devices, the question of how low the thresholds can be reduced must be addressed. The limit is set by the requirement to retain

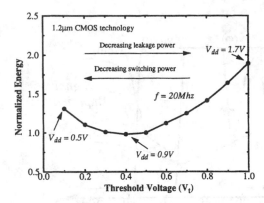

Fig. 9. Compromise between dynamic and leakage power dissipation through V_t variation.

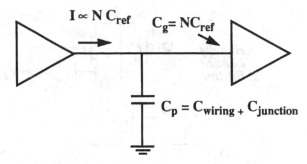

Fig. 10. Circuit model for analyzing the effect of transistor sizing.

adequate noise margins and the increase in subthreshold currents. Noise margins will be relaxed in low power designs because of the reduced currents being switched, however, the subthreshold currents can result in significant static power dissipation.

Fig. 9 shows a plot of energy versus threshold voltages for a fixed throughput for a 16-b datapath ripple carry adder (which essentially represents the power to perform the operation). Here, the power supply voltage is allowed to vary to keep the throughput fixed. For a fixed throughput (e.g., that obtained at a 20 Mhz clock rate), the supply voltage and therefore the switching component of power can be reduced while reducing the threshold voltage. However, at some point, the threshold voltage and supply reduction is offset by an increase in the leakage currents, resulting in an optimal threshold voltage for a given level of logic complexity. That is, the optimum threshold voltage must compromise between improvement of current drive at low supply voltage operation and control of the subthreshold leakage.

IV. PHYSICAL, CIRCUIT, AND LOGIC LEVEL OPTIMIZATIONS

The next level of system design involves optimizing the physical, circuit and logic levels. A few examples which illustrate the tradeoffs that can be made at this level of system design are presented, which include optimizing place and route, transistor sizing, reduced swing logic, logic minimization, and logic level power down. Support circuitry for low voltage operation, which includes voltage level converters and low-voltage dc/dc converters, will be presented.

A. Place and Route Optimization

At the layout level, the place and route should be optimized such that signals that have high switching activity (such as clocks) should be assigned short wires and signals with lower switching activities can be allowed progressively longer wires. Current design tools typically minimize the overall area or wire lengths given a timing constraint, which does not necessarily reduce the overall capacitance switched. An example of the benefits of this optimization is presented in Section VII-A.

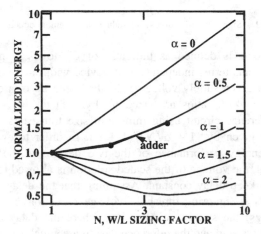

Fig. 11. Plot of energy versus transistor sizing factor for various parasitic contributions.

B. Transistor Sizing

Independent of the choice of logic family or topology, optimized transistor sizing will play an important role in reducing power consumption. For low power, as is true for high speed design, it is important to equalize all delay paths so that a single critical path does not unnecessarily limit the performance of the entire circuit. However, beyond this constraint, there is the issue of what extent the (W/L) ratios should be uniformly raised for all the devices, yielding a uniform decrease in the gate delay and hence allowing for a corresponding reduction in voltage and power. It is shown in this section, that if voltage is allowed to vary, that the optimal sizing for low power operation is quite different from that required for high speed.

In Fig. 10, a simple two-gate circuit is shown, with the first stage driving the gate capacitance of the second, in addition to the parasitic capacitance C_p due to substrate coupling and interconnect. Assuming that the input gate capacitance of both stages is given by NC_{ref}, where C_{ref} represents the gate capacitance of a MOS device with the smallest allowable (W/L), then the delay through the first gate at a supply voltage V_{ref} is given by

$$T_N = K \frac{(C_p + NC_{\text{ref}})}{(NC_{\text{ref}})} \frac{V_{\text{ref}}}{(V_{\text{ref}} - V_t)^2}$$
$$= K(1 + \alpha/N) \frac{V_{\text{ref}}}{(V_{\text{ref}} - V_t)^2} \qquad (11)$$

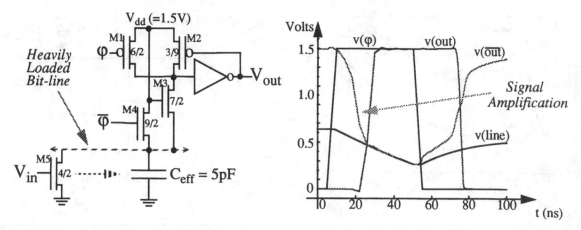

Fig. 12. Signal amplification/swing reduction in memory circuits.

where α is defined as the ratio of C_p to C_{ref}, and K represents terms independent of device width and voltage. For a given supply voltage V_{ref}, the speed up of a circuit whose W/L ratios are sized up by a factor of N over a reference circuit using minimum size transistors ($N = 1$) is given by $(1 + \alpha/N)/(1 + \alpha)$. In order to evaluate the energy performance of the two designs at the same speed, the voltage of the scaled solution is allowed to vary as to keep delay constant. Assuming that the delay scales as $1/V_{dd}$ (ignoring threshold voltage reductions in signal swings) the supply voltage, V_N, where the delay of the scaled design and the reference design are equal is given by

$$V_N = \frac{(1 + \alpha/N)}{(1 + \alpha)} V_{\text{ref}}. \qquad (12)$$

Under these conditions, the energy consumed by the first stage as a function of N is given by

$$\begin{aligned}
\text{Energy}(N) &= (C_p + NC_{\text{ref}})V_N^2 \\
&= \frac{NC_{\text{ref}}(1 + \alpha/N)^3 V_{\text{ref}}^2}{(1 + \alpha)^2}. \qquad (13)
\end{aligned}$$

After normalizing against E_{ref} (the energy for the minimum size case), Fig. 11 shows a plot of $\text{Energy}(N)/\text{Energy}(1)$ versus N for various values of α. When there is no parasitic capacitance contribution (i.e., $\alpha = 0$), the energy increases linearly with respect to N, and the solution utilizing devices with the smallest (W/L) ratios results in the lowest power. At high values of α, when parasitic capacitances begin to dominate over the gate capacitances, the power decreases temporarily with increasing device sizes and then starts to increase, resulting in a optimal value for N. The initial decrease in supply voltage achieved from the reduction in delays more than compensates the increase in capacitance due to increasing N. However, after some point the increase in capacitance dominates the achievable reduction in voltage, since the incremental speed increase with transistor sizing is very small (this can be seen in (11), with the delay becoming independent of α as N goes to infinity). Throughout the analysis we have assumed that the parasitic capacitance is independent of device sizing, which is the common case when interconnect dominates. However, the drain

and source diffusion and perimeter capacitances actually increase with increasing area, favoring smaller size devices and making the above a worst-case analysis. Also plotted in Fig. 11 are simulation results from extracted layouts of an 8-b adder carry chain for three different device (W/L) ratios ($N = 1$, $N = 2$, and $N = 4$). The curve follows the simple first-order model derived very well, and suggests that this example is dominated more by the effect of gate capacitance rather than parasitics. In this case, increasing devices (W/L)'s does not help, and the solution using the smallest possible (W/L) ratios results in the best sizing.

These results indicate that minimum sized devices should be used when the total load capacitance is not dominated by the interconnect. This is indeed the case for circuitry that drive local busses inside datapaths. To drive large capacitances, such as between datapath modules or off-chip, buffers with appropriate strength can be used.

A low-power cell library has been designed which mostly uses minimum sized devices to implement most logic functions and has been used to design the circuits described in Section VII. This cell library typically switches a factor of 2–3 lower capacitance compared to cell libraries optimized for speed.

C. Reduced Swing Logic

Reducing the power supply voltage is clearly a very effective way to reduce the energy per operation since it has a quadratic impact on the power consumption. At a given reduced supply voltage, the output of CMOS logic gates will make rail to rail transitions; an approach to reducing the power consumption further at a fixed reduced supply voltage is to reduce the swing on the output node. For example, using an NMOS device to pull up the output will limit the swing to $V_{dd} - V_t$, rather than rising all the way to the supply voltage. The power consumed for a $0 \rightarrow V_{dd} - V_t$ transition will be $C_L V_{dd}(V_{dd} - V_t)$, and therefore the power consumption reduction over a rail to rail scheme will be $\propto V_{dd}/(V_{dd} - V_t)$. This scheme of using an NMOS device to reduce the swing has two important negative consequences: First, the noise margin for output high (NM_H) is reduced by the amount V_T, which can reduce the margin to 0 V, if the supply voltage is set near the sum of the thresholds.

178

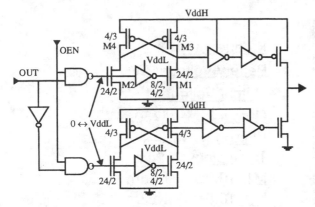

Fig. 13. Level conversion output pad driver.

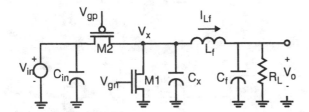

Fig. 14. Low-voltage dc/dc buck converter.

Second, since the output does not rise to the upper rail, a static gate connected to the output can consume static power for a high output voltage (since the PMOS of the next stage will be "on"), increasing the effective energy per transition.

Therefore, to utilize the voltage swing reduction, special gates are needed to restore the noise margin to the signal, and eliminate short-circuit currents. These gates require additional devices that will contribute extra parasitic capacitances. Fig. 12 shows a simplified schematic of such a gate, used in the FIFO memory cells of a low-power cell library [6]. This circuit uses a precharged scheme and the device M3 is used to clip the voltage of the bit-line (which has several transistors similar to M5 connected to it) to $V_{dd} - V_t$, where $V_t > V_{t0}$ due to the body effect. The devices M1 and M4 are used to precharge the internal node (the input of the inverter) to V_{dd}, and the bit-line to $V_{dd} - V_T$. During evaluation ($\varphi = $ "1"), if V_{in} is high, the bit-line will begin to drop, as shown in the SPICE output next to the schematic. Because the capacitance ratio of the bit-line to the internal node is very large, once the bit-line has dropped roughly 200 mV to sufficiently turn on M3, the internal node quickly drops to the potential of the bit-line, providing signal amplification. Thus this circuit greatly reduces the voltage swing on the high-capacitance bit-line, which reduces the energy, and provides signal amplification, which reduces the delay, as well. If the load on the bit-line was on the order of just a few gates, the energy savings would be marginal, due to the extra parasitic capacitances and therefore the signal swing reduction technique presented above is most useful for high-capacitance nodes.

D. Low-Voltage Support Circuitry: Level Converting Pad

One important type of circuit required for low-voltage operation is a level-shifter that can convert low-voltage

signal swings from the core of the low-power chips (e.g., 1.5 V) to the high-voltage signal swings required by I/O devices (e.g., 5 V required by displays) or vice-versa. Fig. 13 shows the schematic of a level converting output pad driver.

This tri-stateable output buffer uses two supply voltages, the low-voltage supply, $V_{dd}L$, that is tied to the pad-ring and the high-voltage supply, $V_{dd}H$, coming in through another unbuffered pad. The low-to-high conversion circuit is a PMOS cross coupled pair (M3, M4) connected to the high supply voltage, driven differentially (via M1, M2) by the low-voltage signal from the core. The N-device pulldowns, M1 and M2, are dc ratioed against the cross coupled P-device pullups, M3 and M4, so that a low-swing input ($V_{dd}L = 1$ V) guarantees a correct output transition ($V_{dd}H = 5$ V). That is, the PMOS widths are sized so that the drive capability of the NMOS can overpower the drive of the PMOS, and reverse the state of the latch. This level-converting pad consumes power only during transitions and it consumes no dc power. The remaining buffer stages and output driver are supplied by $V_{dd}H$. This level-conversion pad will work only with an NWELL process since it requires isolated wells for the PMOS devices that are connected to the high voltage and the ones that are connected to the low-voltage devices.

E. Low-Voltage Support Circuitry: High Efficiency Low-Voltage dc/dc Conversion

The architecture driven voltage scaling strategy which will be presented later assumes that the supply voltage is a free variable. In order to realize such portable systems in which different parts of the system could operate at their own "optimum" supply voltage and communicate with each other using the level-conversion circuitry described in Section IV-D, the design of high efficiency low-voltage (in which the voltage can be made programmable) switching regulators must be considered. Fig. 14 shows the schematic of a Buck regulator that can be used to generate low-supply voltages with high efficiency. The converter works by chopping the input voltage to reduce the average voltage. This produces a square-wave of duty cycle D with frequency f_s at the inverter output node V_x. The chopped signal is filtered by the second-order low-pass filter (L_f and C_f) to reduce the ac component to an acceptable ripple value. Buck converters are capable of a 0% to 100% duty factor and the output voltage is approximately equal to the input voltage multiplied by duty factor. For example, with $V_{in} = 6$ V and $D = 25\%$, the output voltage is 1.5 V. The output can be set to an arbitrary value by controlling the duty cycle on the inverter output node V_x. There are three main sources of dissipation that cause the conversion efficiency of this circuit to be less than unity: conduction loss (I^2R loss in the power transistors and filter components), switching loss ($C_x V_{in}^2 f_s$ loss due to parasitics), and $C_g V_{in}^2 f_s$ gate drive loss for each FET. Four techniques are presented in [7] to improve the efficiency of the converter at reduced voltages and current levels. Synchronous Rectification: The free-wheeling diode used

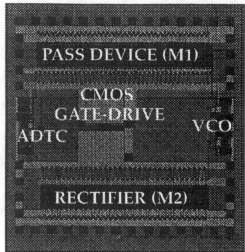

Frequency: 1Mhz

Battery Voltage: 6V

Output Voltage: 1.5V

Output Power: 750mW

Efficiency: 92%

Technology: 1.2μm

M1: 10.2cm / 1.2μm

M2: 10.5cm / 1.2μm

Size: 4.2mm X 4.2mm

Fig. 15. Die photo of a 1.5 V buck converter [7].

in conventional high voltage converters is replaced with a gated NMOS device (M1). For low output voltage levels, the diode's voltage drop becomes a significant fraction of the output voltage, leading to large conduction loss. A gated NMOS device can achieve the same function more efficiently.

1) Soft Switching: Capacitive switching loss is nearly eliminated by using the filter inductor as a current source to charge and discharge the inverter node; therefore, the power transistors are turned on and off at $V_{DS} = 0$ (referred to as zero voltage switching (ZVS)). This factor is especially important for high-frequency converters.

2) Adaptive Dead Time Control: ZVS depends critically on the dead-time, during which neither M1 nor M2 conducts. A fixed dead-time designed assuming an average current load can result in significant losses if operating conditions change. An adaptive dead-time control through a negative feedback loop is used to accommodate varying operating conditions and process variations.

3) Resonant-Transition Gate-drive: A resonant inductor is used to charge and discharge the gate capacitances of the power devices, recycling a significant fraction of the gate charge. Fig. 15 shows the die photo of the low-voltage converter.

F. Logic Minimization and Technology Mapping

Recently some effort has gone into changing the optimization criteria for synthesizing logic structures (both combinational and sequential) to one that addresses low power. Typical reduction in power consumption that can be achieved by optimizing at this level for random logic is on the order of 25%. Techniques have been proposed which choose logic to minimize switching [8], position registers through retiming to reduce glitching activity [9] and to decrease power during the technology mapping phase [10], [11] choosing gates from a library which reduces switching; for example a three input AND gate can be implemented as a single 3-input gate (i.e., NAND followed by an inverter) or two 2-input AND gates with different power results.

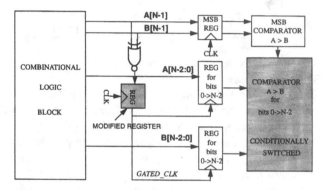

Fig. 16. Using gated clocks to reduce power.

G. Logic Level Power-Down and Gated Clocks

Powering down has traditionally been applied only at the chip and module levels, however, the application to the logic level can also be very beneficial to reduce the switching activity at the expense of some additional control circuitry [12]. Assume a pipelined system for comparing the output of two numbers from a block of combinational logic as shown in Fig. 16; the first pipeline stage is a combinational block and the next pipeline stage is a comparator which performs the function $A > B$, where A and B are generated in the first stage (i.e., from the combinational block). If the most significant bits, $A[N-1]$ and $B[N-1]$, are different then the computation of $A > B$ can be performed strictly from the MSB's and therefore the comparator logic for bits $A[N-1:0]$ and $B[N-2:0]$ is not required (and hence the logic can be powered down). If the data is assumed to be random (i.e., there is a 50% chance that $A[N-1]$ and $B[N-1]$ are different), the power savings can be quite significant. One approach to accomplish this is to gate the clocks as shown in Fig. 16. The XNOR output of the $A[N-1]$ and $B[N-1]$ is latched by a special register to generate a gated clock. This gated clock is then used to clock the lower order registers. Since the latch to the lower bits is conditional, they must be made static. The standard TSPC register [13] is shown in Fig. 17 as modified

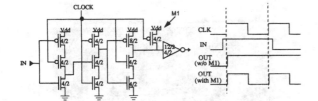

Fig. 17. Schematic of a modified TSPC latch that is used to generate gated clocks.

to support clock gating. As shown in the timing on the right side of Fig. 17, with the extra PMOS device M1, the output can be forced to a ZERO during the low phase of the clock. Without this device, it will not be possible to generate rising edges for the gated clock on two consecutive rising edges of the system clock. Gated clocks have been used extensively in the system presented in Section VII.

V. ARCHITECTURE LEVEL OPTIMIZATION

There is a great deal of freedom in optimizing architectures for low power. An architecture driven voltage scaling strategy is presented in this section in which concurrent architectures are used to retain throughput at reduced supply voltages. This is an extremely effective approach for power consumption reduction and can in many cases, yield more than an order of magnitude savings. Various architectural techniques are also presented to reduce the effective capacitance switched which involves the choice of number representation, exploitation of signal correlations, and resynchronization to minimize glitching.

A. Architecture Driven Voltage Scaling

The simple first order delay analysis presented in (10) is reasonably accurate for long channel devices. However, as feature sizes shrink below 1.0 μm, the delay characteristics as a function of lowering the supply voltage deviate from the first order theory presented since it does not consider carrier velocity saturation under high electric fields. As a result of velocity saturation, the current is no longer a quadratic function of the voltage but linear; hence, the current drive is significantly reduced and is approximately given by $I = WC_{ox}(V_{dd} - V_t)\nu_{max}$ [14]. Given this and assuming the delay of a circuit is given by CV_{dd}/I, we see that the delay for submicron circuits is relatively independent of supply voltages at high electric fields. A "technology" based approach proposes choosing the power supply voltage based on maintaining the speed performance for a given submicron technology [15]. By exploiting the relative independence of delay on supply voltage at high electric fields, the voltage can be dropped to some extent for a velocity-saturated device with very little penalty in speed performance. Because of this effect, there is some movement to a 3.3 V industrial voltage standard since at this level of voltage reduction, to the "critical voltage," there is not a significant loss in circuit speed [15]. This approach can achieve a 60% reduction in power when compared to a 5 V operation [16].

The above mentioned "technology" based approach is focusing on reducing the voltage while maintaining device speed, instead of attempting to maintain computational throughput at a minimum power level. From (1) it is clear that CMOS logic gates achieve lower power-delay products as the supply voltages are reduced. In fact, once a device is in velocity saturation there is a further degradation in the energy per computation, so in minimizing the energy required for computation, the above critical voltage provides an *upper* bound on the supply voltage (whereas in the technology scaling theory it provided a *lower* bound!). It now will be the task of the architecture to compensate for the reduced circuit speed, that comes with operating below the critical voltage. Examples of optimizing architectures for both arithmetic and memory computation are presented below.

1) Arithmetic Computation: A Simple Adder Comparator Datapath [17] (Computation Without Feedback): To illustrate how architectural techniques can be used to compensate for reduced speeds, a simple 8-b datapath consisting of an adder and a comparator is analyzed assuming a 2.0 μm technology. As shown in Fig. 18, inputs A and B are added, and the result compared to the worst-case delay through the adder, comparator and latch is approximately 25 ns at a supply voltage of 5 V, the system in the best case can be clocked with a clock period of $T = 25$ ns. When required to run at this maximum possible throughput, it is clear that the operating voltage cannot be reduced any further since no extra delay can be tolerated, hence yielding no reduction in power. We will use this as the reference datapath for our architectural study and present power improvement numbers with respect to this reference. The power for the reference datapath is given by

$$P_{\text{ref}} = C_{\text{ref}}V_{\text{ref}}^2 f_{\text{ref}} \qquad (14)$$

where C_{ref} is the total effective capacitance being switched per clock cycle. The effective capacitance was determined by averaging the energy over a sequence of input patterns with a uniform distribution. One way to maintain throughput while reducing the supply voltage is to utilize a parallel architecture. As shown in Fig. 19, two identical adder-comparator datapaths are used, allowing each unit to work at half the original rate while maintaining the original throughput. Since the speed requirements for the adder, comparator, and latch have decreased from 25 ns to 50 ns, the voltage can be dropped from 5 V to 2.9 V (the voltage at which the delay doubled, from Fig. 7). While the datapath capacitance has increased by a factor of 2, the operating frequency has correspondingly decreased by a factor of 2. Unfortunately, there is also a slight increase in the total "effective" capacitance introduced due to the extra routing, resulting in an increased capacitance by a factor of 2.15. Thus the power for the parallel datapath is given by

$$P_{\text{par}} = C_{\text{par}}V_{\text{par}}^2 f_{\text{par}}$$
$$= (2.15C_{\text{ref}})(0.58V_{\text{ref}})^2\left(\frac{f_{\text{ref}}}{2}\right) \approx 0.36P_{\text{ref}}. \qquad (15)$$

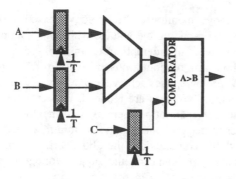

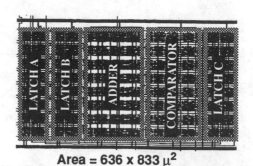

Area = 636 x 833 μ²

Fig. 18. A simple datapath with corresponding layout.

Fig. 19. Parallel implementation of the simple datapath.

Area = 1476 x 1219 μ²

The amount of parallelism can be increased to further reduce the power supply voltage and the power consumption for a fixed throughput. However, as the supply approaches the threshold voltage of the devices, the delays increase significantly with a reduction in supply voltage and therefore the amount of parallelism and corresponding overhead circuitry increase significantly. At some "optimum" voltage, the overhead circuitry due to parallelism dominates and the power starts to increase with further reduction in supply [17].

Another possible approach is to apply pipelining to the architecture, as shown in Fig. 20. With the additional pipeline latch, the critical path becomes the $\max[T_{\text{adder}}, T_{\text{comparator}}]$, allowing the adder and the comparator to operate at a slower rate. For this example, the two delays are equal, allowing the supply voltage to again be reduced from 5 V used in the reference datapath to 2.9 V (the voltage at which the delay doubles) with no loss in throughput. However, there is a much lower area overhead incurred by this technique, as we only need to add pipeline registers. Note that there is again a slight

increase in hardware due to the extra latches, increasing the "effective" capacitance by approximately a factor of 1.15. The power consumed by the pipelined datapath is

$$P_{\text{pipe}} = C_{\text{pipe}} V_{\text{pipe}}^2 f_{\text{pipe}}$$
$$= (1.15 C_{\text{ref}})(0.58 V_{\text{ref}})^2 f_{\text{ref}} \approx 0.39 P_{\text{ref}}. \quad (16)$$

With this architecture, the power reduces by a factor of approximately 2.5, providing approximately the same power reduction as the parallel case with the advantage of lower area overhead. As an added bonus, increasing the level of pipelining also has the effect of reducing logic depth and hence power contributed due to hazards and critical races.

Clearly an even bigger improvement can be obtained by simultaneously exploiting parallelism and pipelining. The summary of all these cases along with the area penalty is presented in Table 2.

2) Memory Access: The same parallelism concept used in the previous section can be used to optimize memory operations for low-power. For example, Fig. 21 shows two alternate schemes for reading 8 b of data from memory

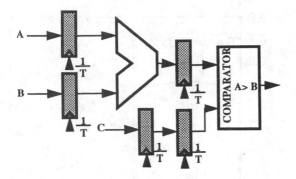

Fig. 20. Pipelined implementation of the simple datapath.

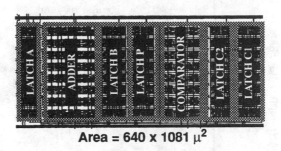

Area = 640 x 1081 μ²

Table 2 Architecture Based Voltage Scaling Results

Architecture	Voltage	Area (normalized)	Power (normalized)
Simple	5 V	1	1
Parallel	2.9 V	3.4	0.36
Pipelined	2.9 V	1.3	0.39
Pipelined-Parallel	2.0	3.7	0.2

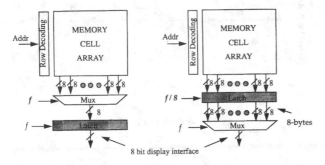

Fig. 21. Parallel memory access reduces circuit speed requirement enabling low-voltage operation.

at throughput f. On the left hand side is the serial access scheme in which the 8-b of data are read in a serial format and the memory is clocked at the throughput rate f. Assume that this implementation can operate at 3 V. Another approach is to read several words from memory and clock the memory at a lower rate for the same throughput. For example, reading 8 bytes in parallel implies that the memory can be clocked at $\frac{1}{8}$ the serial rate. This implies that the time available to read the memory for the parallel implementation is 8 times as long as the serial version and therefore the supply voltage can be dropped for a fixed throughput. The parallel version can run at a supply voltage of 1.1 V (based on the delay versus V_{dd} for the SRAM presented in Section VII) while meeting throughput requirements—note that once again the multiplexor is running at the throughput rate. An important point to note is that this optimization is possible only if the data access pattern is sequential in nature, as is the case for a video frame-buffer.

B. Minimizing Switching Activity by Choice of Number Representation

In most signal processing applications, two's complement is typically chosen to represent numbers since arithmetic

operations (addition and subtraction) are easy to perform. Fig. 22 shows a short segment of a speech signal and the associated transition probabilities (both the $0 \rightarrow 1$ and $1 \rightarrow 0$) assuming two's complement for each bit.

The transition probability represents the average number of transitions per cycle. There are three important regions in the transition graph: the lower-order region (the LSB's), the middle region, and the higher-order region [18]. In the lower-order region, the bits are uncorrelated both temporally and spatially and therefore the transition probability is $\frac{1}{2}$; i.e., there is a 50% probability that the bits will be in the 0 state or 1 state—a reasonable assumption for slowly varying signals. The higher-order bits represent the sign-extension operation performed in two's complement representation; i.e., the transition probability on the higher order bits is determined by the frequency at which the signal changes from positive to negative or from negative to positive. In the middle region, the transition of the bits fall between the transition probabilities of the lower-order bits and higher-order bits. The breakpoints (points of transition between the lower-order bits and the middle region and between the middle region and the higher-order bits) are determined by the mean and variance of the signal. One of the problems with two's complement representation is sign-extension, which causes the MSB sign-bits to switch when a signal transitions from positive to negative or vice-versa (for example, going from -1 to 0 will result in all of the bits toggling). Therefore using a two's complement representation can result in significant switching activity when the signals being processed switch frequently around zero and when they do not utilize the entire bit-width (i.e., the dynamic range is much smaller than the maximum possible value determined from the bit-width) since a lot of the MSB bits will perform sign extension. Even if a signal utilizes the entire bit-width, arithmetic operations such as scaling can reduce the signal dynamic range.

1) Reducing the Number of Transitions on Busses Using Sign Magnitude: One approach to minimizing the switching in the MSB's is to use a sign-magnitude representation, in which only one bit is allocated for the sign and the rest for the magnitude. In this case, if the dynamic range of a signal does not span the entire bitwidth, only one bit will toggle when the signal switches sign, as opposed to the two's complement representation where due to sign

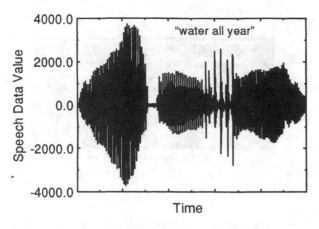

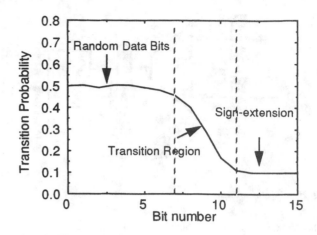

Fig. 22. Data in signal processing applications is often correlated.

extension several of the bits will switch. To illustrate this, consider gaussian data applied to a 16-b data-bus, and the let the signal have a mean of 0 and a $3\sigma = 2^{11}$. Fig. 23(a) shows the transition probabilities versus bit position number for two's complement representation as a function of the first order correlation coefficient $\rho = \mathrm{cov}(X_n, X_{n+1})/\sigma^2$. A $\rho = 0$ indicates that the data is uncorrelated and therefore the transition probability for the MSB's is $\frac{1}{2}$ (i.e., there is a 50% chance the output is going to transition from positive to negative). A large positive correlation coefficient (e.g., +0.99) implies that the signal changes very slowly and therefore switches sign very infrequently (i.e., the transition probability is close to zero). Similarly, a negative correlation coefficient with a large magnitude (e.g., −0.99) implies that the signal changes frequently from positive to negative.

The upper breakpoint (which represents the dynamic range of signal) lies at $\log_2(3\sigma) = 11$ b in this case. Therefore for the two's complement representation, the bits from 11–15 indicate the activity of the sign bit. If a sign-magnitude representation is used (shown in Fig. 23(b)), the bits 11–14 will have a low transition activity, and bit 15, whose transitions represent the sign transition probability, will have the same activity as the two's complement representation. Also, the transition region has lower activity for the sign-magnitude representation. Clearly, there is an advantage in using sign-magnitude representation over two's complement to reduce the switching activity.

From the above example, it is clear that the reduction in the number of transitions for sign-magnitude over two's complement is a function of both the dynamic range of the signal and the signal correlation coefficient. Consider evaluating the reduction in the number of transitions as a function of dynamic range and the correlation coefficient. Let the normalized dynamic range of a signal be defined as $3\sigma/\text{maximum amplitude}$. For a 16 b example, this turns out to be: $3\sigma \, 2^{16}-1$. Let the number of transitions represent the sum of the transitions per clock cycle for all the bits, i.e.,

$$\text{Number of Transistors} = \sum_{i=1}^{16} \alpha_i. \qquad (17)$$

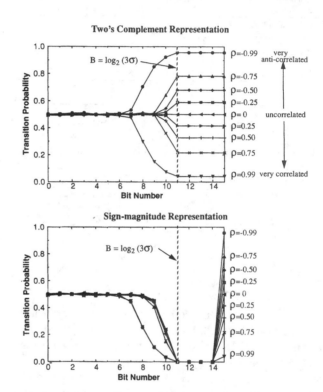

Fig. 23. Transition activity for different number representations.

Fig. 24 shows a plot of the total number of transition as a function of the normalized dynamic range and the correlation factor for two's complement representation. For a correlation factor of $\rho = 0$, the number of transitions is equal to 8 and is independent of the normalized dynamic range since the data is random and each bit has a 50% probability of transitioning (≥ 8 transitions for a 16-b bus). For very correlated data (e.g., $\rho = 0.99$), the MSB's don't switch very often while the LSB's switch 50% of the time. Therefore, for a small normalized dynamic range, the average number of transitions is very small and for a large dynamic range it approaches the value of 8 (since the lower region with activity of 50% extends out to more bits). Similarly, for very anti-correlated data (e.g., $\rho = -0.99$), the MSB switch very frequently. Therefore, a small normalized dynamic range will result in a lot of

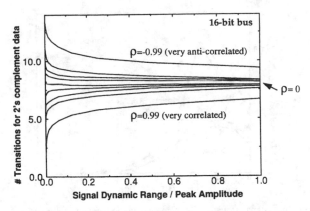

Fig. 24. Transition activity for two's complement representation.

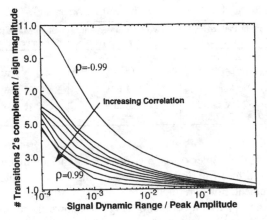

Fig. 25. Activity reduction for sign-magnitude over two's complement.

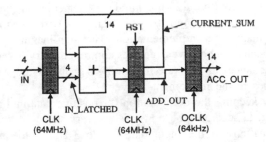

Fig. 26. Two's complement implementation of an accumulator.

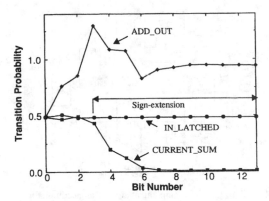

Fig. 27. Signal statistics for two's complement implementation of the accumulator datapath assuming random inputs.

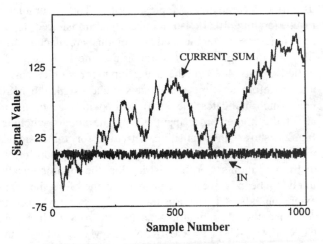

Fig. 28. Signal value for two's complement implementation assuming random inputs.

transitions (since many bits are allocated to sign extension) while the number of transition approaches 8 for high values of the normalized dynamic range.

Fig. 25 shows the ratio of the number of transitions required in the two's complement representation to the sign-magnitude representation as a functions of signal correlation and normalized dynamic range (plotted on a log axis). From this plot, it is clear that the biggest win for sign-magnitude representation is when the dynamic range is smallest and the signal is very anticorrelated.

The above analysis suggests that sign-magnitude has some advantages in terms of the number of the transitions on busses. However, addition and subtraction computation are difficult to implement in sign-magnitude representation.

Sign-magnitude is therefore most useful for cases where large busses have to be driven (for example external memory access), where the overhead for converting back to two's complement is quite insignificant compared to the reduction in capacitance switched in the large busses. That is, a small overhead capacitance is added to the system to reduce the *overall* switched capacitance. The scheme presented here is only one example of coding data for reducing power. There are several other schemes which can be used such as log representation, differential coding, gray-coding, etc. The coding scheme to be used is a function of signal statistics, capacitance overhead introduced, area overhead, delay, latency, etc.

2) Reducing Activity for Arithmetic Computation: As mentioned in the previous section, sign-magnitude has some advantages in terms of reducing the number of the transitions on busses, but addition/subtraction operations are difficult to implement. However, in several applications that require multiple additions inside a sample period (e.g., Multiply Accumulate Units of DSP filters), an architecture optimization can be used that trades silicon area for lower switching activity without altering the throughput. To illustrate this consider a correlator example in which the correlation length is 1024; the samples, whose values range from −7 to +7, are accumulated at 64 MHz. Fig. 26

185

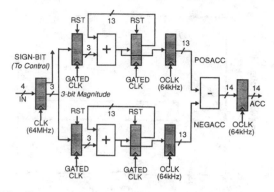

Fig. 29. Sign magnitude implementation of an accumulator.

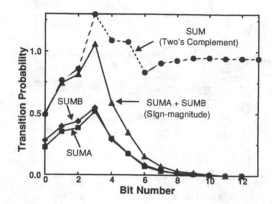

Fig. 30. Signal statistics for sign magnitude implementation of the accumulator datapath assuming random inputs.

Table 3 Number Representation Tradeoff for Arithmetic

Input Pattern (1024 cycles)	Two's Complement Power, 3V	Sign Magnitude Power, 3V
Constant (IN=7)	1.97 mW	2.25 mW
Ramp (-7,-6,...7,-7)	2.13 mW	2.43 mW
Random	3.42 mW	2.51 mW
Min -> Max -> Min (-7, +7,-7,+7,...)	5.28 mW	2.46 mW

shows a conventional architecture for an accumulator that uses two's complement representation. The accumulated result (adding 1024, 4-b numbers) is transferred to an accumulator register at 64 kHz. In this architecture, the MSB of the input, bit 3 (assuming that the LSB is bit 0), is tied to bits 4–13 of the adder input for sign-extension and therefore anytime the input switches sign, the MSB bit (which indicates the sign) will switch, resulting in all of the higher order input bits to the adder switching. Fig. 27 shows the transition activities for three signals in the two's complement datapath assuming uniformly distributed inputs. For this distribution, the input is equally likely to be positive or negative, and therefore the sign-extension bits 3–13 have a transition probability close to $\frac{1}{2}$. Since the accumulator acts as a low-pass filter (i.e., $\text{CURRENT\_SUM}_N = \text{CURRENT\_SUM}_{N-1} + \text{IN}_N$), the higher order bits have little switching activity even when the input is rapidly varying; that is, the accumulator smoothes the input signal. This is shown in Fig. 28 which shows the CURRENT_SUM output and the input value for 1024 samples (here the input is random and varying from -7 to $+7$). Although the CURRENT_SUM has low switching activity, the adder output (before the latch) has significant switching activity due to glitching, as seen from Fig. 27. The glitching activity arises since all of the input bits to the adder switch each time the input changes sign, and this results in high switching activity of the adder (even though the final adder output at the end of the cycle does not change too much relative to its value at the beginning of the cycle). Another approach for implementing the accumulator is to use a sign-magnitude representation whose datapath is shown in Fig. 29. Here two accumulator datapaths are used, one that sums all positive numbers and one that sums all negative numbers. The latched sign-bit from the input register is used to generate gated clocks that enables the positive datapath latch or the negative datapath latch; this ensures that this scheme does not increase effective clock load. At the end of 1024 cycles, the positive and negative accumulated values are transferred to separate registers. A subtract operation is then performed at the lower frequency and therefore is quite negligible in terms of capacitance overhead. The key to low-power is that there is no sign-extension is being performed and therefore the adder has a low switching activity in the higher order bits.

In fact, the higher order bits only need an incrementer (as opposed to a full-adder required by the two's complement implementation), reducing the number of gates that are switched in the accumulator. Fig. 30 shows the transition activities for the output of the adders in both datapaths. Also shown in the figure is the transition activity for the two's complement implementation and the sum of the transition activities for the sign-magnitude implementation; we can see that the glitching activity is significantly reduced in the sign-magnitude implementation. The sign-magnitude implementation, however, requires control circuitry (overhead capacitance) to generate the timing signals for the various latches in the implementation.

By keeping the computation for positive data and negative data separate, the power is not very sensitive to rapid fluctuations in the input data. Table 3 shows the power estimates for various input patterns. Again, the biggest advantage is when the sign toggles frequently. For the case when the input changes very slowly, the sign-magnitude implementation consumes more power (15%) due to the capacitance switched by the control overhead circuitry.

C. Ordering of Input Signals

The switching activity can be reduced by optimizing the ordering of operations in a design. To illustrate this, consider the problem of multiplying a signal with a constant coefficient, which is a very common operation in signal processing applications. Multiplications with constant coefficients are often optimized by decomposing the multiplication into shift-add operations and using the CSD representation. Consider the example in which a multiplication with a constant is decomposed into IN + IN $\gg 7$ + IN $\gg 8$. The shift operations (denoted by $\gg$)

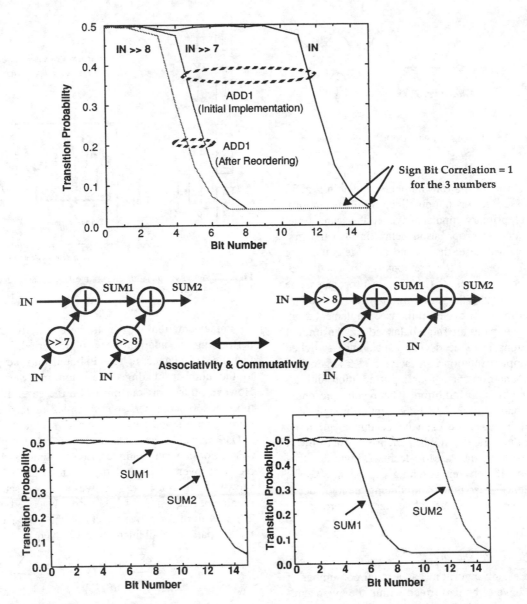

Fig. 31. Reducing activity by reordering inputs.

represent a scaling operation, which has the effect of reducing the dynamic range of the signal. This can be seen from Fig. 31 which shows the transition probability for the 3 signals IN, IN $\gg$ 7 and IN $\gg$ 8. In this example, IN has a large variance and almost occupies the entire bit-width. The shifted signal are scaled and have a much lower dynamic range. Now consider the two alternate topologies of implementing the two required additions. In the first implementation, IN and IN $\gg$ 7 are added in the first adder and the sum (SUM1) is added to IN $\gg$ 8 in the second adder. In this case, the SUM1 transition characteristics is very similar to the characteristics of the input IN since the amplitude of IN $\gg$ 7 is much smaller than IN and the 2 inputs have identical sign-bits (since a shift operation does not change the sign). Similarly SUM2 is very similar to SUM1 since SUM1 is much larger in magnitude than IN $\gg$ 8. In the second implementation (obtained by applying associativity and commutativity), the

two small number IN $\gg$ 7 and IN $\gg$ 8 are summed in the first adder and the output is added to IN in the second adder. In this case, the output of the first adder has a small amplitude (since we are adding 2 scaled number of the same sign) and therefore lower switching activity. The second implementation switched 30% less capacitance than the first implementation. This example demonstrates that ordering of operation can result in reduced switching activity.

D. Minimizing Glitching Activity

As mentioned in Section II-A-4 static designs can exhibit spurious transitions due to finite propagation delays from one logic block to the next (also called critical races and dynamic hazards); that is, a node can have multiple transitions in a single clock cycle before settling to the correct logic level. To minimize the "extra" transitions and power in a design, it is important to balance all signal paths and reduce the logic depth. For example, consider the two

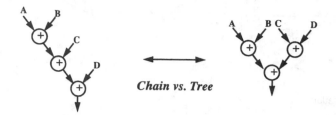

Chain vs. Tree

Fig. 32. Reducing the glitching activity by balancing signal paths.

implementations for adding four numbers shown in Fig. 32 (assuming a cascaded or nonpipelined implementation). Assume that all primary inputs arrive at the same time. Since there is a finite propagation delay through the first adder for the chained case, the second adder is computing with the new C input and the previous output of $A + B$. When the correct value of $A + B$ finally propagates, the second adder recomputes the sum. Similarly, the third adder computes three times per cycle. In the tree implementation, however, the signal paths are more balanced and the amount of extra transitions is reduced. The capacitance switched for a chained implementation is a factor of 1.5 larger than the tree implementation for a four input addition and 2.5 larger for an eight input addition. The results presented above indicate that increasing the logic depth (through more cascading) will increase the capacitance due to glitching while reducing the logic depth will increase register power. Hence the decision to increase or decrease logic depth is based on a tradeoff between glitching capacitance versus register capacitance. Also note that reducing the logic depth can reduce the supply voltage while keeping throughput fixed.

E. Degree of Resource Sharing

A signal processing algorithm has a required number of operations that have to be performed within the given sample period. One strategy for implementing signal processing algorithms is the direct mapping approach, where there is a one to one correspondence between the operations on the signal flow graph and operators in the final implementation. Such an architecture style is conceptually simple and requires a small or no controller. Often, however, due to area constraints or if very high-throughput is not the goal (e.g., speech filtering), time-multiplexed architectures are utilized in which multiple operations on a signal flowgraph can be mapped onto the same functional hardware unit. Given that there is a choice between time-multiplexed and fully parallel architectures, as is the case in low to medium throughput applications, an important question arises which is what is the architecture that will result in the lowest switching activity. To first order, it would seem that the degree of time-multiplexing would not affect the capacitance switched by the logic elements or interconnect. For example, if a data flowgraph has five additions to be performed inside the sample period, it would seem that there is no difference between an implementation in which one physical adder performs all five additions or

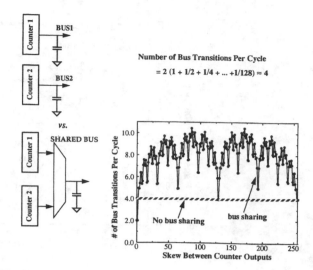

Fig. 33. Activity tradeoff for time-multiplexed hardware: Bus-sharing example.

an implementation in which there are five adders each performing one addition per sample period. It would seem that the capacitance switched should only be proportional to the number of times the additions were performed. However, this turns out not to be the case. To understand this tradeoff, consider two examples of resource sharing: sharing busses and sharing execution units.

1) Example 1: Time-Sharing Busses (Output of Two Counters): Consider an example of two counters whose outputs are sent over parallel and time-multiplexing busses as shown in Fig. 33. In case 1, we have two separate busses running at frequency f while case 2 has one bus running at $2f$. Therefore for a fixed voltage, the power consumption for the parallel implementation is given by

$$P_{\text{no bus-sharing}} = \tfrac{1}{2}(\sum \alpha_i)C_{nbs}V^2f + \tfrac{1}{2}(\sum \alpha_i)C_{nbs}V^2f$$
$$= (\sum \alpha_i)C_{nbs}V^2f \qquad (18)$$

where C_{nbs} is the physical capacitance per bit of BUS1 and BUS2 (assume for simplicity that all bits have the same physical capacitance) and α_i is the transition activity for bit i (for both $0 \rightarrow 1$ and $1 \rightarrow 0$ transitions and therefore there is a factor of $\tfrac{1}{2}$ in (18)). In this case, since the output is coming from a counter, the activity is $\sum \alpha_i = 1 + \tfrac{1}{2} + \tfrac{1}{4} + \cdots + \tfrac{1}{128}$ since the LSB switches every cycle, the 2nd LSB switches every other cycle, etc. This means that each bus will have on the average a total of approximately two transitions per clock cycle. For the time-multiplexed implementation, we have one bus C_{bs} running at twice the frequency. Typically C_{bs} will be smaller than C_{nbs} since with fewer interconnects, it is easier to route the chip and hence the wire lengths are smaller. The goal here is to show that the activity α' is modified due to time-multiplexing and the power consumption for this implementation is given by

$$P_{\text{bus-sharing}} = \tfrac{1}{2}\sum \alpha'C_{bs}V^22f = \sum \alpha'C_{bs}V^2f. \qquad (19)$$

Fig. 33 shows the plot of total number of transitions per cycle (i.e., $\sum \alpha'$) for the time-shared case as a function

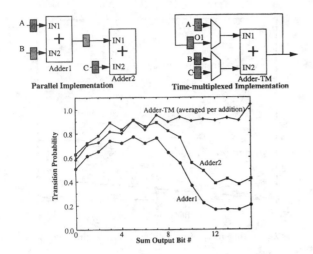

Fig. 34. Activity tradeoff for time-multiplexed hardware: Adder example.

of the skew between the counter outputs. The figure also shows the number of transitions for the nontime-shared implementation which is independent of skew and is equal to 4 (for 2 busses). As seen from this figure, except for one value of the counter skew (when the skew is $= 0$), the "parallel" implementation has the lower switching activity. This example shows that time-sharing can significantly modify the signal characteristics and cause an increase in switching activity.

2) Example 2: Time-Sharing Execution Units: The second example is a simple second order FIR filter which will demonstrate that time-multiplexing of execution units can increase the switched capacitance. The FIR filter is described as follows:

$$Y - a_0 \cdot X + a_1 \cdot X@1 + a_2 \cdot X@2 = A + B + C \quad (20)$$

where @ represents the delay operator and $A = a_0 \cdot X$, $B = a_1 \cdot X@1$ and $C = a_2 \cdot X@2$. Also let $O1 = A + B$. The value of the coefficients are: $a_0 = 0.15625$, $a_1 = 0.015625$, and $a_2 = -0.046875$.

One possible implementation might be to have two physical adders, one performing $A + B = O1$ and the other performing $O1 + C$. An alternate implementation might be to have a single time-multiplexed adder performing both additions. So, in cycle 1, $A + B$ is performed, and in cycle 2, $O1 + C$ is performed. The topologies for the two cases are shown is Fig. 34. In the first implementation, the adders can be chained and therefore the effective critical path can be reduced (chaining two ripple carry adders of N bits has a delay $= (N + 1) \cdot$ Delay of one bit); it is clear that this will result in extra glitching activity but since the objective here is to isolate and illustrate the activity modification only due to time multiplexing, the first implementation is assumed to be registered. The IRSIM simulator was used to determine the switching activity for the adders in the two implementations assuming speech input data. The transition activities for the two adders for the parallel implementation and the average transition activity per addition for the time-multiplexed implementation are

shown in Fig. 34. The results once again indicate that time-multiplexing can increase the overall switching activity. The basic idea is that in the fully parallel implementation, the inputs to the adders change only once every sample period and the source of inputs to the adders are fixed (for example, IN1 of adder1 always comes from the output of the multiplier $a_0 * X$). Therefore, if the input changes very slowly (i.e., the input data is very correlated), the activity on the adders become very low. However, in the time multiplexed implementation, the inputs to the adder change twice during the sample period and more importantly arrive from different sources. For example, during the first cycle, IN2 of the time-multiplexed adder is set to B and during the second cycle, IN2 of the time-multiplexed adder is set to C. Thus even if the input is constant (which implies that the sign of X, $X@1$, and $X@2$ is the same) or slowly varying, the sign of B and C will be different since the sign of the coefficients a_1 and a_2 are different. This will result in the input to adder switching more often and will therefore result in higher switching activity. That is, even if the input to the filter does not change, the adder can still be switching. For this example, the time multiplexed adder switched 50% extra capacitance per addition compared to the parallel case (even without including the input changes to adders or the multiplexor overhead).

VI. ALGORITHM LEVEL OPTIMIZATION

The choice of algorithm is the most highly leveraged decision in meeting the power constraints. The ability for an algorithm to be parallelized is critical and the basic complexity of the computation must be highly optimized. In this section, a few examples illustrating the tradeoffs at the algorithmic level is presented which include exploitation of concurrency and minimizing the number of operations.

A. Voltage Reduction using Algorithmic Transformations

In the Section V-A, an architecture driven voltage scaling strategy was presented in which the computation was parallelized to reduce the supply voltage and power. The examples presented in Section V-A did not have any feedback and the computation was easily parallelizable. In applications that have feedback (e.g., infinite impulse response), the computation cannot be easily parallelized and algorithmic transformations are required to alleviate the recursive bottlenecks. To illustrate the application of speedup transformations to lower power, consider a first order IIR filter, as shown in Fig. 35(a), with a critical path of 2 (assume for simplicity that each operation takes one control cycle) [19]. Due to the recursive bottleneck [20] imposed by the filter structure, it is impossible to reduce the critical path using retiming or pipelining. Also, the simple structure does not provide opportunities for the application of algebraic transformations (associativity, distributivity, etc.) and applying a single transformation is not enough to reduce power in this example. Applying loop unrolling (Fig. 35(b)) does not change the effective critical path or capacitance and therefore the supply voltage cannot be

189

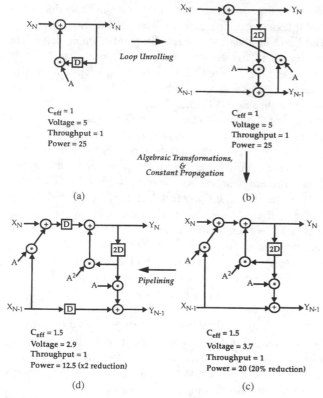

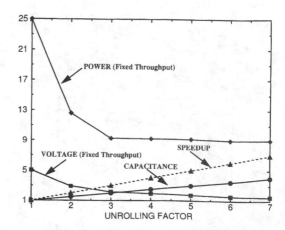

Fig. 36. Optimizing for power is different than optimizing for speed.

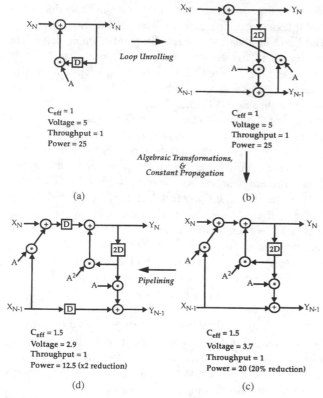

$C_{eff} = 1$
Voltage = 5
Throughput = 1
Power = 25

$C_{eff} = 1$
Voltage = 5
Throughput = 1
Power = 25

Loop Unrolling

*Algebraic Transformations,
&
Constant Propagation*

(a)

(b)

$C_{eff} = 1.5$
Voltage = 2.9
Throughput = 1
Power = 12.5 (x2 reduction)

$C_{eff} = 1.5$
Voltage = 3.7
Throughput = 1
Power = 20 (20% reduction)

Pipelining

(d)

(c)

Fig. 35. Using speedup transformation to reduce power.

altered; as a result, the power cannot be reduced. However, loop unrolling enables several other transformations such as distributivity, constant propagation ($A * A = A^2$), and pipelining which result in a significant reduction in power dissipation. After applying loop unrolling, distributivity, and constant propagation in a systematic way, the output samples can be represented as

$$Y_{N-1} = X_{N-1} + A * Y_{N-2} \qquad (21)$$
$$Y_N = X_N + A * X_{N-1} + A^2 * Y_{N-2}. \qquad (22)$$

The transformed solution has a critical path of 3 (Fig. 35(c)). However, pipelining can now be applied to this structure, reducing the critical path further to 2 cycles (Fig. 35(d)). Since the final transformed block is working at half the original sample rate (since we are processing 2 samples in parallel), and the critical path is same as the original datapath (2 control cycles), the supply voltage can be dropped to 2.9 V (the voltage at which the delays increase by a factor of 2, see Fig. 7). However, note that the effective capacitance increases since the transformed graph requires 3 multiplications and 3 additions for processing 2 samples while the initial graph requires only one multiplication and one addition to process one sample, or effectively a 50% increase in capacitance. The reduction in supply voltage, however, more than compensates for the increase in capacitance resulting in an overall reduction of the power by a factor of 2 (due to the quadratic effect of voltage on power). This simple example can be used to illustrate that optimizing for throughput will result in a *different* solution

than optimizing for power. For this example, arbitrary speedup can be achieved by continuing to apply loop unrolling combined with other transformations (algebraic, constant propagation, and pipelining). The speedup grows linearly with the unrolling factor, as shown in Fig. 36. If the goal is to minimize power consumption while keeping the throughput fixed, the speedup can be used to drop the supply voltage. Unfortunately, the capacitance grows linearly with unrolling factor (since the number of operations per input sample increases) and soon limits the gains from reducing the supply voltage. This results in an "optimum" unrolling factor for power of 3, beyond which the power consumption starts to increase again.

This example brings out two very important points: First, the application of a particular transformation can have conflicting effects on the different components of power consumption. For example, a transformation can reduce the voltage component of power (through a reduction in the critical path) while simultaneously increasing the capacitance component of power. Therefore, while speedup transformations can be used to reduce power by allowing for reduced supply voltages, the "fastest" solution is often NOT the lowest power solution. Second, the application of transformations in a combined fashion almost always results in lower power consumption than the isolated application of a single transformation. In fact, it is often necessary to apply a transformation which may temporary increase the power budget, in order to enable the application of transformations which will result in a more dramatic power reduction.

B. Minimizing the Number of Operations: Vector Quantization Example

Minimizing the number of operations to perform a given function is critical to reducing the overall switching activity. To illustrate the power tradeoffs that can be made at the algorithmic level, consider the problem of compressing a video data stream using vector quantization. Vector quantization (VQ) is a lossy compression technique which exploits the correlation that exists between neighboring samples and quantizes samples together rather than individually. Detailed description of vector quantization can be

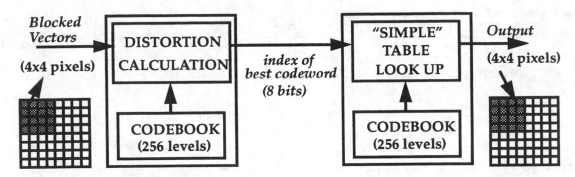

Fig. 37. Video compression/decompression using vector quantization.

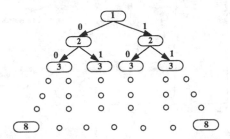

Fig. 38. Tree structured vector quantization.

found in [21]. Fig. 37 shows a block diagram of the VQ encoding/decoding process. On the encoder side, a group of pixels is blocked into a vector and compared (using a metric such as mean square error or absolute error) against a set of predetermined reproduction vectors (a set of possible pixel patterns) and the index of the best match is output. The decoder has a copy of all possible reproduction vectors (codebook) and the index of the best codeword is used to reconstruct the image using a simple table lookup operation. For this example, the image is segmented into 4 × 4 blocks (i.e., the vector size is 16 pixel words) and there are 256 levels in the codebook. In this case, 16:1 compression is achieved since only 8 b are transmitted (choosing 1 out of 256 levels) instead of 16 × 8 b for the true data. In this section, the focus will be on evaluating the computational complexity of encoding algorithms for VQ. Typically, the distortion metric used is mean square error. The distortion metric between an input vector **X** and a codebook vector C_i is computed as follows:

$$D_i = \sum_{j=0}^{15} (X_j - C_{ij})^2. \tag{23}$$

Three VQ encoding algorithms will be evaluated: full search, tree search, and differential codebook tree search.

1) Full Search Vector Quantization: Full search is a brute-force approach in which the distortion between the input vector and every entry in the codebook is computed. The distortion as defined in equation X is computed 256 times (for codebook entries **C0** through **C255**) and the code index that corresponds to the minimum distortion is determined and sent over to the decoder. For each distortion computation, there are 16 8-b memory accesses (to fetch the entries in the codeword), 16 subtractions, and 16 multiplications and 15 additions. In addition to this, the

minimum of the 256 distortion values, which involves 255 comparison operations, must be determined.

2) Tree-Structured Vector Quantization: In order to reduce the computational complexity required by an exhaustive full-search vector quantization scheme, a binary tree-search is typically used. The basic idea is to perform a sequence of binary search instead of one large search. As a result, the computational complexity increases as $\log N$ instead of N, where N is the number of nodes at the bottom of the tree. Fig. 38 shows the structure for the tree search. At each level of the tree, the input vector is compared against two codebook entries. If for example at level 1, the input vector is closer to the left entry, then the right portion of the tree is never compared below level 2 and an index bit 0 is transmitted. This process is repeated till the leaf of the tree is reached. The TSVQ will in general have some degradation over the performance of the full search VQ due to the constraint on the search. However, this is typically not very noticeable [21] and the power reduction relative to the full search VQ is very significant. Here only $2 \times \log_2 256 = 16$ distortion calculations have to be made compared to 256 distortion calculations in the full search VQ. The number of comparison operations is also reduced to 8.

3) Differential Codebook Tree-Structure Vector Quantization: Another option is to use the same search pattern as the previous scheme but perform computational transformations to minimize the number of switching events [22]. In the above scheme, at each level of the tree, the distortion difference between the left node and right node needs to be computed. This is summarized by the following equation:

$$D_{\text{left}-\text{right}} = \sum_{j=0}^{15} (X_j - C_{\text{left}j})^2$$
$$- \sum_{j=0}^{15} (X_j - C_{\text{right}j})^2. \tag{24}$$

This equation can be manipulated to reduce the number of operations.

$$D_{\text{left}-\text{right}} = \sum_{j=0}^{15} ((Xj - C_{\text{left}j})^2$$
$$- (X_j - C_{\text{right}j})^2) \tag{25}$$

Table 4 Computational Complexity of VQ Encoding Algorithms

Algorithm	# of Memory Accesses	# of Multiplications	# of Additions	# of Subtractions
Full Search	4096	4096	3840	4096
Tree Search	256	256	240	264
Differential Search/Tree Search	136	128	128	0

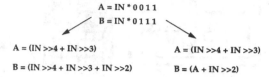

Fig. 39. Simple example demonstrating common subexpression elimination.

$$D_{\text{left}-\text{right}} = \sum_{j=0}^{15} (X_j^2 + C_{\text{left}j}^2 - 2X_j C_{\text{left}j}$$
$$- X_j^2 - C_{\text{right}j}^2 + 2X_j C_{\text{right}}) \quad (26)$$

$$D_{\text{left}-\text{right}} = \sum_{j=0}^{15} (C_{\text{left}j}^2 - C_{\text{right}j}^2)$$
$$+ \sum_{j=0}^{15} 2X_j (C_{\text{right}j} - C_{\text{left}j}). \quad (27)$$

The first term in (27) can be precomputed for each level and stored. By storing and accessing $2 \times (C_{\text{right}j} - C_{\text{left}j})$, the number of memory access operations can be reduced; that is, by changing the contents of the codebook through computational transformations, the number of switching events—number of multiplications, additions/subtractions and memory accesses—can be reduced. Table 4 shows a summary of the computational complexity per input vector (16 pixels).

C. Minimizing Number of Operations: Multiplication with Constants

A powerful technique to reduce the number of operations is conversion of multiplications with constants into shift-add operations. Since multiplications with fixed coefficients are quite common in signal processing applications like DCT, filters, etc., the application scope of this transformation is large. The basic idea is that multiplication with 0 is a NOP and therefore a multiplication with a constant degenerates to shift-add operations corresponding to the 1's in the coefficient. Techniques and tools have been developed to scale coefficients so as to minimize the number of 1's in the coefficients so as to minimize the number of shift-add operations. In addition, if an input is being multiplied with multiple coefficients, some of the shift-add terms can be shared and the number of operations can be further reduced. Fig. 39 shows an example of exploiting multiple coefficients being multiplied with the same input. On the left is a brute force implementation in which each multiply is computed separately. On the right side is another approach which exploits common terms in the coefficients and therefore some of the shift-add terms

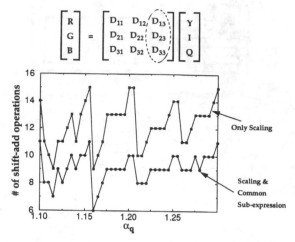

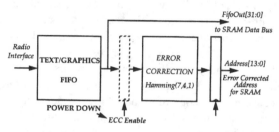

Fig. 40. Scaling and common subexpression can reduce the number of operations.

Fig. 42. Gated clocks to shut down modules when not used.

can be shared. In this implementation, A is first computed and is then used in the computation of B.

Fig. 40 shows an example of converting between the YIQ color space to the RGB color space for the video decompression module to be described later. The translation consists of a matrix multiplication with constant coefficients. The graph in Fig. 40 shows the number of shift-add operations for the three coefficients being multiplied with the Q input as a function of the scaling factor (the constant that is multiplied to the three coefficients). The top curve represents the number of operations for scaling alone and the bottom one represents scaling plus common subexpression elimination. From this figure, it is obvious that there is a great degree of freedom in optimizing the coefficients for minimizing number of operations for low-power.

VII. SYSTEM DESIGN EXAMPLE: A PORTABLE MULTIMEDIA TERMINAL

In this section, the various techniques presented in the previous sections will be applied to the design of a portable multimedia terminal that will allow users to have untethered access to fixed multimedia information servers [23]. The

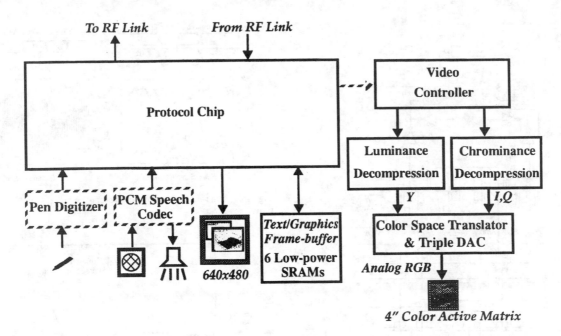

Fig. 41. Overview of the blocks in a multimedia I/O terminal.

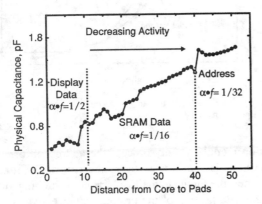

Fig. 43. Optimizing placement for low-power: Example of routing large data/control busses.

portable terminal is designed to transmit audio and pen input from the user to the network on a wireless uplink and will receive audio, text/graphics, and compressed video from the backbone on the downlink. The portability requirement resulted in the primary design focus to be on power reduction. The availability of communications between the terminal and computational resources at a fixed base station at the other end of the wireless link provides a major degree of freedom for optimizing power; i.e., any computation that does not have to be performed on the terminal (such as general purpose computation or application specific tasks such as speech and handwriting recognition) can be removed from the terminal. This resulted in a terminal that only provides the functionality to interface to I/O devices, as shown in Fig. 41. Six chips provide the interface between a high speed digital radio modem and a commercial speech codec, pen input circuitry, and LCD panels for text/graphics and full-motion video display. The chips provide protocol conversion, synchronization, error correction, packetization, buffering, video decompression, and D/A conversion at a

total power consumption of less than 5 mW. A protocol chip is used to communicate between the various I/O devices in the system. On the uplink, 4 Kbps digitized pen data and 64 Kbps, μ-law speech data are buffered using FIFO's, arbitrated and multiplexed, packetized, and transmitted in a serial format to the radio modem. On the down link, serial data from the radio at a 1 Mbps rate is depacketized and demultiplexed, the header information (containing critical information such as data type and length) is error corrected and transferred through FIFO's to one of the three output processing modules: speech, text/graphics, and video decompression. The speech module communicates data serially to a codec from the FIFO at a rate of 64 Kbs. The text/graphics module corrects error sensitive information, and generates the control, timing and buffering for conversion of the 32 b wide data from the text/graphics frame-buffer (implemented using six low-power 64 Kb SRAM chips) to the 4-b at 3 MHz required by a 640 × 480, passive LCD display. The final output module is the video decompression which is realized using four chips. The algorithm used is vector quantization, which involves memory lookup operations from a codebook of 256 4 × 4 pixel patterns. Vector quantization was chosen over the JPEG algorithm to perform the video decompression since VQ requires a very low complexity decoder (and hence lower power). Compressed YIQ video is buffered using a ping-pong scheme (one for Y and one for IQ), providing an asynchronous interface to the radio modem and providing immunity against bursty errors. The amount of RAM required is reduced by a factor of 32 by storing the video in the compressed format. The YIQ decompressed data is sent to another chip which converts this data to digital RGB and then to analog form using a triple DAC which can directly drive a 4″ active matrix color LCD display. A fourth chip performs the video

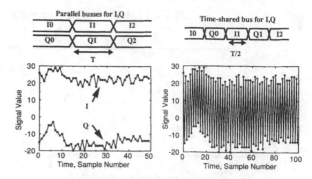

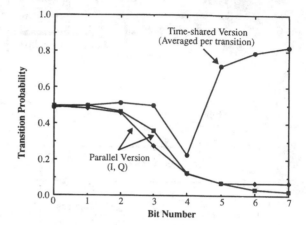

Fig. 45. Time-multiplexed hardware can increase switching activity.

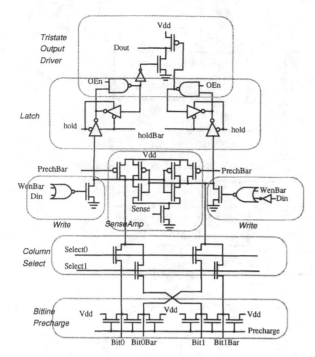

Fig. 44. Precharge, sense-amp, and "glitch" free tri-state buffer.

Fig. 46. Activity tradeoff for time-multiplexed versus non-time-multiplex hardware.

control functions which include the synchronization of the various chips and the LCD display, control of the ping-pong memories and loading of the codebooks, and it uses an addressing scheme which eliminates the need for an output line buffer. As described earlier, the key to low-power design is operating at the lowest supply voltage; which can result in a reduction in computational throughput. However, this can be avoided by the use of parallelism and pipelining. This approach was used extensively for both arithmetic computation and memory access, allowing the supply voltage to be as low as 1.1 V, even with a process which has $V_{tn} = 0.7$ V and $V_{tp} = -0.9$ V. To interface the low-voltage core of the custom chips to I/O devices running at 5 V, level-conversion pads are used (See Fig. 13). The entire chipset is implemented with a low-power cell library that features minimum sized transistors (as described in Section IV-B), optimized layout and logic with minimum switching activity. A few of the specific techniques used in each chip are outlined below.

A. Protocol Chip

At the logic level, gated clocks are used extensively to power down unused modules. For example, consider the error correction module for the text/graphics module. The basic protocol for the text/graphics module that is sent over the radio link and the text/graphics FIFO is address information for the frame-buffer followed by bit-mapped data. While address information is sensitive to channel errors, bit-mapped data is not very sensitive and therefore only the address information is error corrected. Fig. 42 shows a block diagram of a power efficient implementation of this function. A register is introduced at the output of the FIFO and a gated clock is used to enable the error

correction module to only process the address information and the ECC is shut down during the rest of the time. In this manner, the inputs to the ECC are not switching around when the data portion of the protocol is accessed from the FIFO. Since typically, the address in only a small portion of the bandwidth compared to the data, significant power savings is possible.

At the layout level, transition activity can be used to drive the placement and routing. Fig. 43 shows an example of routing large busses from the core to the pads for the protocol chip. The text/graphics module on the protocol chip communicates to both the text/graphics frame-buffer and to the text/graphics display. The display requires 8 b (4 for the top half and 4 for the bottom half) at a rate of 3 MHz, while each SRAM block uses 32 b clocked at 3 MHz/8 (using the parallel access scheme described in Section V-A-2). The address bits are also clocked at 3 MHz/8 and have very little activity since the accesses are mostly sequential. The approach to minimize power is to route the high activity display data (and display clock) which switches every cycle to have the shortest lengths, while the SRAM address, which have an activity factor 16 times lower, are allowed to have the longest wires.

B. Frame-Buffer SRAM

The 64 kb SRAM frame buffer chips are internally organized into eight 8 kb blocks each. These memories

Table 5 Summary of Chipset for the Infopad Terminal

Chip Description	Area (mmxmm)	Minimum Supply Voltage	Power at 1.5V
Protocol	9.4 x 9.1	1.1 V	1.9 mW
Frame-buffer SRAM (with loading)	7.8 x 6.5	1.1 V	500 μW
Video Controller	6.7 x 6.4	1.1 V	150 μW
Luminance Decompression	8.5 x 6.7	1.1 V	115 μW
Chrominance Decompression	8.5 x 9.0	1.1 V	100 μW
Color Space Conversion and Triple DAC	4.1 x 4.7	1.3 V	1.1 mW

Fig. 47. Overview of the IP graphics multimedia I/O terminal.

were synthesized from a parameterized design that can be scaled on the chip, block, and subblock level making it easily retargeted for entire memory chips (as used in the frame buffer) or for on-chip memories for processors (such as in the video decompression chips). The frame buffer conserves power by minimizing both switching activity and voltage swing. At the system level, only one SRAM chip in a bank needs to be active at any given time because each chip has a full 32 b data bus. Thus by dividing the words among the chips by address (rather than by bit) only one chip consumes control overhead. The chip level architecture also minimizes switching by first decoding at the block level (using three address lines), and then activating only one of the eight memory blocks. Since each block has a 32 b data bus, minimal column decoding (in this case 2 to 1 required for sense amp pitchmatching) is needed, so only 64 bitlines are charged and discharged in the whole bank each cycle. Internally, the blocks themselves conserve power by using low-voltage-swing bitlines (as described for the FIFO in Section IV-C). The bitlines are precharged to an NMOS threshold voltage level below the supply voltage. For example, operating with a 1.1 V supply, the bitlines only swing between 0.35 V and 0.0 V.

The blocks also conserve power by eliminating glitches on the data bus. In section, it was shown that balanced paths can be used to reduce glitching for static circuits; another approach is to use self-timing as used in the SRAM design. Basically, when a block is activated to read, its output is initially tristated; only as the sense-amp reads the data is the output enabled, so there can be no spurious transitions on the data bus (Fig. 44).

C. Video Decompression and Display

The video decompression and frame-buffer are very memory intensive and the parallelism technique described in Section V-A-2 was used to access the ping-pong frame-buffers and lookup table at very low rates and run the circuits at 1.1 V. For example, the video frame-buffer was clocked at 156 kHZ while meeting the throughput rate of 2.5 MHz. In the YIQ to RGB translation, which involves multiplication with constant coefficients, the switching events are minimized at the algorithmic level by substituting multiplications with hardwired shift-add operations (in which the shift operations degenerated to wiring) and by optimally scaling coefficients. In this way, the 3 ×

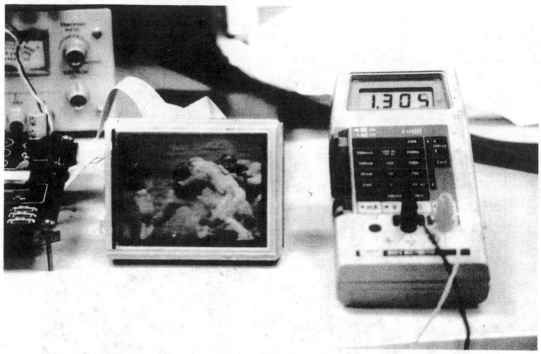

Fig. 48. Video decompression hardware output.

3 matrix conversion operation degenerated to 8 addition. The implementation was fully parallel and therefore there was no controller. For I/O communication (between the decompression chips and the color space chip) and in the matrix computation, sign-magnitude representation is chosen over two's complement to reduce the toggle activity in the sign bits. At the architecture level, time-multiplexing was avoided as it can destroy signal correlations, increasing the activity. Fig. 45 shows two alternate schemes for transmitting the I and Q data from the decompression chips to the color space converter chip. On the left is a fully parallel version in which I and Q have separate data busses. Also shown is the data for I and Q for a short segment in time. As seen, the data is slowly varying and therefore has low switching activity in the higher order bits. On the right is a time-multiplexed version in which there is a single time-shared bus in which the I and Q samples are interleaved. As seen from the signal value on the data bus, there is high switching on the data bus resulting in high activity and power. Fig. 46 shows the transition probability for both the parallel and time-multiplexed version. A low-voltage 6-b DAC based on a current switched array is used to drive the LCD. It operates down to 1.3 V, uses minimum sized devices for digital decoding, a single ended architecture to reduce the average power by a factor of 2 and a low-voltage current reference, and consumes on average 440 μW at 2.5 Mhz and 0.7 V_{pp} output amplitude when continuously operational. The actual power measured in the system was lower since the DAC was shut down during the horizontal and vertical blanking periods. The specifications of the low power chipset which was fabricated in 1.2 μm technology is given in Table 5. A PCB board containing the protocol chip, 6 SRAM chips, a speech codec, and pen interface logic has

been fabricated and tested. Various power supply voltages needed for the design including -17 V (adjustable using a trim-pot) for display drive, 12 V for dc to ac inverter for the backlight, 1.1 V for the custom chips, and -5 V for the speech codec have been realized using commercial chips. This board is integrated with the Plessey radio modem to realize a complete I/O terminal with a 1 Mb/s wireless channel. Fig. 47 shows the photograph of the first generation InfoPad terminal (IPGraphics). The next version of InfoPad will provide support for one-way full motion video. Fig. 48 shows the output of the decompression chips (running at 1.3 V) on a 4″ SHARP active matrix display.

VIII. CONCLUSIONS

A variety of approaches to reduce power consumption in CMOS circuits have been presented which involve optimization at all levels of the system design. At the technology level, the key low-power consideration is threshold reduction which allows supply voltage scaling to below 1 V. An optimum threshold voltage selection based on trading leakage and switching currents was presented which indicates that the threshold voltage should be scaled to the 0.3–0.4 V range as opposed to current day technology which uses a threshold voltage on the order of 0.7 V–1 V. At next level, the physical, circuit, and logic levels are optimized for low power. This includes activity driven place and route, the use of minimum sized devices in cell design, use of reduced swing logic, and logic level optimization and power down. Optimization at the architectural level can have a major impact on power. An architecture driven voltage scaling strategy was presented in which parallel and pipelined architectures were used to scale the supply down into the 1–1.5 V range without loss in functionality.

This strategy combined with threshold reduction allows voltage reduction to about 0.5 V, which would result in two orders of magnitude of power reduction compared to conventional designs running at 5 V. Various strategies were also presented for minimizing switched capacitance at the architectural level which includes choice of number representation, exploitation of signal correlations, and minimizing glitching transitions. Finally, the choice of algorithm is the most highly leveraged decision in meeting the power constraints. The ability for an algorithm to be parallelized is critical and the basic complexity of the computation must be highly optimized. A number of the low-power techniques that were presented were applied to the design of a chipset for a portable multimedia terminal that supports pen input, speech I/O and full-motion video. The entire chipset that performs protocol conversion, synchronization, error correction, packetization, buffering, video decompression and D/A conversion operates from a 1.1 V supply and consumes less than 5 mW—which is more than three orders of magnitude lower power compared to equivalent commercial solutions.

ACKNOWLEDGMENT

The authors thank R. Allmon, T. Burd, A. Burstein, Prof. Rabaey, and A. Stratakos for their invaluable contributions.

REFERENCES

[1] N. Weste and K. Eshragian, *Principles of CMOS VLSI Design: A Systems Perspective.* Reading, MA: Addison-Wesley, 1988.
[2] H. J. M. Veendrick, "Short-circuit dissipation of static CMOS circuitry and its impact on the design of buffer circuits," *IEEE J. Solid-State Circ.,* vol. SC-19, pp. 468–473, Aug. 1984.
[3] S. Sze, *Physics of Semiconductor Devices.* New York: Wiley, 1981.
[4] J. Colinge, *Silicon-on-Insulator Technology: Materials to VLSI.* Amsterdam: Kluwer, 1991.
[5] R. Swansan and J. Meindl, "Ion-implanted complementary MOS transistors in low-voltage circuits," *IEEE J. Solid-state Circ.,* vol. SC-7, pp. 146–153, Apr. 1972.
[6] T. Burd, "Low-power CMOS library design methodology," M.S. thesis, ERL Univ. Calif. Berkeley, Aug. 1994.
[7] A. Stratakos, S. Sanders, and R. W. Brodersen, "A low-voltage CMOS DC–DC converter for a portable low-powered battery-operated system," in *IEEE Power Electron. Specialists Conf.,* 1994.
[8] A. Shen, A. Ghosh, S. Devedas, and K. Keutzer, "On average power dissipation and random pattern testability of CMOS combinational logic networks," in *IEEE Int. Conf. on Computer-Aided Design,* pp. 402–407, 1992.
[9] J. Monteiro, S. Devadas, and A. Ghosh, "Retiming sequential circuits for low power," in *IEEE Int. Conf. on Computer-Aided Design,* pp. 398–402.
[10] V. Tiwari, P. Ashar, and S. Malik, "Technology mapping for low power," in *Proc. 1993 Design Automation Conference,* pp. 74–79.
[11] C. Tsui, M. Pedram, and A. Despain, "Technology decomposition and mapping targeting low power dissipation," in *Proc. 1993 Design Automation Conf.,* pp. 68–73.
[12] M. Alidina, J. Monterio, S. Devadas, A. Ghosh, and M. Papaefthymiou, "Precomputation-based sequential logic optimization for low power," in *1994 International Workshop in Low Power Design.*
[13] J. Yuan and C. Svensson, "High-speed CMOS circuit technique," *IEEE J. of Solid-state Circ.,* pp. 62–70, Feb. 1989.
[14] H. B. Bakoglu, *Circuits, Interconnections, and Packaging for VLSI.* Menlo Park, CA: Addison-Wesley, 1990.
[15] M. Kakumu and M Kinugawa, "Power-supply voltage impact on circuit performance for half and lower submicrometer CMOS LSI," *IEEE Trans. Electron Devices,* vol. 37, pp. 1902–1908, Aug. 1990.
[16] D. Dahle, "Designing high performance systems to run from 3.3 V or lower sources," in *Silicon Valley Personal Computer Conf.,* pp. 685–691, 1991.
[17] A. P. Chandrakasan, S. Sheng, and R.W. Brodersen, "Low-power CMOS digital design," *IEEE J. Solid-State Circ.,* vol. 27, no. 4, pp. 473–484, Apr. 1992.
[18] P. E. Landman and J. M. Rabaey, "Power estimation for high level synthesis," in *Proc. EDAC '93,* Paris, Feb. 1993, pp. 361–366.
[19] A. Chandrakasan, M. Potkonjak, R. Mehra, J. Rabaey, and R. W. Brodersen, "Minimizing power using transformations," *IEEE Trans. Comp.-Aided Design,* Jan. 1995.
[20] K. K. Parhi, "Algorithm transformation techniques for concurrent processors," *Proc. IEEE,* vol. 77, pp. 1879–1895, Dec. 1989.
[21] A. Gersho and R. Gray, *Vector Quantization and Signal Compression.* New York: Kluwer, 1992.
[22] W. C. Fang, C. Y. Chang, and B. J. Sheu,"A systolic tree-searched vector quantizer for real-time image compression," in *VLSI Signal Processing IV.* New York: IEEE Press, 1990.
[23] A. P. Chandrakasan, A. Burstein, and R. W. Brodersen, "A low-power chipset for portable multimedia applications," in *Proc. IEEE Int. Solid State Circ. Conf.,* pp. 82–83, Feb. 1994.

Overview of Low-Power ULSI Circuit Techniques

TADAHIRO KURODA† AND TAKAYASU SAKURAI†, MEMBERS

SUMMARY This paper surveys low-power circuit techniques for CMOS ULSIs. For many years a power supply voltage of 5 V was employed. During this period power dissipation of CMOS ICs as a whole increased four-fold every three years. It is predicted that by the year 2000 the power dissipation of high-end ICs will exceed the practical limits of ceramic packages, even if the supply voltage can be feasibly reduced. CMOS ULSIs now face a power dissipation crisis. A new philosophy of circuit design is required. The power dissipation can be minimized by reducing: 1) supply voltage, 2) load capacitance, or 3) switching activity. Reducing the supply voltage brings a quadratic improvement in power dissipation. This simple solution, however, comes at a cost in processing speed. We investigate the proposed methods of compensating for the increased delay at low voltage. Reducing the load capacitance is the principal area of interest because it contributes to the improvement of both power dissipation and circuit speed. Pass-transistor logic is attracting attention as it requires fewer transistors and exhibits less stray capacitance than conventional CMOS static circuits. Variations in its circuit topology as well as a logic synthesis method are presented and studied. A great deal of research effort has been directed towards studying every portion of LSI circuits. The research achievements are categorized in this paper by parameters associated with the source of CMOS power dissipation and power use in a chip.

key words: LSI, CMOS, low-power, low-voltage, power-delay product, energy-delay product, pass-transistor logic

1. Introduction

"CMOS circuits dissipate little power by nature." So believed circuit designers. But in reality, CMOS power dissipation as a whole has increased by 4 times every 3 years, which is the same pace as increase in bit density of state-of-the-art DRAMs! Figure 1 plots power dissipation of MPUs and DSPs presented in the ISSCC for the past 15 years. The power dissipation has increased by 1000 times over the 15 years and is exceeding 10 watts. Designers are to be looking at CMOS circuits in a new light.

Why is CMOS power dissipation becoming so large? The reason can be found in scaling principles shown in Table 1. A "constant field scaling" theory formulated by Dennard et al. [1] assumes that device voltages as well as device dimensions are scaled by a scaling factor $\varkappa(>1)$, resulting in a constant electric field in the device. This brings a desirable effect that,

while power density remains constant, circuit performance can be improved in terms of density ($\varkappa^2$), speed ($\varkappa$), power ($1/\varkappa^2$), and the power-delay product ($1/\varkappa^3$). Almost limitless progress in CMOS ICs is promised with this scaling scenario. But in practice, neither a supply voltage, V_{DD}, nor a threshold voltage of MOSFETs, V_{th}, had long been scaled till 1990. The resultant effect is better explained with a scaling theory called "constant voltage scaling." With the constant voltage scaling, circuit speed is further improved ($\varkappa^2$), while power density increases very rapidly by $\varkappa^3$. For more precise analysis in the submicron region [2], the power density increases by $\varkappa^{\alpha+1}$, where α represents velocity saturation effects and is typically 1.3 for 0.5

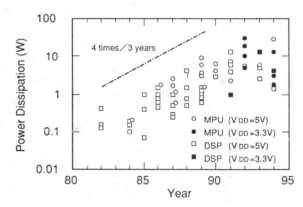

Fig. 1 Power dissipation of MPUs and DSPs presented in the ISSCC for the past 15 years.

Table 1 Influence of scaling on MOS device characteristics.

PARAMETER		SCALING MODEL		
		Constant field	Constant voltage	$\propto 1/\kappa^{0.5}$ voltage
Device size		$1/\kappa$	$1/\kappa$	$1/\kappa$
Gate-oxide thickness	tox	$1/\kappa$	$1/\kappa$	$1/\kappa^{0.5}$
Substrate doping		κ	κ^2	$\kappa^{1.5}$
Supply voltage	V	$1/\kappa$	1	$1/\kappa^{0.5}$
Electric field	E	1	κ	1
Current	I	$1/\kappa$	κ $(\kappa^{\alpha-1})$	$1/\kappa$
Area	A	$1/\kappa^2$	$1/\kappa^2$	$1/\kappa^2$
Capacitance	$C=\varepsilon A/tox$	$1/\kappa$	$1/\kappa$	$1/\kappa^{1.5}$
Gate delay	VC /I	$1/\kappa$	$1/\kappa^2$ $(1/\kappa^\alpha)$	$1/\kappa$
Power dissipation	VI	$1/\kappa^2$	κ $(\kappa^{\alpha-1})$	$1/\kappa^{1.5}$
Power density	VI /A	1	κ^3 $(\kappa^{\alpha+1})$	$\kappa^{0.5}$
Power-delay product	CV^2	$1/\kappa^3$	$1/\kappa$	$1/\kappa^{2.5}$

Manuscript received November 7, 1994.
Manuscript revised December 2, 1994.
† The authors are with Toshiba Corporation, Semiconductor Device Engineering Lab., Kawasaki-shi, 210 Japan.

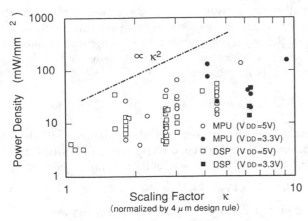

Fig. 2 Power density dependence on scaling factor normalized by 4 μm design rules.

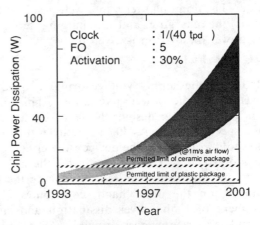

Fig. 3 Prediction of CMOS power dissipation by the year 2001.

μm MOSFETs. This constant voltage scaling causes the rapid increase in CMOS power dissipation. This inference is verified by Fig. 2 where the power density dependence on the scaling factor is plotted using the data in Fig. 1.

A real move to 3.3V or even lower supply voltage is being seen. One of the driving forces is increasing packaging and cooling cost of CMOS LSIs. The permitted limit of chip power dissipation in an inexpensive plastic package is a little over 1 W. Above the criterion, an expensive ceramic package is necessary which cannot meet a tight budget for consumer LSIs at all. Another motivation is emerging battery-operated applications for multimedia that demand intensive computation in portable environments. Furthermore, TTL interface is reaching its speed limit and an alternative is being developed where people don't have to stick to the 5 V supply voltage framework. This is also one of the background reasons. As shown in Fig. 1, the power dissipation is still increasing even under the reduced supply voltage.

How is the CMOS power dissipation changing in future? The $\chi^{\alpha+1}$ increase in power density with the constant voltage scaling cannot be accepted. The constant field scaling is not realistic due to auxiliary factors in MOSFETs such as a subthreshold current slope. Many alternative scaling approaches have been proposed. Scaling V_{DD} by $\chi^{0.5}$ is reported to satisfy both device reliability and circuit performance [3]. The resultant effect is summarized in Table 1. With this scaling approach the increase in power density is held down to $\chi^{0.5}$. In Fig. 3, CMOS power dissipation by the year 2001 is predicted, assuming that the $\chi^{0.5}$ supply voltage scaling is carried out, or assuming that a little bit gradual scaling scenario of $\chi^{0.33}$ is applied. χ is assumed to be 1.26 per year, which is the same scaling factor as has been applied so far. It is predicted that by the year 2000 the power dissipation would exceed that of ECL ICs, even if the supply voltage is appropriately scaled down. CMOS ULSIs are facing a power dissipation crisis. A new philosophy of circuit design is required.

The CMOS power dissipation is analyzed in Sect. 2. The power dissipation can be minimized by reducing: 1) the supply voltage, 2) the load capacitance, 3) switching activity. Reducing the supply voltage is studied in Sect. 3. The capacitance reduction is discussed in Sect. 4. Pass-transistor logic is attracting attention as it comprises fewer transistors and exhibits smaller stray capacitance than conventional CMOS static circuits. Variations in its circuit topology as well as a logic synthesis method are presented and studied in this section. In Sect. 5, an overview of low-power circuit techniques is given as summary.

2. Analysis of CMOS Power Dissipation

CMOS power dissipation is given by

$$P = p_t \cdot (C_L \cdot V_S + \overline{I_{SC}} \cdot \Delta t_{SC}) \cdot V_{DD} \cdot f_{CLK}$$
$$+ (I_{DC} + I_{LEAK}) \cdot V_{DD}. \qquad (1)$$

The first term, $p_t \cdot C_L \cdot V_S \cdot V_{DD} \cdot f_{CLK}$, represents dynamic dissipation due to charging and discharging of the load capacitance, where p_t is the switching probability, C_L is the load capacitance, V_S is the voltage swing, and f_{CLK} is the clock frequency. In most cases, V_S is the same as V_{DD}, but in some logic circuits V_S may be smaller than V_{DD} for high-speed and/or low-power operation. The second term, $p_t \cdot \overline{I_{SC}} \cdot \Delta t_{SC} \cdot V_{DD} \cdot f_{CLK}$, is dynamic dissipation due to switching transient current, where $\overline{I_{SC}}$ is the mean value of the switching transient current, and Δt_{SC} is time while the switching transient current draws. This dissipation can be held down by careful design [4]. The third term, $I_{DC} \cdot V_{DD}$, is static dissipation in such a circuit as a current mirror sense amplifier where current is designed to draw continuously from the power supply. The last term, $I_{LEAK} \cdot V_{DD}$, is due to the subthreshold current and the reverse bias leakage between the source-drain diffusions and the substrate. The dominant term in a well-designed logic circuit is the charging and discharging term, and CMOS power dissipation is given by

$$P \approx p_t \cdot C_L \cdot V_{DD}^2 \cdot f_{CLK}. \tag{2}$$

Reducing the supply voltage brings a quadratic improvement in the power dissipation. This simple solution to low-power designs, however, comes at the cost of a speed penalty. As V_{DD} approaches the sum of the threshold voltages of the devices, the circuit delay increases drastically. Compensation for the increased delay at low voltage is required. Reducing the load capacitance, on the other hand, contributes to the improvement of both power dissipation and circuit speed, and is therefore most principle. The physical capacitance can be minimized through the utilization of certain circuit styles, as well as device miniaturization. Reducing the switching activity has recently been paid attention. A problem to calculate the switching probability of a series connected logic circuits is found to be very time consuming (i.e. NP-Complete). A fast power estimation tool is expected for low-power LSI design. Power loss by glitching is reported to amount to 15% to 20% of the total power dissipation [5]. Optimum logic structures to reduce hazards should be investigated. This paper focuses the discussion associated with voltage and capacitance reduction in the following sections.

3. Voltage Scaling

Lowering the supply voltage is the most attractive choice due to the quadratic dependence. However, as the supply voltage becomes lower, the circuit delay increases and the LSI throughput degrades. There are three means to maintain the throughput: 1) reduce V_{th} to improve circuit speed, 2) introduce parallel and pipelined architecture while using slower device speeds, 3) prepare multiple supply voltages and for each cluster of circuits choose the lowest supply voltage that satisfies the speed requirements.

The second approach is discussed in [6] in detail. The idea here is to utilize the increasing transistor density to provide additional circuits to parallelize the computation, and trade off silicon area against power consumption. This idea is basically based on almost "limitless" number of transistors, which has not yet been obtained. In reality, chip sizes are still increasing even though the transistor density is increasing by 60% per year. It should also be pointed out that the algorithm that is being implemented may be sequential in nature and/or have feedback, which will limit the degree of parallelism.

The problem in the third approach is the necessity of level conversion every time signals interface the circuit clusters in different supply voltages. A good level converter should be developed which exhibits small delay, consumes little power, and occupies small pattern area. How efficiently can circuits be clustered with the minimum number of the interface is another discussion issue. In this section, lowering both V_{DD} and V_{th} is investigated.

3. 1 Optimizing V_{DD} and V_{th}

The I_D-V_{GS} characteristic of short-channel MOSFETs has an α-law dependence (typically α is 1.3 for 0.5 μm MOSFETs) [7] due to velocity saturation effects:

$$I_{DS} = \beta \cdot (V_{GS} - V_{th})^\alpha. \tag{3}$$

The circuit delay D is then approximated as

$$D = \gamma \cdot \frac{Q}{I_{DSO}} = \frac{\gamma \cdot C_L \cdot V_{DD}}{\beta \cdot (V_{DD} - V_{th})^\alpha}. \tag{4}$$

Taking the charging and discharging current and the leak current into account, the CMOS power dissipation is given by

$$P = p_t \cdot C_L \cdot V_{DD}^2 \cdot f_{CLK} + I_{LEAK} \cdot V_{DD}, \tag{5}$$

where

$$f_{CLK} = \frac{1}{nD},$$

$$I_{LEAK} = \lambda \cdot W \cdot \exp\left\{\frac{-V_{th}}{S}\right\},$$

$$S = \frac{kT}{q} \ln 10 \left(1 + \frac{C_D}{C_{OX}}\right). \tag{6}$$

n is the logic depth, k is Boltzmann's constant, T is temperature, C_D is the depletion-layer capacitance, and C_{OX} is the gate capacitance. For with $W_P = 10$ μm, $W_n = 5$ μm, $p_t = 0.3$, $n = 30$, fanout $= 2$, and other parameters for a 0.4 μm CMOS device, the power dissipation and the circuit delay are calculated in various V_{DD} and V_{th}, and the results are plotted in Figs. 4(a) and (b), respectively. Figure 4(b) indicates that the increasing delay with low V_{DD} can be compensated by lowering V_{th}. The delay contour lines are provided on the $V_{DD} - V_{th}$ plane in Fig. 4(b).

The power-delay (PD) product is calculated and plotted in Fig. 4(c). The PD product can be interpreted as the amount of energy expended in each switching event. The dimension is joule (watt × second), or watt/MIPS. By ignoring the subthreshold leak current in (5), the PD product is given by

$$P \cdot D = \frac{p_t}{n} \cdot C_L \cdot V_{DD}^2. \tag{7}$$

This suggests that the energy expended in every circuit transition is independent of the circuit speed (and V_{th} when $V_{th} \geq 0.1$ V). In order to save the energy, it is desirable to operate a circuit at the slowest possible speed with the lowest possible supply voltage. The PD product is considered to be a good index for those who

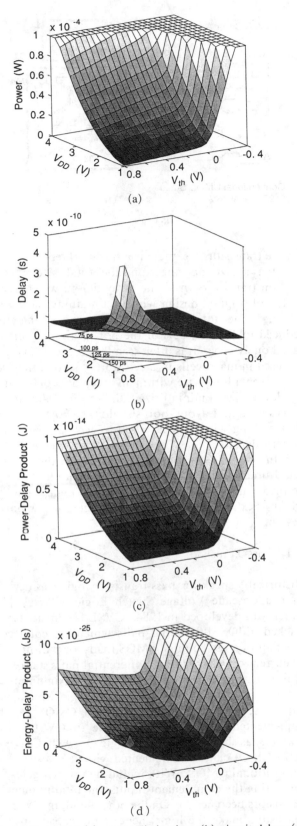

(a)

(b)

(c)

(d)

Fig. 4 Simulated (a) power dissipation, (b) circuit delay, (c) power-delay product, (d) energy-delay product, dependence on supply voltage and threshold voltage. Assumption: $W_p = 10 \ \mu$m, $W_n = 5 \ \mu$m, $V_{thp} = V_{thn}$, $p_t = 0.3$, fanout$=2$, logic depth$=30$ stages, and other parameters for a 0.4 μm CMOS.

care mostly about battery consumption. However, most of the applications of today requires near-peak performance throughput. The PD product gives no information about the trade-off between energy and speed.

In discussing the trade-off between energy and speed, the energy-delay space is explored to find the most energy saving solution under a given speed constraint. The energy-delay (ED) product, that is PD², can be an important index for those who care about speed as much as energy. Figure 4(d) shows the calculation results which suggest V_{th} be lowered to about 0.1 V to minimize the ED product. For example, if V_{DD} is reduced from 3.3 V to 1.0 V, the delay of CMOS circuit in $V_{th}=0.7$ V is increased by a factor of 5. However, if V_{th} is reduced to 0.1 V, the delay increase is held down to only 20%. Consequently, the power dissipation can be reduced to below 1/10, while almost maintaining the circuit speed.

3.2 Controlling V_{th} Fluctuation

V_{th} fluctuation due to process variation is presently around ± 0.15 V, but this value should be scaled down in the low voltage operation. Figure 5 shows how much the V_{th} fluctuation affects the circuit delay in various V_{DD}. It is seen from the figure that the V_{th} fluctuation of ± 0.15 V gives less than 5% speed variation in 5 V V_{DD}, while the same amount of the V_{th} fluctuation doubles the delay in 1 V V_{DD}. For example, if it is specified that 0.4 V is the minimum V_{th} to keep a total leakage current of a VLSI within a specification, the center value of V_{th} should be set 0.55 V. The worst chips then show the V_{th} of 0.7 V. The V_{th} distribution of this case is illustrated by a shaded region in Fig. 5. If the V_{th} fluctuation can be reduced to ± 0.05 V, the worst V_{th} becomes 0.5 V. This case is indicated by another shaded region in the figure. The worst case speed difference between the two V_{th} fluctuation cases is a factor of 1.3 for 1.5 V V_{DD} and a factor of 3 for 1 V V_{DD}.

A new circuit technique was developed to reduce

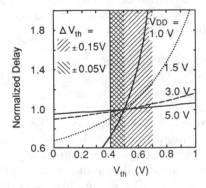

Fig. 5 Calculated circuit delay dependence on V_{th} in various V_{DD}.

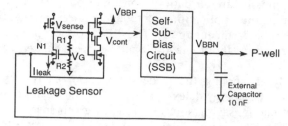

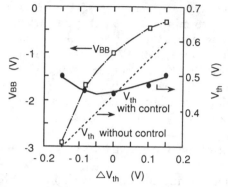

Fig. 6 Circuit diagram of self-adjusting V_{th} scheme.

Fig. 7 Measured V_{th} controllability dependence on process fluctuation ΔV_{th}.

the V_{th} fluctuation by using self-substrate-biasing [8]. Figure 6 shows a circuit diagram. A leakage sensor senses leakage current of a representative MOSFET, N1, amplifies it by a load pMOS, and outputs a control signal, Vcont, to a Self-Substrate-Bias (SSB) circuit. The leakage current can vary by a factor of 10 when V_{th} is changed by 0.1 V. V_G is set around 0.2 V to enhance the leakage current. Vcont is controlled such that it triggers the SSB only when the leakage is higher than a certain level. In an nMOS case, the SSB, when triggered, draws charge from P-wells, and lowers a substrate voltage. V_{th} is then increased to reduce the leakage currents. Thus, the substrate bias is controlled such that leakage current of the MOSFETs is adjusted constant. Figure 7 shows a measured V_{th} static controllability which is found to be less than ± 0.025 V. The delay of the sensor introduces dynamic controllability. The overall V_{th} controllability including static and dynamic effects is ± 0.05 V.

4. Capacitance Reduction

Total load capacitance is the sum of gate capacitance, diffusion capacitance, and routing capacitance. The ratio is case-by-case, but may often be almost even. Using small number of transistors or small size of transistors contributes to the reduction in the gate capacitance and the diffusion capacitance. Pass-transistor logic may have this advantage because it comprises fewer transistors and exhibits smaller stray capacitance than conventional CMOS static logic. As

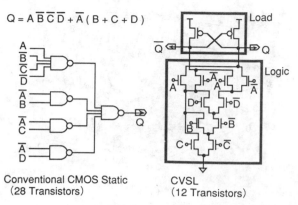

Fig. 8 CVSL.

for the transistor size, there is a study [9] reporting that through a size optimization, the total size of one million transistors in a gate array design was reduced to 1/8 of original design while maintaining the circuit speed. The total load capacitance was therefore reduced to 1/3 and 55% of the power dissipation was saved on average. It is often seen that bigger transistors are used in macrocells in a cell library so that they can drive even a long wire within an acceptable delay time. As device size is miniaturized, the routing capacitance places weight. Layout tools should take power dissipation of long wires into account as well as speed and pattern density.

In this way, the whole design system including a cell library and CAD environments should be reexamined. In this section, pass-transistor logic is studied which is expected as a post CMOS logic for low power design.

4.1 Pass-Transistor Logic

Historically speaking, pass-transistor logic was derived from a Cascade Voltage Switch Logic (CVSL) [10] which was developed in 1984. CVSL is constructed of stacked nMOS differential pairs which are connected to a pair of cross-coupled pMOS loads for pull-up. No dc current draws in static. Differential pairs stacked in N-level can implement a logic with a maximum of $2^N - 1$ inputs. Therefore, complicated logic which may require several gates in conventional CMOS can be implemented in a single stage gate in CVSL. For example, as shown in Fig. 8, a logic $Q = A \cdot \bar{B} \cdot \bar{C} \cdot \bar{D} + \bar{A} \cdot (B + C + C)$ can be implemented with 28 transistors in conventional CMOS. With including inverter gates for generating the complement inputs, the total number of transistors becomes 36. On the other hand, in CVSL 12 transistors make the logic; hence stray capacitance in gate and drain of MOSFETs can be reduced to 1/3.

In 1985 further refinement lead to a Differential Split-Level Logic (DSL) [11] depicted in Fig. 9. In DSL the "H" level of the nMOS logic is clamped by the n-transistors whose gates are biased to $(V_{DD}/2)$

$+ V_{thn}$. This reduces the signal swing to half in the stacked nMOS circuits to gain speed.

CVSL and DSL were closely examined in [12]. Simulated results of full adders in CMOS, CVSL and DSL are summarized in Table 2. Contrary to the expectation, CVSL is slower than CMOS. This is because during the switching action, the nMOS pull-down trees have to "fight" the cross-coupled pMOS loads which are holding the previous data. Reducing the size of the cross-coupled pMOS transistors helps to gain the switching speed at the cost of degradation in drive capability. Both CVSL and DSL consume almost twice as big dynamic power as CMOS. Different current paths through the nMOS stacked trees for the compliment inputs can cause signal skew on the output. If the two compliment outputs in the nMOS

pull-down trees become "L" simultaneously, the cross-coupled pMOS loads don't flip and draw large current. DSL is fast, but consumes larger static power. This is because the clamped "H" level of $V_{DD}/2$ yields leak current through one of the pMOS loads which should be turned off. The PD product and the ED product indicate that neither CVSL nor DSL outperforms CMOS in spite of the reduced stray capacitance.

The nMOS logic circuits in CVSL and DSL can drive only "L" level, and therefore the cross-coupled pMOS loads are necessary for driving "H" level. Pass-transistor circuits, on the other hand, can drive both "H" and "L" levels so that they don't need the cross-coupled pMOS loads. Modifications and derivations of pass-transistor logic have been proposed such as CPL, DPL, DCVSPG, SRPL, and SAPL. Their

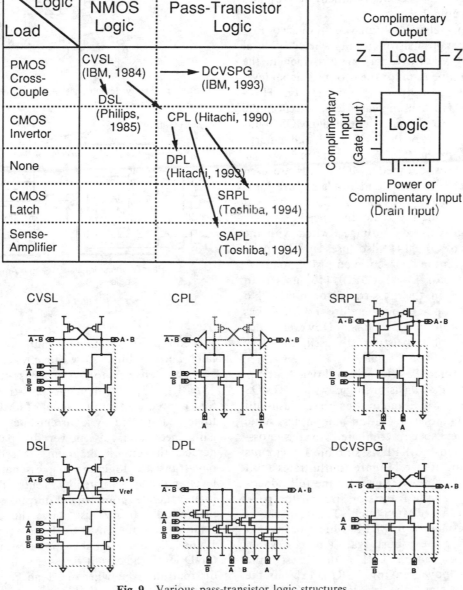

Fig. 9 Various pass-transistor logic structures.

Table 2 Simulated performance comparison of full adder in various pass-transistor structures.

3 μm Device (Full Adder)

Performance / Circuit	Tr. Number	Speed (ns)	Power (mW/25MHz)	P·D (normalized)	E·D (normalized)
CMOS	40	20	0.29	1.00	1.00
CVSL	22	22	0.61	2.31	2.55
DSL	26	14	0.48	1.15	0.80

0.4 μm Device (Full Adder)

Performance / Circuit	Tr. Number	Speed (ns)	Power (mW/100MHz)	P·D (normalized)	E·D (normalized)
CMOS	40	0.82	0.52	1.00	1.00
CPL	28	0.44	0.42	0.43	0.23
DPL	48	0.63	0.58	0.86	0.66
DCVSPG	24	0.53	0.30	0.37	0.24
SRPL	28	0.48	0.19	0.21	0.13

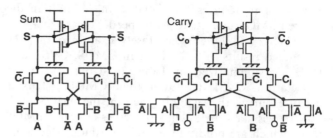

Fig. 10 Full adder circuit in SRPL.

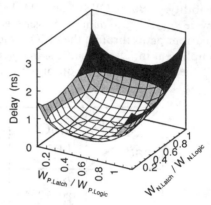

Fig. 11 Delay dependence on transistor width in SRPL.

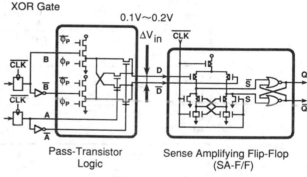

Fig. 12 SAPL.

circuit diagrams are depicted in Fig. 9 together with an illustration which explains a history of their development. Their performance was studied in [16] and summarized in Table 2.

A Complementary Pass-transistor Logic (CPL) [13] uses nMOS pass-transistor circuits where "H" level drops by V_{thn}. CMOS inverters are provided in the output stage to compensate for the dropped signal level as well as to increase output drive capability. However, the lowered "H" level increases leak current in the CMOS inverters. So the cross-coupled pMOS loads can be added to recover the "H" level and enlarge operation margin of the CMOS inverters in low V_{DD}. In this case, the cross-coupled pMOS loads are used only for the level correction and don't require large drive capability. Therefore, small pMOS can be used in order not to degrade the switching speed.

A Differential Cascade Voltage Switch with the Pass-Gate (DCVSPG) [14] also uses nMOS pass-transistor logic with the cross-coupled pMOS load. A Double Pass-transistor Logic (DPL) [15] uses both nMOS and pMOS in the pass-transistor logic. No signal drop is taken place and therefore the circuit has big operation margin in low V_{DD}. However, since twice the number of transistors are used, the stray capacitance doesn't become small.

A Swing Restored Pass-transistor Logic (SRPL) [16] uses nMOS pass-transistor logic with a CMOS latch. Since the CMOS latch flips in a push-pull manner, it exhibits larger operation margin, less static current, and faster speed, compared to the cross-coupled pMOS loads. SRPL is suitable for circuits with light load capacitance. Figure 10 illustrates a full adder in SRPL. Figure 11 illustrates the full adder's delay dependence on the transistor sizes in the pass-transistor logic and the CMOS latch. The figure shows substantial design margin in SRPL which means that SRPL circuits are quite robust against process variations. As shown in Table 2, CPL is the fastest while SRPL is the most power saving. SRPL exhibits the smallest PD product and the smallest ED product in

this comparison. The push-pull action of the CMOS latch and the small stray capacitance bring this result.

A Sense-Amplifying Pass-transistor Logic (SAPL) [17] is a dynamic pass-transistor logic. In SAPL a reduced output signal of nMOS pass-transistor logic is amplified by a current latch sense-amplifier to gain speed and save power dissipation. Figure 12 depicts the circuit diagram. All the nodes in the pass-transistor logic are first discharged to the GND level and then evaluated by inputs. The pass-transistor logic generates complement outputs with small signals of around 100 mV just above the GND level. The small signals are sensed by the sense-amplifier in about 1.6 ns. Since the signal swings are small just above the GND level, the circuit runs very fast with small power dissipation, even when the load capacitance is large. SAPL therefore is suitable for circuits with large load

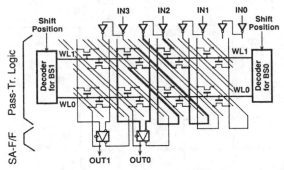

Fig. 13 4-to-2 barrel shifter in SAPL.

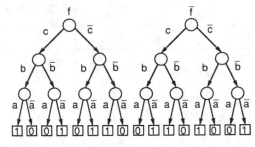

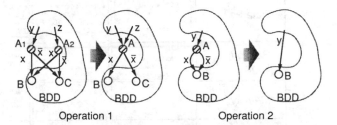

Fig. 14 Logic binary trees of function $f = abc + a\bar{b}\bar{c} + \bar{a}b\bar{c} + \bar{a}\bar{b}c$.

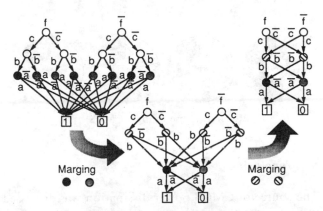

Fig. 15 Two operations to reduce logic binary trees.

capacitance. By adding a cross-coupled NOR latch, the sensed data can be latched so that the SAPL circuit can be used as a pipeline register. An application example is a barrel shifter shown in Fig. 13 where multi-stage logic can be constructed just by concatenating the pass-transistors without inserting an amplification stage.

It should also be noted that test patterns can be generated automatically by using D-algorithm [18] for pass-transistor logic as well as for conventional CMOS static logic.

4.2 Pass-Transistor Logic Synthesis

A synthesis method of pass-transistor network is described in [19]. It is based on Binary Decision Diagram (BDD) [20]. The synthesis begins by generating logic binary trees for separate logic functions which are then merged and reduced to a smaller graph. Lastly the graph is mapped to transistor circuits.

Let's consider a carry generation function in an adder. The function is expressed as $f = abc + a\bar{b}\bar{c} + \bar{a}b\bar{c} + \bar{a}\bar{b}c$. The logic binary trees are instantly generated as shown in Fig. 14 from a truth table of the function f. For example, the path from the source node (f) through edges "$\bar{c}$," "$\bar{b}$," and "a" to the sink node (1) corresponds to the case when $f = 1$ with $c = b = 0$ and $a = 1$.

The trees can be reduced by applying in sequence two operations illustrated in Fig. 15 from the sink node. Operation 1 merges two nodes whose corresponding outgoing complement edges reach the same node. Operation 2 removes from the graph a node with two outgoing complement edges to the same node. In this particular example, a case where the second operation can be applied is not found. Figure 16 illustrates the reduction procedure of the logic binary trees in Fig. 15. The reduced graph is mapped to transistor circuits as shown in Fig. 17. All the edges are replaced with n-transistors whose gates are provided with the variables marked on the edges. The sink nodes (0) and (1) are replaced with V_{ss} and V_{DD}. If a edge "x" reaches the sink node (1) and the compliment edge "$\bar{x}$" reaches the sink node (0), "x" can be fed to the

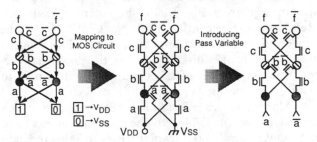

Fig. 16 Logic binary tree reduction.

Fig. 17 Transistor mapping.

node as a pass variable. In this example two transistors are reduced by this rule. Lastly, appropriate buffer circuits should be connected to the output nodes (f) and ($\bar{f}$).

It is always guaranteed that a correct logic circuit can be synthesized in this method. Detail discussion can be found in [19].

Table 3 Low-power CMOS LSI circuit techniques.

	p_t	C_L	V_S	V_{DD}	f_{CLK}	I_{SC}	I_{DC}	I_{LEAK}
General		• device scaling	*Small Signal*	Low V_{DD} • DC-DC converter [22] • 0.25V QuadRail [23]		*Careful Design* • design verification by CAD		• control V_{th} fluctuation by SSB [8]
Clock	• gated clock	• floorplan to reduce wire length • F/F sizing *Charge Recycling* • C stacking [24]	• 1/2 swing [24]					
Bus	*Glitch Suppress* • 3-state-buffer activated after data fix [25]	• C stacking [26] • exclusive bus	• 1/2 swing [26]					
Data Path	• latch insertion to deskew data-in [27]	*Tr. Reduction* • pass-transistor (CPL [13], SRPL [16], SAPL [17], DPL [15], DCVSPG [14])	• pass-transistor (SAPL [17])		• parallelism [6]			
Random Logic	*CAD* • permutation of series-connected transistor order [28]	• library & CAD for pass-tr. logic [29] • tr. sizing [9,30]	• current switch logic (MCML [31])					*Sleep Mode* • 2 type V_{th} (MT-CMOS [32])
Memory		• hierarchy	• reduced swing WL, BL				*Cut Current* • latch S/A [33]	• switched-source-impedance [34]
I / O		• MCM [35] • area pad [35]	• reduced swing I/O (GTL [36], LVDS)		• phase modulation [37]		• dynamic termination [38]	

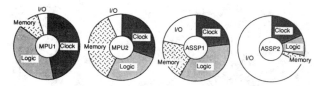

Fig. 18 Power use in logic CMOS LSIs.

5. Conclusions

The source of CMOS power dissipation was discussed in Sect. 2 and the parameters associated with the power dissipation were studied. Another concern is power use in a chip. Figure 18 shows some examples. Reflecting the versatility of logic CMOS LSIs, the power use is different from chip to chip. This suggests that all kinds of efforts in reducing the power parameters in every portion of LSI circuits should be paid to reduce the total power dissipation of a chip.

Many research efforts have been reported which are categorized in Table 3 by the power parameters and the power use. Brief explanations of the research achievements are given in [21]. Even though most of the research fields seem to have been explored, the low-power research has just come to the most interesting stage. In order to achieve power reduction by three figures, various techniques are required in each design domain from architectural level to algorithmic, logical, circuit, layout, and device levels. Wide spread of research activities will be expected involving many researchers working on system design, circuit design, CAD tools, and device design.

Acknowledgment

The authors would like to acknowledge the encouragement of A. Kanuma, K. Maeguchi, and Y. Unno throughout the work.

References

[1] Dennard, R. H., Gaensslen, F. H., Yu, H.-N., Leo Rideout, V., Bassous, E. and Leblank, A. R., "Design of ion-implanted MOSFET's with very small physical dimensions," *IEEE J. Solid-State Circuits*, vol. 9, pp. 256 -268, 1974.

[2] Sakurai, T., "High-Speed/High-Density Logic Circuit Design," in *Proc. IEEE 1993 VLSITSA*, pp. 222-226, 1993.

[3] Kakumu, M., "Process and Device Technologies of CMOS Devices for Low-Voltage Operation," *IEICE Trans. Electron.*, vol. E76-C, no. 5, pp. 672-680, May 1993.

[4] Veendrick, H. J. M., "Short-Circuit Dissipation of Static CMOS Circuitry and Its Impact on the Design of Buffer Circuits," *IEEE J. Solid-State Circuits*, vol. 19, no. 4, pp. 468-473, Aug. 1984.

[5] Benini, L., Favalli, M. and Riccò, B., "Analysis of Hazard Contributions to Power Dissipation in CMOS ICs," in *Proc. IEEE 1994 IWLPD*, pp. 27-32, Apr. 1994.

[6] Chandrakasan, A. P., Sheng, S. and Brodersen, R. W., "Low-Power CMOS Digital Design," *IEEE J. Solid-State Circuits*, vol. 27, no. 4, pp. 473-484, Apr. 1992.

[7] Sakurai, T. and Newton, A. R., "Alpha-Power Law MOSFET Model and its Applications to CMOS Inverter Delay and Other Formulas," *IEEE J. Solid-State Circuits*, vol. 25, no. 2, pp. 584-594, Apr. 1990.

[8] Kobayashi, T. and Sakurai, T., "Self-Adjusting Threshold-Voltage Scheme (SATS) for Low-Voltage

High-Speed Operation," in *Proc. IEEE 1994 CICC*, pp. 271-274, May 1994.

[9] Yamada, M., Kurosawa, S., Nojima, R., Kojima, N., Mitsuhashi, T. and Goto, N., "Synergistic Power/Area Optimization with Transistor Sizing and Write Length Minimization," to be published in the 1994 Symposium on Low Power Electronics, Oct. 1994.

[10] Heller, L. G. and Griffin, W. R., "Cascade Voltage Switch Logic: A Differential CMOS Logic Family," in *ISSCC Dig. Tech. Papers*, pp. 16-17, Feb. 1984.

[11] Pfennings, L. C. M. G., Mol, W. G. J., Bastiaens, J. J. J. and Van Dijik, J. M. F., "Differential Split-Level CMOS Logic for Sub-Nanosecond Speeds," in *ISSCC Dig. Tech. Papers*, pp. 212-213, Feb. 1985.

[12] Chu, K. M. and Pulfrey, D. L., "A Comparison of CMOS Circuit Techniques: Differential Cascade Voltage Switch Logic Versus Conventional Logic," *IEEE J. Solid-State Circuits*, vol. 22, no. 4, pp. 528-532, Aug. 1987.

[13] Yano, K., Yamanaka, T., Nishida, T., Saito, M., Shimohigashi, K. and Shimizu, A., "A 3.8-ns CMOS 16×16-b Multiplier Using Complementary Pass-Transistor Logic," *IEEE J. Solid-State Circuits*, vol. 25, no. 2, pp. 388-395, Apr. 1990.

[14] Lai, F. S. and Hwang, W., "Differential Cascade Voltage Switch with the Pass-Gate (DCVSPG) Logic Tree for High Performance CMOS Digital Systems," in *Proc. IEEE 1993 VLSITSA*, pp. 358-362, 1993.

[15] Suzuki, M., Ohkubo, N., Yamanaka, T., Shimizu, A. and Sasaki, K., "A 1.5ns 32b CMOS ALU in Double Pass-Transistor Logic," in *ISSCC Dig. Tech. Papers*, pp. 90-91, Feb. 1993.

[16] Parameswar, A., Hara, H. and Sakurai, T., "A High Speed, Low Power, Swing Restored Pass-Transistor Logic Based Multiply and Accumulate Circuit for Multimedia Applications," in *Proc. IEEE 1994 CICC*, pp. 278-281, May 1994.

[17] Matsui, M., Hara, H., Seta, K., Uetani, Y., Kim, L.-S., Nagamatsu, T., Shimazawa, T., Mita, S., Otomo, G., Ohto, T., Watanabe, Y., Sano, F., Chiba, A., Matsuda, K. and Sakurai, T., "200 MHz Video Compression Macrocells Using Low-Swing Differential Logic," in *ISSCC Dig. Tech. Papers*, pp. 76-77, Feb. 1994.

[18] Roth, J. P., Oklobdzija, V. G. and Beetem, J. F., "Test Generation for FET Switching Circuits," in *Proc. IEEE Int'l Test Conference*, pp. 59-62, Oct. 1984.

[19] Sakurai, T., Lin, B. and Newton, A. R., "Multiple-Output Shared Transistor Logic (MOSTL) Family Synthesized Using Binary Decision Diagram," *Dept. EECS*, Univ. of Calif., Berkeley, ERL Memo M90/21, Mar. 1990.

[20] Akers, S. B., "Binary Decision Diagrams," *IEEE Trans. on Computers*, vol. C-27, no. 6, pp. 509-516, Jun. 1978.

[21] Kuroda, T. and Sakurai, T., "A Technical White Paper on Low-Power LSI," in *Nikkei Microdevices*, pp. 78-104, Oct. 1994.

[22] Stratakos, A. J., Brodersen, R. W. and Sanders, S. R., "High-Efficiency Low-Voltage DC-DC Conversion for Portable Applications," in *Proc. IEEE 1994 IWLPD*, pp. 105-110, Apr. 1994.

[23] Carley, L. R. and Lys, I., "QuadRail: A Design Methodology for Ultra-Low Power ICs," in *Proc. IEEE 1994 IWLPD*, pp. 225-230, Apr. 1994.

[24] Kojima, H., Tanaka, S. and Sasaki, K., "Half-Swing Clocking Scheme for 75% Power Saving in Clocking Circuitry," in *Proc. IEEE 1994 Symposium on VLSI Circuits*, pp. 23-24, Jun. 1994.

[25] Chandrakasan, A., Burstein, A. and Brodersen, R. W., "A Low Power Chipset for Portable Multimedia Applications," in *ISSCC Dig. Tech. Papers*, pp. 82-83, Feb. 1994.

[26] Yamauchi, H., Akamatsu, H. and Fujita, T., "A Low Power Complete Charge-Recycling Bus Architecture for Ultra-High Data Rate ULSI's," in *Proc. IEEE 1994 Symposium on VLSI Circuits*, pp. 21-22, Jun. 1994.

[27] Lemonds, C. and Shetti, S. S. M., "A Low Power 16 by 16 Multiplier Using Transition Reduction Circuitry," in *Proc. IEEE 1994 IWLPD*, pp. 139-142, Apr. 1994.

[28] Prasad, S. C. and Roy, K., "Circuit Optimization for Minimization of Power Consumption under Delay Constraint," in *Proc. IEEE 1994 IWLPD*, pp. 15-20, Apr. 1994.

[29] Yano, K., Sasaki, Y., Rikino K. and Seki, K., "Lean Integration: Achieving a Quantum Leap in Performance and Cost of Logic LSIs," in *Proc. IEEE 1994 CICC*, pp. 603-606, May 1994.

[30] Tan, C. H. and Allen, J., "Minimization of Power in VLSI Circuits Using Transistor Sizing, Input Ordering, and Statistical Power Estimation," in *Proc. IEEE 1994 IWLPD*, pp. 75-80, Apr. 1994.

[31] Mizuno, M., Yamashina, M., Furuta, K., Igura, H., Abiko, H., Okabe, K., Ono A. and Yamada, H., "A GHz MOS Adaptive Pipeline Technique Using Variable Delay Circuits," in *Proc. IEEE 1994 Symposium on VLSI Circuits*, pp. 27-28, May 1994.

[32] Mutoh, S., Douseki, T., Matsuya, Y., Aoki, T. and Yamada, J., "1 V High-Speed Digital Circuit Technology with 0.5 μm Multi-Threshold CMOS," in *Proc. IEEE 1993 ASIC Conf.*, pp. 186-189, 1993.

[33] Sakurai, T., "High-Speed Circuit Design with Scaled-Down MOSFET's and Low Supply Voltage," in *Proc. IEEE 1993 ISCAS*, pp. 1487-1490, May 1993.

[34] Horiguchi, M., Sakata, T. and Itoh, K., "Switched-Source-Impedance CMOS Circuit for Low Standby Subthreshold Current Giga-Scale LSI's," in *Proc. IEEE 1993 Symposium on VLSI Circuits*, pp. 47-48, May 1993.

[35] Zhu, Q., Xi, J. G., Dai, W. W.-M. and Shukla R., "Low Power Clock Distribution Based on Area Pad Interconnect for Multichip Modules," in *Proc. IEEE 1994 IWLPD*, pp. 87-92, Apr. 1994.

[36] Gunning, B., Yuan, L., Nguyen T. and Wong, T., "A CMOS Low-Voltage-Swing Transmission-Line Transceiver," in *ISSCC Dig. Tech. Papers*, pp. 58-59, Feb. 1992.

[37] Nogami, K. and Gamal, A. E., "A CMOS 160Mb/s Phase Modulation I/O Interface Circuit," in *ISSCC Dig. Tech. Papers*, pp. 108-109, Feb. 1994.

[38] Kawahara, T., Horiguchi, M., Etoh, J., Sekiguchi T. and Aoki, M., "Low Power Chip Interconnection by Dynamic Termination," in *Proc. IEEE 1994 Symposium on VLSI Circuits*, pp. 45-46, Jun. 1994.

Multi-Level Pass-Transistor Logic
for Low-Power ULSIs

Yasuhiko Sasaki, Kazuo Yano, Shunzo Yamashita, Hidetoshi Chikata*,
Kunihito Rikino**, Kunio Uchiyama and Koichi Seki

Central Research Laboratory, Hitachi, Ltd., Kokubunji, Tokyo 185, Japan
*Hitachi ULSI Engineering Corp., Kodaira, Tokyo 187, Japan
**Hitachi Device Engineering Ltd., Mobara, Chiba 297, Japan

Abstract

Multi-level pass-transistor logic (MPL) removes the redundancy in conventional single-level pass-transistor circuits to improve both power and delay. MPL is synthesizable based on the multi-level binary decision diagram, a new logic representation, and it has the potential to replace CMOS in any synthesized control block of an MPU. Overall improvement in the product of power, delay, and area of 42 % over CMOS is confirmed in actual microprocessor benchmark tests.

Introduction

Pass-transistor logic needs fewer transistors than conventional CMOS logic to perform the same function. Decreasing the number of transistors reduces not only power but also delay. However due to the difficulty of the synthesis, pass-transistor logic has not been used in random logic circuits. To overcome this difficulty, we previously developed an integration technique that features a new logic synthesizer with an original cell library as shown in Fig. 1 [1].

The pass-transistor logic we developed was a single-level logic. In CMOS circuits, however multi-level logic plays an important role in power and delay reduction, so it is conjectured that multi-level logic is essential to improve pass transistor circuits. But a good synthesis method combined with multi-level logic is needed to develogp multi-level pass-transistor circuits. For single-level pass transistor logic, we developed a technique that uses a graph representation called a binary decision diagram (BDD)[2], because it conforms well with pass transistor circuits in that a node in a BDD corresponds to a pass-transistor multiplexer as shown in Fig. 2. In CMOSs, multi-level logic optimization using Boolean manipulation is well established, so we tried to apply this technique to the BDD-based pass-transistor synthesis. But due to the wide difference between the Boolean equation and BDD, it was very difficult to make use of the CMOS factorization technique for pass-transistor logic synthesis. So we developed a new optimization technique based on the BDD that made possible the use of MPL.

Multi-Level Pass-Transistor Logic

MPL is a hierarchical logic unlike single-level pass-transistor logic. The hierarchy is expanded in the direction of "gate inputs" (Fig. 3). In the conventional logic the source-drain inputs are connected to each other and gate inputs are driven only by the primary inputs. On the other hand. in MPL gate inputs are driven by either primary inputs or the outputs of other pass transistor circuits.

The advantage of MPL in terms of power and small area comes from the high sharing capacity of the circuits. MPL shares circuits in the upper level which are generated separately when using single-level pass-transistor logic. The more a circuit is shared , the more redundant transistors can be eliminated. Another advantage lies in its parallel operation in contrast to the sequential operation of single-level logic. Parallel operation changes the order of delay time from $O(n)$ to $O(log\ n)$, where n is the number of inputs. This change results in a conspicuous improvement in circuits with large delay.

How multi-level circuits can be constructed is the key to MPL. Instead of using the conventional Boolean technique, we developed a technique based on a new extension of BDD. Multi-level pass transistor circuits are derived from this extended diagram by replacing all edges with multiplexers using pass-transistors and by inserting buffers where needed. We call this extension multi-level BDD and it is obtained through an algorithm which consists of the following three steps as shown in Fig. 4.

1) Extract partial diagrams from the original BDD, each of which share a logic (Fig. 4. (a)).
2) Replace extracted diagrams with new diagrams each of which have the same number of leaves as the original partial diagrams (Fig. 4. (b)).
3) Construct the upper-level logics for the control inputs of the replaced nodes. These new logics are adjusted to be the same as the original ones (Fig. 4. (c)).

An example of the effect of sharing logic is shown in Fig. 5. In this circuit a logic is shared by two outputs in the second level and two pass-transistors are eliminated.

The reduction in the delay time is illustrated in Fig. 6. Pass-transistor circuits for 8-NAND logic using conventional pass transistor logic and MPL are shown. In the conventional circuit, the critical path consists of seven cascaded pass transistors and three buffers. On the other hand, in MPL the critical path consists of only three cascaded pass transistors and one buffer.

Experimental

We applied MPL to 27 randomly selected circuits. These were mainly random logic circuits in a microprocessor. The delay time using MPL was compared with that of single-level pass-transistor logic as shown in Fig. 7. The effect of the multi-level optimization is clearly confirmed in circuits which have large delay. Also the power, the area, and the delay when using MPL were compared with those of conventional CMOS logic as shown in Fig. 8. On average, power was reduced by 23 % and area was reduced by 15 %. The delay time was also reduced by 12 %. The overall improvement with MPL was 42 %.

Acknowledgments

We thank H. Inayoshi, and E. Takeda for their steady encouragement; S. Narita, O. Nishii and J. Nishimoto for providing the sample circuits and for their advice on the logic circuits; I. Kudoh for providing the data of the device an for his advice; and T. Shonai for his valuable advice.

Reference

[1] K. Yano, Y. Sasaki, K. Rikino, K. Seki, 1994 CICC, pp603-606.
[2] R. Bryant IEEE Trans. Comp., Vol. c-35, pp677-691, Aug. 1986

Reprinted from *Proceedings of the 1995 Low-Power Symposium*, pp. 14-15, 1995.

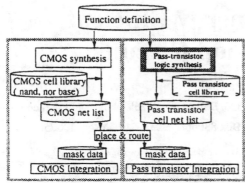

Fig. 1 CMOS integration vs. pass transistor integration

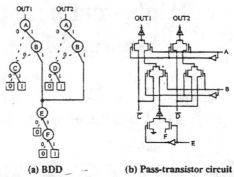

(a) BDD   (b) Pass-transistor circuit

Fig. 2 Conventional pass-transistor logic synthesis using BDD

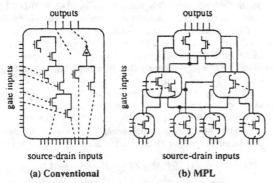

(a) Conventional   (b) MPL

Fig. 3 Conventional pass-transistor logic vs. MPL

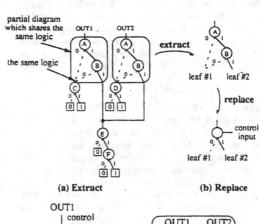

(a) Extract   (b) Replace

(c) Construct the upper-level BDD   (d) Constructed multi-level BDD

Fig. 4 Construction of multi-level BDD

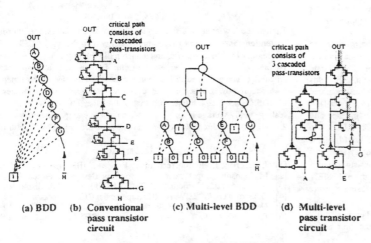

(a) Multi-level BDD   (b) Multi-level pass-transistor circuit

Fig. 5 Transformation into a multi-level pass-transistor circuit

(a) BDD   (b) Conventional pass transistor circuit   (c) Multi-level BDD   (d) Multi-level pass transistor circuit

Fig. 6 Delay time reduction (8-NAND logic)

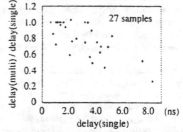

Fig. 7 Single-level pass-transisor logic vs. MPL

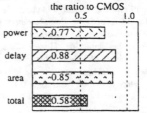

Fig. 8 CMOS vs. MPL in benchmark tests   15

209

High Performance, Energy Efficient Master-Slave Flip-Flop Circuits

Uming Ko[1], and Poras T. Balsara[2]

[1]Texas Instruments Incorporated, P. O. Box 655303, M/S 8316, Dallas, TX 75265, uko@daldd.sc.ti.com

[2] Dept. of Elect. Engg., University of Texas at Dallas, P.O. Box 830668, EC33, Richardson, TX 75083

Abstract

This paper investigates performance, power and energy efficiency of several CMOS master-slave D-flip-flops (DFFs). To improve performance and energy efficiency, a push-pull DFF and a push-pull isolation DFF are proposed. Among the five DFFs compared, the proposed push-pull isolation circuit is found to be the fastest with the highest energy efficiency and a minimum data pulse width property. Effects of using DPL circuit and tri-state push-pull driver are studied. The impact of scaling supply voltage alone and scaling transistor threshold voltage with supply voltage on speed and power consumption of these circuits is also examined.

I. Introduction

DFFs are one of major functions in finite state machines (FSM) which in turn is the critical part of control logic. It has been reported in [1] that the control logic of a microprocessor can occupy 20% of the processor's power. As more advanced architecture concepts, such as register renaming and out-of-order execution in a superscalar microprocessor [2], continue to prevail, the control logic will likely be more complicated and it power dissipation will likely grow beyond this current level. In addition, to boost processor clock frequency, modern processors typically adopt superpipelined execution [2] which uses DFFs. Enhancing DFFs' speed can either lead to a higher clock rate or allow more logic depths between two pipeline registers. In this paper, we compare area, speed, and power of five different DFF implementations: a regular low-risk DFF [3], a low-area DFF, a low-power DFF [4], a proposed push-pull DFF for performance, and another proposed push-pull isolation DFF (PPI-DFF) for performance and energy efficiency. Discussion is then extended to the use of double pass-transistor logic (DPL) for speed [5] and tri-stated circuit for reducing short-circuit power dissipation. Lastly, effects of scaling supply voltage at constant threshold voltage, as well as scaling threshold voltage with supply voltage are examined.

II. Design Techniques and Comparison of Energy Efficiency

A conventional negative edge-triggered DFF consists of two level-sensitive latches or 16 MOSFETs is illustrated in Fig. 1(a). The speed of this regular DFF is limited by two-gate delay (245 ps, Table I) after the clock signal, C, transitions from logic 1 to 0. The advantage of this DFF design is that it involves minimum design risk. A common approach to reduce area overhead of the regular DFF is to remove the two feedback transmission gates. This low-area DFF is depicted in Fig. 1(b), and it uses 25% fewer transistors. Although the strength of feedback inverters has been weakened to minimize the short-circuit power dissipation due to voltage contention, this low-area DFF still consumes 18% more total power and is 42% slower (or 76% more energy, Table I) than that of the regular DFF.

One approach to optimize for power dissipation is to replace the inverter and transmission gate in the feedback path of Fig. 1(a) with a single tri-state inverter, which is referred to as a low-power DFF [4] as shown in Fig. 1(c). The tri-state inverter avoids short-circuit power dissipation in the feedback path, and yields only 1% reduction in total power and 3% (Table I) slower speed when compared to the regular DFF. Considering area and energy efficiency, the low-power DFF is comparable to the regular DFF. To optimize for speed, an inverter and transmission gate are added between outputs of the master and slave latches to accomplish a push-pull effect at the slave latch, as depicted in Fig. 1(d). This adds four MOSFETs, but reduces the clock-to-output (C-to-Q) delay from two gates in a regular DFF to one gate. One method to reduce the transistor count is to use nMOSFET for latches' input [6]. However, the output of the nMOSFET can only reach a voltage level of $V_{dd}-V_t$ when it is at logic 1, causing a power overhead up to 50% [5]. A second issue of the nMOSFET input is the speed degradation due to a slow transition from logic 0 to logic 1. Therefore, a full transmission gate is kept in the push-pull DFF. To offset the four added

MOSFETs for push-pull, the two transmission gates in feedback paths are eliminated. Compared to the regular DFF, this push-pull DFF is 31% faster but with a 22% power overhead.

To optimize for energy usage in the push-pull DFF, two pMOSFETs are added to isolate the feedback path. This push-pull isolation DFF (PPI-DFF) increases the transistor count to 18, but achieves 16% reduction in total power and a speedup of 25% (Table I) relative to the previous push-pull DFF. Compared to the regular DFF, PPI-DFF improves speed by 56% at an expense of 6% more power. Energy efficiency of this PPI-DFF is enhanced by 45%–122% when compared to the previous four DFFs. Applying a DPL [5] input (Fig. 1(f)) to the PPI-DFF can result in a 20% reduction in the setup time. However, when D is at logic 1 and C switches from logic 1 to 0, a DC-path exists (*INV2-P2-P1-C*) leading to a 60% power overhead. Another option is to use a tri-state inverter to replace the push-pull driver of PPI-DFF, as shown in Fig. 1(g). Though this approach reduces the short-circuit power of the push-pull driver, it weakens the drive strength due to stacked MOSFETs and is 10% less efficient in energy compared to the PPI-DFF.

III. Effects of Scaling V_t and V_{dd}

SPICE simulation results for various supply voltages (V_{dd}) using the same device models with constant threshold voltages (V_t) are summarized in Fig.2–Fig. 5. In Fig. 2, speed of the low-area DFF degrades faster than others under low voltages, as it takes longer to resolve the logic value. In Fig. 3, the push-pull and low-area DFFs consistently dissipate higher power due to voltage contention in feedback loops. For a V_{dd} range of 1.5–3.5V, PPI-DFF's energy efficiency is 42–51% higher than that of the regular, low-power, and push-pull DFFs, and is 218–272% higher than that of the low-area DFF (Fig. 4). From 3.5V to 1.5V of V_{dd}, on an average, energy efficiency is improved by a factor of 2. At 1.5V, the minimum data pulse width of low-area and push-pull DFFs degrades to 3ns while the other three DFFs maintain at 0.7ns, which strongly suggests the former should be avoided in low-voltage, high-performance applications (Fig. 5). Scaling V_t with V_{dd} can maintain a proper signal-to-noise ratio [5]. Assuming V_t can be kept at 1/5 of V_{dd}, the effects of scaling both simultaneously down to a V_{dd} of 1.0V are summarized in Fig. 6–Fig. 9. In contrast to Fig. 4, Fig. 8 indicates that energy efficiency is improved by a factor of 5.7 when V_{dd} is scaled from 3.5V down to 1.0V. From Fig. 5 to Fig. 9 at a V_{dd} of 1.5V, minimum data pulse of low-area and push-pull DFFs improves by a factor of 5.8 as the V_t scaling reduces effects of voltage contention, while that of the other DFFs improve only by a factor of 1.8.

IV. Conclusions

Though the low-area DFF uses up to 33% fewer transistors, the internal voltage contention consumes up to 122% more energy than the rest of DFFs. Compared to a regular DFF, a low-power and a push-pull DFF improve power dissipation by 1% and delay by 31%, respectively, but end up with a comparable energy efficiency. The proposed PPI-DFF improves speed by 56% at the expense of only 6% of more power, when compared to a regular DFF. Energy efficiency of this PPI-DFF is 45–122% higher than that of the other DFFs. On an average, while scaling supply voltage from 3.5 to 1.xV enhances energy efficiency by a factor of 2, scaling it with threshold voltage can boost the efficiency by a factor of 5.7.

References

[1] IBM, "Blue Lightning Technology Preview," IBM, 1993.
[2] M. Johnson, "Superscalar Microprocessor Design," Prentice Hall, NJ, 1991.
[3] N. Weste *et al.*, "Principles of CMOS VLSI Design," Addison-Wesley, 1993.
[4] G. Gerosa *et al.*, "2.2W, 80MHz Superscalar RISC Processor," JSSC, 12/1994.
[5] U. Ko *et al.*, "Low Power Techniques for HP Adder," Trans. on VLSI, 6/1995.
[6] R. Hossain *et al.*, "Low Power Design with DET FF," Trans. on VLSI, 6/1994.

Reprinted from *Proceedings of the 1995 Low-Power Symposium*, pp. 16-17, 1995.

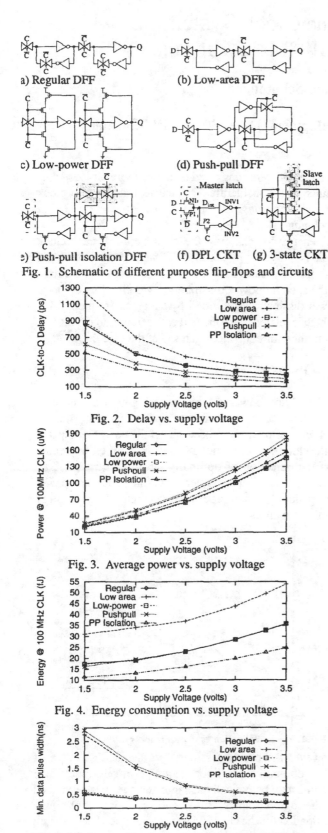

a) Regular DFF

(b) Low-area DFF

c) Low-power DFF

(d) Push-pull DFF

e) Push-pull isolation DFF

(f) DPL CKT (g) 3-state CKT

Fig. 1. Schematic of different purposes flip-flops and circuits

Fig. 2. Delay vs. supply voltage

Fig. 3. Average power vs. supply voltage

Fig. 4. Energy consumption vs. supply voltage

Fig. 5. Minimum pulse (=t_{setup}+t_{hold}) vs. supply voltage

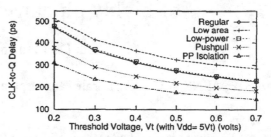

Fig. 6. Dependency of delay on V_t & V_{dd} scaling

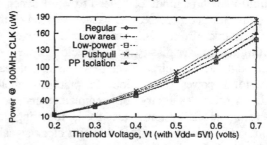

Fig. 7. Dependency of power on V_t & V_{dd} scaling

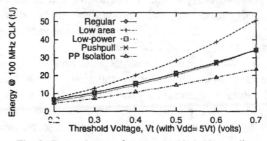

Fig. 8. Dependency of energy on V_t & V_{dd} scaling

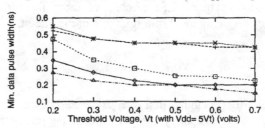

Fig. 9. Dependency of min. pulse (=t_{setup}+t_{hold}) on V_t & V_{dd} scaling

TABLE I. Comparison of power, delay, & energy for various DFFs

Parameters 3.3V,100MHz	Regular	Low-area	Low-power	Push-pull	Push-pull Isolation	unit
# of transistors	16	12	16	16	18	tr.
Total tr. width	45.8	40.2	46.8	43	48	μm
Power, avg.	122.9	146.9	121.7	152.6	131.4	μW
Percentage	94	112	93	116	100	%
Delay,C-to-Q	245.0	311.5	250.0	195.5	157.0	ps
Percentage	156	198	159	125	100	%
Energy	30.11	45.75	30.41	29.83	20.63	fJ
Percentage	146	222	147	145	100	%

Power Dissipation in the Clock System
of highly pipelined ULSI CMOS Circuits

Erik De Man and Matthias Schöbinger

SIEMENS A.G., Corporate Research and Development, D-81730 Munich, Germany

Abstract: Pipelining is an efficient way to increase the computational throughput of a feed-forward synchronous data-path but increases the capacitive load of the clock network. Even with an optimized level of pipelining, the power dissipated in the clock system of high-throughput ULSI circuits can be a considerable part of the total power dissipation. In practical implementations complementary or complementary non-overlapping clock systems are used, requiring the distribution of two or even four different clock signals. True-single-phase registers requiring only a single clock signal enable a significant reduction of the capacitive load of the clock network and will be analyzed with respect to their potential for reduction of the total power dissipation of pipelined ULSI circuits.

Basically, the power dissipation depends quadratically on the voltage swing, so reducing the swing of the clock signals offers a potential for power dissipation savings. A novel clock driver circuit for on-chip generation of clock signals with a reduced swing will be presented, allowing a reduction of the power dissipated in the clock system of high performance ULSI circuits of more than 60%, without reducing the throughput rate of the circuit.

1. Introduction

Introducing pipelining in a data path based circuit, is an efficient and customary way to increase the throughput rate of the circuit and thus increasing the computational power of the data path. Circuits featuring a high degree of pipelining inherently have short and simple critical paths. The short paths are essential for achieving a high throughput rate, but they also allow the circuits to be realized with minimum sized devices, which is a prerequisite for low power dissipation [1]. Moreover the inherent locality of the logical paths between the pipeline registers considerably reduces the power dissipation of the logic circuits due to glitches [2].

So the principle of pipelining, applied to increase the throughput rate, may at the same time reduce the power dissipated in the logic circuitry resulting in a reduction of the total power dissipation. However, the large number of pipeline registers controlled by one or more clock signals results in a considerable capacitive load in the clock network requiring an optimization of the degree of pipelining [3].

The energy for charging the clock network each clock cycle is delivered by the clock system of the circuit. An appropriate measure for the power efficiency of a circuit is the specific power dissipation p, defined as the total power dissipation P divided by the clock frequency 1/T (which is related to the critical path delay) and divided by the number of equivalent gates G of the circuit

$$p = P \cdot T / G \qquad (1)$$

In Fig. 1 the power dissipation per gate vs. the clock frequency for a number of chips realized in 1.5-μm CMOS is compared. It shows that if the power dissipated in the clock network is not taken into account, the specific power dissipation of optimally pipelined circuits is an order of magnitude lower than the specific power dissipation of non-pipelined circuits or gate array implementations.

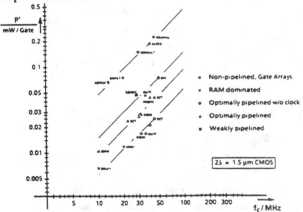

Fig. 1. Power dissipation vs clock Frequency.

However the power dissipated in the clock system of even optimally pipelined [3] high-throughput ULSI circuits may be typically as large as the power dissipated in the logic circuitry. But even if the additional power dissipation for the clock system is also taken into account (Fig. 1), the specific power dissipation of highly pipelined circuits is still about a factor 5 better than of non-pipelined circuits. Nevertheless, the clock system of such circuits still holds a great potential to reduce the total power dissipation.

2. The Clock Network

2.1. Clocking strategies

The complexity of the clock network of pipelined circuits depends on the clocking and pipelining strategy used. Because of the finite transition time of the clock signals, strategies to prevent races have to be considered in the clocking concept [4].

Complementary non-overlapping clock systems basically avoid races by the interdependent timing characteristics of the clock signals but they require the generation and routing of four different signals. But the dynamic pipeline registers of a non-overlapping clock system can be distributed over the logic blocks (Fig. 2) and can be efficiently realized as transmission gate latches exploiting the input capacitances of the preceeding logic block as dynamic storage elements. Moreover, if the latches are distributed in such a way that the delay time between two successive latches is longer than the non-overlap time of the clock phases, the introduction of a non-overlap time in the clock system will not reduce the computation time of the logical circuits. The complexity in terms of capacitive clock load for non-overlapping clock systems will be analyzed in section 2.2.

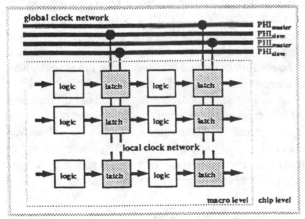

Fig.2 The clock network.

In a strict complementary clock system races must be avoided by a careful design of the clock network and pipeline registers. The registers must be optimized for delay time and clock transition requirements. The clock network must be thoroughly laid out to minimize signal delays mainly resulting from the RC-nature of the clock wiring. The advantage of strict complementary clock systems is that only two clock signals have to be routed around the chip. However this advantage can only be fully exploited if the registers are not distributed as latches over the logic blocks. The pipeline registers can be based on e.g. C^2MOS or tranmission-gate inverter latches which are more complex than the transmission based latches applicable with non-overlaping clock systems.

True single-phase clock systems [5] originally developed for high-speed CMOS circuits have the most simple clock network since only one signal has to be routed around the chip. However even more complex registers are required, which also demand a carefully designed clocking network in respect to the transition time of the clock signal. In section 3. the potential of such simple single-phase clock systems to reduce the capacitive clock load and power dissipation is evaluated. Especially the impact of the more complex register circuits on the total power dissipation will be discussed.

2.2. Capacitive load of the clock network

The expansive clock network is charged with the gate capacitances (C_g) of the clocked transistors of the pipeline registers. Another capacitive part of the clock network, is the junction capacitance of the source-drain regions of the output nodes of the clock drivers (C_j). The wiring capacitance of the clock network can be separated into the capacitance of the global clock network (C_{wg}) and the capacitance of the local clock network (C_{wl}). This is illustrated in Fig.2. The total capacitive load C_L of the clock network can be written as :

$$C_L = C_{wg} + C_{wl} + C_g + C_j \qquad (2)$$

The power needed for switching the clock network 1 / T times per second is then given by :

$$P_{cn} = 1 / T . Vdd^2 . C_L \qquad (3)$$

macro	gate capacitance (C_g)	local wiring capacitance (C_{wg})	global wiring capacitance (C_{wg})	junction capacitamce (C_j)
FIR macro	50pF	105pF	19pF	5pF
correlator macro	16pF	38pF	12pF	2pF

Table 1 Capacitive load C_L of the clock Network

In Table 1 the relative portion of the different capacitive parts of the clock network is given for two realized examples of pipelined carry-save macros realized in a 1.0-µm technology [6]. The macros are designed for a complementary non-overlapping clock system and the registers are distributed as latches over the logic of the data path as descussed above (Fig.2).Apparently, the dominant part of the capacitive load is the capacitance of the local clock network (C_{wl}). Also for complementary clock systems for which a 50% reduction of the wiring capacitance can be expected, the local wiring capacitance still represents a major part of the clock load.

3. Single-phase registers

C^2MOS or transmission gate registers (Fig 3a) have traditionally been used as hardware efficient dynamic pipeline registers. They require complementary clock phases for the controlling of the p- and n-channel devices so that the local clock network will require the wiring of at least two different clock signals.

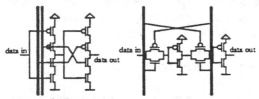

Fig. 3a. C²MOS register and transmission-gate register.

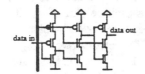

Fig. 3b. Single-phase register.

True-single-phase registers [5], originally developed for high-speed circuits, require only one clock phase to be distributed (Fig. 3b). As already mentioned, the capacitance of the global and local clock wiring can be reduced by about 50% compared to a complementary clock system . The gate capacitance will not change very much since single-phase registers also have four transistors connected to the clock signal. Since the local wiring capacitance C_{wl} represents a major part of the total capacitace of the clock network C_L , a significant saving can be expected for the power dissipation of the clock system. However, the circuits for the single-phase registers are more complex than simple C^2MOS register or pass-gate registers (Fig. 3).

In Table 2 some characteristic features of both C^2MOS and single-phase registers based on 1.0-μm CMOS designᶜ are summarized. The capacitive load of the local clock network is reduced by 53% (from 64fF to 30fF) as expected. But for a typical input transition probability of 25% (σ=0.25), the power dissipation of the logic itself has increased by 182% (from 32μW to 102μW). The main reasons for this increase in power dissipation of the logic is that the single-phase register has more internal nodes than the C^2MOS register and it also contains a

precharge inverter (central part of the register) which depending on the transition statistic of the input signal can have an unfavourable power dissipation balance.

	C²MOS Register (Fig. 3a)	Single-phase Register (Fig. 3c)
complexity (transistors / register)	8	9
number of internal nodes	2	3
area (1.0–μm CMOS)	618 μm²	570 μm²
minimum clock transition time	4 ns	2 ns
clock load (local wiring and gate cap.)	64 fF	30 fF
power dissipation (@ 100 MHz) clock network logic (σ = 0.25) total	160 μW 36 μW 196 μW	75 μW 102 μW 177 μW

Table 2. Comparison C²MOS registers and single-phase registers.

From the results of Table 2 it can be concluded, that the saving in power dissipation in the local clock network is almost completely compensated by an increase of the power dissipation of the logic of the registers. Only the saving in the global clock network remains but this is only a minor part of the complete power dissipation of the clock network (Table 1.). So applying the single-phase registers shown in Fig. 3b as pipeline registers, will bring no significant over-all saving of power dissipation of the complete circuit. Moreover, using the single-phase registers as shown in Fig. 3b results in undesirable higher demands for the clock drivers compared to conventional C^2MOS registers since faster transition times of the clock signals must be guaranteed (Table 2). However, especially for high-speed applications, the advantage of the simple clock network for circuits applying single-phase registers, of course remains unaffected.

4. The Clock Drivers

Since the power P_{cs} delivered by the clock drivers of the clock system can be a considerable part of the total power dissipation P of the circuit, the design of the drivers must also be optimized for minimum power dissipation.

CMOS clock drivers consist of a chain of cascaded drivers with increasing driving capability. An important design parameter for such driver chains is the stage ratio f defined by the ratio of the widths of the transistors of successive driver stages :

$$f = W_p^i / W_p^{i-1} = W_n^i / W_n^{i-1} \qquad (4)$$

W_p^i is the width of the p-channel transistor of the driver of stage i in the chain and W_n^i is the width of the associated n-channel transistor .

In a first approximation the wiring capacitances between successive driver stages are neglected. Then the parasitic capacitance of an intermediate node of the driver chain can be taken proportional to the widths of the transistors interacting with this node. Thus the capacitances of two successive intermediate nodes will also have a ratio equal to the stage ratio f and an approximate value for the power dissipation of the complete clock system can be given:

$$P_{cs} = 1 / T \cdot Vdd^2 \cdot C_L \cdot (1+1/f+1/f^2+..+1/f^N) \qquad (5)$$

C_L is the capacitive load for the clock driver, Vdd is the supply voltage, T is the clock period and N is the total number of stages in the driver chain. N is detemined by the desired input capacitance C_I of the first stage of the driver chain:

$$N = \log(C_L / C_I) / \log(f) \qquad (6)$$

The power efficiency of the clock system, can then be defined as :

$$\eta_P = P_{cn} / P_{cs} = 1 / (1+1/f+1/f^2+..+1/f^N), \qquad (7)$$

which for a given ratio of C_L / C_I is only a function of f. The factor $1-\eta_P$ characterizes the additional power dissipation overhead needed in the clock driver circuits for charging and discharging the clock network.

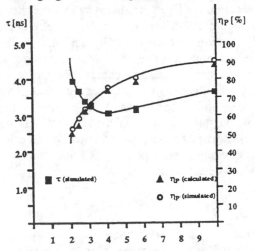

Fig. 4. Power efficiency and delay of clock drivers.

Fig. 4 shows the power efficiency η_P calculated from (6) and (7) for $C_L / C = 1000$. For comparision also the results obtained from SPICE simulations of driver chains designed in a 1.5-μm CMOS technology, are shown (for $C_L = 100p$ and $C_I = 100fF$). The small deviation between the calculated and simulated results show that the approximation of neglecting the wiring capacitance between the driver stages in (5) is justified. Moreover this also shows that the power dissipation due to short-circuit currents during clock signal transitions is negligable.

In Fig. 4 also simulation results for the delay t as a function of f are shown. This well known relation [4] between t and f has a minimum for $f \cong 3$. But from (7) and also from Fig. 4 it is clear that in order to miminize the power dissipation of the clock system, the stage ratio f of the drivers should be chosen as large as possible. This means that the number of drivers in the chain should be minimized. However another design parameter is the transition time of the clock signals. For complementary and single-phase clock systems requiring smaller clock transition times than non-overlapping clock systems, it may not be possible to apply stage ratios much larger than $f = 3$. Also the delay time of the complete driver chain must be taken into account, but it is well known that the total delay increases only slightly with increasing stage ratio f (Fig. 4.). So depending on the relative importance of power dissipation and delay time of the clock drivers, the optimum for the stage ratio will be in the range $f = 3..10$. For example, the clock drivers for the QAM-processor of [6], driving a total clock capacitance of almost 2nF, are designed with a stage ratio of $f = 8$, reducing the power dissipation of the predriver stages by almost a factor of 4 with only a major delay penalty and a minor degradation of the clock signal transition time. This is also a reasonable choice considering the degradation of the transition time throughout the expansive clock network due to RC-effects

5. Reduced Clock swing

The power needed for charging and discharging the clock network 1 / T times per seconds is given by (3). If the clock swing V_{clock} can be chosen different from the supply voltage V_{dd}

$$V_{clock} = r \cdot V_{dd} \qquad \text{with } r < 1 \qquad (8)$$

then the power dissipation of a clock system supplied from V_{dd} would be

$$P_{cs}' = 1 / T \cdot C_L \cdot V_{dd} \cdot V_{clock} \qquad (9)$$

or with (8)

$$P_{cs}' = 1/T \cdot C_L \cdot r \cdot V_{dd}^2 \qquad (10)$$

A reduction of V_{clock} results in a proportional reduction of P_{cs}'. The reduced clock swing however reduces the saturation current of the transistors connected to the clock phases :

$$I_{DS} \cong W \cdot (V_{clock}-V_{th})^\alpha \qquad \text{with } 1 < \alpha < 2 \qquad (11)$$

(with W and V_{th} the channel width and treshold voltage of the transistor involved; $\alpha = 2$ for conventional CMOS and $\alpha = 1$ for sub-micrometer CMOS).

As a concequence of the reduced saturation current, the delay time of the pipeline registers will increase. In order to retain the original throughput rate, the channel width W of the clocked transistors will have to be resized according to :

$$W' \cong W \cdot (V_{dd}-V_{th})^\alpha / (V_{clock}-V_{th})^\alpha \qquad (12)$$

This will increase the gate capacitance portion of C_L. But since this gate capacitance is only a smaller part of C_L, already for $\alpha = 2$ a reduction of the power dissipation can be obtained. For $\alpha = 1$ the required increase of the channel width and gate capacitance is smaller and the power dissipation saving correspondingly higher.

If the clock drivers could be supplied with V_{clock} instead of V_{dd} (e.g. from an external source) then the power dissipation in the clock system is :

$$P_{cs}'' = 1/T \cdot C_L \cdot V_{clock}^2 \qquad (13)$$

or with (8)

$$P_{cs}'' = 1/T \cdot C_L \cdot r^2 \cdot V_{dd}^2 \qquad (14)$$

In this case, due to the quadratic dependence of P_{cs}'' on r , a more significant saving of power dissipation of the clock system on the chip could be obtained. However the power dissipated in the additional external source or level conversion circuit is equal to

$$P_{su} = (1/T) \cdot C_L \cdot (V_{dd}-V_{clock}) \cdot V_{clock} \qquad (15)$$

or with (8)

$$P_{su} = (1/T) \cdot C_L \cdot (1-r) \cdot r \cdot V_{dd}^2 \qquad (16)$$

is only removed off-chip but not removed from the total system power dissipation which for e.g. portable systems is an important cost factor. So even if the clock system is supplied from a reduced supply voltage, the total power dissipation on system level is still given by (10).

A novel circuit for an on-chip generation of reduced-swing clock signals, requiring no extra (external nor internal) supply voltage is shown in Fig. 5, which enables to achieve a true reduction of the power dissipation as described in (14). The circuit exploits the principle of charge charing between the capacitances of complementary clock phases and additionally reduces the swing of the clock signals to approximately half the supply voltage. In this way both principles proposed in [1] to reduce the power dissipation in the clock system were combined in an efficient circuit realization. The circuit is used in a complementary non-overlapping clock system for which the principle of distributed registers (Fig.2) can be exploited, allowing a hardware efficient implementation of the pipeline latches as simple transmission gates as described before.

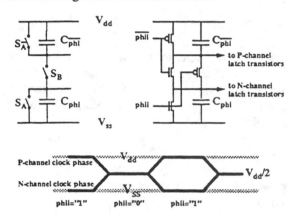

Fig. 5 Principle and CMOS implementation of complementary-phase clock driver.

The circuit of Fig. 5 requires no additional overhead compared to the standard solution. Also the widths of the clock driver transistors can be retained so that the silicon area remains unaffected. Whereas the power dissipation of the driver circuit of Fig. 5 is reduced by 75% compared to the standard clock driver circuits. The penalty in the throughput rate can be compensated by resizing the width of the clocked transisitors of the latches according to (12). This will increase the gate capacitance part of the clock load, but will only slightly increase the total clock capacitance. Simulations showed that even when maintaining the throughput rate of the circuit by resizing of the clocked transistors, a power dissipation reduction of the clock system of higly pipelined ULSI circuits of at least 60% can be achieved.

6. Conclusion

Complementary non-overlapping clock systems, complementary and true-single-phase clock systems were analyzed considering clock network complexity and register circuit complexity.

The comparison of power dissipation between C^2MOS registers and an implementation of true-single-phase registers shows that the straightforward application of single-phase registers as pipeline registers in highly pipelined ULSI circuits will yield no significant saving in power dissipation compared to complementary driven C^2MOS registers. Since the power dissipation savings due to the simpler clock networks are compensated by the increased power dissipation in the more complex register circuits, a more substantial saving of power dissipation can be obtained with a reduction of the swing of the clock signals. A special driver circuit for the on-chip generation of these reduced-swing clock signals is presented. For the conventional complementary non-overlapping clock systems, it allows a reduction of the power dissipation in the clock system of more than 60%, while maintaining the throughput rate of the circuit.

7. Acknowlegement

The ideas presented here are the result of extensive discussions with the co-workers of our staff. The authors would like to thank especially Stefan Meier, Heinz Söldner and Tobias Noll and for their valuable suggestions.

8. References

[1] T.G. Noll and E. De Man, "Pushing the Performance Limits due to Power Dissipation of future ULSI Chips", Proceedings of the ISCAS-92, pp. 1652 - 1655, San Diego, Mai 1992.

[2] C. M. Huizer, "Power Dissipation Analysis of CMOS VLSI Circuits by means of Switch-level Simulation", Proceedings of the European Solid-State Circuit Conference, pp. 61-64, Grenoble, Sept. 1990.

[3] T.G. Noll, "Carry-Save Architectures for high-speed Digital Signal Processing", Journal of VLSI Signal Processing, No. 3, pp. 121-140, 1991.

[4] C. Mead and L. Conway, "Introduction to VLSI Systems", Reading, MA: Addison-Wesley, 1980.

[5] J. Yuan and C. Svenson, "High-speed CMOS circuit techniques", IEEE Journal of solid-state circuits, Vol SC-24, No 1, pp. 62 - 70, Feb. 1989.

[6] E. De Man, M. Schöbinger, T.G. Noll and G. Sebald, "A 60-MBaud Single-Chip QAM Processor for the complete Base-Band Signal Processing of QAM Demodulators", Proceedings ISCAS-94, London, June 1994.

Power-Delay Characteristics
of CMOS Adders

CHETANA NAGENDRA, ROBERT MICHAEL OWENS, AND MARY JANE IRWIN

Abstract—An approach to designing CMOS adders for both high speed and low power is presented by analyzing the performance of three types of adders-linear time adders, $\log N$ time adders and constant time adders. The representative adders used are a ripple carry adder, a blocked carry lookahead adder and several signed-digit adders, respectively. Some of the tradeoffs that are possible during the logic design of an adder to improve its power-delay product are identified. An effective way of improving the speed of a circuit is by transistor sizing which unfortunately increases power dissipation to a large extent. It is shown that by sizing transistors judiciously it is possible to gain significant speed improvements at the cost of only a slight increase in power and hence a better power-delay product. Perflex, an in-house performance driven layout generator, is used to systematically generate sized layouts.

Index Terms— Power-delay product, static CMOS adders, transistor sizing.

I. INTRODUCTION

The three most widely accepted metrics for measuring the quality of a circuit are area, delay and power. Minimizing area and delay has always been considered important, but reducing power consumption has been gaining prominence recently [8], [7], [6], [5]. This can be attributed to the increasing popularity of mobile communication systems. Since dramatic improvements in battery technology are not foreseen, low-power designs are crucial not only to lighten the overall weight, but also to reduce the time between recharges. On the other hand, with advances in CMOS technology and reduction of the feature size, area is no longer as scarce a resource as it once used to be. Portability imposes a strict limitation on power dissipation while still demanding high computational speeds as required by real-time tasks. Hence we use the **power-delay product** as the metric of performance in this paper and downplay the importance of the area occupied by the circuit.

Chandrakasan et. al. describe the *four* degrees of freedom available in the design of low-power circuits and systems, namely technology, circuit design styles, architecture and algorithms [8]. One popular technology family is static CMOS since it has large noise margins and consumes the least power in its class [9]. The dynamic power consumption of a CMOS gate is given by $P = p_f C_L V_{dd}^2 f$, where p_f is the activity factor, C_L is the load capacitance, V_{dd} is the supply voltage and f is the clock frequency. In this paper, we assume that the supply voltage is fixed at 5 V and clock all adders at the same frequency. We attempt to lower the load capacitance and the activity factor of the circuit by playing with the circuit design styles (transistor sizing) and addition algorithms, respectively. We compare the power consumed and the delay of adders using three different addition algorithms, namely, ripple carry addition, blocked carry lookahead addition and signed-digit addition. The choice of the addition algorithm significantly affects p_f, the activity factor of a circuit [8].

Reference [6] compares the power consumption of some CMOS adders. But they do not size transistors whereas practical circuits usually employ transistor sizing to improve speed. Further, estimating the power consumption of signed-digit adders has not been done before.

The remainder of this paper is organized as follows. Section II presents a brief description of the three adder topologies. An overview of the layout generation of the adder circuits using *Perflex*, a performance driven module generator, is given in Section III. In Section IV, the experiments conducted to estimate the energy and the delay of the adders are described along with the performance numbers and we conclude in Section V.

II. ADDER TOPOLOGIES

We have chosen adders with a wide spectrum of timing and complexity which makes it interesting to compare their performance in terms of power and power-delay product. The adders range from the simple but slow (linear time) ripple carry adder to the fairly complex but extremely fast (constant time) signed-digit adders. The third type of adder is the $O(\log N)$ time blocked carry lookahead adder. The N-bit operands in the ripple carry and blocked carry lookahead adders are represented in 2's complement format and the output is generated in the same format. The operands in the signed-digit adder are, of course, signed-digit.

A. Ripple Carry Adder

The basic unit of a ripple carry adder (RCA) is a Full Adder (FA) which computes a sum bit and a carry bit: $s_i = x_i \oplus y_i \oplus c_i$, $c_{i+1} = x_i y_i + x_i c_i + y_i c_i$. That is, it adds the two operand bits with the incoming carry bit to produce a sum bit and an outgoing carry bit. Since, in the worst case, the carry can propagate from the least significant bit position to the most significant bit position, the addition time of an N-bit RCA is $O(N)$.

B. Blocked Carry Lookahead Adder

We use the structure of the "ELM" blocked carry lookahead adder (BCLA) [14]. Compared to the Brent and Kung adder [4], the ELM adder occupies the same area ($O(N \log N)$), but uses fewer

Manuscript received December 7, 1993; revised February 24, 1994 and May 10, 1994. This work was supported in part by NSF Grant MIP-9102500.

The authors are with the Department of Computer Science and Engineering, The Pennsylvania State University, University Park, PA 16802 USA.

IEEE Log Number 9403170.

Reprinted from *IEEE Transactions on VLSI Systems*, Vol. 2, No. 3, pp. 377-381, September 1984.

TABLE I
ALGORITHM: SIGNED DIGIT ADDITION

For $0 \leq i \leq k$,
Step 1. Compute the interim sum digit u_i and
carry digit c_i: $u_i = x_i + y_i - rc_i$
such that $|u_i| \leq (r-2)/2$ and $|c_i| \leq 1$.
Step 2. Compute the final sum digit :
$s_i = u_i + c_{i-1}$.

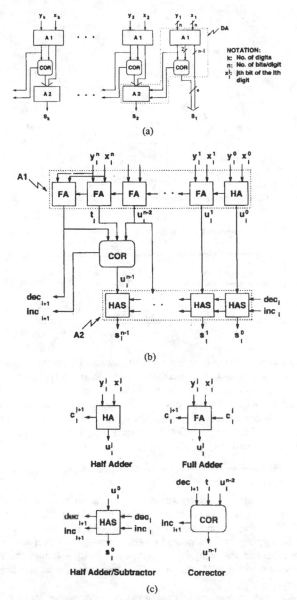

Fig. 1. Signed digit adder.

interconnects ($2N \log N + N$ as opposed to $3N \log N$). The ELM adder makes use of a binary tree of simple processors to perform addition in $O(\log N)$ time. The leaves of the tree compute partial sums and pass them on to the next level. The nodes at the higher levels of the tree receive partial sums as well as the generate and propagate information necessary to update the partial sums at that level of the tree.

C. Signed Digit Adder

Signed-digit (SD) representations [1] are positional number representations with a constant radix $r \geq 3$ in which the individual SD's $\in \{-a, \ldots, -1, 0, 1, \ldots, a\}$ where:

$$\frac{1}{2}(r_0 + 1) \leq a \leq r_0 - 1, \quad \text{for odd indices } r_0 \geq 3$$

$$\frac{1}{2}r_e + 1 \leq a \leq r_e - 1, \quad \text{for even indices } r_e \geq 4$$

In this paper we consider only those bases which are powers of 2. The use of SD numbers allows addition to be performed in constant time by restricting carry propagation to at most one digit position [2]. Each SD is encoded using a 2's complement encoding, where each SD occupies $n = \lceil \log r \rceil + 1$ bits. To represent an N-bit number, $k = \lceil N/(n-1) \rceil$ SD's are required. The SD addition algorithm is given in Table I [2].

Fig. 1 shows the general structure of a base-2^{n-1} SD adder (denoted as SDA-2^{n-1}) and its modules. Step 1 of the SD addition algorithm involves the sign extended addition of two n-bit SD's. The advantage of using a 2's complement encoding is that the digit addition can be done using a simple n-bit RCA, namely module **A1**. The output at this stage lies between $-2r+2$ and $2r-2$ and module **COR** is used to correct the output and generate the interim sum digit and the 2-b carry such that the conditions in step 1 are satisfied. Since the carry can only be $+1, 0$ or -1, it is encoded using two bits *inc* and *dec* (increment and decrement respectively). Thus step 2 of the SD addition algorithm calls for just an increment/decrement circuit, namely module **A2**. The longest path in the SDA-2^{n-1} is marked DA (Digit Adder) in Fig. 1(a).

III. LAYOUT STYLES AND ADDER LAYOUT GENERATION

A few alternatives were considered to produce the adder layouts. Hand design usually results in smaller layouts which can be optimized for performance but can be very tedious and time consuming. On the other hand, automatic layout generation tools result in a faster turn-around time but seldom match the hand layouts in area or performance. But a comparison based on a level 'playing field' was a prime consideration for a controlled set of experiments and hence we decided to use the latter option. Among the automatic layout generation tools, we had to choose between standard cell place and route tools (for example, Timberwolf-SC4 [18]) and *Perflex*, an in-house performance driven module generator [15, 16]. Experimental results showed that *Perflex* can significantly improve the timing of a circuit while the layout area is kept comparable to that obtained when the layout is optimized for area. A 286-transistor 8-b adder

description based on gates from the *MisII* standard-cell library [3] was given as input to both Timberwolf-SC4 and *Perflex*. HSPICE [17] simulations showed that the layouts generated by *Perflex* were better in all respects of maximum signal delay, layout area and power consumption. Hence we decided to use *Perflex*.

Perflex uses a new flexible CMOS layout style that supports sizing of individual transistors to generate fast static combinational CMOS circuit modules. The techniques used for optimizing timing are transistor sizing, transistor reordering and reduction of wiring capacitances along the critical paths. All the three are performed in close interaction with a simulated annealing layout process and transistors are sized only when necessary for improving the speed and just enough to satisfy the slack constraints. This increases the load capacitance to a minimum extent and therefore the increase in power consumption is much less than in the case of standard cells.

A. Transistor Reordering

Transistor reordering is a simple, yet effective timing optimization technique which is achieved by arranging transistors connected in

219

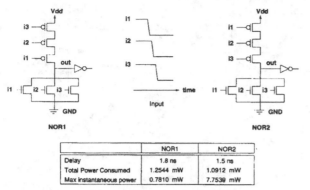

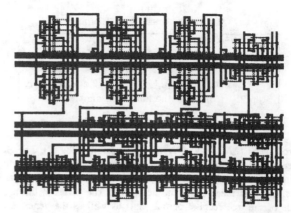

Fig. 3. Layout of a SD-4 DA built by connecting *Perflex* generated modules.

	NOR1	NOR2
Delay	1.8 ns	1.5 ns
Total Power Consumed	1.2544 mW	1.0912 mW
Max instantaneous power	0.7810 mW	7.7539 mW

Fig. 2. Effect of reordering on a 3-input NOR gate ($V_{dd} = 5$ V).

TABLE II
A COMPARISON OF TWO 8-b SDA-4s (UNIT SIZED INVERTER LOAD).

	Sized full DA	Sized parts DA
Max transistor size (nfet/pfet)	$35/21 \lambda$	$5/5 \lambda$
Bounding box area	$279 \times 2338 \lambda^2$	$288 \times 1670 \lambda^2$
Average power dissipation	109.904 mW	70.066 mW
Max instantaneous dissipation	33.729 mW	29.418 mW
Worst case delay	12.585 ns	12.625 ns
Power × Delay	1.383142×10^{-9}	0.884578×10^{-9}

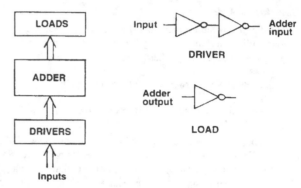

Fig. 4. Adder with driver and load.

series in the order of their input arrival times so that the latest arriving input is assigned to the transistor closest to the output node. Our experiments showed that reordering can not only improve the speed of a gate, but can decrease power consumption as well. Reordering in *Perflex* is carried out once prior to the layout and whenever transistors are sized. The 3-input NOR gate example in Fig. 2 illustrates the effects of reordering.

B. Transistor Sizing

The transistor sizing algorithm used in *Perflex* is the posynomial programming approach used in AT&T's TILOS [12]. But in the TILOS heuristic, transistors are sized only as a preprocessing step to the layout process whereas in *Perflex* sizing is done repeatedly as the simulated annealing layout optimizer slowly solidifies the gate placement to the final one. This allows transistors to be sized more optimally.

Although increasing the transistor size improves the speed of the circuit, it also increases power dissipation since the load capacitance increases. Using *Perflex*, we generated two types of sized layouts of an 8-b SDA-4 and observed their performance. In the first type of adder, a logical description of a single DA of SDA-4 was given as input to *Perflex*. Since the critical path of a SDA lies entirely within a single DA, optimizing the performance of the DA is sufficient to optimize the performance of the entire adder. This "sized full DA" was then used to construct an 8-b SDA-4.

In the second type of adder, each basic module comprising the DA, namely, **HA, FA, HAS** and **COR**, was sized individually using *Perflex*. These modules were then interconnected with hand layout to generate "sized parts DA". The results in Table II show that the 8-b SDA-4 built using the latter DA exhibited almost no loss in speed and consumed far less power giving a superior power-delay product. This is because the maximum size of a transistor in the "sized full DA" is much larger than that in the "sized parts DA". However this extra sizing produces negligible speed improvement (0.003%) which does not justify the cost of increasing power dissipation by 56%.

Hence our layout strategy for optimizing the power-delay product of the adder is to optimize the individual modules separately using

Perflex which sizes the transistors only as much as necessary. Then the modules are interconnected by hand for a compact layout. Since the modules are automatically generated, most of the tedium of hand layout is avoided. Fig. 3 shows an example of a layout generated in this manner.

IV. EXPERIMENTAL RESULTS

Layouts of 32-b versions of the adders presented in Section II were generated in 1.2μ static CMOS technology as described in Section III. Table III lists their salient features. These layouts were then used to extract circuit equivalents for use in a detailed circuit simulation using HSPICE v. 93a [17] to obtain the power and delay measures. The transistor parameters used were from a recent 1.2μ MOSIS fabrication run and the simulations were carried out at 27°C with an input frequency of 10 MHz. All measurements were taken with each input supplied through a driver consisting of two inverters in series, and each output node driving a unit sized inverter load (see Fig. 4).

The delay of each adder was measured directly from the output waveforms by presenting the adder with inputs that caused the maximum carry ripple. It can be seen from Table III that the RCA suffers a greater delay due to its long carry chain. Among the SDA's, SDA-4 has the least delay and the delay of the SDA increases with the base since the digit addition carry chains become progressively longer.

The estimation of power dissipation of a circuit is a difficult problem and has received a lot of attention [5], [10], [11], [13]. HSPICE can measure the power consumed by a circuit given a set of inputs. But the power dissipation is a strong function of the inputs.

TABLE III
NUMBER OF TRANSISTORS, AREA AND DELAY OF 32-b ADDERS

Adder Type	No. of transistors	Area (λ^2)	Worst case delay (ns)
RCA	884	183 × 3145	63.4
BCLA	2228	792 × 1200	32.7
SDA-64	2184	288 × 4856	26.4
SDA-32	2236	288 × 4955	23.1
SDA-16	2184	288 × 4853	19.7
SDA-8	2500	288 × 5568	16.2
SDA-4	2904	288 × 6489	12.6

TABLE IV
POWER MEASURES FOR 32-b ADDERS (95% CONFIDENCE INTERVAL, 3% ERROR).

Adder	Mean power dissipation per addition (mW)	Power × Delay (×10⁻⁹)	Max instantaneous power (mW)
RCA	149.866 ± 2.148	9.502 ± 0.1362	72.727
BCLA	300.162 ± 3.293	9.830 ± 0.1078	74.739
SDA-64	298.145 ± 3.824	7.871 ± 0.1009	88.645
SDA-32	310.312 ± 3.791	7.168 ± 0.0875	87.425
SDA-16	306.772 ± 3.756	6.043 ± 0.0740	82.170
SDA-8	355.003 ± 3.730	5.751 ± 0.0604	97.810
SDA-4	409.695 ± 3.677	5.162 ± 0.0463	103.21

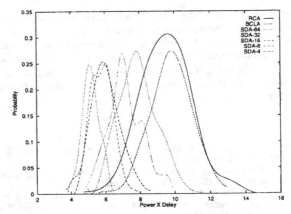

Fig. 6. Power delay histogram.

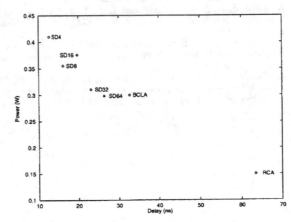

Fig. 7. Power versus delay.

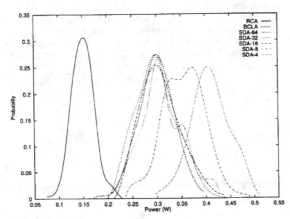

Fig. 5. Power histogram.

The node voltages of each adder were initialized by supplying the adder with zero inputs. Then each adder was presented with 500 independent, pseudorandom inputs and the power consumed was monitored. The power dissipated during the initialization phase was not considered. Since the propagation delay of the slowest adder is approximately 63 ns, each addition was allowed a time of 100 ns for the voltages to stabilize before the next addition. The power dissipation measures include the power consumed by the drivers and the loads. The average power consumed by a single driver driving a single load is about 0.8165 mW. For example, a 32-b SDA-4 has 48 pairs of inputs and 48 outputs and therefore requires 96 drivers and 48 loads.

The height of the power curve was recorded as the maximum instantaneous power dissipation of the circuit. Since the inputs are independent, power can be approximated to be *normally distributed* [5]. Hence the mean power dissipation of the circuit is given by $\bar{x} \pm t_c \frac{s}{\sqrt{N}}$, where $\bar{x}$ is the sample mean, s is the standard deviation, N is the number of samples and t_c is obtained from the t-distribution for c% confidence interval [19]. (For 95% confidence and $N = 500, t_c = 1.9647$.) The power-delay product, i.e. (power consumed per addition × worst case delay), is computed in a similar way. The power measures for the seven adders are shown in Table IV. Figs. 5 and 6 show the probability distribution of the power dissipation and the power-delay product, respectively.

The RCA has approximately one third the number of transistors contained in either a BCLA or a SDA. Because of its simplicity, it is not surprising that it exhibits the lowest average and maximum instantaneous power dissipations. The SDA's may be very fast; but unfortunately their speed advantage does not come for free. They are logically more complicated than the RCA's and consume a lot more power. But they are clearly superior in the power-delay product since their speed advantage over the RCA is sufficient to overcome even their higher power consumption. However, the BCLA fails to achieve the same performance level.

Among the SDA's, the larger the base, the fewer the number of digit adders. Thus with larger bases there is a saving in the logic since fewer sign bits and corrections have to be handled. Therefore we observe a reduction in the power consumption with increasing base. It can be observed from Table IV as well as the power-delay histogram of Fig. 6 that the reduction in power is not enough to overcome the increase in delay. Finally, Fig. 7 shows a plot of the sample mean power dissipation versus the delay of the adders.

V. CONCLUSION

In this paper, we have compared the power-delay product of three types of adders, namely, RCA, BCLA and SDA. We have also shown a design methodology wherein, by sizing the transistors only to the extent necessary, we can derive the speed benefits of sizing without increasing power dissipation to a large extent. The SDA's exhibit very good power-delay products compared to the RCA and BCLA. Although higher bases reduce power consumption, the delay of the adder is more due to longer ripple carry chains in the digit adders. The reduction in power dissipation is not sufficient to balance the increase in delay when the digit addition is done using RCA's. Hence we observe the phenomenon of increasing power-delay product with the base of the SDA. We are investigating the design of faster DA's by using simple BCLA's instead of RCA's to perform signed-digit addition.

REFERENCES

[1] D. E. Atkins, "Introduction to the role of redundancy in computer arithmetic," *Computer*, pp. 74–77, Jun. 1975.
[2] A. Avizienis, "Signed-digit number representation for fast parallel arithmetic," *IRE Trans. Electronic Comput.*, p. 389, 1961.
[3] R. G. Brayton and R. Rudell *et al.*, "MIS: A multiple-level logic optimization system," *IEEE Trans. Computer Aided Design*, vol. 6, no. 6, pp. 1062–1081, Nov. 1987.
[4] R. P. Brent and H. T. Kung, "A regular layout for parallel adders," *IEEE Trans. Comp.*, vol. C-31, no. 3, pp. 260–264, Mar. 1982.
[5] R. Burch, F. N. Najm, P. Yang, and T. N. Trick, "A Monte Carlo approach for power estimation," *IEEE Trans. VLSI Syst.*, vol. 1, no. 1, pp. 63–71, 1993.
[6] T. K. Callaway and E. E. Swartzlander Jr., "Estimating the power consumption of CMOS adders," in *Proc. 11th Symp. on Comp. Arithmetic*, June 1993, pp. 210–219.
[7] A. Chandrakasan, M. Potkonjak, J. Rabaey, and R. Brodersen, "An approach for power minimization using transformations," in *IEEE VLSI Signal Process. Workshop*, pp. 41–50, June 1992.
[8] A. Chandrakasan, S. Sheng, and R. Brodersen, "Low power CMOS digital design," *IEEE J. Solid-state Circ.*, pp. 685–691, Apr. 1992.
[9] K. M. Chu and D. L. Pulfrey, "A comparison of CMOS' circuit techniques: Differential cascade switch logic versus conventional logic," *IEEE J. Solid-state Circ.*, vol. 22, pp. 528–532, Aug. 1987.
[10] M. Cirit, "Estimating dynamic power consumption of CMOS circuits," in *ICCAD*, pp. 534–537, 1987.
[11] S. Devadas, K. Keutzer, and J. White, "Estimation of power dissipation in CMOS combinational circuits using Boolean function manipulation," *IEEE Trans. Computer Aided Design*, vol. 11, no. 3, pp. 373–383, Mar. 1992.
[12] J. P. Fushburn and A. E. Dunlop, "TILOS: A posynomial programming approach to transistor sizing," in *Proc. ICCAD*, pp. 326–328, 1985.
[13] S. M. Kang, "Accurate simulation of power dissipation in VLSI circuits," *IEEE J. Solid-State Circ.*, vol. SC-21, no. 5, pp. 889–891, Oct. 1986.
[14] T. P. Kelliher, R. M. Owens, M. J. Irwin, and T.-T. Hwang, "ELM—A fast addition algorithm discovered by a program," *IEEE Trans. Comput.*, vol. 41, no. 9, Sept. 1992.
[15] S. Kim, "CMOS VLSI layout synthesis for circuit performance," Ph.D. dissertation, The Penn State Univ., Univ. Park, 1992.
[16] S. Kim, R. M. Owens, and M. J. Irwin, "Experiments with a performance driven module generator," in *Proc. DAC*, June 1992, pp. 687–690.
[17] Meta-Software, *HSPICE User's Manual version H92*. Campbell, CA: Meta-Software, 1992, 1300 White Oaks Road, Campbell, CA 95008.
[18] C. Sechen and A. L. Sangiovanni-Vincentelli, "The Timberwolf routing package," in *Proc. Custom Integrat. Circuit Conf.*, May 1984.
[19] G. W. Summers, W. S. Peters, and C. P. Armstrong, *Basic Statistics in Business and Economics*. Belmont CA: Wadsworth, 1981, ch. 8.

Delay Balanced Multipliers for Low Power/Low Voltage DSP Core

Toshiyuki Sakuta*, Wai Lee and Poras T. Balsara**

Integrated Systems Laboratory, Texas Instruments Inc., P.O.Box 655474, MS446, Dallas, TX 75265
* Permanent address: Research & Development Division, Hitachi America, Ltd., 50 Prospect Avenue, Tarrytown, NY 10591
** Permanent address: Department of Electrical Engineering, University of Texas at Dallas, P.O.Box 830688, EC33, Richardson, TX 75083
e-mail addresses : sakuta@hc.ti.com, lee@hc.ti.com, poras@utdallas.edu

1. Introduction

The choice between an array and a Wallace-tree multiplier architecture depends on the tradeoffs between area and speed. Wallace-tree multipliers offer higher speed but consume larger areas than array multipliers. As for the power dissipation, recent research on signal transition activity, which directly influences the dynamic power dissipation, indicated that the array multiplier had an architectural disadvantage [1]. Delay imbalance of the signals in a full adder array causes many spurious transitions and consequently unnecessary power dissipation. In this work, the potentials for power reduction of both Wallace-tree multiplier and array multiplier were explored. By implementing a few delay buffer circuits appropriately, the delays in the full adder array can be balanced and the majority of the spurious transitions are suppressed. Thus, significant reduction in power dissipation of the array multiplier can be achieved.

2. Delay balanced array multiplier

As shown in the lower half of Fig. 1, multiplicand bits are simultaneously input to all the partial product generators at every stage in the conventional array multiplier. All full adders start computing at the same time without waiting for the propagation of sum and carry signals from the previous stage. This results in spurious transitions at the output and wastes power [2]. Furthermore, as shown in Fig. 2, since these spurious transitions are propagated to the next stage continuously, their numbers grow stage by stage like a snow ball. This causes a significant increase in power dissipation.

The following two countermeasures, shown in Fig. 1, were taken to solve this problem. One is that the delay circuit which has the same delay property as that of a full adder circuit was inserted in the multiplicand signal path at each full adder stage. Another is that a register was placed at the input of booth encoder so that the timing of multiplier encoding could be controlled by the clock. The same delay circuit as the one used in the first measure was inserted in the clock signal path at each full adder stage. By these measures, the output transition of a partial product generator at each full adder stage and the transitions of the sum and the carry output of a full adder at the previous stage are synchronized. Therefore, the significant reduction of spurious transitions was achieved with virtually no penalty in performance. The area increase for these additional circuits in a 24 bit x 24 bit Booth multiplier is about 8.5%.

3. Delay balanced Wallace-tree multiplier

A Wallace-tree multiplier occupies about 40% more area than a corresponding array multiplier. It also has a problem of high wiring density of internal signals. On the other hand, it has low probability of occurrence of spurious transitions since most inputs to full adders at each stage are naturally synchronized due to its inherent parallel structure. In addition, fast multiplication is achieved because of the small number of necessary full adder stage [3, 4]. Figure 3 shows a Wallace-tree structure for a 13 bit column compressor required for a 24 bit x 24 bit Booth multiplier.

Signals A and B, which jump over one full adder stage and are the inputs to the full adder of the next stage, cause spurious transitions inside this multiplier. A delay balanced Wallace-tree multiplier is shown in Fig. 4. As shown in this figure, delay circuits which have the same delay as a full adder circuit were inserted in the signal paths of A and B so that A and B are synchronized with the other inputs of the corresponding adders. The penalty in performance is also negligible as in the array multiplier's case.

4. SPICE simulation results

SPICE simulations of both multipliers were performed utilizing random input patterns as multiplicands and multipliers. Table 1 shows the device technology used for this study. The number of spurious transitions (S) and valid transitions (V) generated at the output of each full adder and the carry lookahead adder (CLA) in 64 clock cycles were extracted from simulation results. S/V represents the ratio of wasted power to necessary power. 2D contour map of the values of S/V was plotted at the corresponding position on the floor plan as shown in Fig. 5 (a) through (d). Figure 5 (a) shows that a large number of spurious transitions occurs all over the conventional array multiplier. S/V increases rapidly as the number of stages increases. The average value of S/V over the entire multiplier is 2. The remarkable reduction in spurious transitions by the techniques stated in section 2 can be seen in Fig. 5 (b). The average value of S/V in this case is only 0.3. In Fig. 5 (c), many spurious transitions can also be observed at the stage 2.3 and the stage 5 of Wallace-tree multiplier where the inputs to full adders are not synchronized. The effect of the delay circuit for suppression of spurious transitions in this multiplier is shown in Fig. 5 (d). Since a Wallace-tree multiplier consumes relatively less power by nature, the effect of matching delay in power saving is only 6.5% while 36% of power saving was attained in the delay balanced array multiplier. The SPICE simulation results at different operating voltages are summarized in Fig. 6 (a) and (b). Finally, a comparison of layout area, delay and power-delay product among three multiplier architectures is shown in Table 2.

5. Conclusion

A simple but effective technique, which synchronizes the propagation of signals at each full adder stage, has cut the power dissipation of an array multiplier down to equal to or less than that of a Wallace-tree multiplier with a minimal penalty in performance and layout area. This delay balanced array multiplier is a strong candidate for low power and small area DSP core for portable equipment.

Reference

[1] J.Leijten, et al., "Analysis and Reduction of Glitches in Synchronous Networks", European Design & Test Conf., Dig. Tech. papers, pp.398-403, Mar. 1995.

[2] C. Lemonds, et al., "A Low Power 16 by 16 Multiplier Using Transition Reduction Circuitry", Intl. Workshop on L/P Design, Dig. Tech. papers, pp.13-142, Apr. 1994.

[3] C. S. Wallace, "A Suggestion for a Fast Multiplier", IEEE Trans. Electron. Computer, vol.EC-13, pp.14-17, 1964.

[4] L. Dadda, "Some Schemes For Parallel Multipliers", Alta Freq., vol.34, p.349-356, 1965.

Reprinted from *Proceedings of the 1995 Low-Power Symposium*, pp. 36–37, 1995.

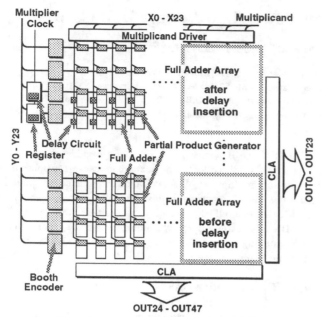

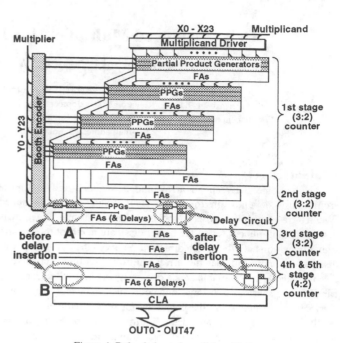

Figure 1 Delay balanced array multiplier (top half) and conventional array multiplier (bottom half)

Figure 4 Delay balanced parallel multiplier

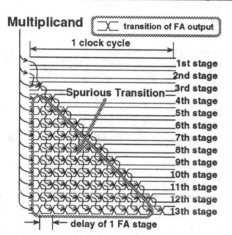

Figure 2 Spurious transition in a conventional array multiplier

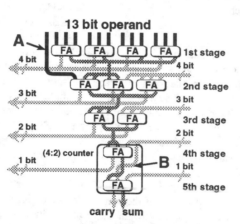

Figure 3 Wallace tree structure for 13 bit column compressor

Table 1 Device technology

Lg	0.35 μm
Vthn/Vthp	+0.5 V / -0.5 V (+0.3 V / -0.3 V @Vdd = 0.9 V)
tox	6 nm

Table 2 Comparison of three different architectures (normalized with conventional array multiplier's values)

Type \ Item	Conventional Array Multiplier	Delay Balanced Array Multiplier	Delay Balanced Parallel Multiplier
Layout Area	1.00	1.09	1.40
Delay	1.00	1.02	0.75
PxD product	1.00	0.64	0.66

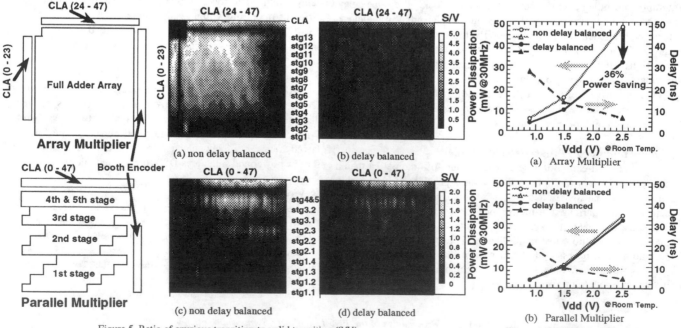

Figure 5 Ratio of spurious transition to valid transition (S/V)

(a) non delay balanced

(b) delay balanced

(c) non delay balanced

(d) delay balanced

(a) Array Multiplier

(b) Parallel Multiplier

Figure 6 SPICE simulation result

224

Minimization of Power in VLSI Circuits Using Transistor Sizing, Input Ordering, and Statistical Power Estimation

Chin Hwee Tan and Jonathan Allen

Research Laboratory of Electronics and Department of EECS
Massachusetts Institue of Technology, Cambridge, MA 02139

Abstract

Low-power design is becoming increasingly important in today's technology as wireless communication and mobility of equipment become increasingly desirable. In this work, a fast and efficient low power design method using cell libraries is developed. This optimization routine utilizes accurate and efficient statistical power estimation methods, transistor sizing, and input ordering. A delay model which takes into account input transition time is developed. An augmented cell library that contains cells that have been designed and sized to give good power and delay trade-offs is constructed, keeping area and input ordering in mind. Finally, an algorithm that selects the best sized versions to use for the circuit so that a given delay constraint is satisfied with minimal power dissipation is developed. Options that use the switching probabilities of a node, input ordering, and critical path analysis, are also provided to enhance the basic algorithm.

1 Introduction and Background

As wireless communication and mobility of equipment become increasingly desirable, power dissipation of circuits has become a major concern in circuit synthesis. In performance driven synthesis of VLSI circuits, low-power design has joined the ranks of area and delay as major motivations in optimization.

On the circuit level, transistor sizing has been well established as a good way to achieve reductions in delay of circuits, as shown by programs such as TILOS (Fishburn and Dunlop) [1] and the Tailor Layout System (Marple) [2]. With special layout techniques, the increase in rectangular area from transistor sizing can be minimized, hence rendering transistor sizing an even better tool for delay reduction. The trade-offs between area, power and delay have been examined by Berkelaar [5,6], who optimized the circuits to adhere to a prescribed delay value, while minimizing power consumption. He used linear programming techniques with transistor widths as decision variables. However, no techniques for accurately estimating power were given.

Recent developments of probabilistic techniques have produced a fast and efficient way of estimating power [4], which is estimated as proportional to the average switching probability of a node. The power dissipation of a gate is approximated by the change in energy for charging and discharging the output capacitance of the gate. Since a gate does not necessarily switch at every clock cycle, the frequency of switching is estimated by the clock frequency multiplied by the expected number of switches per cycle.

Average power is given by:

$$P_{avg} = (\tfrac{1}{2} \times C_{load} \times V_{dd}^2) \times \left(\frac{E\,(transitions/cycle)}{T_{cyc}} \right) \qquad \text{(EQ 1)}$$

where P_{avg} denotes average power, C_{load} is the load capacitance, V_{dd} is the supply voltage, T_{cyc} is the global clock period, and E(transitions) is the expected value of the number of gate output transitions per global clock cycle [4].

This probabilistic method of power estimation provides a simple way of examining power dissipation in terms of sizing, since gate capacitance is proportional to the width of the transistor. Furthermore, the switching probabilities of a node can be used to determine how effectively sizing can be used for power and delay optimization. A node with a low switching probability can be sized larger to achieve better speed, but the increase in power dissipation as a consequence will not be large due to the low probability of switching.

Due to differences in pin delays and the order in which inputs arrive at a gate, different ordering of inputs to a gate can affect the final delay in the output. This has been examined and shown to achieve significant reduction in the delay of an individual gate [8]. Differences in pin delays are larger when transistors in a gate are sized differently. Input ordering does not cause any increase in power dissipation, and may actually decrease power due to shorter delay times and hence possible reduction in glitches. It provides yet another means of delay and power optimization.

Previous work has attempted to size individual transistors in custom design circuits. However, commercially, product to market times must be small, and hence much circuit design is done with standard cells as the target technology. Since standard cell libraries are widely used, it is feasible to have a library of gates with transistors that are previously sized to give good power and delay trade-offs, and then the problem of optimization is to choose the best version of each cell to use. This provides a method with greater granularity, and because of the early binding of transistor widths, a computationally simple method of optimization.

In many available standard cell libraries, cell transistors are not minimum-sized, and only one version of each gate is available. In this paper, we describe the construction of an augmented standard cell library where several sized versions of a gate are available. These gates are sized with area minimization and input ordering in mind such that they give good delay response without high power dissipation. Next we describe optimization routines

This work was supported by the Singapore Economic Development Board and IBM.

that select the version that is best suited for each node so that a given delay constraint is satisfied with minimum power. Since the aim is to have low power dissipation, circuits are first mapped with minimum-sized gates and then changed to larger gates as necessary to satisfy delay constraints. In this way, we start with minimum power, and gates that are not dominant in determining the delay of the circuit remain minimum-sized, thereby ensuring low-power dissipation.

2   Delay modeling

An accurate delay model is a necessity for speed optimization. Often, in optimization routines, gate delays are treated as a fixed quantity, regardless of input slope and output load. There is also a tendency to associate a gate with only one delay, ignoring the difference between output rise (pull-up) and fall (pull-down) times. These issues, however, have been shown to affect the speed of a gate [3, 7], and hence need to be taken into account in delay estimation.

Furthermore, relative differences in input arrival times can change the delay of a gate [10]. This gives rise to the need for correct input ordering, which can achieve up to a two fold increase in speed [8].

In this paper, a "pin delay" model for delay is used. This model is built upon that provided by the Sequential Interactive Synthesis (SIS) package [9]. The output delay is modelled for each pin of a gate when it is the last one to change, or the one that will cause a switching event. This delay model introduces block and drive values for each input pin into the cell library characterization. Separate block and drive values are derived for rise and fall delay. Delay is estimated by:

$$Gate\_delay = block\_delay + \frac{Output_{load}}{Output_{drive}} \quad \text{(EQ 2)}$$

This delay model was verified by simulating cells from a standard cell library with different output loads. Fig. 1 shows the falling output delay times for an or-and-invert gate. It can be seen that the output delay is proportional to the output load, but on examining the output delay times for different input transition times, it was discovered that the block delay varied with input transition times.

Delay is plotted against the input load, or the number of inverters the previous stage drives, in Fig.2. It can be seen that the delay is proportional to the input load too. Delay values are taken from the point where the input reaches 50% of its final value (close to threshold) to when the output reaches 50% of its final value. This is to ensure that no overlap of delay calculation occurs between levels of gates.

Recognizing the effect of input slope on output delay, a new parameter, the input drive, is added into the delay model. Delay is now estimated by:

$$Gate\_delay = \frac{Input_{load}}{Input_{drive}} + block\_delay + \frac{Output_{load}}{Output_{drive}} \quad \text{(EQ 3)}$$

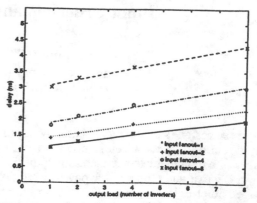

Fig. 1. Fall delay of oai21 gate versus output load with different input transition times.

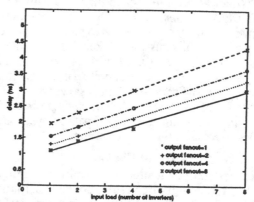

Fig. 2. Fall delay of oai21 gate versus input load with different output loads.

The input drive, block and output drive parameters are obtained by simulating the cells from extracted layout and curve fitting the delay values. This ensures that we have realistic and accurate parameters to work with.

Having established the delay model, we can proceed to describe the input ordering process. At first glance, input ordering seems to be a simple shuffle of input nodes among pins that have the same logic function, arranging them in the order of their arrival times. However, by reordering the inputs, the arrival times of the inputs have changed due to the change in load seen by the input nodes. The delay calculations must take this change into account. Inputs are ordered such that the delay caused by the last input to change is the shortest.

The delay of a circuit is defined by the delay of its longest path from input to output. This is the critical path. Decreasing the critical path delay will therefore also decrease the delay of the circuit. The critical path is obtained by first running a static timing analysis on the circuit, then tracing backwards from the latest arriving primary output towards the primary inputs. Attention is paid to whether it is the rise time or fall time of a node that is contributing to the critical path.

When the version of the gate at a node is changed, not only is the delay through this gate altered, but the arrival times of its inputs are changed too. This is because the input load seen from the input has changed due to the different sizing of transistors. Hence delay update of a gate involves calculating the new arrival times of the inputs as well as the new delay times through the gate. Again, rise and fall times are calculated separately.

3 Cell Library Approach

An augmented standard cell library has been constructed. The aim is to obtain a library where each gate is available in different sizes to cater to different output loads, with each version sized to give good delay response without high power dissipation. The gates (or their logic function) to be used in our library are based on standard cell libraries available in the Sequential Interactive Synthesis (SIS) package.

Cell versions of each logic function are chosen by careful experiment and analysis of the power and delay curves for each sized version. This involves laying out the circuit, sizing the transistors, simulating the circuit, plotting the power and delay curves for each logic function, selecting the versions and finally extracting the delay parameters (block and drive values) by simulation.

When laying out the cells, two things are taken into account. Firstly, the rectangular area should be minimal, hence diffusion breaks are avoided. Furthermore, the increase in rectangular area should be minimal when the transistor sizes are increased. Secondly, relative arrival times of inputs and input ordering are taken into account. In selecting which transistor to size, the inputs are expected to be ordered such that the latest arriving input is placed at the transistor that has the shortest delay to output. This is, in most cases, the transistor closest to the output. Taking into account that the latest arriving input will be placed at this transistor, the transistors in the gate are sized such that the load on the latest arriving input is minimal, but the drive for the gate output is maximal. By sizing a transistor, the drive of the next stage is increased, but the load on the previous stage is also increased. This process is illustrated in Fig. 3.

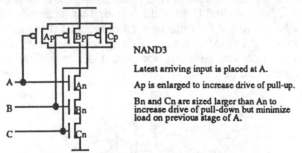

NAND3

Latest arriving input is placed at A.

Ap is enlarged to increase drive of pull-up.

Bn and Cn are sized larger than An to increase drive of pull-down but minimize load on previous stage of A.

Fig. 3. Sizing of transistors in a three-input nand gate.

The layouts of the different sized versions are extracted into SPICE decks for simulation. This is done to ensure that parasitic capacitances are taken into account. The power and delay values of these cells are plotted. As shown by Fig. 4, the shape of the curves show that by sizing a cell, delay can be reduced from that of the minimum sized version by a significant amount without much increase in power. Cell versions that yield good power and delay trade-offs, such as those on the power

and delay curve where the slope is changing the fastest, are chosen. Three versions of each gate, each suitable for a different range of output load, were selected. The advantage of having more than three versions is not apparent as computation time for the optimization algorithm would increase, and for most mapped circuits, the three versions can cater to the range of output load.

Once the cell versions are chosen, the cell is laid out with different numbers of inverters as its output load and also with different input loads, or the load the input stages drive. The layouts are then extracted into SPICE decks, simulated, and the delay values plotted as shown in Fig. 1 and Fig. 2. Finally curve fitting is done to find the input drive, block and output drive parameters.

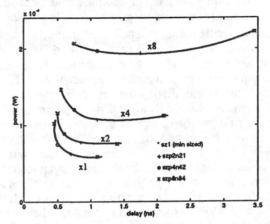

Fig. 4 Power versus Delay curve for 3-input nand gate with output loads of x1, x2, x4 and x8.

Transistor widths were treated as continuous variables in previous work by Berkelaar [7] and in sizing algorithms such as TILOS [1]. In this case, we want to take into account sizing methods that take into account relative delays of input arrivals, input ordering and area minimization. In addition, because of our use of cells as individual design elements, the sizes of the cells are fixed before they are actually mapped onto a circuit, which means that the output load they will drive is not known at the time of sizing. Hence heuristic methods of sizing are required. Other optimization formalizations remain to be explored.

4 Optimization strategy

Since the purpose of the algorithm is to minimize power, the circuit is first mapped with minimum-sized gates. Power is proportional to the load capacitance of the gate, so by using minimum-sized gates we begin with minimal power. Delay constraints are then satisfied by using bigger gate versions as necessary to reduce the delay.

Nodes with low input switching probabilities but large output loads have good potential for delay reduction, and given the low switching probability, increased transistor sizes may not increase power by much. The ratio of load to switching probability of a node, R_{lp}, is used to determine the priority with which the gate will be optimized.

A desirable property of an optimization routine is convexity, so that any local improvement is guaranteed to correspond to a global improvement. It would be desirable to have a provable optimal choice, but no appropriate theory is currently available for this problem. Given all the factors that must be utilized, it is not clear how this problem can be modelled for linear or nonlinear programming. Therefore, the approach taken in this work is heuristic. By using a heuristic approach, a fast and efficient algorithm that takes little computation time to optimize a circuit is developed.

4.1 Basic algorithm

One of the main aims of our algorithm is to optimize circuits quickly. In order to achieve this, updating the delay of the whole circuit at each change of gate version is not feasible. Hence the algorithm needs to be able to utilize the limited amount of information at local nodes to make intelligent choices for gate versions to change.

The circuit is first mapped with minimum-sized cell versions and the critical path is found. If the delay constraint is not satisfied, the algorithm traverses this path from input to output. At each node, each version of a gate is tried out by calculating the new delay values (taking into account change in arrival times of inputs due to changes in input loads, as well as change in output arrival due to new block and drive values) and the change in power from the original gate using the switching probabilities multiplied by the change in input load seen at the pins. The version that gives the best ratio of reduction in delay to increase in power, R_{dp}, is selected. Optimization on a critical path is stopped when the slack of the circuit becomes positive. This procedure is done until all paths satisfy the delay constraint, or if all gate versions have been examined.

While this algorithm makes use of all available information at a given node to make its decisions, it is nonetheless limited by the sequential scan of gates on the path, and by the limited information at each node. To overcome these shortcomings, several options are implemented, and the user can choose to utilize one or more of them in addition to the basic algorithm.

4.2 Threshold

To reduce computation complexity, the circuit is traversed sequentially along the critical path. However, certain gates are better candidates than others for optimization, namely those with large R_{lp} ratios. A threshold value of this ratio is used to determine which gates to change first. The threshold is determined by an estimated percentage of gates on the critical path that will need to be changed in order to achieve the delay constraint. This is obtained by dividing the magnitude of the slack of the circuit (difference between the required arrival time and the actual arrival time) by the expected reduction in delay by changing a cell version, and dividing this by the number of gates on the critical path. The ratio that will allow this fraction of gates on the critical path to be changed is then used as the threshold. Only nodes with ratios higher than this threshold will be analyzed and changed. If the delay constraint is still not satisfied after completing the traversal of the critical path, the threshold is lowered to allow more gates to be changed. If, however, the critical path has changed, then this threshold value is recalculated as before.

4.3 Input ordering

The fact that different inputs of a gate often arrive at different times means that the inputs can be ordered such that the worst case arrival time at the gate output is minimal. In addition, since the gate version is not changed, the delay of the gate can be reduced without any increase in power dissipation. This makes input ordering an attractive option.

One point to note about input ordering is, as mentioned above, the fact that if the order of inputs is changed, the load seen into the pins by the previous stages would change too. Such changes need to be taken into account when determining the delay of the output for a given order of inputs. The order that gives the earliest arrival time at the output is chosen.

There are two input ordering options. The first is input ordering of all nodes before delay optimization is carried out. Input ordering is performed on all nodes. After the critical path is obtained, it is done again on all nodes on this path, this time taking into account whether it is the rise or fall time that needs to be minimized. The second is input ordering at each stage along the critical path. When the gate of a node is changed to a bigger version, the delay is decreased, and hence for the next stage, input ordering can be done again to achieve the minimal delay for this new set of fanins and input arrival times. Referring to figure 5, A was originally on the critical path, and hence was placed at the fastest pin of gate c. If gate a was changed such that the arrival time of B is now slower than A, delay could be shortened if B is now placed at the fastest pin of gate c.

4.4 Check critical path

When the gate version at a node is changed, the delay of the next stage could be caused by a different input than before. Referring to Fig. 5, the critical path is originally from node A to C. When a faster cell version of gate a is used, the arrival time at A is reduced, and the critical path may have shifted from node A to B. This means that the delay at output C could be further reduced if the arrival time at B is made smaller. This option checks if such a shift in critical path has occurred. If so, optimization is done on the new critical path.

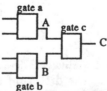

Fig 5. Illustration of change in critical path. When arrival time at A is reduced, arrival time at B could be large enough for B to become the critical input to gate c.

4.5 Maxtime

A gate has two arrival times, rise and fall. The arrival time that arrives the latest, for gates on the critical path, is the critical slope delay. To ensure a reduction in delay when a gate version is changed, a version is accepted only if the critical slope delay (rise or fall) decreases while the other remains the same or decreases. While this guarantees a reduction in final delay, there could be gate versions where the critical slope delay could be reduced further, but the delay of the non-critical slope would increase. If the slack for this non-critical slope is big enough to absorb the increase in delay time, the final delay will be smaller if this gate version is used. This option allows the non-critical slope delay at a node to increase.

5 Results

The basic algorithm, combined with one or more of the options, was tested on several circuits, including a simple circuit with just 11 gates (ex1), another combinational circuit consisting of 101 gates (ex2), a four-bit adder, a multiplexer and x4, a large circuit with 503 nodes. These circuits were chosen to reflect different circuit configurations. Circuits like the four-bit adder have just one dominant critical path, while the multiplexer has a number of dominant critical paths.

Results are shown in Tables 1 to 5. Table 1 shows the number of gates in each circuit, and the delay and power dissipation of the circuits when initially mapped with minimum-sized versions.

The circuits were then optimized under a delay constraint using the basic algorithm and one or more options. The delay constraints are chosen to be mid-range between the initial delay using minimum-sized gates and the minimum delay achievable using the augmented library. This gives the algorithm sufficient flexibility for optimization. If the constraints are too close to either of the end values mentioned above, then the number of possible combinations of the circuits that have delay values close to the constraint becomes limited and hence the effectiveness of the routines will not be apparent.

Table 2 shows the results from running the basic algorithm, first using only the reduction in delay as a guideline for the selection of versions (the gate version that gave the best delay reduction is used), and secondly utilizing the switching probabilities by using the R_{dp} ratio of decrease in delay to rise in power (estimated with the probability estimates). This is done to test the effectiveness of using R_{dp}. Comparing the two methods in Tables 2a and 2b, power dissipation reduces by up to 9% by using R_{dp}, showing that this method is effective in minimizing power increase.

circuit	required time (ns)	a) Basic algorithm using delay only			b) basic algorithm using switching probabilities		
		delay (ns)	power (uW)	cpu time (s)	delay (ns)	power (uW)	cpu time (s)
ex1	7.8	7.71	25.9	<0.1	7.71	25.9	<0.1
ex2	16.8	16.58	237.5	0.1	16.62	232.0	0.1
adder4	14.0	13.92	185.8	0.1	14.11	180.1	0.2
mux	15.7	15.66	390.2	0.3	15.91	357.9	0.6
x4	14.6	14.67	1597.8	6.4	14.59	1548.6	5.9

Table 2: Results for basic algorithm

circuit	required time (ns)	a) threshold			b) threshold, maxtime		
		delay (ns)	power (uW)	cpu time (s)	delay (ns)	power (uW)	cpu time (s)
ex1	7.8	7.71	25.9	<0.1	7.71	25.9	<0.1
ex2	16.8	16.74	225.5	0.2	16.74	225.5	0.2
adder4	14.0	14.11	180.1	0.2	13.98	180.8	0.1
mux	15.7	15.91	357.7	0.9	15.68	342.6	0.5
x4	14.6	14.58	1545.8	9.1	14.59	1528.6	7.0

Table 3: Results for using threshold with and without maxtime

circuit	required time (ns)	a) threshold, input ordering at all nodes			b) threshold, maxtime, input ordering at all nodes		
		delay (ns)	power (uW)	cpu time (s)	delay (ns)	power (uW)	cpu time (s)
ex1	7.8	7.57	25.9	<0.1	7.57	25.9	<0.1
ex2	16.8	16.63	221.4	0.2	16.63	221.4	0.2
adder4	14.0	13.81	181.9	0.2	13.98	170.3	0.2
mux	15.7	15.68	343.2	1.2	15.62	334.3	0.8
x4	14.6	14.60	1515.5	5.9	14.60	1506.8	4.9

Table 4: Results of using input ordering and threshold with and without maxtime

circuit	required time (ns)	a) threshold, input ordering at all nodes and at each stage			b) threshold, input ordering at all nodes and at each stage, check critical path		
		delay (ns)	power (uW)	cpu time (s)	delay (ns)	power (uW)	cpu time (s)
ex1	7.8	7.47	25.9	<0.1	7.47	25.9	<0.1
ex2	16.8	16.70	222.5	0.2	15.90	230.6	0.5
adder4	14.0	13.82	172.9	0.3	14.10	167.1	0.3
mux	15.7	15.70	304.8	0.6	15.53	325.0	1.2
x4	14.6	14.59	1538.3	8.2	14.92	1520.3	6.2

Table 5: Results for using input ordering at each stage and check_critical_path

Next the circuits were tested with the option "threshold". (Table 3a) The "threshold" routine worked well in giving better power values when compared to just using the basic algorithm. Since the basic algorithm traverses the critical path sequentially, gates were optimized according to their order on the critical path. The "threshold" routine avoids this sequential traversal to select gate versions according to their potential in giving good delay and power trade-offs. However, the effectiveness of this option is dependent on the slack of the circuit for the minimum-sized mapping. If this slack is large, then all gates on the critical path would need to be changed, and the traversal would be the same as without using a threshold. The effect of using a threshold is felt only as the magnitude of the circuit slack reduces with iterations of the algorithm.

"Maxtime" returned better results in most cases because of the extra flexibility it provided in selecting what versions to use. Comparing Tables 3a with 3b and Tables 4a with 4b, maxtime reduced power dissipation by up to 7%.

Table 4 shows the results of running "input ordering" with "threshold" and "maxtime". "Input ordering of all nodes", as expected, gave us better delay values without

circuit	number of gates	initial mapping with minimum-sized gates	
		delay (ns)	power (uW)
ex1	11	9.13	25.2
ex2	101	21.06	218.5
adder4	55	17.98	155.0
mux	93	22.40	279.4
x4	503	18.19	1469.5

Table 1: Results for minimum-sized mapping

increase in power. Since inputs were ordered to achieve the best gate output delay, the gate size was not changed, thereby keeping power dissipation constant. The reduction in delay due to input ordering meant fewer gates needed to be changed for the constraint to be satisfied and hence the power dissipation decreased.

The dual use of "Input ordering at all nodes" and "Input ordering at all stages" reduced power dissipation by as much as 17% (in the case of the mux circuit) when compared to not using the input ordering options (comparing Tables 3a and 5a). Comparing Tables 4a and 5a, it can be seen that input ordering at each stage is effective in further reducing power dissipation after input ordering is done at each node.

Finally, the results of running "check critical path" were somewhat varied. This option is useful in determining the smallest delay available as it searches for all possible gates for optimization. However, by switching critical paths more often redundant optimization could occur. For example, in Fig. 5, gate a was sized to be the largest version to reduce the delay, and in doing so, the gate is off the critical path and gate b is now on the critical path. If at the end of the optimization gate b is still on the critical path, a smaller version of gate a could have been used without changing the delay of gate c.

The results showed that the effectiveness of each of the options depended on the topography of the circuit being optimized and on the delay constraint. Comparing the minimum-sized mapping with the best solution from the different combinations of options, delay was reduced by up to 43% with only up to 8% increase in power. For most circuits, delay was reduced by more than 22% with power increase less than 4%. This shows that the augmented library cells were indeed effective in giving good delay trade-offs. Combinations of options produced up to a 28% decrease in power dissipation when compared to the basic algorithm which did not take into account switching probabilities. All routines took little computation time to run. The worst case scenario in traversing a critical path is to change all gates on the path, and the complexity is simply the number of gates on the critical path multiplied by the number of versions of each gate.

In conclusion, the basic algorithm, with the various options, provided a fast and efficient way to determine the best combination of gate versions to use. Together with the usage of mainly minimum-sized cell versions coupled with cells sized to give good delay and power trade-offs, this package provides a fast and convenient way to design standard cell circuits for low-power.

6   Unique Contributions

This optimization routine utilizes an accurate model of power dissipation, using probability estimates. With these probability estimates, intelligent choices can be made about what gate sizes to use to get the best results in terms of both power and delay. Accurate delay models are also used, with parameters extracted form layout. In addition to

sizing techniques, input ordering is also used in our optimization.

The construction of an augmented cell library with cells that give good power and delay trade-offs provide an easy and efficient way to design circuits using standard cell libraries for low power dissipation.

7   Future directions

Possible extensions to the algorithm includes a similar but separate algorithm to optimize delay under a power constraint. This work is currently underway. Other possible improvements include developing a technique to relax the cells that are not on the critical path, and not in the dependency chains for that path, so as to reduce power further. This can also eliminate redundant optimization. Critical path analysis could also include false path detection. Theoretical characterization of the algorithm so that some provable results may be derived could also be looked into, though it may be difficult to characterize such heuristic methods.

Other possible future work could involve establishing a more extensive library to include latches for sequential circuits, and finding other optimization formalizations for determining the sizes of the cells. Circuit forms other than static CMOS could also be examined and the influence of circuit styles could also be studied.

8   Bibliography

[1] J.P. Fishburn and A.E. Dunlop, "TILOS: A Posynomial Programming Approach to Transistor Sizing," In IEEE International Conference on Computer Aided Design, pg. 326-328, November 1985.

[2] D. Marple, "Transistor Size Optimization in the Tailor Layout System," In IEEE Design Automation Conference, Pg. 43-48, 1989.

[3] L. Brocco, S. McCormick and J. Allen, "Macromodeling CMOS Circuits for Timing Simulation", IEEE Transactions on Computer-Aided Design, Vol. 7, No. 12, December 1988.

[4] A. Ghosh, S. Devadas, K. Keutzer and J. White, "Estimation of Average Switching Activity in Combinational and Sequential Circuits," Proceedings of the 29th Design Automation Conference, pg. 253-259, 1992.

[5] M. Berkelaar and J. Jess, "Gate Sizing in MOS Digital Circuits with Linear Programming," Proceedings of the European Design Automation Conference 1990, pg. 217-221.

[6] M. Berkelaar, "Area-Power-Delay Trade-off in Logic Synthesis," Ph.D. dissertation, Technische Universiteit Eindhoven, September 1992.

[7] A. Kayssi, K. Sakallah and T. Mudge, "The Impact of Signal Transition Time on Path Delay Computation," IEEE Transactions on Circuits and Systems II: Analog and Digital Signal Processing, Vol. 40, No. 5, May 1993.

[8] B. Carlson and C. Chen, "Performance Enhancement of CMOS VLSI Circuits by Transistor Reordering," Design Automation Conference, pg. 361-366, 1993.

[9] E. Sentovich, K. Sing, L. Lavagno, C. Moon, R. Murgai, "SIS: A System for Sequential Circuit Synthesis," Electronics Research Laboratory Memorandum No. UCB/ERL M92/41, University of California at Berkeley, 1992.

[10] C. T. Gray, W. Liu and R. Cavin, *Wave Pipelining: Theory and CMOS Implementation*, Kluwer Academic Publishers, 1993. Section 4.2.2, "Delay Models", pg. 65ff.

Low-Power Digital Design

Mark Horowitz, Thomas Indermaur, and Ricardo Gonzalez

Center for Integrated Systems, Stanford University, Stanford, CA 94305
{horowitz,tni,ricardog}@chroma.stanford.edu

Introduction

Recently there has been a surge of interest in low-power devices and design techniques. While many papers have been published describing power-saving techniques for use in digital systems, trade-offs between the methods are rarely discussed. We address this issue by using an energy-delay metric to compare many of the proposed techniques. Using this metric also provides insight into some of the basic trade-offs in low-power design.

The next section describes the energy-loss mechanisms that are present in CMOS circuits, which provides the parameters that must be changed to lower the power dissipation. With these factors in mind, the rest of the paper reviews the energy saving techniques that have been proposed. These proposals fall into one of three main strategies: trade speed for power, don't waste power, and find a lower power problem.

CMOS Power Dissipation

Power dissipation in CMOS circuits arises from two different mechanisms: static power, which results from resistive paths from the power supply to ground, and dynamic power, which results from switching capacitive loads between two different voltage states. Dynamic power is frequency dependent, since no power is dissipated if the node values don't change, while static power is independent of frequency and exists whenever the chip is powered on. For uses where the electronics will be inactive for much of the time (most portable applications), the static power must be made very low in the inactive state.

Even if there are no explicit circuits using static current, the chip will dissipate some static power. This power is the result of leakage current through nominally off transistors. The leakage is set by the sub-threshold current of the transistor,

$$I_{ds} \sim \frac{W}{L} I_s \times \exp\left(\frac{V_{gs} - V_{th}}{av_T}\right) \qquad (1)$$

where v_T (kT/q) is around 26mV at room temperature, 'a' is a constant slightly larger than 1, and I_s is roughly $\mu C_{ox}(av_T)^2$ or $0.3\mu A/\mu$. This leakage current and the allowable static power limit how low one can make the threshold voltage. The situation is made worse by the fact that the threshold voltage is not perfectly controlled, and thus the nominal value must guarantee that the leakage is acceptable in the worst-case situation.

Some numbers will make this limit clearer. A 10mm square chip will generally contain a few meters of transistor width. If the static current limit for this chip is 100μA, then the leakage current of an off transistor must be under 0.1nA/μ. To achieve this leakage requires Vth to be around 8 av_T in the worst-case situation, which would be high temperature and a low-threshold fabrication run. If the fab control on Vth is ±100mV, the nominal value of the threshold would be around 0.35V.

With small static power, charging and discharging capacitors generally consumes most of the power on a CMOS circuit.[*] In charging a load capacitor C up ΔV volts, and then discharging it to its original voltage, a gate pulls C ΔV from the Vdd supply to charge up the capacitor, and then sinks this charge to Gnd to discharge the node. So at the end of a cycle, the gate / capacitor combination has moved C ΔV of charge from Vdd to Gnd, which uses C ΔV Vdd of energy and is independent of the cycle time. The dynamic power of this node is the energy per cycle, times the number of cycles it makes a second, or

$$P = C \Delta V \ Vdd \ \alpha F \qquad (2)$$

where α is the number of times this node cycles each clock cycle and is usually called the activity ratio. The dynamic power for the whole chip is the sum of (eq. 2) over all the nodes in the circuit.

From this formula it is clear what we need to do to reduce the dynamic power. We can either reduce the capacitance being switched, the voltage swing, the power-supply voltage, the activity ratio, or the operating frequency. The power-saving techniques described in the following sections provide a number of ways to reduce these parameters.

Low-Power Design Techniques

Until relatively recently, power was an afterthought in the design process. Designers would optimize their design to meet performance and area constraints, and then talk with the packaging and system designers to figure out how they were going to deal with the power of the chip. Probably the most important low-power design method is simply to make low power a key objective in the design process. Once this is done, a lot of power can be saved by not doing "stupid" things – by simply not wasting power. For example, lowering the power supply from 5V to 3.3V, rather than using internal

Funding for this research was provided by ARPA under contract J-FBI-92-194.

[*] Shunt current that occurs when both devices are on is usually a small percent of the dynamic power (5-10%) and will be ignored in this paper.

Reprinted from *Proceedings of the 1994 IEEE Symposium on Low-Power Electronics*, pp. 8-11, 1994.

voltage regulation, is an obvious design decision if low power is an objective. Removing circuits that dissipate static power and powering down inactive blocks are other examples of how wasted power can be saved.

To help find wasted power we need a metric that allows us to compare two designs to see which is more efficient. The obvious choices for a low-power metric, power and energy, turn out to have serious flaws. Using power as the metric has the problem that CMOS circuits use energy mostly when they switch their outputs. One can always reduce the power by reducing the operating frequency, which is not a useful result.

An alternative metric is the energy needed to complete an operation. This is an improvement over power because running the part slower does not directly change the energy used in an operation, it simply spreads the same energy use over a longer time. The problem with this metric is that the energy an operation requires can be made smaller by reducing the supply voltage since the energy is roughly nCV^2, where nC is the sum of capacitance times transitions that are needed to complete the operation. However, the lower supply voltage also affects performance, and dramatically increases the delay of the operation. Thus the lowest energy solution also will run very slowly.

To avoid these problems, we use the metric of delay/op x energy/op. Smaller energy-delay values imply a lower energy solution at the same level of performance – a more energy-efficient design. The following sections will discuss various low-power design techniques, and show how they affect the energy-delay product. The first three methods (voltage scaling, transistor sizing, and adiabatic circuits) only have a small effect on the energy-delay product and are really methods for trading speed for power. The next two sections describe ways of not using energy needlessly. Finally, the last two sections describe how reformulating the problem at the system level can yield large improvements in the energy-delay product.

Voltage Scaling

In a given technology the energy per operation can be reduced by lowering the power-supply voltage. However, since both capacitance and threshold voltage are constant, the speed of the basic gates will also decrease with this voltage scaling. We can use a charge control model to estimate the delay of a gate by dividing the charge needed to transition the node by the transistor current. As other researchers have shown [3], using a quadratic model of a transistor leads to:

$$t_d = k \frac{CV}{(V - V_{th})^2} \qquad (3)$$

Figure 1 plots energy / operation, delay and energy-delay as the supply voltage is scaled. At large voltages, reducing the supply reduces the energy for a modest change in delay

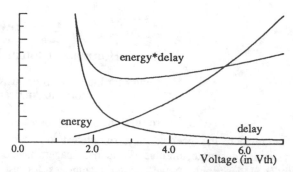

Figure 1. Energy and Delay vs. Voltage

(especially in the velocity saturated case, where the delay change is even less than shown in the figure). At voltages near the device threshold, small supply changes cause a large change in delay for a modest change in energy. While there is a minima at Vdd = $3V_{th}$, it is pretty flat. Around this point changing the supply voltage does not strongly affect the energy-delay product, allowing one to trade delay for energy. From the 3 V_{th} point, there is a factor of about 4 in energy in either direction (from 1.5 V_{th} to 6 V_{th}) that can be traded for delay without greatly changing the energy-delay product. Below 1.5 V_{th} the surplus performance would be better spent in some other way, like reducing the transistor sizes.

Transistor Sizing

Like supply voltage, sizing gates mostly presents the designer an opportunity to trade speed for power, rather than reducing their product. Since some of the load capacitance is caused by the gate capacitance of other transistors, one can reduce the energy of an operation by making all the transistors smaller. However, decreasing the size of the transistors also decreases their current drive, and thus makes the gates slower. This trade-off can be easily seen using a chain of uniformly loaded inverters, which are shown in Figure 2.

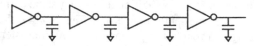

Figure 2. Simple Inverter Chain

Figure 3 graphs the delay, energy, and energy-delay of a stage as a function of the transistor's capacitance contribution to the total load. The load will be mostly load capacitance for small transistor, and will be mostly gate for large devices. For very small transistors, energy is dominated by switching the load capacitance, while the delay is inversely proportional to the transistor width, so increasing the transistors improves the energy-delay product. For large transistors, the gates are limited by self loading, so decreasing the transistor size improves the metric. The optimal operating point is when the transistor loading is the same as the wire loading.

Obviously, real circuits are more complex. The gate and wire capacitance is different for different gates, nodes transition at

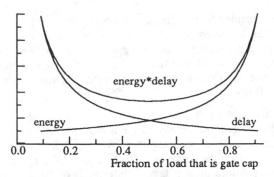

Figure 3. Energy, Delay vs. Transistor Width

different frequencies, and not all gates are on the critical path. While this problem is difficult to solve precisely, the structure of the solution remains roughly the same as the simple inverter chain: making the critical path transistors much smaller than their loads will greatly increase the delay without reducing the power, and making the transistors much larger than their loads will greatly increase the energy without having a large effect on the delay.

The energy-delay product is roughly constant as the percentage of gate loading changes from 20% to 80%, which is roughly a factor of 5 in speed and power. While using minimum-sized devices can lead to lower power solutions [3], they do not lead to more energy efficient solutions.

Adiabatic Circuits

Adiabatic or charge-recovery circuits, are another method that allow a designer to explicitly trade performance for lower energy requirements [11][7]. These circuits resonate the load capacitance with an inductor, which recovers some of the energy needed to change the capacitor's voltage. The energy loss in switching the load can be reduced to $\tau/T\ CV^2$, where τ is the intrinsic delay of the gate, and T is the delay set by the LC circuit. While this ease in trading energy for delay is attractive, the energy-delay product for these circuits is much worse than normal CMOS gates[6]. Thus adiabatic circuits become attractive only when you need to operate at delays beyond the range viable by voltage scaling and transistor sizing standard CMOS.

Technology Scaling

One way to greatly improve the energy-delay product, and thus save energy, is to improve the technology. In ideal scaling as first described by Dennard[5], all voltages and linear dimensions are reduced by a scale factor, γ (<1). Since the E-fields in the devices and wires remain constant, the device current* and device and wire capacitance all scale as γ. Since the voltage also scales by γ, the energy of an operation scales as γ^3. The delay of each gate also improves by γ, since the delay is roughly $t_d = CV/i$. The energy-delay

product decreases by γ^4, implying a 0.7 shrink of a chip can be run at the same performance for roughly 1/4 the power.

The difficulty with ideal scaling is the requirement for Vth to scale along with the supply voltage. As was mentioned earlier, static power caused by leakage current through the off transistors will limit how low the threshold voltage can be scaled.[†] Even with constant voltage scaling, the reduced capacitance improves both the energy and the delay, so their product scales at least as γ^2.

Transition Reduction

Another way to improve the energy-delay product is to avoid wasting energy – avoid causing node transitions that are not needed. One common approach to solve this problem is to make sure that idle blocks do not use any power. The key to selective activation is to control objects that dissipate a significant amount of power. From our work analyzing power of digital systems, around 70% of the power comes from high-transition count, high-capacitance nodes – like clocks and buses – which comprise less than 20% of the nodes in a given design. While doing selective activation of a set of 64 bus lines might make sense, trying to reduce the number of transitions in the adder that drives the bus does not.

As long as the static power is small, the circuit only uses power when a node switches. Thus an idle section can be powered down simply by preventing its outputs from switching (generally by keeping its inputs stable). At the block level on a chip, the activation is usually done by gating the clock to the function blocks[10]. When the clock is turned off, none of the latch outputs change state, and thus the logic outputs are also stable. Gating the clock has the added advantage that it reduces the clock load that toggles each cycle, since the clocks in the inactive blocks are effectively turned off. On low-power processors, the caches, FPU, and integer unit can all be independently controlled [1]. Generally the performance impact of the clock gating is small, so the energy-delay product decreases by the energy saving.

Reducing unnecessary toggles will reduce the energy-delay product, but it rarely changes it by more than a small integer factor (2 or 3). To get more significant reductions requires examining the problem from the system level.

Parallelism

One can improve the energy-delay product by reducing either the energy or the delay. Voltage and transistor scaling allow a designer to trade excess performance for lower energy operations. The ability to trade delay for energy points out the strong connection between high-speed and low-power designs. One wants to start with a solution with a large

---

* This relations holds independent of whether the devices are velocity saturated or not.

† There has been some work to allow larger leakage currents and switch the power supply off to these sections using lower leakage (higher threshold) transistors. This might allow slightly lower threshold transistors in the active circuits but requires a sophisticated power management system on chip [8].

amount of excess performance that can then be traded for reduced power. A way of generating this performance is by exploiting parallelism.

When an application has parallelism, one can build N functional units instead of one, and solve N problems at the same time. Doing this increases the performance by nearly N (there is some time needed to distribute the operands, and collect the results), and increases the power by slightly over N (again because of overhead). Thus using parallelism increases the energy/op by only the overhead while the effective delay/op drops by N minus the delay overhead. The energy-delay product of the parallel solution is much lower (roughly N times lower) than the original sequential approach. This argument is independent of how the parallelism is extracted (pipelining, parallel machine, etc.), although the overhead factors will be different. For DSP applications with a large amount of parallelism, the performance gains allow the resulting systems to run at very low power supply voltage, use small transistor sizes, and still meet their performance targets [4].

In some applications, the available parallelism is smaller and harder to extract. In processors the cost of issuing multiple instructions is not small, and does not yield a performance gain for all code sequences. As a result, as shown in Table 1, parallel execution neither helps or hurts a processor's energy-delay product (Watts/SPEC^2). Fabrication technology seems more important for the energy-delay product than whether the machine is superscalar (21064, PPC604) or not.

Table 1   Energy-Delay for some Recent Processors

μP	DEC 21064	MIPS R4200	IDT R4600	PPC 604	PPC 603
SPECavg	155	42.5	64	162.5	80
Power	30W	1.8W	3W	13W	3W
SPEC^2/W	800	1000	1400	2000	2100
Min L	0.75μ	0.64μ	0.64μ	0.5μ	0.5μ

Redefine the Problem

So far we have looked at ways to more efficiently implement the tasks needed to complete some operation. Yet this discussion missed the most important method of reducing system energy – reduce the number/complexity of tasks that the operation requires. It is at this level that the designer can make the largest changes to the energy-delay product, since simplifications often reduce both the energy and the delay of the operation. The key point to realize is that the energy-delay product measures the energy to complete some user operation and the delay to complete that operation. If we can simplify the operation, we reduce the number of primitive steps required, and thus reduce both the energy and the delay.

As a simple example of the saving that is possible, consider a operation that is implemented as a program on a micro-controller. The initial code for this operation takes N micro

instructions to execute, so the energy for the operation is N times the instruction energy, and the delay is N times the instruction delay. If another approach can perform the same operation in M instruction, the energy-delay product will change by $(M/N)^2$, since both the delay and energy decrease by (M/N).

This strategy works for hardware designs as well, with similar quadratic gains. Often a reformulation of a problem can lead to a solution that requires less computation to accomplish the same task [2][9]. Orders of magnitude gains are possible at this level. Unfortunately the optimizations used tend to be tied to the specific application that is being optimized. The good news is that this process is similar to the ones used to increase system performance. The bad news is that these system level optimizations generally require some creative insight.

Conclusions

Good design has always required one to make careful trade-offs, and low-power design simply means one needs to consider energy dissipation in addition to the normal concerns of speed, area, and design-time. The energy-delay product is a useful guide for making these trade-offs. It allows a designer to find optimizations that provide the largest reduction in energy for the smallest change in performance. It also makes clear the strong coupling between performance and power which is the reason that many high-performance techniques are useful for low-power design.

References

[1] R. Bechade, et al., "A 32b 66MHz 1.8W Microprocessor, ISSCC, Feb 1994, pg 208-209.

[2] B. Brandt, B. Wooley, "A Low Power, Area-Efficient Digital Filter for Decimation and Interpolation," IEEE Journal of Solid State Circuits, SC29, June 1994.

[3] A. Chandrakasan, et al. "Low-power CMOS digital design." IEEE Journal of Solid-state Circuits Vol 27 pg 473-484.

[4] A. Chandrakasan, et al, "A Low Power Chipset for Portable Multimedia Applications," ISSCC, Feb 1994, pg 82-83.

[5] R. Dennard et al., "Design of Ion Implanted MOSFET's with Very Small Dimensions," IEEE Journal of Solid State Circuits, SC9, pg 256-267, 1974.

[6] T. Indermaur, et al., "Evaluation of Charge Recovery Circuits and Adiabatic Switching for Low Power CMOS Design," Symposium on Low-Power Electronics, Oct 1994.

[7] J. Koller, W. Athas, "Adiabatic Switching, Low Energy Computing, and the Physics of Storing and Erasing Information," Proceedings of Physics of Computation Workshop, Oct. 1992.

[8] D. Takashima, et al., "Standby/Active Model Logic for Sub-1V Operating ULSI Memory," IEEE Journal of Solid State Circuits, Vol 29, pg 441-447, 1994.

[9] E. Tsern, et. al., "Video Compression For Portable Communication Using Pyramid Vector Quantization of Subband Coefficients," IEEE Workshop on VLSI Signal Processing, Oct 1993.

[10] N. Yeung et al., "The Design of a 55 SPECint92 RISC Processor under 2W," ISSCC, Feb 1994, pg 206-207.

[11] S. Younis, T. Knight, "Practical Implementation of Charge Recovering Asymptotically Zero Power CMOS," Proceedings of the 1993 Symp. on Integrated Sys., MIT Press, pg. 234-250, 1993.

Power Analysis of a Programmable DSP for Architecture/Program Optimization

Hirotsugu Kojima, Douglas J. Gorny, Kenichi Nitta, and Katsuro Sasaki

Research and Development Division, Hitachi America, Ltd.
201 E. Tasman Dr., San Jose, CA 95134, U. S. A.

Introduction

Power consumption has become one of the primary metrics in CMOS LSI design. A high level power etimation model will be indispensable to evaluate architectures and programming styles for performance and power consumption optimization. A model was proposed in which a constant energy was used for every module but data dependency was not taken into account[1,2]. The purpose of this paper is to quantify the module break down and the data dependency of the power consumption and to find a key for high level power estimation. We analyzed power consumption of a 24bit fixed point DSP, HX24, which we developed previously[3]. We have found that the buses don't consume as much power as we originally expected while the data operation modules consume much power and the data dependency caused about 30% variation in worst case chip power. This is the first paper that describes how large the data dependency of data operation is and how low the bus power consumption is in a DSP of an extended Harvard architecture.

Switch level and Cell based power simulation

We analyzed power using a switch level and cell based simulation. The simulation causes errors by (a) ignoring power caused by short circuit current, (b) ignoring power caused by switching capacitance inside cells, and (c) regarding intermediate swings as full swings. The extreme case of (a) is signal conflict, but we assume that signal conflict never occurs in a correct design. We estimate that an error of -7 to -10% is caused by ignoring the short circuit current. We compensate error (b) by increasing the simulated power by a factor that is derived from the load capacitance inside cells. With this compensation, the error is between -10% and -20%, while it is -30% to -40% without.

Module breakdown power

Fig. 1 shows the block diagram of HX24 and Fig. 2 shows a module break down of the power consumption with many test programs. Power consumption is normalized by the worst case chip power. We focus on the power of the data operation, clocking and buses in this paper.

'Data operation' consists of a register file, an ALU and a multiplier, and it consumes 3-33%. The data dependency is as large as 30% of the worst case power. Thus, ignoring the data dependency causes an error of as much as 30% in high level power estimation. Clock and bus circuits are dominant power consuming components in general. We separately evaluated the power consumed by clocking and bus driving, though they are already included in the modules. Clocking power is calculated by summing up all the load capacitance on the clock line from the source to a control gate that enables/disables the clock. The total is 17%. Bus power is calculated by multiplying the number of transitions and load capacitance on each bit of the buses. The bus power consumption is less than 5%, while it was reported that the bus power was 9% in a microprocessor[4].

Bus transition activity

In order to confirm the unexpected result of bus power, we evaluated the bus activity with two speech CODEC programs. Fig. 3 shows the activity on each bit of the buses. The activity here is defined as the average number of transitions per cycle. The activity of the address buses shown in (a), (b), and (c) is significantly small because successive accesses often refer successive memory locations. The activity of the data buses shown in (d), (e), and (f) is at most 2 because the data buses are precharged. If the data is white noise, the activity is 1. Fig. 3 demonstrates that the activity on the X and Y data buses is less than white noise, but the instruction activity is higher. The bus power is not dominant in an extended Harvard architecture because of less activity on address buses, while it is higher in von Neumann architecture because the bus is multiplexed.

Data dependency of data operation

Since the data operation modules consumed as much as 33% of the worst case chip power, we investigated more about the data path power. Fig. 4 and 5 show the data dependency of the ALU power for addition and the register file power respectively. The ALU power is in proportion to the number of transitions of the input data. The register file power is in proportion to the number of zero's, which is because the register file employs precharged circuit. Fig. 6 and 7 show the data dependency of the multiplier power. Two random data series are fed in to each input of the multiplier and there is no correlation between the power and the number of transitions of the input nor output(Fig. 6). When one of the inputs is fed by a constant value, we observe a good correlation between the power and the number of transitions of the other input(Fig. 7). In Summary, we found noteworthy hints for data path power etimation: 1) the register file power can be estimated by observing number of zero's of the input and output data, 2) the ALU power can be estimated by observing the number of transitions of the input, and 3) a constant value can be used to estimate the multiplier power as long as the both input change.

Conclusion

We demonstrated a module break down of a DSP power. We have found that the bus power is significantly small while the data path power is large. We obtained some noteworthy hints for high level power estimated of DSPs, which we believe useful for architecture and program optimization of DSP's in terms of power consumption.

Reprinted from *Proceedings of the 1995 Low-Power Symposium*, pp. 26–27, 1995.

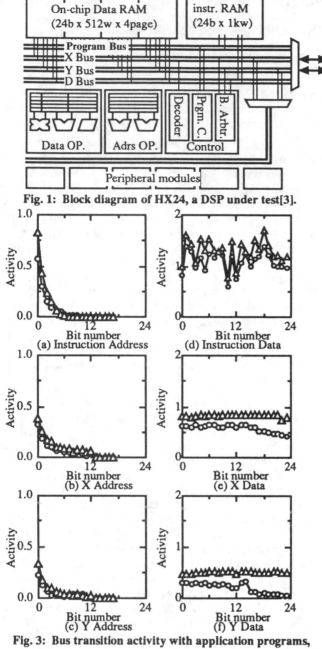

Fig. 1: Block diagram of HX24, a DSP under test[3].

Fig. 3: Bus transition activity with application programs, calculated as the average number of transitions on each bit of the buses.

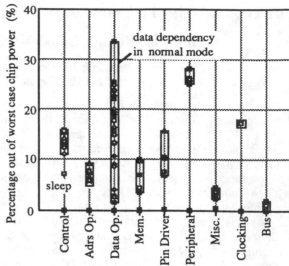

Fig. 2: Module break down of power analysis. Each plot represents a test program, and a white square on each category represents the range in use.

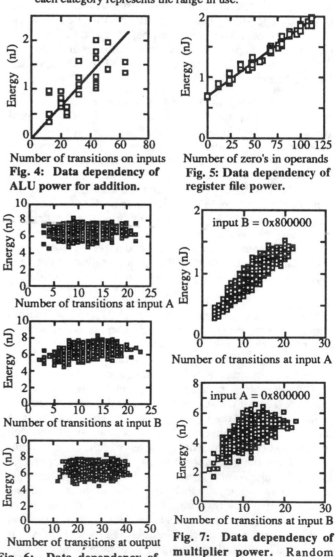

Fig. 4: Data dependency of ALU power for addition.

Fig. 5: Data dependency of register file power.

Fig. 6: Data dependency of multiplier power. Random data was fed to both inputs.

Fig. 7: Data dependency of multiplier power. Random data was fed to one of the inputs and the other input was constant.

References

[1] T. Sato, M. Nagamatsu, and H. Tago, "Power and performance simulator:ESP and its application for 100MIPS/W class RISC design," Symposium on Low Power Electronics, pp. 46-47, Oct. 1994.

[2] Ping-Wen Ong and Ran-Hong Yan, "Power-conscious Software design - a framework for modeling software on hardware," Symposium on Low Power Electronics, pp. 36-37, Oct. 1994.

[3] T. Baji, et al., "HX24 24-bit Fixed Point Digital Signal Processor," the International Conference on Signal Processing Applications and Technology, pp. 622-629, Oct. 1993.

[4] S. Kawasaki, "SH2: A low power RISC micro for consumer applications," Hot Chips IV, pp. 97-103, Aug. 1994.

A Review of Adiabatic Computing

John S. Denker

AT&T Bell Laboratories
Holmdel, NJ 07733

Abstract

We explain

a) why people want a *low-energy* computer.

b) under what conditions there is — or is *not* — an irreducible energy per computation for CMOS circuits.

c) partial versus full adiabatic computation, and their relationship to logically reversible computation.

d) various schemes for achieving adiabatic operation.

Motivations and Objectives

Low-energy computing is important for many reasons, including:

- According to estimates by the Environmental Protection Agency, computer equipment is responsible for 5–10% of the electrical power consumption in the United States. Building more-efficient computers makes more sense than building new power plants and rewiring our homes and offices to accomodate inefficient computers.

- It is becoming harder and harder to remove the heat produced by high-performance chips. Typical workstations rely on forced-air cooling, but that tends to create unwelcome amounts of noise. Supercomputers rely on liquid cooling, but that is exceedingly expensive and inconvenient.

- In many portable electronic devices, there is an unpleasant tradeoff: autonomy-time versus battery size and weight. A reduction in energy consumption (even at the cost of some slight increase in circuit complexity) would be quite advantageous.

Consider trying to perform a certain difficult computation on a typical battery-powered computer — without completely discharging the batteries. We will consider the energy budget of the logic circuits, omitting for now the display and other power-hungry components.

It is important to distinguish between *energy* and *power*. Power is energy per unit time. For an ordinary CMOS logic circuit, if we simply reduce its clock rate, its power consumption will be reduced in the same proportion. Unfortunately, the time required to complete the computation will be increased, so the energy per computation is unchanged. Therefore, at the end of the computation the batteries will be just as dead as if the computation had been performed at full speed.

If we wanted a low-power computer, we could just reduce the clock rate — or turn it off entirely :-) — but that's not what we want. What we really want is a *low energy* computer.

Also, given two computers with equally low energy per computation, we would prefer the faster one. We conclude that high speed is at least as important as low power.

Energy Theorem

To change the logic state of a typical node in a CMOS circuit requires transferring a certain amount of charge $Q = CV_{dd}$, where C is the capacitance of the node and V_{dd} is the operating voltage. This implies transferring a certain amount of energy (the switching energy), $\frac{1}{2}CV_{dd}^2$. The energy *dissipated* during this transfer need not be related to the energy *transferred*, but in ordinary CMOS logic circuits both quantities happen to be comparable. Dissipation on the order of $\frac{1}{2}CV_{dd}^2$ is unavoidable *if* all the needed charge is extracted from the V_{dd} terminal of the power supply and returned to the ground terminal.

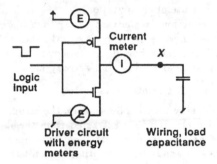

Fig. 1. Example Circuit: Vanilla CMOS

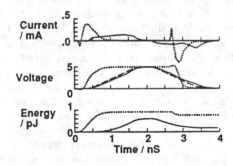

Fig. 2. Example Circuit Waveforms

We illustrate this argument by applying it to the node X in figure 1. The dotted waveforms in figure 2 apply to this circuit. The bottom panel shows, as a function of time, the amount of energy extracted from the power supply up to that time; the slope of the curve is the power. An energy $E = QV_{dd} = CV_{dd}^2 \approx 1\text{pJ}$ is taken from the from the V_{dd} terminal as the node is charged up. No energy is carried by current returning via the ground terminal, since the voltage is zero. (The tiny energy recovery seen at time 2.75nS is due to charge injected to the V_{dd} terminal via the

Reprinted from *1994 IEEE Symposium on Low Power Electronics*, pp. 94–97, 1994.

gate-drain capacitance of the PFET.) The energy curve at time 2nS includes the energy stored in the capacitance, plus the energy dissipated during charging (in the form of I^2R heating in the PFET channel, wiring, etcetera). The stored energy is dissipated during discharge (in the NFET channel, etcetera).

During the two operations (charge and discharge), an energy CV_{dd}^2 disappears from the supply. Therefore all circuits of this type must dissipate at least $\frac{1}{2}CV_{dd}^2$ per operation. This is a powerful result. It is independent of the internal details of the circuit, and a similar result holds even if the node's charge-versus-voltage characteristic is nonlinear.

The dissipated energy can easily exceed this lower bound. For instance, if the NFET and PFET are ever turned on simultaneously, a "crowbar current" will flow directly from V_{dd} to ground, dissipating energy without contributing to the desired charging or discharging of the node X.

The conventional approaches to reducing the energy per computation are:
- Reducing the operating voltage V_{dd}.
- Reducing the capacitance C.
- Reducing the activity factor (i.e. the number of node transitions per useful computation).

Obviously such reductions are advantageous, but there are limits; in any case for present purposes we take such reductions for granted and show how dissipation can be *further* reduced at any particular C, V_{dd}, and activity factor.

An Adiabatic Example

The foregoing energy theorem depends on the assumption that all the needed electrons are extracted from the supply via the V_{dd} terminal and returned via the ground terminal. The essential idea of adiabatic computing is to lift this assumption. That is:
- We extract charge from the supply at the lowest feasible voltage, and return it at the highest feasible voltage.

We now show how this can be done; figure 3 is an example. (This circuit is similar to the vanilla CMOS inverter in figure 1, and performs a similar function — but is not quite as practical, for reasons to be discussed below.) The solid waveforms in figure 2 apply to this circuit. The transistors remain on while the power supply is ramped up and down. As we shall see later, this time-dependent power supply can take over the timing functions of a traditional clock signal, so we will call it a *power-clock*. The rise and fall times of the power-clock are several-fold longer than the natural RC time of the node (where R includes the resistance of the transistor channels, wiring resistance, etcetera).

The adiabatic circuit dissipates less energy during charging: the energy trace at time 2nS is mostly *transferred* energy. During discharge (2.25 – 2.75 nS) most of this energy is returned to the supply.

The adiabatic circuit charges node X to the same voltage as the vanilla circuit. This can be seen in the voltage waveform at time 2nS. However, the charge is delivered over a longer time. This circuit's peak currents are several-fold smaller. This is crucial. Remember, the dissipated power, I^2R, is a nonlinear function of I. If we slow down the power-clock's risetime by a factor of N,

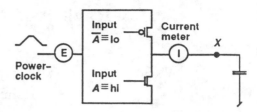

Fig. 3. Toy Example: Adiabatic CMOS

- The time required increases by a factor of N.
- The current decreases by a factor of N.
- The power decreases by a factor of N^2.
- The dissipated energy per operation decreases by a factor of N.
- The transferred charge and energy are unchanged.

This is in sharp contrast to vanilla CMOS, where simply reducing the clock rate reduces the power by only one factor of N and reduces the energy not at all.

The voltage at node X closely tracks the power-clock voltage (the dashed line in the figure). This is crucial: at no time does a current flow across a large potential drop. This is in contrast to vanilla CMOS, where currents routinely flow across drops on the order of V_{dd}.

Types of Adiabatic Circuitry

Circuits can be categorized on the basis of their energy performance as follows:

1 — Fully adiabatic circuits (e.g. figure 3) — which, if operated arbitrarily slowly, would dissipate arbitrarily little energy per operation. Remember, "full" adiabaticity is a statement about the asymptotic behavior at very low frequencies.

2 — Partially adiabatic circuits — in which charge is transferred across reduced potential drops, and some energy is recovered. However, some energy is lost due to operations that are irreversible in principle.

3 — Utterly non-adiabatic circuits (e.g. vanilla CMOS, figure 1) — in which no attempt is made to minimize potential drops, or to recover the transferred energy.

This field is rooted in discussions of "logical reversibility" and the thermodynamics of computation[1; 2]. A logically reversible device is one where if you tell me the output I can tell you what the input must have been; an inverter is a perfect example. In contrast, an adder is not logically reversible, because if you tell me the sum I cannot tell you what the addends must have been.

Figure 4 shows a mechanical adder. Even though it is not logically reversible, it is thermodynamically and mechanically reversible in the usual sense. If operated slowly, it would dissipate arbitrarily little energy per operation. As will become clearer below, logical reversibility is neither necessary nor sufficient for thermodynamic reversibility (or adiabaticity).

We classify the mechanical adder as a *push-through* logic device. The output changes almost as soon as the input is changed. The bad news is that the device provides no

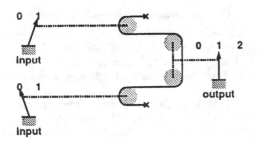

Fig. 4. Mechanical Adder

power gain; the force and energy required to move the output pointer must be supplied by whomever is moving the input pointer. Furthermore, there is no logic-level restoration (i.e. a degraded input produces a degraded output).

The circuit of figure 3 can be extended to perform a NAND function. (It needs another NFET in series and another PFET in parallel.) We classify this as an *escalator logic device*, which is very different from a push-through device. The output is carried up and down by attaching it to the power-clock. The output energy and the output timing come from the power-clock, not from the inputs. The device provides power gain, and performs logic-level restoration. It is not logically reversible, but it is fully adiabatic.

In a push-through device, the outputs necessarily and immediately follow the inputs, so an adiabatic input generally guarantees an adiabatic output. In an escalator device, however, some restrictions must be enforced to prevent an "oops" — an accidental non-adiabatic transition. Specifically, an output that is low must not be connected to a power-clock that is already high; similarly, an output that is high (perhaps as the result of a previous calculation) must not be connected to a power-clock that is already low. Consequently, most escalator devices are *pulse-mode* escalator devices. The circuit in figure 3 illustrates this. The output has a predetermined "resting" level (ground in this case). Whenever the output makes a transition away from the resting level, it must be returned ("recharged") to the resting level before the start of the next calculation. This recharge step carries a terrible price.

The problem is that recharge must be performed if *and only if* it was necessary. Suppose (as is diagrammed) the input is asserted (i.e. A = high, $\overline{A}$ = low) while the power-clock ramps up. Then the input must remain asserted while the power-clock ramps down, so that the output may return to its resting level.

On the other hand, now suppose that (contrary to what is diagrammed) the input had been unasserted (i.e. A = low, $\overline{A}$ = high) while the power-clock ramped up. The output, node X, would have remained at the resting level (low). The input must not become asserted until the power-clock has returned to the resting level — lest an oops occur.

Adiabatic Recharge Schemes

There are three ways to deal with the recharge problem:
- Retractile cascade schemes.
- Memory schemes.
- Regenerative schemes.

Retractile Cascade — The most direct way to guarantee that recharge will be performed if and only if necessary is to require not just (a) that the input be valid while the desired output is being computed (the evaluate phase) but also (b) that it remain valid until the recharge phase is completed.

Let's see what this implies for a complex calculation consisting of M stages. Let's suppose the final output, the output of the Mth stage, must be valid for one clock phase. The input to that stage must be valid for one phase before and after, for a total of 3 phases. The input to stage 1 must be valid for $2M + 1$ phases. This is illustrated in figure 5 for the case $M = 3$.

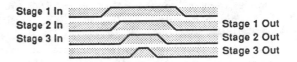

Fig. 5. Retractile Cascade Timing

This staggered timing diagram is the signature of the retractile cascade scheme. The basic idea[2] has several electronic embodiments[6; 8].

One undesirable aspect is that the latency (the interval between accepting one input and accepting the next) is increased by a factor of two, because time is needed for the retractions. What's worse is that the throughput is reduced by a factor of M or so, since no pipelining is possible.

Memory Schemes — To permit pipelining, the first thing that comes to mind is to recharge the output node *without* having valid inputs available at recharge time. The circuit of figure 6 uses this scheme.

This approach carries its own price. If the output remains valid after the input has gone invalid, then the device is performing a *memory* function, whether we intended it or not. The laws of physics[1] tell us that erasing one bit of memory must cost some energy, no matter how slowly it is done. (Retraction, unlike erasure, can be dissipation-free since the still-valid input allows us to know the prior state of the node we are trying to recharge.) The fundamental physical limit is on the order or kT; the practical limit appears to be on the order of CV_t^2, where V_t is a threshold voltage — several orders of magnitude larger than kT, but still much less than CV_{dd}^2.

The memory function need not be explicit or complex. Some logic families[4; 5; 9] leave the output node in a high impedance state until recharge time; charge trapped on the node constitutes the memory.

Because of this erasure energy, so memory schemes are *not* asymptotically adiabatic. However, for the applications of most interest to us, the operating frequency is well above the asymptotic regime anyway. At any given nonzero frequency, all we care about is the actual dissipation at that frequency.

Regenerative schemes — As mentioned above, the recharge energy can be arbitrarily small if we know the prior state of the node. If the gate at stage m implements a *logically reversible* function, a tantalizing possibility arises: we can use the stage-m outputs to control the recharge of the stage-m inputs[10; 11]. This idea can be applied without difficulty to a chain of inverters and/or buffers (i.e. a shift register). Unfortunately, for typical logic functions F, it is prohibitively difficult to implement the functional inverse F^{-1}.

Clockability

In any digital circuit, dissipation in the clock generator is an important part of the overall energy budget. An adiabatic system can use a N constant-voltage sources (capacitors and/or batteries) and a N switches, to synthesize an N-step approximation to the required ramplike waveforms[7]. This works fine at low frequencies, but at high frequencies the energy required to operate the switches is prohibitive.

We prefer resonant supplies; that is, an RLC circuit, where R is the parallel combination of all the transistor channel resistance and wiring resistance on the chip, and C is the parallel combination of all the gate capacitance and wiring capacitance. We need one or two inductors per chip (certainly not per logic gate). Resonant schemes, alas, require that the clock driver see a relatively constant load capacitance — independent of data patterns.

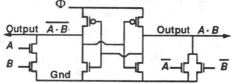

Fig. 6. Adiabatic Logic Gate

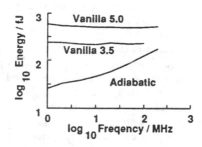

Fig. 7. Energy versus Frequency

A Working System

Figure 6 shows a high-performance adiabatic logic gate. It has excellent noise immunity compared to previous logic families[9]. The transistor area required for an inverter in in this family is large compared to vanilla CMOS, but is actually smaller for many-input gates. Two-wire (differential) signalling ensures a constant load to the power-clock driver — but doubles the wiring area.

Figure 7 shows the energy per operation for a chain of inverters, comparing the adiabatic family to vanilla CMOS. At 200MHz, the adiabatic circuit dissipates fourfold less energy. At lower frequencies, the advantage is even larger. The dissipation of the vanilla CMOS can be reduced by lowering V_{dd} from 5 volts to 3.5 volts, but then the operating speed is compromised.

Conclusions + Acknowledgements

We will not see fully adiabatic micrprocessor circuits on the market any time soon. In the short run, there are easier ways to reduce system energy requirements. On the other hand, these are novel ideas that will have important applications eventually.

Many of the ideas presented here were developed in collaboration with Alan Kramer. The Bell Labs adiabatic computing effort has benefitted from the contributions of Steve Avery, Bryan Ackland, Alex Dickinson, Al Dunlop, Thad Gabara, Yann leCun, Larry Jackel, Tom Wik, and many others.

References

[1] Rolf Landauer, "Irreversibility and Heat Generation in the Computing Process," *IBM J. Res. Devel.* **5** (1961).

[2] C. H. Bennett, "Logical Reversibility of Computation," *IBM J. Res. Devel.* **17**, 525–532 (1973).

[3] C. Seitz et al., "Hot Clock nMOS," *Proceedings of the 1985 Chapel Hill Conference on VLSI.* Computer Science Press (1985).

[4] Roderick T. Hinman and Martin F. Schlecht, "Power Dissipation Measurements on Recovered Energy Logic," *1994 Symposium on VLSI Circuits / Digest of Technical Papers*, 19. IEEE (June 1994); also R. T. Hinman and M. F. Schlect, "Recovered Energy Logic ...," *Proceedings of the IEEE Power Electronics Specialists Conference.* (1993).

[5] A. G. Dickinson and J. S. Denker, "Adiabatic Dynamic Logic," *Proceedings of the Custom Integrated Circuits Conference.* IEEE (1994).

[6] J. G. Koller and W.C. Athas, "Adiabatic Switching, Low Energy Computing, and the Physics of Storing and Erasing Information," *PhysComp '92: Proc. of the Workshop on Physics and Computation.* IEEE (1993).

[7] W. C. Athas, L. "J." Svensson, J. G. Koller, N. Tzartzanis, Y-C Chou, "A Framework for Practical Low-Power Digital Cmos Systems Using Adiabatic Switching Principles," *Int'l Workshop on Low-Power Design*, (unpublished, 1994).

[8] Ralph C. Merkle, "Reversible Electronic Logic using Switches," *Nanotechnology* **4** 21–40. (1993).

[9] Alan Kramer, John S. Denker, Stephen C. Avery, Alex G. Dickinson, and Thomas R. Wik, "Adiabatic Computing with the 2N-2N2D Logic Family," *1994 Symposium on VLSI Circuits / Digest of Technical Papers*, 25. IEEE (June 1994).

[10] J. S. Hall, "An Electroid Switching Model for Feversible Computer Architectures," *PhysComp '92: Proc. of the Workshop on Physics and Computation.* IEEE (1993).

[11] S. G. Younis and T. Knight, "Practical Implementation of Charge Recovering Asymptotically Zero Power CMOS," *Proc. of 1993 Symposium on Integrated Systems*, 234–250. MIT Press (1993).

[12] T.J. Gabara, "Pulsed Low Power CMOS," *Inter. J. of High Speed Elec. and Systems*, **5** 2, (1994).

2ND ORDER ADIABATIC COMPUTATION
WITH 2N-2P AND 2N-2N2P LOGIC CIRCUITS

A. Kramer, J. S. Denker, B. Flower, J. Moroney

AT&T Bell Laboratories
Holmdel, NJ 07733

ABSTRACT

Recent advances in compact, practical adiabatic computing circuits which demonstrate significant energy savings have renewed interest in using such techniques in low-power systems. Several recently introduced circuits for adiabatic computing make use of diodes in a way which reduces switching energy from $O(CVdd^2)$ in the non-adiabatic (ie: standard CMOS) case, to $O(CVddVt)$. These circuits provide an energy savings of at most one order of Vdd/Vt. This paper introduces a new class of adiabatic computing circuits which offer several advantages over existing approaches, the primary one being that, because no diodes are used, switching energy can be reduced to an energy floor of $O(CVt^2)$. These *second order* adiabatic computing circuits provide an energy savings of as much as $O(Vdd/Vt^2)$ over conventional CMOS. Additional advantages of the proposed circuits include the fact that, in comparison to most compact adiabatic circuits which have floating output levels over the entire data valid time, these new circuits have non-floating output levels over most of the data valid time. This is important for restoring logic levels and minimizing problems with crosstalk. The proposed circuits have been simulated and demonstrate adiabatic power savings compared to standard CMOS circuits over an operating frequency range from 1MHz to 100MHz of as much as a factor of 3. One circuit topology has been fabricated and tested and operates properly at up to 100MHz, the maximum speed which could be tested. Power measurements on the functioning circuit are in progress and preliminary results demonstrate adiabatic power-vs-frequency behavior. These *second order* adiabatic computing circuits provide an attractive alternative to achieve adiabatic power savings without suffering from many of the limitations of alternative approaches and without costing much more either in terms of complexity or size.

1. INTRODUCTION - ENERGETICS AND ADIABATIC CHARGING

The energetics of standard CMOS (or any other switching system based on a single fixed DC power rail) are straightforward: when the charging switch is closed to charge the load C up to the rail voltage V, a charge $Q = CV$ is pulled out of the positive power rail. When the discharging switch is closed to discharge the load C to ground, the same charge $Q = CV$ is transferred to the ground terminal of the power supply. Over an entire charge/discharge cycle, a total charge of $Q = CV$ was taken *from* the positive rail of the power supply and returned *to* the ground terminal, and thus the total energy dissipated over the entire cycle corresponds to $Ed_{total} = Ed_{charge} + Ed_{discharge} = QV = CV^2$. Since there were two switching events involved, the average energy dissipated during charging and discharging is one half of this total dissipated energy, $Ed_{average} = Ed_{total}/2 = 1/2CV^2$. In the case where all switch resistors and load capacitors are linear, the energetics are symmetric and the energy dissipated during charging or discharging is exactly equal to this average switching energy: $Ed_{charge} = Ed_{discharge} = Ed_{average} = 1/2CV^2$. In the case of nonideal or nonlinear circuit elements, the energy dissipated during charging and discharging need not be equal, but because all of the charge was taken from the positive rail and returned to ground, total energy dissipated over the charge/discharge cycle must *always* equal twice the average switching energy: $Ed_{total} = Ed_{charge} + Ed_{discharge} = 2Ed_{average} = CV^2$.

This energy is dissipated by the integrated I^2R loss of the charging and discharging currents through the effective resistance of the circuit (switch resistances (transistor channels) and parasitic resistances from the power rail to the load C and from C to the ground node):

$$Ed_{total} = \int I(t)^2 R(I, t)dt$$

where we have used R(I,t) to include all changes in effective current path resistance either as a function of time (ie: from switching) or as a function of current (from nonlinearities). The key point is that *independent* of the sizes and/or function of this effective resistance R, the integration of I^2R over the entire charging/discharging cycle is always the same and is equal to $Ed_{total} = Ed_{charge} + Ed_{discharge} = QV = CV^2$. The reason for this is that in a fixed DC-powered switching system, the circuit elements and the switching current are related: the only way to change the switching current is to change the circuit elements (the linearity of R or C, or the size of R for example), but the dependency between the two will result in no change in the average dissipated energy: $Ed_{total} = CV^2$.

For standard single supply-rail switching systems, the only way to reduce energy consumption is to reduce the supply voltage V, or the load capacitance C. Of course, architectural approaches can also be employed at the system level to reduce the number of switching events in the system. The essential point is that for systems of this type, if a particular load must be switched to a particular voltage with a particular average frequency, there is nothing that can be done to reduce energy consumption.

Adiabatic computing is compatible with the energy savings that can be achieved through reductions in V or C, yet achieves additional reductions in dissipated energy by avoiding the single-rail DC power supply architecture. If a single non-DC power supply rail is used both to charge and discharge a switching node, the energetics change considerably. In this case, total dissipated energy over the charging/discharging cycle need not be related to transferred charge and can in fact be made arbitrarily small.

While for the DC power supply case analyzed above, where nodes are charged from the DC power supply rail and discharged into the ground node, the total dissipated energy *must* be related to the transferred charge: $Ed_{total} = Ed_{charge} + Ed_{discharge} = QV$, in the case that the power supply rail charges the switching node by ramping up and the same power supply rail later discharges the node by ramping down, this dependency between transferred charge and dissipated energy need no longer be true because charge transferred from the power supply to charge the node can be recovered by the same power supply when it later discharges the node. By taking advantage of adiabatic charging principles and charge recovery, this approach to switching breaks the dependency between the switching current and the circuit elements so that the energy dissipated as I^2R losses during the charging/discharging cycle can be made arbitrarily small. This is accomplished by making use of periodic ramp-like clocked power supplies.

How this is done can most easily seen by considering the I^2R dissipation losses in the adiabatic charging case. For a given ramp time T, the transferred charge in the adiabatic and non-adiabatic cases must be the same: $Q = CV$. The difference between the two in terms of energy dissipated is that, while in the non-adiabatic case the current is highly nonuniform, in the adiabatic case, because of the ramp, it can be made much more uniform over the ramp time T, and in fact ideally constant ($I = Q/T$). By slowing down the ramp (increasing T), the charging current can be made arbitrarily small. The energy dissipated during the charging cycle is $Ed_{charge} = I^2RT = IRQ$. Increasing time T by a factor of α will decrease current I by a factor of α (transferred charge $Q = IT$ will remain the same), but because I^2 is not linear in I, dissipated energy $Ed_{charge} = I^2RT$

will decrease by a factor of α. Adiabatic charging principles allow dissipated energy to be an arbitrarily small percentage of transferred energy by transferring charge at a constant and arbitrarily slow rate.

In the nonadiabatic case, maximum switching current typically flows when the voltage difference between the load C and the voltage rail V or ground are greatest, leading to energy dissipation spikes. While it might be possible to devise a nonadiabatic circuit which had a uniform current flow, perhaps even equal to that of the adiabatic circuit, this would only be possible with a highly nonuniform resistor which had greatest resistance when the voltage across it was greatest. Because of charge loss, the resistor needed for uniform current would also lead to the same total dissipated energy:

$$Ed_{total} = \int I(t)^2 R(I, t) dt = QV.$$

2. RECENT WORK

While the concepts of adiabatic charging have been known for some time [1,2], early circuit proposals, while very interesting from a theoretical standpoint, were not practical for large-scale implementation due to the unwieldiness and complexity of the circuits, the large overheads involved, the complexity of the timing and power supply/clock generation, and the relatively slow speeds at which they would operate.

Recent interest has resulted in several much more practical circuit implementations of adiabatic computing circuits [4,5,6,7,8,9,10,11]. These circuits, rather than aiming to achieve energy dissipation floors approaching the theoretical minimum, achieve much more practical implementations by aiming for energy floors which are "only" a factor of 2 - 20 less than that of conventional static CMOS logic (called "vanilla CMOS" from now on).

Several recently-proposed implementations make use of the fact that diodes can be used to provide very compact and efficient adiabatic charging elements [4,5,8,11]. The resulting circuits are fast (>100MHz), compact (no larger than vanilla CMOS), and compatible with existing fabrication technologies. These circuits exhibit adiabatic energy savings, but the use of diodes for adiabatic charging in any circuit limits this saving to a factor of V/Vt over that of conventional circuits. The reason for this is that a diode will have a voltage drop which is to first order constant and equal to Vt for any positive current driven through it. This means that for a diode, $IR = V = Vt$ (it is a nonlinear current-dependant resistor), and energy dissipated in adiabatic charging through a diode cannot be less than $Ed = I^2RT = IT * IR = QVt = CVVt$. The maximum energy savings possible though any diode-based adiabatic charging circuits is thus limited to $1/(QV/QVt) = 1/(V/Vt)$, no matter how slowly the charging occurs.

3. ORDER OF ADIABATIC DISSIPATION

We have found this factor of V/Vt to be a useful reference in analyzing energy dissipation in adiabatic circuits as compared to conventional switching circuits. *First order adiabatic losses* correspond to losses which have a floor of $O(CVVt) = O(QV/[V/Vt])$, such as the diode charging losses described above. *Second order adiabatic losses* correspond to losses which have a floor of $O(CVt^2) = O(QV/[V/Vt]^2)$, such as a nonadiabatic switching event from Vt to ground. By this convention, theoretical energy floors which are independent of Vdd and Vt such as kT would be called N^{th} *order adiabatic losses*. Practical adiabatic computing circuits typically contain first and/or second order loss terms, and thus have energy floors which are high compared to the theoretical minimum.

While it would seem that first order losses are more important than second order losses, V/Vt is typically not very large and any loss term has a scaling factor in front of it, so it does not take many second order loss terms to equal a first order loss term. In fact, while diode-based adiabatic charging systems must have a first order energy loss, they often have one more second order losses which may dominate the actual energy floor.

4. 2ND ORDER ADIABATIC CIRCUITS

This work introduces a new class of circuits with a low energy floor. The essential energy advantage of these circuits comes from the fact that they have been "adiabatically designed" to eliminate all first order energy losses and to minimize second order losses as much as possible. This is accomplished primarily by charging nodes through minimal switch resistances rather than diodes. The circuits we describe realize this advantage with minimal additional overhead and complexity over diode-based circuits, either at the circuit or system level. These circuits are at most two times the size of the smallest diode-based adiabatic circuits, making them about the same size as vanilla CMOS, and they operate at similar frequencies of greater than 100MHz. Another advantage of these new circuits is that, while most diode-based adiabatic computing circuits have output levels which are floating during their output valid time, these new circuits provide outputs which are clamped during their output valid time (like vanilla CMOS). This is important at the system level in terms of reducing crosstalk and restoring logic levels.

5. BASIC OPERATION OF THE 2N-2P FAMILY

The first of these circuits we will introduced is called 2N-2P. The name is based on our convention of using the number of transistors in a gate because the cost for each input in terms of transistors is 2 Nfets and the overhead for each complete gate is 2 Pfets. The circuit uses differential logic, so each gate computes both a logic function and its complement, and each input to a gate requires both polarities to be represented. The basic circuit for a inverter-buffer is shown in fig 1. Each Nfet input gets the corresponding positive and negative polarity inputs and the cross-coupled Pfets are connected to the clock-supply. The timing and logical operation of the gate is as follows (fig 2):

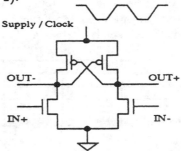

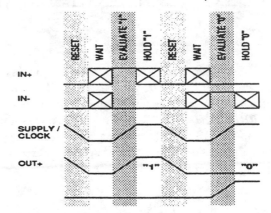

FIGURE 1: Basic 2N-2P differential buffer/inverter.

FIGURE 2: Timing for 2N-2N2P buffer/inverter.

In the *RESET* (first) phase the inputs are low, the outputs are complementary (one high, the other low), and the power supply ramps down. The high output, because its Pfet is held on by the low output, will "ride" the ramp down so that at the end of the first phase both outputs will be low. In the *WAIT* (second) phase the power-supply stays low, maintaining the outputs low (the necessary condition for the next logical gate, which is delayed by a quarter cycle, to perform its *RESET* phase) and the inputs are evaluated. Note that because the gate is "powered down", the evaluation of the inputs will have no effect on the state of the gate. In the *EVALUATE* (third) phase, the power supply ramps up and the outputs will evaluate to a complementary state. The half-gate with its input high will have its output held low while the half-gate with its input low will "ride" the ramp up. At the end of *EVALUATE* the outputs will always be complementary. This condition is guaranteed by the inverse logic of the two half gates and their cross coupled Pfets (this is the reason that 2N-2P logic *must* be differential). In the *HOLD* (fourth) phase the power supply clock stays high while the inputs ramp down to low. Gate outputs remain valid for the entire phase.

243

6. COMPLEX GATES AND SEQUENCES OF GATES

Because there are four phases to the timing, there must be four quadrature clocks in a complete system, each clock 90 degrees in advance of the previous clock. In this way, each logic phase in the system holds its outputs valid while its successor is evaluating (ramping up) and its predecessor is resetting (ramping down) and waits with its outputs both low while its successor is resetting (down) and its successor is evaluating (up).

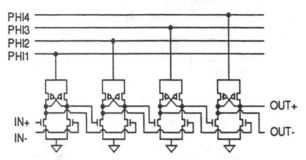

FIGURE 3: 4-Phase shift register bit.

A shift register can be constructed by making a sequence of buffer/inverter gates connected sequentially and in the proper phase relationship, that is: *PHI1, PHI2, PHI3, PHI4, PHI1,* ... (fig 3). We have simulated an 0.8 um CMOS implementation of such shift registers using minimum size transistors at speeds in excess of 250 MHz.

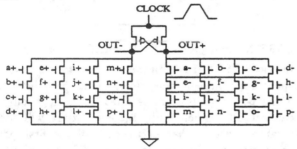

FIGURE 4: Complex gate.

More complex gates can be constructed by replacing the single Nfets used in the inverter/buffer with an arbitrary Nfet-based logic tree and its inverse (fig 4). Because the differential logic provides both negative and positive polarity signals, providing both positive and negative logic trees using only Nfets is straightforward: while in vanilla CMOS the positive logic tree is created by connecting a single input polarity to Nfets and the negative logic tree is created by connecting the same input polarity to the Pfet-based inverse tree, in the case of differential logic both the logic tree and its inverse can be Nfet-based as every Nfet connected to an input in the logic tree has a corresponding Nfet connected to the inverted input in the inverse logic tree. We have simulated logic gates with up to 4x4=16 inputs at speeds up to 100MHz.

7. 2N-2N2P AND SYSTEM ISSUES

A variant on the 2N-2P logic family described above is that of the 2N-2N2P family, the only difference being that 2N-2N2P has a pair of cross-coupled Nfets in addition to the cross-coupled Pfets common to both families (fig 5). 2N-2N2P thus has cross-coupled full inverters and thus is very similar to a standard SRAM cell. The timing and logical operation of 2N-2N2P is identical to that of 2N-2P.

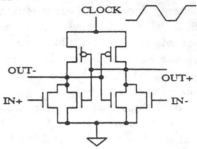

FIGURE 5: Basic 2N-2N2P Inverter/Buffer Gate.

Fully-static logic such as vanilla CMOS has outputs which offer two important advantages at the system level. The first of these is that its outputs are always clamped to either Vdd or Gnd. This is important to restore logic levels and reduce the effects of crosstalk. The second advantage is that fully-static logic has static outputs which are always valid; if the inputs do not change neither do the outputs. This is important for simplifying timing and system design. Dynamic logic such as domino CMOS enjoys neither of these advantages. During the output valid time, outputs may be floating (if the evaluation did not change them from the precharged state). Also, the outputs are pulse-mode: they are only valid following an evaluation which may change their state from that of being precharged. Unchanging inputs can result in a gate whose output "pulses" back and forth between high and low values following precharge and evaluate phases.

Most diode-based adiabatic logic is similar to dynamic logic in that it has floating pulse-mode outputs. 2N-2P and 2N-2N2P also have pulse-mode outputs, but they enjoy an important advantage in that, like vanilla CMOS, their outputs are not floating during the output valid time. The primary advantage of 2N-2N2P over 2N-2P is in fact that the addition of the cross-coupled Nfets results in non-floating data valid over 100% of the *HOLD* phase, as opposed to 2N-2P where, because the inputs are ramping down during the *HOLD* phase, a gates low output is only clamped for the first 50% of the *HOLD* phase. (Note that, from the point of view of the evaluation of the successive gate, the first 50% of the *HOLD* time is the most important period in which to maintain a good logic level as it is in this period that the next gate will make its decision.)

The rough analysis of the energetics of the 2N-2P and 2N-2N2P adiabatic logic families are identical. The analysis requires a more electrical description of the timing. As already described in the logical timing description, during the *RESET* phase, when the clock is ramping down and the inputs are held low, one output is already low and the other output "rides" the clock down. The high output will ride down only to Vt, rather than gnd, because at that point the Pfet ceases to conduct. Following *RESET* then, both outputs are not low but rather the low output is low while the high output is floating at Vt. If during the *EVALUATE* phase, the logical state of the gate has not changed (the high output should continue to be high), the high output which is floating at Vt will ride the clock up, beginning its conduction when the rising clock has again reached a voltage of Vt. These details do not really change the analysis of the energetics in the case when the logical state of the gate has not changed. Because the upward and downward ramp on the output is fully adiabatic, energy loss can be made arbitrarily small by making the ramp time arbitrarily long. During the *HOLD* phase, when the outputs are floating, there is no energy loss.

When the gate output state makes a transition from one logical state to the other, the fact that the old high output was floating at Vt becomes critical however. During the *HOLD* phase the inputs become valid and the logical state of the gate will change. This means that the floating output, which in the previous state was high and thus had a nonconducting logical Nfet tree, will now have a conducting tree. These valid inputs will thus connect the floating output, which is at a voltage of Vt, to ground and the result is a *nonadiabatic* charge transfer of $Ed_{discharge} = O(CVt^2)$. The same is true for the old low output which now must ride the ramp up; it will make a nonadiabatic transition from Gnd to Vt when the ramp reaches Vt and this will dissipate energy of $Ed_{charge} = O(CVt^2)$. Because charge of $Q = CVt$ was supplied from the ramp to the load when the ramp was at a voltage of Vt and this charge was later transferred to Gnd, the total energy dissipation for the charge/discharge cycle can be determined as before: $Ed_{total} = Ed_{charge} + Ed_{discharge} = CVt^2$. Because this energy loss is nonadiabatic, there is no way to reduce it; it is independent of clock speed.

These 2 logical families will thus lose some arbitrarily small energy at each clock cycle corresponding to adiabatic I^2R losses in the Pfets and will lose CVt^2 at each gate transition cycle. This can be compared to vanilla CMOS, which will lose some small energy as leakage at all times and will lose CV^2 for each transition cycle.

We have made simulations comparing the energetics of 2N-2N2P logic with vanilla CMOS. The circuits have been chosen to give the best possible advantage to CMOS (for 5V switching): despite the fact that both adiabatic logic families provide both a logical operation and a quarter-latch with each gate, comparison is being made with a CMOS circuit which provides only the equivalent number of logical operations per clock period. The circuits are essentially clocked "Mobius" circuits which provide a close approximation of an arbitrarily long inverter chain (fig 6).

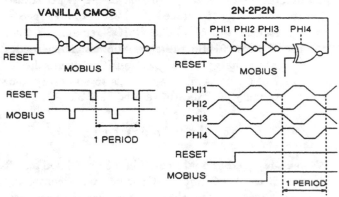

FIGURE 6: Vanilla CMOS and 2N-2N2P Mobius Circuits.

These circuits have been designed in an 0.8 um CMOS process and simulated by a detailed circuit simulator (ADVICE) at Vdd=5V from speeds of 1MHz up to 250MHz, where CMOS circuit operation begins to fail. Note that, because there are four gates per period, 250MHz corresponds to a gate speed of 1GHz. At each simulation speed, correct logical circuit operation was checked and energy per gate transition was extracted. This energy includes only the energy dissipated by the gate itself; in a real system the energy consumed by the clock driver providing the energy-recovery ramp-like clocks must of course also be considered. The results are plotted in figure 7.

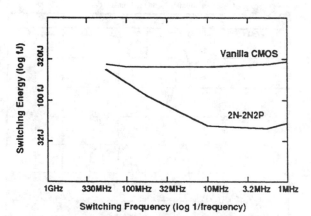

FIGURE 7: Simulated switching energy-vs-frequency curves.

The results for the vanilla CMOS circuit show an essentially flat transition energy of 0.8pJ. At speeds above

100 MHz (2.5ns/gate), the energy begins to climb because the gates no longer have enough time to fully complete their transition and crowbar currents begin to dominate. At speeds below 1 MHz, energy lost to leakage becomes significant and so total energy begins to rise.

The energy curve for 2N-2N2P demonstrates nice adiabatic characteristics. At high speeds above 100MHz, the energy used by the 2N-2N2P circuits is roughly equivalent to that of CMOS. As both circuits are based on the same underlying circuit elements (transistors), their top speeds and their energetics in this regime, where they are operating so fast that they are never in equilibrium, are similar. From 100 MHz to 9MHz, the adiabatic circuit demonstrates a steady decrease in energy per transition corresponding to reduced I^2R losses at lower speeds, while the energetics of vanilla CMOS remain constant. At speeds below 9MHz the adiabatic circuit energetics level out at an energy of 0.25 pJ/transition due to the energy floor of CVt^2. This corresponds to an energy floor which is roughly 30% that of vanilla CMOS at Vdd=5V.

10. TEST RESULTS

We have designed and fabricated in 0.8 um CMOS a shift register in the 2N-2P adiabatic logic family containing 1000 shift stages. The shift register has been successfully tested at frequencies up to 100MHz (the maximum frequency of our test setup). The measured waveforms at 100MHz (fig 8) show clearly the correct operation of the shift register.

We are currently engaged in energy measurements of this circuit and a vanilla CMOS equivalent. Our goal is to make energy measurements corresponding to the simulated transition-energy-vs-frequency curves shown in figure 7. We have preliminary measurement results from our 2N-2P circuit indicating adiabatic charging behavior; over an operating frequency range from 5-100MHz we have seen qualitatively that doubling the frequency more than doubles the energy per transition.

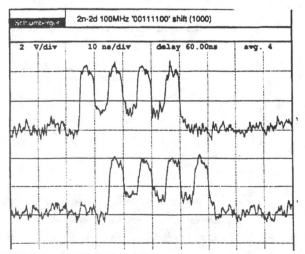

FIGURE 8: 2N-2P shift register measured waveforms at 100 MHz. Measurement was made with e-beam tester.

11. CONCLUSIONS

2N-2P and 2N-2N2P are novel adiabatic logic families which demonstrate lower energy floors than diode-based adiabatic logic families. The two families are related and have another advantage over most diode-based circuits because they provide non-floating outputs over their output valid times. Size is comparable to conventional circuits, complexity and timing are similar to that of other practical adiabatic computing circuits. Simulations of 2N-2N2P show expected adiabatic charging behavior and an energy savings compared to standard static CMOS of as much as a factor of three in worst cases (worst for 2N-2N2P that is). Inverter chains (built in a standard 0.8 um CMOS process) operate at > 100MHz; higher speeds appear possible (though dissipation increases). Preliminary efforts to measure energy consumption qualitatively indicate adiabatic charging behavio.

12. REFERENCES

1. Rolf Landauer, "Irreversibility and Heat Generation in the Computing Process", IBM J. Res. Devel. vol. 5 pp. 183-191 (1961).

2. C. H. Bennett, "Logical Reversibility of Computation", IBM J. Res. Devel. vol. 17, pp525-532 (1973).

3. C. Seitz et al., "Hot Clock nMOS", Proceedings of the 1985 Chapel Hill Conference on VLSI. Computer Science Press (1985).

4. Roderick T. Hinman and Martin F. Schlecht, "Power Dissipation Measurements on Recovered Energy Logic", 1994 Symposium on VLSI Circuits / Digest of Technical Papers, 19. IEEE (June 1994).

5. A. G. Dickinson and J. S. Denker, "Adiabatic Dynamic Logic", Proceedings of the Custom Integrated Circuits Conference. IEEE (1994).

6. J. G. Koller and W.C. Athas, "Adiabatic Switching, Low Energy Computing, and the Physics of Storing and Erasing Information", PhysComp '92: Proc. of the Workshop on Physics and Computation. IEEE (1993).

7. Ralph C. Merkle, "Reversible Electronic Logic using Switches", Nanotechnology Vol. 4 21–40. (1993).

8. Alan Kramer, John S. Denker, Stephen C. Avery, Alex G. Dickinson, and Thomas R. Wik, "Adiabatic Computing with the 2N-2N2D Logic Family", 1994 Symposium on VLSI Circuits / Digest of Technical Papers, 25. IEEE (June 1994).

9. J. S. Hall, "An Electroid Switching Model for Reversible Computer Architectures", PhysComp '92: Proc. of the Workshop on Physics and Computation. IEEE (1993).

10. S. G. Younis and T. Knight, "Practical Implementation of Charge Recovering Asymptotically Zero Power CMOS", Proc. of 1993 Symposium on Integrated Systems, 234–250. MIT Press (1993).

11. T.J. Gabara, "Pulsed Low Power CMOS", Inter. J. of High Speed Elec. and Systems, Vol. 5 2, (1994).

A Low-Power Microprocessor
Based on Resonant Energy

William C. Athas, Nestoras Tzartzanis, *Student Member, IEEE*, Lars "J." Svensson, *Member, IEEE*,
and Lena Peterson, *Member, IEEE*

Abstract— We describe AC-1, a CMOS microprocessor that derives most of its operating power from the clock signals rather than from dc supplies. Clock-powered circuit elements are selectively used to drive high-fan-out nodes. An inductor-based, all-resonant clock-power generator allows us to recover 85% of the clock-drive energy. The measured top frequency for the microprocessor was 58.8 MHz at 26.2 mW. The resulting overall decrease in dissipation ranges from four to five times at clock frequencies from 35 to 54 MHz. We also compare the performance of the processor to a reimplementation in static logic.

Index Terms—Adiabatic switching, energy recovery, low-power digital CMOS, microprocessor design, resonant charging.

I. INTRODUCTION

TO maximize computing performance while minimizing power dissipation is a common goal when designing VLSI CMOS systems. The well-known starting assumption for assessing dissipation in voltage-mode digital CMOS circuits is the fCV^2 component to dynamic power dissipation. The CV^2 energy quantity equals twice the energy required to represent the signal (the *signal energy*). The energy dissipation per signal transition (the *switching energy*) is, under this assumption, at least equal to the signal energy. This basic premise for the relationship between switching energy and signal energy for all the switching activity inside the CMOS system firmly establishes the two available means for minimizing power: namely, to minimize the signal energy and to minimize the number of transitions required to perform the computation. Often these two strategies are at odds with each other. For example, transition count can sometimes be reduced by increasing signal fan-out, which requires signal capacitance (C) and its associated energy to increase.

This relationship between switching energy and signal energy for voltage-mode digital CMOS circuits is not a fundamental one. From computational thermodynamics, the speed of energy transport determines the absolute switching-energy dissipation, which can be but a small fraction of the signal energy [1]. A technique that has come to be known as *adiabatic charging* [2] uses this basic result to reduce switching energy well below the signal energy by explicitly controlling the transition time. Watkins [3] first investigated the technique and applied it to a MOS shift register. More recently, the idea

has surfaced in the form of hot-clock nMOS [4] for building low-power, high-speed nMOS circuits. In both investigations, the practical key to exploiting adiabatic charging was to tap the clock lines for ac power. In contrast, in conventional practice, all operating power is derived from dc voltage supplies.

In this paper, we describe the architecture and circuit implementation of the AC-1 microprocessor, which systematically applies adiabatic charging in the form of clock-powered logic. Furthermore, we describe a compatible resonant clock-driver circuit that can efficiently deliver energy to the circuit nodes and then recover and reuse circuit energy which would otherwise have been dissipated as heat. The principal result is that power dissipation can be reduced well below fCV^2.

AC-1 has been fabricated as a 0.5-μm n-well CMOS chip. We present laboratory measurements of the chip which clearly show that significant amounts of energy can be recovered and reused with this approach. Our paper concludes with performance predictions, based on chip-level simulations, of how well this new approach compares to the supply-voltage scaling of fully restored, static CMOS logic.

II. SYSTEM APPROACH

An effective low-power system that utilizes energy recovery demands the combined solution of two interrelated engineering problems. The first is to devise an efficient source of pulsed power for extrinsically controlling signal transition times. The second is to develop a style of logic circuitry that can efficiently harness the power pulses to execute the computation. The attainable efficiency depends on the nature of the circuit function. For the simple case of the output of one logic gate driving another nearby gate, the circuitry overhead for supporting energy recovery outweighs the benefits of recycling circuit energies. Under such circumstances, it is better to use fully restored static logic and minimize the energy dissipation by conventional means (viz., voltage scaling). However, when the output signal energy is large compared to the internal switching energy of the gate, the overhead becomes relatively less significant. Such differences are common in microprocessors, where the application of regular structures and distributed control signals result in drive transistors with high fan-out.

For the design of the pulsed-power source, there are two general approaches: resonant circuits and stepwise-charging circuits [5]. The resonant approach uses inductances to commute circuit energy between electric and magnetic form. The stepwise approach uses banks of switched capacitances to incrementally deliver and recover circuit energy. The stepwise

Manuscript received April 10, 1997; revised June 16, 1997. This work was supported by DARPA Contracts DABT63-92-C0052 and DAAL01-95-K3528.

The authors are with the Information Sciences Institute, University of Southern California, Marina del Rey, CA 90292 USA.

Publisher Item Identifier S 0018-9200(97)08034-7.

Reprinted from *IEEE Journal of Solid-State Circuits*, Vol. 32, No. 11, pp. 1693-1700, November 1997.

247

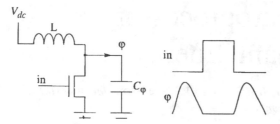

Fig. 1. A single-rail resonant clock driver.

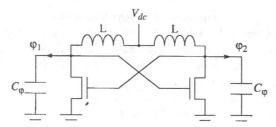

Fig. 2. The blip circuit, an all-resonant dual-rail *LC* oscillator, used as a clock driver.

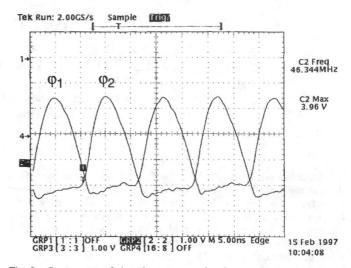

Fig. 3. Scope trace of the *almost*-nonoverlapping two-phase clock waveforms of AC-1's blip circuit.

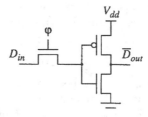

Fig. 4. Three-transistor dynamic latch.

approach is simpler to control and integrate into a single-chip design, but the resonant approach offers inherently higher efficiency.

There are myriad ways to generate resonant-energy pulses for driving on-chip capacitance. A simple and sufficient scheme is the flyback circuit shown in Fig. 1. When the nFET is on, the load is clamped to ground while a current builds up quasi-linearly in the inductor. When the nFET is turned off, the built-up inductor energy transfers to the load as a positive-going sinusoidal pulse. At the 50% time point of the pulse, the output voltage reaches its peak and the current will then reverse direction. During the latter half of the pulse, charge is returned from the load back through the inductor. If the nFET is turned on again when the load voltage returns to zero, the process can be repeated.

Two such circuits working in tandem can generate two nonoverlapping pulse trains which are readily usable as system clocks. The overall power dissipation for such a configuration will be determined by the gate-drive energy for the nFET's: the devices should be wide to cause little loss during current build-up, but wider devices require more gate energy.

For maximum overall power efficiency, it is possible to cross-couple two such circuits, as shown in Fig. 2. We then arrive at the *blip circuit* [6]: a simple, dual-rail *LC* oscillator whose active element is a cross-coupled pair of nFET's. Similar circuit topologies have been used to generate antiphase small-signal oscillations [7]. The blip circuit is unique by virtue of its large loop gain: the oscillation amplitude stabilizes only when the nFET's severely limit the signal during the second half of the period. The resulting waveforms (Fig. 3) are very similar to those generated by two single-ended circuits working 180° out of phase. The most important difference is that some clock-phase overlap is unavoidable. The pulse amplitude is approximately three times the dc supply voltage (V_{dc}).

The key to the high energy efficiency of the blip circuit is its all-resonant topology. The nFET gate-drive energy is supplied resonantly by the circuit itself. The nFET's can

therefore be made wider without undue power penalty. In this configuration, the blip circuit is free-running, with the pulse widths determined by the inductances and capacitances. For a constant capacitive load, the frequency will be stable and can be locked to a specific frequency with a varactor-based phase-locked loop. To impose frequency and phase stability when the capacitance varies from cycle to cycle (e.g., conditionally on data signals), the simple phase-locked loop approach is unlikely to be workable. For such cases, the gate drive must be synchronized to an external timebase. The cost of the frequency and phase stability will invariably be some additional power dissipation in the resonant-pulse generation circuitry. An alternative which avoids this contribution to dissipation is to treat the free-running blip circuit and its clock-powered-logic as an asynchronous subsystem. Synchronizers, such as self-timed first-in–first-out (FIFO) memories, would then be necessary to buffer data at the interfaces.

To use the waveform shapes of the blip circuit for system clocks imposes constraints on the choice of latch circuitry and clocking scheme. The latches cannot require very fast clock edges for correct operation, nor can they require the complements of the clock signals. A simple circuit that meets these requirements is the three-transistor dynamic latch that consists of a pass gate followed by an inverter (Fig. 4). The clock voltage must swing above the logic supply voltage to ensure a full-swing input to the inverter. The cost of overdriving the clock voltage is increased energy stored in the clock lines. However, the resonant driver efficiently recycles a high percentage of this energy.

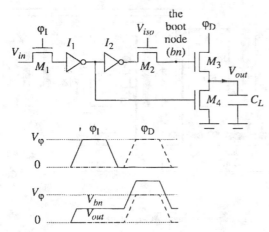

Fig. 5. Energy-recovery latch (E-R latch) and a sketch of its signal timing when V_{in} is high.

A resonant clock driver, such as the blip circuit, can readily be utilized as an energy-efficient clock generator to recycle clock-line energy. The clock lines are then a *de facto* source of system-wide ac power. This suggests that clock power might be useful for other large capacitive loads, such as data buses and control lines. The design problem for clock powering these additional nodes is that the energy transfer is *conditional* on an input datum.

Our solution is the *energy-recovery latch* (E-R latch) which serves two purposes: to latch the input data, and, conditionally on the latched datum, to transfer charge from a clock line to a load capacitance and back again. In CMOS technology, charge can be transferred adiabatically [2] and conditionally to a load capacitance by means of a charge-controlled switch. The switch implementation is therefore the critical aspect of the E-R latch design.

The circuit diagram for the E-R latch is shown in Fig. 5. The all-important charge-controlled switch is implemented as a bootstrapped nFET which is used in a clocked-buffer configuration [8]. The input-latch stage is a dynamic three-transistor latch. The *bootstrap transistor* M_3 transfers clock-line charge to and from the load capacitance C_L during clock phase φ_D. The *clamp transistor* M_4 clamps the output to ground when, during phase φ_D, it is not driven by the bootstrap transistor. Note that the load capacitance is always discharged through the bootstrap transistor, never through the clamp transistor. Inverters I_1 and I_2 are powered from a dc power supply with a voltage V_{dd} that is approximately one threshold voltage lower than the clock-phase amplitude V_φ. The dc power supply connected to inverter I_2 supplies the control energy for the switch; this energy is not recovered. The *isolation transistor* M_2 isolates the boot node bn from the output of the inverter, thus allowing the boot node to rise above V_{dd}. The isolation-transistor gate is connected to a dc supply at voltage V_{iso}. We use a dc voltage, instead of phase φ_I, to ensure that when the input is low, the boot node is actively held low also during φ_D. Otherwise, it could bootstrap enough, when φ_D goes high, to cause short-circuit current to flow from the clock driver through M_3 and M_4.

The rationale for a bootstrapped circuit topology is to minimize the dissipation in the switch so that as much as possible of the energy supplied to the load capacitance can be recovered. The total dissipation required to operate the switch has two terms: one for the control charge and one for the controlled charge. For all CMOS switch topologies we have investigated, we have found an inverse dependency between the control energy and the loss in the switch. System and circuit parameters, such as voltage swings and device ratios, determine the quantitative relationship. The best we can do to limit the total E-R-latch dissipation is to choose the switch topology that requires the smallest control energy for a specified loss. We found the bootstrapped nFET to be the most suitable switch implementation in the CMOS technology. Effective bootstrapping makes the switch-transistor gate rise high enough above the highest applied clock voltage to keep the channel conductance high, and consequently the instantaneous loss low, even for the maximum clock voltage. The output swing is thus fully restored to the clock amplitude. Alternatives include applying a logic supply voltage higher than the clock amplitude and using a transmission gate (e.g., a pFET and nFET in parallel). In both these solutions, more energy is dissipated to control the switch than is needed for a bootstrapped device with the same loss.

III. THE AC-1 MICROPROCESSOR EXPERIMENT

The objective of the AC-1 microprocessor experiment was to determine the net effect that the E-R latches and resonant clocking would have on clock and logic power dissipation for a small microsystem. Since the main goal of the experiment was to evaluate circuit techniques, the exact microarchitecture was unimportant except that it contain a representative cross section of circuit elements such as registers, data buses, function units, and control logic.

The genesis for the AC-1 instruction set architecture (ISA) was a variant of DLX [9] in which the 32-b, three-operand instruction format was replaced with a 16-b, two-operand instruction format [10]. A C compiler, assembler, and architecture-level simulator were readily available. Additionally, the 16-b instruction format helped to reduce implementation cost since the chip size was determined by the pad ring. The shorter instructions did impose certain limitations compared to their 32-b counterparts: shorter immediate and offset fields, a smaller register file, and more complicated instruction decoding.

A. Chip Microarchitecture

AC-1 is a pipelined microprocessor which can complete one instruction per clock cycle. It has separate instruction and data buses. The major function units are (Fig. 6): a control unit that generates all the control signals; a PC unit that fetches and sequences instructions; a 16-word, three-port register file (RF); a shifter; a compare unit; an ALU; and a load/store unit that handles data memory accesses. The ALU computes the effective memory addresses which are then used by the load/store unit.

To maximize throughput (i.e., instructions per second) in pipelined microprocessors, the ISA and the allowable circuit delays determine the assignment of microtasks to pipeline

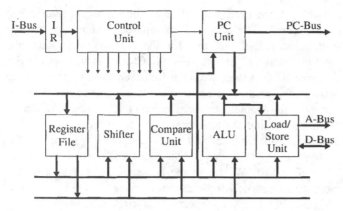

Fig. 6. AC-1 function-unit organization.

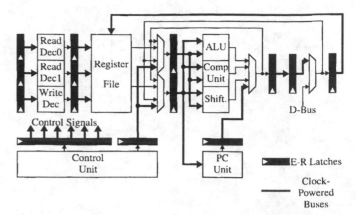

Fig. 7. E-R-latch placement in AC-1 datapath.

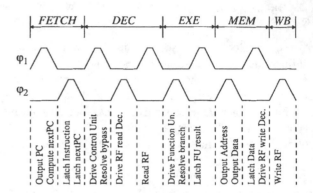

Fig. 8. AC-1 pipeline timing. Note the separate phase for RF decoder access which is required for clock-powering the RF word lines.

stages. The fraction of instructions that complete per clock cycle and the operating frequency establish the throughput. For clock-powered logic, a third consideration is the placement of the E-R latches to exploit energy recovery. Energy is recovered mainly along the paths from the resonant clock driver to the inputs of the logic blocks. The energy for transitions triggered by these pulses in circuit nodes that are not part of the clock-powered logic nets is not recoverable. For this reason, partitioning of logic blocks into smaller blocks to increase the pervasiveness of clock-powered logic nets is advantageous. However, as the pipeline becomes deeper from the insertion of the additional E-R latches, the typical circuit fan-out decreases since less logic is driven per clock phase. This fan-out reduction diminishes the energy-recovery advantage.

With this rudimentary understanding of the tradeoff in exploiting clock-powered logic, we identified the nodes inside the design where the circuit fan-out would necessarily be high. These nodes included the RF address buses, the RF word and write bit lines, the source operand buses, the PC-unit adder inputs, and the control-unit outputs. These high-fan-out nodes were clock powered from E-R latches. In some cases, the pipeline structure was affected. For example, to make it possible to recover the energy from the high-capacitance RF word lines, its decoders and its array operate in different phases. Fortunately, the extra pipeline latency did not decrease the instruction throughput because of other factors related to the instruction decoding (discussed below).

The resulting E-R-latch placement for the datapath is shown in Fig. 7. The only high-fan-out nodes that were not clock powered were in the two bypass buses. These nodes were not clock powered because of pipeline constraints, but could have been if the function units had been designed to operate in one phase.

The pipeline contains five stages (Fig. 8): FETCH (instruction fetch), DEC (instruction decode), EXE (instruction execute), MEM (data-memory access), and WB (write back). The main difference between this pipeline and that of a simple single-instruction-issue microprocessor is that the RF access is delayed until after the instruction opcode has been decoded. This delay is necessary because of the dense instruction encoding. Some instructions implicitly define their operand registers, which requires the instruction to be decoded before

the operand register addresses can be formed. A benefit is that RF read operations occur only when necessary, i.e., when the instruction uses data from the RF which would not be bypassed from the EXE or MEM pipeline stage. The drawback is that one extra clock cycle of latency is introduced after conditional and register-indexed branch instructions. To reduce the latency penalty of these instructions, the first instruction of the fall-through branch path is fetched and is input to the pipeline. If the branch path is taken, the fall-through instruction is squashed in a later pipeline stage before it alters the program semantics.

B. Circuit-Level Design

The logic circuits are predominantly of the precharged [Fig. 9(a)] and pass transistor [Fig. 9(b)] styles. Both design styles work straightforwardly with clock-powered signals. The E-R-latch outputs are valid during one clock phase and low during the other phase. Therefore, they can drive gates that are precharged during the other phase. Precharged gates are arranged in domino style; the outputs of the final stage are stored in E-R latches. Precharged gates driven by E-R latches do not need protection nFET's in their pull-down stacks since the input signals are low during precharging. The precharged gates and the E-R-latch inverters are powered from the same dc supply (V_{dd}). Precharging with pFET's is problematic because the blip circuit does not provide the clock complements for output. To solve this problem, we exploited the higher-than-

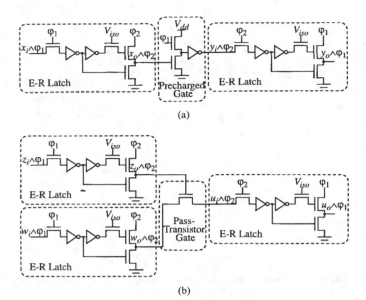

(a)

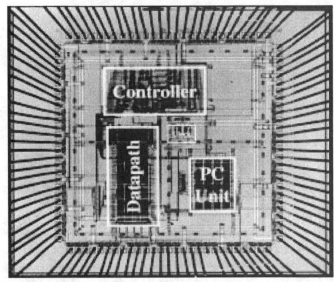

Fig. 9. E-R latches used with (a) precharged logic and (b) pass-transistor logic.

(b)

Fig. 10. Chip micrograph. Note: this is a photo of a 0.8-μm part. We used the same layout for the 0.5-μm part. Only the pad frame was changed. Core size is 2.63 mm × 2.63 mm for the 0.5-μm part; chip size is 3.88 mm × 3.88 mm for both parts; the core contains 12.7k transistors.

V_{dd} voltage swing of the clock phases to drive nFET pull-up transistors for the precharged gates.

An advantage of the E-R-latch design is that it converts the logic-voltage levels into clock-voltage levels. We used this feature in pass-gate circuitry by having clock-powered signals drive the pass-transistor gates while the logic-level signals are steered through these transistors. The higher voltage swing of the clock-powered signals compared to the voltage swing of the logic signals allowed passing the logic-level signals at their full swing. Furthermore, some energy along the pass-transistor path could be recovered, although HSPICE simulations indicate that most of the injected energy would be trapped in the path.

For low-power operation of the E-R latch, the rise and fall times of the clocks must be longer than the practically obtainable minimum transition times. The consequence of stretching the rise time is that, within a clock cycle, the logic will activate later than it would from a minimal-transition-time input signal. A consequence of stretching the fall time is that the input cutoff voltage for the pass-gate section of an E-R latch will occur earlier in the clock cycle. The net result is that for a fixed cycle time, the amount of computation that can be done during a phase is traded off for power dissipation. Serendipitously, the slight overlap in the clock phases produced by the all-resonant blip circuit serves to mitigate the time lost in the cycle due to the longer transition times.

The chip (Fig. 10) includes four major blocks: the clock driver, the datapath, the control unit, and the PC unit. It was fabricated in the Hewlett-Packard CMOS14B process which is a 0.5-μm, 3.3-V, 3-metal-layer, n-well CMOS process offered through MOSIS. The chip was custom designed with Magic and simulated with IRSIM and PowerMill. Small circuit sections were simulated with HSPICE. The core size is 2.63 × 2.63 mm. It contains 12.7k transistors. Approximately 17% of the transistors are pFET's which are located in the RF cells

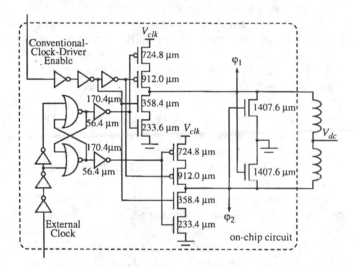

Fig. 11. Clock-driver schematic.

and the inverters of the precharged gates and the E-R latches. The chip was packaged in a 108-pin PGA.

For experimentation purposes, the AC-1 chip contains two independent clock circuits: a conventional NOR-based two-phase generator and a resonant clock driver (Fig. 11). When enabled by an external control signal, the clock driver generates the two nonoverlapping phases from an external clock source. A separate dc supply V_{clk} powers the conventional clock driver for measurement purposes. In resonant mode, we disable the conventional clock driver and externally attach two inductors to the clock lines to generate the two phases. The resonant driver (i.e., the blip circuit) nFET's are on chip.

Clock phases are distributed inside the chip through a global clock grid. To minimize clock skew, we placed the conventional clock driver close to the center of the grid. The calculated resistance through the clock grid is less than 4 Ω. Each of the two large transistors of the blip circuit was

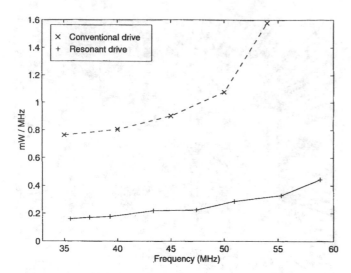

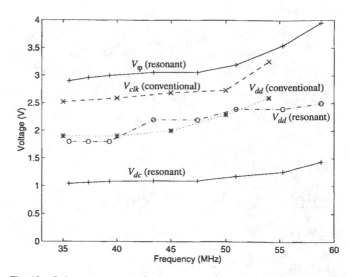

Fig. 12. Lab measurements of AC-1 combined clock and core energy dissipation (mW/MHz) per clock cycle as function of frequency. Supply voltages are set to their minimum values for each frequency point (see Fig. 13).

Fig. 13. Lab measurements of minimum supply and clock voltages required for correct operation. V_φ is the clock swing generated from V_{dc}.

partitioned into 153 smaller transistors which were connected in parallel around the clock grid.

The 16-b datapath contains the RF decoders and array, the bypass multiplexors, the function units, and several pipeline registers which were implemented with E-R latches. The RF cell is based on the six-transistor SRAM cell with two read ports and one write port. The RF decoders consist of precharged NAND gates. All function units share common, clock-powered, dual-rail source-operand buses. The ALU adder is a carry-select design with four 4-b stages. The logarithmic-stage shifter is based on pass-transistor logic. The compare unit uses the ALU adder and includes a precharged equal/nonequal comparator.

The control unit includes several AND–OR domino PLA's and random logic. Its output latches are clock powered, as are the internal pipeline latches that drive high-capacitance nodes. Pipeline latches that only drive the latches of the next phase are implemented with three-transistor dynamic latches. The PC unit contains two adders of the same design as that used for the ALU.

C. Lab Results

The chip was tested on a wire-wrap board which included 7-ns SRAM chips for an external instruction store. We downloaded small test programs and measured power for both resonant and conventional modes while the programs executed. We observed no significant dissipation difference among the different programs. The blip circuit was set up in its free-running configuration in which the clock-cycle period varied by several percent from cycle to cycle. The SRAM chips were sufficiently fast to work correctly for the shortest attainable clock cycle in the test setup.

To test for correct program execution, the external address bus was permanently enabled. This bus carries the function-unit result of the instruction that is in the *MEM* pipeline stage. We recorded the outputs of the function units by storing them in the memory buffer of a 16-channel storage oscilloscope

while each test program executed. We then verified that the recorded data sequences were correct.

Our power measurement procedure was, first, to find the top operating frequency for the maximum clock voltage allowed by the technology (i.e., 3.3 V). For each lower frequency point, the power-minimization method was to find the lowest voltages for the two dc supplies for which the output was correct. We gave precedence to reducing the clock voltage over reducing the core voltage since the clock lines drive most of the switching capacitance. Therefore, the clock voltage was first reduced until the minimum voltage was reached for which the outputs were correct. The same procedure was carried out for the core voltage.[1] The power was then measured. The results are plotted in Figs. 12 and 13.

In resonant mode, we varied the frequency from 35.5 to 58.8 MHz by connecting external inductors that ranged from 290 down to 99 nH. The voltages for increasing frequencies ranged from 1.0 to 1.4 V for V_{dc} (the blip-circuit supply), which corresponded to a resonant-clock voltage swing (V_φ) from 2.9 to 4.0 V, and 1.8 to 2.5 V for V_{dd} (the core supply). The combined power dissipation ranged from 5.7 to 26.2 mW.

Under conventional drive, we adjusted the external clock frequency from 35 to 54 MHz and repeated the power measurement procedure. The voltages for increasing frequencies ranged from 2.5 to 3.3 V for the supply voltage of the conventional clock driver (V_{clk}) and from 1.9 to 2.6 V for V_{dd}. The combined power dissipation ranged from 26.7 to 85.3 mW.

The results show that in resonant mode, the dissipation is a factor of four to five less than conventional mode. The clock power is approximately 90% of the total power under conventional drive and 60–70% under resonant drive. The core supply V_{dd} is about the same for both resonant and conventional modes. Since clock power dominated the dissipation, we adjusted the clock voltage by 10-mV steps and the core voltage by 100-mV steps. The 100-mV steps

---

[1] Although varying the voltages individually may not find the overall global minimum which maintains correct outputs, we believe it will be quite close.

caused the core voltage not to change for some successive measurements (Fig. 13).

IV. SIMULATION COMPARISON

The laboratory experiments have validated that AC-1 can exploit energy recovery at levels of practical significance: based on the power consumed by the clock drivers, an average of 85% of the energy injected into the clock nets was recovered by the resonant clock driver and reused. The next level of evaluation is to determine how well energy recovery compares to a fully static approach which relies on voltage scaling for low-power operation. The overall power savings from the lab measurements indicate an encouraging trend; however, AC-1 was specifically designed for clock-powered logic. Thus, there is an obvious bias against the conventional clock-drive mode.

Direct comparisons with commercially available microprocessor cores are possible, but fraught with difficulty. Even if we restrict ourselves to 16-b architectures, the instruction sets, the semiconductor processes, and the benchmarks are likely to differ in the different cases. Some of these difficulties can be resolved (for example, by using industry-standard benchmarks such as Dhrystone), but the information to be gleaned from such comparisons is still rather limited.

We chose instead to compare AC-1 to a complete reimplementation of the same ISA using a reasonable, industrial-style approach. This design, called AC-1/c, was developed independently from AC-1, by a different designer, but at approximately the same level of architectural ambition. For example, the pipelines are very similar, both designs use register files with two read ports and one write port, and the bypass logic is of similar sophistication.

As mentioned earlier, AC-1 was a custom layout done with Magic. For AC-1/c, the approach was different: the layout was generated with a commercial place-and-route system (the Epoch System from Cascade Design Automation). This reference design uses static-CMOS libraries included with the layout system, with no modifications. The implementation allows the clock signal to be gated away from unused blocks on an instruction-by-instruction basis. According to simulation, this feature saves approximately 40% of the processor core's dissipation [11].

In terms of transistor count, AC-1/c is bigger than AC-1, comprising 30k devices as compared with 13k. Precharged and especially pass-transistor logic uses fewer devices than static logic for the same functionality. In terms of area, the AC-1/c core is considerably smaller. Both designs are I/O-limited. For AC-1, no effort was spent on compacting its layout; for AC-1/c, this task was handled well by the place-and-route system. In terms of capacitance, the total extracted capacitance for all the switching nodes in AC-1/c was 500 pF. In contrast, the corresponding capacitance for AC-1 was 260 pF.

AC-1/c has not yet been fabricated. However, we have generated a complete layout of the core (Fig. 14), using the same design rules used for AC-1. Circuit extraction (using Magic) and simulation (with PowerMill) allow us to predict the AC-1/c dissipation for small benchmark programs. The laboratory results for AC-1 are not directly comparable to these

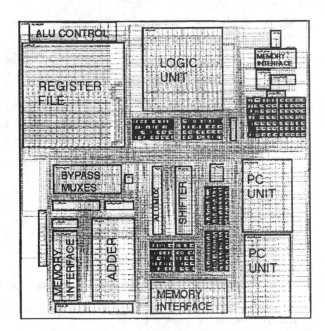

Fig. 14. Layout checkplot for AC-1/c core. Size is 1.59 mm × 1.56 mm; transistor count is 30k.

simulations, since the laboratory setup includes parasitic packaging capacitances which are not modeled in the simulations. For comparison, we carried out the corresponding extraction and simulations for AC-1 as well. The power-minimization procedure was the same as that used for laboratory measurements: for each frequency, we found the lowest voltages at which the designs would execute the test programs correctly and measured the corresponding power level.

Figs. 15 and 16 show the simulation results. AC-1 was only simulated in conventional-clock mode since problems were encountered with simulation accuracy for the resonant mode. The third curve in Fig. 15 shows the simulated AC-1 dissipation with the clock power prorated by a factor of 6.5. The factor is based on the lab-measured energy-recovery performance of AC-1 in resonant mode. At 50 MHz, AC-1 dissipation in conventional mode is 2.4 times higher than that of AC-1/c. In energy-recovery mode, the projected dissipation of AC-1 is lower by a factor of 1.9.

V. SUMMARY

In this paper, we have presented a new approach to low-power CMOS microprocessors and have described a prototype implementation. The approach is based on using the clock lines as a source of ac power for the high-fan-out circuit nodes.

The rationale that we used for applying clock power was rudimentary. The goal of the experiment was to produce and measure the energy-recovery effect at a significant level of circuit complexity. The result of the experiment was that clock-powered logic, when used with an energy-efficient clock driver, can make a dramatic difference in the *net* power dissipation. The overall, lab-measured net power reduction was a factor of 4.5 in resonant mode versus conventional mode for the prototype chips. From simulation and laboratory measurements for conventional mode, 90% of the power

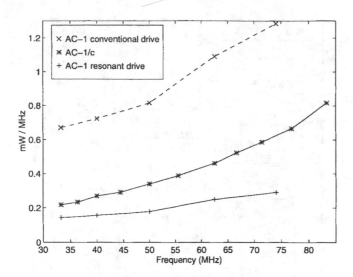

Fig. 15. Simulated energy dissipation (mW/MHz) per clock cycle as a function of frequency. Results were obtained with PowerMill. Values for AC-1 under resonant drive are compensated for the projected energy-recovery efficiency. Supply voltages are shown in Fig. 16.

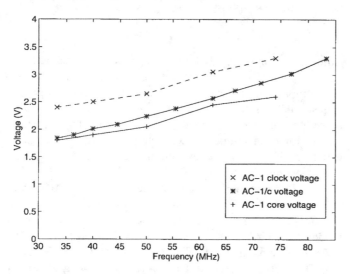

Fig. 16. Minimum supply and clock voltages required for correct PowerMill simulation of AC-1 and AC-1/c.

dissipation was due to the 10% of the nodes that were clock powered.

The simulations of AC-1 and AC-1/c, its fully static, conventional counterpart, offer a fair comparison in that the microprocessors both implement the same ISA, are based on the same CMOS process, were evaluated using the same extraction and simulation tools, and share similar microarchitectures. The simulation results indicate that naïvely the cost of clock-powered logic is high, but with the high efficiency of the blip circuit taken into consideration, AC-1 power should be approximately 50% that of AC-1/c.

This result is encouraging, though far from conclusive. AC-1 and AC-1/c were designed quite differently. AC-1/c benefited from the technique of gating away the clock to unused function blocks, extensive supply-voltage scaling, and area and buffer sizing optimization performed by the CAD synthesis tool. The influence upon low power of the standard-cell library and the implementation choices made by the synthesis tool is unclear. In contrast, AC-1 did not use gated clocks, and no optimization was carried out for area and buffer sizing. Also, AC-1 demonstrated less supply-voltage scalability in simulation. Because it was a custom design for low power, dissipation at the logic-block level was a principal consideration of the designers.

Theoretically, it should be possible to reduce power dissipation with energy recovery. The AC-1 experiment has demonstrated that energy recovery is practically feasible. The initial comparison to supply-voltage scaling at the small-system level is favorable. Further research into both approaches will determine the extent to which the relative strengths and weaknesses of the two will alter these initial findings.

ACKNOWLEDGMENT

The authors wish to thank H. "Deadman" Li, P. Wang, and X. Jiang for their important contributions to the physical design of the AC-1 microprocessor, A. Ekelund for his vital work in the design and implementation of AC-1/c, J. Jones of Synopys, Inc. and S. Sugiyama of Cascade Design Automation.

REFERENCES

[1] R. P. Feynmann, *Feynmann Lectures on Computation*, A. J. G. Hey and R. W. Allen, Eds. Reading, MA: Addison-Wesley, 1996.
[2] W. C. Athas, "Low-power design methodologies," in *Energy-Recovery CMOS*, J. Rabaey and M. Pedram, Eds. Norwell, MA: Kluwer, 1996, pp. 65–100.
[3] B. G. Watkins, "A low-power multiphase circuit technique," *IEEE J. Solid-State Circuits*, vol. SC-2, p. 215, Dec. 1967.
[4] C. L. Seitz, A. H. Frey, S. Mattisson, S. D. Rabin, D. A. Speck, and J. L. A. van de Snepscheut, "Hot-clock NMOS," in *Proc. 1985 Chapel Hill Conf. VLSI*, pp. 1–17.
[5] L. Svensson, "Low power digital CMOS design," in *Adiabatic Switching*, A. Chandrakasan and R. Brodersen, Eds. Norwell, MA: Kluwer, 1996, pp. 181–217.
[6] W. C. Athas, L. J. Svensson, and N. Tzartzanis, "A resonant signal driver for two-phase, almost-nonoverlapping clocks," in *Proc. 1996 Int. Symp. Circuits and Systems*, Atlanta, GA, May 12–15, 1996, vol. 4, pp. 129–132.
[7] A. Rofougaran, J. Rael, M. Rofougaran, and A. Abidi, "A 900 MHz CMOS LC-oscillator with quadrature outputs," in *ISSCC Dig. Tech. Papers*, San Francisco, CA, Feb. 1996, pp. 392–393.
[8] L. A. Glasser and D. W. Dobberpuhl, in *The Design and Analysis of VLSI Circuits*. Reading, MA: Addison-Wesley, 1985.
[9] D. Patterson and H. Hennessy, *Computer Architecture: A Quantitative Approach*. San Mateo, CA: Morgan Kaufman, 1989.
[10] J. D. Bunda, "Instruction-processing optimization techniques for VLSI microprocessors," Ph.D. dissertation, The University of Texas at Austin, TX, 1993.
[11] A. M. Ekelund, "Low power microprocessor design," Master's thesis, Lund Institute of Technology, Mar. 1997.

Clocked CMOS Adiabatic Logic with Integrated Single-Phase Power-Clock Supply: Experimental Results

Dragan Maksimović[1], Vojin G. Oklobdžija[2], Borivoje Nikolić[2], K. Wayne Current[2]

[1]Department of Electrical and Computer Engineering, University of Colorado, Boulder, CO 80309-0425

[2]Department of Electrical and Computer Engineering, University of California, Davis, CA 95616

ABSTRACT

In this paper we describe the design and experimental evaluation of a clocked CMOS adiabatic logic (CAL). CAL is a dual-rail logic that operates from a single-phase AC power-clock supply in the 'adiabatic' mode, or from a DC power supply in the 'non-adiabatic' mode. In the adiabatic mode, the power-clock supply waveform is generated using an on-chip switching transistor and a small external inductor between the chip and a low-voltage DC supply. Circuit operation and performance are evaluated using a chain of inverters realized in 1.2μm technology. Experimental results show energy savings in the adiabatic mode versus the non-adiabatic mode at clock frequencies up to about 40MHz.

1 INTRODUCTION

The potential for energy savings using energy-recovery (or 'adiabatic') circuits has been examined using various circuit implementations [1-11]. Weaknesses of the previously proposed approaches include the need for multi-phase AC power-clock supplies for proper interfacing between stages, and correspondingly high complexity of both the logic and the required power-clock generator [1-9].

This paper describes results of experimental evaluation of the clocked adiabatic logic (CAL) [10,11] operated from a single-phase power-clock generator integrated with logic [11]. The measurements are focused on verification of operation of the logic over frequency and supply voltage ranges and a comparison of CAL energy consumption to the case when the logic is operated from a DC power supply.

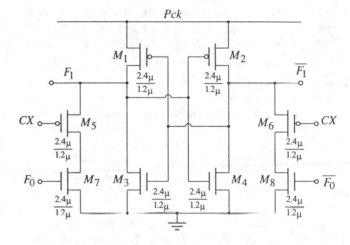

Figure 1: CAL inverter.

CAL circuit configuration and operation are reviewed in Section 2. Implementation issues and the test chip are discussed in Section 3. Measurement results are presented in Section 4. Section 5 concludes the paper.

2 CIRCUIT OPERATION

The basic CAL gate, the inverter, is shown in Fig. 1. Cross-coupled CMOS inverters, transistors M_1-M_4, provide the memory function. In general, the devices M_7 and M_8 can be replaced with NMOS logic trees to perform switching involved in the evaluation of an arbitrary binary function. As an example, implementation of a 2:1 MUX stage is shown in Fig. 2. The CAL topology is similar to the logic proposed by Denker [6]. Our innovation is the inclusion of path control switches, devices M_5 and M_6, added in series with the logic trees. This modification allows operation of the circuit with a single-phase power-clock supply, Pck, as opposed to the four-phase power clock required by the logic proposed in [6].

Reprinted with permission from *Proceedings of 1997 International Symposium on Low Power Electronics and Design,*, D. Maksimovic, V. G. Oklobdzija, B. Nikolic, and K. W. Current, "Clocked CMOS Adiabatic Logic with Integrated Single-Phase Power-Clock Supply: Experimental Results," pp. 323 - 327, 1997. © 1997 ACM.

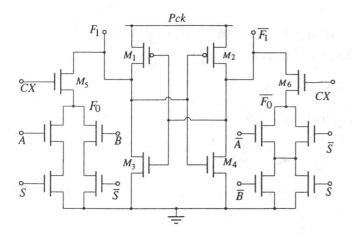

Figure 2: 2:1 MUX implemented with CAL.

Idealized CAL timing waveforms for the inverter are shown in Fig. 3. The power clock Pck is shown as a trapezoidal waveform. Logic evaluation is enabled by the auxiliary clock CX. For $F_0 = 0$, M_7 is off, M_8 is on, the complementary output $\overline{F_1}$ goes to 0, and the true output F_1 follows the power-clock waveform. In the next clock period, the auxiliary clock $CX = 0$ disables the logic evaluation and the outputs repeat the result stored during the evaluation in the previous clock period. As a result, the CAL logic states are represented by presence or absence of a pair of pulses. When controlling a chain of logic stages, the same power clock Pck supplies all CAL stages. The logic evaluation is enabled in alternate logic stages by the auxiliary clock CX and its complement $\overline{CX}$. To reduce power consumption in the auxiliary clock distribution, the swing of the auxiliary clocks can be reduced without affecting the signal levels in the logic. It is interesting to note that CAL can also be operated as a conventional clocked logic with a DC power supply connected to Pck.

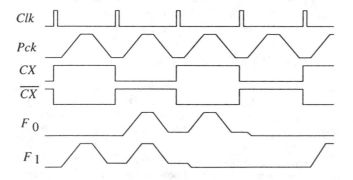

Figure 3: Idealized CAL timing waveforms.

3 IMPLEMENTATION

Fig. 4 is a block diagram of the experimental CAL chip built as a chain of $n = 736$ dual-rail inverters in 1.2μm CMOS technology. Fig. 5 shows the test chip die photo.

The device sizes in the CAL logic stages are indicated in Fig. 1 and Fig. 4. For testing purposes, twelve of the inverters have both outputs connected to the output pins, via conventional output buffers that serve as voltage comparators. The conventional output buffers and the circuits used to generate the auxiliary clocks are supplied from a separate DC supply V_{DD}. The inductor L and the low-voltage DC source V_B are used for energy-recovery (adiabatic) operation of the CAL chip. The AC power-clock waveform Pck is generated using a single NMOS device Q in parallel with the CAL logic. The device Q is turned on during a small fraction of the clock period at the point when Pck is approximately zero. During this time, energy is added to sustain oscillation in the resonant circuit formed by the external inductance L and the equivalent logic capacitance C_{eq}. When Q is clocked close to the resonant frequency, Pck swings between 0 and a peak value approximately equal to $2 \cdot V_B$ [11]. The switching transistor Q takes a small fraction of the total chip area, as shown in Fig. 5, where Q is the area immediately to the left from the CAL label.

Given a desired power-clock frequency f, the required inductance L can be found from

$$f = \frac{1}{2\pi\sqrt{LC_{eq}}},$$

where $C_{eq} = 101\text{pF}$ is the measured equivalent chip capacitance at the Pck node. The equivalent chip capacitance was measured by connecting a resistor R between V_B and Pck and by measuring the time constant RC_{eq} of the charge-up transient after the device Q is turned off. The logic can also be operated from a DC supply connected directly to Pck, while the power-clock device Q is disabled. This non-adiabatic mode of operation is used to evaluate how much energy can be recovered by the power clock.

4 EXPERIMENTAL RESULTS

Operation of the circuit is illustrated by the waveforms shown in Fig. 6 for an input sequence of 1100 and two power-clock frequencies. The pair of quasi-sinusoidal pulses that corresponds to the logic high output is observed as a pair of rectangular pulses through the conventional, DC-supplied output buffers. The maximum operating frequency of the circuit in this test configuration was 50MHz, due to signal source limitation. Higher operating frequencies are predicted by simulation [10].

Energy consumption of the CAL chip (excluding the consumption of the output buffers) was measured as a function of frequency for three cases: (1) for adiabatic operation when 3V peak-to-peak sinusoidal Pck was supplied from an external function generator and the on-chip power-clock device Q was disabled; (2) for adiabatic operation with quasi-sinusoidal Pck generated using $V_B = 1.5$V and an external L, and by clocking Q as shown in Fig.3; and (3) for non-adiabatic operation with the minimum DC supply voltage $V_B = 2.5$V for which the DC-

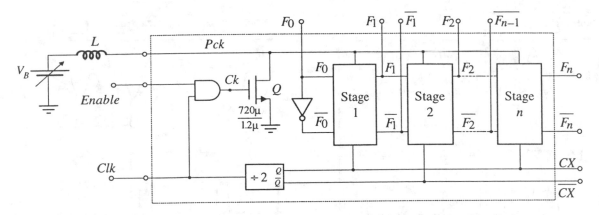

Figure 4: A chain of CAL logic gates with on-chip power-clock generation. For adiabatic operation, *Enable* = 1, and inductor *L* is connected between *Pck* and the DC supply V_B; for non-adiabatic operation, *Enable* = 0 and *Pck* = V_B.

supplied logic was found to function properly. In all three cases, the activity factor (defined as the normalized number of input transitions) is equal to 1. The results are shown in Figs. 7 and 8. The non-adiabatic energy consumption is approximately constant at about 0.58pJ per inverter per cycle. The theoretical non-adiabatic energy consumption, based on the measured C_{eq} = 101pF is 0.61pJ per inverter per cycle. The small difference comes from the fact that C_{eq} includes capacitance of the *Pck* distribution, which is not switched during non-adiabatic operation. These results show that the energy consumption of the CAL operated from the DC supply is indeed equal to CV^2 losses. Therefore, the results of Fig. 7 show how much of the CV^2 can be recovered through the power clock during adiabatic operation of the chip with external or internal power clock. Energy savings are very large at low operating frequencies and diminish as the frequency approaches f = 30MHz. The results with the externally supplied power-clock waveform are significantly better than the results with the internally generated *Pck*. We believe that significantly better performance of the internal power-clock generator can be obtained by increasing the size and reducing the on-resistance of the switching transistor Q.

By changing the activity factor at a constant power-clock frequency it was found that the CAL power consumption at the activity factor equal to 0 is approximately one half of the power consumption at the activity factor equal to 1, as shown in Fig. 8. This confirms that at low activity factors non-adiabatic operation from a DC supply can be significantly more efficient.

The CAL ability to operate from either a single-phase AC power-clock supply or from a DC supply opens interesting possibilities to combine adiabatic and non-adiabatic modes of operation to achieve energy-efficient operation for a very wide range of throughput rates and activity factors.

The energy consumption measurements shown in Figs. 7 and 8 included only the part of the circuit operated from the supply V_B: the logic and the switching transistor Q. The measured energy consumption per cycle of the circuits supplied from the constant DC voltage V_{DD} = 3V (auxiliary clocks and the driver for the switching transistor Q) is about 30pJ, or about 8% of the total non-adiabatic energy consumption.

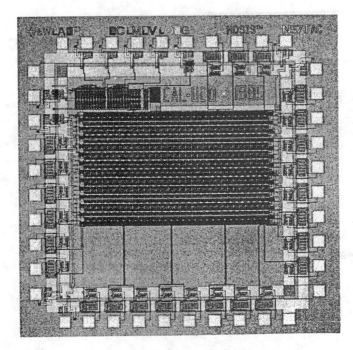

Figure 5: Test chip die photo.

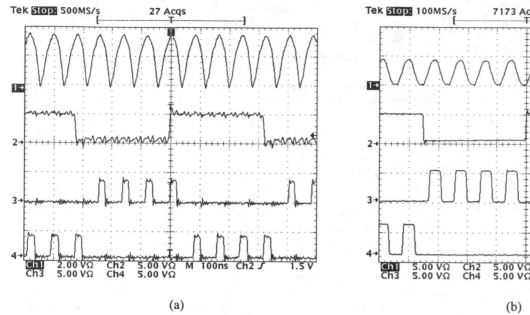

(a)

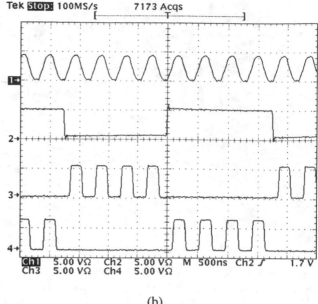

(b)

Figure 6: Measured CAL waveforms: Ch1: power clock *Pck*, Ch2: logic input F_0, Ch3: buffered logic output F_1, Ch4: buffered complementary logic output $\overline{F_1}$ for (a) $f = 12.5$MHz, the logic is supplied from $V_B = 1.8$V, while the DC supply for the conventional output buffers is $V_{DD} = 4$V and (b) $f = 2.36$MHz, the logic is supplied from $V_B = 1.5$V, the DC supply for the output buffers is $V_{DD} = 5$V.

5 CONCLUSION

The proposed clocked adiabatic logic (CAL) operates from a single-phase power-clock supply. The test chip, a chain of inverters, is implemented in a 1.2µm CMOS technology. Operation of the logic and its energy consumption are measured for adiabatic operation using an external power-clock generator or using a simple on-chip power-clock generator. These results are compared to the energy consumption measured in non-adiabatic operation when the chip is supplied from a DC voltage source. Experimental results show significant energy savings at relatively low clock rates. The CAL ability to operate from either AC power-clock supply or from a conventional DC supply opens further possibilities for energy-efficient operation in a very wide range of throughput rates by combining adiabatic and non-adiabatic modes of operation.

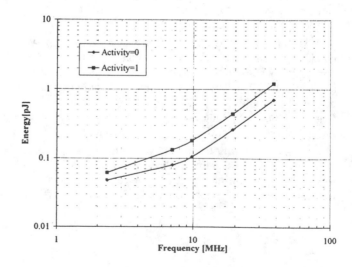

Figure 7: Energy/inverter per cycle vs. frequency.

Figure 8: Energy/inverter vs. frequency for different activity factors.

258

ACKNOWLEDGMENTS

The authors would like to thank Fengchen Lin and Olivier Greneche for their help in layout and testing of the CAL chip.

REFERENCES

[1] A.G. Dickinson, J.S. Denker, "Adiabatic Dynamic Logic," *IEEE Journal of Solid-State Circuits*, vol. 30, no. 3, pp. 311-315, March 1995.

[2] T. Gabara, "Pulsed Low Power CMOS," *International Journal of High Speed Electronics and Systems*, vol. 5, no. 2, pp. 159-177, June 1994.

[3] J.G. Koller, W.C. Athas, "Adiabatic Switching, Low Energy Computing and Physics of Storing and Reading of Information," *Proceedings PhysComp* '92, Dallas, Texas, Oct. 2-4, 1992.

[4] W.C. Athas, L.J. Svensson, J.G. Koller, N. Tzartzanis, E. Y.-C. Chou, "Low-Power Digital Systems Based on Adiabatic-Switching Principles," *IEEE Transactions on VLSI Systems*, vol. 2, no. 4, December 1994.

[5] S.G. Younis, T. Knight, "Practical Implementation of Charge Recovering Asymptotically Zero Power CMOS," *Proceedings of 1993 Symposium on Integrated Systems*, pp. 234-250, MIT Press, 1993.

[6] J.S. Denker, "A Review of Adiabatic Computing," *Proceedings of the 1994 Symposium on Low Power Electronics*, San Diego, October 1994.

[7] A. Kramer, J.S. Denker, S.C. Avery, A.G. Dickinson, T.R. Wik, "Adiabatic Computing with the 2N-2N2D Logic Family," *IEEE Symposium on VLSI Circuits, Digest of Technical Papers*, pp. 25-26, 1994.

[8] R.T. Hinman, M.F. Schlecht, "Recovered Energy Logic - a Highly Efficient Alternative to Today's Logic Circuits," *IEEE Power Electronics Specialists Conference*, 1993.

[9] Y. Moon, D-K. Jeong, "An Efficient Charge Recovery Logic Circuit", *IEEE Journal of Solid-State Circuits*, vol. 31, no. 4, pp. 514-522, April 1996.

[10] D. Maksimović, V.G. Oklobdžija, "Clocked CMOS Adiabatic Logic with Single AC Power Supply," 21st *European Solid State Circuits Conference, ESSCIRC'95*, Lille, France, September 1995.

[11] D. Maksimović, V.G. Oklobdžija, "Integrated Power Clock Generators for Low-Energy Logic," 26th *Annual IEEE Power Electronics Specialists Conference*, Atlanta, June 1995.

Chapter 4

Clock Subsystem

Clock Subsystem

Proper timing and clock system design are two of the most critical components of digital systems, as has been summarized by Professor Steven Unger: "Despite the deceptively simple outward appearance of the clocking system, it is often a source of considerable trouble in actual systems." This chapter is dedicated to the subject of clocking in digital systems.

The function of the clock in the digital system can be compared to that of a metronome in music. The metronome designates the beginning of a musical score and the exact moment when certain notes are to be played by particular instruments in orchestra; designates the end of the part or section; that is, it provides synchronization for various instruments in the orchestra during various parts and periods of the score that is being performed. Similarly, in the digital system the clock designates the exact moment when the signal is to change as well as when its final value is to be captured, when the logic is active or inactive. Finally, all the logic operations have to finish before the tick of the clock and the final values of the signals are being captured at the tick of the clock. Therefore, the clock provides the time reference point, which determines the movement of data in the digital system. This definition fits the description of the synchronous systems, which will be the subject of this chapter.

Asynchronous and self-timed systems are not covered here, even though they have captured much research and academic interest for quite some time. Self-timed systems are attracting attention because of the increasing difficulty of controlling the clock skew and distribution as the operating frequency of today's systems keeps increasing, reaching 600 MHz or even 1 GHz and beyond, for the next generation of systems currently under development. The trend in clocking speed is shown in Fig. 4.1, indicating the exponential growth in clock frequency over the years.

Clock Signals

Clocks are defined as pulsed, synchronizing signals that provide the time reference for the movement of data in the synchronous digital system. The clocking in a digital system can be either single-phase, multiphase (usually two-phase), or edge-triggered, as illustrated in Fig. 4.2.

The dark rectangles in the figure represent the interval during which the bi-stable element samples its data input. Figure 4.2 shows the possible types of clocking techniques and corresponding general finite-state machine structures:

a. Single-phase clocking and single-phase latch machine (Fig. 4.2a).
b. Edge-triggered clocking and flip-flop machine (Fig. 4.2b).
c. Two-phase clocking and two-phase latch machine with single latch (Fig. 4.2c).
d. Two-phase clocking and two-phase latch machine with double latch (Fig. 4.2d).

In Fig. 4.2 c and d, W_j is the pulse width of the phase j and g_{ij} is the interphase gap from phase i to phase j; if $g_{ij} > 0 \Rightarrow$ two-phase, nonoverlapping, if $g_{ij} < 0 \Rightarrow$ two-phase, overlapping clocking scheme.

The multiphase design typically extends to three, but not more than four, nonoverlapping phases.

Multiphase clocking was used in the early dynamic MOS circuits from the very beginning of VLSI, as well as in the systems where this is dictated by the nature of the computation. In the mainframe computer systems such as IBM's, the two-phase clocking was almost exclusively used, and it was incorporated in the Level-Sensitive Scan Design (LSSD) discipline. The reason for that is that the two nonoverlapping clocks provide a most reliable and robust clocking system that fits well into the design for testability methodology that was incorporated in LSSD.

A single-phase clock was used in simpler systems with less stringent requirements such as those of LSSD. Some of the high-performance systems used single-phase clocking, (CRAY). Because of the increased demand for performance and increasing difficulties with handling clock-skews, the single-phase clock has seen its resurgence. This is due to the in-creasing penalties from the clock-skew, which does not scale down proportionally as the technology permits ever faster

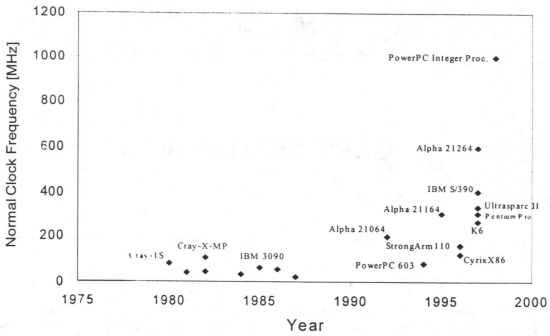

Fig. 4.1. The trend in Clock frequency over the years.

transistors. This means that the percentage of the cycle that is lost due to the clock-skew has increased to the point where it cannot be tolerated. Therefore, some sacrifice in reliability has to be made as a compromise to performance. Given that the machine cycle may take one or several clock cycles, as illustrated in Fig. 4.3 (especially in the micro-programmed systems), the overhead for the clock-skew directly influences the machine cycle and therefore the performance of the overall system.

Timing analysis strongly depends on the type of clocking used in the system. Single-phase systems and multiphase overlapping systems require more extensive timing analysis than multiphase nonoverlapping and edge-triggered systems. The single-phase and multiphase overlapping timing requirements are bounded by both *short* and *long paths*. This constraint is illustrated in Fig. 4.4 for the simplified case where *setup* and *hold* times are set to zero. The advantage of these systems over their nonoverlapping counterparts is their so-called cycle stealing feature, which effectively reduces the clock cycle and boosts the performance at the price of more difficult timing analysis and less robust operation.

Figure 4.4*a* shows the timing constraints for a single-phase system:

1. LS data are available at t_1.
2. LS data must arrive at LD after t_2 (or be latched up in Cycle 1 $\Rightarrow$ *short path*).
3. LS data arrive at LD by t_3 (or reduce the path length available in Cycle 2).
4. LS data must arrive at LD before t_4 (or be latched up in Cycle 3 $\Rightarrow$ *long path*)

Figure 4.4*b* presents the timing constraints for two-phase overlapping systems:

1. L_2S data are available at t_1.
2. L_2S data must arrive at L_2D after t_2 (or be latched up in Cycle 1 $\Rightarrow$ *short path*).
3. L_2S data must arrive at L_2D by t_4 (or violate the system cycle time requirement).
4. L_2S data must arrive at L_1D before t_5 (or be latched up in Cycle 3 $\Rightarrow$ *long path*).

Bi-stable Elements

In order to define the clocking of the system, the nature and behavior of the bi-stable element, often referred as a "latch" or flip-flop, need to be specified precisely. Quite commonly, both the terms flip-flop and latch are used indiscriminately for the bi-stable element seen in synchronous systems. Later in the text we will make a distinction between the flip-flop and latch.

Latch. Latch is a device capable of storing the value of the input D in conjunction with the clock C and providing it at its output Q. The latch has the following relationship between input D and clock C: While $C = 0$, output Q remains constant regardless of the value of D (the latch is "opaque"). While $C = 1$, the output $Q = D$, and it reflects all the changes of D (the latch is "transparent.") Often we describe the behavior of a latch as "level-sensitive." (The behavior of the latch is dependent on the value of the clock—not the changes.) The behavior of an ideal latch is illustrated in Fig. 4.5.

"Flip-Flop". Flip-flop is defined as a bi-stable memory element with the same inputs and outputs as a "latch" (Fig. 4.6). However, the output Q responds to the changes of D only at the moment that clock C is making transitions. We define this as being "edge triggered." The internal mechanisms of the flip-flop and latch are entirely different. We further define a flip-flop as a "leading edge triggered" if output Q assumes a value of input

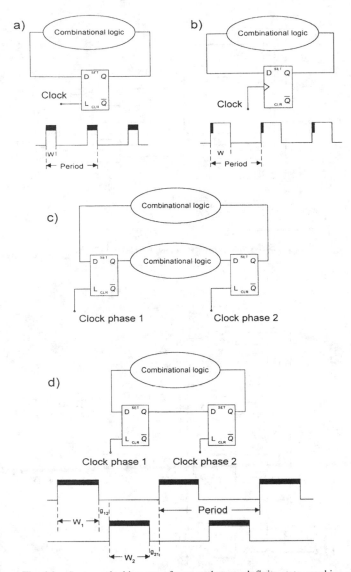

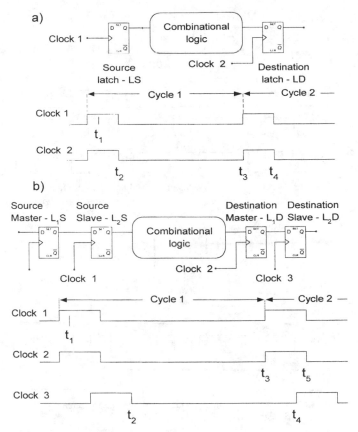

Fig. 4.4. Timing constraints in single-phase and two-phase overlapping clocking techniques [Wagner].

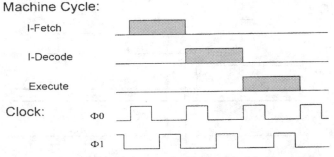

Fig. 4.2. System clocking waveforms and general finite-state machine structures [Wagner].

Machine Cycle:

Fig. 4.3. Relationship between the Clock and the Machine cycle.

D as a result of the transition of the clock *C* from 0 to 1. Conversely, in a "negative edge-triggered" flip-flop, output *Q* assumes a value of input *D* as a result of the transition of the clock *C* from 1 to 0. It is also possible to build a "double-edge triggered" flip-flop that responds to both the leading and trailing edge of the clock *C*. Such flip-flop implementations, first published in 1981 by Unger, are starting to gain attention given the increasing demand for performance.

Difference between a Master-Slave Latch and a Flip-Flop. The difference between a "master-slave latch" and a "flip-flop" is often not understood, and there is a widespread misconception in which a *master-slave latch* combination is referred to as a *flip-flop*. It is important to understand a fundamental difference between a *master-slave latch* and a *flip-flop* which exists in their latching (triggering) mechanisms. In a master-slave latch a latching mechanism occurs as a result of the clock reaching a logic value of one (level) allowing the data to be locked. This mechanism is fundamentally different in a flip-flop. What causes the data to be latched in a flip-flop is not the value (level) of the clock but the transition from 0 to 1 (or 1 to 0—negative edge triggered flip-flop). The rate of this transition causes the latching to occur in a *flip-flop*. This dependency of the transition rate, in order to make a reliable latching, is precisely why use of *flip-flops* represents a hazard. For the very same reason use of *flip-flop* has been forbidden in some design disciplines such as IBM's LSSD.

In order to picture the "latching" mechanism in flip-flop structures, we took the SN 7474 flip-flop as an example. A schematic diagram of a D-type flip-flop (SN 7474) is presented in Fig. 4.7, and the analysis of the "latching" mechanism in Figs. 4.8 and 4.9. It is a useful exercise to analyze what happened in this *flip-flop* when the clock changes from 0 to 1.

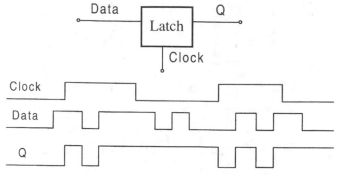

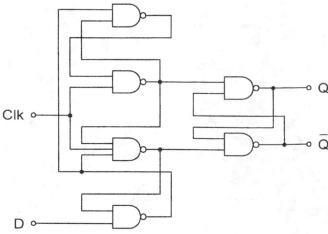

Fig. 4.5. Signal relationship of an ideal latch [Unger, Tan].

Fig. 4.7. SN 7474, leading-edge-triggered flip-flop.

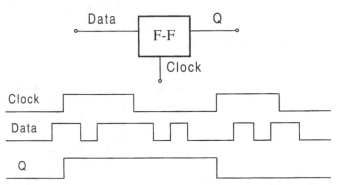

Fig. 4.6. Signal relationship of an ideal leading-edge-triggered flip-flop [Unger, Tan].

The whole scheme has two functional segments. The left part of the scheme (with output nodes $\overline{S}$ and $\overline{R}$, in Figs. 4.8 and 4.9) is multiplexing one of the two clocked NAND gates depending on the state of D. The state of D before the rising edge of the clock determines which clocked NAND gate is to be isolated. The output of the isolated NAND gate remains high. The output of the selected clocked NAND gate goes low at the rising edge of the clock, causing the regenerative process in S-R latch.

On the leading edge of the clock, depending on the state of input D, a flow of signals occurs and causes the latching to a new state. If we assume that the delay of all the NAND gates is the same, τ, and that the leading edge of the clock occurs at 0 time point, we will have changes in different nodes of the circuit in moments of τ, 2τ, and 3τ, as the signal progresses through the circuit. That is illustrated in Fig. 4.8 for a 0 to 1 transition of Q and in Fig. 4.9 for a 1 to 0 transition.

For the sake of clarity, only the first part of the clock cycle is shown in the diagrams. In the remaining part of the cycle, the low level on the Clk node forces both nodes $\overline{S}$ and $\overline{R}$ to a high level, making them ready for another cycle.

The key point in the edge-triggered behavior of this flip-flop is that once the clock has made a transition from 0 to 1, one of the nodes $\overline{S}$ or $\overline{R}$ (depending on the state of D before the leading edge of the clock) makes a $1 \rightarrow 0$ transition, thus disabling the further impact of D on the value of the nodes $\overline{S}$ and $\overline{R}$.

In Fig. 4.8, node $\overline{S}$ makes a transition $1 \rightarrow 0$, thus disabling the further changes of the nodes A and $\overline{R}$. In this way, input D is isolated and cannot influence the state of nodes $\overline{S}$ and $\overline{R}$ while the clock is 1.

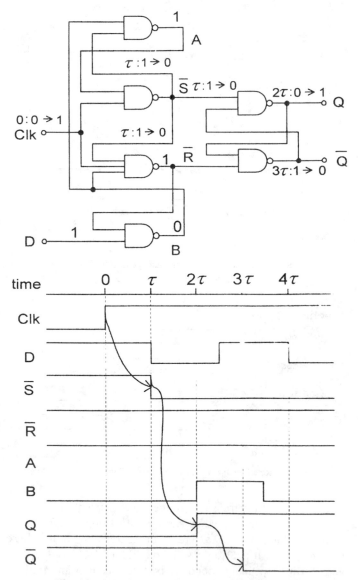

Fig. 4.8. SN 7474, Signal flow for **0** to **1** transition of Q.

264

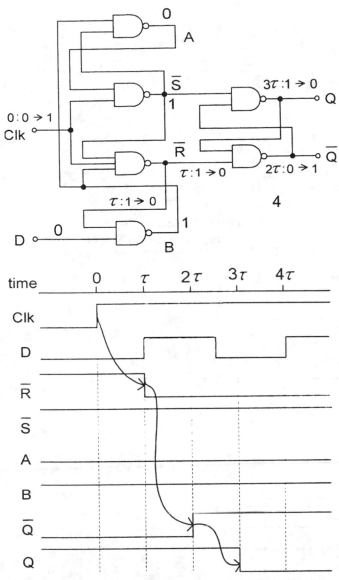

Fig. 4.9. SN 7474, Signal flow for **1** to **0** transition of Q.

In Fig. 4.9, node $\bar{R}$ makes a transition $1 \rightarrow 0$, thus disabling the further changes of the nodes B and $\bar{S}$. In this way, input D is isolated and cannot influence the state of nodes $\bar{S}$ and $\bar{R}$ while the clock is 1.

The issue of races in this structure should be mentioned here. Races are caused only if data input changes in the window either around the leading edge of the clock (for data: 0 to 1) or around one gate delay before the leading edge of the clock (for data: 1 to 0). The width of the critical window is determined by the parameters of the NAND gates in the first (multiplexing) stage of a flip-flop.

Figures 4.10 and 4.11 present a discussion of the races. If data change in the first part of the critical window, the change will have a chance to win the race. If data change in the second part of the window, the change will cause the race, but it will not win and will not be remembered. In the ideal case of the equal delays of the NAND gates N1 and N4 for race 01 and N2 and N3 for race 10, the first part and the last part of the window will be

equal. Any change in parameters of the NAND gates will cause one or the other part of the window to be bigger. The races are very hard to present on the logic level and in the way of changes in discrete time intervals of NAND gate delays because they occur between these discrete points. This is why we have chosen to present the signals in the fashion of windows.

For the case in Fig. 4.10, the race occurs in the N1-N4 loop between the signals in the nodes B and $\bar{R}$. Both nodes are initially high. If both D and clock go high in the time window of two gate delays, the positive-loop feedback is enabled and the race begins. The race ends when either B or $\bar{R}$ goes low, depending on the relative position of data and clock transition within the window.

A similar situation occurs in the N2-N3 loop for the transition of D from 1 to 0 within the critical window. The difference is that the window is shifted one delay earlier because signals B and Clk enable the positive-feedback loop. The race is between the signals on node A and $\bar{S}$.

Parameters of Bi-Stable Elements. Without intimate knowledge of the bi-stable element used in the system, it is not possible to establish any requirements on the clock. The clocking of a digital system and the choice and parameters of a bi-stable elements are closely interrelated. The clocking strategy often implies and determines the choice of the bi-stable element used in the system and vice versa. For the start we should establish some basic parameters:

Parameters presented in this chapter define the conventionally accepted timing parameters of bi-stable elements (Fig. 4.12). The inadequacies of this approach will be discussed later in this chapter.

Setup time U is defined as the minimum amount of time that the input signal D needs to be stable before the latching mechanism takes place in order to ensure reliable latching.

Hold time H is defined as the minimal amount of time data D needs to remain stable after the latching mechanism occurred in order to ensure reliable latching of data.

In addition, we define delay of the bi-stable element, which is added to the signal: In a case of flip-flop, this delay D_{CQ} is defined as a time difference between the time the triggering event C takes place (leading or trailing edge of the clock) and the time the value of the input D propagates to the output Q. (As stated by Unger and Tan, we can make this concept more precise by requiring that the change in D occurs sufficiently early that making it appear any earlier would have no effect on when Q changes.)

In case of a latch, two delays are to be defined:

1. *Delay D_{CQ}*: from the time clock C changes from 0 to 1, and the time output Q reflects the value of the signal at input D.
2. Delay D_{DQ}: from the time input D changes until the time output Q reflects this change (while clock $C = 1$ the latch being in *"transparent"* mode).

The *latch delay* is usually taken as the worse of the two D_{CQ} or D_{DQ}—whichever happens to be greater.

265

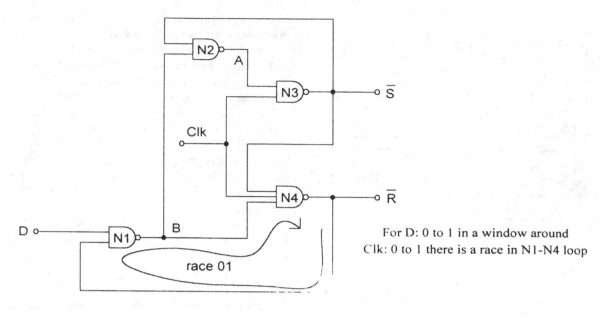

For D: 0 to 1 in a window around
Clk: 0 to 1 there is a race in N1-N4 loop

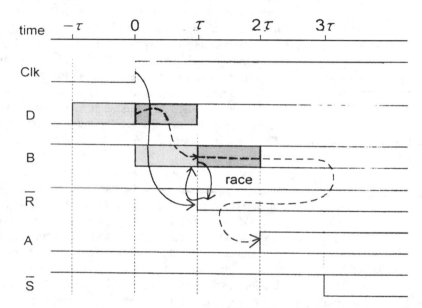

Fig. 4.10. SN7474, The race for 0 to 1 transition of D.

Design of the High-Performance Latch

The availability of a latch that can support requirements for high clock rates and a shallow pipeline is an essential factor in high-performance systems. A high-performance latch has to exhibit design flexibility, immunity to noise, and immunity to race-through. Given that the latch delay is directly subtracted from the time available to perform useful computation in the clock period, the latch has to introduce a minimal amount of delay in the critical path. Various latch designs were developed to satisfy this last requirement. One of the significant advancements made was the so-called Earle's Latch developed by Earle in 1965 and Halin and Flynn in 1972, during the course of design of IBM 360/91 mainframe computer. Earl's Latch is shown in Fig. 4.13.

Earl's Latch represents a logical AND-OR combination, which is a Sum-of-Products (SOP) form providing a general expression for representing any logic function that can be implemented with this latch. The delay of *Earl's Latch* is two gates, which is one gate delay less compared to the commonly used cross-coupled gate combination.

Minimizing latch delay has been the subject of work by Yuan and Svensson and later Afghahi and Svensson. They developed very fast latch and flip-flops better known as "True-Single-Phase-clock" (TSPC) Latch (shown in Fig. 4.14). The same latch (with some improvements) has been used in the first implementation of Digital's "Alpha" processor achieving a 200 MHz clock rate, as shown in Fig. 4.15. An interesting development of a hybrid latch flip-flop ele-

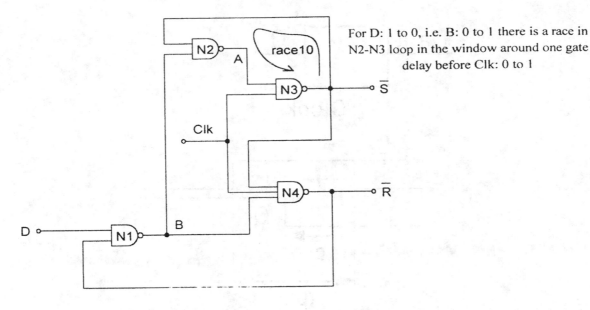

For D: 1 to 0, i.e. B: 0 to 1 there is a race in N2-N3 loop in the window around one gate delay before Clk: 0 to 1

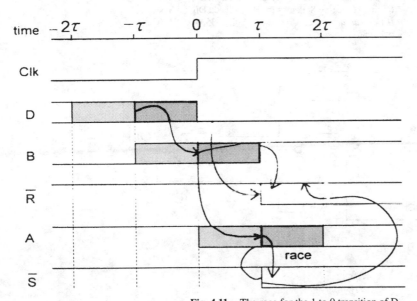

Fig. 4.11. The race for the 1 to 0 transition of D.

ment is presented by Partovi et al. Development and modifications of the clocking strategy, including the selection and development of an appropriate latch structure, is best illustrated in the papers describing Digital's "Alpha" processor development.

An attempt to reduce power dissipated by the clock subsystem is described in Kojima et al. and Kawaguchi and Sakurai which the clock signal swing is reduced. In the paper by Kawaguchi and Sakurai, a differential sense-amplifier structure is used for latching the signal. Such a latch, developed by Toshiba (see Fig. 4.16), was later used in Digtal's 21264 "Alpha" processor, which runs at a 600 MHz clock rate.

Finally, the synthesis of the clock tree in order to reduce the clock skews is described by Minami and Takano.

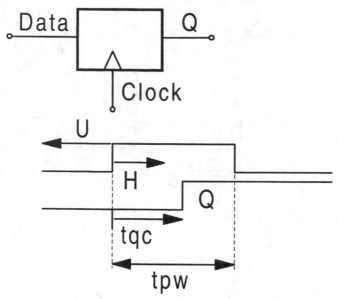

Fig. 4.12. Parameters of a bi-stable element. Based on S. H. Unger, C. Tan, "Clocking Schemes for High-Speed Digital Systems," *IEEE Transactions on Computers,* Vol. C-35, pp. 881, (October 1986).

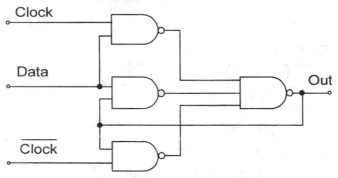

Fig. 4.13. Earl's Latch.

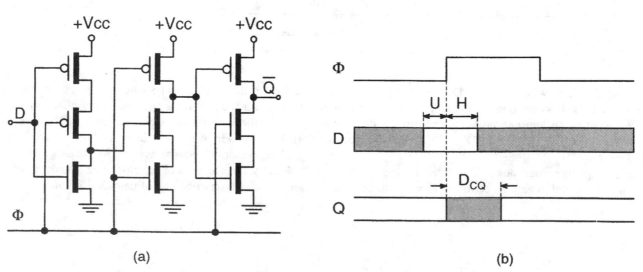

(a)

(b)

Fig. 4.14. True-single-phase-clock (TSPC) flip-flop: (*a*) schematic diagram (*b*) timing. Adapted from M. Afghahi and C. Svensson, "A Unified Single-Phase Clocking Scheme for VLSI Systems," *IEEE Journal of Solid-State Circuits,* Vol. 25, pp. 226, (February 1990).

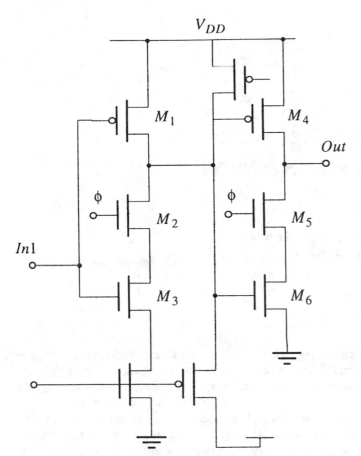

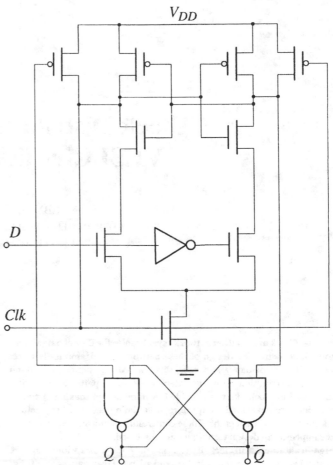

Fig. 4.15. Modified TSPC latch as used in the first-generation 21064 "Alpha" processor from Digital. Adapted from D. Dobberpuhl et. al., "A 200MHz 64-b Dual-Issue CMOS Microprocessor," *IEEE Journal of Solid-State Circuits,* Vol. 27, pp. 1560 (November 1992).

Fig. 4.16. Toshiba latch used in the third-generation 212064"Alpha" processor from Digital 21264. The latch is differential. Adapted from P. E. Gronowski et al, "High-Performance Microprocessor Design," *IEEE Journal of Solid-State Circuits,* Vol. 33, pp. 683, (May 1998).

Clock Distribution Networks in VLSI Circuits and Systems

EBY G. FRIEDMAN
DEPARTMENT OF ELECTRICAL ENGINEERING
UNIVERSITY OF ROCHESTER

Abstract—Clock distribution networks synchronize the flow of data signals among data paths. The design of these networks can dramatically affect system wide performance and reliability. In order to better understand how clock distribution networks interact with data paths, a theoretical background of clock skew is provided. Minimum and maximum timing constraints are developed from the relative timing between the localized clock skew and the data paths. These constraint relationships are reviewed and compensating design techniques are discussed.

Significant interest in clock distribution networks exists within both the industrial and academic communities, and a diverse spectrum of results has been developed. These results, representing the field of clock distribution network design and analysis, can be grouped into a number of subtopics: 1) circuit and layout techniques for structured custom VLSI systems; 2) the automated synthesis of clock distribution networks with application to automated placement and routing of gate arrays, standard cells, and larger block-oriented circuits; 3) the analysis and modeling of the timing and power dissipation characteristics of clock distribution networks; 4) the specification of the optimal timing characteristics of clock distribution networks based on architectural and functional performance requirements; and 5) the design of clock distribution networks for specific architectures, such as systolic arrays and wafer scale integration. Each of these areas is described, the clock distribution networks of specific example circuits are surveyed, and future trends discussed.

1. INTRODUCTION

In a synchronous digital system, the clock signal is used to define a time reference for the movement of data within that system. Since this function is vital to the operation of synchronous systems, much attention has been given to the characteristics of these clock signals and the networks used in their distribution. Clock signals are often regarded as simple control signals; however, these signals have some very special characteristics and attributes. Clock signals are typically loaded with the greatest fanout, travel over the greatest distances, and operate at the highest speeds of any signal (either control or data), within the entire system. Since the data signals are provided with a temporal reference by the clock signals, the clock waveforms must be particularly clean and sharp. Furthermore, these clock signals are particularly affected by technology scaling, in that long global interconnect lines become much more highly resistive as line dimensions are decreased. This increased line resistance is one of the primary reasons for the increasing significance of clock distribution networks on synchronous performance. Finally, the control of any differences in the delay of the clock signals can severely limit the maximum performance of the entire system and create catastrophic race conditions in which an incorrect data signal may latch within a register.

Most synchronous digital systems consist of cascaded banks of sequential registers with combinatorial logic between each set of registers. The functional requirements of the digital system are satisfied by the logic stages. The global performance and local timing requirements are satisfied by the careful insertion of pipeline registers into equally spaced time windows to satisfy critical worst case timing constraints. The proper design of the clock distribution network further ensures that these critical timing requirements are satisfied and that no race conditions exist [B1–B25, 1, 2]. With the careful design of the clock distribution network, system-level synchronous performance can actually increase, surpassing the performance advantages of asynchronous systems by permitting synchronous performance to be based on average path delays rather than worst case path delays, without incurring the handshaking protocol delay penalties required in most asynchronous systems. Thus, the upper limit to system clock frequency is strongly tied to the synchronization and clock distribution strategies used in the design and implementation of the particular system.

In a synchronous system, each data signal is typically stored in a latched state within a bistable register [3] awaiting the incoming clock signal, which determines when the data signal leaves the register. Once the enabling clock signal reaches the register, the data signal leaves the bistable register, propagates through the combinatorial network and, for a properly working system, enters the next register, and is fully latched into that register before the next clock signal appears. Thus, the delay components that make up a general synchronous system are composed of the following three individual subsystems [B128, B129, 4]:

1) the memory storage elements,
2) the logic elements, and
3) the clocking circuitry and distribution network.

Interrelationships among these three subsystems of a synchronous digital system are critical to achieving maximum levels of performance and reliability, and represent the primary and most significant material within this introductory chapter. The important area of clock generation, as compared to clock distribution (which is the primary topic of this book), is only briefly mentioned and bears separate focus.

The introductory chapter is organized as follows. In Section 2, fundamental definitions and the timing characteristics of clock skew are discussed. The timing relationships between a local data path and the clock skew of that path are described in Section 3, which also contains a review of globally asynchronous, locally synchronous systems. The interplay among the aforementioned three subsystems making up a synchronous digital system is described in Section 4, particularly, how the timing characteristics of the memory and logic elements constrain the design and synthesis of clock distribution networks. Different forms of clock distribution networks, such as buffered trees and H-trees, are discussed along with compensation techniques, microwave frequency clock distribution networks, clock distribution networks optimized for low power, and the testing and evaluation of clock distribution networks. The automated layout and synthesis of clock distribution networks are described in Section 5. Various models exist to calculate the timing characteristics of clock distribution networks. Some examples of these models, both deterministic and probabilistic, are presented in Section 6, as well as techniques for designing process insensitive clock distribution networks. The information describing the localized scheduling of the clock delays is useful in optimizing the performance of high speed synchronous circuits. The specification of the optimal timing characteristics of clock distribution networks is reviewed in Section 7. In VLSI systems, certain architectures exist that place unusual constraints on the design of the clock distribution network. The design of clock distribution networks for these specific VLSI-based architectures are discussed in Section 8. The application of clock distribution networks to high speed circuits has existed for many years. The design of the clock distribution networks of certain important VLSI-based systems has been described in the literature, and some examples of these circuits are described in Section 9. In an effort to provide some insight into future and evolving areas of research relevant to

high performance clock distribution networks, some potentially important topics for future research are discussed in Section 10. Finally, a summary with some concluding remarks is provided in Section 11. Note that each section is independent and referenced within other sections as appropriate.

2. THEORETICAL BACKGROUND OF CLOCK SKEW

A schematic of a generalized synchronous data path is presented in Figure 1, where C_i and C_f represent the clock signals driving a sequentially-adjacent pair of registers, specifically the initial register R_i and the final register R_f of a data path, respectively. Both clock signals originate from the same clock signal source and a pair of registers are sequentially-adjacent if only combinatorial logic (no sequential elements) exists between the two registers. The propagation delay from the clock source to the jth clocked register is the **clock delay**, T_{Cj}. The clock delays to the initial clock signal T_{Ci} and the final clock signal T_{Cf} define the time reference when the data signals begin to leave their respective registers, R_i and R_f. These clock signals originate from a clock distribution network designed to generate a specific clock signal waveform, used to synchronize each register. This standard clock distribution network structure is based on *equipotential clocking*, where the entire network is considered a surface that must be brought to a specific voltage (clock signal polarity) at each half of the clock cycle. Ideally, clocking events occur at all registers simultaneously. Given this global clocking strategy, the clock signal arrival times (at each register) are defined with respect to a universal time reference.

Definition of Clock Skew

The difference in clock signal arrival times between *two sequentially-adjacent registers*, as shown in (1), is the **clock skew**, T_{Skew}. A sequentially-adjacent pair of registers is a path with only combinational logic and/or interconnect between two registers. If the clock signals C_i and C_f are in complete synchronism (i.e., the clock signals arrive at their respective registers at exactly the same time), the clock skew is zero. A definition of clock skew is provided below:

DEFINITION 1: *Given two sequentially-adjacent registers, R_i and R_j, and an equipotential clock distribution network, the clock skew between these two registers is defined as*

$$T_{Skew i,j} = T_{Ci} - T_{Cj}, \qquad (1)$$

where T_{Ci} and T_{Cj} are the clock delays from the clock source to the registers R_i and R_j, respectively.

It is important to observe that the temporal skew between the arrival time of different clock signals is only relevant in sequentially-adjacent registers making up a single data path, as shown in Figure 1. Thus, system-wide (or chip-wide) clock

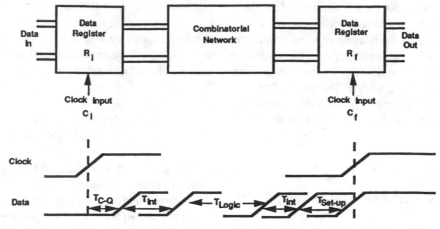

Fig. 1. Timing Diagram of Clocked Data Path.

skew between nonsequentially connected registers, from an analysis viewpoint, has no effect on the performance and reliability of the synchronous system and is essentially meaningless. However, from a design perspective, system-wide global clock skew places constraints on the permissible local clock skew. It should be noted that in [B11, 5], Hatamian designates the lead/lag clock skew polarity (positive/negative clock skew) notation as the opposite of that used here.

Different clock signal paths can have different delays for a variety of reasons. Wann and Franklin [B3] present the following causes of clock skew:

1) differences in line lengths from the clock source to the clocked register;

2) differences in delays of any active buffers (e.g., distributed buffers) within the clock distribution network (due to 3 and 4 below);

3) differences in passive interconnect parameters, such as line resistivity, dielectric constant and thickness, via/contact resistance, line and fringing capacitance, and line dimensions; and

4) differences in active device parameters, such as MOS threshold voltages and channel mobilities, which affect the delay of the active buffers.

It should be noted that for a well designed and balanced clock distribution network, the distributed clock buffers are the principal source of clock skew.

Delay Components of Data Path

The minimum allowable clock period $T_{CP}(min)$ between any two registers in a sequential data path is given by

$$\frac{1}{f_{Clkmax}} = T_{CP}(min) \geq T_{PD} + T_{Skew} , \qquad (2)$$

where

$$T_{PD} = T_{C-Q} + T_{Logic} + T_{Int} + T_{Set-up} , \qquad (3)$$

and the total path delay of a data path, T_{PD}, is the sum of the time required for the data to leave the initial register once the clock signal C_i arrives, T_{C-Q}, the time necessary to propagate through the logic and interconnect, $T_{Logic} + T_{Int}$, and the time required to successfully propagate to and latch within the final register of the data path, T_{Set-up}. Observe that the latest arrival time is given by $T_{Logic(max)}$ and the earliest arrival time is given by $T_{Logic(min)}$, since data is latched into each register within the same clock period.

The sum of the delay components in (3) must satisfy the timing constraint of (2) in order to attain the clock period $T_{CP}(min)$, which is the inverse of the maximum possible clock frequency, f_{Clkmax}. Note that in (1), the clock skew $T_{Skewi,j}$ can be positive or negative depending on whether C_j leads or lags C_i, respectively. The clock period is chosen such that the latest data signal generated by the initial register is latched by the final register with the next clock edge after the clock edge that activated the initial register. Furthermore, in order to avoid race conditions, the local path delay must be chosen such that for any two sequentially-adjacent registers in a multistage data path, the latest data signal must arrive and be latched within the final register before the earliest data signal generated with the next clock pulse at the output of the initial register arrives. The waveforms in Figure 1 show the timing requirement of (2) being barely satisfied (i.e., the data signal arrives at R_f just before the clock signal arrives at R_f).

To determine the clock delay from the clock source to each register, it is important to investigate the relationships among the clock skews of the sequentially-adjacent registers occurring within a global data path. Furthermore, it is necessary to consider the effects of feedback within global data paths on the clock skew.

The path between two sequentially-adjacent registers is described in this introductory chapter as a local data path, in contrast to a global data path, where a local data path is composed of two sequentially-adjacent registers and a global data path can consist of one or more local data paths. The relationship between the clock skew of sequentially-adjacent registers in a global data path is called **conservation of clock skew** and is formalized below:

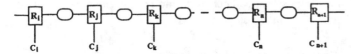

Fig. 2. Global Data Path Composed of Multiple Local Data Paths.

THEOREM 1: *For any given global data path, clock skew is conserved. Alternatively, the clock skew between any two registers in a global data path that are not necessarily sequentially-adjacent is the sum of the clock skews between each pair of registers along the global data path between those same two registers.*

PROOF: For a global data path with only two registers, the clock skew is defined by Definition 1. Now add an extra register to the global data path as illustrated in Figure 2. The clock skew between registers R_i and R_j is $T_{Skewi,j} = T_{CDi} - T_{CDj}$, and the clock skew between registers R_j and R_k is $T_{Skewj,k} = T_{CDj} - T_{CDk}$. Adding $T_{Skewi,j}$ to $T_{Skewj,k}$, $T_{Skewi,j} + T_{Skewj,k} = T_{CDi} - T_{CDk}$ is obtained, which is the clock skew between registers R_i and R_k. By adding n registers to the global data path, the clock skew between the first and the last register is $T_{Skewi,n} = T_{Skewi,j} + \ldots + T_{Skewn-1,n} = T_{CDi} - T_{CDn}$. If one register is added to the global data path, the clock skew between registers R_n and R_{n+1} is $T_{Skewn,n+1} = T_{CDn} - T_{CDn+1}$. Substituting $T_{Skewn,n+1}$ into $T_{Skewi,n}$, $T_{Skewi,n+1} = T_{Skewi,j} + \ldots + T_{Skewn,n+1}$ is obtained. Since n can be any value greater than two, this theorem is proved. □

Although clock skew is defined between two sequentially-adjacent registers, Theorem 1 shows that clock skew can exist between any two registers in a global data path. Therefore, it extends the definition of clock skew introduced by Definition 1 to any two nonsequentially-adjacent registers belonging to the same global data path. It also illustrates that the clock skew between any two nonsequentially-adjacent registers that do not belong to the same global data path has no physical meaning, since no functional data transfer between these registers occurs.

A typical sequential circuit may contain sequential feedback paths, as illustrated in Figure 3. It is possible to establish a relationship between the clock skew in the forward path and the clock skew in the feedback path because the initial and final registers in the feedback path are also registers in the forward path. As shown in Figure 3, the initial and final registers in the feed-

back path R_l–R_j are the final and initial registers of the forward path R_j–R_k–R_l. This relationship is formalized in the following:

THEOREM 2: *For any given global data path containing feedback paths, the clock skew in a feedback path between any two registers, say R_l and R_j, is related to the clock skew of the forward path by the following relationship,*

$$T_{Skewfeedbackl,j} = -T_{Skewforwardj,l} . \qquad (4)$$

PROOF: Applying Theorem 1 to the forward path between registers R_l and R_j, $T_{Skewl,j} = T_{CDl} - T_{CDj}$ is obtained. Likewise, applying Theorem 1 to the feedback path between registers R_j and R_l, $T_{Skewj,l} = T_{CDj} - T_{CDl}$ is obtained, which is the negative of $T_{Skewl,j}$, validating the theorem. □

Both Theorems 1 and 2 are useful for determining the optimal clock skew schedule within a synchronous digital system, specifically, the set of local clock skew values that maximizes system performance and reliability. The process for determining these clock skew values is discussed in subsection 7.1.

3. TIMING CONSTRAINTS DUE TO CLOCK SKEW

The magnitude and polarity of the clock skew have a two-sided effect on system performance and reliability. Depending upon whether C_i leads or lags C_f and upon the magnitude of T_{Skew} with respect to T_{PD}, system performance and reliability can either be degraded or enhanced. These cases are discussed below:

3.1 MAXIMUM DATA PATH/CLOCK SKEW CONSTRAINT RELATIONSHIP

For a design to meet its specified timing requirements, the greatest propagation delay of any data path between a pair of data registers, R_i and R_f, being synchronized by a clock distribution network must be less than the minimum clock period (the inverse of the maximum clock frequency) of the circuit as shown in (2) [B7, B8, B11, B16, B28, B126, 4–7]. If the time of arrival of the clock signal at the final register of a data path T_{Cf} *leads* that of the time of arrival of the clock signal at the initial register of the same sequential data path T_{Ci} (see Figure 4A), the clock skew is referred to as **positive clock skew** and, under this condition, the maximum attainable operating frequency is decreased. Positive clock skew is the additional amount of time that must be added to the minimum clock period to reliably apply a new clock signal at the final register. Reliable operation implies that the system will function correctly at both low and high frequencies (assuming fully static logic). It should be noted that positive clock skew only affects the maximum frequency of a system and cannot create race conditions.

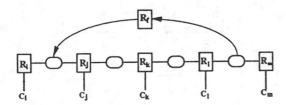

Fig. 3. Data Path with Feedback Paths.

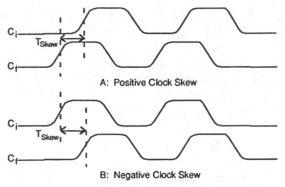

Fig. 4. Clock Timing Diagrams.

In the positive clock skew case, the clock signal arrives at R_f before it reaches R_i. From (2) and (3), the maximum permissible positive clock skew can be expressed as [B7, B8, B11, B16, B28, B126, 4–7]

$$T_{Skewi,f} \leq T_{CP} - T_{PD(max)} =$$
$$T_{CP} - (T_{C-Qi} + T_{Logic(max)} + T_{Int} + T_{Set-upf}) \quad for\ T_{Ci} > T_{Cf}. \tag{5}$$

where $T_{PD(max)}$ is the maximum path delay between two sequentially-adjacent registers. This situation is the typical critical path timing analysis requirement commonly seen in most high performance synchronous digital systems. If (5) is not satisfied, the system will not operate correctly at that specific clock period (or clock frequency). Therefore, T_{CP} must be increased for the circuit to operate correctly, thereby decreasing the system performance. In circuits where the tolerance for positive clock skew is small [T_{Skew} in (5) is small], the clock and data signals should be run in the same direction, thereby forcing C_f to lag C_i and making the clock skew negative.

3.2 MINIMUM DATA PATH/CLOCK SKEW CONSTRAINT RELATIONSHIP

If the clock signal arrives at R_i before it reaches R_f (see Figure 4B), the clock skew is defined as being negative. **Negative clock skew** can be used to improve the maximum performance of a synchronous system by decreasing the delay of a critical path; however, a potential minimum constraint can occur, creating a race condition [B11, B12, B31, B118, B119, B126, 4, 5, 7]. In this case, when C_f *lags* C_i, the clock skew must be less than the time required for the data to leave the initial register, propagate through the interconnect and combinatorial logic, and latch in the final register (see Figure 1). If this condition is not met, the data signal stored in register R_f is overwritten by the data signal that had been stored in register R_i and has propagated through the combinatorial logic. Furthermore, a circuit operating close to this condition might pass system diagnostics but malfunction at unpredictable times due to fluctuations in ambient temperature or power supply voltage [B126]. Correct operation requires that R_f latches the data signal corresponding

to the data signal R_i latched during the previous clock period. This constraint on clock skew is

$$|T_{Skewi,f}| \leq T_{PD(min)} = T_{C-Qi} + T_{Logic(min)} + T_{Int} - T_{Holdf}$$
$$for\ T_{Cf} > T_{Ci}, \tag{6}$$

where $T_{PD(min)}$ is the minimum path delay between two sequentially-adjacent registers and T_{Hold} is the amount of time the input data signal must be stable once the clock signal changes state.

An important example in which this minimum constraint can occur is in those designs using cascaded registers, such as a serial shift register or a k-bit counter, as shown in Figure 5. Note that a distributed *RC* impedance is between C_i and C_f. In cascaded register circuits, $T_{Logic(min)}$ is zero and T_{Int} approaches zero (since cascaded registers are typically designed, at the geometric level, to abut). If $T_{Cf} < T_{Ci}$ (i.e., negative clock skew), then the minimum constraint becomes

$$|T_{Skewi,f}| \leq T_{C-Qi} - T_{Holdf} \quad for\ T_{Cf} > T_{Ci}, \tag{7}$$

and all that is necessary for the system to malfunction is a poor relative placement of the flip flops creating a highly resistive connection between C_i and C_f. In a circuit configuration such as a shift register or counter, where negative clock skew is a more serious problem than positive clock skew, provisions should be made to force C_f to lead C_i, as shown in Figure 5.

As higher levels of integration are achieved in high complexity VLSI circuits, on-chip testability [8] becomes necessary. Data registers, configured in the form of serial set/scan chains when operating in the test mode, are a common example of a design for testability (DFT) technique. The placement of these circuits is typically optimized around the functional flow of the data. When the system is reconfigured to use the registers in the role of the set/scan function, different local data path delays are possible. In particular, the clock skew of the reconfigured local data path can be negative and greater in magnitude than the local register delays. Therefore, with increased negative clock skew, (7) may no longer be satisfied and incorrect data may latch into the final register of the reconfigured local data path. Therefore, it is imperative that attention be placed on the clock distribution of those paths that have nonstandard modes of operation.

In ideal scaling of MOS devices, all linear dimensions and voltages are multiplied by the factor $1/S$, where $S > 1$ [B16,

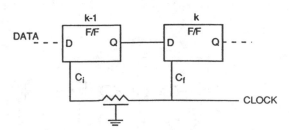

Fig. 5. K-Bit Shift Register with Positive Clock Skew.

9–11]. Device dependent delays, such as T_{C-Q}, T_{Set-up}, and T_{Logic}, scale as $1/S$ while interconnect dominated delays such as T_{Skew} remain constant to first order, and if fringing capacitance and electromigration are considered, actually increase with decreasing dimensions. Therefore, when examining the effects of dimensional scaling on system reliability, (6) and (7) should be considered carefully [12]. One straightforward method to avoid the effect of technology scaling on those data paths particularly susceptible to negative clock skew is to not scale the clock distribution lines. Svensson and Afghahi [B116] show that, by using coarser than ordinary lines for global clock distribution, 20-millimeter-wide chip sizes with CMOS circuits scaled to 0.3-μm polysilicon lines would have comparable logic and cross-chip interconnect delays (on the order of 0.5 ns), making possible synchronous clock frequencies of up to 1 GHz. Therefore, the scaling of device technologies can severely affect the design and operation of clock distribution networks, necessitating specialized strategies and compensation techniques.

3.3 Enhancing Synchronous Performance by Applying Localized Clock Skew

Localized clock skew can be used to improve synchronous performance by providing more time for the critical worst case data paths [B122, B126, B128, B129, 4, 7]. By forcing C_i to *lead* C_f at each critical local data path, excess time is shifted from the neighboring less critical local data paths to the critical local data paths. This negative clock skew represents the additional amount of time that the data signal at R_i has to propagate through the logic stages and interconnect sections, and into the final register. Negative clock skew subtracts from the logic path delay, thereby decreasing the minimum clock period. Thus, applying negative clock skew, in effect, increases the total time that a given critical data path has to accomplish its functional requirements by giving the data signal released from R_i more time to propagate through the logic and interconnect stages and latch into R_f. Thus, the differences in delay between each local data path are minimized, thereby compensating for any inefficient partitioning of the global data path into local data paths that may have occurred, a common situation in many practical systems. Different terms have been used in the literature to describe negative clock skew, such as "double-clocking" [B126], "deskewing data pulses" [B31], "cycle stealing" [B121, B122], "useful clock skew" [11], and "prescribed skew" [B72].

The maximum permissible negative clock skew of a data path, however, is dependent upon the clock period itself as well as the time delay of the previous data paths. This dependence results from the temporal nature of the serially cascaded local data paths making up the global data path. Since a particular clock signal synchronizes a register that functions in a dual role, as the initial register of the next local data path and as the final register of the previous data path, the earlier C_i is for a given data path, the earlier that same clock signal (now C_f) is for the previous data path. Thus, the use of negative clock skew in the ith path results in a more positive clock skew for the preceding path, which may then establish the new upper limit for the system clock frequency.

An Example of Applying Localized Negative Clock Skew to Synchronous Circuits

Consider the nonrecursive synchronous circuit shown in Figure 6, where the horizontal ovals represent logic elements with logic delays and the vertical ovals represent clock delays. Since the local data path from R_2 to R_3 represents the worst case path (assuming the register delays are equal), by delaying C_3 with respect to C_2, negative clock skew is added to the R_2–R_3 local data path. If C_1 is synchronized with C_3, then the R_1–R_2 local data path receives some positive clock skew. Thus, assuming the register delays are both 2 ns, C_2 should be designed to lead C_3 by 1.5 ns, forcing both paths to have the same total local data path delay, $T_{PD} + T_{Skew} = 7.5$ ns. The delay of the critical path of the synchronous circuit is temporally refined to the precision of the clock distribution network, and the entire system (for this simple example) could operate at a clock frequency of 133.3 MHz, rather than 111.1 MHz if no localized clock skew is applied. The performance characteristics of the system, both with and without the application of localized clock skew, are summarized in Table 1.

Note that $|T_{Skew}| < T_{PD}$ ($|-1.5$ ns$| < 9$ ns) for the R_2–R_3 local data path; ensuring that the correct data signal is successfully latched into R_3 and no minimum data path/clock skew constraint relationship exists. This design technique of applying localized clock skew is particularly effective in sequentially-adjacent, temporally irregular local data paths; however, it is applicable to any type of synchronous sequential system. For certain architectures, a significant improvement in performance is both possible and likely.

The limiting condition for applying localized negative clock skew is determined by the control of the clock skew variations

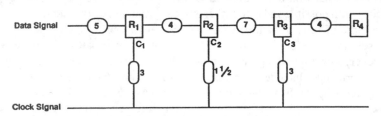

Fig. 6. Example of Applying Localized Negative Clock Skew to Synchronous Circuit.

275

TABLE 1. PERFORMANCE CHARACTERISTICS OF CIRCUIT OF FIGURE 6
WITHOUT AND WITH LOCALIZED CLOCK SKEW.

Local Data Path	$T_{PD(min)}$ − zero skew	T_{Ci}	T_{Cf}	T_{Skew}	$T_{PD(min)}$ − with non-zero skew
R_1 to R_2	$4 + 2 + 0 = 6$	3	1.5	1.5	$4 + 2 + 1.5 = 7.5$
R_2 to R_3	$7 + 2 + 0 = 9$	1.5	3	−1.5	$7 + 2 − 1.5 = 7.5$
f_{Max}	111.1 MHz				133.3 MHz

(All time units are in nanoseconds.)

and by the differences in path delays among neighboring local data paths. These clock skew variations are due to power supply variations, to process tolerances (where process parameters may vary over a specified range), and to environmental effects, such as temperature or radiation, which, for example, can affect both MOS threshold voltages and channel mobilities.

3.4 GLOBALLY ASYNCHRONOUS AND LOCALLY SYNCHRONOUS SYSTEMS

In order to minimize the difficulty in distributing global clock signals across large systems, specifically, the creation of large amounts of localized clock skew, asynchronous approaches have been developed to constrain intermodule timing [B16]. These different asynchronous approaches may be divided into two general classes: fully asynchronous systems (including self-timed systems requiring handshaking protocols and wavefront processors not requiring specialized intermodule communication [13]) and hybrid globally asynchronous, locally synchronous systems [14].

The concept of globally synchronous, locally asynchronous timing does not exist, since it is impossible for a globally synchronous system to synchronously control the timing of a locally asynchronous system. Moving data with no information describing the local temporal data flow to another module requiring some information about the local temporal data flow is not possible without providing some form of temporal reference point. If a temporal reference point is provided, the initial module is no longer fully asynchronous and contains some level of synchronous information. Therefore, a globally synchronous, locally asynchronous system can not effectively exist.

However, a *globally asynchronous, locally synchronous* system is possible, since a system with some temporal information (the locally synchronous system) can interact globally at the asynchronous level since no additional timing information is necessary. Thus, this approach is both possible and practical and is briefly reviewed here. The topic of fully asynchronous systems represents a field unto itself and therefore is not discussed in any great detail here. However, a useful and highly interesting comparison between the performance characteristics of synchronous and asynchronous systems by Afghahi and Svensson [B16] is included in this book.

In a globally asynchronous, locally synchronous system, individual modules operate completely synchronously, using self-contained local clocks within the module and distributing these clock signals via clock distribution networks. These synchronous modules, for example, can occur at the IC level or at the board level, and communicate asynchronously at the system level. The global asynchronous signals are converted to local fully synchronous signals by using local arbiters or synchronizers [15, 16]. Arbiters accept asynchronous signals and synchronize them to a temporal reference, while minimizing the probability of a synchronous register becoming metastable [3]. **Metastability** is the state of a register in which neither binary state (a "0" or a "1") appears at the output within a time period consistent with the normal operation of the register. The register will remain in this tenuous state of equilibrium until some circuit parameter varies sufficiently so as to drive the state of the register into one of the two binary states. This state of metastability is categorized as a type of reliability problem. The register could remain in this state for milliseconds, thereby forcing the overall system to function incorrectly. Therefore, it is important to ensure that the local synchronous clock signals are correctly received at the global asynchronous level.

An approach to minimizing the occurrence of metastability is to increase the clock period so as to allow sufficient time for the synchronous register to return to its proper state. This method, however, assumes that the register will return to its correct digital state (instead of the reverse polarity of the proper state) and will return in a sufficiently short amount of time. This concept can be applied to pausible or extendible clock schemes [B16], where the local synchronous clock is stopped until the output state of the register is properly defined, whereupon a new clock signal is generated once a proper register output is achieved. This strategy requires that the arbiters are monitored to guarantee that the register output is both stable and logically well defined. Since there is no absolute bound on the register resolution time, the clock period can become unacceptably long. If a specified maximum delay is exceeded, the system fails. Therefore, the primary performance penalty of globally asynchronous, locally synchronous systems is the delay due to signal arbitration and any required metastable resolving time. Note how this disadvantage compares with fully synchronous systems, in which the primary performance penalty is due to positive clock skew.

An important attribute of globally asynchronous, locally synchronous timing is that different modules can operate at different locally optimal clock frequencies. The system communication is performed using some form of handshaking protocol among the independent synchronous modules. The delay of the handshaking protocol can significantly affect the overall data rate of the system. In general, it has been shown that local data paths perform faster when operated fully synchronously, while larger modules tend to perform faster when operated asynchronously [B16], where the speed of an asynchronous system is approximately the average speed of the modules plus some handshaking delay.

4. CLOCK DISTRIBUTION DESIGN OF STRUCTURED CUSTOM VLSI CIRCUITS

Many different approaches, from ad hoc to algorithmic, have been developed for designing clock distribution networks in VLSI circuits. The requirement of distributing a tightly controlled clock signal to each synchronous register on a large nonredundant hierarchically structured VLSI circuit (an example floorplan is shown in Figure 7) within specific temporal bounds is difficult and problematic. Furthermore, the tradeoffs that exist among system speed, physical die area, and power dissipation are greatly affected by the clock distribution network. The design methodology and structural topology of the clock distribution network should be considered in the development of a system for distributing the clock signals. Therefore, various clock distribution strategies have been developed. The most common and general approach to equipotential clock distribution is the use of buffered trees, discussed in subsection 4.1. In contrast to these asymmetric structures, symmetric trees such as H-trees are used to distribute high speed clock signals. This topic is described in subsection 4.2. In developing structured custom VLSI circuits, such as exemplified by the floorplan pictured in Figure 7, specific circuit design techniques are used

to control the delays within the clock distribution network. One important compensation technique is described in subsection 4.3. Furthermore, additional design issues exist if the VLSI circuit operates at extremely high frequencies. Some discussion on distributing microwave clock frequencies is provided in subsection 4.4. Low power design is an area of significant currency and importance. Some very recent efforts to reduce the power dissipated within the clock distribution network are reviewed in subsection 4.5. Finally, only minimal material has been published on the testing and evaluation of high performance clock distribution networks. This topic is briefly discussed in subsection 4.6.

4.1 BUFFERED CLOCK DISTRIBUTION TREES

The most common strategy for distributing clock signals in VLSI-based systems is to insert buffers at the clock source and/or along a clock path, forming a tree structure. Thus, the clock source can be described as the root of the tree, the initial portion of the tree as the trunk, individual paths driving each register as the branches, and the driven registers as the leaves. This metaphor for describing a clock distribution network is commonly accepted and used throughout the literature, and is illustrated in Figure 8. Occasionally a mesh version of the clock tree structure is used, in which shunt paths farther down the clock distribution network are placed to minimize the interconnect resistance within the clock tree. This mesh structure effectively places the branch resistances in parallel, minimizing both the clock delay and the clock skew. An example of this mesh structure is described and illustrated in subsection 9.2. The mesh version of the clock tree is considered in this introductory chapter as an extension of the standard, more commonly used, clock tree depicted in Figure 8.

If the interconnect resistance of the buffer at the clock source is small compared to the buffer output resistance, often a single buffer is used to drive the entire clock distribution network. This strategy may be appropriate if the clock is distributed entirely on metal, making load balancing of the network less critical. The primary requirement of a single buffer system is that the buffer provide enough current to drive the network capacitance (both interconnect and fanout) while maintaining high quality waveform shapes (i.e., short transition times) and minimizing the effects of the interconnect resistance. The output resistance of the buffer must be designed to

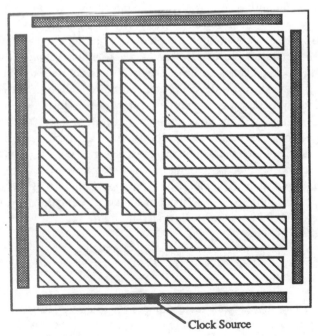

Fig. 7. Floorplan of Structured Custom VLSI Circuit Requiring Synchronous Clock Distribution.

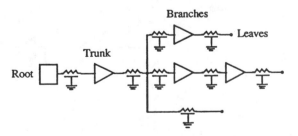

Fig. 8. Tree Structure of Clock Distribution Network.

277

be much greater than the resistance of the interconnect section being driven.

An alternative approach to using only a single buffer at the clock source is to distribute buffers throughout the clock distribution network, as shown in Figure 8. This approach requires additional area but greatly improves the precision and control of the clock signal waveforms and is necessary if the resistance of the interconnect lines is not negligible. The distributed buffers serve the double function of amplifying the clock signals degraded by the distributed interconnect impedances and isolating the local clock nets from upstream load impedances [B28]. A three-level buffer clock distribution network utilizing this strategy is shown in Figure 9. In this approach a single buffer drives multiple clock paths (and buffers). The number of buffer stages between the clock source and each clocked register depends on the total load capacitance, in the form of registers and interconnect, and the permissible clock skew [B39]. It is worth noting that the buffers are a primary source of the total clock skew within a well-balanced clock distribution network since the active device characteristics vary much more greatly than the passive device characteristics. The maximum number of buffers driven by a single buffer is determined by the current drive of the source buffer and the capacitive load (assuming an MOS technology) of the destination buffers. The final buffer along each clock path provides the control signal driving the register.

Historically, the primary goal in designing clock distribution networks has been to ensure that a clock signal arrives at every register within the entire synchronous system at precisely the same time. This concept of zero clock skew design has been extended, as is explained in subsection 3.3, to provide either a positive or a negative clock skew at a magnitude depending upon the temporal characteristics of each local data path to improve system performance and enhance system reliability.

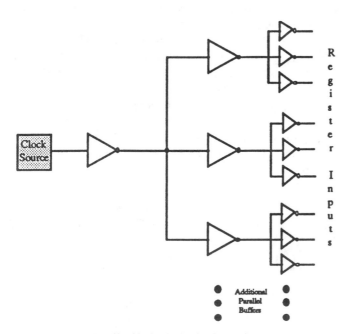

Fig. 9. Three-Level Buffer Clock Distribution Network.

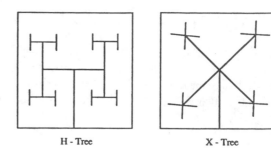

Fig. 10. Symmetric H-Tree and X-Tree Clock Distribution Networks.

4.2 SYMMETRIC H-TREE CLOCK DISTRIBUTION NETWORKS

Another approach for distributing clock signals, a subset of the distributed buffer approach depicted in Figure 8, utilizes a hierarchy of planar symmetric H-tree or X-tree structures (see Figure 10) [B30, B44, 11]. Zero clock skew is ensured by maintaining identical paths from the clock signal source to the clocked register of each clock path. In the H-tree network, the primary clock driver is connected to the center of the main "H" structure. The clock signal is transmitted to the four corners of the main "H." These four close-to-identical clock signals provide the inputs to the next level of the H-tree hierarchy, represented by the four smaller "H" structures. The distribution process then continues through several levels of progressively smaller "H" structures. The final destination points of the H-tree are used to drive the local registers or are amplified by local buffers that drive the local registers. Thus, each clock path from the clock source to a clocked register has practically the same delay. The primary delay difference among the clock signal paths is due to variations in process parameters that affect the interconnect impedance and, in particular, any active distributed buffer amplifiers. As described in Section 8, the amount of clock skew within an H-tree structured clock distribution network is strongly dependent upon the physical size, the control of the semiconductor process, and the degree to which active buffers and clocked latches are distributed within the H-tree structure.

The conductor widths in H-tree structures are designed to decrease progressively as the signal propagates to lower levels of the hierarchy. This strategy minimizes reflections of the high speed clock signals at the branching points. Specifically, the impedance of the conductor leaving each branch point Z_{K+1} must be twice the impedance of the conductor providing the signal to the branch point Z_K for an H-tree structure [B30, B44, B142, 11] and four times the impedance for an X-tree structure. This tapered H-tree structure is illustrated in Figure 11.

$$Z_k = \frac{Z_{k+1}}{2} \quad \textit{for an H-tree structure} \qquad (8)$$

The planar H-tree structure places constraints on the physical layout of the clock distribution network as well as on the design methodology used in the development of the VLSI system. For example, in an H-tree network, clock lines must be routed in

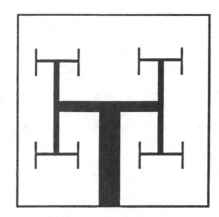

Fig. 11. Tapered H-Tree Clock Distribution Network.

both the vertical and horizontal directions. For a standard two-level metal CMOS process, this requirement creates added difficulty in routing the clock lines without using either resistive interconnect or multiple high resistance vias between the two metal lines. This difficulty is a primary reason for the development of three or more layers of metal in logic-based CMOS processes. Furthermore, the interconnect capacitance (and therefore the power dissipation) is much greater for the H-tree as compared with the standard clock tree, since the total wire length tends to be much greater [B170]. This increased capacitance of the H-tree structure exemplifies an important tradeoff between clock delay and clock skew in the design of high speed clock distribution networks. Symmetric structures are used to minimize clock skew; however, an increase in clock signal delay is incurred. Therefore, the increased clock delay must be considered when choosing between buffered tree and H-tree clock distribution networks. Also, since clock skew only affects sequentially-adjacent registers, the obvious advantages to using highly symmetric structures to distribute clock signals are significantly degraded. There may be, however, certain sequentially-adjacent registers distributed across the integrated circuit. For this situation, a symmetric H-tree structure may be appropriate.

Another consideration in choosing a clock distribution topology is that the H-tree and X-tree clock distribution networks are difficult to implement in those VLSI-based systems that are irregular in nature, such as pictured in Figures 7 and 12. In these types of systems, buffered tree topologies integrated with structured custom design methodologies [17] should be used in the

design of the clock distribution networks in order to maximize system clock frequency, minimize clock delay, and control any deleterious effects of local (particularly negative) clock skew.

4.3 COMPENSATION TECHNIQUES FOR CONTROLLING CLOCK SKEW

One structured custom approach, oriented to hierarchical VLSI-based circuits, utilizes compensation techniques to minimize the variation of interconnect impedances and capacitive loads between clock signal paths [B28, B31, B34, B36, 18, 19]. A general schematic of a clock distribution network is shown in Figure 13, in which the nodes i, j, and k represent different clock signal destinations (i.e., clocked registers). Different clock paths could conceivably have different levels of buffering, where each buffer drives a localized distributed RC impedance. The location of these buffers is often chosen so that the active buffer output impedance is comparable to or greater than the interconnect resistance seen at the buffer output. The buffer location ensures that the locally distributed RC interconnect section can be modeled accurately as being primarily capacitive. The use of distributed buffers in this manner is described as buffer repeaters [11, 20]. However, in general the interconnect impedance should be modeled as a distributed resistive-capacitive section of interconnect.

The difficulty with applying symmetric clock distribution strategies is that they do not easily support the ability to partition large VLSI systems into hierarchically structured functional blocks. Preferably, each large functional block would contain its own locally optimized clock distribution network to satisfy the local timing and loading of that particular functional block. For a globally synchronous system, however, local optimization within a functional element does not necessarily lead to global optimization of the overall on-chip clock distribution system.

If the interconnect resistance of the global clock distribution network is relatively small, a chip-level centralized clock buffer circuit can be used to satisfy the synchronization requirements of a VLSI circuit. However, in most large VLSI circuits the physical distances are such that line resistances, coupled with any via/contact resistances and the typically significant line and coupling capacitances, will create large interconnect impedances. Therefore, even with a centrally located clock generation and distribution circuit, additional techniques are

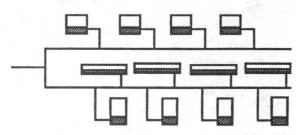

Fig. 12. Clock Distribution Network for Structured Custom VLSI Circuit.

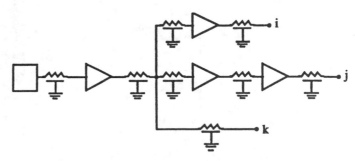

Fig. 13. Clock Distribution Network.

279

required to compensate for variations in interconnect and register loading.

In order to control the delay of each clock signal path and to minimize the skew between these paths, passive RC delay elements [B31] or geometrically sized transistor widths [B28] are used to compensate for the variation of the delay of each clock signal path. These delay variations are caused by different on-chip locations (i.e., different path-dependent interconnect impedances) and capacitive loading of the clock destinations (i.e., the number and load of the clocked registers per clock signal path). Clock buffers are placed along the clock path such that the highly resistive interconnect lines (typically long lines) drive loads with low capacitance, while the low resistance interconnect lines (typically short lines) drive loads with high capacitance. Thus, either a centralized module of clock buffer drivers can be used or those clock buffers driving large capacitive loads can be placed close to the registers, thereby decreasing the interconnect resistance. This design strategy of using compensation techniques to control the local clock skew is depicted graphically in Figure 14. The variation of clock delay between each of the functional ele-ments is compensated for by parameterizing the current drive of each of the functional block clock buffers resident in the centrally located clock buffering circuit (see Figure 14). If feedback circuitry is being used to further control the delays and skews within the clock distribution network, as in on-chip phase lock loops (PLLs), taps are placed close to the register, which are fed back to maintain lock.

In order to ensure that the clock distribution network is successfully designed, the following practices should be followed: 1) the number of stages of clock buffering within each of the functional blocks should be the same to maintain equal polarity, 2) the maximum clock signal rise and fall times within each functional block should be specified and controlled, and 3) the internal functional block clock skew should be specified and controlled using the same hierarchical clock distribution strategy as is used at the global VLSI system level [B28].

Advantages and Disadvantages of Compensation Technique

The primary advantage of using a compensation technique is controlling (and reducing) on-chip clock skew. Also, the clock delay from the clock source to the clocked registers is reduced, due to improved partitioning of the RC loads. Since the inverters, located within each functional block, drive large capacitive loads, the interconnect impedance, and in particular the interconnect resistance driven by any specific clock buffer, is small in comparison to the buffer output impedance. The fairly long distances of the intra-block clock signal paths are fairly resistive. These paths, however, are isolated from the highly capacitive loads. Thus, the RC time constants are reduced, reducing the overall clock delay. Another important advantage of this

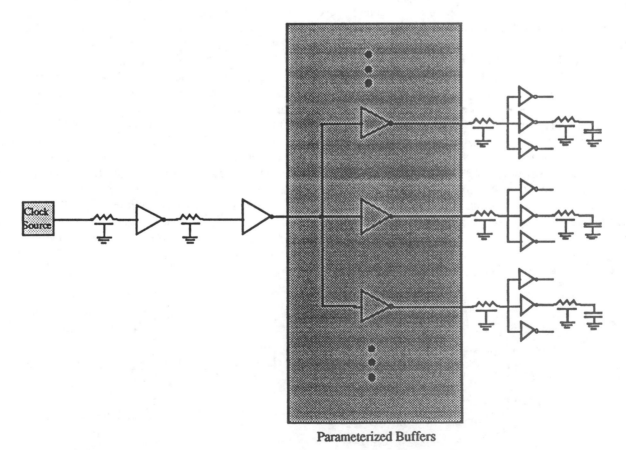

Parameterized Buffers

Fig. 14. Parameterized Buffers Integrated into a Clock Distribution Network to Control Local Clock Skew [B28].

design technique is the ease of partitioning the clock distribution problems among a team of VLSI circuit designers. The overall VLSI system design can be partitioned hierarchically into a manageable domain of information while still providing a strategy for implementing optimal clock distribution networks. The usefulness of this compensation technique is dependent upon the ability to characterize the device and interconnect impedances within the VLSI circuit. With an accurate estimate of these impedances, the parameterized buffers can be designed to satisfy a specific clock skew schedule.

It is important to note a disadvantage of this compensation technique. Unlike interconnect impedances, transistor conductances tend to be very sensitive to variations in supply voltage, and to process and environmental conditions (e.g., temperature, radiation). A clock signal path whose delay is dominated by interconnect impedances may vary differently from a clock signal path whose delay is dominated by device impedances [B28].

Several specific examples of clock distribution networks are discussed in the literature [e.g., B1, B11, B28, B35, B36, B39, B161]. Some examples of clock distribution networks applied to high speed circuits are described in Section 9. In each of these clock distribution networks, significant effort has been placed on accurately estimating the magnitude of the resistive and capacitive interconnect impedances to determine the effect of these RC loads on the shape of the clock signal waveform. This information is typically back-annotated into a SPICE-like circuit simulator to adjust the clock delays for minimum clock skew [21]. Minimal work exists, however, in developing circuit procedures and algorithms for automating the circuit design of clock distribution networks in structured custom VLSI circuits [B95–B98, B174]. One primary requirement for developing these algorithms is a physical model for estimating the delay of the clock signal path. A distributed RC interconnect delay model must be integrated with a model of the delay of a distributed buffer in order to estimate the local clock skews. This topic is further discussed in Section 6. An important observation is that the accuracy required to calculate delay differences (as in clock skew) is much greater than that required when calculating absolute delay values (as in the delay of a clock path).

4.4 CLOCK DISTRIBUTION TECHNIQUES FOR MICROWAVE CLOCK FREQUENCIES

Those systems with signals propagating at very high speeds over long lengths may induce reflections into the propagating signal that could seriously degrade the signal waveform shape. Therefore, the design of clock distribution networks where clock signals are distributed over a large area, as is common in VLSI systems, may become increasingly more difficult as clock frequencies approach or exceed microwave frequencies ($>10^9$ Hz or 1 GHz). A simple convention for estimating whether a transmission line model is necessary to analyze the waveform characteristics rather than a distributed RC impedance is to compare the length of the line over which the signal is propagating $l_{Clockpath}$ to the speed of light in the dielectric material c_{Di} (where $c = 3 \times 10^8$ meters per second in a vacuum) multiplied by the waveform transition time τ_t, as shown below [11].

$$\tau_t c_{Di} = l_{Clockpath} \qquad (9)$$

If $l_{Clockpath}$ is greater than $\tau_t c_{Di}$, a transmission model for the clock signal path becomes necessary; otherwise the distributed RC impedance model is satisfactory.

A transmission line model to analyze clock distribution networks is described in [B46] and is based on a finite Fourier Series expansion of an approximate square wave clock signal assuming finite rise and fall times. Complex transmission and reflection coefficients are determined for each harmonic at each termination (clocked register) and branch point within the clock distribution network. Assuming TEM mode propagation, accurate results up to 2 GHz can be achieved. If corrections are made to the dielectric constant, higher frequency results could accurately be attained. Load and source impedances lead to reflection coefficients at all net terminations and tapping points within the clock tree, while the interconnect is treated as sections of linear transmission lines with a fixed characteristic impedance. A time-domain waveform is reconstructed from steady-state solutions of the reflections at each harmonic frequency for any desired point within the clock distribution network.

With this high frequency model, important characteristics of the clock distribution network can be determined, such as clock skew, waveshape, and ringing. Specifically, the magnitude and location of the terminating resistors can be determined and any ringing or false clock edges can be anticipated [B46].

Bußmann and Langmann [B47] describe an alternative approach to standard tree-structured equipotential clock distribution networks. Recognizing the difficulty in accurately and efficiently transmitting microwave clock signals, sine waves rather than square waves typically are used. The authors apply damping compensation techniques to the design of high frequency clock distribution networks. Active compensation with negative impedance converters is used to improve the performance of a clock distribution network without significantly increasing power dissipation or clock signal delay. The idea behind applying active compensation to clock distribution networks is to change the propagation characteristics of the microstrip, reducing the attenuation constant of the transmission line. An on-chip differential microstrip is used to distribute the clock signals, thereby minimizing adjacent line coupling, resulting in well defined signal propagation characteristics. Differential current switch circuits operating into the Gigabit-per-second range are used to complement the high speed differential microstrip.

The capacitive loading of the registers is considered to be part of the transmission line structure itself. This loading dominates the transmission line characteristics in integrated circuits, affecting the characteristic impedance and propagation constant of the transmission line. The distributed load model is applied to the periodically loaded transmission lines since the distance between loads is chosen not to exceed the limit set by the operating frequency [B47]. This frequency-domain approach provides insight into the basic performance characteristics of clock distribution networks operating at microwave frequencies.

In distributing sinusiodal clock signals across the backplane of a large massively parallel computer operating at a frequency of 160 MHz, Chi [B17, 18] presents a novel strategy for distributing clock signals over long distances (up to several meters). This strategy is denoted by Chi as "salphasic" clock distribution, where the word **salphasic** is intended to connote the concept of sudden transitions in phase between regions of relative constancy. Clock skew is minimized with this approach by exploiting the spatial properties of standing waves, particularly the property that the amplitude and frequency of the forwoard and reverse traveling waves are identical. This strategy has been demonstrated on a computing system, producing a clock skew of only 175 ps, effectively an order-of-magnitude improvement over conventional approaches (~2ns). The computing system being synchronized by the salphasic clock distribution network, after an initial tuning procedure, "has performed continuously without attention since August 1990" [B17, B18]. Optimizing salphasic networks, however, does require selecting specific reactive terminations to control the position of the standing waves and, if the system frequency is changed, the values of the termination impedances must also be changed. In general, salphasic networks must operate in relatively lossless transmission lines; therefore, the applicability of this strategy to VLSI systems is the most appropriate for synchronizing systems at the board level rather than on the VLSI circuit itself, due to the highly resistive, lossy nature of on-chip interconnect lines.

4.5 Design of Low Power Clock Distribution Networks [B52–B54]

In a modern VLSI system, the clock distribution network may drive thousands of registers, creating a large capacitive load that must be sourced efficiently. Furthermore, each transition of the clock signal changes the state of each capacitive node within the clock distribution network, in contrast to the switching activity in combinational logic blocks, where the change of logic state is dependent on the logic function. The combination of large capacitive loads and a continuous demand for higher clock frequencies has led to an increasingly larger proportion of the total power of a system being dissipated within the clock distribtuion network, in some applications much greater than 25% of the total power [B53, B166].

The primary component of power dissipation in most CMOS-based digital circuits is dynamic power. It is possible to reduce CV^2f dynamic power by lowering the clock frequency, the power supply, and/or the capacitive load of the clock distribution network. Lowering the clock frequency, however, conflicts with the primary goal of developing high speed VLSI systems. Therefore, for a given circuit implementation, low dynamic power dissapation is best achieved by employing certain design techniques that minimize the power supply and/or the capacitive load.

Recently, De Man [B52] introduced a technique for designing clock buffers and pipeline registers such that the clock distribution network operates at half the power supply swing, reducing the power dissipated in the clock tree by 60% without compromising the clock frequency of the circuit. Kojima, Tanaka, and Sasaki [B53] describe a similar strategy in which the clock signals operate also only over half of the power supply rail, reducing the power dissipated in the clock tree by ideally 75%. The degradation in system speed is very small since, unlike the clock signals, the data signals operate over the full power supply rail. Thus the voltage is reduced only in the clocking circuitry, resulting in significantly reduced power with a minimal degradation in system speed. Experimentally derived savings of 67% were demonstrated on a test circuit (a 16-stage shift register) fabricated in a 0.5-μm CMOS technology with only a 0.5 ns degradation in speed using this half-swing clocking scheme.

Other approaches exist for reducing the power dissipated within a clock distribution network. These approaches reduce power by decreasing the total effective capacitance required to implement a clock tree. Reductions of 10% to 25% in power dissipated within the clock tree are reported with no degradation in clock frequency [B54]. As described in Section 10, the development of design strategies for minimizing the power dissipated both internal to the clock tree as well as the overall system being synchronized is a research topic of great relevance to a variety of important applications.

4.6 Test and Evaluation of High Speed Clock Distribution Networks

Following the design of clock distribution networks, these circuits must also be tested. Deol [B41] and Keezer [B42] describe different functional test systems for performing prototype testing in which the time differences within the clock distribution networks are evaluated. In [B41] an analysis system is described specifically tailored for evaluating tester skew and clock distribution networks, permitting both enhanced test programs and improved circuit implementations. Non-invasive electron beam testing is demonstrated in [B42] for both characterizing and optimizing clock distribution networks within VLSI circuits. Experimental results are described in which race conditions exist due to negative clock skew [see (6)]. These nonfunctional clock distribution networks are detected using electron beam probing. Both of these papers exemplify the importance of verifying and evaluating the design and implementation of the clock distribution networks when building high speed and highly reliable VLSI systems.

5. AUTOMATED SYNTHESIS AND LAYOUT OF CLOCK DISTRIBUTION NETWORKS

Different approaches have been taken in the automated synthesis and layout of clock distribution networks, ranging from procedural behavioral synthesis of pipelined registers [B100–B106] to the automated layout of clock distribution nets for application to gate array and standard cell-based integrated circuits [B35, B55–B85, B91, 11]. In the area of automated layout, two paths have been taken, although with time these approaches should converge. These two paths, commercial ad hoc and algorithmic, are described below in subsections 5.1 and 5.2, respectively. The integration of the effects of clock distribution into behavioral synthesis methodologies is described in subsection 5.3.

5.1 Commercial Strategies for Implementing Clock Distribution Networks in In-House Design Systems

This research path is oriented to commercial semiconductor foundries and supporting design tools [B35, B55–B58, B61, B68, B69] in which a variety of strategies are in use. Each of these strategies is aimed at increasing the routing prioritization of the clock signal nets over the data signal nets and connecting these clock nets to previously placed distributed local buffers. These buffers are used for amplifying the clock signals, since the clock signals traverse long interconnect sections. Empirical delay models coupled with back annotation are typically used to calculate the clock path delays. Either the clock skews are estimated for inclusion in timing analysis or the clock paths are compensated for, thereby forcing the clock skew to be of negligible magnitude.

An interesting example of this type of strategy, developed for gate arrays, is described in [B58]. The clock distribution system consists of a single chip-wide clock conditioner and, for each clock net, a clock ring, a primary clock buffer, and a set of secondary clock buffers. The clock ring is a global clock net that surrounds the entire integrated circuit just inside the I/O buffers. The clock conditioner is located physically inside the ring; it is sourced by the off-chip input clock signal and drives a primary clock buffer. The primary clock buffer drives the clock ring, which in turn drives the secondary clock buffers and any registers within the I/O buffers. Each secondary buffer drives a portion of the on-chip registers. For reasons of symmetry (i.e., to minimize the clock skew), the secondary buffers are located along the vertical sides of the integrated circuit, and the primary buffers are centered on one horizontal row. All connections between buffers are in metal for low resistance and the number of registers clocked by a particular buffer is limited by load balancing requirements. A primary source of delay within the clock path is the *RC* delay within the clock ring; however, internal clock skews of less than 500 ps are maintained. Due to area concerns, internal strapping of the clock lines is not done. A design methodology is described that extends commercial software by performing register pin distribution (physical location, capacitive load, and number), partitioning of registers by chip region for load balancing, additional balancing of capacitive loads, clock net routing, and back-annotation for precise timing analysis.

5.2 Automated Layout of Clock Distribution Networks

A second research path in the area of automated layout has been the development of algorithms that carefully control the variations in delay between clock signal net length so as to minimize clock skew [e.g., B59, B62–B67, B72, B78]. The strategy used is to construct binary tree-like structures with the clock pins at the leaf nodes. Minimal skew clock distribution networks are created using a recursive bottom-up approach. At each of the clock pins of the registers, which represent the leaves of the clock distribution tree, a clock net is defined. The point where two zero skew clock nets connect is chosen such that the effective delay from that point to each clocked register is identical (within the accuracy of the delay model). This process continues up the clock distribution tree, the point of connection of each new branch being chosen to satisfy the zero skew design goal. The layout process terminates when the root (or source) of the clock tree is reached. The schematic diagram of this geometric matching process is illustrated in Figure 15. Thus, the automated layout algorithm attempts to balance the delay of each clock branch in a recursive manner, moving from the leaves to the root of the tree. The appropriate branching points of the zero skew subtree are chosen to maintain equal delay. If the zero skew tapping point falls directly on an unroutable location, such as an existing macrocell, a non-zero clock skew would be realized [B91–B93].

Some early clock routing algorithms [B59, B65, B66] define the delay as a measure of the total wire length along a path. These algorithms attempt to equalize the lengths of each net from the root of the clock tree to each of the leaf nodes. Thus, the clock skew is minimized during the routing phase of the layout process. No attempt is made to postprocess the layout database to further improve the skew characteristics.

In [B91–B93] the automated layout of the clock tree is composed of a two-phase process. The clock net is initially routed in a binary-tree manner with the clock pins as leaf nodes and the clock buffer as the root. This layout phase is followed by a post-

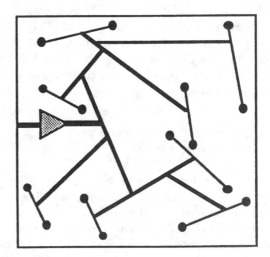

Fig. 15. Geometric Matching for Zero Clock Skew.

layout phase in which the clock nets are widened according to the zero skew specifications, thereby giving the clock layout system additional flexibility in routing around possible blockages. The choice of which clock net to widen is determined by analyzing the sensitivity of the clock net impedance. These sensitivities provide a means of choosing those nets that will decrease the average delay of the RC trees as near as possible to a specified target delay. The nets that are widened are less sensitive to increasing capacitance and whose clock path delay must be increased. However, if all the clock nets are relatively thin, statistical variations in the widths of those wires closest to the clock driver may affect the actual clock skew the greatest [B 91]. This effect occurs since the section of interconnect closest to the driver sees the greatest portion of the distributed RC impedance of the interconnect line. Therefore, the greatest change in delay will occur in those clock paths in which the interconnect impedance closest to the clock source varies significantly.

These automated clock layout algorithms tend to use simplified delay models, such as linear delay, where the delay is linearly related to the path length, or the Elmore delay [22–24], where the delay along a path is the summation of the products of the branch resistance and the downstream capacitance of every branch on the path from the root to the downstream node i (the clock pin of the register). The Elmore delay is

$$T_{Di} = \sum_k R_{ki} C_k , \qquad (10)$$

where C_k is the capacitance at node k, and R_{ki} is the resistance of the portion of the (unique) path between the input and the output node i that is common with the (unique) path between the input and node k. The Elmore delay is a first-order step response approximation of the delay through a distributed resistive-capacitive interconnect section. For slow input waveforms, the Elmore delay approximation can become highly inaccurate, since the shape and magnitude of the clock waveforms are not considered.

The fundamental difficulty with these delay models, however, is the inability to consider accurately the effects of active devices, such as distributed buffers, when estimating delay as well as more subtle considerations, such as bias-dependent loading and varying waveform shapes. The primary focus of the existing research into the automatic layout of clock distribution networks has been placed on minimizing total wirelength, metal-to-metal contacts and crossovers, as well as attaining zero system-wide clock skew (assuming nominal process and environmental conditions).

5.3 INTEGRATION OF CLOCK DISTRIBUTION INTO BEHAVIORAL SYNTHESIS

Localized clock distribution has been considered only minimally in automated layout or physical synthesis. However, early work in applying local clock skew to behavioral synthesis is described in [B95–B106, B174]. These papers represent early efforts to develop strategies that consider the effects of clock distribution networks in the behavioral synthesis process, rather than after the circuit has been partitioned into logic (or register transfer level) blocks. This capability will improve high level exploratory design techniques as well as optimize the performance of circuits implemented with high level synthesis tools.

As described in [B95–B97], the automated synthesis of clock distribution networks can be broken up into four phases: 1) optimal clock scheduling, 2) topological design, 3) circuit design, and 4) physical layout. Optimal scheduling represents a primary research activity in clock distribution networks and is discussed further in subsection 7.1. The area of topological design, in which the structure of the clock distribution network is derived from the scheduling of the local clock skew schedule, is discussed in [B95]. Algorithms are developed for converting the clock skew information into path-specific clock delays. With this information and some information describing the hierarchy of the circuit function, a clock distribution tree is developed with delay values assigned to each branch of the tree. With the topological structure and delay information determined, circuit delay elements are synthesized that satisfy the individual branch delays. Circuit techniques to implement the clock tree delay elements are discussed further in [B96, B97]. Finally, a variety of techniques exist to lay out the clock distribution trees. Some of these layout techniques are discussed in subsections 5.1 and 5.2. This work represents early research in the development of a systematic methodology for synthesizing tree-structured clock distribution networks that contain distributed cascaded buffers and, furthermore, exploit non-zero localized clock skew.

In [B101–B103], a delay model characterizing the timing components of a local data path, similar to (2) and (3), is used to incorporate the effects of local clock distribution delays into the retiming process. This improvement in model accuracy is accomplished by assuming that physical regions of similar clock delay exist throughout an integrated circuit. Retiming is an automated synthesis process for relocating pipeline registers such that the critical worst case path delay is minimized, creating a synchronous system with the highest possible clock frequency while maintaining the function and latency of the original system. Previous work in the area of retiming ignored clock skew in the calculation of the minimum clock period. In the algorithm presented in [B102, B103], clock delays are attached to individual paths between logic gates. As a register is placed on a new path, it assumes the clock delay of that path. Since each sequentially-adjacent pair of registers defines a local data path, as registers are moved from one local data path to another local data path during the retiming process, the displaced registers assume the clock delay of the new physical region. Thus, the local clock skews of each data path are determined at each iteration of the retiming process, permitting both increased accuracy in estimating the maximum clock frequency and detection and elimination of any catastrophic race conditions. If a choice of register locations does not satisfy a particular clock period or if a race condition is created, that specific register instantiation is disallowed. This algorithm, therefore, integrates the effects of clock skew (and variable register and interconnect delays) directly into the synchronous retiming process.

It is interesting to note that adding clock delay to a clock path (applying localized clock skew) has an effect similar to retiming, where the register crosses logic boundaries. Thus, time can be shifted by moving the registers or changing the local clock delays, where retiming is discrete in time and localized clock skew is continuous in time. In general, the two methods complement each other [B126]. As Fishburn mentions in [B126], since both methods are linear, "it is likely that efficient procedures could be given for optimizing systems by jointly considering both sets of variables."

6. ANALYSIS AND MODELING OF THE TIMING CHARACTERISTICS OF CLOCK DISTRIBUTION NETWORKS

This research area is composed of a number of disparate topics, all of which have in common the attributes of modeling the general performance characteristics of clock distribution networks. An important and active area of research in clock distribution networks is the design of circuits less sensitive to variations in process parameters. This topic is discussed in subsection 6.1. Deterministic approaches for calculating clock skew are summarized in subsection 6.2. In contrast to the use of deterministic approaches, probabilistic and statistical estimates of clock skew are possible, particularly for highly regular circuit structures. Some probabilistic and statistical models of clock skew are described in subsection 6.3.

6.1 DESIGN OF PROCESS INSENSITIVE CLOCK DISTRIBUTION NETWORKS

A primary disadvantage of clock distribution networks is that the delay of each of the elements of a clock path (the distributed buffers and the interconnect impedances) is highly sensitive to geometric, material, and environmental variations that exist in an implementing technology. Thus, as device and interconnect parameters vary from process lot to process lot, the specific performance characteristics of the clock distribution network may change. This phenomenon can have a disastrous effect on both the performance and the reliability of a synchronous system, thereby limiting the precision and the design methodology of the clock distribution network. It is essential for a robust clock distribution network to exhibit a certain degree of tolerance to variations in process parameters and environmental conditions. In an effort to overcome this problem, various approaches have been developed that mitigate the effects of process tolerances on the design of clock distribution networks while maintaining an effective methodology for designing these networks.

Threshold Tracking to Control Clock Skew

An important circuit design technique for making clock distribution networks less process sensitive is described by Shoji [B107]. The technique uses the MOS circuit characteristic that N-channel and P-channel parameters tend not to track each other

as a process varies. Interestingly, the response times of these devices tend to move in opposite directions, since the cause of a positive threshold voltage shift in one type of MOS transistor (e.g., an N-channel device) will typically cause the P-channel threshold voltage to shift in the opposite direction (due to charge build-up in the thin oxide between the polysilicon gate and the conducting channel). Shoji quantitatively describes how the delays of the P-channel and N-channel transistors within the distributed buffers of a clock distribution network should be individually matched to ensure that, as the process varies, the path delays between different clock paths will continue to track each other.

The primary objective of this process insensitive circuit design technique is to match the two clock edges (of either a P-channel or an N-channel transistor) as the process parameters vary. Shoji presents two rules to minimize the effects of process variations on clock skew. The rules are:

1) match the sum of the pull-up delays of the P-channel MOSFET with the pull-up delays of any related clock signal paths, and
2) match the sum of the pull-down delays of the N-channel MOSFET with the pull-down delays of any related clock signal paths.

Although process variations may change the total clock delay along a given path, the difference in delay between paths will track each other, keeping the skew small.

A circuit utilizing this technique is shown in Figure 16. Delay times T_1, T_3, and T_A are directly related to the conductances of the N-channel devices, N_1, N_3, and N_A, respectively. Delay times T_2 and T_B are directly related to the conductances of the P-channel devices, P_2 and P_B, respectively. The conductance of

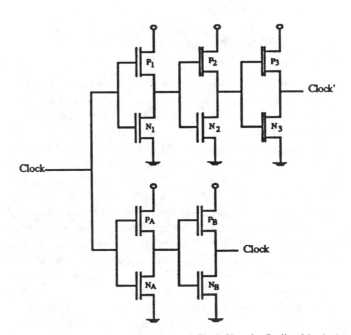

Fig. 16. Elimination of Process Induced Clock Skew by Scaling Matched Transistor Types.

each of these devices is proportional to the size of th MOSFET geometric width. In conventional CMOS circuit design, the transistor widths are adjusted to satisfy

$$T_A + T_B = T_1 + T_2 + T_3, \quad (11)$$

thereby ensuring that the skew at the output of each clock branch is close to zero. If, for example, either the N-channel or the P-channel threshold voltage varies significantly from its assumed value, (11) would no longer be satisfied and a large skew would develop between the outputs, Clock and Clock', shown in Figure 16. Instead of designing the circuit in Figure 16 to satisfy (11), the circuit is designed to satisfy the two equations, (12) and (13). Thus, the primary objective of (11) (zero clock skew) is maintained, while the added more stringent design constraint makes the entire clock distribution network more tolerant to process variations. This increased tolerance occurs, since by satisfying both (12) and (13), the N-channel and P-channel transistors of both branches individually track each other, making the system more tolerant to variations in the N-channel and P-channel transistor characteristics.

$$T_A = T_1 + T_3, \quad (12)$$

$$T_B = T_2. \quad (13)$$

This design technique can be used to make circuits less sensitive to process variations and environmental conditions even if the circuits are not inverters but are more general forms of logic gates. The technique also ensures similar behavior when interconnect impedances are included within the circuit. Simulated worst case clock skews of circuits using this technique exhibit skews that are 10% less than that of conventionally designed circuits [B107].

Interconnect Widening to Minimize
Clock Skew Sensitivity

As described in subsection 5.2, one approach for the automated layout of clock nets is to lengthen specific clock nets to equalize the length of every clock line, thereby keeping the clock skew close to zero. A disadvantage of this approach is that these minimum width lines are very susceptible to variations in the etch rate of the metal lines, as well as to mask misalignment or local spot defects. Therefore, the effective interconnect impedance (and the delay) of these long thin clock nets can vary greatly from wafer to wafer as these line widths vary. In order to design these clock nets to be less sensitive to process variations, Pullela, Menezes, and Pillage [B91–B93] have developed an automated layout algorithm that widens rather than lengthens the clock nets while equalizing the line delays. These nets are therefore less sensitive to both under- and over-etching during the metal patterning process. By widening the clock lines, the interconnect resistance is decreased; however, the interconnect capacitance increases. It is interesting to note that increasing the line width of those branches closer to the root of the

RC tree has a greater effect on the clock path delay than increasing the widths closer to the leaf nodes (the clocked registers). Thus, decreasing the resistance at the source by increasing the line width affects the total path delay more significantly than decreasing the resistance at the leaf node, since more capacitance is seen by the large source resistance than if the resistance is greater near the leaf. Therefore, the clock skew is particularly sensitive to changes in line width close to the clock source. One approach to making the clock lines more tolerant of process variations is to make the width of the clock interconnect lines widest near the clock source and thinner as the leaf nodes are approached. This strategy would provide a reasonable tradeoff between controlling the effect of process variations (particularly, metal etch rates) on the clock skew and minimizing line dimensions for process yield and circuit layout efficiency. The relative sensitivities of each net to changes in the capacitive and resistive interconnect impedances are analyzed and integrated into the Elmore delay model [22–24]. One of the primary advantages of this approach is that the process of automatically laying out the clock nets is separate from the clock skew reduction process. Thus, local layout techniques such as widening the clock nets can be used to make the overall circuit less sensitive to variations in process parameters.

6.2 Deterministic Models for Estimating Clock Skew

A clock signal path within a clock distribution network has a single input and, although paths branch off from the trunk of the tree, a single path (or branch) exists from the clock source to the clock input of a register. This branch is typically composed of distributed buffers and interconnect sections, as shown in Figure 13. In order to simplify the calculation of the path delay and to provide simple closed form delay models of the path, it is typically assumed that the buffer on-resistance is much greater than the interconnect resistance that the buffer is driving. This assumption permits the distributed RC interconnect section to be modeled as a simple lumped capacitor. Percent errors reflecting this assumption are provided in [25], where the errors are dependent upon the ratio of the load resistance to the output buffer on-resistance. However, if the line resistance is not significantly smaller than the buffer output resistance, repeaters [11, 20] are often inserted at a point within the clock line to ensure that the output resistance of the repeater buffer is much larger than the local line resistance of the interconnect section between the repeaters [B16].

In order to calculate the clock path delay and skew, a simple model of a CMOS inverter driving another inverter with line resistance and capacitance between the two inverters is often used. A well-known empirical estimate of the rise (or fall) time of a single CMOS inverter driving an RC interconnect section with a capacitive load (representing the following CMOS inverter) is [10, 11, 25]

$$T_{R/F} = 1.02 R_{Int} C_{Int} + 2.21 (R_{Tr} C_{Int} + R_{Tr} C_{Tr} + R_{Int} C_{Tr}), \quad (14)$$

where R_{Int} and C_{Int} are the resistance and capacitance of the interconnect section, respectively, and R_{Tr} and C_{Tr} are the output on-resistance of the driving buffer and the input load capacitance ($= C_{OX} W L$) of the following buffer, respectively. Note that C_{OX} is the oxide capacitance per unit area, and W and L are the width and length, respectively, of the following buffer. An approximate estimate of the output resistance of the buffer may be obtained from [10, 11, 25]

$$R_O \approx \frac{L/W}{\mu C_{ox}(V_{DD} - V_T)}, \qquad (15)$$

where μ is the channel mobility, V_{DD} is the power supply voltage, and V_T is the device threshold voltage. Equation (15) is derived from the large signal I-V equation of a MOSFET operating in the saturation region close to the linear region and is accurate for small channel geometries, since velocity saturation decreases the quadratic behavior of the MOS device operating in the saturation region. The physical delay model represented by (14) and (15) is a fairly simple approximation of the delay of a CMOS inverter driving a distributed RC impedance. More complex and accurate delay models exist. This area of inquiry represents an important topic of intensive research unto itself and is discussed in great detail throughout the literature.

An important research area in VLSI circuits is timing analysis, where simplified RC models are used to estimate the delay through a CMOS circuit. Clock characteristics are provided to a timing analyzer to define application-specific temporal constraints, such as the minimum clock period or hold time, on the functional timing of a specific synchronous system [B113]. In [B121, B122], Tsay and Lin continue this approach by describing an innovative timing analyzer which exploits negative clock skew (i.e., time is "stolen" from adjacent data paths to increase system performance). Therefore, the descriptive term "cycle stealing" is used to describe this process. In [B114, B115], Dagenais and Rumin present a timing analysis system that determines important clocking parameters from a circuit specification of the system, such as the minimum clock period and hold time. This approach is useful for top-down design when performing exploratory estimation of system performance.

6.3 PROBABILISTIC/STATISTICAL MODELS FOR ESTIMATING CLOCK SKEW

Clock skew is caused by variations in clock signal delay between sequentially-adjacent registers. Differences in process parameters across an integrated circuit die are a primary source of this delay variation. A strategy for analyzing these effects is to use a statistical approach to model the process parameter variations in order to estimate changes in the clock path delay. This method of estimating clock skew is quite different from classical deterministic techniques used within industry and described throughout the literature. The primary application of statistical

models of clock skew is highly regular architectures, such as parallel processors.

In developing a statistical model of clock skew for parallel processors, three basic assumptions regarding the clock distribution network are made [B111]: 1) the clock paths from the clock source to the processing elements are identical; 2) the clock arrival times are considered to be random variables and are the sums of uncertain independent delays due to the distributed interconnect and buffers along the clock path; and 3) to make the results independent of topology, any processor can communicate with any other. Since the delay of each cascaded buffer is modeled as a real random variable, the entire clock path delay is modeled as the sum of these random variables. As described by Afghahi and Svensson in [B16, B117] and Kugelmass and Steiglitz in [B110, B111], the variance of the clock path is equal to the sum of the variances of each buffer along the path (see Figure 13). Therefore, according to the Central Limit Theorem [26], the distribution of the clock path delay tends to a normal Gaussian distribution, regardless of the distribution of the individual buffer delays. Kugelmass and Steiglitz provide upper bounds on clock skew, assuming a Gaussian distributed clock delay with a variance proportional to the wire length. Afghahi and Svensson analyze the individual device parameters characterizing the on-resistance of a transistor operating in the saturation region. The authors discern an interesting aspect: long highly resistive interconnects are often broken up into equally spaced interconnect sections by inserting individual buffers (i.e., repeaters [11, 20]), thereby making the line resistance negligible as compared with the buffer resistance (since each interconnect section is small in comparison to the output on-resistance of the buffer). This strategy effectively makes the interconnect delay linear with line length. Furthermore, the use of repeaters actually reduces the standard deviation of the interconnect delay, and the line delay becomes proportional to the square root of the interconnect length. Thus, the use of repeaters minimizes the effect of scaling the interconnect lines on the standard deviation of the clock signal paths.

Kugelmass and Steiglitz show that the expected range of variation of a set of related random variables is no larger than the expected range of the corresponding set of independent random variables. As shown in [B16, B110], for the set of normally distributed independent paths, the expected value and variance of the difference between the maximum and minimum delay (clock skew) is

$$T_{Skew} = \sigma \frac{4 \ln R - \ln \ln R - \ln 4\pi + 2C}{\sqrt{2 \ln R}}, \quad (16)$$

$$Variance\ (T_{Skew}) \approx \frac{\pi^2 \sigma^2}{6 \ln R}, \qquad (17)$$

where $C \approx 0.5772$ (Euler's constant), R is the number of clock signal paths, and σ is the standard deviation of a single path delay. Equation (16) is an asymptotic upper bound on the expected clock skew for a clock distribution network with R signal paths. Note that the clock skew grows with the square root of the logarithm of the size of the system (i.e., the number of clock paths). This statistical approach assumes a highly regular structure, thereby assuring that a balanced clock distribution network is possible. The asymptotic upper bound of the expected clock skew, using this model and assuming an H-tree clock distribution network, is $\Theta(N^{0.25}\sqrt{\log N})$, where N is the number of leaves (or destination points) in the H-tree.

A statistical approach has also been used for estimating the distribution of the delays of the leaf nodes of a clock tree with application to the automated layout problem [B88–B90]. The extremal points of the distribution determine the clock skew while the average of the distribution determines the delay. Attention is placed on reducing the variance of the nominal delay distribution as well as the variances due to any process variations [B91].

One major weakness in applying a statistical approach to the design and analysis of clock skew in VLSI systems is that a statistical methodology ignores the localized deterministic nature of clock skew in affecting system performance and reliability, and instead considers clock skew as a system-wide global parameter. The local data path dependent clock skew directly determines the reliability and performance of the entire system, rather than an average system-wide clock skew. Any individual local data path can limit the maximum speed or create a catastrophic race condition within the system. Therefore, the probabilistic models described in this subsection are appropriate only if applied to highly regular architectures, such as systolic parallel processors [B110, B111]. The primary reason that a probabilistic clock skew model is applicable to a parallel processor architecture is that the local data paths within these types of circuit structures tend to have similar path delays; therefore, any local clock skew is close in magnitude to an average global clock skew. The distribution of clock signals within parallel processors is presented in much greater detail in subsection 8.1.

7. SPECIFICATION OF THE OPTIMAL TIMING CHARACTERISTICS OF CLOCK DISTRIBUTION NETWORKS

An important element in the design of clock distribution networks is choosing the minimum local clock skews that increase circuit performance by reducing the maximum clock period while ensuring that no race conditions exist. This design process is called *optimal clock skew scheduling;* it has been extensively studied in [B126–B138] and is described in subsection 7.1. Starting with the timing characteristics of the circuit, such as the minimum and maximum delay of each combinational logic block and register, it is possible to obtain the localized clock skews and the minimum clock period. This process is accomplished by formulating the optimal clock scheduling problem as a linear programming problem and solving with linear programming techniques [B126, B133]. In subsection 7.2 a graphical technique for investigating architectural effects on pipelining is described. Specifically, the performance tradeoff between latency and clock frequency for increasing pipelining is presented.

7.1 OPTIMAL CLOCK SKEW SCHEDULING

The concept of scheduling the system-wide clock skews for improved performance while minimizing the likelihood of race conditions was first presented by Fishburn in 1990 [B126], although the application of localized clock skew to increase the clock frequency and to eliminate race conditions was known previously [7]. Fishburn presents a methodology in which a set of linear inequalities are solved using standard linear programming techniques in order to determine each clock signal path delay from the clock source to every clocked register. Two clocking hazards, identical to the constraint relationships described in Section 3, are eliminated by solving the set of linear inequalities derived from (5) and (6) for each local data path. The deleterious effect of positive clock skew (the maximum data path/clock skew constraint relationship) is described as *zero-clocking,* while the deleterious effect of negative clock skew (the minimum data path/clock skew constraint relationship) is described as *double-clocking.* This approach is demonstrated on a 4-bit ripple-carry adder with accumulation and input register in a 1.25-μm CMOS technology. The minimum clock period is decreased from 9.5 ns with zero clock skew to 7.5 ns with localized clock skew [B126]. Szymanski improves this methodology for determining an optimal clock skew schedule by selectively generating the short path constraints, permitting the inequalities describing the timing characteristics of each local data path to be solved more efficiently [B133].

In order to describe the process for determining an optimal clock skew schedule, a system timing model is presented. A block diagram of a multistage synchronous digital system is depicted in Figure 17. Between each stage of registers there is, typically, a block of combinational logic with possible feedback paths between the registers. Each register is either a multi- or single-bit register, and all the inputs are assumed to change at the same time point of the transition of the clock signal. Only one single-phase clock signal source in the circuit is assumed. The registers are composed of edge-triggered flip flops and assume a single value for each clock cycle. The combinational logic block is described in terms of the maximum and minimum delay values, $T_{Logic(max)}$ and $T_{Logic(min)}$. These logic delay values are obtained by considering the delay of all possible input to output paths within each combinational logic block. For simplicity and without loss of generality, the block diagram in Figure 17 only considers a single-input, single-output circuit. In this figure, Δ_1 and Δ_2 are off-chip clock delays outside the VLSI-based system and δ is that portion of the on-chip clock delay that is shared by each of the clock paths (the initial trunk of the clock tree). The registers R_{in} and R_{out} make up one set of registers placed at the input and output, respectively, of the VLSI-based

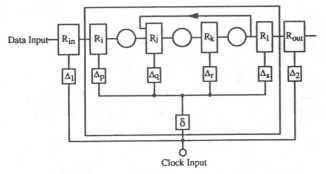

Fig. 17. Multistage Block Diagram of a Synchronous Digital System.

system. The circuit is composed of the registers, R_i and R_l, where the logic blocks are between the registers and the on-chip signal clock delays are Δ_p to Δ_s ($+\delta$).

Off-Chip Clock Skew

This circuit model, pictured in Figure 17, also considers the relationship between off-chip and on-chip clock skew. The registers R_{in} and R_{out} symbolize off-chip registers and are controlled by the off-chip clock source, which also provides the on-chip clock signals, since the circuit is assumed to be a fully synchronous system. This relationship is represented by

$$T_{Skewin,out} = T_{Skewin,i} + T_{Skewi,j} + \ldots + T_{Skewl,out} = 0. \tag{18}$$

Therefore, to satisfy (18), in Figure 17, $\Delta_1 = \Delta_2$.

Although it is possible to have off-chip non-zero clock skew, it is desirable to ensure that the clock skew between VLSI I/O approaches zero, in order to avoid complicating the design of a circuit board or the specification of the interface of the circuit with other components also controlled by the same clock source. For example, a circuit with intentional non-zero clock skew requires that the clock distribution network of any other synchronous circuit sharing the same global clock be offset by the same amount of temporal skew; otherwise, race conditions such as described in subsection 3.2 may occur at the board level. This strategy of minimizing clock skew as the level of design complexity shifts should be applied at each higher level of design, such as from the board level to the multiboard level.

A fully synchronous circuit must generate data at a rate defined by the clock period of the clock source; otherwise, race conditions may occur at the interface of the circuit with other parts of the system. These race conditions occur when there is a negative clock skew, intentionally or unintentionally introduced into the circuit. Observe that every circuit has unintentional clock skew caused by several factors, one common cause being variations in process parameters. This type of clock skew must be considered during the design of the circuit, and should be less than the intentional on-chip clock skew introduced to increase the performance of the circuit. Furthermore, the magnitude of the intentional clock skew at each register I/O may vary sub-

stantially, according to the optimization applied to the data path. Therefore, clock skew at the system level of a VLSI circuit should be constrained to approach zero, in order to allow the circuit to communicate correctly with other board-level circuits. For example, a symmetric zero clock skew distribution system should be used for the external registers (symmetric networks are discussed in subsection 4.2). Observe that restricting the off-chip clock skew to zero does not preclude the circuit from being optimized with localized clock skew. The primary effect is that the performance improvement is less than that obtained without this constraint.

Observe that (18) is valid only if the interface circuitry is controlled by the same clock source. The restriction does not apply to asynchronous circuits or synchronous circuits that communicate asynchronously (i.e., globally asynchronous, locally synchronous systems) (see subsection 3.4).

Global and Local Timing Constraints

As described in Section 3, in order to avoid either type of clock hazard (either a maximum or minimum data path/clock skew constraint relationship), a set of inequalities must be satisfied for each local data path in terms of the system clock period T_{CP} and the individual delay components within the data path. To avoid limiting the maximum clock rate between two sequentially-adjacent registers, R_i and R_j, (5) must be satisfied. To avoid race conditions between two sequentially-adjacent registers, R_i and R_j, (6) must be satisfied.

The system-wide clock period is minimized by finding a set of clock skew values that satisfy (5) and (6) for each local data path and (18) for each global data path. These relationships are sufficient conditions to determine the optimal clock skew schedule such that the overall circuit performance is maximized while eliminating any race conditions.

The timing characteristics of each local data path are asumed to be known. The minimum clock period is obtained when the problem is formalized as a linear programming problem, such as

Minimize

$$T_{CP}$$

subject to the local and global timing constraints:

$$T_{Skewi,j} \leq T_{CP} - T_{PD(max)}$$
$$= T_{CP} - (T_{C-Qi} + T_{Logic(max)} + T_{Int} + T_{Set-upj})$$
$$for \ T_{Ci} > T_{Cj}$$

$$|T_{Skewi,j}| \leq T_{PD(min)}$$
$$= T_{C-Qi} + T_{Logic(min)} + T_{Int} - T_{Holdj} \quad for \ T_{Cj} > T_{Ci}$$

$$T_{CP} \geq T_{PD(max)} + T_{Skewi,j}$$

$$T_{PD(max)} = T_{C-Qi} + T_{Logic(max)} + T_{Int} + T_{Set-upj}$$

$$T_{Skewin,i} + T_{Skewi,j} + T_{Skewj,k} + \ldots + T_{Skewn,out} = 0.$$

289

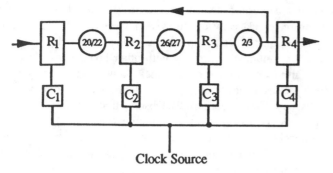

Fig. 18. Circuit Example with Feedback.

An Example of Determining the Optimal Clock Skew Schedule of a Pipelined System

An example of determining the minimum clock period of a multistage system with feedback paths is the circuit illustrated in Figure 18. This example is similar to that used in [27], adapted to consider zero clock skew between off-chip registers. The numbers inside the logic blocks are the minimum and maximum delays of each block, respectively. Similar to the approach taken in [27], for simplicity all the register and interconnect timing parameters are assumed to be zero.

The linear program that gives the minimum clock period and the optimal clock skew schedule for the circuit shown in Figure 18 is

Minimize

$$T_{CP}$$

subject to:

$$R_1 - R_2: \quad C_1 - C_2 = T_{Skew1,2} \geq -20 \ ns$$

$$T_{Skew1,2} - T_{CP} \leq -22 \ ns$$

$$R_2 - R_3: \quad C_2 - C_3 = T_{Skew2,3} \geq -26 \ ns$$

$$T_{Skew2,3} - T_{CP} \leq -27 \ ns$$

$$R_3 - R_4: \quad C_3 - C_4 = T_{Skew3,4} \geq -2 \ ns$$

$$T_{Skew3,4} - T_{CP} \leq -3 \ ns$$

$$R_3 - R_2: \quad C_3 - C_2 = T_{Skew3,2} \geq -2 \ ns$$

$$T_{Skew3,2} - T_{CP} \leq -3 \ ns$$

$$R_1 - R_4: \quad T_{Skew1,2} + T_{Skew2,3} + T_{Skew3,4} = 0$$

where the optimal clock schedule and minimum clock period are

$$T_{Skew1,2} = -3 \ ns$$

$$T_{Skew2,3} = -12 \ ns$$

$$T_{Skew3,4} = 15 \ ns$$

$$T_{CP} = 19 \ ns$$

If zero clock skew between off-chip registers is not considered, the minimum clock period is $T_{CP} = 15$ ns. Although the restriction of zero clock skew increases the clock period, there is still an improvement in performance by applying intentional localized non-zero clock skew to the circuit. With zero clock skew, the minimum period is $T_{CP} = 27$ ns due to the worst case path delay of the local data path between registers R_2 and R_3.

7.2 EFFECTS OF CLOCK DISTRIBUTION DESIGN ON ARCHITECTURAL TRADEOFFS

Before the design effort of a clock distribution network can commence, certain timing constraints and goals must be specified. These timing characteristics are typically application-specific and depend greatly on the architectural and circuit tradeoffs of a given system implementation. A few papers exist that consider the effects of clock distribution on different aspects of these architectural tradeoffs. For example, Friedman and Mulligan [B128, B129, 4] describe the tradeoff between latency and clock frequency when pipelining a synchronous digital system. The authors provide equations and a graphical technique for determining the optimal level of pipelining.

As described earlier, registers are inserted into global data paths in order to increase the clock frequency of a digital system, albeit with an increase in latency. This tradeoff between clock frequency and latency is illustrated graphically in Figure 19. In this figure, both the latency and the clock period are shown as a function of the number of pipeline registers M inserted into a global data path. Thus, as M increases, the latency increases by the delay of each inserted register and any local clock skew while the maximum possible clock frequency increases, because the critical path is shortened (since there is less logic and interconnect delay per local data path).

If no registers are inserted into a data path, the minimum latency L_{Min} is the summation of the individual logic delays, NT_{fN}, where N is the number of logic stages per global data path and T_{fN} is the average delay per logic stage. For each register inserted into the global data path, L increases by $T_{Reg} + T_{Skew}$, assuming a constant register delay and clock skew per local data path, where $T_{Reg} = T_{C-Q} + T_{Set-up}$. Thus, L increases linearly with M, as shown in Figure 19.

The local data path having the largest value of T_{PD}, defined as the critical data path, establishes the maximum clock frequency f_{Clkmax} for the system. This argument assumes that the circuit is not a simple shift register and logical operations are being performed between the registers of each local data path. If the global data path is a simple shift register, the maximum clock frequency is limited by the maximum path delay T_{PD} of a local data path. The MAX subscript in Figure 19 is used to emphasize that the critical local data path limits the minimum

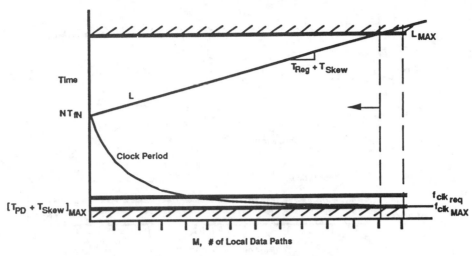

Fig. 19. Design Paradigm for Pipelined Synchronous Systems [B128, B129, 4].

clock period (maximum clock frequency) of the total global data path.

Most design requirements must satisfy some specified maximum time for latency while satisfying or surpassing a required clock frequency. The design constraints due to an application-specific limitation on the maximum permissible latency L_{Max} and the maximum possible clock frequency f_{Clkmax} are shown in Figure 19 by the vertical dashed lines. Thus, for a given L_{Max}, an appropriate maximum clock frequency and level of pipelining M are defined by the intersection of the L curve and the L_{Max} line. If L_{Max} is not specified or is very large and the desire is to make the clock frequency as high as possible, then an appropriate f_{Clk} is defined by the intersection of the clock period curve and the f_{Clkmax} line. Thus, for a particular L and f_{Clk}, the extent of the possible design space is indicated by the horizontal arrow. If L and f_{Clk} are both of importance and no L_{Max} or f_{Clkmax} is specified or constrains the design space, then some optimal level of pipelining is required to provide a "reasonably high" clock fre-

quency while maintaining a "reasonable" latency. This design choice is represented by a particular value of M, defining an application-specific f_{Clk} and L.

Effect of Positive Clock Skew on Pipelined Architecture

The effects of clock skew, technology, and logic architecture on clock period and latency are demonstrated graphically in Figures 20 and 21. If the clock skew is positive or if a poorer (i.e., slower) technology is used, then, as shown in Figure 20, L reaches L_{Max} at an increasingly smaller value of M. Also, the minimum clock period increases, which decreases the maximum clock frequency. Furthermore, large positive clock skew or a very poor technology eliminates any possibility of satisfying a specified clock frequency f_{Clkreq} and limits the entire design space as defined by the intersection of L and L_{Max}. For a poorer technology or logic architecture, the intersection between either the clock period or the latency curve and the ordinate shifts up-

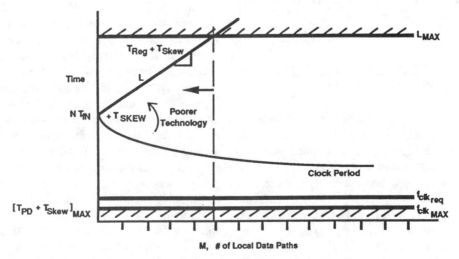

Fig. 20. Effect of Positive Clock Skew and Technology on Design Paradigm.

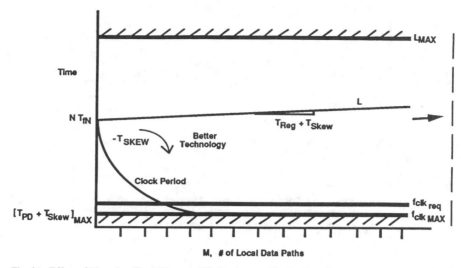

Fig. 21. Effect of Negative Clock Skew and Technology on Design Paradigm.

ward, since T_{fN} increases due to the slower technology and N increases for the less optimal architecture.

Effect of Negative Clock Skew on Pipelined Architecture

If the clock skew is negative or if a better (i.e., faster) technology is used, as shown in Figure 21, L reaches L_{Max} at a larger value of M. Also, the minimum clock period decreases, satisfying f_{Clkreq} and f_{Clkmax} with less pipelining. The possible design space, represented by the intersection of L and L_{Max}, is greatly increased, permitting higher levels of pipelining if very high clock rates are desired. In addition, for a better technology or logic architecture, the intersection between either the clock period or the latency curve and the ordinate shifts downward, since T_{fN} decreases due to the faster technology and N decreases for a more optimal architecture.

Thus, this graphical paradigm can be used to explore the tradeoff between latency and clock frequency in synchronous digital systems. Furthermore, it provides insight into how the design of the clock distribution network affects system-wide performance and pipelining decisions.

8. CLOCK DISTRIBUTION NETWORKS FOR TARGETED VLSI/WSI ARCHITECTURES

Unlike the general problem of designing clock distribution networks for structured custom VLSI circuits, certain circuit structures and architectures have been developed that lend themselves to specific synchronzation strategies and can create special problems and constraints. Some examples of these architectures are parallel processors, neural networks, multi-chip modules (MCMs) or hybrid cirsuits [B171], and wafer scale integrated (WSI) circuits. Interestingly, there exist several examples in which the research areas of parallel processing and wafer scale integration have been merged in order to build

highly dense, fine-grained parallel processors (in certain cases, performing neural processing) at the wafer level. The main reasons for this symbiotic relationship are that parallel processors become much more effective as the number of individual processors increase, and wafer scale integration is most appropriate when implemented with those circuit structures that are naturally redundant, thereby permitting localized wafer yield loss. This strategy permits those WSI processors that are faulty to be replaced via software or hardware techniques. In this section, clock distribution networks targeted to these architectures are described. Specifically, clock distribution in parallel processors and in wafer scale integrated circuits are discussed in subsections 8.1 and 8.2, respectively.

8.1 CLOCK DISTRIBUTION IN PARALLEL PROCESSORS

Many different parallel processing architectures exist. The clock distribution networks utilize a global clock to synchronize the data flow from one processor to the next, such as in a systolic array [13].

Much of the existing literature analyzing clock distribution networks in processor arrays uses graph theory to investigate the connectivity characteristics of these array structures. Fisher and Kung investigated clock distribution issues in large two-dimensional arrays clocked by a planar H-tree clock structure [B145, B146]. Two deterministic clock skew models are considered, one in which clock skew is proportional to the difference in path length from two nodes to the nearest common upstream tapping point of the clock distribution tree (*difference model*), and the other in which the clock skew is proportional to the sum of those lengths (*summation model*). Note that because the summation model is weaker and less constraining than the difference model, any clocking strategy that operates correctly with the summation model will operate correctly with the difference model. Since all sources of clock skew are treated collectively as a distance metric, the authors show that if small variations in delay are considered, clock

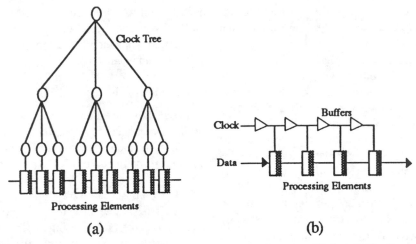

Fig. 22. Clocking Schemes Used in Systolic Arrays. a) Tree Structure and b) Straight-Line Structure [B150].

skew tends to grow with system size (the summation model). The summation model of Fisher and Kung [B146] predicts both an upper and lower bound of $\Theta\ (\sqrt{N})$ for the clock skew of an H-tree clock distribution network, where N is the number of processing elements.

One of the limiting factors in implementing long systolic arrays is the difficulty in providing high speed and reliably synchronized signals. The simplest form of a systolic array is a linear array, composed of a one-dimensional pipeline of processing elements (PEs). These one-dimensional arrays are particularly important in practical systems due to the large number of applications, and the performance is typically bounded by the speed of the I/O circuitry [B146]. In the synchronization of a linear systolic array, two types of clock distribution strategies are possible.

Tree-Structured Clocking Scheme

One equipotential clocking approach, similar to that shown in Figure 9, is the use of a buffered tree structure, as depicted in Figure 22a [B149, B150, B153]. For these networks, the clock delay becomes more severe as the size of the network increases. Note that in Figure 22a the circles represent distributed buffers and the edges (i.e., the connecting wires) represent the wires of the clock tree, while the rectangles are individual processing elements. Thus, the larger the systolic array, the greater the number of branches in the clock distribution network. Assuming that the interconnect resistance and capacitance of the clock distribution network is significantly greater than the driver source resistance and register input capacitance, the clock signal waveform will degrade as the array size increases [B148]. This degradation of the clock signal for increasing array size does not necessarily introduce clock skew, but may limit the maximum clock frequency of the system [28].

As additional PEs are added to the string (or the processor is scaled), the clock distribution network must be completely redesigned, since different loading appears at different points within the clock tree. If the PEs are on a single integrated circuit, then adding PEs to the string requires redesigning the clock distribution network. If, however, the PEs are on different integrated circuits, then the difference in clock delays driving different PEs will be more pronounced.

These issues of scalability represent some of the primary problems in applying global clocking strategies to ever expanding parallel processors. Dikaiakos and Steiglitz [B150], however, show that as the length of a one-dimensional systolic array synchronized by a tree-structured clock distribution network increases by N (the number of processing elements or the pipeline length), the clock period must be increased in proportion to $\log N$, to guarantee that the probability of a negative clock skew constraint relationship not being satisfied, as described by (6), remains negligible.

Buffered Straight-Line Clocking Scheme

A second strategy for distributing clock signals in systolic arrays is shown in Figure 22b. This approach is called *buffered clocking* (also straight-line or pipelined clocking [B146]) and uses a straight-line data-driven structure. The clock distribution network is composed of a series of distributed buffers generating successive clock pulses. These pulses originate from the same global clock signal, allowing clocking events to be pipelined through the network, so that several clock signals are simultaneously active within the system [B150, B153, 28]. Consecutive events are separated by the buffer delay as well as any local interconnect delay, which is the effective clock delay. Note that with respect to clocking large networks, buffered clocking is less limiting. Note that in Figure 22b each PE is synchronized by the clock signal from its left and a new buffered clock signal is used in communication with the PE to its right. Thus, data transfers are resynchronized with the buffered clock at each PE. Each buffered clock delay is chosen to be equal to the logic delay T_{PD}. This strategy, however, ignores the data dependency of the path delays. The output from one logic function is synchronized and passed to the successive logic function with minimum delay [28]. Buffered clocking, however, is only prac-

293

tical if the clock transmission between processors is sufficiently uniform to gain a performance advantage over the buffered tree structure [B146].

Dikaiakos and Steiglitz [B150] show that differences in delay between the rising and falling clock signal edges may create synchronization failure. The failure mode of the buffered clocking strategy is different from that commonly seen in tree-structured clock distribution networks as described in (5) and (6). For buffered clocking, clock skew should be redefined to be the deviation of the actual clock buffer delay from the desired clock buffer delay [28]. Careful control of the buffer delay characteristics is therefore required to ensure that each PE is synchronized properly. Also, as N increases, the clock period increases proportionally in order to ensure that the systolic pipeline will function properly. Thus, as systolic arrays increase in length, tree-structured clock distribution networks become preferable to straight-line clock distribution networks [B150], despite the added requirement of redesigning the clock distribution network due to the increased asymmetric loading.

The straight-line clocking structure is also difficult to generalize to two-dimensional arrays. Therefore, a hybrid globally asynchronous, locally synchronous clocking approach (see subsection 3.4) is recommended for two-dimensional processor arrays [B146]. In noting the self-timed nature of the synchronization pulses when using straight-line or buffered clocking, one readily sees why asynchronous strategies are commonly used in parallel processors, probably best exemplified by wavefront processors [13].

Both deterministic and probabilistic clock skew models have been developed to analyze clock distribution networks in parallel processors. One of the primary applications of a probabilistic clock skew model is highly regular architectures, such as parallel processors. Probabilistic clock skew models are further discussed in subsection 6.3.

8.2 CLOCK DISTRIBUTION NETWORKS IN WAFER SCALE INTEGRATED CIRCUITS

As system complexity, circuit densities, and clock frequencies evolve to more challenging levels, specialized circuit architectures become more enticing. One such architecture is wafer scale integration (WSI) [28]. In WSI, an entire multi-inch wafer is kept undiced and is packaged as a single wafer, utilizing only those circuits that function correctly. Redundancy techniques, either in hardware or software, are used to insulate the system from any non-yielding circuits on the wafer by functionally removing those circuits that do not behave correctly. Since wafer scale distances are measured in terms of several centimeters rather than VLSI signal lengths, typically on the order of a few millimeters, speed-of-light limitations may become significant when clocking high speed WSI circuits. Furthermore, as circuits are removed from the WSI system, loading within the clock distribution network changes. Therefore, it is necessary to design a clock distribution network that is robust enough to handle this varying load environment. These issues can severely affect the performance of high speed clock distribution networks in WSI systems. Therefore, this type of high density architecture requires a synchronization strategy that can satisfy these difficult and unusual requirements.

Application of H-Trees to WSI

In developing strategies for distributing clock signals in constrained architectures, specific solutions have arisen that are similar to the more general problem of distributing clock signals in largely unstructured VLSI systems, but that incorporate aspects peculiar to the WSI architecture. For example, Keezer and Jain [B141–B143] apply both passive and active (with distributed buffers) H-tree structures, as shown in Figures 10 and 11, to the problem of distributing synchronous clock signals across a large multi-inch wafer, implementing a large neural network processor. The authors integrate distributed active buffers into an H-tree structure (see Figure 23) in order to offset the significant waveform degradation that would arise due to the great distances over which the global clock signals must propagate across the wafer (a 7-cm line for a 4-in wafer [B143]). The final buffer at the end of each H-tree path drives a specific VLSI circuit die located on the wafer. In [B141, B143], the authors note that the distributed active buffers provide the greatest portion of the total variation in delay caused by any process variations across the wafer. In the example cited in [B141, B143], assuming each of the process parameters vary by $\pm 20\%$ of nominal value, approximately 60% of the clock skew induced by process tolerances are due to the variation of the delay of the internal buffers placed within the H-tree clock distribution network. Clock skews of over 5 ns across a poorly processed wafer are not uncommon. Therefore, process variations across a wafer represent one of the primary limitations to wafer scale circuits operating at high frequencies.

In contrast, Bakoglu [11] describes a completely passive (no active buffers are distributed within the H-tree structure) 3-in wafer scale implementation of an H-tree exhibiting clock skews less than 30 ps. Thus, *passive* H-tree structures provide significantly less variation in clock skew than *active* H-tree structures; however, this advantage must be weighed against the increased signal degradation that occurs in completely passive networks. A summary of some of the salient features of the H-tree structure used by Bakoglu are 1) a consistent fanout of two at each breakpoint, 2) minimal induced crosstalk from differently phased clock signals, and 3) controlled impedance of tapered transmission lines to minimize the reflections in the high frequency clock signals.

Clocking Issues in WSI

In [29], Johnstone and Butcher describe the design of power and clock distribution networks for a wafer scale-based systolic processor. The authors note the interrelationship between the two distribution systems and describe how to minimize power transients within the power distribution system by exploiting localized clock skew to shift the delays of each of the local data paths within the system. Thus, the current being drawn from the

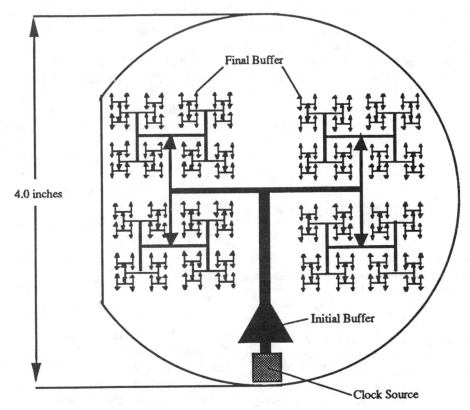

Fig. 23. Wafer Scale Four-Level H-Tree Clock Distribution Network [B141, B143].

power distribution circuitry is offset in time by applying localized clock skew, effectively "smoothing out" the transient current requirements of the power distribution over time.

In addition to the challenging requirements imposed on the design of a clock distribution network of a standard VLSI circuit, wafer scale integration places completely different and new requirements for synchronizing these large physical areas. Specifically, there is additional complexity due to the load impedance being variable; therefore, the buffers must be restructured. This variability in loading occurs since wafer yield is not expected to reach 100%. Redundancy techniques, both through programmable switches (soft restructuring) or through selective metal connections or fuses (hard restructuring), are therefore necessary [B143, 30]. In either case, the effects of these programmable circuit paths can significantly degrade or at best change the characteristics of the on-wafer clock distribution circuitry. High resistance elements placed near the driving source contribute more to the path delay than if placed near the receiving end (since the load resistance sees most of the load capacitance). One interesting result, described in [B143], is that high resistance fuses, when used in clock distribution networks, should be placed close to the clocked register element. Since it is preferable not to place high impedance links directly within the clock path, another strategy is to disable the clock drivers by building controllable tristate drivers within the clock buffers [30]. This procedure maintains the low impedance nature of the high speed clock lines while supporting the use of redundant drivers, important to wafer scale circuits.

Another option in synchronizing wafer scale circuits is to use some degree of asynchronous self-timing. Many synchronization possibilities exist, ranging from data flow architectures such as wavefront processors to the aforementioned globally clocked systolic arrays [13], including globally asynchronous, locally synchronous systems. One type of approach targeted for a fine-grained parallel processor utilizing wafer scale technology is summarized in [B140]. Two strategies are presented: either distribute a fast global clock signal and generate a slower clock locally, or distribute slower global out-of-phase clock signals and multiply them together to generate a high speed clock signal locally. Distributing a slow clock signal has the advantage of being less dependent on power supply variations across a wafer, although inaccuracies may develop during the out-of-phase multiplying process. The preferred synchronization strategy presented in [B140] is to distribute a slow clock signal globally and to multiply the clock signal further down the clock tree as the lower levels are reached. It should be noted, however, that the skew of the slow clock should be much less than the clock period of the fast clock.

In closing this topic on synchronizing WSI circuits, a highly interesting and provocative clock distribution strategy should be mentioned, specifically, distributing the clock signals across a large wafer (or printed circuit board) using optical transmitters to generate the clock signal and local electro-optical receivers (or detectors) to receive the clock signals. The clock signals are then distributed locally via standard electronic means. This strategy would have negligible clock skew and signal latency

295

(i.e., clock delay). Chou and Franklin [B144] discuss using silicon detectors placed advantageously across a wafer to provide system-wide clock distribution. These detectors would receive the transmitted global clock signal, convert it to a current (or voltage), and distribute locally the clock signal to neighboring local data paths (or circuit modules), thereby ameliorating the clock distribution problem across the entire system. Thus, as the requisite technologies mature, electro-optical techniques could potentially be used to distribute global clock signals across a wafer or system.

9. EXAMPLE IMPLEMENTATIONS OF CLOCK DISTRIBUTION NETWORKS

A number of interesting and innovative examples of high performance fully synchronous clock distribution networks have been developed for highly specialized and high performance commercial processors and have been described in the literature. These VLSI-based systems required an unusual combination of methodologies and practices commensurate with large design teams while maintaining the localized circuit optimization requirements important to high speed VLSI circuit design. The design of the clock distribution network used to synchronize some well known industrial circuit examples are discussed in this section.

9.1 THE BELL TELEPHONE BELLMAC-32A AND WE32100 32-BIT MICROPROCESSORS [B157, B158]

In the early 1980s, a family of 32-bit microprocessors was developed at Bell Laboratories using a variety of advanced CMOS circuit design techniques and methodologies. Since performance was of fundamental importance, significant attention was placed on synchronization, particularly the design of the global clock distribution network. In 1982, Shoji [B157] described the clock distribution of the BELLMAC-32A, a 146,000-transistor CPU operating at 8 MHz and built using a 2.5-μm single-level metal silicide CMOS technology. The clock delay and maximum tolerance of the clock distribution network were specified at 15 ns ± 3.5 ns, defining the maximum permissible clock skew. A four-phase clocking strategy was utilized; each phase synchronizes in order: the slave latches, the slave latch logic, the master latches, and the master latch logic. Each time a clock signal crosses a power buss, a silicide crossunder is used to route the clock signal. In order to equalize the series resistance, each clock path is routed with an identical number of three power buss crossunders from either of the two primary clock lines around the chip periphery. Buffers are placed strategically after each crossunder to amplify the degraded clock signal. With this clock distribution strategy, the circuit satisfies the clock frequency specification of 8 MHz at 70°C with the clock skew not exceeding ±3.5 ns. It is worth noting that, due to the significantly increased complexity encountered when distributing four separate clock signals, a four-phase clocking strategy is

not particularly appropriate for higher density, higher speed VLSI-based systems. This perspective is consistent with the processor described next.

In 1986, Shoji [B158] reported on the electrical design of the WE32100 CPU built using a 1.75-μm CMOS technology. The approach used in designing the WE32100 synchronizing clock system is described; local clock skew was optimized for a small number of the critical paths by applying negative clock skew. This strategy is consistent with the customized design methodology used in the design of the CPU. The clock distribution network utilizes a standard tree structure where the input clock signal is buffered by the clock driver and distributed over the entire circuit. Buffers are again placed after each crossunder. A strategy very similar to the approach presented in subsection 4.3 [B28] and depicted in Figure 14 is used to compensate for the variation in interconnect impedance and register load of each clock line. Clock edges at each register are further synchronized by adjusting the MOSFET transistor geometries of the distributed buffers within the clock distribution network.

Another circuit technique used in the WE32100 to minimize the dependence of the clock skew on any process variations is discussed in subsection 6.1. This technique minimizes process-induced clock skew caused by asymmetric variations of the device parameters of the N-channel and P-channel MOSFETs. Clock distribution networks with skews an order of magnitude less than conventionally designed circuits have been simulated with this technique [B107].

The issue of chip-to-chip synchronization is important in the design of integrated circuits because these circuits make up the components of a larger computing system. Since individual integrated circuits are processed in different wafer lots, the device parameters may differ, and therefore any internal delays, such as the clock line delays, will also differ. If the clock delay is T_{CD} for the slowest integrated circuit, as much as $T_{CD}/2$ clock skew may be developed when the fastest and slowest chips communicate synchronously. Therefore, a preferable strategy is to reduce the on-chip clock delay T_{CD} to a minimum and to preselect chips for similar clock delays [B158].

9.2 THE DEC 64-BIT ALPHA MICROPROCESSOR [B166]

An important application area for high speed clock distribution networks is the development of high speed microprocessors. The performance of these circuits is often limited by the clocking strategy used in their implementation. The DEC Alpha chip currently represents a significant milestone in microprocessor technology. The VLSI circuit operates above 200 MHz with a 3.3-V power supply implemented in 0.75-μm CMOS three-level metal technology. A clock period of 5 ns must be satisfied for each local data path. Therefore, the clock skew should be of very small magnitude (e.g., less than 0.5 ns for a 10% positive clock skew requirement). This strategy assumes the clock skew to be a global effect rather than a local effect. Thus, careful attention to modeling the circuits and interconnects is required in order to design and analyze this type of high speed system. The Alpha microprocessor contains 1.68 million transistors and supports a fully pipelined 64-bit data structure. The functional attributes of the microprocessor are described in greater detail in [B166], since the focus herein is on the clocking strategy used within the circuit.

In designing this high speed microprocessor, significant attention has been placed on the circuit implementation. The single-phase clock signal is distributed globally on the topmost level of the trilevel metal process, as is the power distribution, since the third layer of metal is thicker and wider (7.5-μm pitch with contacts as compared to 2.625 μm and 2.25 μm for the second and first layers of metal, respectively). Therefore, the resistivity per unit length of the third layer of metal and the metal-to-substrate capacitance is less. A number of inherent difficulties exist within the clock distribution requirements of the Alpha chip. For example, a substantial capacitive load must be driven at high speed by the clock distribution network, 3250 pF (3.25 nF). Also, for both latch design and power dissipation reasons, so as to minimize short-circuit current [31], a fast clock edge rate (<0.5 ns) must be maintained throughout the clock distribution network. The huge capacitative load is due to the 63,000 transistor gates being driven by the clock distribution system. The distribution of the loads is non-symmetric, necessitating a specialized strategy for distributing the clock.

The single 200-MHz clock signal is distributed through five levels of buffering, where the total network consists of 145 separate elements. Each of the elements contains four levels of buffering with a final output stage locally driving the clocked registers. These distributed buffers are configured as a mesh as shown in Figure 24 [B43]. Vertical straps are placed on the second level of metal (M2) to minimize any skew that may develop within the initial four-stage portion of the buffer tree. The primary signal-wide distribution is on the third level of metal (M3), designed to be particularly thick to minimize any line resistance as well as to improve process yield.

The approach used in designing the clock distribution of the Alpha chip is to permit only positive clock skew, thereby assuring that no catastrophic race conditions induced by negative clock skew can occur. Thus, only a graceful degradation in maximum clock rate caused by localized positive clock skew is possible. This strategy is accomplished by locating the clock generation circuitry centrally within the integrated circuit. The clock signal is then distributed radially from the center of the chip die to its periphery. By carefully monitoring the design of this clock distribution methodology, the likelihood of developing a catastrophic amount of negative clock skew (i.e., $|T_{Skew}| > T_{PD(min)}$) is minimized.

9.3 Pipelined Multiplier [B11, B139, B159, B160, B170, 32]

Another application area that requires sophisticated clock distribution is heavily pipelined digital signal processors (DSP), such as FIR/IIR digital filters, multiply-adders, multiply-accumulators, and frequency synthesizers. These types of circuits repeatedly use similar circuit elements, such as multipliers, adders, and registers. Careful attention is placed on developing high performance customized versions of these circuit elements, which are used repeatedly within larger DSP systems. The primary difference between different versions of these DSP circuit components is typically the size of the bit slice (e.g., 8-bit versus 64-bit) and the degree of pipelining. Substantial pipelining is applied to these circuits to increase system clock frequency (see [B128, B129, 4] and subsection 7.2).

The multiplier function is a good example of a complex circuit element capable of significant improvements in clock rate with high levels of pipelining. Since the data flow is nonrecursive, fewer pipeline registers are required as compared to those structures that contain substantial feedback. Furthermore, the multiplier tends to be the critical element (in terms of speed, area, and power) in most DSP circuits. Heavily pipelined multipliers requiring sophisticated clock distribution networks are the focus of considerable research. In this subsection, specific circuit examples of clock distribution networks in highly pipelined DSP-based multipliers implemented in VLSI technologies are described.

A common feature of these VLSI-based multipliers (and many VLSI-based systems) is the repetitive organization of the physical layout. Repetitive parallel arrays of abutted adder cells, pipelined at each bit (a register placed between each adder cell), provide worst case path delays of only a single adder and a register delay (T_{C-Q} and T_{Set-up}), permitting very high multiplication throughput. Specialized architectures beyond the scope of this book, such as carry-save addition, are used to further improve the throughput of these VLSI-based multipliers [see e.g., B11, B139, B159, B160, B170, 32].

In highly arrayed structures, clock skew can appear both horizontally (serial skew), in the direction of the data flow, and vertically (parallel skew), orthogonal to the data flow. As described by Hatamian and Cash in [B139, B159] and pictured in Figure 25, assuming the clock source originates from point A, the clock skew between points D and E at the cell inputs is quite close to the clock skew between points B and C. As long as this horizontal clock skew is less than the local data path delay between cells, no negative clock skew condition will occur [see subsection 3.2 and (6)], and the multiplier array will operate properly. Furthermore, additional cells can be added to the array without

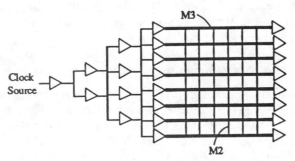

Fig. 24. Clock Distribution Network of DEC Alpha Microprocessor.

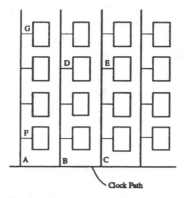

Fig. 25. Clock Distribution Network of Multiplier Array [B139, B159].

creating any race conditions as long as the same constraint is maintained. Unlike the horizontal skew, however, the vertical skew is cumulative. The clock skew increases as the signal propagates vertically from point A. Thus, the cumulative skew between points F and G dominates over the horizontal skew, again assuming the clock source originates from point A. It is worth noting that the highly arrayed structure of the multiplier, which makes it particularly amenable to a VLSI implementation, also constrains and limits the topology and layout of the clock distribution network used to synchronize the circuit.

In the 8-bit X 8-bit multiplier described in [B139, B159], which is implemented in a 2.5-μm CMOS technology and operates up to 70 MHz, the clock signals are distributed entirely on metal, except where short polysilicon crossunders are used to transverse the power lines. A two-level buffer clock distribution network is used, where the clock load is balanced at the output of the second buffer, and the path before the second buffer is kept symmetric with the other clock paths. This clock distribution strategy is similar to that described in [B160]. Also an 8-bit X 8-bit multiplier, the multiplier described in [B160] is implemented in a 1.0-μm NMOS technology and operates up to 330 MHz at room temperature and up to 600 MHz with liquid nitrogen cooling. Pipelining after each 1-bit addition, the multiplier architecture is very similar to that described in [B139]. The clock signals are distributed in metal outside the multiplier array and in polycide (strapped polysilicon with silicide to minimize the line resistance) inside the multiplier array. Two-phase clocking is used with a total master and slave register fanout of 8 pF. No special circuitry to prevent overlap of the two-phase clock is used, since this practice degrades the active-high portion of the clock signal.

Maximal pipelining of multiplying architectures is taken one step further by pipelining each half-bit of an 8-bit X 8-bit multiplier [B170]. Operating at 230 MHz and implemented in a 1.6-μm double-level metal CMOS process, the multiplier architecture is similar to that used in [B139, B159, B160]. The circuit uses a single-phase clocking scheme [33]. A standard three-level buffer network is used to distribute the clock signal. Each row of the multiplier provides a capacitive load of 5.3 pF. A common clock line runs horizontally in metal one (M1) and is driven by a large sized buffer. The complete buffer tree is composed of fourteen buffers, where ten of these buffers drive the registers (i.e., the leaves of the tree). The maximum skew measured between two clock lines is 150 ps [B170].

These three multiplier examples are intended to provide some insight into distributing clock signals within highly arrayed VLSI-based DSP systems. The primary attributes of the multiplier—repetitive circuit elements, abutted or closely spaced layouts, and extremely high throughput (i.e., many hundreds of MHz) due to heavy pipelining—make the multiplier an important class of VLSI system that requires highly specialized clock distribution networks.

10. DIRECTIONS FOR FUTURE RESEARCH IN THE DESIGN OF CLOCK DISTRIBUTION NETWORKS

Significant research opportunities remain in the design of clock distribution networks. Some examples of these research areas are briefly described in this section.

Automated Synthesis of Clock Distribution Networks

Much of the current research focuses on automating the synthesis of clock distribution networks to support higher performance requirements. The optimal placement of locally distributed buffers, improved delay models that account for nonlinear active transistor behavior, the use of localized clock skew to increase circuit speed, and the integration of RC interconnect and buffer physical delay models for more accurate delay analysis must be considered in the automated design and layout of clock distribution networks. The effects of clock skew, both positive and negative, must also be integrated into behavioral synthesis and RC timing analyzers to detect race conditions as well as to satisfy local and global performance constraints. Synchronous timing constraints must be integrated into high level behavioral synthesis algorithms, thereby improving the accuracy and generality of these synthesis (and simulation) tools.

Most clock distribution networks are tree-structured; however, in many customized VLSI circuits, certain portions of the network are strapped in a mesh configuration to minimize interconnect resistance. These mesh networks decrease clock skew as well as improve circuit reliability. Therefore, one area of future research is the automated layout of clock meshes [B91]. Both timing models and physical layout must be integrated to handle this important extension of tree-structured clock distribution networks.

Design of Process Insensitive Clock Distribution Networks

A research area of primary importance to the practical design of high speed clock distribution networks is improving the tolerance of these networks to both process and environmental variations. As functional clock periods approach 1 ns (i.e., 1 GHz clock frequency), variations in delay of tens of picoseconds could degrade the performance and reliability of these high speed synchronous systems. Variations of this magnitude are quite common in modern semiconductor technologies. This topic is of immediate importance to the existing high speed processor community.

Design of Microwave Frequency Clock Distribution Networks

As system-wide clock frequencies increase beyond Gigahertz levels, transmission line effects will begin to influence the performance characteristics of clock distribution networks. Models of interconnect sections will require the inclusion of accurate inductive elements. These and related microwave effects will become more apparent as ultra-high speed digital technologies become more readily available. For example, the superconductive digital electronic technology, Single Flux Quantum (SFQ) logic [B51, 34], can operate well into the 10- to 100-GHz frequency range.

System Issues Affecting the Design of Clock Distribution Networks

System issues in the design of clock distribution networks also necessitate focused research. Important and developing architectural structures, such as parallel processors, neural networks, supercomputers, hybrid circuits or monolithic multichip modules [B171], and wafer scale integration all require specialized synchronization strategies. Some approaches, such as WSI-based optical clock distribution, have been discussed briefly within this introductory chapter and represent important and challenging areas of research. Related systems issues, such as ultra-low power circuits requiring ultra-low power clock distribution networks [B52–B54], are becoming increasingly important. Improving the power dissipation characteristics of clock distribution networks is particularly important, since these networks dissipate a large portion of the total system-wide power budget. Topics such as distributing small differential signals and task monitoring (or power management) strategies are important areas of research in low power and low voltage clock distribution network design.

Debug and Production Test of Clock Distribution Networks

A necessary requirement in developing a product is evaluating the quality of that product. Both debug and production test of high performance clock distribution networks are of fundamental importance. As exemplified in subsection 4.6, minimal research exists that describes how to debug high speed clock distribution networks as well as how to test these networks in a production environment.

A testimony of the importance of the design and analysis of clock distribution networks to future VLSI systems is the large number of recent PhD dissertations on this topic. A few are referenced here [7, 14, 27, 32, 35–39].

11. SUMMARY AND CONCLUSIONS

All electronic systems are fundamentally asynchronous in nature; by the careful insertion of precise localized timing relationships and storage elements, an asynchronous system can be adapted to appear to behave synchronously. As long as the specific local timing and functional relationships are satisfied, synchronous systems can be used, easing the timing constraints on data flow, albeit requiring a clock distribution network to provide the synchronizing reference signal. By synchronizing with distributed clock signals, clock frequency (a measure of how often new data appear at the output of a system) will remain the primary performance metric in synchronous systems. Furthermore, systems can be developed by careful design of the clock distribution network that operate at performance levels otherwise unattainable without significant architectural or technological improvements.

It is often noted that the design of the clock distribution network represents the fundamental circuit limitation to performance in high speed synchronous digital systems. The local data path dependent nature of clock skew, rather than its global characteristics, requires extreme care in the design, analysis, and evaluation of high speed clock distribution networks. The design complexity and difficulty in scaling these networks to finer geometries are the primary reasons for the recent emphasis placed on asynchronous systems. Clearly, however, synchronous systems will be commonplace for a long time to come, necessitating improved techniques for designing and implementing high speed, highly reliable clock distribution networks. Furthermore, as tighter control of the clocking parameters improves, approaches such as localized clock skew will be applied more generally to the design of clock distribution networks to further enhance system performance.

A singular commentary on the current immaturity of the research area of clock distribution design is the complete lack of an agreed upon terminology and notation defining the primary concepts and terms. This problem is evidenced by the large variety of terms used to describe such issues as race conditions, negative clock skew, and T_{Skew}.

In this book on synchronous VLSI-based clock distribution networks, timing relationships are examined and are used to constrain the timing characteristics of the overall system. Various architectures and applications are considered, and circuit strategies for distributing the clock signals are offered. It is the intention of this introductory chapter to integrate the various subtopics and to provide some sense of cohesiveness to the field of clocking and, specifically, clock distribution networks.

ACKNOWLEDGMENT

I would like to acknowledge J.L.C. Neves for his significant assistance with Sections 2 and 7.

REFERENCES

Note: references beginning with a "B" are placed in the bibliography following this immediate list of references. All other references are listed below in the order of their citation.

[1] K. Wagner and E. McCluskey, "Tuning, Clock Distribution, and Communication in VLSI High-Speed Chips," Stanford University, Palo Alto, California, CRC Technical Report 84-5, June 1984.

[2] K. D. Wagner, "A Survey of Clock Distribution Techniques in High-Speed Computer Systems," Stanford University, Palo Alto, California, CRC Technical Report No. 86-20, December 1986.

[3] E. G. Friedman, "Latching Characteristics of a CMOS Bistable Register," *IEEE Transactions on Circuits and Systems I: Fundamental Theory and Applications,* Vol. CAS-40, No. 12, pp. 902–908, December 1993.

[4] E. G. Friedman and J. H. Mulligan, Jr., "Pipelining and Clocking of High Performance Synchronous Digital Systems," *VLSI Signal Processing Technology,* M. A. Bayoumi and E. E. Swartzlander, Jr., eds., pp. 97–133, Kluwer Academic Publishers, 1994.

[5] M. Hatamian, "Understanding Clock Skew in Synchronous Systems," *Concurrent Computations (Algorithms, Architecture and Technology),* S. K. Tewksbury, B. W. Dickinson, and S. C. Schwartz, eds., pp. 87–96, Plenum Publishing, 1988.

[6] J. Alves Marques and A. Cuhna, "Clocking of VLSI Circuits," *VLSI Architecture,* Randell and Treleaven, eds., pp. 165–178, Prentice-Hall, 1983.

[7] E. G. Friedman, *Performance Limitations in Synchronous Digital Systems,* PhD dissertation, University of California, Irvine, California, June 1989.

[8] T. W. Williams and K. P. Parker, "Design for Testability—A Survey," *Proceedings of the IEEE,* Vol. 71, No. 1, pp. 98–112, January 1983.

[9] R. H. Dennard, F. H. Gaensslen, H-N. Yu, V. L. Rideout, E. Bassous, and A. R. LeBlanc, "Design of Ion-Implanted MOSFET's with Very Small Physical Dimensions," *IEEE Journal of Solid-State Circuits,* Vol. SC-9, No. 5, pp. 256–268, October 1974.

[10] H. B. Bakoglu and J. D. Meindl, "Optimal Interconnection Circuits for VLSI," *IEEE Transactions on Electron Devices,* Vol. ED-32, No. 5, pp. 903–909, May 1985.

[11] H. B. Bakoglu, *Circuits, Interconnections, and Packaging for VLSI,* Addison Wesley, 1990.

[12] C. V. Gura, "Analysis of Clock Skew in Distributed Resistive-Capacitive Interconnects," University of Illinois, Urbana, Illinois, SRC Technical Report No. T87053, June 1987.

[13] S. Y. Kung, *VLSI Array Processors,* Prentice-Hall, 1988.

[14] D. M. Chapiro, *Globally-Asynchronous Locally-Synchronous Systems,* PhD dissertation, Stanford University, Palo Alto, California, October, 1984.

[15] D. J. Kinniment and J. V. Woods, "Synchronization and Arbitration Circuits in Digital Systems," *Proceedings of the IEE,* pp. 10–14, 1st Quarter 1980.

[16] T. Sakurai, "Optimization of CMOS Arbiter and Synchronization Circuits with Submicrometer MOSFET's," *IEEE Journal of Solid-State Circuits,* Vol. SC-23, No. 4, pp. 901–906, August 1988.

[17] D. D. Gajski, *Silicon Compilation,* Addison Wesley, 1988.

[18] R. H. Krambeck and M. Shoji, "Skew-Free Clock Circuit for Integrated Circuit Chip," Patent #4,479,216, AT&T Bell Laboratories, October 23, 1984.

[19] C. M. Lee and B. T. Murphy, "Trimmable Loading Elements to Control Clock Skew," Patent #4,639,615, AT&T Bell Laboratories, January 27, 1987; *IEEE Journal of Solid-State Circuits,* Vol. SC-22, No. 6, pp. 1220, December 1987.

[20] S. Dhar and M. A. Franklin, "Optimum Buffer Circuits for Driving Long Uniform Lines," *IEEE Journal of Solid-State Circuits,* Vol. SC-26, No. 1, pp. 32–40, January 1991.

[21] G. Yacoub, H. Pham, M. Ma, and E. G. Friedman, "A System for Critical Path Analysis Based on Back Annotation and Distributed Interconnect Impedance Models," *Microelectronics Journal,* Vol. 19, No. 3, pp. 21–30, May/June 1988.

[22] W. C. Elmore, "The Transient Response of Damped Linear Networks with Particular Regard to Wideband Amplifiers," *Journal of Applied Physics,* Volume 19, pp. 55–63, January 1948.

[23] J. Rubinstein, P. Penfield, Jr., and M. A. Horowitz, "Signal Delay in RC Tree Networks," *IEEE Transactions on Computer-Aided Design,* Vol. CAD-2, No. 3, pp. 202–211, July 1983.

[24] J. Vlach, J. A. Barby, A. Vannelli, T. Talkhan, and C. J. Shin, "Group Delay as an Estimate of Delay on Logic," *IEEE Transactions on Computer-Aided Design,* Vol. CAD-10, No. 7, pp. 949–953, July 1991.

[25] T. Sakurai, "Approximation of Wiring Delay in MOSFET LSI," *IEEE Journal of Solid-State Circuits,* Vol. SC-18, No. 4, pp. 418–426, August 1983.

[26] A. Papoulis, *Probability, Random Variables, and Stochastic Processes,* McGraw-Hill, 1965.

[27] C. T. Gray, *Optimal Clocking of Wave Pipelined Systems and CMOS Applications,* PhD dissertation, North Carolina State University, Raleigh, North Carolina, July 1993.

[28] S. K. Tewksbury, *Wafer-Level Integrated Systems: Implementation Issues,* Kluwer Academic Publishers, 1989.

[29] K. K. Johnstone and J. B. Butcher, "Power Distribution for Highly Parallel WSI Architectures," *Proceedings of the IEEE International Conference on Wafer Scale Integration,* pp. 203–214, January 1989.

[30] J. Fried, "An Analysis of Power and Clock Distribution for WSI Systems," *Wafer Scale Integration,* G. Saucier and J. Trilhe, eds., pp. 127–142, Elsevier Science Publishers B.V. (North-Holland), 1986.

[31] H.J.M. Veendrick, "Short-Circuit Dissipation of Static CMOS Circuitry and its Impact on the Design of Buffer Circuits," *IEEE Journal of Solid-State Circuits,* Vol. SC-19, No. 4, pp. 468–473, August 1984.

[32] M. R. Santoro, *Design and Clocking of VLSI Multipliers,* PhD dissertation, Stanford University, Palo Alto, California, October 1989.

[33] Y. Jiren, I. Karlsson, and G. Svensson, "A True Single-Phase-Clock Dynamic CMOS Circuit Technique," *IEEE Journal of Solid-State Circuits,* Vol. SC-22, No. 4, pp. 899–901, October 1987.

[34] K. K. Likharev and V. K. Semenov, "RSFQ Logic/Memory Family: A New Josephson-Junction Technology for Sub-Terahertz Clock-Frequency Digital Systems," *IEEE Transactions on Applied Superconductivity,* Vol. 1, No. 1, pp. 3–28, March 1991.

[35] D. C. Noice, *A Clocking Discipline for Two-Phase Digital Integrated Circuits,* PhD dissertation, Stanford University, Palo Alto, California, January 1983.

[36] Y. H. Kim, *Accurate Timing Verification for Digital VLSI Designs,* PhD dissertation, University of California, Berkeley, California, January 1989.

[37] P. R. Mukund, *Optimal Clock Distribution in VLSI Systems,* PhD dissertation, University of Tennessee, Knoxville, Tennessee, December 1990.

[38] D. Wong, *Techniques for Designing High-Performance Digital Circuits using Wave-Pipelining,* PhD dissertation, Stanford University, Palo Alto, California, September 1991.

[39] D. A. Joy, *Clock Period Minimization with Wave Pipelining,* PhD dissertation, University of Massachusetts, Amherst, Massachusetts 1991.

BIBLIOGRAPHY

Reference numbers with asterisks refer to papers included in the book. Citations beginning with a "B" refer to this list of publications.

1. BASIC CONCEPTS AND ANALYSIS

[1]* F. Anceau, "A Synchronous Approach for Clocking VLSI Systems," *IEEE Journal of Solid-State Circuits,* SC-17, No. 1, pp. 51–56, February 1982.

[2] M. A. Franklin and D. F. Wann, "Asynchronous and Clocked Control Structures for VLSI Based Interconnection Networks," *Proceedings of 9th Annual Symposium on Computer Architecture,* pp. 50–59, April 1982.

[3]* D. Wann and M. Franklin, "Asynchronous and Clocked Control Structures for VLSI Based Interconnection Networks," *IEEE Transactions on Computers,* Vol. C-32, No. 3, pp. 284–293, March 1983.

[4]* S. Dhar, M. Franklin, and D. Wann, "Reduction of Clock Delays in VLSI Structures," *Proceedings of IEEE International Conference on Computer Design,* pp. 778–783, October 1984.

[5] S. Unger and C-J. Tan, "Optimal Clocking Schemes for High Speed Digital Systems," *Proceedings of IEEE International Conference on Computer Design,* pp. 366–369, October 1983.

[6] J. Beausang and A. Albicki, "A Method to Obtain an Optimal Clocking Scheme for a Digital System," *Proceedings of IEEE International Conference on Computer Design,* pp. 68–72, October 1983.

[7]* S. H. Unger and C-J. Tan, "Clocking Schemes for High-Speed Digital Systems," *IEEE Transactions on Computers,* Vol. C-35, No. 10, pp. 880–895, October 1986.

[8] D. Noice, R. Mathews, and J. Newkirk, "A Clocking Discipline for Two-Phase Digital Systems," *Proceedings of IEEE International Conference on Circuits and Computers,* pp. 108–111, September 1982.

[9] C. Svensson, "Signal Resynchronization in VLSI System," *Integration, the VLSI Journal,* Vol. 4, No. 1, pp. 75–80, March 1986.

[10] M. S. McGregor, P. B. Denyer, and A. F. Murray, "A Single-Phase Clocking Scheme for CMOS VLSI," *Proceedings of the Stanford Conference on Advanced Research in VLSI,* pp. 257–271, March 1987.

[11]* M. Hatamian and G. L. Cash, "Parallel Bit-Level Pipelined VLSI Designs for High-Speed Signal Processing," *Proceedings of the IEEE,* Vol. 75, No. 9, pp. 1192–1202, September 1987.

[12] M. Hatamian, L. A. Hornak, T. E. Little, S. T. Tewksbury, and P. Franzon, "Fundamental Interconnection Issues," *AT&T Technical Journal,* Volume 66, Issue 4, pp. 13–30, July/August 1987.

[13]* K. D. Wagner, "Clock System Design," *IEEE Design & Test of Computers,* pp. 9–27, October 1988.

[14]* A. F. Champernowne, L. B. Bushard, J. T. Rusterholz, and J. R. Schomburg, "Latch-to-Latch Timing Rules," *IEEE Transactions on Computers,* Vol. C-39, No. 6, pp. 798–808, June 1990.

[15]* D. G. Messerschmitt, "Synchronization in Digital System Design," *IEEE Journal on Selected Areas in Communications,* Vol. 8, No. 8, pp. 1404–1419, October 1990.

[16]* M. Afghahi and C. Svensson, "Performance of Synchronous and Asynchronous Schemes for VLSI Systems," *IEEE Transactions on Computers,* Vol. C-41, No. 7, pp. 858–872, July 1992.

[17] V. Chi, "Designing Salphasic Clock Distribution Systems," *Proceedings of Symposium on Integrated Systems,* pp. 219–233, March 1993.

[18]* V. L. Chi, "Salphasic Distribution of Clock Signals for Synchronous Systems," *IEEE Transactions on Computers,* Vol. C-43, No. 5, pp. 597–602, May 1994.

[19] M. C. Papaefthymiou and K. H. Randall, "Edge-Triggering vs. Two-Phase Level-Clocking," *Proceedings of Symposium on Integrated Systems,* pp. 201–218, March 1993.

[20] W.K.C. Lam, R. K. Brayton, and A. L. Sangiovanni-Vincentelli, "Valid Clocking in Wavepipelined Circuits," *Proceedings of IEEE International Conference on Computer-Aided Design,* pp. 124–131, November 1992.

[21] D. A. Joy and M. J. Ciesielski, "Clock Period Minimization with Wave Pipelining," *IEEE Transactions on Computer-Aided Design,* Vol. CAD-12, No. 4, pp. 461–472, April 1993.

[22]* C. T. Gray, W. Liu, and R. K. Cavin, III, "Timing Constraints for Wave-Pipelined Systems," *IEEE Transactions on Computer-Aided Design,* Vol. CAD-13, No. 8, pp. 987–1004, August 1994.

[23] X. Zhang and R. Sridhar, "Synchronization of Wave-Pipelined Circuits," *Proceedings of IEEE International Conference on Computer Design,* pp. 164–167, October 1994.

[24] E. G. Friedman, "Clock Distribution Design in VLSI Circuits—an Overview," *Proceedings of IEEE International Symposium on Circuits and Systems,* pp. 1475–1478, May 1993.

[25]* K. A. Sakallah, T. N. Mudge, T. M. Burks, and E. S. Davidson, "Synchronization of Pipelines," *IEEE Transactions on Computer-Aided Design,* Vol. CAD-12, No. 8, pp. 1132–1146, August 1993.

[26] R. Peset Llopis, L. Ribas Xirgo, and J. Carrabina Bordoll, "Short Destabilizing Paths in Timing Verification," *Proceedings of IEEE International Conference on Computer Design,* pp. 160–163, October 1994.

[27] S.-Z. Sun, D.H.C. Du, Y.-C. Hsu, and H.-C. Chen, "On Valid Clocking for Combinational Circuits," *Proceedings of IEEE International Conference on Computer Design,* pp. 381–384, October 1994.

2. CLOCK DISTRIBUTION DESIGN OF STRUCTURED CUSTOM VLSI CIRCUITS

[28]* E. G. Friedman and S. Powell, "Design and Analysis of a Hierarchical Clock Distribution System for Synchronous Standard Cell/Macrocell VLSI," *IEEE Journal of Solid-State Circuits,* Vol. SC-21, No. 2, pp. 240–246, April 1986.

[29] E. Friedman, "A Partitionable Clock Distribution System for Sequential VLSI Circuits," *Proceedings of IEEE International Symposium on Circuits and Systems,* pp. 743–746, May 1986.

[30]* H. B. Bakoglu, J. T. Walker, and J. D. Meindl, "A Symmetric Clock-Distribution Tree and Optimized High-Speed Interconnections for Reduced Clock Skew in ULSI and WSI Circuits," *Proceedings of IEEE International Conference on Computer Design,* pp. 118–122, October 1986.

[31] "Method of Deskewing Data Pulses," *IBM Technical Disclosure Bulletin,* Vol. 28, No. 6, pp. 2658–2659, November 1985.

[32] M. Afghahi and C. Svensson, "A Scalable Synchronous System," *Proceedings of IEEE International Symposium on Circuits and Systems,* pp. 471–474, May 1988.

[33]* M. Afghahi and C. Svensson, "A Unified Single-Phase Clocking Scheme for VLSI Systems," *IEEE Journal of Solid-State Circuits,* Vol. SC-25, No. 1, pp. 225–233, February 1990.

[34] B. Wu and N. A. Sherwani, "Effective Buffer Insertion of Clock Tree for High-Speed VLSI Circuits," *Microelectronics Journal,* Vol. 23, No. 4, pp. 291–300, July 1992.

[35]* G. M. Blair, "Skew-Free Clock Distribution for Standard-Cell VLSI Designs," *IEE Proceedings-G,* Vol. 139, No. 2, pp. 265–267, April 1992.

[36]* S. Padin, "Scheme for Distributing High-Speed Clock Signals in a Large Digital System," *Electronics Letters,* Vol. 25, No. 2, pp. 92–93, January 1989.

[37] M. A. Cirit, "Clock Skew Elimination in CMOS VLSI," *Proceedings of IEEE International Symposium on Circuits and Systems,* pp. 861–864, May 1990.

[38]* F. Minami and M. Takano, "Clock Tree Synthesis Based on RC Delay Balancing," *Proceedings of IEEE Custom Integrated Circuits Conference,* pp. 28.3.1–28.3.4, May 1992.

[39] D. Mijuskovic, "Clock Distribution in Application Specific Integrated Circuits," *Microelectronics Journal,* Vol. 18, pp. 15–27, July/August 1987.

[40] D. Renshaw and C. H. Lau, "Race-Free Clocking of CMOS Pipelines Using a Single Global Clock," *IEEE Journal of Solid-State Circuits,* Vol. SC-25, No. 3, pp. 766–769, June 1990.

[41]* I. Deol, "Automatic Analysis of Circuits for Tester Skew and Clock Distribution for VLSI Circuits," *Proceedings of the IEEE International Conference on Computer-Aided Design,* pp. 350–353, October 1987.

[42]* D. C. Keezer, "Design and Verification of Clock Distribution in VLSI," *Proceedings of IEEE International Conference on Communications,* pp. 811–816, April 1990.

[43] M. Horowitz, "Clocking Strategies in High Performance Processors," *Proceedings of the IEEE Symposium on VLSI Circuits,* pp. 50–53, June 1992.

[44] M. Nekili, Y. Savaria, G. Bois, and M. Bennani, "Logic-Based H-Trees for Large VLSI Processor Arrays: A Novel Skew Modeling and High-Speed Clocking Method," *Proceedings of the 5th International Conference on Microelectronics,* pp. 144–147, December 1993.

[45] W. Chuang, S. S. Sapatnekar, and I. N. Hajj, "A Unified Algorithm for Gate Sizing and Clock Skew Optimization to Minimize Sequential Circuit Area," *Proceedings of IEEE International Conference on Computer-Aided Design,* pp. 220–223, November 1993.

[46]* C. Kraft, "Harmonic Series Analysis of Digital Clock Distribution Circuits," *Proceedings of 32nd IEEE Midwest Symposium on Circuits and Systems,* pp. 206–211, August 1989.

[47]* M. Bußmann and U. Langmann, "Active Compensation of Interconnect Losses for Multi-GHz Clock Distribution Networks," *IEEE Transactions on Circuits and Systems-II: Analog and Digital Signal Processing,* Vol. CAS-39, No. 11, pp. 790–798, November 1992.

[48] Q. Zhu, W.W.-M. Dai, and J. G. Xi, "Optimal Sizing of High Speed Clock Networks Based on Distributed RC and Lossy Transmission Line Models," *Proceedings of IEEE International Conference on Computer-Aided Design,* pp. 628–633, November 1993.

[49] P. Ramanathan, A. J. Dupont, and K. G. Shin, "Clock Distribution in General VLSI Circuits," *IEEE Transactions on Circuits and Systems-I: Fundamental Theory and Applications,* Vol. CASI-41, No. 5, pp. 395–404, May 1994.

[50] W. D. Grover, "A New Method for Clock Distribution," *IEEE Transactions on Circuits and Systems-I: Fundamental Theory and Applications,* Vol. CASI-41, No. 2, pp. 149–160, February 1994.

[51] K. Gaj, E. G. Friedman, M. J. Feldman, and A. Krasniewski, "A Clock Distribution Scheme for Large RSFQ Circuits," *IEEE Transactions on Applied Superconductivity,* 1995.

[52]* E. De Man and M. Schobinger, "Power Dissipation in the Clock System of Highly Pipelined ULSI CMOS Circuits," *Proceedings of the International Workshop on Low Power Design,* pp. 133–138, April 1994.

[53] H. Kojima, S. Tanaka, and K. Sasaki, "Half-Swing Clocking Scheme for 75% Power Saving in Clocking Circuitry," *Proceedings of the IEEE Symposium on VLSI Circuits,* pp. 23–24, June 1994.

[54] J. L. Neves and E. G. Friedman, "Minimizing Power Dissipation in Non-Zero Skew-based Clock Distribution Networks," *Proceedings of the IEEE International Symposium on Circuits and Systems,* May 1995.

3. AUTOMATED LAYOUT AND SYNTHESIS OF CLOCK DISTRIBUTION NETWORKS

[55]* Y. Ogawa, T. Ishii, Y. Shiraishi, H. Terai, T. Kozawa, K. Yuyama, and K. Chiba, "Efficient Placement Algorithms Optimizing Delay for High-Speed ECL Masterslice LSI's," *Proceedings of ACM/IEEE 23rd Design Automation Conference,* pp. 404–410, June 1986.

[56]* S. Boon, S. Butler, R. Byrne, B. Setering, M. Casalanda, and A. Scherf, "High Performance Clock Distribution for CMOS ASICs," *Proceedings of IEEE Custom Integrated Circuits Conference,* pp. 15.4.1–15.4.5, May 1989.

[57] A. Chao, "Clock Tree Synthesis for Large Gate Arrays," *High Performance Systems,* p. 32, 1989.

[58]* D. Y. Montuno and R.C.S. Ma, "A Layout Methodology for the Synthesis of High Speed Global Clock Nets," *Proceedings of IEEE Custom Integrated Circuits Conference,* pp. 28.4.1–28.4.4, May 1992.

[59]* P. Ramanathan and K. G. Shin, "A Clock Distribution Scheme for Non-Symmetric VLSI Circuits," *Proceedings of the IEEE International Conference on Computer-Aided Design,* pp. 398–401, November 1989.

[60] K. D. Boese and A. B. Kahng, "Zero-Skew Clock Routing Trees with Minimum Wirelength," *Proceedings of IEEE International Conference on ASICs,* pp. 17–21, September 1992.

[61] J. Burkis, "Clock Tree Synthesis for High Performance ASICs," *Proceedings of IEEE International Conference on ASICs,* pp. 9.8.1–9.8.4, September 1991.

[62] T.-H. Chao, Y.-C. Hsu, and J.-M. Ho, "Zero Skew Clock Net Routing," *Proceedings of ACM/IEEE Design Automation Conference,* pp. 518–523, June 1992.

[63]* J. Cong, A. B. Kahng, and G. Robins, "Matching-Based Methods for High-Performance Clock Routing," *IEEE Transactions on Computer-Aided Design,* Vol. CAD-12, No. 8, pp. 1157–1169, August 1993.

[64] M. Edahiro, "A Clock Net Reassignment Algorithm Using Voronoi Diagrams," *Proceedings of IEEE International Conference on Computer-Aided Design,* pp. 420–423, November 1990.

[65]* M.A.B. Jackson, A. Srinivasan, and E. S. Kuh, "Clock Routing for High Performance ICs," *Proceedings of ACM/IEEE Design Automation Conference,* pp. 573–579, June 1990.

[66] A. B. Kahng, J. Cong, and G. Robins, "High-Performance Clock Routing Based on Recursive Geometric Matching," *Proceedings of ACM/IEEE Design Automation Conference,* pp. 322–327, June 1991.

[67] R. S. Tsay, "Exact Zero Skew," *Proceedings of IEEE International Conference on Computer-Aided Design,* pp. 336–339, November 1991.

[68] T. Saigo, S. Watanabe, Y. Ichikawa, S. Takayama, T. Umetsu, K. Mima, T. Yamamoto, J. Santos, and J. Buurma, "Clock Skew Reduction Approach for Standard Cell," *Proceedings of IEEE Custom Integrated Circuits Conference,* pp. 16.4.1–16.4.4, May 1990.

[69] P. D. Ta and K. Do, "A Low Power Clock Distribution Scheme for Complex IC System," *Proceedings of IEEE International Conference on ASICs,* pp. 1-5.1–1-5.4, September 1991.

[70] P. R. Mukund and D. W. Bouldin, "A Graph Theoretic Approach to the Clock Distribution Problem," *Proceedings of IEEE International Conference on ASICs,* pp. 7-5.1–7-5.4, September 1991.

[71] J. Cong, A. Kahng, and G. Robins, "On Clock Routing for General Cell Layouts," *Proceedings of IEEE International Conference on ASICs,* pp. 14.5.1–14.5.4, September 1991.

[72]* T.-H. Chao, Y.-C. Hsu, J.-M. Ho, K. D. Boese, and A. B. Kahng, "Zero Skew Clock Routing with Minimum Wirelength," *IEEE Transactions on Circuits and Systems-II: Analog and Digital Signal Processing,* Vol. CAS-39, No. 11, pp. 799–814, November 1992.

[73] N. A. Sherwani and B. Wu, "Clock Layout for High-Performance ASIC Based on Weighted Center Algorithm," *Proceedings of IEEE International Conference on ASICs,* pp. 15.5.1–15.5.4, September 1991.

[74] D. A. Joy and M. J. Ciesielski, "Placement for Clock Period Minimization with Multiple Wave Propagation," *Proceedings of ACM/IEEE 28th Design Automation Conference,* pp. 640–643, June 1991.

[75] Y-M. Li and M. A. Jabri, "A Zero-Skew Clock Routing Scheme for VLSI Circuits," *Proceedings of the IEEE International Conference on Computer-Aided Design,* pp. 458–463, November 1992.

[76] W. Khan, M. Hossain, and N. Sherwani, "Zero Skew Clock Routing in Multiple-Clock Synchronous Systems," *Proceedings of IEEE International Conference on Computer-Aided Design,* pp. 464–467, November 1992.

[77] Q. Zhu and W.W-M. Dai, "Perfect-Balance Planar Clock Routing with Minimal Path-Length," *Proceedings of IEEE International Conference on Computer-Aided Design,* pp. 473–476, November 1992.

[78]* R-S. Tsay, "An Exact Zero-Skew Clock Routing Algorithm," *IEEE Transactions on Computer-Aided Design,* Vol. CAD-12, No. 2, pp. 242–249, February 1993.

[79] W. Khan and N. Sherwani, "Zero Skew Clock Routing Algorithm for High Performance ASIC Systems," *Proceedings of IEEE International Conference on ASICs,* pp. 79–82, September 1993.

[80] W. Khan, S. Madhwapathy, and N. Sherwani, "An Hierarchical Approach to Clock Routing in High Performance Systems," *Proceedings of IEEE International Symposium on Circuits and Systems,* pp. 1.467–1.470, May/June 1994.

[81] S. Lin and C. K. Wong, "Process-Variation-Tolerant Zero Skew Clock Routing," *Proceedings of IEEE International Conference on ASICs,* pp. 83–86, September 1993.

[82] M. Edahiro, "A Clustering-Based Optimization Algorithm in Zero-Skew Routings," *Proceedings of ACM/IEEE Design Automation Conference*, pp. 612–616, June 1993.

[83] N.-C. Chou and C.-K. Cheng, "Wire Length and Delay Minimization in General Clock Net Routings," *Proceedings of the IEEE International Conference on Computer-Aided Design*, pp. 552–555, November 1993.

[84]* M. Edahiro, "Delay Minimization for Zero-Skew Routing," *Proceedings of the IEEE International Conference on Computer-Aided Design*, pp. 563–566, November 1993.

[85] M. Edahiro, "An Efficient Zero-Skew Routing Algorithm," *Proceedings of ACM/IEEE 31st Design Automation Conference*, pp. 375–380, June 1994.

[86] J. D. Cho and M. Sarrafzadeh, "A Buffer Distribution Algorithm for High-Speed Clock Routing," *Proceedings of ACM/IEEE Design Automation Conference*, pp. 537–543, June 1993.

[87] G. E. Tellez and M. Sarrafzadeh, "Clock-Period Constrained Minimal Buffer Insertion in Clock Trees," *Proceedings of the IEEE International Conference on Computer-Aided Design*, pp. 219–223, November 1994.

[88] C.W.A. Tsao and A. B. Kahng, "Planar-DME: Improved Planar Zero-Skew Clock Routing with Minimum Pathlength Delay," *Proceedings of the European Design Automation Conference*, pp. 440–445, September 1994.

[89] A. B. Kahng and C.-W.A. Tsao, "Low-Cost Single Layer Clock Trees with Exact Zero Elmore Delay Skew," *Proceedings of the IEEE International Conference on Computer-Aided Design*, pp. 213–218, November 1994.

[90] M. Seki, T. Inoue, K. Kato, K. Tsurusaki, S. Fukasawa, H. Sasaki, and M. Aizawa, "A Specified Delay Accomplishing Clock Router Using Multiple Layers," *Proceedings of the IEEE International Conference on Computer-Aided Design*, pp. 289–292, November 1994.

[91]* S. Pullela, N. Menezes, and L. T. Pillage, "Reliable Non-Zero Clock Trees Using Wire Width Optimization," *Proceedings of ACM/IEEE Design Automation Conference*, pp. 165–170, June 1993.

[92] N. Menezes, A. Balivada, S. Pullela, and L. T. Pillage, "Skew Reduction in Clock Trees Using Wire Width Optimization," *Proceedings of IEEE Custom Integrated Circuits Conference*, pp. 9.6.1–9.6.4, May 1993.

[93] S. Pullela, N. Menezes, J. Omar, and L. T. Pillage, "Skew and Delay Optimization for Reliable Buffered Clock Trees," *Proceedings of IEEE International Conference on Computer-Aided Design*, pp. 556–562, November 1993.

[94] K. Zhu and D. F. Wong, "Clock Skew Minimizing During FPGA Placement," *Proceedings of ACM/IEEE 31st Design Automation Conference*, pp. 232–237, June 1994.

[95]* J. L. Neves and E. G. Friedman, "Topological Design of Clock Distribution Networks Based on Non-Zero Clock Skew Specifications," *Proceedings of 36th IEEE Midwest Symposium on Circuits and Systems*, pp. 468–471, August 1993.

[96]* J. L. Neves and E. G. Friedman, "Circuit Synthesis of Clock Distribution Networks Based on Non-Zero Clock Skew," *Proceedings of IEEE International Symposium on Circuits and Systems*, pp. 4.175–4.178, May/June 1994.

[97] J. L. Neves and E. G. Friedman, "Synthesizing Distributed Buffer Clock Trees for High Performance ASICs," *Proceedings of IEEE International Conference on ASICs*, pp. 126–129, September 1994.

[98] J. Chung and C.-K. Cheng, "Optimal Buffered Clock Tree Synthesis," *Proceedings of IEEE International Conference on ASICs*, pp. 130–133, September 1994.

[99] J. Chung and C.-K. Cheng, "Skew Sensitivity Minimization of Buffered Clock Tree," *Proceedings of the IEEE International Conference on Computer-Aided Design*, pp. 280–283, November 1994.

[100] N. Park and A. Parker, "Synthesis of Optimal Clocking Schemes," *Proceedings of ACM/IEEE 22nd Design Automation Conference*, pp. 489–495, June 1985.

[101]* E. G. Friedman, "The Application of Localized Clock Distribution Design to Improving the Performance of Retimed Sequential Circuits," *Proceedings of IEEE Asia-Pacific Conference on Circuits and Systems*, pp. 12–17, December 1992.

[102] T. Soyata, E. G. Friedman, and J. H. Mulligan, Jr., "Integration of Clock Skew and Register Delays into a Retiming Algorithm," *Proceedings of IEEE International Symposium on Circuits and Systems*, pp. 1483–1486, May 1993.

[103]* T. Soyata and E. G. Friedman, "Retiming with Non-Zero Clock Skew, Variable Register, and Interconnect Delay," *Proceedings of the IEEE International Conference on Computer-Aided Design*, pp. 234–241, November 1994.

[104]* N. V. Shenoy, R. K. Brayton, and A. L. Sangiovanni-Vincentelli, "Resynthesis of Multi-Phase Pipelines," *Proceedings of ACM/IEEE Design Automation Conference*, pp. 490–496, June 1993.

[105] L.-F. Chao and E.H.-M. Sha, "Retiming and Clock Skew for Synchronous Systems," *Proceedings of IEEE International Symposium on Circuits and Systems*, pp. 1.283–1.286, May/June 1994.

[106] B. Lockyear and C. Ebeling, "The Practical Application of Retiming to the Design of High-Performance Systems," *Proceedings of IEEE International Conference on Computer-Aided Design*, pp. 288–295, November 1993.

4. ANALYSIS AND MODELING OF THE TIMING AND POWER CHARACTERISTICS OF CLOCK DISTRIBUTION NETWORKS

[107]* M. Shoji, "Elimination of Process-Dependent Clock Skew in CMOS VLSI," *IEEE Journal of Solid-State Circuits*, Vol. SC-21, No. 5, pp. 875–880, October 1986.

[108] S. Lin and C. K. Wong, "Process-Variation-Tolerant Clock Skew Minimization," *Proceedings of the IEEE International Conference on Computer-Aided Design*, pp. 284–288, November 1994.

[109] S. R. Kunkel and J. E. Smith, "Optimal Pipelining in Supercomputers," *Proceedings of IEEE International Symposium on Computer Architecture*, pp. 404–411, June 1986.

[110] S. D. Kugelmass and K. Steiglitz, "A Probabilistic Model for Clock Skew," *Proceedings of IEEE International Conference on Systolic Arrays*, pp. 545–554, May 1988.

[111]* S. D. Kugelmass and K. Steiglitz, "An Upper Bound on Expected Clock Skew in Synchronous Systems," *IEEE Transactions on Computers*, Vol. C-39, No. 12, pp. 1475–1477, December 1990.

[112] C.-S. Li and D. G. Messerschmitt, "Statistical Analysis of Timing Rules for High-Speed Synchronous Interconnects," *Proceedings of IEEE International Symposium on Circuits and Systems*, pp. 37–40, May 1992.

[113] E. Vanden Meersch, L. Claesen, and H. De Man, "Automated Analysis of Timing Faults in Synchronous MOS Circuits," *Proceedings of IEEE International Symposium on Circuits and Systems*, pp. 487–490, May 1988.

[114] M. R. Dagenais and N. C. Rumin, "Automatic Determination of Optimal Clocking Parameters in Synchronous MOS VLSI Circuits," *Proceedings of the 1988 Stanford Conference on Advanced Research in VLSI*, pp. 19–33, March 1988.

[115]* M. R. Dagenais and N. C. Rumin, "On the Calculation of Optimal Clocking Parameters in Synchronous Circuits with Level-Sensitive Latches," *IEEE Transactions on Computer-Aided Design*, Vol. CAD-8, No. 3, pp. 268–278, March 1989.

[116]* C. Svensson and M. Afghahi, "On *RC* Line Delays and Scaling in VLSI Systems," *Electronics Letters*, Vol. 24, No. 9, pp. 562–563, April 1988.

[117] M. Afghahi and C. Svensson, "Calculation of Clock Path Delay and Skew in VLSI Synchronous Systems," *Proceedings of IEEE European Conference on Circuit Theory and Design*, pp. 265–269, September 1989.

[118] S. C. Menon and K. A. Sakallah, "Clock Qualification Algorithm for Timing Analysis of Custom CMOS VLSI Circuits with Overlapping Clocking Disciplines and On-section Clock Derivation," *Proceedings of the First International Conference on Systems Integration*, pp. 550–558, April 1990.

[119] K. A. Sakallah, T. N. Mudge, and O. A. Olukoton, "Analysis and Design of Latch-Controlled Synchronous Digital Circuits," *Proceedings of ACM/IEEE Design Automation Conference*, pp. 111–117, June 1990.

[120]* K. A. Sakallah, T. N. Mudge, and O. A. Olukoton, "Analysis and Design of Latch-Controlled Synchronous Digital Circuits," *IEEE Transactions on Computer-Aided Design,* Vol. CAD-11, No. 3, pp. 322–333, March 1992.

[121]* R.-S. Tsay and I. Lin, "Robin Hood: A System Timing Verifier for Multi-Phase Level-Sensitive Clock Designs," *Proceedings of IEEE International Conference on ASICs,* pp. 516–519, September 1992.

[122]* I. Lin, J. A. Ludwig, and K. Eng, "Analyzing Cycle Stealing on Synchronous Circuits with Level-Sensitive Latches," *Proceedings of ACM/IEEE Design Automation Conference,* pp. 393–398, June 1992.

[123] S. Narayan and D. D. Gajski, "System Clock Estimation based on Clock Wastage Minimization," *Proceedings of the European Design Automation Conference,* pp. 66–71, September 1992.

[124] M. C. Papaefthymiou and K. H. Randall, "TIM: A Timing Package for Two-Phase, Level-Clocked Circuitry," *Proceedings of ACM/IEEE Design Automation Conference,* pp. 497–502, June 1993.

[125] P. V. Argade, "Sizing an Inverter with a Precise Delay: Generation of Complementary Signals with Minimal Skew and Pulsewidth Distortion in CMOS," *IEEE Transactions on Computer-Aided Design,* Vol. CAD-8, No. 1, pp. 33–40, January 1989.

5. Specification of the Optimal Timing Characteristics of Clock Distribution Networks

[126]* J. P. Fishburn, "Clock Skew Optimization," *IEEE Transactions on Computers,* Vol. C-39, No. 7, pp. 945–951, July 1990.

[127] K. A. Sakallah, T. N. Mudge, T. M. Burks, and E. S. Davidson, "Optimal Clocking of Circular Pipelines," *Proceedings of IEEE International Conference on Computer Design,* pp. 642–646, October 1991.

[128]* E. G. Friedman and J. H. Mulligan, Jr., "Clock Frequency and Latency in Synchronous Digital Systems," *IEEE Transactions on Signal Processing,* Vol. SP-39, No. 4, pp. 930–934, April 1991.

[129] E. G. Friedman and J. H. Mulligan, Jr., "Pipelining of High Performance Synchronous Digital Systems," *International Journal of Electronics,* Vol. 70, No. 5, pp. 917–935, May 1991.

[130]* K. A. Sakallah, T. N. Mudge, and O. A. Olukoton, "*checkTc* and *minTc*: Timing Verification and Optimal Clocking of Synchronous Digital Circuits," *Proceedings of IEEE International Conference on Computer-Aided Design,* pp. 552–555, November 1990.

[131] K. A. Sakallah, T. N. Mudge, and O. A. Olukoton, "Optimal Clocking of Synchronous Systems," *Proceedings of TAU90 ACM International Workshop on Timing Issues in the Specification and Synthesis of Digital Systems,* August 1990.

[132] T. M. Burks, K. A. Sakallah, K. Bartlett, and G. Borriello, "Performance Improvement through Optimal Clocking and Retiming," *Proceedings of International Workshop on Logic Synthesis,* May 1991.

[133]* T. G. Szymanski, "Computing Optimal Clock Schedules," *Proceedings of ACM/IEEE Design Automation Conference,* pp. 399–404, June 1992.

[134]* T. G. Szymanski and N. Shenoy, "Verifying Clock Schedules," *Proceedings of IEEE International Conference on Computer-Aided Design,* pp. 124–131, November 1992.

[135] N. Shenoy and R. K. Brayton, "Graph Algorithms for Clock Schedule Optimization," *Proceedings of IEEE International Conference on Computer-Aided Design,* pp. 132–136, November 1992.

[136] R. B. Deokar and S. S. Sapatnekar, "A Graph-theoretic Approach to Clock Skew Optimization," *Proceedings of IEEE International Symposium on Circuits and Systems,* pp. 1.407–1.410, May/June 1994.

[137] W. Chuang, S. S. Sapatnekar, and I. N. Hajj, "A Unified Algorithm for Gate Sizing and Clock Skew Optimization to Minimize Sequential Circuit Areas," *Proceedings of IEEE International Conference on Computer-Aided Design,* pp. 220–223, November 1993.

[138] T. M. Burks, K. A. Sakallah, and T. N. Mudge, "Identification of Critical Paths in Circuits with Level-Sensitive Latches," *Proceedings of IEEE International Conference on Computer-Aided Design,* pp. 137–141, November 1992.

6. Clock Distribution Networks for Targeted VLSI/WSI Architectures

[139] M. Hatamian and G. Cash, "A 70-MHz 8-bit × 8-bit Parallel Pipelined Multiplier in 2.5 μm CMOS," *IEEE Journal of Solid-State Circuits,* Vol. SC-21, No. 4, pp. 505–513, August 1986.

[140] J. N. Coleman and R. M. Lea, "Clock Distribution Techniques for Wafer-Scale Integration," *Proceedings of Southhampton Workshop on Wafer Scale Integration,* pp. 1–8, 1985.

[141] D. C. Keezer and V. K. Jain, "Neural Network Clock Distribution," *Proceedings of IFIP Workshop on Silicon Architecture for Neural Nets,* pp. 101–111, November 1990.

[142] D. C. Keezer and V. K. Jain, "Clock Distribution Strategies for WSI: A Critical Survey," *Proceedings of IEEE International Conference on Wafer Scale Integration,* pp. 277–283, January 1991.

[143]* D. C. Keezer and V. K. Jain, "Design and Evaluation of Wafer Scale Clock Distribution," *Proceedings of IEEE International Conference on Wafer Scale Integration,* pp. 168–175, January 1992.

[144]* H. U. Chou and M. A. Franklin, "Optical Distribution of Clock Signals in Wafer Scale Digital Circuits," *Proceedings of IEEE International Conference on Computer Design,* pp. 117–121, October 1987.

[145] A. L. Fisher and H. T. Kung, "Synchronizing Large VLSI Processor Arrays," *Proceedings of 10th Annual International Symposium on Computer Architecture,* pp. 54–58, June 1983.

[146]* A. L. Fisher and H. T. Kung, "Synchronizing Large VLSI Processor Arrays," *IEEE Transactions on Computers,* Vol. C-34, No. 8, pp. 734–740, August 1985.

[147] M. Hatamian and G. Cash, "High Speed Signal Processing, Pipelining, and VLSI," *Proceedings of IEEE Conference on Acoustics, Speech, and Signal Processing,* pp. 1173–1176, April 1986.

[148]* S. Y. Kung and R. J. Gal-Ezer, "Synchronous Versus Asynchronous Computation in Very Large Scale Integrated (VLSI) Array Processors," *Proceedings of SPIE,* Vol. 341, Real Time Signal Processing V, pp. 53–65, May 1982.

[149] M. D. Dikaiakos and K. Steiglitz, "Comparison of Tree and Straight-Line Clocking for Long Systolic Arrays," *Proceedings of IEEE Conference on Acoustics, Speech, and Signal Processing,* pp. 1177–1180, April 1991.

[150]* M. D. Dikaiakos and K. Steiglitz, "Comparison of Tree and Straight-Line Clocking for Long Systolic Arrays," *Journal of VLSI Signal Processing,* Vol. 2, pp. 287–299, 1991.

[151] M. Roumeliotis, J. McKeeman, and G. Gray, "A Distributed Fault Tolerant Clocking Scheme for Systolic Array Architectures," *Proceedings of 6th Annual International Phoenix Conference on Computers and Communications,* pp. 105–109, February 1987.

[152] A. El-Amawy, "Branch-and-Combine Clocking of Arbitrarily Large Computer Networks," *Proceedings of IEEE International Conference on Parallel Processing,* pp. 409–417, August 1991.

[153]* A. El-Amawy, "Clocking Arbitrarily Large Computing Structures under Constant Skew Bound," *IEEE Transactions on Parallel and Distributed Systems,* Vol. PDS-4, No. 3, pp. 241–255, March 1993.

[154] N. Nigam and D. C. Keezer, "A Comparative Study of Clock Distribution Approaches for WSI," *Proceedings of IEEE International Conference on Wafer Scale Integration,* pp. 243–251, January 1993.

[155] N. G. Sheridan, C. M. Habiger, and R. M. Lea, "WSI Clock & Signal Distribution: a Novel Approach," *Proceedings of IEEE International Conference on Wafer Scale Integration,* pp. 252–261, January 1993.

[156] A. El-Amawy and U. Maheshwar, "Synchronous Clocking Schemes for Large VLSI Systems," *Proceedings of 27th Asilomar Conference on Signals, Systems, & Computers,* pp. 761–765, November 1993.

7. Example Implementations of Clock Distribution Networks

[157]* M. Shoji, "Electrical Design of BELLMAC-32A Microprocessor," *Proceedings of IEEE International Conference on Circuits and Computers,* pp. 112–115, September 1982.

[158]* M. Shoji, "Reliable Chip Design Method in High Performance CMOS VLSI," *Proceedings of IEEE International Conference on Computer-Aided Design,* pp. 389–392, October 1986.

[159] M. Hatamian and G. L. Cash, "High Speed Signal Processing, Pipelining, and VLSI," *Proceedings of IEEE International Conference on Acoustics, Speech, and Signal Processing,* pp. 1173–1176, April 1986.

[160] T. G. Noll, D. Schmitt-Landsiedel, H. Klar, and G. Enders, "A Pipelined 330-MHz Multiplier," *IEEE Journal of Solid-State Circuits,* Vol. SC-21, No. 3, pp. 411–416, June 1986.

[161] R. Maini, J. McDonald, and L. Spangler, "A Clock Distribution Circuit with a 100 PS Skew Window," *Proceedings of IEEE Bipolar Circuits and Technology Meeting,* pp. 41–43, September 1987.

[162] D. Chengson, L. Costantino, A. Khan, D. Le, and L. Yue, "A Dynamically Tracking Clock Distribution Chip with Skew Control," *Proceedings of IEEE Custom Integrated Circuits Conference,* pp. 15.6.1–15.6.4, May 1990.

[163] H. Itoh, N. Masuda, S. Kawashima, B. Fujita, S. Ishii, and M. Usami, "A Novel Design Concept for Small-Skew Clock LSIs with the Self-Delay-Adjustment," *Proceedings of IEEE Bipolar Circuits and Technology Meeting,* pp. 130–133, September 1991.

[164] M. Usami, S. Ishii, S. Kawashima, B. Fujita, N. Masuda, and H. Itoh, "An Automatic 5ps Skew-time Control Clock-pulse Adjustment LSI for High-speed Computers," *Proceedings of IEEE Symposium on VLSI Circuits,* pp. 53–54, May/June 1991.

[165] G. M. Blair, "Bit-Serial Correlator with Novel Clocking Scheme," *Proceedings of European Solid-State Circuits Conference,* pp. 157–160, 1991.

[166]* D. W. Dobberpuhl, *et al.,* "A 200-MHz 64-b Dual Issue CMOS Microprocessor," *IEEE Journal of Solid-State Circuits,* Vol. SC-27, No. 11, pp. 1555–1565, November 1992.

[167] R. B. Watson, Jr., H. A. Collins, and R. Iknaian, "Clock Buffer Chip with Absolute Delay Regulation Over Process and Environmental Variations," *Proceedings of IEEE Custom Integrated Circuits Conference,* pp. 25.2.1–25.2.5, May 1992.

[168] A. Ishibashi, A. Maeda, T. Arakawa, K. Higashitani, and M. Tatsuki, "High-Speed Clock Distribution Architecture Employing PLL for $0.6\mu m$ CMOS SOG," *Proceedings of IEEE Custom Integrated Circuits Conference,* pp. 27.6.1–27.6.4, May 1992.

[169] K. Ishibashi, T. Hayashi, T. Doi, N. Masuda, A. Yamagiwa, and T. Okabe, "A Novel Clock Distribution System for CMOS VLSI," *Proceedings of IEEE International Conference on Computer-Aided Design,* pp. 289–292, October 1993.

[170] D. Somasekhar and V. Visvanathan, "A 230-MHz Half-Bit Level Pipelined Multiplier Using True Single-Phase Clocking," *IEEE Transactions on VLSI Systems,* Vol. VLSI-1, No. 4, pp. 415–422, December 1993.

[171]* R. Reinschmidt and D. Leuthold, "Clocking Considerations for a Pentium Based CPU Module with a 512K Byte Secondary Cache," *Proceedings of IEEE Multi-Chip Module Conference,* pp. 26–31, March 1994.

8. RECENT PAPERS

[172] Y.-M. Li, S. Ashtaputre, J. Greidinger, M. Hartoog, M. Hossain, and S.-T. Hui, "Skew Controllable Buffered Clock Tree," *Proceedings of 1st International Conference on ASIC,* pp. 29–32, October 1994.

[173] P. Popescu, C. Kurowski, G. Thomsen, and W. Jager, "High Speed Clocking Methodology for ASIC Design in Fault Tolerant Systems," *Proceedings of 1st International Conference on ASIC,* pp. 271–274, October 1994.

[174] A. Balboni, C. Costi, A. Pellencin, M. Quadrini, and D. Sciuto, "Automatic Clock Tree Generation in ASIC Designs," *Proceedings of European Design and Test Conference,* March 1995.

[175] A. Erdal, M. Yue, L. Hiramoto, and J. Stahler, "An Implementation of a Clock-Tree Distribution Scheme for High-Performance ASICs," *Proceedings of IEEE International ASIC Conference,* pp. 26–29, September 1992.

[176] W. Khan, M. Hossain, and N. Sherwani, "Minimum Skew Multiple Clock Routing in Synchronous ASIC Systems," *Proceedings of IEEE International ASIC Conference,* pp. 22–25, September 1992.

[177] W. Chuang, S. S. Sapatnekar, and I. N. Hajj, "Timing and Area Optimization for Standard-Cell VLSI Circuit Design," *IEEE Transactions on Computer-Aided Design,* vol. CAD-14, no. 3, pp. 308–320, March 1995.

[178] J.-D. Cho and M. Sarrafzadeh, "A Buffer Distribution Algorithm for High-Performance Clock Net Optimization," *IEEE Transactions on VLSI Systems,* vol. VLSI-3, no. 1, pp. 84–98, March 1995.

[179] N.-C. Chou and C.-K. Cheng, "On General Zero-Skew Clock Net Construction," *IEEE Transactions on VLSI Systems,* vol. VLSI-3, no. 1, pp. 141–146, March 1995.

[180] T. Soyata, E. G. Friedman, and J. H. Mulligan, Jr., "Monotonicity Constraints on Path Delays for Efficient Retiming with Localized Clock Skew and Variable Register Delay," *Proceedings of the International Symposium on Circuits and Systems,* May 1995.

Clock System Design

KENNETH D. WAGNER
IBM CORPORATION

A well-designed clock system is a fundamental requirement in high-speed computers. In this tutorial, the author provides a framework for understanding system timing and then describes how the clock system executes the timing specifications. The tutorial examines clock generation and the construction of clock-distribution networks, which are integral to any clock system. Examples from contemporary high-speed systems highlight several common methods of clock generation, distribution, and tuning. Tight control of system clock skew is essential to an effective clock system.

The careful design of clock systems is often neglected. Part of the reason is that older, slower computers had higher tolerances to variations in the clock signal and had less exacting timing requirements. Today, however, as the demand for high-speed computers grows, the design of their clock systems should become a major concern not only in achieving high performance, but also in reducing assembly and maintenance costs.

A well-planned and well-built clock system is a prerequisite to reliable long-term computer operation. Conversely, a badly designed clock system can plague a computer throughout its lifetime, affecting its operation at any speed. To make such systems function, components often have to be tuned individually at several stages of manufacturing.

Despite these costs and performance penalties, timing design is still overlooked in many systems. Although significant decisions that must be made early in computer design include such issues as clocking scheme and type of memory element, designers seldom participate. Instead, system architects may simply repeat a previously successful set of choices, despite significant changes in design specifications, technology, and environment. Of course, these systems will eventually be functional, but they will require much more maintenance and tuning—costs not always reflected back to the developers.

These attitudes prevail in part because timing design problems are rarely reported in the literature. Also, design teams tend to be secretive about their clock systems, either because they believe they are doing something new or because they are doing nothing new and are afraid to be associated with an older technique. Either way, the result is a scarcity of information on how to avoid timing problems through proper design of the clock system.

THE CLOCK SYSTEM

System timing specifications are executed using a clock system. The clock system has two main functions, clock generation and clock distribution. We use clock-generation circuitry to form highly accurate timing signals, which we then use to synchronize

Reprinted from *IEEE Design & Test of Computers*, pp. 9-27, October 1988.

Two types of clocked bistable elements are important in contemporary high-speed computers: the latch and the edge-triggered flip-flop.

changes in the system state. These pulsed, synchronizing signals are known as clocks. We use clock distribution to deliver the clocks to their destinations at precisely specified instants. A network, called the *clock-distribution network*, propagates clocks formed by clock generation to clocked memory elements.

Most logic design texts, such as that by McCluskey (see "Additional Reading" at the end of this article), describe bistable ele-synchronized by the clock signals. A system oscillator is the source for these periodic signals. We generate and manipulate the clock signals and precisely place clock pulses to meet the system timing requirements. We may also tune the clocks to compensate for inaccuracies in the clock pulsewidth or pulse position.

BISTABLE ELEMENTS

The focus of this article is on the timing design of systems that use static bistable elements. The techniques described can also be used in the timing design for other types of clocked memory elements, such as arrays and dynamic latches, or for precharging circuitry.

Most logic design texts, such as that by McCluskey (see "Additional Reading" at the end of this article), describe bistable elements and their characteristics in great detail. Two types of clocked bistable elements are important in contemporary high-speed computers: the latch and the edge-triggered flip-flop. The latch is transparent while its clock (control) input is active. By transparent, we mean that its outputs reflect any of its data inputs. Edge-triggered elements, such as the D flip-flop, respond to their data inputs only at either the rising or falling transition of their clock input. They do not have the transparency property of the latch.

We can describe the time-dependent behavior of a bistable element using the following parameters:

- *setup time*, the minimum time that the data input of the bistable element must be held stable before the active edge or latching level of the clock pulse occurs
- *hold time*, the minimum time that the data input of the bistable element must be held stable after the active edge or latching level of the clock pulse disappears
- *propagation delay*, the time between a change on the clock or data input of the bistable element and the corresponding change on its output

For system operation to be correct, the setup time, hold time, and minimum clock pulsewidth must be satisfied for each bistable element. Signals whose propagation delay is so long that it violates the setup time are called long-path signals. Signals whose propagation delay is so short that it violates the hold time are called short-path signals. Both conditions result in incorrect data being stored.

SYSTEM CLOCKING SCHEMES

System clocking is either single-phase, multiphase (usually two-phase), or edge-triggered. Figure 1 illustrates. The dark rectangles in the figure represent the interval during which a bistable element samples its data input. Each scheme requires a minimum clock pulsewidth.

The most widely used scheme is multiphase clocking. The multiphase clocking scheme in Figure 1b is two-phase, nonoverlapping. In this scheme, two distinct clock phases are distributed within the system, and each bistable element receives one of these two clocks. Systems that have adopted two-phase clocking include microprocessors such as the Intel 80x86 series and Motorola MC68000 family, micro-mainframes such as the HP-9000, and mainframes such as the IBM 3090 and the Univac 1100/90.

Figure 2 shows a finite-state machine, a machine that realizes sequential logic functions, with each clocking scheme. (For more on finite-state machines, see McCluskey's text.) For simplicity, primary I/O is not shown. The Amdahl 580 mainframe and Cray-1 vector processor are single-phase latch machines, such as that shown in Figure 2a. Modern high-speed microprocessors like the Bellmac-32A are two-phase latch machines with a single-latch design using nonoverlapping clock phases, such as that shown in Figure 2b. Figure 2c shows a two-phase latch machine with a double-latch design. This type of machine supports scan-path testing, since it can use LSSD latch pairs, which are hazard-free master-slave latches with a scan input port. Most contemporary IBM products, including IBM 3090 mainframes, incorporate design for testability using this structure. Systems built with catalog parts are usually flip-flop machines, such as that shown in Figure 2d, because clocked bistable elements commonly offered in bipolar and CMOS MSI chips are edge-triggered.

THE CLOCK CYCLE

System designers characterize a computer's functionality in terms of its clock cycle, also called its machine cycle. The average number of clock cycles required per machine instruction is a measure of computer performance. Table 1 gives clock rates for some well-known systems. The designer focuses on the clock cycle because it determines the standard work interval for internal machine functions. The system state is the set of values in system memory elements at the end of a clock cycle.

A clock cycle has the following properties:

1. It consists of a sequence of one or more clock pulses.
2. The sequence of clocks generated in each cycle is identical to every other cycle.
3. No partial clock sequences can occur: clocks can only stop and start at cycle boundaries.
4. Each bistable element can be updated at most once per cycle.

These properties ensure that the transition to the next state of the system is predictable and correct. This deterministic system

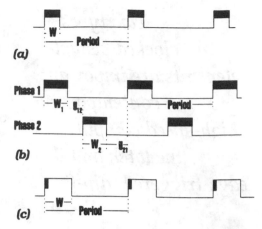

Figure 1. *System clocking waveforms; single-phase (a), two-phase (b), and edge-triggered (c). W_i=pulsewidth of phase j and g_{ij}= interphase gap from phase i to phase j; if $g_{ij} > 0 \Rightarrow$ two-phase, nonoverlapping, if $g_{ij} < 0 \Rightarrow$ two-phase overlapping.*

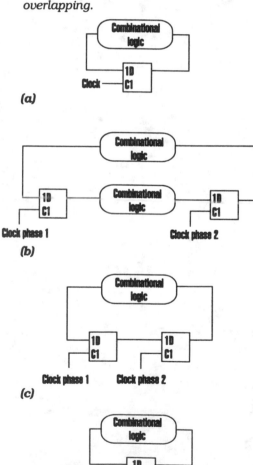

Figure 2. *General finite-state machine structures: one-phase latch machine (a), two-phase latch machine with single-latch (b) and double latch (c), and flip-flop machine (d).*

behavior will hold whether clock cycles occur at the system operating rate or one at a time. We can reproduce system behavior at the operating rate by issuing single clock cycles or bursts of clock cycles, which makes system debugging much simpler.

TIMING ANALYSIS

Programs for timing analysis are used routinely to verify system timing. They can identify long or short paths, and the designer can interact with them to get estimates of signal-path delays in parts of the system. Designers can also run them after layout to get more accurate results. The delay models used for system elements are validated by circuit simulation.

Single-phase systems and multiphase overlapping systems require more extensive timing analysis than multiphase nonoverlapping and edge-triggered systems. The timing constraints of single-phase and multiphase overlapping systems are two-sided, bounded by both short paths and long paths. Figure 3 illustrates these constraints in a simplified example, where setup time and hold time are set to 0. The advantage of these systems is that they operate more quickly than their nonoverlapping counterparts.

CLOCK SIGNALS

In a conventional computer system, one source generates the system clock signal. Multiple processors operating synchronously may also share one signal. We can manipulate this clock signal in many ways before it reaches its destinations. We can divide it, delay it, shape it, buffer it, and gate it. Clocked bistable elements, either latches or flip-flops, use the signal that results from such manipulations.

Table 1. *System clock rates.*

System	Intro Date	Technology	Class	Nominal Clock Period (ns)	Nominal Clock Frequency (MHz)
Cray-X-MP	1982	MSI ECL	Vector processor	9.5	105.3
Cray-1S,-1M	1980	MSI ECL	Vector processor	12.5	80.0
CDC Cyber 180/990	1985	ECL	Mainframe	16.0	62.5
IBM 3090	1986	ECL	Mainframe	18.5	54.1
Amdahl 58	1982	LSI ECL	Mainframe	23.0	43.5
IBM 308X	1981	LSI TTL	Mainframe	24.5,26.0	40.8,38.5
Univac 1100/90	1984	LSI ECL	Mainframe	30.0	33.3
MIPS-X	1987	VLSI CMOS	Microprocessor	50.0	20.0
HP-900	1982	VLSI NMOS	Micro-mainframe	55.6	18.0
Motorola 68020	1985	VLSI CMOS	Microprocessor	60.0	16.7
Bellmac-32A	1982	VLSI CMOS	Microprocessor	125.0	8.0

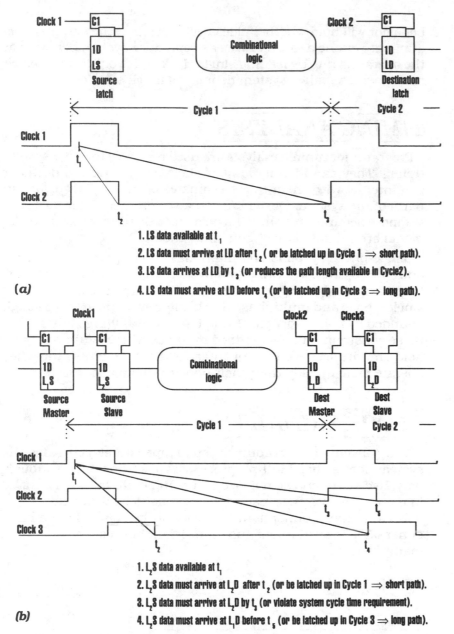

(a)

1. LS data available at t_1
2. LS data must arrive at LD after t_2 (or be latched up in Cycle 1 $\Rightarrow$ short path).
3. LS data arrives at LD by t_3 (or reduces the path length available in Cycle2).
4. LS data must arrive at LD before t_4 (or be latched up in Cycle 3 $\Rightarrow$ long path).

(b)

1. L_2S data available at t_1
2. L_2S data must arrive at L_2D after t_2 (or be latched up in Cycle 1 $\Rightarrow$ short path).
3. L_2S data must arrive at L_2D by t_3 (or violate system cycle time requirement).
4. L_2S data must arrive at L_2D before t_5 (or be latched up in Cycle 3 $\Rightarrow$ long path).

Figure 3. *Path requirements in a single-phase machine* **(a)** *and in a two-phase overlapping latch machine with a double latch* **(b)**.

For all systems, we must correctly place the leading- or trailing-edge positions of the distributed clock pulses to ensure that bistable elements switch at the correct times.

SIGNAL CHARACTERISTICS

Clocked sequential logic responds to several characteristics of the clock signal: the clock period, the pulsewidth, and the leading-edge or trailing-edge position of the clock pulse. The *clock period* is the interval before the signal pattern repeats. The ideal clock signal for a bistable element is a sequence of regularly repeating pulses. Ideal pulses are rectangular with sufficient duration and amplitude to ensure the reliable operation of the bistable element. The duration of the pulse, or pulsewidth (W), can be any fraction of the clock period, but is usually less than or equal to half of it. An accurate model of a real clock pulse includes actual voltage levels and the shapes of the pulse edges.

For all systems, we must correctly place the leading- or trailing-edge positions of the distributed clock pulses to ensure that bistable elements switch at the correct times. Also, distributed clock pulses must be wide enough or they will either be filtered out in transmission or be unable to switch a bistable element because they lack the energy. *Clock-manipulation elements* reposition clock pulses and change their pulsewidths. They consist of delay elements and elements that manipulate the pulsewidth. *Delay elements* either delay a pulse, or, in a timing chain, produce a sequence of delayed pulses in response to a single pulse input. *Pulsewidth-manipulation elements* require both delay elements and logic gates.

Delay elements are available as both analog and digital circuits and are chosen according to the accuracy, flexibility, and range of signal delay required. Analog delay elements vary from simple printed or discrete wire interconnections to delay lines. Delay lines, packaged in hybrid chips, consist of lumped LC elements or distributed printed wire, which provides more accurate control. Digital delay elements include logic gates and counters. Logic gates are relatively inaccurate because of their wide delay ranges, while the time resolution of counters depends on their operating frequency.

Some delay elements are programmable, providing a range of delays. To select a particular delay, we can either connect to a particular chip output pin or tap, or control the configuration electronically by a multiplexer. A typical integrated delay line provides delays from 1 to 10 ns in 1-ns increments with a ±0.5-ns tolerance.

Pulsewidth-manipulation elements have three functions: chop, shrink, and stretch. Figure 4b shows the effect of a chopper, shrinker, and stretcher on a positive pulse. The effect of each manipulation element differs for positive and negative clock pulses. Thus, for each pulse polarity, only three of the four elements are useful. The other element has only a delay effect. Table 2 shows the values for the signal characteristics after chopping, shrinking, and stretching. AND gates have delay d_a, OR gates have delay d_o, inverters have delay d_i, delay elements have delay D, and interconnections have no delay. The signal input is a pulse of width W whose leading edge occurs at time $t=0$. For an element to have an effect during the pulse, the sum of d_i and D must be less than W.

Table 2. *Effect of elements that manipulate the clock pulsewidth.*

Element	Positive Pulse			Negative Pulse		
	Leading Edge	Pulse-width	Function	Leading Edge	Pulse-width	Function
A	d_a	$D+d_i$	Chopper	—	—	—
B	—	—	—	d_o	$D+d_i$	Chopper
C	$D+d_a$	$W-D$	Shrinker	d_a	$W+D$	Stretcher
D	d_a	$W+D$	Stretcher	$D+d_a$	$W-D$	Shrinker

CLOCK GENERATION

We can derive all clock signals in a synchronous machine from the system clock signal. The system clock is often a rectangular pulse train with a 50% duty cycle, called a *square wave*. The circuit that generates the system clock is at the base of the clock-distribution network. Its input is from either a voltage-controlled oscillator (VCO), a crystal oscillator (XO), or a voltage-controlled crystal oscillator (VCXO). All three sources produce a sinusoidal (single-frequency) output, which is then clamped or divided to generate the rectangular system clock. Excluding the quartz crystal, the oscillator circuit is usually packaged on a single hybrid IC.

A simple oscillator consists of an LC circuit, which we tune by carefully selecting component values that allow the circuit to resonate at the desired frequency. When we need extreme frequency stability over a wide temperature range, we use an XO. An XO consists of a tuned circuit with an embedded quartz crystal in the feedback loop. The crystal stabilizes the resonant frequency of the oscillator circuit.

When we need a larger range of selectable frequencies, we use either a VCO or a VCXO, because the XO has a very limited tunable range. A DC voltage input controls both the VCO and VCXO. The VCO could be an emitter-coupled multivibrator that produces a square wave that we can tune over a 10:1 frequency range up to 20 MHz. It could also be a capacitance-controlled oscillator that produces a sine wave tunable over a 2:1 frequency range up to microwave frequencies. If we modify the resonant frequency of an XO, we get a tuning accuracy of a few hundred parts per million in the VCXO. Thus, the XO has the most frequency stability but the least tuning flexibility, the VCXO is in the middle on both, and the VCO has the least frequency stability and the most tuning flexibility. Frequency instability in the oscillator can cause clock jitter, requiring us to assign a tolerance to the clock-edge placement in timing analysis.

From the system clock we derive the full set of clocks and clock phases that the system requires. We can generate multiphase clocks from a square-wave input in many different ways. These methods include one shots, clock choppers or shrinkers, shift-register latches, and frequency dividers, depending on the precision and flexibility required. To prevent the overlap of adjacent clock phases in a nonoverlapping clocking scheme, we use output feedback or clock choppers. If there is uneven loading on each clock phase, the relative pulse-edge positions may change, which might cause some of the clock phases to overlap. Another cause of overlap is the asymmetric rising and falling delays of contemporary devices.

Figure 5 shows two simplified circuits that create two-phase clocks. The techniques are applicable to general multiphase clock generation. The first circuit is used in the Univac 1100/90 for four-phase clock generation. It requires a fast-running square-wave clock input and a ring counter. Each stage of the ring counter enables one clock phase, and the single clock chopper

From the system clock we derive the full set of clocks and clock phases that the system requires.

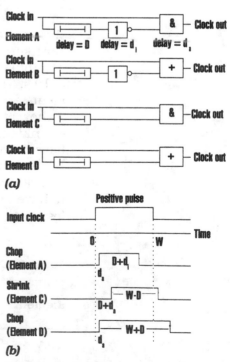

Figure 4. *Elements that manipulate the clock pulsewidth (a) and their effect on a positive pulse (b).*

For developing, diagnosing, and producing high-speed systems, we ideally want a wide-bandwidth oscillator source that is highly accurate.

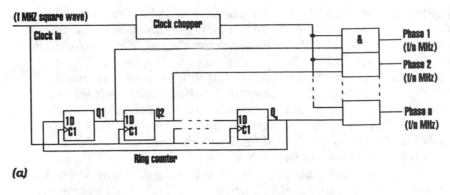

(a)

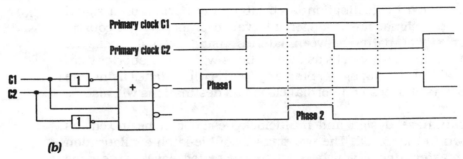

(b)

Figure 5. *Creating a two-phase clock: selecting the pulses of a fast-running clock (a) and decoding the primary clocks (b).*

determines pulsewidth. The second circuit is used in the Bellmac-32A. It generates two-phase clocks by decoding primary clock signals. We can use a gray-code counter to produce these primary clocks, or we can use *clock shaping*. Clock shaping allows us to generate clock phases from a system clock with fixed gaps between phases (forcing pulsewidths to vary with frequency).

CLOCK SEQUENCES

The three schemes for system clocking we have looked at—single-phase, multiphase, and edge-triggered—determine the basic data flow in latch and flip-flop machines during each clock cycle. Complicating these requirements, though, are special timing considerations. For example, subsystems may require dfferent clock-arrival times so that they can communicate with each other across interfaces with large delays. Also, paths within subsystems may be too long for normal system timing. We can accommodate irregular interfaces and paths without affecting the clock cycle, although system timing becomes more complex. To handle these cases, we generate a sequence of clocks during each clock cycle and do not use normal data-path timing.

There are two timing design styles for handling the clock sequences generated during a clock cycle: multiphase design and multiclock design. Figure 6 illustrates. The dashed vertical lines represent the boundaries of the clock cycle. The solid vertical lines represent active clock edges. Time proceeds left to right across each diagram and only paths originating from the earliest (leftmost) cycle are shown. In a normal multiphase (*k*-phase) design

313

(Figure 6a), latches clocked by phase 1 feed latches clocked by phase 2, and so on. Only the latch clocked on the last phase feeds the phase-1 latch of the succeeding cycle. All data movement proceeds phase i to phase $i+1$ modulo k.

In contrast, the multiclock design (Figure 6c) ensures that bistable elements clocked at any time T_i during one cycle feed only bistable elements clocked in the succeeding cycle. For instance, the three cycle $n-1$ clocks are early, normal and late, which correspond to the times T_0, T_1 and T_2. Each can feed any of the T_0, T_1 or T_2 bistable elements in cycle n.

In the Amdahl 580, early clocks prevent long paths between the remote channel frame and I/O processor. If we clock the source latch earlier or destination latch later than normal on a signal path, the signal has a longer interval to propagate. Of course, other signal paths between latches using normal clocks as sources and early clocks as destinations will have a shorter than normal time to propagate. Similarly, paths with latches using late clocks as sources and normal clocks as destinations will also be shorter.

Multiphase design and multiclock design can be mixed, as shown in Figure 6d. The two-phase, double-latch configuration has master latches, which feed their associated slaves in the same cycle. Each master latch is clocked at one of three timings: T_0, T_1 or T_2. The slave latch of each pair communicates with any of the master latches in the next cycle. The IBM 3033, 308X, and 3090 mainframes use similar techniques.

Figures 6b, 6e, and 6f show examples of more complex paths. Figure 6b shows the possibility of paths that skip adjacent phases in a three-phase system. The Univac 1100/90 is an example of a design with nonadjacent phase paths. Note that any phase-i-to-phase-i path in the succeeding cycle would require identical analysis to a single-phase system. Figures 6e and 6f show fractional cycle and multicycle paths. Such paths are typical of a performance-oriented design that uses two-phase latch machines.

Systems can also generate clocks that operate at several distinct cycle times, usually integer multiples of a base cycle time. We can use clocks with lower rates for parts of the system that do not need faster clocks. All clocking between subsystems must be synchronous, or else we must use techniques to reduce metastable behavior at subsystem interfaces.

THE SYSTEM CLOCK SOURCE

For developing, diagnosing, and producing high-speed systems, we ideally want a wide-bandwidth, highly accurate oscillator source. Most systems have both a crystal-oscillator source input for production systems and a tunable source input for prototype development and AC diagnosis. During development of a multiphase system, we may need to vary the pulsewidth of any clock phase as well as to vary the relative pulse positions.

To detect marginal path-delay problems, the Amdahl 580 selects any one of three crystal oscillators as the clock source in production machines, lengthening or shortening its clock cycle.

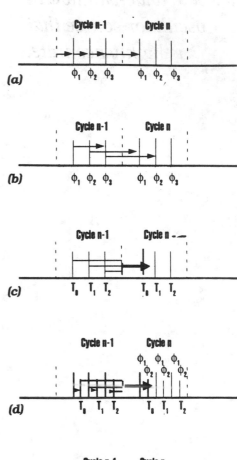

Figure 6. Placing clock pulses; three-phase, adjacent paths (a); three-phase, nonadjacent paths (b); multiclock (three clocks) (c); multiclock, two-phase (d); multiclock, two-phase with fractional cycle paths (e); and multiclock, two-phase with multicycle paths (f).

The goal of clock distribution is to organize clocks so that the delays from the source of each clock or clock phase to its bistable elements are identical.

Operating modes are called normal, fast margin, and slow margin. These correspond to nominal clock frequency, 5% faster than nominal, and 5% slower than nominal. An external oscillator input is also available, bypassing the internal oscillators during diagnosis and development.

To detect marginal timing problems in the IBM 3090, a two-phase double-latch machine, designers made it possible to lengthen the delay between the leading edge of the slave clock and the trailing edge of the master latch clock for a selected system region (see Figure 3b). In addition, lengthening the clock cycle allows us to verify the slave-latch-to-slave-latch path delay.

We can choose between distributed or centralized clock sources to control multiprocessors synchronously. In distributed control, we let each processor or processor group in the complex use its own local oscillator, with some form of enforced synchronization between oscillators, like a phase-locked loop. Alternatively, in centralized control, we designate one oscillator as the master oscillator and have each system select this master through a local/remote switch. The second method is simpler and is common in mainframe multiprocessor models such as the Amdahl 580, IBM 3033, and IBM 370/168. Although the IBM multiprocessors use a master oscillator, other standby oscillators are phase-locked to the master oscillator and can be selected if it fails.

CLOCK DISTRIBUTION

The goal of clock distribution is to organize clocks so that the delays from the source of each clock or clock phase to its bistable elements (its destinations) are identical. In reality, however, no matter how each clock path is constructed, any two clock paths in the same machine or any two corresponding paths in different machines will always have a delay difference. Every computer operates in a different temperature, power supply, and radiation environment, and duplicate components will differ in subtle ways between computers. We must build in tolerance to these variations in any system timing design.

The most common approaches to ensure correct and reliable machine timing are worst-case analysis and statistical analysis. In worst-case analysis, we assume that all component parameters lie within some range, and the cumulative worst-case effect is still within the timing tolerance of the machine. In statistical analysis, the intent is that most machines have tolerable timing characteristics, and so we can rely on the cumulative statistical variations of component parameters to remain within the timing tolerance.

CLOCK SKEW

We specify system timing such that every system memory element has an expected arrival time for the active edge of its clock signal. Clock-edge inaccuracy is the difference between the actual and expected arrival time of this clock edge. For every pair of system memory elements that communicate, we define *path clock skew* as the sum of the clock-edge inaccuracies of the pair's source and destination. *System clock skew* is the largest path clock skew in the system. It is the value of the worst-case timing inaccuracy among all paths. We can break it into interboard skew, on-board interchip skew, and so on to the smallest timed component.

The challenge to designers of clock-distribution networks is how to control system clock skew so that it becomes an acceptably small fraction of the system clock period. As a rule, most systems cannot tolerate a clock skew of more than 10% of the system clock period. If system clock skew goes beyond the design limit, system behavior can be affected. Setup and hold times are missed, which results in long and short paths. No scheme is immune from these problems—even flip-flop machines can malfunction when clock skew is present.

Clock-powering trees, such as the one in Figure 7, are a source of clock skew. These trees are used to produce multiple copies of the clock signal for distribution. Each gate of the tree has some uncertainty associated with its delay, which is the difference between its best-case and worst-case delays. This difference is called *gate skew*. Using a worst-case timing analysis, the clock skew caused by a powering tree equals the arithmetic sum of the gate (and interconnection) skews on the path from the tree root to an output. In other words, clock skew has a cumulative effect by tree level. We can minimize this clock skew by placing all gates at a given tree level, or even the entire tree, on the same chip. In addition, we can realize elements at each tree level by using electrically matched devices and careful wiring.

DISTRIBUTION TECHNIQUES

We must efficiently distribute the rectangular clock pulses produced through the interaction of the oscillator and clock-generation circuitry. Critical to efficient distribution is the clock-network layout—the physical placement of the network. It must conform to design rules that ensure the integrity of the clock signal by minimizing electrical coupling, switching currents, and impedance discontinuities. Other rules must prevent excessive clock skew by equalizing path delays and maintaining the quality of the signal edge. Symmetry and balanced loading at many levels of packaging, such as on the chip or on the board, are characteristic of effective clock-network layouts. To achieve these qualities, we can prearrange positions of the clock pins and make clock paths as short as possible.

It is sometimes difficult to coordinate the relative lengths of two paths that originate from a common source. To match any two paths or path segments in the clock system, we may need identical lengths of cable, wire, and interconnections; balanced loading; and equal numbers of buffer gates. A technique called *time-domain reflectometry* helps in this process by accurately measuring the line delays of long cables. In this process, line sig-

> *Critical to efficient distribution is the clock-network layout—the physical placement of the network.*

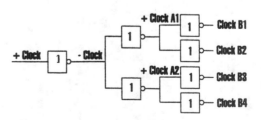

Figure 7. *Clock-powering tree.*

In practice, large systems distribute a small number of clock signals to each board or module.

nals are generated, and the signal reflections from line terminations are detected in real time. Once we measure the delays, we can equalize them by adjusting the lengths of the cables.

Duplicating the composition of two paths is not the only method of ensuring that two paths have equal delay. Another technique for matching different paths is called *padding*. In padding, we add delay elements to one or both paths. The Cray-1 uses extra interconnections and spare IC packages as padding, for example.

System designs often use a mixture of strategies. Designers might use duplication in subsections of the clock-distribution network with intermediate padding. Component screening also ensures that the performance of each system component is acceptable. Despite these techniques, some system clock skew is inevitable. However, as long as the timing analysis includes the effect of clock skew and determines that the system will function correctly, no other precautions are needed. If we detect that some system paths are failing because of excessive clock skew, then the clocks have to be tuned, or some part of the network has to be redesigned.

High-speed systems use a hierarchical structure to distribute clocks efficiently. A model for such a structure consists of *logic islands*. A logic island is a partition in the system, like a printed circuit board. Each island has a single point for clock entry, and all islands have the same line delay from their clock source, outside the island, to their clock entry point. (Tunable delay lines may precede the clock entry points.) We apply the same technique recursively to the islands themselves and to the subislands, such as chips, until we reach the individual clocked elements. Figure 8 is a diagram of the resulting structure, which resembles a star with the clock source in the center and the islands on the periphery. This model is remarkably similar to the physical organization of the Cray-1. In systems that need multiphase clocks, clock phases are generated at some convenient level in the hierarchy, and clock entry points at subsequent levels have one input for each clock phase.

In practice, large systems distribute a small number of clock signals to each board or module. Either the leading- or trailing-edge position of these clocks is very tightly controlled. When these signals reach board or module distribution, phase-generation circuits and clock choppers produce the final, well-controlled pulses. With this strategy, we can simplify clock distribution and reduce clock skew because only one clock edge has to follow a predetermined relationship to a reference clock. Using a single edge also minimizes clock skew by exploiting *common-mode action*, in which changes in power and temperature have similar effects across clock circuitry.

Clock choppers should be close to the bistable elements that are the signal's final destination. Otherwise, asymmetric rising and falling delays in the buffer gates downstream of the clock chopper may shrink or stretch the pulse excessively. For instance, in the IBM 3090's two-phase double-latch design, the leading edge of the slave clock and trailing edge of the master clock are formed from a common input clock edge. These edges control the

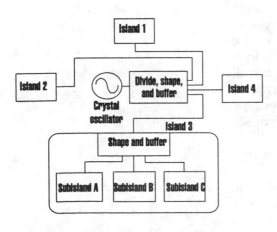

Figure 8. *Logic islands.*

critical system path, so they must have minimal skew. Exploiting common-mode action in this way is also known as edge-tracking.

There are a number of simple guidelines for on-chip clock distribution to minimize asymmetries in clock-path delays and keep clock edges sharp. We balance and limit on-chip clock loading, using clock-powering trees, special buffers, symmetric layouts, and careful buffer placement. In very high speed ICs, we can select H-trees, which are symmetric, controlled-impedance clock-distribution trees composed entirely of metal. We can also reduce process-dependent clock skew in CMOS chips by adjusting clock-buffer FET parameters appropriately.

DESIGN DECISIONS

One of the first decisions in the design of the clock-distribution network is where to put square-wave production—the function that will produce the system clock—and where to put the function that will generate additional clocks. If a system requires many different clocks, we should distribute only a small number of clocks globally, and then generate the necessary clocks for each system section locally by manipulating (including decoding) these primary clocks.

In systems constructed from gate arrays or other semicustom chips, we cannot manipulate device characteristics individually. Most mainframes use gate arrays because the designers need to trade off chip turnaround time with the large number of different chips required. Both MOS and bipolar gate arrays have limited fanout of clock driver circuits. Whole chips or portions of chips are dedicated to controlling clock signals and buffering clock signals through powering trees (see Figure 7). The designers' intent is to produce as many copies of the clock signal as necessary to satisfy the loading requirements of the system. The more system bistable elements and associated chips and the smaller the amount of fanout allowed for each clock copy, the larger the number of required tree levels.

Figure 9 shows a simplified example of the clock-distribution network in the Amdahl 580. In this mainframe, which is based on bipolar gate arrays, the circuitry that produces the system clock is on the same card as the oscillator. Special chips on the console board then receive the system clock. These chips do clock gating and powering to provide the primary clock input for all boards. On the boards, tunable clock-distribution chips receive the primary clock and do clock chopping; derive the early, normal, and late clocks; and power these clocks sufficiently to satisfy loading requirements. The Amdahl 580 uses two ECL LSI clock-distribution chips to drive about 118 ECL LSI logic chips and static RAMs on each of its boards.

Custom systems and those constructed from catalog parts also use separate clock buffer chips when clock signals must drive large capacitive loads. Dedicated chips produce the clock copies that the system needs. Buffers on these chips must supply large

Most mainframes use gate arrays because designers need to trade off chip turnaround time with the large number of different chips required.

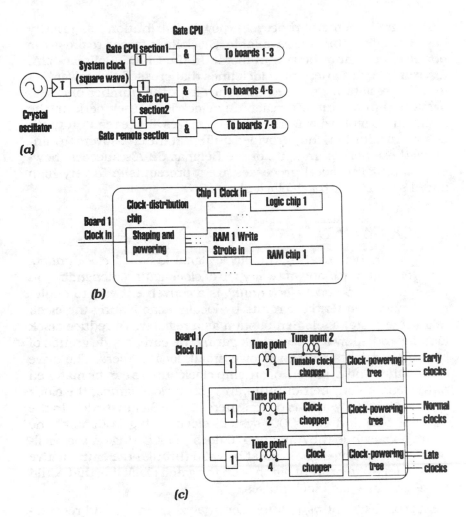

Figure 9. *Clock-distribution network with system clock gating; system-wide clock distribution (a), board-clock distribution (b), and clock-distribution chip (c).*

currents, and accordingly, each buffer occupies a large area of each buffer chip. Buffer chips in the HP-9000 also generate clock phases.

In MOS custom chips or chip sets like the MC68000, we can produce clock phases on the chip and then buffer them immediately to provide adequate fanout. Most MOS single-chip systems use a single or cascaded clock buffer as a source for each distributed clock phase. Each buffer drives a very wide metal line (large load). The line composition and dimensions minimize the clock signal's power loss and voltage drop, while maximizing the speed of signal propagation. The wide metal line branches into groups of narrower metal lines which themselves may branch off, and so on. Nonmetal segments with a higher resistance, like polysilicon, should be avoided in clock-distribution lines. The distortion of the clock signal increases when it is propagated across such impedance mismatches. In these regions, clock phases may overlap and violate system timing constraints.

In single-level metal processes, power distribution has priority on the single metal layer available. Clock lines forced to cross the power lines cannot be run in metal. In the CMOS Bellmac-32A, for example, designers ran clock lines that crossed the power bus in a low-resistance silicide, and they kept the number of cross-overs to a minimum. To reach any clock load, they perform the same number of crossovers, which equalizes all the resulting path delays. In addition, they provided buffers at the crossovers to minimize these delays. In spite of the Bellmac 32-A's success, however, multilevel metal processes are a prerequisite to very high speed systems with low clock skew.

CLOCK GATING

Selectively deactivating the clock signal is called *clock gating*. Designers can use one of two types of clock gating, depending on the application. *Local clock gating* is a convenient way to implement many sequential circuits by locally deactivating the clock to a set of bistable elements, such as a register. To reduce clock loading and allow for local clock gating, we can provide groups of bistable elements with their own small, local buffers. The drive capabilities and numbers of on-chip clock buffers can be matched to the loading on their outputs. In *system clock gating*, the clock to an entire subsystem is deactivated. Figure 9a illustrates. In the Amdahl 580 and HP-3000, system-clock gating is done at the board level, while the Amdahl 470 does clock gating on the oscillator card. Gating signals must be valid through the entire active clock interval to prevent glitches on the gated clock line that could be sensed as valid clock pulses.

System clock gating is a tool for analyzing errors and recovering from them. It allows us to test the machine in a deterministic fashion by ensuring that a predetermined minimum number of clock cycles occurs after some machine stop condition. This type of testing makes it easier to isolate faults. For example, in the IBM 308X, the operator console deactivates the channel subsystem and logs out its contenfs through scan-out. It then scans in to reset the failed section. Amdahl 580 mainframes have separate console clocks, instruction and execution unit clocks, and I/O processor clocks, all of which are separately gated versions of a common, ungated system clock (also called a free-running clock).

CLOCK TUNING

High-speed computer systems with multiple boards and many chips on each board often require clock tuning after assembly. Clock tuning is calibrating the signals of the clock-distribution network. Some designers manage to avoid clock tuning by carefully designing and routing the clock network. Clock tuning during assembly and in the field is an expensive process, both in time and in the cost of the technical expertise. Because of this, designers must minimize the number of clock-tuning operations.

Some designers manage to avoid clock tuning by carefully designing and routing the clock network.

Tuning proceeds down the clock-distribution tree from clock source towards the clock destination.

We can determine how much clock tuning is needed by comparing clock signals probed at specified observation points. Modification of delays in the clock-distribution path then compensates for any significant inaccuracy in the clock-edge position or pulsewidth. Tuning can be manual, automatic, adaptively automatic, or a combination.

REFERENCE CLOCKS

To specify the placement of clock edges in a system, we designate one or more of the system's clocks as reference clocks. We use transitions of these timing references, or reference-clock edges, for comparison with other clock edges. We can specify the arrival time of a clock signal at any particular point in the clock-distribution network relative to these reference clocks.

TUNE POINTS

Tuning proceeds down the clock-distribution tree from clock source towards the clock destination. By tuning in this direction, we have the fewest number of tuning operations to calibrate the system because there is no backtracking. We can tune large components, such as printed circuit boards, separately to reduce the tuning requirements of the fully assembled system.

Tune points are the observation points in the clock-distribution tree where we can change the delay. Tuning methods to reposition clock edges include modifying wire lengths, selecting different taps in delay lines, or selecting one delay element from a set by controlling a multiplexer. We can tune clock choppers by adjusting their internal delay elements to control pulsewidth as well.

A tune-point hierarchy is embedded within the clock-distribution network. Figure 10 shows how we can use a tree structure to model this hierarchy. The level of the tune point in the tree is referred to as the depth in the tune-point hierarchy—the deeper the tune point, the farther away it is from the system oscillator. Chip or module primary I/O are usually the deepest accessible tune points.

Designing accessible and effective tune points may be difficult and their tuning resolution and range may be limited. One ap-

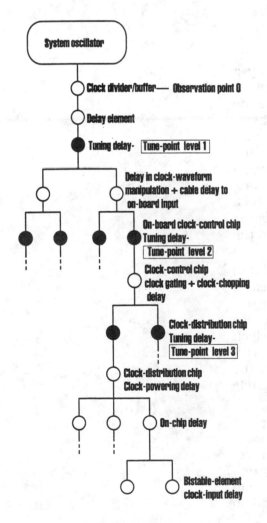

Figure 10. Tune points.

Table 3. *Sample clock-specification plan; all clock times are in nanoseconds.*

| Parameters | Observation Point Level 0 | Observation Point | | | |
		Tune-Point Level 1	Tune-Point Level 2	Tune-Point Level 3	Tune-Point Level 4
Local reference clock edge	0	+3.0	+7.0	+13.0	+20.0
Local tolerance	—	±0.5	±0.5	±0.5	±0.5
Effective clock arrival time	0	3.0±0.5	+7.0±1.0	+13.0±1.5	+20.0±2.0

321

proach to the problem is to determine the worst-case delay in a clock path and pad other clock paths to match this delay. For instance, in subsections of the clock-distribution network, one path may be significantly longer than all others and not have any tune points. By adding delay elements at tune points, we can pad all other comparable paths, either to have the same or shorter delay, which is fixed by design. A sophisticated physical design system can automatically design in this type of tuning by adding wire and capacitive elements as needed.

System timing is based on a clock-specification plan. The plan details the allowable ranges for clocks at each tune point, relative to a reference clock. We need to place tune points in such a way that the sum of the clock skew at the deepest tune point plus the additional skew for the signal to reach bistable elements beyond that tune point does not exceed the acceptable limit on system clock skew. Table 3 shows a sample clock-specification plan for Figure 10. The uncertainty is ±0.5 ns for each of 13 required tuning operations. In this plan, system clock skew equals 3.0 ns plus the maximum of the clock skews on the paths between the 12 tune points at Level 3 and the bistable elements that their clocks control.

TUNING SCHEMES

Tuning schemes vary in sophistication. Manual tuning by a trained technician using an oscilloscope is common. The technician calibrates the tune points sequentially, starting from the one closest to the system clock source. We can provide extra observation points for clock tuning by distributing supplementary clock signals for use as precise reference clocks. No powering trees or other skew-increasing elements are allowed in the signal paths of these clocks, so they have little or no need for tuning themselves. We can thus use them safely and tune all other clock signals relative to the these references. This strategy increases tuning accuracy and decreases the number of tune points in the system.

Knowing the clock signal internal to the chip is often helpful in tuning. For MOS systems with multiple chips, on-chip clock buffering creates uncertain delays, so we must observe a representative internal clock signal. We can use this internal reference clock, which is output at a chip pin, as both a functional clock signal (for small loads), and as a reference to be compared with the corresponding references of other chips. We can then use the relative edge positions of the internal reference clocks to guide tuning.

In the clock-tuning schemes of the ETA 10 supercomputer and IBM 4341, designers provided two separate tuning resolutions, or tuning levels: rough tuning and fine tuning. Fine tuning is required for minute adjustments deep in the clock tree, while rough tuning provides the coarse adjustment earlier in the distribution and (in the ETA machine) before the system is immersed in coolant.

We can provide extra observation points for clock tuning by distributing supplementary clock signals for use as precise reference clocks.

*Untuned systems
—designed with
attention to component
variations and to
equalizing wire lengths
and clock loading
—eventually proved less
expensive and
entirely adequate.*

A typical automatic tuning technique, devised for the discontinued STC CMOS mainframe, uses clock-distribution chips with many degrees of time-shifted clocks available through a large crossbar switch. In this case, a logic chip has four internal reference clocks. Each internal reference clock is a representative internal clock produced after chopping and powering one of four chip clock inputs. These representative clocks are compared with a precise reference clock. The correct skew-minimizing clock for each chip clock's primary input is then selected automatically from the crossbar switch. Registers on the clock-distribution chips are loaded to properly configure each crosspoint of the switch.

Automatic tuning often consists of closing a special feedback connection in a clock network that has an odd number of inversions in the signal path. When this connection is closed, we get oscillations with a period proportional to the total delay of the path. If necessary, we can adjust one or more tune points automatically through control signals determined by diagnostic code. Feedback tuning techniques seem attractive, and numerous patents exist for them, but their sensitivity to the clock duty cycle and the signal transition time make them complex and usually impractical.

Thus, more sophisticated tuning schemes are not necessarily better ones. Proposals for system clocking of the IBM 308X and 3090 mainframes required complex automatic tuning techniques. Eventually, untuned systems—designed with careful attention to component variations and to equalizing wire lengths and clock loading—proved less expensive and entirely adequate.

The clock system is an integral part of synchronous computers, yet it is not a widely studied aspect of their design. Early attention to system timing issues can provide benefits in system performance as well as product development time. The goal of the clock system designer is to control system clock skew at the system operating frequency as well as to minimize electrical hazards that may add undesirable components to the clock signals. Familiarity with clock-generation and clock-distribution techniques suitable for high-speed systems is essential as cycle times decrease. ⬛

ACKNOWLEDGMENTS

This tutorial was supported in part by the National Sciences and Engineering Research Council of the Government of Canada under its postgraduate scholarship program, in part by the National Science

Foundation under grant DCR-8200129, and in part by IBM Corp. The work was performed at Stanford University's Center for Reliable Computing and IBM.

I thank Edward McCluskey and Mark Horowitz of Stanford University, as well as Glen Langdon, Jr., and John DeFazio of IBM Corp., and Ron Kreuzenstein of Amdahl Corp. for their many helpful comments and suggestions.

ADDITIONAL READING

Bakoglu, H.B., J.T. Walker, and J.D. Meindl, "A Symmetric Clock Distribution Tree and Optimized High-Speed Interconnections for Reduced Clock Skew in ULSI and WSI Circuits," *Proc. IEEE Int'l Conf. on Computer Design*, 1986.

Domenik, S., "On-Chip Clock Buffers," *Lambda*, 1st qtr., 1981.

Friedman, E., and S. Powell, "Design and Analysis of a Hierarchical Clock Distribution System for Synchronous Standard Cell/Macrocell VLSI, *IEEE J. Solid-State Circuits*, Vol. SC-21, No. 2, 1986.

Glasser, L., and D. Dobberpuhl, *The Design and Analysis of VLSI Circuits*, chapt. 6, Addison-Wesley, Reading, Mass., 1985.

Hitchcock, R., Sr., "Timing Verification and the Timing Analysis Program," *Proc. Design Automation Conf.*, 1982.

IBM 3033 Processor Complex TO/DM, IBM Corp., Mechanicsburg, Pa., 1981.

Kogge, P., "Hardware Design and Stage Cascading," *The Architecture of Pipelined Computers*, chapt. 2, McGraw-Hill, New York, 1981.

Langdon, G., Jr., *Computer Design*, appendices C and D, Computeach Press, 1982.

Lob, C., and A. Elkins, "HP-9000: 18-Mhz Clock Distribution System," *HP J.*, Aug. 1983.

Maini, J., J. McDonald, and L. Spangler, "A Clock Distribution Circuit with a 100-ps Skew Window," *Proc. Bipolar Circuits and Technology Meeting*, 1987.

McCluskey, E.J., *Logic Design Principles: With Emphasis on Testable Semicustom Circuits*, chapts. 7-8, Prentice-Hall, Englewood Cliffs, N.J., 1986.

Seitz, C., "System Timing," *Introduction to VLSI Systems*, chapt. 7, C. Mead and L. Conway, eds., Addison-Wesley, Reading, Mass., 1980.

Shoji, M., "Electrical Design of the BELLMAC-32A Microprocessor," *Proc. Circuits and Computers Conf.*, 1982.

Shoji, M., "Elimination of Process-Dependent Clock Skew in CMOS VLSI," *IEEE J. Solid-State Circuits*, Vol. SC-21, No. 5, 1986.

Unger, S., and C.J. Tan, "Clocking Schemes for High-Speed Digital Systems," *IEEE Trans. Computers*, Vol. C-35, No. 10, 1986.

Wagner, K.D., *A Survey of Clock Distribution Techniques in High-Speed Computer Systems*, tech. rpt. 86-309, CSL Stanford Electronics Lab., CRC 86-20, Stanford Univ., Stanford, Calif., 1986.

Clocking Schemes for High-Speed Digital Systems

STEPHEN H. UNGER, FELLOW, IEEE, AND CHUNG-JEN TAN, MEMBER, IEEE

Abstract—A key element (one is tempted to say the *heart*) of most digital systems is the clock. Its period determines the rate at which data are processed, and so should be made as small as possible, consistent with reliable operation.

Based on a worst case analysis, clocking schemes for high-performance systems are analyzed. These are 1- and 2-phase systems using simple clocked latches, and 1-phase systems using edge-triggered *D*-flip-flops. Within these categories (any of which may be preferable in a given situation), it is shown how optimal tradeoffs can be made by appropriately choosing the parameters of the clocking system as a function of the technology parameters. The tradeoffs involve the clock period (which of course determines the data rate) and the tolerances that must be enforced on the propagation delays through the logic. Clock-pulse edge tolerances are shown to be an important factor. It is shown that, for systems using latches, their detrimental effects on the clock period can be converted to tighter bounds on the short-path delays by allowing *D* changes to lag behind the leading edges of the clock pulses and by using wider clock pulses or, in the case of 2-phase systems, by overlapping the clock pulses.

Index Terms—Clocking, clock pulses, delays, digital systems, edge-triggered flip-flops, edge tolerances, latches, one-phase clocking, skew, synchronous circuits, timing.

I. INTRODUCTION

VIRTUALLY all contemporary computers and other digital systems rely on clock pulses to control the execution of sequential functions. A number of different general schemes are used, along with several different types of flip-flops or similar storage elements. Despite the deceptively simple outward appearance of the clocking system, it is often a source of considerable trouble in actual systems. The number of parameters involved, particularly in 2-phase systems, is large, and a close analysis reveals a surprising degree of conceptual complexity.

If one is not particularly interested in maximizing performance, then a 2-phase system with nonoverlapping clocks, or a 1-phase system with edge-triggered FF's is not difficult to design. However, if minimizing the clock period is a prime issue, then the problem becomes far more complex. However, significant performance gains are possible by carefully choosing the clocking parameters (period, pulse-widths, overlap), and further gains may be achieved by using well-designed latches.

In this study we develop sets of relations for three basic

Manuscript received December 14, 1982; revised April 5, 1985.

S. H. Unger is with the Department of Computer Science, Columbia University, New York, NY 10027.

C.-J. Tan is with the IBM T. J. Watson Research Center, Yorktown Heights, NY 10598.

IEEE Log Number 8610485.

types of systems that make possible intelligent tradeoffs between speed maximization (period minimization) and the difficulty of satisfying constraints on the logic path delays. We begin with discussions of the state devices considered, the nature of imprecision in clock-pulse generation and distribution systems, logic block delays, and the design goals. We then analyze the simple case of the 1-phase system using edge-triggered FF's. After this warm-up, we proceed to treat the 1-phase system using latches, a considerably more complicated case. An extension of the methodology used in that section is then applied to the case of 2-phase systems using latches. Some overall conclusions are then presented in the final section.

A. State Devices and Their Parameters

The state devices (or storage elements) treated here are:

The latch [2], [6], [1] (sometimes referred to as the *polarity hold* latch. This is a device with inputs C and D, and output Q (often Q', the complement of Q is also generated), such that, ideally, while $C = 0$, Q remains constant (regardless of the value of D), and while $C = 1$, $Q = D$, changing whenever D changes (see Fig. 1). (For real latches, as is explained below, there are nonzero delays in the response times, and there must be constraints on the behavior of the inputs.) The C and D inputs are usually referred to as the clock and data inputs, respectively. Although it is not, in general, necessary to do so, in the applications treated here, the system clock signals are indeed fed to the C inputs of the latches. A variety of implementations of latches are known, differing in such factors as suitability for various technologies, load driving ability, and relative values of the parameters to be discussed subsequently. Latches with logic hazards have been used in some systems. In order to eliminate the possibility of malfunction due to those hazards, the *complement* of the C signal is distributed independently to the latches with its edges carefully controlled relative to the corresponding edges of the C signals. We do not discuss such systems here, where it is assumed that the latches are free of hazards.

The *edge-triggered D-flip-flop* (ETDFF) [2], [6] has the same inputs and outputs as the latch, but Q responds to changes in D only on one edge of the C pulse (see Fig. 2). That is, Q can change only at the time that C changes from 0 to 1 (the rising edge of the C signal), and then only if necessary to assume the same value that D has at that time. (There are also ETDFF's that change state on the negative-going edge of the C signal. Furthermore, it is possible to build a double-edge-triggered D-FF [9] that will respond on *both* edges of the C pulse.)

1) Latch Parameters: The significant parameters for a latch

Reprinted from *IEEE Transactions on Computers*, Vol. C-35, No. 10, pp. 880–895, October 1986.

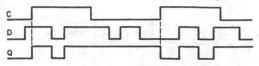

Fig. 1. Behavior of an ideal latch.

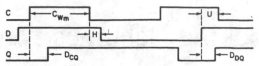

Fig. 3. Latch parameters.

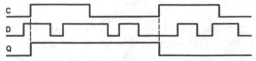

Fig. 2. Behavior of ideal positive triggered ETDFF.

are listed below, with rough definitions (illustrated in Fig. 3). These definitions are then refined to take into account dependencies that exist among the parameters.

C_{Wm}: Minimum clock-pulse width, the minimum width of the clock pulse such that the latch will operate properly even under worst case conditions, and such that widening the C pulse further by making its leading edge occur earlier will not affect the values of D_{DQ}, U, or H, as defined below.

D_{CQ}: Propagation delay from the C terminal to the Q terminal, assuming that the D signal has been set early enough relative to the leading edge of the C pulse.

D_{DQ}: Propagation delay from the D terminal to the Q terminal, assuming that the C signal has been turned on early enough relative to the D change.

U: The *setup time*, the minimum time between a D change and the trailing edge of the C pulse such that, even under worst case conditions, the Q output will be guaranteed to change so as to become equal to the new D value, assuming that the C pulse is sufficiently wide.

H: The *hold time*, the minimum time that the D signal must be held constant *after* the trailing edge of the C signal so that, even under worst case conditions, and assuming that the most recent D change occurred no later than U prior to the trailing edge of C, the Q output will remain stable after the end of the clock pulse. (It is not unusual for the value of this parameter to be negative.)

Note that D_{DQ}, for example, may vary significantly depending on whether the latch output is being changed from 0 to 1 or vice versa. A similar situation exists for D_{CQ}. Where appropriate it is useful to add subscripts R or F to these parameters to distinguish between the rising and falling output cases. This will not be done here. Instead, we shall confine ourselves to using overall maximum and minimum values, as indicated below.

The addition to the subscripts of D_{DQ} or D_{CQ} of an M or m make these parameters the maximum or minimum values, respectively. These are the extremes with respect to variations in the parameters of the components from which the latches are constructed, the directions of signal changes, and the destinations (Q or Q') of the signals.

In the definition of D_{CQ}, it is assumed that D has assumed its proper value early enough. We can make this concept more precise by requiring that the change in D occurs sufficiently early so that making it appear any earlier would have no effect on when Q changes. For any real latch it is always possible to define such an interval. Similarly, when defining D_{DQ}, it is assumed that the leading edge of C appears sufficiently early so that turning C on any earlier would not make Q change any sooner. Again this is possible for any real latch.

Now we state an important postulate regarding propagation delays:

Suppose that C goes on at time t_C, and that D changes, making D different from Q, at time t_D. Then we postulate that the time t_Q at which Q changes is, at the latest:

$$t_Q = \max \ [t_C + D_{CQM}, \ t_D + D_{DQM}]. \tag{1}$$

Although for some latches there are higher order effects, depending on the technology, that may cause t_Q to be larger when the difference between the arguments of the max is small, the error is small enough to justify our postulate for most practical purposes. Refining the model to take such effects into account is left for further research.

A related assumption about latch behavior is that, provided that the setup, hold-time, and minimum pulse-width constraints are observed, the propagation delay will not be affected by the clock-pulse going off before the output changes in response to a D change. An examination of a variety of latch designs appears to justify this assumption.

There are other possibilities for refining our results, by using more complex definitions of latch parameters. If we define the actual interval between the occurrence of a D change and the trailing edge of C as u (note that proper operation requires that $u \geq U$), then, for many latch designs it will be found that the hold time H is, over some range of values of u, a decreasing function of u. There are also possibilities for reducing the clock-pulse width below C_{Wm} (within limits), usually at a cost of increasing propagation delays and/or setup and hold times. For the sake of making the analysis more tractable, we shall not consider these alternatives, but instead shall assume that there is a fixed, consistent, set of latch parameters, as described above.

In summary, we assume that the minimum clock-pulse width is large enough so that further increases cannot reduce any of the other latch parameters, that U is minimal, that H is minimal given U, and that the postulate stated above regarding propagation delays is valid.

2) Edge-Triggered-D-FF Parameters: The significant parameters for an ETDFF are defined below (see also Fig. 4).

U: The *setup time*, the minimum time that the D signal must be stable prior to the triggering edge of the C pulse.

H: The *hold time*, the minimum time that the D signal must be held constant *after* the triggering edge of the C pulse. (The value of H may be 0 or even negative for some ETDFF's.)

C_{Wm}: Minimum clock-pulse width, the minimum width

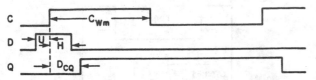

Fig. 4. Parameters of a positive-edge-triggered-D-FF.

of the clock pulse such that the ETDFF will operate properly even under worst case conditions.

D_{CQ}: *Propagation delay* from the C terminal to the Q terminal, assuming that the D signal has been set up sufficiently far in advance as specified by the setup time constraint.

B. Clock-Pulse Edge Deviation

In any real-world system there are limits to the precision with which events can be timed. Our concern here is with synchronous systems with clock-pulses distributed to a multitude of devices for the purpose of coordinating events. The intent is to have certain clock-pulse edges occur simultaneously at all devices (in some cases fixed displacements may be specified for corresponding signals at different devices). In designing clocking schemes, it is necessary to take into account the extent to which this goal cannot be fully attained.

The approach taken here is to assume that, at each significant clock-pulse edge, there is a specified tolerance range, within which we can assume the errors will be confined. This is, essentially, a "worst case" approach. No attempt will be made to exploit statistical information that could make possible more precise estimates of errors, nor will any effort be made to consider the effects of correlations between errors or between delays.

The most elaborate situation that we deal with is that of 2-phase systems using latches as storage elements. Here both the leading and trailing edges of both clock-pulses are of interest (although the analysis makes it clear that certain edges are more significant than others). We define tolerances for all 4 edges, designating them as T_{1L}, T_{1T}, T_{2L}, and T_{2T}, corresponding to the leading and trailing edges of $C1$ and $C2$, respectively. Assume that, for example (see Fig. 13), the leading edge of the $C1$ pulse for some period would have arrived at every latch at time t (which we refer to as its *nominal* arrival time) if there were no inaccuracies in timing. Then, in the actual system, this edge is received at every latch somewhere in the time interval, $(t - T_{1L}, t + T_{1L})$. Corresponding assumptions of course apply for the other three edges. Our goal is to design our systems so that if this assumption, and corresponding assumptions about the precision of the components used, are valid, then there will be no failures due to timing, even if some malicious demon is, in each case, permitted to choose the extreme deviations most likely to cause trouble. Of course in 1-phase systems we need only define two edge tolerances, T_L and T_T.

We are lumping together in these edge tolerances all sources of imprecision in clock timing and distribution. These are principally in the circuits used to determine the clock-pulse widths, often called "shapers," and in the networks used to distribute the pulses to the individual latches (or other similar devices). This latter factor is generally referred to as clock-pulse skew. In the case of 2-phase systems, it is also necessary to consider the circuits that determine the phase relationship between the $C1$- and $C2$-clocks.

Relative to other sources of error, the precision with which the clock frequency can be maintained, at least in high-performance systems, is so great (due to the use of crystal controlled oscillators) that we can safely neglect this factor. (If this assumption is not justified in any particular case, it is not difficult to introduce a tolerance factor on the clock period, which can be superposed on our basic results.)

By representing all of the timing deviations in terms of the edge tolerances, we simplify our analysis, making it easier to treat, as a separate issue, the mechanisms whereby precision is lost.

The precision with which clock-pulse widths can be controlled is generally a function of how precisely delay elements can be specified. The same factor usually is involved in controlling the phase between the $C1$ and $C2$ pulses of a 2-phase system. The ratio of 2 delays on the same chip can be specified with much greater precision than is the case for delays on different chips. Usually one edge of the output of a shaper can be controlled more precisely than the other. In the 2-phase case, there are techniques for minimizing the edge tolerances for particular pairs of edges. As is shown in the sequel, T_{2L} and T_{1T} are usually more significant. They should therefore be kept smaller, relative to the other two-edge tolerances.

Several factors contribute to clock-pulse skew. Despite all efforts to equalize conduction path lengths between the clock source and each clock-pulse "consumer," differences inevitably occur in both off-chip wiring and in paths on chips. Since it is usually necessary to provide amplifiers in the distribution paths, variations in the delays encountered in such devices along different paths produce significant amounts of skew.

Another contribution to skew results from the fact that pulse edges are never vertical as shown in our idealized diagrams, and that there is variability among individual latches, even on the same chip, with respect to the voltage thresholds that effectively distinguish 1's from 0's. Thus even if a pulse edge should arrive simultaneously at the inputs to two different latches, its effect might be felt at different times due to a difference in thresholds. The result is the same as if the delays in the paths leading to the two latches differed. Hence, such effects are considered as part of the skew. Note that, unlike the factor due to varying length conduction paths, this effect could result in the delayed sensing of a *positive*-going edge at a latch that is relatively quick in sensing a *negative*-going edge. (This would occur if the device involved had a relatively high threshold.)

C. Logic Block Delays

In addition to the various parameters associated with the clocking system and with the latches or FF's, a very important pair of parameters is that associated with the logic circuitry: the *maximum* and *minimum* delays in any path through the logic block, designated as D_{LM} and D_{Lm}, respectively. As is made evident in our analysis, large variations among logic path delays are clearly detrimental. That is, for a given value of D_{LM}, it is desirable to keep the *smallest* path delay as close to D_{LM} as possible.

It is frequently the case, when choosing the clocking parameters, that the value of D_{LM}, the *long-path* delay is given; it is a function of the maximum number of stages of logic, the amount of fan-in and fan-out associated with gates in the longest paths, and of the technology, which determines propagation delay through individual gates. The lower bound on the *short-path* delay D_{Lm}, on the other hand, can often be dictated, within limits, by the clock system designer, using such means as adding delay pads to increase the delays in the shortest paths, or adjusting the power levels of certain key gates.

The ultimate limits on how tightly the short-path delays can be controlled, that is, on how high a lower bound D_{LmB} on them is feasible, depends on the tolerances with which gate delays can be specified, as well as on how well wire lengths, both on and off chip can be predicted at design time. It is these factors that determine, for a given value of D_{LM}, what the largest feasible value of D_{LmB} is.

D. Goals for Design of Clocking Schemes

It is assumed here that a principal goal in the specification of a clocking scheme is to make the period as small as possible, which is tantamount to maximizing the speed of the system. But of course this must be done within the confines of a design that results in a system that can be made to operate reliably.

It is obvious that minimizing D_{LM} is basic to minimizing the clock period. But, as pointed out above, it is also important to keep the *smallest* path delay as large as possible. But it is by no means easy to make the logic path delays uniform in value. For this reason, we have developed procedures for finding the minimum possible value of P given the maximum achievable lower bound D_{LmB} on the short-path delays.

II. Optimum Parameters for 1-Phase Clocking with ETDFF's

For 1-phase systems using ETDFF's, the clocking parameters to be determined, (see Fig. 4) are the period P and the clock-pulse width W. A block diagram of the systems under consideration is shown as Fig. 5.

We develop a set of constraints, such that if all are satisfied, and if the D signals arrive on time for the first cycle, then they will also arrive on time for the next cycle and will remain stable long enough to ensure that the FF's react properly. By induction, it follows that, for all succeeding cycles, the FF inputs are also stable over the appropriate intervals, so that the system will behave according to specifications.

For any clock-pulse period, proper operation requires that the D signals become stable at least U prior to the earliest possible occurrence of the triggering edge. (It is assumed here that this is the positive-going edge. Precisely the same arguments apply where the triggering edge is negative going— or even if the FF's trigger on both edges.) If we assume that $t = 0$ coincides with the nominal time of the leading edge of the current clock pulse, then the earliest possible occurrence time

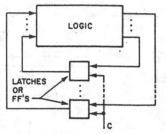

Fig. 5. Block diagram of a 1-phase system.

of that edge is $-T_L$. (See Fig. 6(a).) Hence, the latest possible arrival time, under worst case conditions, of the D signals for the current clock pulse t_{DLArr} must meet the constraint

$$t_{DLArr} \le -T_L - U.$$

Defining the latest possible arrival time, under worst case conditions, of the D signals for the *next* clock-pulse as t_{DLArrN}, it follows that "on time arrival" of D for the next cycle means

$$t_{DLArrN} \le P - T_L - U. \tag{2}$$

Since the *latest* possible occurrence of the leading edge of the current clock-pulse is at T_L, it follows that the latest arrival time of the D signals for the next cycle is

$$t_{DLArrN} = T_L + D_{CQM} + D_{LM}. \tag{3}$$

(see Fig. 6(b)).

Replacing t_{DLArrN} in (2) by its value from (3), we have the required constraint to ensure that D signals are not late

$$P - T_L - U \ge T_L + D_{CQM} + D_{LM}.$$

Solving for P converts it to the following more meaningful form:

$$P \ge 2T_L + U + D_{CQM} + D_{LM}. \tag{4}$$

Next it is necessary to constrain the system so as to ensure that the *earliest* arrival time of a D signal for the *next* cycle does not arrive so early as to violate the hold-time constraint for the current cycle. (See Fig. 7.)

Given that the latest occurrence time of the leading edge of a clock pulse is T_L, the hold-time constraint mandates that the earliest occurrence time of a D signal for the next cycle, t_{DEArrN}, satisfy

$$t_{DEArrN} > T_L + H. \tag{5}$$

Since the *earliest* occurrence of a leading edge of a clock pulse is at $-T_L$, we can express t_{DEArrN} in terms of the FF propagation delay and the logic delay as

$$t_{DEArrN} = -T_L + D_{CQm} + D_{Lm}.$$

Inserting the value of t_{DEArrN} from the above equation into (5) gives us a relation, the satisfaction of which is a necessary and sufficient condition for preventing, under worst case assumptions, premature changes in D signals:

$$-T_L + D_{CQm} + D_{Lm} > T_L + H.$$

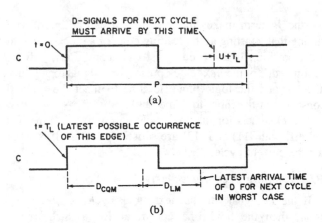

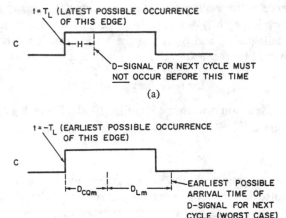

Fig. 6. Ensuring that D signals do not arrive too late in ETDFF systems. (a) Specification of latest permissible arrival time of D signal for next cycle. (b) Arrival time (worst case) for D signal for next cycle.

Fig. 7. Ensuring that D signals in ETDFF systems do not arrive too early. (a) Specification of earliest permissible arrival time of D signal for next cycle. (b) Computation of earliest possible arrival time of D signal for next cycle.

Simplifying and rearranging terms yields the basic constraint that defines D_{LmB}, the lower bound on the short-path delays:

$$D_{Lm} > D_{LmB} = 2T_L + H - D_{CQm}. \tag{6}$$

In addition to constraints (4) and (6) on the period and short-path delays, it is necessary to add a third constraint to ensure that the minimum pulse-width specification for the FF's is satisfied. Since, under worst case assumptions skew might make the leading edge late and the trailing edge early, the minimum width specification for the clock pulses is

$$W \geq T_L + T_T + C_{Wm}. \tag{7}$$

The procedure for choosing optimum clocking parameters for 1-phase systems using ETDFF's is usually very straightforward. We simply set W at any convenient value satisfying constraint (7) and set P to satisfy constraint (4) with equality. In most cases, it will be found that the constraint on the short-path bound given by (6) is not difficult to meet. In the unlikely event that this is not the case, it may be necessary to insert delay pads at the outputs of the FF's. The procedure for doing this is the same as that for the 1-phase case with latches, treated in Section III-D.

III. OPTIMUM PARAMETERS FOR 1-PHASE CLOCKING WITH LATCHES

Fig. 5 is a block diagram of the 1-phase systems treated here. Clock signals with parameters noted are shown in Fig. 8. We shall develop a set of constraints, involving the various parameters we have discussed, such that if and only if they are all respected, the system will operate properly in the sense that the D inputs to all the latches will arrive on time for each clock cycle (as specified by the setup time parameter), and will remain stable for a sufficient interval (as specified by the hold-time parameter).

The argument is in the form of induction on the clock periods. It is assumed at the outset that the D signals arrive on time for the first clock cycle. Constraints are developed to ensure that, given this assumption, the D signals will arrive on time for the next cycle. Additional constraints are then found to ensure that the D signals remain stable for an adequate interval during the first cycle. It is then obvious by induction that the same will be true for all subsequent clock cycles.

More specifically, our initial assumption is that, under worst case conditions (of delay values, edge tolerances, etc.), every D signal must arrive (at a latch input terminal) no later than U prior to the trailing edge of the clock pulse. Taking $t = 0$ as the *nominal* time of occurrence of the *leading* edge of the clock pulse for the current cycle (i.e., the time this edge would arrive if the tolerance on this edge T_L were 0), the earliest possible occurrence time of the *trailing* edge would be $W - T_T$.

Since the D signal must arrive at least U prior to this edge, we have for the latest permissible arrival time for D, t_{DLArr}:

$$t_{DLArr} \leq W - T_T - U. \tag{8}$$

Assume now that the above constraint is satisfied for the first clock cycle.

A. Preventing Late Arrivals of D Signals

The latest (under worst case conditions) arrival time of D signals for the next cycle is designated as t_{DLArrN}. The maximum permitted value of t_{DLArrN} is found by simply adding P to the right side of (8)

$$t_{DLArrN} \leq W - T_T - U + P \tag{9}$$

(see Fig. 9(a)).

The worst case value of t_{DLArrN} is the latest time at which the output of a latch could respond to a D signal, plus the maximum delay through the logic. Designating the latest occurrence time of a leading edge of a clock pulse as t_{CLL}, and using postulate (1) for determining the latest time at which the output of a latch could change, we obtain

$$t_{DLArrN} = \max~[t_{CLL} + D_{CQM},~t_{DLArr} + D_{DQM}] + D_{LM}.$$

(The discussions pertaining to the left and right parts, respectively, of the max expression are illustrated by Fig. 9(b)

329

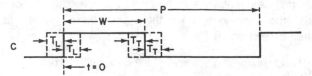

Fig. 8. Parameters for 1-phase systems.

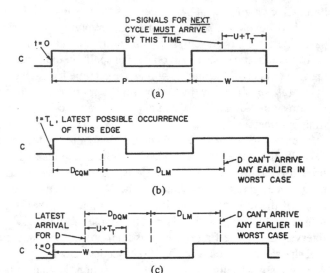

Fig. 9. Ensuring that D arrives sufficiently early. (a) Deadline for arrival of D signals. (b) Bound on D arrival due to leading edge of C. (c) Bound on next-cycle D arrival due to arrival time of D for current cycle.

and (c).) The value of t_{CLL} is clearly T_L, and the value of t_{DLArr} is given by (8), so replacing those variables in the above relation gives us

$$t_{DLArrN} = \max \ [T_L + D_{CQM}, \ W - T_T - U + D_{DQM}] + D_{LM}.$$

(10)

Combining (9) with (10) produces

$$\max \ [T_L + D_{CQM}, \ W - T_T - U + D_{DQM}]$$
$$+ D_{LM} \leq W - T_T - U + P.$$

Solving for P yields

$$P \geq \max \ [T_L + T_T + U + D_{CQM} - W, \ D_{DQM}] + D_{LM}.$$

This expression can be decomposed into 2 constraints that, in combination, are equivalent to it:

$$P \geq D_{CQM} + D_{LM} + U + T_L + T_T - W \qquad (11)$$

and

$$P \geq D_{DQM} + D_{LM}. \qquad (12)$$

The constraint (12) can be intuitively justified by noting that it represents the total time for a signal to traverse a complete loop, under worst case conditions. If the period were any less, then, if the worst case conditions were actually realized, a signal following a sequence of such maximum delay paths would fall increasingly far behind the clock pulses until it eventually violated a setup time constraint.

Constraint (11) can also be justified intuitively. (Transpos-

ing the W term makes this clearer.) It can be interpreted as stating that, starting at the leading edge of a clock pulse, there must be time, under even worst case conditions, before the trailing edge of the *next* clock pulse, for a signal to get through a latch, and the logic block in time to meet the setup time constraint at the input to some latch.

The D signals for the next cycle will arrive on time if, and only if, both (11) and (12) are satisfied, and if (8) is satisfied for the current cycle.

B. Preventing Premature Arrivals of D Signals

If the D signal for the next clock cycle is generated too soon, then the hold-time constraint for a latch might be violated. This is where the short-path delays become important. In order to prevent the possibility of a hold-time violation, it is necessary that, in the worst case, a D change for the next cycle not occur until at least H after the latest possible occurrence of the trailing edge of the clock-pulse defining the current cycle. With t_{CLT} as the latest occurrence of a clock-pulse trailing edge, and t_{DEArrN} as the earliest possible arrival of a D signal for the next cycle, this constraint is expressed as

$$t_{DEArrN} > t_{CLT} + H.$$

(This discussion is illustrated by Fig. 10(a).) Replacing t_{CLT} by its value, $W + T_T$, we obtain

$$t_{DEArrN} > W + T_T + H. \qquad (13)$$

Letting t_{CEL} represent the earliest possible arrival time of a clock-pulse leading edge, and t_{DEArr} represent the earliest arrival time of a D signal for the *current* cycle, we again utilize (1) to obtain the following:

$$t_{DEArrN} = \max \ [t_{CEL} + D_{CQm}, \ t_{DEArr} + D_{DQm}] + D_{Lm}.$$

(The discussion involving the left part of the max is illustrated in Fig. 10(b).) Replacing t_{CEL} by its value $-T_L$ and bringing D_{Lm} inside the max, yields

$$t_{DEArrN} = \max \ [-T_L + D_{CQm} + D_{Lm}, \ t_{DEArr} + D_{DQm} + D_{Lm}].$$

(14)

Inserting the above value of t_{DEArrN} in (13) yields

$$\max \ [-T_L + D_{CQm} + D_{Lm}, \ t_{DEArr} + D_{DQm} + D_{Lm}]$$
$$> W + T_T + H. \qquad (15)$$

Now we show that, for a system that operates properly even under worst case conditions, (15) is satisfied if, and only if, the left part of the max in (15) exceeds the right side of the inequality. The "if" part of this assertion is obviously true.

To prove necessity (the "only if" part) let us assume that (15) is valid but that the left part of the max does *not* exceed the right part of the inequality. Then it follows that the *right* part of the max must satisfy the inequality, and hence must exceed the left part of the max. In that case, (14) is reduced to

$$t_{DEArrN} = t_{DEArr} + D_{DQm} + D_{Lm}. \qquad (16)$$

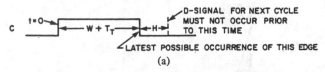

(a)

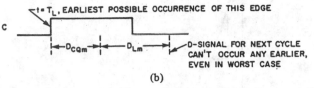

(b)

Fig. 10. Ensuring that D does not arrive too early. (a) Earliest permissible arrival time of D. (b) Lower bound on D arrival for next cycle due to leading edge of C.

But from (12) it is clear that

$$P > D_{DQm} + D_{Lm}.$$

Adding t_{DEArr} to both sides yields

$$t_{DEArr} + P > t_{DEArr} + D_{DQm} + D_{Lm}.$$

From the above and from (16) we then obtain

$$t_{DEArr} + P > t_{DEArrN}.$$

But this means that, for each cycle (in the worst case), D arrives earlier and earlier relative to the trailing edge of C. Therefore, even if t_{DEArr} is comfortably above the minimum for the first cycle, it will eventually violate the hold-time constraint, and hence the system would not operate properly under worst case conditions. Hence, by contradiction, we have completed the argument that (15) is equivalent to:

$$-T_L + D_{CQm} + D_{Lm} > W + T_T + H$$

or, solving for D_{Lm}:

$$D_{Lm} > D_{LmB} = T_L + T_T + H + W - D_{CQm}. \qquad (17)$$

The above expression gives us the lower bound D_{LmB} on the short-path delay. Satisfying this bound is necessary and sufficient to ensure against the premature arrival of a D signal.

C. Consequences of the Constraints

The basic constraints derived in the previous subsections are reproduced below.

$$P \geq D_{CQM} + D_{LM} + U + T_L + T_T - W \qquad (11)$$

$$P \geq D_{DQM} + D_{LM} \qquad (12)$$

$$D_{Lm} > D_{LmB} = T_L + T_T + H + W - D_{CQm}. \qquad (17)$$

To these we must add one more to ensure that, even under worst case conditions, the clock-pulse width at any latch input meets the minimum clock pulse width specifications of the latches. This is:

$$W \geq C_{Wm} + T_L + T_T. \qquad (18)$$

W in (11) cannot usefully be increased beyond the point where the right side of (11) would, if equality held, violate

(12), which of course also represents a lower bound on P. Note that it is undesirable to increase W gratuitously, since this would, as indicated by (17), raise the lower bound on the short-path delays. To find the maximum useful value of W, treat (11) and (12) as equalities and solve them simultaneously (eliminating P) to obtain

$$W = U + T_L + T_T + D_{CQM} - D_{DQM}. \qquad (19)$$

When W is less than the above value, (11), with equality, specifies the minimum value of P. When W equals that value, the minimum value of P is given by (12). The maximum useful value of D_{LmB} is found by substituting into (17) the maximum useful value of W. This gives us

$$D_{LmB} = H + U + 2(T_L + T_T) + D_{CQM} - D_{CQm} - D_{DQM}. \qquad (20)$$

If the value of the lower bound on the short-path delays given by the above relation is attainable, then the minimum P value of (12) is attainable. If not, then, to find the minimum P value as a function of an achievable value of D_{LmB}, solve (17) and (11) (as equations) simultaneously for P, eliminating W. This results in

$$P = H + U + 2(T_L + T_T) + D_{CQM} - D_{CQm} + D_{LM} - D_{LmB}. \qquad (21)$$

Since W must also satisfy constraint (18), there is a corresponding lower bound on D_{LmB}, which is found by substituting into (17) the right side of (18) for W to obtain

$$D_{LmB} = 2(T_L + T_T) + H - D_{CQm} + C_{Wm}. \qquad (22)$$

The relations developed here are the basis for the optimization procedure of the next subsection. First, however we must consider a possible variation of the development thus far.[1] The initial assumption in the discussion of 1-phase systems was that the D signals must appear at latch inputs no later than U prior to the trailing edges of the clock pulses. In what followed, this constraint was consistently observed. But what if we had made a stronger assumption, i.e., that the D changes must appear even earlier, say at $U + r$ ($r > 0$) prior to the trailing edges of the clock pulses? Is it possible that there might be some advantages to this?

The key to analyzing this question is to observe that the proposal is exactly equivalent to assuming a larger value of the setup time U. The effect of this can be determined by looking at those constraints and derived relations that involve U, namely (11), (19), (20), and (21). The value of D_{LmB} necessary to achieve the minimum P increases with U. So does the minimum value of P for any value of D_{LmB} in the range for which (21) is valid. Thus there are clear disadvantages to this alternative of effectively increasing U, and no apparent advantages to compensate for them. It follows then that any 1-phase clocking scheme that violates any of our constraints will, under the worst case assumption, either be vulnerable to failure, or will be suboptimum in that either P or D_{LmB} would be reducible without increasing the other.

---

[1] The necessity for considering this possibility was pointed out by V. Pitchimani and G. Smith.

D. When the Short-Path Bounds Cannot be Met

Now observe that neither the basic constraint (17) on D_{LmB}, nor either of the derived extremes of D_{LmB} given by (20) and (22) involve D_{LM}. Thus, there is no inherent reason why the range found for D_{LmB} (in terms of the afore-noted extremes) should be much *below* or indeed not *above* D_{LM}. If, despite all efforts, including the use of delay pads in critical paths, it is still not possible to satisfy the lower bound on the short-path delays represented by (22), then (assuming that the relevant latch or other parameters cannot be favorably altered so as to remedy this situation), it is necessary to resort to more drastic measures.

The most practical technique appears to be to introduce uniform delay elements into *all* logic paths so as to increase the minimum path delays by an amount sufficient to get us into the desired range. Suppose, for example that the largest value of D_{LmB} that can be reliably guaranteed, is less than the bound of (22) by the amount d_x. Then we could add delay pads with *minimum* values d_x to the outputs of all latches. The effect would be to increase the attainable D_{LmB} to the desired minimum, and to increase D_{LM} by the amount corresponding to the *maximum* value of delay elements with minimum values d_x. If we define T_d as the delay element tolerance ratio d_M/d_m then the addition to D_{LM} is $T_d d_x$. Note that P increases by $T_d d_x$ over the value obtained for it if the D_{LmB} from (22) is used in (21). The graph of Fig. 11 illustrates how P varies with the maximum attainable value of D_{LmB}. It is piece-wise linear, with the left part corresponding to the region where uniform pads must be added as just indicated, and with the right part generated directly from (21). The value P_1 corresponds to the value given by (12).

E. Procedure for Optimizing the Clocking Parameters

We are now in position to describe a procedure for finding the minimum clock period, given D_{MLmB}, the maximum lower bound we can enforce on the short-path delays. The corresponding value of W is also determined.

A complicating factor is the possibility that the lower bound on W given by (18), might exceed what we have called the "maximum useful value of W," given by (19). In that event, the W value is given by (18), and D_{LmB} is given by (22). Note that, when D_{MLmB} is less than the required value of D_{LmB}, it is necessary to pad the outputs of all latches with delay elements whose *minimum* values make up the difference. This adds to the period an amount T_d times this minimum value.

The procedure is as follows.

IF the right side of (18) $\leq$ the right side of (19)
THEN
   IF $D_{MLmB} \geq$ right side of (20)
   THEN
      $D_{LmB} \Leftarrow$ right side of (20)
      $W \Leftarrow$ right side of (19)
      $P \Leftarrow$ right side of (12)
   ELSE
      $D_{LmB} \Leftarrow D_{MLmB}$
      IF $D_{MLmB} \geq$ right side of (22)

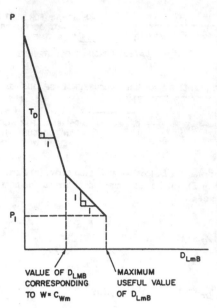

Fig. 11. *P* as a function of the largest achievable lower bound on short-path delay.

      THEN
         $P \Leftarrow$ right side of (21)
         Solve (17) to determine W
      ELSE
         $d \Leftarrow$ right side of (22) $- D_{MLmB}$
         In all latch outputs put delay pads with minimum value d
         $W \Leftarrow$ right side of (18)
         $P \Leftarrow$ right side of (11) $+ T_d d$
ELSE
   $W \Leftarrow$ right side of (18)
   $D_{LmB} \Leftarrow$ right side of (22)
   IF $D_{MLmB} \geq$ right side of (22)
   THEN
      $P \Leftarrow$ right side of (12)
   ELSE
      $d \Leftarrow$ right side of (22) $- D_{MLmB}$
      In all latch outputs put delay pads with minimum value d
      $P \Leftarrow$ right side of (12) $+ T_d d$.

Other procedures based on the constraints developed here may be useful under special circumstances.

IV. OPTIMUM PARAMETERS FOR 2-PHASE CLOCKING WITH LATCHES

Fig. 12 is a general block diagram of the 2-phase clocked systems treated here. Clock signals (shown in Fig. 13) go directly to the C inputs of the latches. Facilities for scan-in and scan-out are not included as they do not affect the basic arguments.

The strategy to be followed is based on the assumption that if the D inputs to all of the latches are valid in the intervals specified by the U and H parameters, then the system will operate as specified. A set of constraints will be derived, such that if the D inputs to all of the $L1$ latches arrive early enough for the first clock cycle, then if, and only if, all of the constraints are satisfied, the inputs to the $L2$ latches will arrive on time for the first $C2$ clock interval, and the D inputs to the

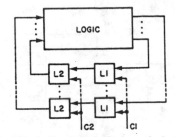

Fig. 12. Block diagram of a 2-phase clocked system.

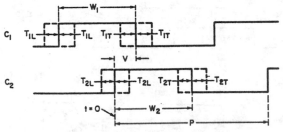

Fig. 13. Parameters for 2-phase systems.

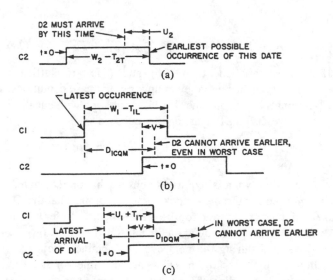

Fig. 14. $D2$ arrival time. (a) Deadline for arrival of $D2$ signals. (b) Lower bound on $D2$ arrival due to $C1$ leading edge. (c) Lower bound on $D2$ arrival due to arrival of $D1$ signal.

$L1$-latches will arrive early enough for the next $C1$-clock interval. Also, the D inputs to the $L1$ latches will remain valid long enough during the first $C1$ interval, and the D inputs to the $L2$ latches will remain valid sufficiently long during the first $C2$ interval. By induction, it then follows that, for all subsequent clock periods, the latches will all have valid inputs during the prescribed intervals.

Throughout the following discussion it is assumed that $t = 0$ at the *nominal* time (by "nominal time" we mean what the time would be if the edge tolerances were 0) of the leading edge of the $C2$ clock. (The *actual* arrival time of this edge at any $L2$ latch may be anywhere between $-T_{2L}$ and $+T_{2L}$). It follows then that the earliest arrival time of the trailing edge of the $C1$ pulse is $V - T_{1T}$. To ensure that the $L1$ latch setup time constraints are met, even under worst case conditions, t_{D1LArr}, the latest arrival time for $D1$ signals during the current clock cycle, must satisfy

$$t_{D1LArr} \leq V - T_{1T} - U_1. \tag{23}$$

In all that follows, it is assumed that, for the first clock period, all $D1$ signal arrival times satisfy (23).

The argument that the constraints developed here are necessary as well as sufficient is dependent on the assumption that, in the worst case, (23) is satisfied with equality. Since this is not actually necessary, it follows that the constraints are not strictly necessary. However, enforcing a more stringent constraint on arrival times of $D1$ signals, namely that they be required to be earlier by some additional amount, is equivalent to assuming that U_1 has increased by this same amount. The effect of this is considered at the end of this section, where it is shown that, as compared to the disadvantages, there is very little to be gained by increasing U_1 (or U_2, which is equivalent to insisting that the $D2$ signals arrive at a time earlier than required by the setup time requirements).

A. Latest Arrival Times of D2-Signals for First Clock Interval

First we develop constraints to ensure that, if the $D1$ signals arrive on time, the $D2$ signals will also arrive on time. (Refer here to Fig. 14(a).) In this case, "on time" means that in order to respect the setup time constraint for the $L2$ latches, the $D2$ signals must arrive no later than U_2 prior to the trailing edge of the $C2$ pulses. At the earliest, the trailing edge of a $C2$ pulse might occur at $W_2 - T_{2T}$.

So, the latest arrival time t_{D2LArr} of the $D2$ signals must satisfy

$$t_{D2Larr} \leq W_2 - T_{2T} - U_2. \tag{24}$$

Let t_{C1LL} be the latest arrival time of the leading edge of a $C1$ pulse. Then, recalling (1) about latch propagation delays, the latest time when the output of an $L1$ latch changes (an alternate description of t_{D2LArr}) is as follows (the left side of the max is illustrated by Fig. 14(c) and the right side by part Fig. 14(b)):

$$t_{D2LArr} = \max \ [t_{D1LArr} + D_{1DQM}, \ t_{C1LL} + D_{1CQM}].$$

Replacing t_{C1LL} by its value $V - W_1 + T_{1L}$ and t_{D1LArr} by the value given in (23) (assuming that (23) is satisfied with equality) gives us

$$t_{D2LArr} = \max \ [V - U_1 - T_{1T} + D_{1DQM},$$
$$V - W_1 + T_{1L} + D_{1CQM}]. \tag{25}$$

Combining (24) with (25) we obtain

$$\max \ [V - U_1 - T_{1T} + D_{1DQM}, \ V - W_1 + T_{1L} + D_{1CQM}]$$
$$\leq W_2 - T_{2T} - U_2.$$

This can be expressed as two separate constraints:

$$V - U_1 - T_{1T} + D_{1DQM} \leq W_2 - T_{2T} - U_2$$

and

$$V - W_1 + T_{1L} + D_{1CQM} \leq W_2 - T_{2T} - U_2$$

which can be rewritten, respectively, as

$$W_2 \geq V + U_2 - U_1 + D_{1DQM} + T_{2T} - T_{1T} \tag{26}$$

333

and

$$W_1 + W_2 \geq V + U_2 + D_{1CQM} + T_{1L} + T_{2T}. \qquad (27)$$

If (23) is satisfied, then (26) and (27) are sufficient conditions for ensuring that even under worst case conditions, the $D2$ signals arrive on time. If (23) is satisfied with equality, then they are also sufficient for this purpose.

A. Latest Arrival Times of D1 Signals During the Next Cycle

Now consider what is required to ensure that the $D1$ signals arrive on time for the *next* clock cycle, assuming that the $D1$ and $D2$ signals are on time for the present cycle. (Refer here to Fig. 15(a).) The upper bound on the latest arrival time $t_{D1LArrN}$ of a $D1$ signal during the next cycle is obtained from (23), which gives the latest permissible arrival time for the first cycle by simply adding the period P to the right side. This gives us

$$t_{D1LArrN} \leq P + V - U_1 - T_{1T}. \qquad (28)$$

Now consider how long it might take a signal to get through an $L1$ latch, through the following $L2$ latch, and through the logic to reach an $L1$ latch input in time for the next $C1$ pulse. (See Fig. 12). In terms of the latest arrival time at an $L2$ input t_{D2LArr} and the latest possible occurrence of a $C2$ leading edge t_{C2LL}, (1) gives us for the latest arrival time t_{Q2LArr} for a signal at an $L2$ output

$$t_{Q2LArr} = \max\ [t_{D2LArr} + D_{2DQM},\ t_{C2LL} + D_{2CQM}].$$

Adding the maximum delay through the logic D_{LM} gives us the latest arrival time, $t_{D1LArrN}$ for a signal at an $L1$ input during the *next* cycle

$$t_{D1LArrN} = \max\ [t_{D2LArr} + D_{2DQM},\ t_{C2LL} + D_{2CQM}] + D_{LM}.$$

Equation (25) gives us t_{D2LArr}, and t_{C2LL} is simply T_{2L}. Substituting in the above relation yields

$$t_{D1LArrN} = \max\ [\max\ [V - U_1 - T_{1T} + D_{1DQM},$$
$$V - W_1 + T_{1L} + D_{1CQM}] + D_{2DQM},$$
$$T_{2L} + D_{2CQM}] + D_{LM}.$$

Expanding the inner max yields

$$t_{D1LArrN} = \max\ [V - U_1 - T_{1T} + D_{1DQM} + D_{2DQM},$$
$$V - W_1 + T_{1L} + D_{1CQM} + D_{2DQM},$$
$$T_{2L} + D_{2CQM}] + D_{LM}. \qquad (29)$$

There are three factors restricting the propagation of signals thru the two latches: propagation thru the D inputs of both $L1$ and $L2$ latches, propagation from the C inputs of the $L1$ latches (involving the location of the $C1$ leading edge) through the D inputs of $L2$ latches, and propagation from the C inputs of the $L2$ latches (involving the location of the $C2$ leading edge). These are all accounted for in the above expression. They are illustrated in Fig. 15(b), (c), and (d), respectively.

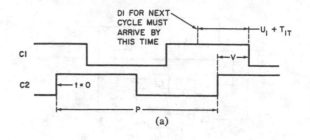

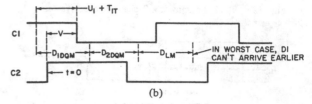

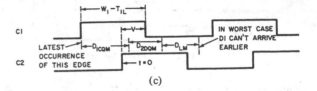

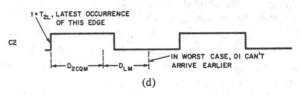

Fig. 15. $D1$ arrival time. (a) Deadline for $D1$ arrival during next cycle. (b) Lower bound on next-cycle $D1$ arrival due to propagation delays through D inputs of latches. (c) Lower bound on next cycle $D1$ arrival due to $C1$ leading edge. (d) Lower bound on $D1$ arrival for next cycle due to $C2$ leading edge.

Replacing $t_{D1LArrN}$ in (28) by the value found in (29) gives us

$$\max\ [V - U_1 - T_{1T} + D_{1DQM} + D_{2DQM},$$
$$V - W_1 + T_{1L} + D_{1CQM} + D_{2DQM},$$
$$T_{2L} + D_{2CQM}] + D_{LM} \leq P + V - U_1 - T_{1T}.$$

Solving for P and simplifying yields

$$P \geq \max\ [D_{1DQM} + D_{2DQM},$$
$$- W_1 + D_{1CQM} + D_{2DQM} + U_1 + T_{1L} + T_{1T},$$
$$T_{1T} + T_{2L} - V + D_{2CQM} + U_1] + D_{LM}. \qquad (30)$$

Relation (30) can be decomposed into the following three equivalent constraints which, taken together, are equivalent to it.

$$P \geq D_{1DQM} + D_{2DQM} + D_{LM} \qquad (31)$$

$$P \geq - W_1 + D_{1CQM} + D_{2DQM} + U_1 + D_{LM} + T_{1L} + T_{1T}$$

or, solving for W_1

$$W_1 \geq - P + D_{1CQM} + D_{2DQM} + U_1 + D_{LM} + T_{1L} + T_{1T} \qquad (32)$$

$$P \geq - V + D_{2CQM} + U_1 + D_{LM} + T_{1T} + T_{2L}. \qquad (33)$$

Each of the above constraints can be justified intuitively.

- Constraint (31) indicates that the period cannot be less than the total time it would take a signal, under worst case conditions, to propagate around a loop (i.e., thru an $L1 - L2$ latch pair and the logic).

- Constraint (33) (when the $-V$ is transposed) states that, starting at the leading edge of a $C2$ pulse, there must be time, prior to the end of the *next* $C1$ pulse, for signals to get through $L2$ latches and the logic to the inputs of $L1$ latches prior to the setup times for those latches, under worst case conditions of logic delay, latch delay and edge tolerances.

- Similarly, (32) states (transposing the $-P$ term helps make this clearer) that a similar relation holds with respect to starts made at the leading edge of $C1$ pulses and ending at the trailing edges of $C1$ pulses during the next cycle.

Note that if (26) is satisfied with equality, and if (27) is satisfied, then, it is not difficult to show, with the aid of (31), that (32) is implied. Alternatively, satisfying both (32) with equality and (27) ensures that (26) is satisfied.

C. Premature Changes of D1 Signals

Next we ensure that changes in $D1$ signals do not propagate through the $L1$ and $L2$ latches and the logic so fast that they cause some $D1$ inputs to change to their values for the *next* cycle prematurely, i.e., before the hold times for the current cycle have expired. (Refer here to Fig. 16(a).) The earliest arrival time $t_{D1EArrN}$ of such "short-path" signals for the next cycle must be later than H_1 after the latest possible occurrence of a $C1$ trailing-edge; that is

$$t_{D1EArrN} > V + T_{1T} + H_1. \tag{34}$$

The earliest time that a $D1$ signal can change as a result of signal changes generated during the same clock period getting around the loop is arrived at analogously to the way (29) was produced; the same three categories of constraints must be considered. Now, however, since we seek the *minimum* delays, we use *minimum* values for the delays within the max expressions, and the *earliest* times for the critical clock-pulse edges.

With t_{C2EL} as the earliest occurrence time of a $C2$ pulse leading edge, and with t_{D2EArr} as the earliest arrival time of a $D2$ input change, postulate (1) indicates that the earliest output from an $L2$ latch can occur at t_{Q2E}, given by

$$t_{Q2E} = \max \ [t_{C2EL} + D_{2CQm}, \ t_{D2EArr} + D_{2DQm}]$$

Adding D_{Lm} to each component of the max of the right side of the above relation, and replacing t_{C2EL} by its value $- T_{2L}$ gives us $t_{D1EArrN}$, the earliest arrival time of a $D1$-change for the *next* clock cycle

$$t_{D1EArrN} = \max \ [- T_{2L} + D_{2CQm} + D_{Lm},$$
$$t_{D2EArr} + D_{2DQm} + D_{Lm}]. \tag{35}$$

To find t_{D2EArr} is the same as finding the earliest output of an $L1$ latch. If we represent the earliest occurrence time of a $C1$ pulse leading edge by t_{C1EL}, and the earliest arrival of a $D1$

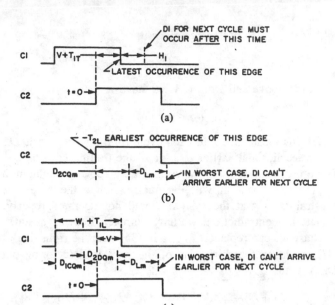

Fig. 16. Premature $D1$ changes. (a) Lower bound for occurrence time of $D1$ for next cycle. (b) Lower bound on next cycle $D1$ arrival due to leading edge of $C2$. (c) Lower bound on $D1$ arrival for next cycle due to leading edge of $C1$.

input for the current cycle as t_{D1EArr}, then we have

$$t_{D2EArr} = \max \ [t_{C1EL} + D_{1CQm}, \ t_{D1EArr} + D_{1DQm}]. \tag{36}$$

Replacing t_{C1EL} in the above equation by $V - W_1 - T_{1L}$, and inserting the resulting expression for t_{D2EArr} in (35), yields

$$t_{D1EArrN} = \max \ [- T_{2L} + D_{2CQm} + D_{Lm}, \ \max \ [V - W_1 - T_{1L}$$
$$+ D_{1CQm}, \ t_{D1EArr} + D_{1DQm}] + D_{2DQm} + D_{Lm}].$$

Expanding the inside max in the above equation gives us

$$t_{D1EArrN} = \max \ [- T_{2L} + D_{2CQm} + D_{Lm},$$
$$V - W_1 - T_{1L} + D_{1CQm} + D_{2DQm} + D_{Lm},$$
$$t_{D1EArr} + D_{1DQm} + D_{2DQm} + D_{Lm}]. \tag{37}$$

(The first 2 parts of the max are illustrated in Fig. 16(b) and (c), respectively.)

Now we show that, for a system that operates properly even under worst case conditions, (34) is valid if, and only if, it is valid when the value used for $t_{D1EArrN}$ is that of (37) with the third part of the max deleted. The "if" part of this assertion is obviously true.

To prove necessity (the "only if" part), let us assume the contrary, namely that (34) is valid and that neither of the first 2 parts of the max of (37) exceeds the right side of (34).

Then, since $t_{D1EArrN}$ *must* satisfy (34), it follows that the *third* part of the max must do so. Therefore, it must exceed each of the first two parts, both of which can therefore be deleted from (37), reducing it to

$$t_{D1EArrN} = t_{D1EArr} + D_{1DQm} + D_{2DQm} + D_{Lm}. \tag{38}$$

But, from (31) it is clear that

$$P > D_{1DQm} + D_{2DQm} + D_{Lm}.$$

Adding t_{D1EArr} to both sides gives us

$$t_{D1EArr} + P > t_{D1EArr} + D_{1DQm} + D_{2DQm} + D_{Lm}.$$

From the above and from (38) we have

$$t_{D1EArrN} < t_{D1EArr} + P.$$

But this means that, for each cycle (in the worst case), $D1$ arrives earlier and earlier relative to the trailing edge of $C1$. Therefore, even if t_{D1EArr} is comfortably above the minimum for the first cycle it will eventually violate the hold-time constraint, so that the system would not operate properly. Hence, by contradiction, we have completed our argument.

Thus, we can replace $t_{D1EArrN}$ in (34) with the right side of (37), omitting the third part of the max (and factoring out D_{Lm}), which gives us

$$\max\ [-T_{2L} + D_{2CQm},\ V - W_1 - T_{1L} + D_{1CQm} + D_{2DQm}]$$
$$+ D_{Lm} > V + T_{1T} + H_1.$$

Solving for D_{Lm} produces

$$D_{Lm} > D_{LmB} = \min\ [V + H_1 + T_{1T} + T_{2L} - D_{2CQm},$$
$$W_1 + H_1 + T_{1L} + T_{1T} - D_{1CQm} - D_{2DQm}].$$

The above expression can be partitioned into two relations, at least one of which must be satisfied.

$$D_{Lm} > D_{LmB} = V + H_1 + T_{1T} + T_{2L} - D_{2CQm} \quad (39)$$

$$D_{Lm} > D_{LmB} = W_1 + H_1 + T_{1T} + T_{1L} - D_{1CQm} - D_{2DQm}. \quad (40)$$

While it is conceivable that a system might exist for which the right side of (40) is less than the right side of (39), an examination of the 2 expressions suggests that this is very unlikely. Hence, in most cases it is constraint (39) that should be relied upon.

D. Premature Changes of D2 Signals

Now consider how to ensure that the $D2$ signals, once on, remain stable long enough for proper operation, i.e., that the hold-time constraints for the $L2$ latches are satisfied. It is necessary to ensure that $t_{D2EArrN}$ the time of the earliest change in a $D2$ signal resulting from a signal passed by the *next* $C1$ pulse satisfies the following relation where t_{C2LT} is the latest occurrence time of the trailing edge of $C2$.

$$t_{D2EArrN} > t_{C2LT} + H_2. \quad (41)$$

The latest appearance of the trailing edge of $C2$, C_{2LT}, occurs at $W_2 + T_{2T}$. (Refer now to Fig. 17(a).) Replacing t_{C2LT} in (41) by this value, we obtain

$$t_{D2EArrN} > W_2 + T_{2T} + H_2. \quad (42)$$

Noting that the earliest time that any $D1$ signal is permitted to change as a result of a previous $D1$ change during the same cycle is $V + H_1 + T_{1T}$ [see (34)], and that the leading edge of the next $C1$ pulse occurs no earlier than $P + V - W_1 - T_{1L}$,

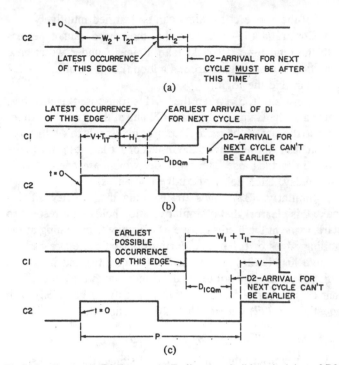

Fig. 17. Premature $D2$ changes. (a) Earliest permissible arrival time of $D2$. (b) Lower bound on $D2$ arrival time for next cycle due to $D1$ arrival time. (c) Lower bound on $D2$ arrival for next cycle due to the leading edge of $C1$ for the next cycle.

we can compute $t_{D2EArrN}$ as follows:

$$t_{D2EArrN} = \max\ [V + H_1 + T_{1T} + D_{1DQm},$$
$$P + V - W_1 - T_{1L} + D_{1CQm}]. \quad (43)$$

Combining (42) and (43) yields

$$\max\ [V + H_1 + T_{1T} + D_{1DQm},\ P + V - W_1 - T_{1L} + D_{1CQm}]$$
$$> W_2 + H_2 + T_{2T}. \quad (44)$$

The left and right parts of the max of (44) are illustrated in Fig. 17(b) and (c), respectively.

Relation (44) can be expressed as the following pair of relations, *at least one of which* must be satisfied:

$$V + H_1 + T_{1T} + D_{1DQm} > W_2 + H_2 + T_{2T}$$

$$P + V - W_1 - T_{1L} + D_{1CQm} > W_2 + H_2 + T_{2T}.$$

These may be more conveniently expressed, respectively, as

$$W_2 < H_1 - H_2 + D_{1DQm} + V + T_{1T} - T_{2T} \quad (45)$$

and

$$W_1 + W_2 < D_{1CQm} + V + P - H_2 - T_{1L} - T_{2T}. \quad (46)$$

They constitute necessary and (along with the other constraints developed above) sufficient conditions for ensuring that the inputs to the $L2$ latches will remain on for a sufficiently long time relative to the trailing edges of the $C2$ pulses. Under most circumstances, it would appear that (46) is much more likely to be satisfied than is (45).

E. Intervals During Which Output Signals are Valid

(Since the material in this subsection is not essential to what follows, it may be skipped at first reading.)

If outputs are taken from the logic block, and are thereafter sent to external receivers instead of to $L1$ latches, then it is clear that those signals will be stable and valid at least over the interval during which we have ensured that the $D1$ signals are valid, namely

$$(V - U_1 - T_{1T}, \quad V + H_1 + T_{1T}).$$

If the outputs are taken directly from $L2$ latches, then we can compute the stable output interval as follows.

The *unstable* interval begins at the earliest time at which a $Q2$ signal can change (i.e., the earliest time an $L2$ latch output can change). This time t_{StUn} can be found in terms of the time of occurrence of the earliest leading edge of a $C2$ pulse, which is $-T_{2L}$, and t_{D2EArr}, the earliest time at which a $D2$ input can change.

$$t_{StUn} = \max \ [-T_{2L} + D_{2CQm}, \ t_{D2EArr} + D_{2DQm}].$$

We have already found an expression for t_{D2EArr} in (36), which we can insert in the above expression. Let us do so, also replacing the t_{C1EL} term by its value as indicated by $V - W_1 - T_{1L}$. This gives us

$$t_{StUn} = \max \ [-T_{2L} + D_{2CQm},$$
$$\max \ [V - W_1 - T_{1L} + D_{1CQm}, \ t_{D1EArr} + D_{1DQm}] + D_{2DQm}].$$

Expanding the inner max yields

$$t_{StUn} = \max \ [-T_{2L} + D_{2CQm},$$
$$V - W_1 - T_{1L} + D_{1CQm} + D_{2DQm},$$
$$t_{D1EArr} + D_{1DQm} + D_{2DQm}]. \quad (47)$$

As was shown earlier [see (34)] the earliest change of $D1$ permitted for the next cycle is at time

$$H_1 + V + T_{2T}.$$

Therefore, the earliest time we can expect $D1$ to change for the *current* cycle, i.e., the value of t_{D1EArr} is P less than that amount, or

$$t_{D1EArr} = -P + H_1 + V + T_{1T}.$$

Substituting this value into (47) gives us

$$t_{StUn} = \max \ [-T_{2L} + D_{2CQm},$$
$$V - W_1 - T_{1L} + D_{1CQm} + D_{2DQm},$$
$$-P + H_1 + V + T_{1T} + D_{1DQm} + D_{2DQm}]. \quad (48)$$

The Q_2 signals become stable again after the latest $D2$ change prior to the setup time propagates to the latch outputs. Using the value for the latest $D2$ change given in (24) we get for $t_{End\ Un}$, the latest time that the unstable period can end

$$t_{End\ Un} = \max \ [W_2 - T_{2T} - U_2 + D_{2DQM}, \ T_{2L} + D_{2CQM}].$$

$$(49)$$

At all other times, the Q_2 signals are guaranteed to be stable and valid.

F. Consequences of the Constraints

The necessary and sufficient constraints derived above are reproduced below.

$$W_2 \geq V + U_2 - U_1 + D_{1DQM} + T_{2T} - T_{1T} \quad (26)$$

$$W_1 + W_2 \geq V + U_2 + D_{1CQM} + T_{1L} + T_{2T} \quad (27)$$

$$P \geq D_{1DQM} + D_{2DQM} + D_{LM} \quad (31)$$

$$W_1 \geq -P + D_{1CQM} + D_{2DQM} + U_1 + D_{LM} + T_{1L} + T_{1T} \quad (32)$$

$$P \geq -V + D_{2CQM} + U_1 + D_{LM} + T_{1T} + T_{2L}. \quad (33)$$

At least one of the following 2 constraints on D_{Lm} must be satisfied. In most cases, (39) is less stringent (its right side is smaller) and so determines D_{LmB}, the lower bound on D_{Lm}.

$$D_{Lm} > D_{LmB} = V + H_1 + T_{1T} + T_{2L} - D_{2CQm} \quad (39)$$

$$D_{Lm} > D_{LmB} = W_1 + H_1 + T_{1T} + T_{1L} - D_{1CQm} - D_{2DQm}. \quad (40)$$

At least one of the following 2 constraints must be satisfied. In most cases, this will be (46).

$$W_2 < H_1 - H_2 + D_{1DQm} + V + T_{1T} - T_{2T} \quad (45)$$

$$W_1 + W_2 < D_{1CQm} + V + P - H_2 - T_{1L} - T_{2T}. \quad (46)$$

In addition to the above constraints, two more are necessary to ensure that the clock-pulse widths satisfy the minimum requirements of the latches themselves. These are:

$$W_1 \geq C_{W1m} + T_{1L} + T_{1T} \quad (50)$$

and

$$W_2 \geq C_{W2m} + T_{2L} + T_{2T}. \quad (51)$$

Our objective is to choose the clock parameters (widths, period, and overlap) so as to maximize the speed of the system (clearly this is achieved when the period P is minimized), while making it as insensitive as possible to parameter variations. That is, we would like to make the tolerances as large as possible. We often start out with a desired value for the maximum logic delay D_{LM} in a logic path (the *long-path delay*) as this is largely determined by the given technology and the desired maximum number of stages of logic. The crucial factor determining feasibility with known tolerances for delay per logic stage is then the *minimum* delay in a logic path D_{Lm} or *short-path delay*. If the required lower bound on the short-path delay is too large compared to the long-path delay, then the system may be difficult or impossible to realize reliably.

We therefore define the problem as that of finding the minimum value of P such that the lower bound on the short-path delay (D_{LmB}) is acceptable (not too large). It is assumed that we are given all of the latch parameters, the clock-pulse edge tolerances, and the long-path delay D_{LM}.

The key constraint on D_{Lm} is almost always (39). Hence, we set D_{LmB} equal to the right side of that constraint and solve for

V:

$$V = D_{LmB} - H_1 - T_{1T} - T_{2L} + D_{2CQm}. \qquad (52)$$

Now substitute the above right side for V in (33), which is the key constraint on P, to obtain an expression for the minimum value of P as a function of the short-path delay:

$$P = H_1 + U_1 + D_{2CQM} - D_{2CQm} + D_{LM} - D_{LmB} + 2(T_{1T} + T_{2L}). \qquad (53)$$

This expression is valid provided that the value of P obtained does not violate (31). Thus, to find the maximum value of D_{LmB} beyond which no further reductions in P are possible, we must first find the maximum value of V for which (33) is valid (i.e., the value for which (31) is not violated). We do this by substituting the right side of (31) for P in (33) and, treating the resulting expression as an equality, solving for V:

$$V = T_{1T} + T_{2L} + D_{2CQM} + U_1 - D_{1DQM} - D_{2DQM}. \qquad (54)$$

There is clearly nothing to be gained by making the overlap any larger than the value given in (54), since the effect would be to increase the lower bound on the short-path delay without reducing P beyond the absolute minimum given by (31).

Now we can compute the maximum useful value of D_{LmB} by substituting into (39) the above value of V:

$$D_{LmB} = 2(T_{1T} + T_{2L}) + H_1 + U_1$$
$$- D_{1DQM} - D_{2DQM} + D_{2CQM} - D_{2CQm}. \qquad (55)$$

Now we are in position to discuss the question mentioned at the beginning of this section as to the consequences of forcing the $D1$ and/or the $D2$ signals to appear earlier than the minimum bounds dictated by the setup times for the latches. The effect of doing this is the same as if the values of the setup times (the U_i's) were increased. Let us examine the relations derived here to see what effects such increases would have.

First observe that U_1 appears in (26), (32), and (33), as well as in (54) for the maximum useful overlap, in (55) for the value of D_{LmB} corresponding to the absolute minimum bound on P, and in (53) for the minimum value of P as a function of the lower bound on the short-path delay. The direct effects of increasing U_1 are detrimental in all cases except that corresponding to (26). That is, the period would have to be increased and/or D_{LmB} would have to be increased (various tradeoffs are possible), both of which are bad, but the lower bound on the width of the $C2$ pulse would be relaxed, a benefit, but seldom one that is needed.

The U_2 term appears only in (26) and (27), and in (49) for the end of the unstable period for the outputs of $L2$. In the first two cases it tightens (by increasing) the lower bounds on the pulse widths, which is mildly bad, and in the last case it increases the interval during which the $Q2$ signals are stable, which might conceivably be advantageous in some situation.

It therefore does not seem useful to consider requiring the D inputs to the latches to arrive earlier than necessary, unless a very special circumstance should make important one of the factors discussed above. An interesting and perhaps useful

added conclusion from the above discussion is that the setup time for the $L2$ latches is of less importance with respect to speed and tolerances than is the set-up time for the $L1$ latches.

G. Computing Optimum Clock Parameters

Let $D_{\max LmB}$ be the largest lower bound that we can enforce on the short-path delays. To compute optimum clock parameters, proceed as follows.

IF $D_{\max LmB} \geq$ right side of (55)
THEN
    $D_{LmB} \Leftarrow$ right side of (55)
    $P \Leftarrow$ right side of (31)
    $V \Leftarrow$ right side of (54)
ELSE
    $D_{LmB} \Leftarrow D_{\max LmB}$
    $P \Leftarrow$ right side of (53)
    Compute V from relation (39)
$W_2 \Leftarrow$ max[right side of (26), right side of (51)]
Compute W_1 from (27) (use equality)
Increase W_1 if necessary to satisfy (50)
IF $W_1 + W_2 >$ right side of (46) (Not likely.)
THEN
    IF W_2 violates (45) (It probably will.)
    THEN increase P to satisfy (46)
IF $D_{LmB} >$ right side of (40) (Not likely.)
THEN decrease D_{LmB} until (40) is satisfied with equality.

The procedure given above is intended as a general guide to the use of the constraints developed here. In particular cases alternative procedures may be more appropriate.

V. CONCLUSIONS

As is evident from the length of the corresponding section, the task of determining optimum clocking parameters for systems using ETDFF's is relatively simple. The clock-pulse width is not critical, and the constraint on the short-path delays is seldom stringent. The price paid for this is that the minimum clock period is the sum, not only of the maximum delays through the logic and the FF's, but also of the setup time and twice the edge tolerance. No tradeoffs are possible to reduce this quantity.

For 1-phase systems using latches, it may be possible to make the period as small as the sum of the maximum delays through a latch (from the D input) and the logic. In order to do this, the clock-pulse width must be made sufficiently wide (usually past the point where the leading edge of the clock-pulse precedes the appearance of the D signals). Wider clock pulses imply increased values of D_{LmB}, the lower bound on the short-path delays. If this bound is not to become unreasonably high, it is necessary to keep the edge tolerances small. It is also helpful if the difference between the maximum and minimum values of the propagation delays from the C inputs of the latches are small.

The 2-phase system with latches is inherently more complex in that more variables are involved. As in the previous case, tradeoffs are possible between P and D_{LmB}. Here the intermediate variable is V, the amount of overlap between the

338

$C1$ and $C2$ pulses. In very conservative designs there is a negative overlap and D_{LmB} is zero. If positive overlaps are permitted, P can be decreased, but at the cost of making D_{LmB} nonzero. A continuum of tradeoffs exists to the point where P is reduced to the sum of the maximum propagation delays through the $L1$ and $L2$ latches (from the D inputs) and the logic. Again it is possible to absorb the effect of edge tolerances in terms of short-path rather than long-path problems.

An important advantage of 2-phase over 1-phase systems is that, for every 2-phase system, simply by varying the overlap (i.e., the phasing between the $C1$ and $C2$ clock pulses), D_{LmB} can be varied continuously from zero to the highest useful value (with the minimum P of course changing in the opposite direction). On the other hand, for 1-phase systems, the range of variation of D_{LmB} possible by varying the clock-pulse width is often much smaller, particularly at the low end. As illustrated in the graph of Fig. 11, there may be a significant range of values of D_{LmB} that is attainable only by adding delay pads at the outputs of all latches.

In 1-phase systems, if the designer is overly aggressive and it becomes apparent during the test phase that the short-path bound cannot be met, then it is usually necessary to add delay pads at the latch outputs as well as to increase the clock period. This usually means very extensive changes, affecting many chips. Should the same situation arise in connection with a 2-phase system, in addition to increasing the clock period, all that need be done is to reduce the amount of overlap, adjustments that affect only the clocking system, usually a much simpler process affecting far fewer chips. Hence designers of 2-phase systems can afford to be bolder in choosing the clock period since the penalty for over-reaching is less severe.

With only one latch in each feedback path, the lower limit on the clock period is lower for 1-phase systems, although this factor is somewhat attenuated by the fact that some latches in 1-phase systems will have both inputs from sources that fan out to other latches, and outputs that fan-out to many gate inputs. Both of these are factors that reduce speed. But in 2-phase systems each $L1$ latch feeds only one other device (an $L2$ latch), and each $L2$ latch receives its D input from a source (an $L1$ latch) feeding no other device. Hence, all other things being equal, we would expect the delays through the two latches in the feedback paths of 2-phase systems to have less than twice the delays of the one latch in the feedback path of a 1-phase system.

An advantage of 2-phase systems over both of the other types considered here is that they are somewhat more compatible with the LSSD concept for system testing [1], [2].

It appears that all three types of systems have their places. Where there is a willingness to exert great efforts to suppress skew (e.g., by hand-tuning the delays in clock distribution paths), and to control other related factors very precisely, the 1-phase system may be the best choice, as in the case of the CRAY 1 machine. In other cases of high-performance machines, 2-phase clocking may be more suitable. Use of ETDFF's seems to have advantages for less aggressive designs.

The results presented here in such precise looking relations obviously depend heavily on the precision with which the parameters of those relations can be determined. Realistic figures must be obtained that take into account such matters as power supply and temperature variations, as well as data sensitive loading considerations.

The relations developed here may be useful in determining what latches to use in certain situations and to determine how to modify latch designs so as to improve system performance. For example, an examination of the constraints developed in Section III-C for 1-phase systems with latches suggests that the *minimum* value of D_{DQ} is of no importance, whereas the minimum value of D_{CQ} *is* important in that the larger it is, the less stringent is the constraint on short-path delays.

In the 2-phase case, minimizing $(D_{2CQM} - D_{2CQm})$ is clearly helpful. It relaxes the requirement on D_{LmB} imposed by (55), which, if it can be satisfied, allows P to be set to the minimum value given by (31). If (55) cannot be satisfied, then P is given by (53), and will therefore vary directly with $(D_{2CQM} - D_{2CQm})$.

On the other hand, neither D_{1CQM}, D_{1CQm}, D_{1DQm}, nor D_{2DQm} seem to be of primary importance. As was pointed out on page 893, the setup and hold-time requirements for the $L1$-latches are much more important than are the corresponding parameters for the $L2$ latches. It is clear that there are different optimum requirements for $L1$ and $L2$ latches. Furthermore, different choices may be appropriate depending upon whether or not an effort is being made to attain the minimum period corresponding to the maximum loop delay.

It is clear from the results developed here that minimizing clock edge tolerances is of considerable importance in high-performance digital systems. In 2-phase systems, a special effort is warranted to minimize T_{1T} and T_{2L}, which appear in key several constraints. Unfortunately, technology trends are such as to emphasize factors that cause skew. For example, as the dimensions of logic elements on chips shrink, the ratio of wiring delays to gate delays grows. A high priority must therefore be given in wiring algorithms to the clock distribution system. Off-chip wiring forming part of the clock distribution network must be carefully controlled. In some cases, the insertion of adjustable delays in these paths may be warranted. It is quite likely that the continuation of the trends that exacerbate the skew problem will soon make it worthwhile to consider systems that do not use clock pulses or that use clock pulses only locally. Discussions of such asynchronous, self-timed, or speed-independent systems are in [4] and [8].

Logic designers and those developing computer aids for logic design customarily pay a great deal of attention to minimizing long-path delays. It is also important to consider techniques for increasing short-path delays. In line with this there is a need for circuit designers to develop techniques for introducing *precisely controlled* delay elements where needed. At present, in many technologies, logic designers are forced to cascade inverters to produce delays. This is wasteful in terms of both chip area and power. In general, the idea that greater speed may result from better delay elements should be conveyed to those developing digital technology.

Further developments along the lines developed here would

include the use of statistical rather than worst-case analyses, which would allow us to choose clocking parameters such that the likelihood of a timing failure is very small, but not zero. This usually implies shorter clocking periods. In using this approach it is important to be able to take into account correlations among delay values, skew etc., in various parts of the system [5], [7].

It is also possible to speed up systems by exploiting detailed knowledge of the logic paths. There may be, for example, constraints on the sequencing of signals through certain combinations of paths that allow us to consider consecutive pairs, triples, etc., of cycles together and thereby realize that shorter periods are feasible than would be the case if each period were considered separately. Research along this line is being conducted by K. Maling [3].

An earlier presentation of the work discussed here, in a different form with different notation was issued by the authors several years ago [10], [11]. The idea that clocked systems could be speeded up by permitting the D-inputs to latches to lag behing the leading edges of the clock-pulses and by allowing the C1- and C2-clock pulses to overlap is not new. These ideas are included in the very interesting book on digital systems design by Langdon [2], and have been pointed out by D. Chang of IBM's Poughkeepsie Laboratories a number of years ago in at least one internal memorandum. Other pertinent work, in connection with pipelining, is by Kogge [12] and Cotten [13].

Acknowledgments

Substantial contributions to the form and substance of this work were made by R. Risch. The comments and criticisms of K. Maling were very helpful, and we also profited from discussions with V. Kumar, W. Donath, G. Buchanan, D. Kurshner, J. Shelly, G. Maley, P. Meehan, and D. Lee.

References

[1] E. B. Eichelberger and T. W. Williams, "A logic design structure for LSI testability," in *Proc. 14th Design Automat. Conf.*, June, 1977, pp. 462–468.

[2] G. Langdon, Jr., *Computer Design*. San Jose, CA: Computeach, 1982.

[3] K. Maling, "Automatic clock optimization," IBM internal memorandum, 1982.

[4] C. Mead and L. Conway, *Introduction to VLSI Systems*. Reading, MA: Addison-Wesley, 1980.

[5] J. Shelly and D. Tryon, "Statistical techniques of timing verification," in *Proc. 20th Design Automation Conf.*, 1983, pp. 396–402.

[6] *The TTL Data Book for Design Engineers*, Texas Instruments, Inc., Dallas, TX, 1981.

[7] D. Tryon, F. M. Armstrong, and M. R. Reiter, "Statistical failure analysis of system timing," *IBM J. Res. Develop.*, vol. 28, pp. 340–355, July, 1984.

[8] S. H. Unger, *Asynchronous Sequential Switching Circuits*. New York: Wiley, 1969; 3rd printing issued by R. E. Krieger, Malabar, FL, 1983.

[9] ——, "Double-edge-triggered flip-flops," *IEEE Trans. Comput.*, vol. C-30, pp. 447–451, June, 1981.

[10] S. H. Unger and C.-J. Tan, "Clocking schemes for high-speed digital systems," IBM T. J. Watson Res. Center, Yorktown Heights, NY, Tech. Rep. RC9754, Dec. 1982.

[11] ——, "Clocking schemes for high-speed digital systems," in *Proc. IEEE Int. Conf. Comput. Des.: VLSI Comput.*, Oct.–Nov., 1983, pp. 366–369.

[12] P. Kogge, *The Architecture of Pipelined Computers*. New York: McGraw-Hill, 1981.

[13] L. W. Cotten, "Circuit implementation of high-speed pipelined systems," in *Proc. AFIPS*, vol. 27, pt. 1, pp. 489–504, 1965.

Clock Tree Synthesis Based on RC Delay Balancing

FUMIHIRO MINAMI AND MIDORI TAKANO

ULSI RESEARCH CENTER, TOSHIBA CORPORATION

Abstract

This paper presents a novel clock tree synthesis method based on top down binary tree construction, optimal buffer insertion, and bottom up wiring with precise RC delay balancing. The proposed method has achieved near-zero clock skews with minimal clock delays while achieving similar total wire lengths in comparison with previous heuristics.

1 Introduction

The circuit speed in a synchronous VLSI design is limited by the clock skew which is defined as the maximum difference in the delays from the clock source to the synchronous components. The system speed is also limited by the inter-chip clock skew which is the clock delay difference among chips. In general, most systems require a clock skew of less than 10% of the system clock period. Thus, clock skew minimization and clock delay reduction are needed for high performance chips.

Several approaches for clock routing using a tree structure have been reported as follows: the H-tree method [1], the Method of Means and Medians (MMM) [2], and the Path Length Balancing method (PLB) [3]. However, the H-tree structure is not applicable for asymmetric distributions of synchronous components, and MMM and PLB are not aimed at path delay balancing which is the primary objective. Therefore, these techniques are insufficient in minimizing the clock skew.

This paper presents a novel approach to clock routing that minimizes the clock skew by constructing a binary tree with RC delay balancing. The proposed Path Delay Balancing method (PDB) always yields near-zero clock skews. The key point of PDB is that the branching point of a binary tree is so determined as to equilibrate the RC delay for each subtree calculated by Elmore's formula [4]. The proposed method uses a hierarchical tree structure divided by buffer cells to prevent the delay increase caused by the quadratic term of the wire length for a circuit with too many synchronous components. The synchronous components are so clustered in the lowest tree level as to balance the load capacitance within each cluster where the wire resistance is negligible, and thus the clock skew among clusters is minimized. As a result, the clock skew and clock delay are perfectly minimized with the above mentioned techniques.

This paper is organized as follows: in Section 2, the keys to minimize the skew and delay are presented. Section 3 presents the definition of the problem. The algorithm for clock tree synthesis is presented in Section 4, and practical considerations are presented in Section 5. The experimental results are presented in Section 6, and the paper is concluded in Section 7.

2 Skew and Delay Minimization

In this section, a delay calculation model is presented and then the keys to minimize the skew and delay are presented.

2.1 Delay calculation model

For a lumped RC tree network, Elmore's formula [4][6] is commonly used for delay calculation and the authors have also adopted it in this paper. According to this formula, the delay

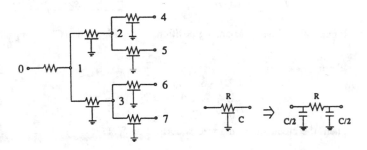

Figure 1: Binary clock tree   Figure 2: π-model subcircuit

from a root to a leaf in a RC tree is defined as the sum of the products of both the resistances R_i and its downstream capacitances C_i along the path, which is defined as follows:

$$Delay = w \sum R_i C_i$$

where w is a constant coefficient.

In a binary tree for clock routing, Elmore's delay is calculated in the following bottom up manner.

Let the highest branch node of the tree be labeled 1 and two child nodes of node k be labeled $2k$ and $2k+1$, and the driver output pin be labeled 0, as shown in Figure 1. Let $C(k)$ be the downstream capacitance from node k and $L(k)$ be the wire length between node k and its parent node. Let r and c be the unit length resistance and capacitance, and let R_{on} and dt be the on-resistance and the intrinsic delay of a driver cell. Then, the delay from a driver input pin to leaf node e is expressed as follows.

$$Delay(e) = dt + \sum d(k) \qquad (1)$$
$$d(k) = w\, r\, L(k) \cdot (\, c\, L(k) / 2 + C(k)\,) \qquad \text{if } k > 1$$
$$d(1) = w\, R_{on}\, C(1)$$
$$C(k) = C(2k) + C(2k+1) + c\,(\, L(2k) + L(2k+1)\,)$$
$$C(e) = \text{input pin capacitance of leaf } e$$

In this expression, each wire segment which is expressed by a distributed RC line is assumed to be approximately equivalent to a π-model subcircuit as shown in Figure 2.

2.2 Skew minimization

The skew between two different paths is equal to the difference of the sum of RC products from the last common branch to each leaf because the RC products along a common path is the same. According to this, the local skew $S(k)$ at each node k can be defined as the maximum delay difference for all paths from node k to its descendant leaves. Let $T_L(k)$ and $T_S(k)$ be the maximum and minimum delays of all paths from node k to its descendant leaves, then $S(k)$ is expressed as follows:

$$
\begin{aligned}
S(k) &= T_L(k) - T_S(k) \\
&= \max (\, S(2k),\, S(2k+1), \\
&\qquad T_L(2k) + d(2k) - T_S(2k+1) - d(2k+1), \\
&\qquad T_L(2k+1) + d(2k+1) - T_S(2k) - d(2k)\,) \\
&= \max (\, S(2k),\, S(2k+1),\, \mu(k) + |\lambda(k)|\,) \qquad (2)
\end{aligned}
$$

Reprinted from *Proceedings of IEEE Custom Integrated Circuits Conference*, pp. 28.3.1-28.3.4, May 1992.

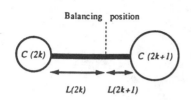

Figure 3: RC delay balancing position

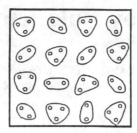

Figure 4: Clustering

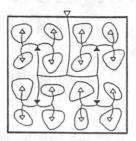

Figure 5: Tree construction

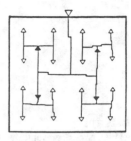

Figure 6: Balanced routing

where

$$\mu(k) = (S(2k) + S(2k+1)) / 2$$
$$\lambda(k) = d(2k) + \tau(2k) - d(2k+1) - \tau(2k+1)$$
$$\tau(j) = (TL(j) + Ts(j)) / 2 .$$

Then, the clock skew in a tree is defined as $S(1)$.

$\lambda(k)$ means, the skew between the delay median of the left side descendants and that of the right side descendants. Because $\mu(k)$ is less than either $S(2k)$ or $S(2k+1)$, if $\lambda(k)$ is forced to be equal to zero, then $S(k)$ is rewritten as follows:

$$S(k) = \max (S(2k), S(2k+1)).$$

This equation suggests that the local clock skews do not exceed the descendant clock skews. Consequently, if a position of each node k is so determined that $\lambda(k)$ is zero in a bottom up manner, the clock skew will be equal to the maximum skew at the leaves and will normally be zero. This is the key to minimize the clock skew in this paper.

The condition that $\lambda(k)$ is zero is as follows:

$$d(2k) + \tau(2k) = d(2k+1) + \tau(2k+1)$$

or

$$r L(2k) \cdot (c L(2k) / 2 + C(2k)) + \varepsilon(k)$$
$$= r L(2k+1) \cdot (c L(2k+1) / 2 + C(2k+1)) \qquad (3)$$

where

$$\varepsilon(k) = (\tau(2k) - \tau(2k+1)) / w .$$

As shown in this equation, delay balancing is analogous to moment balancing of a mobile where the capacitance corresponds to the weight and the resistance corresponds to the arm length. Figure 3 shows the image.

2.3 Delay minimization

For a circuit with too many synchronous components, the clock delay increases in proportion to the square of the wire length. In such a case, a hierarchical clock tree structure by multistage buffering is commonly used to reduce the clock delay.

For multistage buffering, the delay is influenced by the positions of the inserted buffer cells. Buffer insertion close to the primary driver cell causes a large delay in the buffer stage, and buffer insertion close to the leaf nodes causes a large delay in the primary driver stage, so there exists an optimal position for buffer cells to be inserted.

In the proposed method, buffer cells are so inserted as to minimize the total clock delay by a greedy method.

In addition, the clock pins in the lowest trees are routed by the Steiner-tree method which minimizes the wire length and thus reduces the delay. This is because the wire resistances in the lowest tree are very small in comparison with the on-resistances of the buffers and RC delay balanced routing is unnecessary.

3 Problem definition

Given a circuit placement, the clock pin positions, the number of buffering stages and the performance data for buffer cell types, the clock layout optimization problem can be defined as follows: construct a hierarchical clock tree by multi-stage buffering which minimizes the clock skew, delay, and wire length subject to the clock pin positions. In this paper, it is assumed that two layers are available for clock layout.

4 Algorithm

The proposed method consists of three steps, which are synchronous component clustering, hierarchical tree topology construction, and RC delay balanced routing. In tree topology construction, the optimal combination of the buffer types and levels to be inserted into the tree is obtained by a greedy method. The main process flow is shown as follows.

```
procedure clock_tree_synthesis
Begin
    B : set of sequentially inserted buffer types and
        their insertion depths in the tree
    for b ∈ B , do
        synchronous component clustering
        hierarchical tree topology construction
        /* RC delay balanced routing */
        for each level of hierarchical tree, do
            branch position determination for each subtree
            buffer positions adjustment for delay equalization
            tree level decrement
        end
        saving the routing result if the delay is the smallest
    end
End
```

4.1 Synchronous component clustering

First, synchronous components are clustered by recursive bisectioning until the wire resistance within each cluster becomes negligibly small, as shown in Figure 4. At each division, the dividing line is so determined as to balance the load capacitance of two regions which is defined as the sum of the estimated wire capacitance [5] and the input gate capacitance, or

$$LoadCap = c \, h \, f(fanout) + \Sigma C_g \qquad (4)$$

where

c	: capacitance per unit wire length
h	: half perimeter of the minimal bounding box
$f(fanout)$	: statistical coefficient function of wire length
C_g	: input pin capacitance for each gate .

After clustering, the clock pins in each cluster are routed by the Steiner-tree method to minimize the wire length.

342

By this clustering, the inter-cluster skew is minimized because of the uniform load capacitance and negligible wire resistance for each cluster.

4.2 Hierarchical tree topology construction

Secondly, the binary tree topology whose root is a primary clock driver and whose leaves are clusters is constructed by top down bi-sectioning [2], and then buffer cells are inserted to construct the hierarchical tree. In each buffering stage, buffer cells of the same type are topologically inserted at the same level of the binary tree, as shown in Figure 5. The lowest buffer cells are placed at the center of each cluster region, and middle buffer cells are placed at or near the highest branch points of the subtrees, which are determined by the process explained in Section 4.3.

4.3 RC delay balanced routing

Lastly, each subtree of the hierarchical tree is recursively routed in a bottom up manner, as shown in Figure 6. In the subtree which is driven by a middle buffer cell or a primary driver cell, the branch position determination and the buffer position adjustment are so performed as to minimize skew as follows.

1) Branch position determination

In the subtree, the branch position for each tree node is so determined as to minimize its local skew $S(k)$, or to satisfy Equation (3).

Let l be $L(2k) + L(2k+1)$ and x be the ratio of $L(2k)$ to l, then x is solved as follows:

$$x = \frac{C(2k+1) + c\,l\,/\,2 - \varepsilon(k)\,/\,r\,l}{C(2k) + C(2k+1) + c\,l} \quad (5)$$

To minimize the wire length and clock delay, l is assumed to be the Manhattan length between node $2k$ and $2k+1$. If x is not in the range from 0 to 1, x is set at 0 or 1 so that $S(k)$ is minimized.

After each branch position determination, node k is connected to each child node with a minimum length, and $C(k)$, $T_L(k)$, and $T_S(k)$ are calculated. These are recursively processed by the following depth first procedure, as shown in Figure 7.

```
procedure delay_balance( P )
Begin
   if P is a leaf then
      F.pos = input pin position of leaf
      F.c   = input pin capacitance
      F.TL  = max delay of lower tree
      F.TS  = min delay of lower tree
      return( F )
   else
      PL : left side subtree of current node
      PR : right side subtree of current node
      FL = delay_balance( PL )
      FR = delay_balance( PR )
      decide branch position so as to minimize the λ(k)
      route from the current node to two child nodes
      l : wire length between one child node and another
      F.pos = branch position for current node
      F.c   = FL.c + FR.c + c l
      F.TL = max ( FL.TL + d(2k), FR.TL + d(2k+1) )
      F.TS = min ( FL.TS + d(2k), FR.TS + d(2k+1) )
      return( F )
   endif
End
```

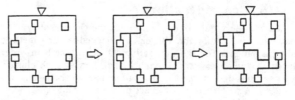

(a) Routing image

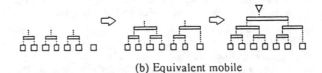

(b) Equivalent mobile

Figure 7: Bottom up routing in a single tree

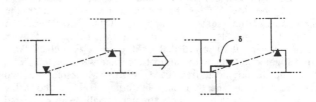

Figure 8: Buffer position adjustment

2) Buffer positions adjustment

After routing at each tree level, the buffer cell positions are so adjusted as to balance the downstream delays by the following technique.

Initially, each buffer cell is placed at the highest branch point of its subtree. Next, the buffer cell is moved to the point with a Manhattan distance δ from the initial position, where the downstream delay is equal to the largest one at the same tree level. A new position is searched along the line which connects the initial position with its brother buffer position of the upper subtree, as shown in Figure 8.

Let Δ be the difference between $\tau(l)$ of the current subtree and the largest one of the subtree at the same level, then δ is required to satisfy the following equation.

$$\Delta = w\,R_{on}\,c\,\delta + w\,r\,\delta\,(\,c\,\delta\,/\,2 + C(l)\,) \quad (6)$$

Such δ is solved as follows:

$$\delta = \frac{\sqrt{b^2 + 2\,rc\Delta/w} - b}{rc} \quad (7)$$

where

$$b = R_{on}\,c + r\,C(l) .$$

In most cases, small δ is sufficient for equalizing the delays at the same level because $C(l)$ is relatively large. And such selection of the moving direction leads to a wire length reduction at an upper level which cancels out the wire length increase at the current level. Consequently, the adjustment for buffer cell positions is so effective that subtree delays of the same level can be balanced without elongating the total wire length.

As mentioned above, RC delay balanced routing for a hierarchical tree results in a near-zero clock skew and minimal clock delay.

5 Practical considerations

In practice, one would like to minimize the skew exactly. In this case, we can modify the RC delay balanced routing method by adding a process for the elongating the wire length in order to balance the delay exactly at each branch and at each cluster. Branch position determination just after routing between two child nodes is also effective for exact delay balancing.

Another practical concern is the many possible combinations for buffer insertion to be checked. To reduce the combinations and the computational time, the authors propose to use the upper driving limit of the tree depth for each driver type.

6 Experimental results

The proposed PDB method was implemented in C on a SUN4 system, and was tested on single and hierarchical tree examples. Table 1 shows the statistics for the test examples and the coefficient *w* was set at 0.7 for delay calculation.

In the single tree case, the routing performance for PDB was compared with that for MMM, and PLB for three industrial data and the MCNC benchmark data *Primary2*, under the condition of the same topology. Table 2 shows the computational results of the clock skew, delay, and wire length. The results showed that PDB yields a zero clock skew with a slight small delay and almost the same wire length.

In the hierarchical tree case, PDB was tested on two examples with one and two stage buffering. Table 3 shows the excellent results which had an extremely small delay compared with the single tree and a sufficiently small skew. The skew was mostly caused by the inter-cluster skew, or the maximum capacitance difference between the clusters which was always less than 10% of the average capacitance.

Figure 9 and 10 show the global and detail routing results of *Data3* by the PDB method.

7 Conclusion

This paper has presented a novel clock tree synthesis method based on top down binary tree construction, optimal buffer insertion, and bottom up wiring with precise RC delay balancing. The proposed method has achieved near-zero clock skews with minimal clock delays for a wide range of clock pin counts and distributions on industrial chips.

Future work will address wire width optimization and give consideration to internal path delay in a large macro with its blockage area during the construction of the clock tree.

References

[1] H. B. Bakoglu, J. T. Walker and J. D. Meindl, "A symmetric clock-distribution tree and optimized high-speed interconnections for reduced clock skew in ULSI and WSI circuits", *Proc. IEEE Int. Conference on Computer Design*, pp. 118-122, 1986.

[2] M. A. B. Jackson, A. Srinivasan, and E. S. Kuh, "Clock Routing for High-Performance ICs", *Proc. 27th Design Automation Conference*, pp. 573-579, 1990.

[3] A. Kahng, J. Cong, and G. Robins, "High-Performance Clock Routing Based on Recursive Geometric Matching", *Proc. 28th Design Automation Conference*, pp. 322-327, 1991.

[4] J. Rubinstein, P. Penfield and M. A. Horowitz, "Signal Delay in RC Tree Networks", *IEEE Trans. on CAD*, CAD-2(3), pp. 202-211, 1983.

[5] S. Boon, S. Butler, R. Byrne, B. Setering, M. Casalanda, and Al Scherf, "High Performance Clock Distribution for CMOS ASICs", *IEEE Custom Integrated Circuits Conference*, pp. 15.4.1-15.4.5, 1989.

[6] H. B. Bakoglu: Circuits, Interconnections, and Packaging for VLSI, Chapter 5, Addison-Wesley Publishing Co., Reading, Mass., 1990.

	dt (nsec)	*Ron* (psec/LU)	*Cg* (LU)
Driver	0.79	3.6	
Buffer1	0.70	31.4	1
Buffer2	0.68	15.7	2
Flipflops			1

(a) Performance data for used cells

	r (psec/LU mm)	*c* (LU/mm)
Horizontal wire	1.22	1.09
Vertical wire	2.14	1.50

(b) Performance data for wires

	Pin counts	Chip width (mm)	Chip height (mm)
Data1	410	6.6	6.6
Primary2	603	12.0	12.0
Data2	1026	14.7	14.7
Data3	1661	14.7	14.7

(c) Test examples

Table 1: Statistics of experiments

		MMM	PLB	PDB
Data1	Skew	0.37	0.33	0.00
	Delay	5.31	5.10	5.07
	Length	178	175	176
Primary2	Skew	0.68	0.39	0.00
	Delay	13.73	13.32	12.95
	Length	432	430	432
Data2	Skew	0.47	0.24	0.00
	Delay	33.33	32.50	32.41
	Length	726	721	723
Data3	Skew	0.71	0.49	0.00
	Delay	48.16	46.96	46.81
	Length	927	918	920

Table 2: Skew (nsec), maximum delay (nsec), and total wire length (mm) comparison in a single tree

		Without buffering	With buffering
Primary2	Skew	0.00	0.07
	Delay	12.95	3.03
	Length	432	232
Data3	Skew	0.00	0.05
	Delay	46.81	3.94
	Length	920	544

Table 3: Skew (nsec), maximum delay (nsec), and total wire length (mm) comparison between without buffering and with buffering

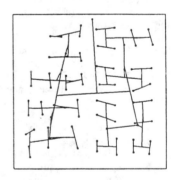

Figure 9: Global routing example

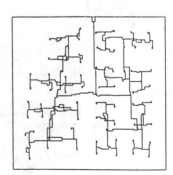

Figure 10: Detail routing example

Half-Swing Clocking Scheme for 75% Power Saving in Clocking Circuitry

Hirotsugu Kojima, *Member, IEEE*, Satoshi Tanaka, *Member, IEEE*, and Katsuro Sasaki, *Member, IEEE*

Abstract—We propose a half-swing clocking scheme that allows us to reduce power consumption of clocking circuitry by as much as 75%, because all the clock signal swings are reduced to half of the LSI supply voltage. The new clocking scheme causes quite small speed degradation, because the random logic circuits in the critical path are still supplied by the full supply voltage. We also propose a clock driver which supplies half-swing clock and generates half V_{DD} by itself. We confirmed that the half-swing clocking scheme provided 67% power saving in a test chip fabricated with 0.5 μm CMOS device, ideally 75%, in the clocking circuitry, and that the degradation in speed was only 0.5 ns by circuit simulation. The key to the proposed clocking scheme is the concept that the voltage swing is reduced only for clocking circuitry, but is retained for all other circuits in the chip. This results in significant power reduction with minimal speed degradation.

I. INTRODUCTION

REDUCING power consumption without sacrificing processing speed is a critical factor in LSI design, especially for hand-held devices. In CMOS circuits, dynamic power consumption is proportional to the transition frequency, capacitance, and square of supply voltage. Consequentially, reducing supply voltage provides significant power savings at the expense of speed. This technique employs high-performance architectures to achieve the specified speed, and is quite effective for ASIC's [1]. In general purpose processors, however, it is more difficult to employ high-performance architectures, because the architecture is already a part of the specifications. It is therefore very important to reduce power consumption without reducing supply voltage or sacrificing performance.

Several power reduction techniques in random logic and clocking have been reported [2]–[7]. Such techniques can be applied for all random logic circuits, but they do not save much power. The clocking circuitry generally consumes a large portion of the total power in digital LSI's. Fig. 1 shows that the clocking circuitry in an adaptive equalizer consumes 33% of the total power [8]. In a microprocessor, 18% of the total chip power is consumed by clocking [9]. This is because the clock frequency is typically several times higher than other signals, such as data and control.

Our proposal is a new clocking scheme in which all the clock signal swings are reduced to half of the LSI supply voltage. This technology allows us to reduce power consump-

Manuscript received October 5, 1994; revised December 5, 1994.

H. Kojima and K. Sasaki are with Hitachi America, Ltd., R&D, San Jose, CA 95134 USA.

S. Tanaka is with the Central Research Laboratory, Hitachi Ltd., Tokyo 185 Japan.

IEEE Log Number 9409857.

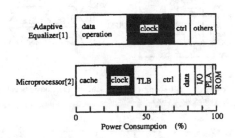

Fig. 1. Power consumption analysis.

tion of clocking circuitry by as much as 75%. The speed degradation caused by the proposed clocking scheme is quite small because the random logic circuits in the critical path are still supplied by the full supply voltage. The key to the proposed clocking scheme is the concept that the voltage swing is reduced only for clocking circuitry, but is retained for all other circuits in the chip. This results in significant power reduction with minimal speed degradation.

II. HALF-SWING CLOCKING SCHEME

Fig. 2 shows the proposed half-swing clocking scheme compared with a conventional scheme. In Fig. 2(a), a conventional latch is gated by two full-swing clocks. To decrease the clocking power, the voltage swing of the clock is reduced to half V_{DD} ('V_{DD}' represents the LSI supply voltage). The proposed scheme, as shown in Fig. 2(b), uses two separate clock signals for NMOS and PMOS transistors, respectively. The clock for NMOS's swings from zero to half V_{DD}, and the clock for PMOS's swings from V_{DD} to half V_{DD}. The power consumed by clocking circuitry is decreased to 25% of conventional clocking circuitry.

We propose to use two stacked inverters to generate half-swing clock signals. Fig. 3 shows the proposed clock driver circuit which supplies the two half-swing clock signals described above and generates a half V_{DD} voltage by itself. Here, C_1 and C_2 represent PMOS loads on drivers, and C_3 and C_4 represent NMOS loads on the other drivers. C_A and C_B are additional capacitors which can be fabricated on-chip or connected externally. The intermediate voltage at the node $H\text{-}V_{dd}$ is given by the following equations:

$$V_{H\text{-}V_{dd}} = \frac{C_1 + C_A}{C_1 + C_4 + C_A + C_B} V_{DD} \quad \text{when CLK is 'low'}$$

$$V_{H\text{-}V_{dd}} = \frac{C_2 + C_A}{C_2 + C_3 + C_A + C_B} V_{DD} \quad \text{when CLK is 'high.'}$$

Reprinted from *IEEE Journal of Solid-State Circuits*, Vol. 30, No. 4, pp. 432–435, April 1995.

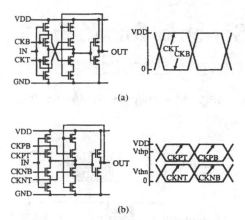

(a)

(b)

Fig. 2. Concept of half-swing clocking scheme. (a) Conventional clocking scheme. (b) Half-swing clocking scheme.

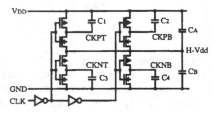

Fig. 3. Proposed clock driver.

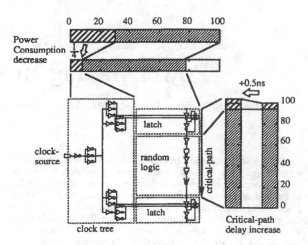

Fig. 4. Summary of tradeoff with half-swing clocking scheme.

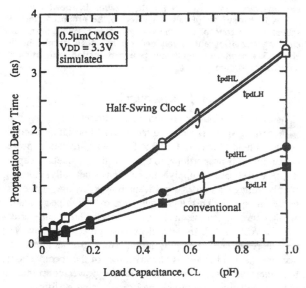

Fig. 5. Simulated load capacitance dependency of the propagation delay in clock drivers.

The H-V_{dd} node is stabilized at $V_{DD}/2$, when C_A and C_B are equal, yet large enough for C_1 through C_4 to be considered negligibly small. If C_1 through C_4 are made equal, C_A and C_B are not needed. In actual LSI's, however, C_A and C_B should be large enough to compensate for production fluctuation of C_1 through C_4.

Fig. 4 shows a summary of the proposed idea by using a simple model of digital circuits. Most digital LSI's are separated into three parts: latches, random logic circuits between the latches, and a clock driver tree which provides clock signals to the latches. The chip performance is determined by the critical path delay from one latch to another. The half-swing clocking scheme causes a two delay increase: in the clock drivers, and in a latch driven by half-swing clock. The delay increase from clock source to the latch does not degrade the performance, because no clock driver is located on the critical path. The performance degradation is caused by the delay increase in the latch that is located on the critical path and is driven by the half-swing clock. Since the clocking scheme never degrades the speed of other random logic circuits on the critical path, the speed degradation is minimal.

III. CHARACTERISTICS

The two important characteristics of our proposed clocking scheme are evidenced through circuit simulation: the delay increase in a half-swing clock driver and the delay increase in a latch driven by a half-swing clock.

Fig. 5 shows the load capacitance dependency of the propagation delay in proposed and conventional clock drivers.

The results were obtained through circuit simulation using 0.5 μm CMOS FET models. The propagation delay of the half-swing clocking driver is approximately twice that of the conventional driver. However, the delay itself does not affect the performance determined by a critical path delay. Since a clock skew is proportional to the delay, deskewing techniques are important in using the half-swing clocking scheme.

The propagation delay of a latch is the interval from when the clock arrives at the latch to the time when the data is output from the latch. Fig. 6 shows the relationship of the simulated propagation delay of a latch driven by conventional and half-swing clocks to the load capacitance. The delay increase caused by the proposed half-swing clocking is at most 0.5 ns. Note that the increase is regardless of the load capacitance. The propagation delay of the latch is the sum of two delays: the delay caused when the clock gated transistor drives the

346

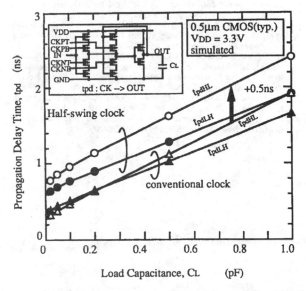

Fig. 6. Simulated load capacitance dependency of the propagation delay of latch.

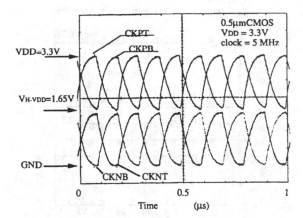

Fig. 7. Observed waveform of half-swing clock signals.

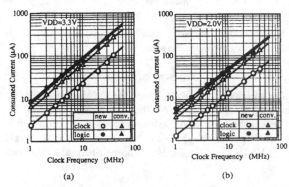

Fig. 8. Measured power consumption. (a) $V_{DD} = 3.3V$. (b) $V_{DD} = 2.0V$.

last stage inverter and the delay caused when the last stage inverter drives the load capacitance of the latch. The former delay increases when the clock gated transistor is driven by a half-swing clock, but the amount of delay is independent of the load capacitance of the latch, i.e., the clocking scheme does not affect the latter delay.

Fig. 4 shows that the critical path delay of an LSI is the sum of the propagation delay of the latch and the random logic. As demonstrated in the simulation, the propagation delay of the latch increases by 0.5 ns. The propagation delay in the random logic is the same as the conventional one, because the random logic circuits are supplied by full V_{DD}. Thus, the critical path delay increases by 0.5 ns by employing the half-swing clocking scheme. The speed degradation of 0.5 ns is acceptable for most LSI's.

IV. EXPERIMENTAL RESULTS

We fabricated a test chip that consisted of two sixteen-stage shift registers: one employing the half-swing clocking scheme and the other employing the conventional clocking scheme. Fig. 7 shows observed waveforms of the four half-swing clocks, which are true and bar clock signals driving PMOS and NMOS. The clock signals driving PMOS swing between 1.65 V and 3.3 V, and the clock signals driving NMOS swing between ground and 1.65 V. The waveforms were observed only at low frequencies, because the pin drivers were designed to be small enough not to affect the internal capacitor balance while the probes had large (10 pF) capacitive loads.

Fig. 8 shows measured power consumption of the test circuits. The random logic circuits in the proposed and conventional shift registers consume the same amount of power. The result confirmed that the proposed clock driver saves 67% of the power of the conventional one throughout a wide clock frequency range of 1 MHz to 40 MHz. The ratio between the power consumed by the proposed and conventional clocking

circuitry was 33% instead of the ideal ratio of 25%. This is because the NMOS transistors that load of the clock driver in the proposed latch are twice that of the conventional latch in order to equalize the PMOS and NMOS loads on each line of the half-swing clocks. The ideal ratio of power saving can be achieved if the latch was designed to have the same capacitive load as the conventional latch.

The intermediate voltage, $V_{H-V_{dd}}$, was measured at various supply voltages and frequencies as shown in Fig. 9. The results confirmed that the intermediate voltage, $V_{H-V_{dd}}$, was successfully stabilized at a V_{DD} of 5 to 1.5 V and at clock frequencies of 1 to 40 MHz.

V. CONCLUSION

We propose the half-swing clocking scheme and clock driver which supplies half-swing clock and generates half V_{DD} by itself. We confirmed that the proposed clock driver successfully generates half-swing clock signals with stable half V_{DD} generation by itself throughout the measurements of the test chip fabricated using 0.5 μm CMOS process. Half-swing clocking scheme allows us to save 67% of power in the test chip, ideally 75%, in the clocking circuitry with only 0.5 ns degradation in speed for a 0.5 μm CMOS device.

The proposed half-swing clocking scheme can achieve a significant power savings for most CMOS digital LSI's

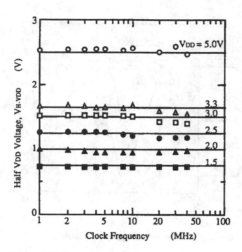

Fig. 9. Measured H-V_{dd} voltage stability.

including microprocessors, DSP's, and other custom chips, with little degradation in speed.

ACKNOWLEDGMENT

The authors would like to thank Mr. A. Masumura, Mr. Y. Ishigami, Mr. H. Misawa, Mr. H. Akama, Mr. S. Katoh, Mr. M. Otsuka, and Mr. T. Akazawa for their great cooperation on layout design, and Mr. K. Nitta, Mr. D. Gorny, Dr. M. Hiraki, Dr. Y. Hatano, and Dr. M. Hotta for their helpful discussions.

REFERENCES

[1] A. P. Chandrakasan, S. Sheng, and R. W. Brodersen, "Low-power CMOS digital design," *IEEE J. Solid-State Circuits*, vol. 27, no. 4, pp. 473–484, Apr. 1992.

[2] K. Yano *et al.*, "A 3.8 ns CMOS 16 × 16 multiplier using complementary pass-transistor logic," *IEEE J. Solid-State Circuits*, vol. 25, no. 2, pp. 388–395, Apr. 1990.

[3] M. Suzuki *et al.*, "An 1.5 ns 32 bit CMOS ALU in double pass-transistor logic," in *1993 IEEE Int. Solid-State Circuit Conf. Dig. Tech. Papers*, Feb. 1993, pp. 90–91.

[4] F. S. Lai and W. Hwang, "Differential cascode voltage switch with pass gate logic tree for high performance digital systems," in *1993 Int. Symp. VLSI Technol.*, June 1993, pp. 358–362.

[5] L. G. Heller, W. R. Griffin, J. W. Davis, and N. G. Thoma, "Cascode voltage switch logic: A differential CMOS logic family," in *1984 IEEE Int. Solid-State Circuit Conf. Dig. Tech. Papers*, Feb. 1984, pp. 16–17.

[6] A. Parameswar, H. Hara, and T. Sakurai, "A high speed, low power, swing restored pass-transistor logic based multiply and accumulate circuit for multimedia applications," in *1994 IEEE Custom Integrated Circuit Conf.*, May 1994, pp. 278–281.

[7] E. De Man and M. Schobinger, "Power dissipation in the clock system of highly pipelined ULSI CMOS circuits," in *Proc. 1994 Int. Workshop Low Power Design*, Apr. 1994, pp. 133–138.

[8] H. Kojima, S. Tanaka, Y. Okada, T. Hikage, F. Nakazawa, H. Matsushige, H. Miyasaka and S. Hanamura, "A multi-cycle operational signal processing core for an adaptive equalizer," *VLSI Signal Process. VI*, Oct. 1993, pp. 150–158.

[9] R. Bechade, R. Flaker, B. Kauffmann, A. Kenyon, C. London, S. Mahin, K. Nguyen, D. Pham, A. Roberts, S. Ventrone, and T. Voreyn, "A 32 b 66 MHz 1.8 W microprocessor," *1994 IEEE Int. Solid-State Circuit Conf., Dig. Tech. Papers*, Feb. 1994, pp. 208–209.

A Reduced Clock-Swing Flip-Flop (RCSFF) for 63% Power Reduction

HIROSHI KAWAGUCHI AND TAKAYASU SAKURAI

Abstract—A reduced clock-swing flip-flop (RCSFF) is proposed, which is composed of a reduced swing clock driver and a special flip-flop which embodies the leak current cutoff mechanism. The RCSFF can reduce the clock system power of a VLSI down to one-third compared to the conventional flip-flop. This power improvement is achieved through the reduced clock swing down to 1 V. The area and the delay of the RCSFF can also be reduced by a factor of about 20% compared to the conventional flip-flop. The RCSFF can also reduce the *RC* delay of a long *RC* interconnect to one-half.

Index Terms—Differential circuit, flip-flops, leak current, low-power CMOS circuit, low-voltage CMOS circuit, *RC* bus, *RC* delay, *RC* interconnect.

I. INTRODUCTION

FOUR pie charts in Fig. 1 show power distributions in various very large scale integrations (VLSI's). As seen from the charts, the power distribution of VLSI's differs from product to product. However, it is interesting to note that a clock system and a logic part itself consume almost the same power in various chips, and the clock system consumes 20–45% of the total chip power. In this clock system power, 90% is consumed by the flip-flops themselves and the last branches of the clock distribution network which directly drives the flip-flops [1].

One of the reasons for this large power consumption of the clock system is that the transition probability of the clock is 100% while that of the ordinary logic is about one-third on average. Consequently, in order to achieve low-power designs, it is important to reduce the clock system power. In order to reduce the clock system power, it is effective to reduce a clock voltage swing. This is because the power consumption of the clock system is proportional either to the clock swing or to the square of the clock swing, depending on the circuit configuration, which is described later.

One idea to reduce the clock voltage swing was pursued in [2], but it required four clock lines, which will increase clock interconnection capacitance. Moreover, routing four clock lines is disadvantageous in area, and the skew adjustment is difficult.

This paper describes a new small-swing clocking scheme which requires only one reduced swing clock line. The RCSFF

Manuscript received August 22, 1997; revised October 17, 1997. This work was supported by a grant from Toshiba Corporation.

The authors are with the Institute of Industrial Science, University of Tokyo, 7-22-1, Roppongi, Minato-ku, Tokyo, 106-8558 Japan (e-mail: kawapy@cc.iis.u-tokyo.ac.jp).

Publisher Item Identifier S 0018-9200(98)02237-9.

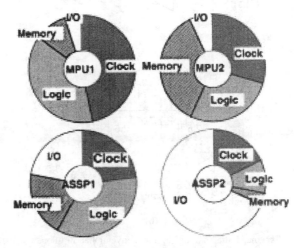

Fig. 1. Power distribution in various VLSI's.

is also beneficial to decrease the capacitance of a clock system by reducing the number of MOSFET's connected to the clock distribution network.

II. REDUCED CLOCK-SWING FLIP-FLOP

Reduced clock-swing flip-flop (RCSFF) is proposed to lower the voltage swing of the clock system. Fig. 2 shows schematic diagrams of the conventional flip-flop and the proposed RCSFF. With the conventional flip-flop, the clock swing cannot be reduced because ϕ and $\overline{\phi}$ are required, and overhead becomes imminent if two clock lines ϕ and $\overline{\phi}$ are to be distributed. On the other hand, if only ϕ is distributed, most of the clock-related MOSFET's operate at full swing, and only minor power improvement is expected.

The RCSFF is composed of a true single-phase master-latch and a cross-coupled NAND slave-latch. The master-latch is a current-latch-type sense-amplifier. The salient feature of the RCSFF is that it can accept a reduced voltage swing due to the single-phase nature of the flip-flop. The voltage swing, V_{Clock}, can be as low as 1 V.

While the MOSFET count of the conventional flip-flop is 24, that of the RCSFF is 20 including an inverter for generating $\overline{D}$. The number of MOSFET's that are related to a clock is also as small as 3, which should be compared to 12, in the conventional flip-flop. Since only three MOSFET's, $P1$, $P2$, and $N1$, are clocked, the capacitance of a clock network can be reduced with the RCSFF, which in turn decreases the power.

The clock swing can be reduced with the RCSFF, but the issue is that when a clock is "high" at the voltage of V_{Clock}, $P1$

Reprinted from *IEEE Journal of Solid-State Circuits*, Vol. 33, No. 5, pp. 807–811, May 1998.

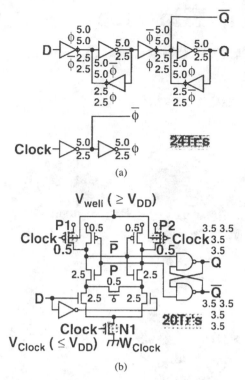

(a)

(b)

Fig. 2. Schematic diagram of (a) the conventional flip-flop and (b) the RCSFF. Numbers in the figure signify MOSFET gate width in microns. Gate length is 0.5 μm for all MOSFET's. W_{Clock} is the gate width of $N1$.

and $P2$ do not switch off completely, leaving leakage current flow through either $P1$ or $P2$. The RCSFF, however, has a leak current cutoff mechanism. By applying backgate bias, V_{well}, to the precharge MOSFET's, $P1$ and $P2$, the threshold voltage of $P1$ and $P2$ can be increased. Then the leakage current can be completely cut off. Although it will be shown afterwards that even without the backgate bias the power can be reduced, the further power improvement is possible by cutting off the leak current. The other way to increase the V_{th} of $P1$ and $P2$ is by an ion-implant, which needs process modification and is usually prohibitive. Thus, this case has not been considered in this paper, but it is one technically promising way if additional ion-implant is allowed. When the clock should be stopped in a standby mode, it should be stopped at V_{SS}. Then there is no leak current even without the backgate bias.

III. REDUCED SWING CLOCK DRIVERS

The RCSFF has a reduced swing clock driver. There are basically two types of clock drivers shown in Fig. 3—type A and type B. In type A, the clock swing, $V_{\text{Clock}} = V_{\text{DD}} - n \cdot V_{th}$ depending on the number of inserted MOSFET's. The power consumption associated with the clock distribution is proportional to V_{Clock} in this case. Type A drivers do not require either dc–dc converters or external voltage supplies, so they are easily implemented.

In type B, on the other hand, V_{Clock} is generated and supplied either from an on-chip dc–dc converter or from an external voltage supply. The power consumption is proportional to the square of V_{Clock}. Thus, it is more efficient than type

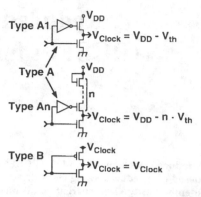

Fig. 3. Types of reduced swing clock drivers. type $A1$ and type An are grouped as type A. In type B, V_{Clock} is supplied externally.

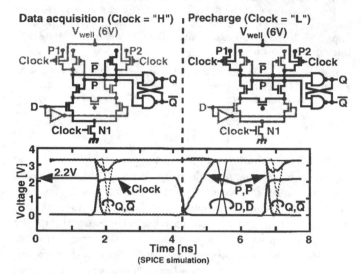

Fig. 4. Operation waveforms of the RCSFF.

A drivers, but more difficult to implement and needs V_{Clock} supply lines to each clock driver.

IV. OPERATION OF RCSFF

Fig. 4 shows the typical behavior of the RCSFF with the type $A1$ driver simulated by SPICE. The left half of the figure is for a data acquisition phase, and the right half shows a precharge phase. It can be seen that the clock goes up only to 2.2 V.

In the figure, the data input D is assumed to be "high" when the clock is asserted. The solid line path turns on and the node $\overline{P}$ goes down to "low" while P remains "high." P and $\overline{P}$ drive a low-active RS flip-flop and an output Q becomes "high." In the precharge phase, MOSFET $P1$ and $P2$ precharge nodes P and $\overline{P}$ to "high." The output Q and $\overline{Q}$ keep the previous state because both P and $\overline{P}$ are "high." The threshold voltage of MOSFET's is 0.6 V, but with the well bias V_{well} of 6 V, the threshold voltage of $P1$ and $P2$ becomes 1.4 V, which is high enough to cut off the leakage path with 2.2 V clock swing.

The RCSFF behaves as an edge-triggered flip-flop because when the clock goes to "high", P and $\overline{P}$ are determined dependent on the input D, and once the data are latched, the change of the input D does not affect P and $\overline{P}$ status thanks to the cross-coupled inverters.

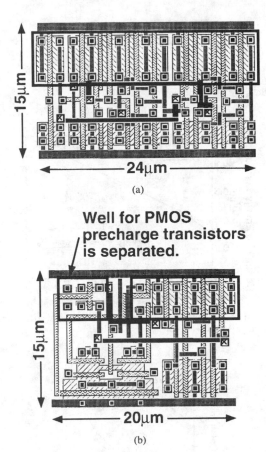

Well for PMOS precharge transistors is separated.

(b)

Fig. 5. Layout of (a) the conventional flip-flop and (b) the RCSFF. W_{Clock} is assumed to be 10 μm and the other value is the same as in Fig. 2.

Let us consider the sizes of MOSFET's here. Numbers in Fig. 2 signify gate width in microns. The nodes P and $\overline{P}$ can be precharged slowly while the clock is "low." Therefore, the size of the precharge PMOSFET's, $P1$ and $P2$, can be minimum—0.5 μm in this case. The width of $N1$ should be large for a faster Clock-to-Q operation. There is a tradeoff between speed and power in choosing the optimum width for $N1$.

V. Performance Comparison

A. Area

Fig. 5(a) is a layout example of the conventional flip-flop, and Fig. 5(b) is the RCSFF case. The well for the precharge PMOSFET's, $P1$ and $P2$, is separated from the normal well for applying the backgate bias. Nevertheless, the area can be reduced by a factor of about 20% compared to the conventional flip-flop. In reality, however, the extra bias lines are needed for the RCSFF case, and this 20% reduction is cancelled out by the bias line overhead. If V_{th} of $P1$ and $P2$ was adjusted by ion-implant, the 20% area reduction could be enjoyed.

B. Delay

A SPICE analysis is carried out assuming typical parameters of a generic 0.5-μm double metal CMOS process. The rise time of V_{Clock} is assumed to be 0.2 ns in the simulations; but even if the rise time is changed from 0.2 to 0.6 ns, the

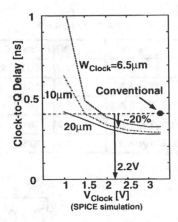

Fig. 6. Clock-to-Q delay characteristic of the RCSFF simulated by SPICE. The delay depends on V_{Clock} but is not affected by V_{well}. The supply voltage is 3.3 V.

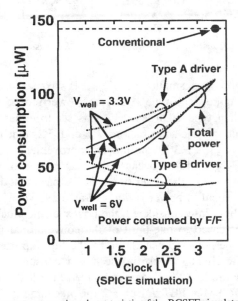

Fig. 7. Power consumption characteristic of the RCSFF simulated by SPICE.

change in Clock-to-Q delay is less than 0.04 ns. Fig. 6 shows Clock-to-Q delay characteristics of RCSFF where the gate width of $N1$, W_{Clock}, is varied as a parameter. Since delay improvement is saturated with W_{Clock} being 10 μm, this value of W_{Clock} is used in the area and power estimation. When V_{Clock} of 2.2 V (type A1 driver) and W_{Clock} of 10 μm are used, the RCSFF is improved by a factor of about 20% over the conventional flip-flops.

Data setup time and hold time in reference to clock are 0.04 and 0 ns, respectively, being independent of V_{Clock}, compared to 0.1 and 0 ns for the conventional flip-flop.

C. Power

Fig. 7 shows power characteristics of the RCSFF. The clock interconnection length is assumed to be 200 μm and transition probability of data is assumed to be 30%. The clock frequency f_{Clock} is assumed to be 100 MHz. These are typical values for low-power processors.

Power consumption per flip-flop is a sum of a clock driver, a flip-flop itself, and interconnection between them. Power

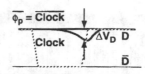

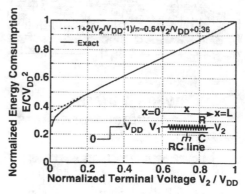

Fig. 9. Behavior of the differential RC bus.

TABLE I
PERFORMANCE COMPARISON

	Driver	V_{Clock} [V]	Power	Delay	Area
Conventional		3.3	100%	100%	100%
RCSFF	Type A1	2.2	59%	82%	83%
V_{well} = 6V	Type A2	1.3	48%	123%	83%
W_{Clock} = 10μm	Type B	2.2	48%	82%	83%
f_{Clock} = 100MHZ	Type B	1.3	37%	123%	83%

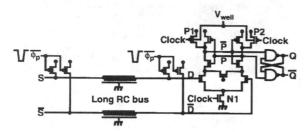

Fig. 8. An application of the RCSFF to a long RC bus.

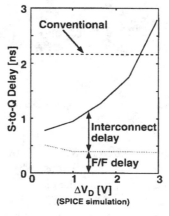

Fig. 10. Normalized energy consumption by distributed RC line if the terminal voltage V_2 is reduced.

consumption gets smaller as V_{Clock} is decreased. As seen from the figure, with type A drivers, power reduction is less efficient than type B drivers. V_{well} is set to either 3.3 or 6 V. Without the backgate bias to $P1$ and $P2$, that is, in the case that V_{well} is 3.3 V, the power improvement is saturated around V_{Clock} of 1.5 V because the leak current increases as V_{Clock} lowers. On the other hand, in the case that V_{well} is 6 V, improvement in power is not saturated even at V_{Clock} of 1 V. In the best case considered, the power of the clock system can be decreased to one-third of the conventional flip-flops. In Fig. 7, the power consumption by the flip-flop itself is also shown. The slight increase in the power consumption of the flip-flop in the low V_{Clock} region is due to the leakage current through the PMOSFET $P1$ or $P2$ for precharge.

Table I summarizes a performance comparison. When the type $A1$ driver, which is easy to implement, is used, the power is reduced to 59% and the Clock-to-Q delay is reduced to 82%. If a dc–dc converter and a type B driver is used, the power consumption can be reduced to 37%, that is, 63% power saving even if the delay increases by 23%. Considering the improvement level and the delay increase, this type $A1$ driver case and this type B driver case can be practical choices.

VI. APPLICATION TO REDUCED SWING BUS

In Fig. 8, an application of the RCSFF to a long RC bus is considered. Since the RCSFF is a differential amplifier in nature, it can be used to amplify a small voltage signal on a differential RC bus, and at the same time it can latch the data.

Behavior of a differential RC bus is shown in Fig. 9. The differential bus is first precharged to V_{DD}, and then, when the voltage difference of D and $\overline{D}$ reaches ΔV_D, the clock is asserted and the amplifier is activated. Since ΔV_D can be as small as less than 1 V, delay reduction of the long RC bus can be achieved. Furthermore, power reduction of logic system can also be realized because D and $\overline{D}$ do not need to be in full swing. Let us consider what amount of power gain is

Fig. 11. Delay improvement of long RC bus. Length of the RC bus is assumed to 10 mm and width is assumed to 0.5 μm. W_{Clock} is 10 μm and a type $A1$ driver is used.

observed when a distributed RC line is driven in full swing at one end and switched off when the other terminal becomes V_2:

$$\frac{V(x, t)}{V_{\text{DD}}} = 1 + \frac{2}{\pi} \sum_{k=1}^{\infty} \frac{(-1)^k}{k - \frac{1}{2}} \cos\left[\left(k - \frac{1}{2}\right)\pi\left(1 - \frac{x}{L}\right)\right]$$
$$\cdot e^{-(k-1/2)^2 \pi^2 (t/RC)}$$

$$Q = \int_0^L CV(x, t)\, dx$$
$$= CV_{\text{DD}}\left[1 - \frac{2}{\pi}\sum_{k=1}^{\infty}\frac{1}{\left(k - \frac{1}{2}\right)^2 \pi^2}\right.$$
$$\left. \cdot e^{-(k-1/2)^2\pi^2(t/RC)}\right].$$

Fig. 10 shows the normalized energy consumption $E (= QV_{\text{DD}})$ by the RC line. If the energy per cycle, $E (= QV_{\text{DD}})$, is expressed in terms of the terminal voltage, $V_2 [= V(L, t)]$, $E \approx 0.36 + 0.64V_2$. This means that about 50%

power saving is possible if an RC interconnect is driven when the voltage swing of V_2 is $0.2V_{DD}$.

Fig. 11 shows the delay dependence of the long RC bus with the RCSFF. The delay is dependent on ΔV_D. Faster operation is possible as ΔV_D is decreased. Compared to the conventional flip-flop, acceleration by a factor of more than two is possible.

VII. CONCLUSION

The RCSFF, which is compatible with the conventional process, is proposed to save up to 63% of the clock system power. With the RCSFF, area can be reduced to 80%, delay can be decreased to 80%, and the power is reduced to one-third of the conventional flip-flop. Leakage current through precharge MOSFET's can be eliminated by backgate bias. As an application of the RCSFF, RC buses are considered. RC delay and power consumed by the interconnect can be reduced to less than one-half compared to the case where the conventional flip-flops are used.

REFERENCES

[1] T. Sakurai and T. Kuroda, "Low-power circuit design for multimedia CMOS VLSI's," in *Proc. Synthesis Sys. Integration Mixed Technol. (SASIMI)*, Nov. 1996, pp. 3–10.
[2] H. Kojima, S. Tanaka, and K. Sasaki, "Half-swing clocking scheme for 75% power saving in clocking circuitry," in *1994 Symp. VLSI Circuits Dig. Tech. Papers*, June 1994, pp. 23–24.
[3] M. Matsui, H. Hara, K. Seta, Y. Uetani, L. Kim, T. Nagamatsu, T. Shimazawa, S. Mita, G. Otomo, T. Oto, Y. Watanabe, F. Sano, A. Chiba, K. Matsuda, and T. Sakurai, "200MHz video compression macrocells using low-swing differential logic," in *ISSCC Dig. Tech. Papers*, Feb. 1994, pp. 76–77.
[4] M. Matsui, H. Hara, Y. Uetani, L. Kim, T. Nagamatsu, Y. Watanabe, A. Chiba, K. Matsuda, and T. Sakurai, "A 200MHz 13mm² 2-D DCT macrocell using sense-amplifying pipeline flip-flop scheme," *IEEE J. Solid-State Circuits*, vol. 29, pp. 1482–1490, Dec. 1994.
[5] J. Montanaro *et al.*, "A 160-MHz, 32-b, 0.5-W CMOS RISC microprocessor," *IEEE J. Solid-State Circuits*, vol. 31, pp. 1703–1714, Nov. 1996.

A Unified Single-Phase Clocking Scheme for VLSI Systems

MORTEZA AFGHAHI, MEMBER, IEEE, AND CHRISTER SVENSSON

Abstract —Two of the main consequences of advances in VLSI technologies are increased cost of design and wiring. In CMOS synchronous systems, this cost is partly due to tedious synchronization of different clock phases and routing of these clock signals. In this paper a single-phase clocking scheme is described that makes the design very compact and simple. It is shown that this scheme is general, simple, and safe. It provides a structure that can contain all components of a digital VLSI system including static, dynamic, and precharged logic as well as memories and PLA's. Clock and data signals are presented in a clean way that makes VLSI circuits and systems well-suited for design compilation.

Index terms —Clocking, compilation, edge-triggered flip-flops, latches, one-phase clocking, skew, synchronous systems, timing, VLSI digital systems.

I. INTRODUCTION

VLSI technologies are continuously being developed towards an increased level of integration and higher speeds. This has several consequences. Two of the most important consequences are an increased design cost and increased cost of wiring.

The increased cost of design is due to the increased complexity of the design. One way to cope with this is to use more systematic approaches to the design, sometimes called structured design [1]. So far structured design has been mainly applied to the structural part of the design (architecture, floorplan, etc.) and not so much to the timing part. In this paper we will discuss a structural timing design. Our goal is to develop a clocking strategy that is general, simple, and safe. Generality and simplicity make design easier and therefore decrease the design cost. Safety decreases the need for careful timing analysis and will also reduce design cost. These factors also make the CMOS circuits and systems well suited for design compilation.

Part of the high wiring cost of CMOS systems is caused by the many clock phases used; often four different clock phases are to be distributed all over a design [2]. The reason for this is that multiphase clocking has been recognized as both safe and transistor-count effective. In this paper we will show that a recently proposed scheme for single-phase clocking [3], [4] is just as safe and transistor-

Manuscript received June 13, 1989; revised September 25, 1989.
The authors are with the LSI Design Centre, Linköping University, S-581 83 Linköping, Sweden.
IEEE Log Number 8932697.

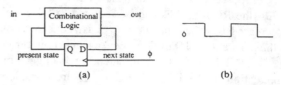

Fig. 1. (a) Single-phase state machine structure and (b) the clock signal.

count effective as the previous four-phase clocking techniques. This technique will be shown to be general enough to handle static, dynamic, precharged, and precharged domino logic as well as array logic like memories or PLA's. Furthermore, it allows considerably higher clock frequencies than the normal multiphase techniques [4].

II. SAFE CLOCKING

In this paper we will limit our discussions on "safety" to timing behavior. We will thus use the term "safe" to denote safety against clock-skew and clock-slope problems, races, hazards, etc. Safe clocking is easily discussed in connection to a finite-state-machine model of logic (see Fig. 1) [1], [5]. The machine in this figure uses a positive edge-triggered D-flip-flop (ETDFF) as storage element. Input signals together with present state signals are inputs to a combinational logic block whose outputs are outputs and next-state signals of the machine. At the positive clock edge the next state becomes the present state and a new state is generated. The circuit is safe if the flip-flop is always nontransparent; that is, a state signal cannot pass the logic block more than once during one clock cycle. The discussion is easily extended to several interconnected finite state machines, a system which can be used as a model for any synchronous digital system. Designing circuits with ETDFF's is rather simple and straightforward and is used in most basic digital design texts [5].

However, if instead of edge-triggered elements data latches are used, it is possible to obtain faster and more transistor-count-effective circuits. But, it is shown that when traditional latches are used, single-phase clocking is neither safe nor fast [6] and a multiphase clock is required. This is due to the lower bound constraint on the minimum short-path delay of the combinational circuits, that could result in a data racethrough problem. For CMOS, when

Reprinted from *IEEE Journal of Solid-State Circuits*, Vol. 25, No. 1, pp. 225-232, February 1990.

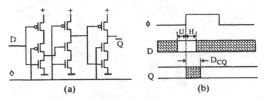

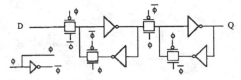

Fig. 3. Static single-phase ETDFF.

Fig. 2. Single-phase positive ETDFF: (a) circuit diagram and (b) parameters.

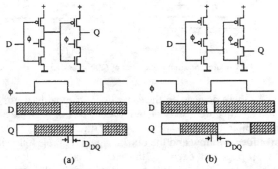

Fig. 4. Circuits and parameters of latches: (a) N-latch and (b) P-latch.

using latches, several schemes exist, in most cases requiring a four-phase clock signal (or pseudo-two-phase, which also means four clock signals) [2]. A simple example in this respect is a circuit according to Fig. 1, where the flip-flop is replaced by two dynamic latches (tristate inverters) clocked by a pseudo-two-phase clock [2]. Some of these schemes are used in connection with other techniques to save transistors, like precharged or precharged domino techniques. In the following, however, it is our intention to show that it is possible to combine the simple and safe single-phase clocking with transistor-count-effective latches. Furthermore, we propose a method that can safely handle up to half a clock period clock skew between two communicating machines without any constraint on the minimum short-path delay of the combinational logic circuit.

In this paper, we divide signals in a system into two classes, namely, clock signal and data signal. The only constraint on the clock signal is that its reference edges must occur only when specified. On the other hand, there are data signals that must obey certain specifications with respect to the clock-signal reference edges. This clear picture of the signals in a system makes the design and verification simpler. To gate the clock signal with a data signal or to delay the primary clock signal destroys this picture. Moreover, gating the clock signal usually results in considerable clock skew. Circuit and bus structures will be considered to selectively transfer data between different parts of the system without gating the clock signal with data signals.

III. BASIC COMPONENTS OF SINGLE-PHASE CLOCKING

A. Storage Elements

In a synchronous system the movement of data is controlled by clocked storage elements. Thus, in this section we consider the timing behavior of storage elements of a single-phase clocking scheme.

A dynamic positive ETDFF [4] is shown in Fig. 2(a). The timing behavior of this device is the same as that of other ETDFF's, i.e., the input data must be valid during the interval given by the setup (U) and hold (H) times of the device and the output data are available and stable D_{CQ} after the active clock edge. The value of the hold time H of this kind of flip-flop is close to zero (see Section V). It is also possible to design a similar circuit which is triggered on the negative clock edge instead of the positive edge [4]. One may note that the proposed flip-flop is inverting. This is normally of little importance. The only limitation of this class of circuits is that the clock rise time must be short enough. This is, however, not a strong limitation, as a rise time of the order of 20 times the logical delay can be accepted [4], [12].

This dynamic flip-flop can thus replace the flip-flop in Fig. 1. It uses nine transistors and one clock line, which is more effective than the standard pseudo-two-phase CMOS solution using eight transistors and four clock lines (two tristate inverters as flip-flop).

If a static flip-flop is needed, a traditional CMOS master–slave flip-flop based on transmission gates [2] fits well into this timing scheme. In Fig. 3 this static ETDFF is demonstrated. This circuit is not really a true single-phase clocked circuit but is safely clocked by a single-phase clock if properly designed. By using a local inverter, as in the figure, the clock skew is under local control so that a properly designed cell will be safe in any environment. This cell uses 18 transistors.

The reason for using a two-phase nonoverlapping clock (pseudo-four-phase in CMOS) when latches are used is mainly to create two different points within each clock period to latch the data. In this way, a holding period is created before each latch that prevents the racethrough problem. This requirement may also be achieved by designing two latches that are transparent on different parts of the same clock signal, i.e., high and low part of the signal. Fig. 4(a) shows a circuit for a latch that is transparent when the clock signal is high and latches the data when the clock signal goes low. Fig. 4(b) shows the counterpart of this circuit, i.e., a circuit that is transparent when the clock signal is low and latches the data when the clock signal goes high [4]. These circuits may be called N- and P-latches, as the clock signal is applied to n and p transis-

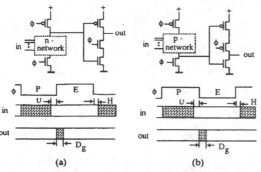

Fig. 6. Block diagram of a single-phase system with ETDFF's.

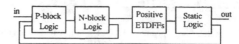

Fig. 7. A mixed technique.

Fig. 5. Circuits and timing diagrams of (a) N-block and (b) P-block.

TABLE I
ALLOWABLE COMBINATIONS OF FLIP-FLOPS
AND PRECHARGE LOGICS

To: ↓	+edge	- edge	N - Block	P - Block
+edge	ok	ok	ok	ok
- edge	ok	ok	ok	ok
N-Block		ok		ok
P-Block	ok		ok	

+edge= Positive ETDFF   - edge= Negative ETDFF

tors in these devices. In the same figure timing diagrams of these circuits are also depicted. Parameters of these latches are similar to other latches. Again, these devices have a forbidden window and the output signal is stable following the propagation delay of the element after the latching edge of the clock signal. As in the case of the dynamic ETDFF above, the clock edge must be fast enough.

B. Precharged Logic and Domino Logic

Precharged dynamic logic utilizes the MOS device characteristics to their best advantages. In fact, another reason for appealing to a multiphase clocking scheme was to enable several levels of nMOS dynamic logic evaluation to be performed per clock cycle [7]. This is now made possible for CMOS circuits even when a single-phase signal is used [3]. Single-phase precharged dynamic logic is based on N- and P-blocks. These blocks, with their respective timing diagrams, are demonstrated in Fig. 5(a) and (b). An N-block is precharging (P) when the clock signal is low and evaluates (E) its stable inputs when the clock signal is high. On the contrary, a P-block is in the precharge phase when the clock is high and in the evaluation phase when the clock signal is low. Therefore, a P-block using the same clock as an N-block can follow an N-block as it is in the precharging phase when the N-block is evaluating; that is, it can change its outputs. Obviously, the output of the N-block cannot be used to drive another N-block as they are in the evaluation phase at the same time. Again, these blocks can be characterized by a forbidden window during which the input data must stay stable and a delay after which the output data are stable.

As in the case of multiphase precharged logic, several levels of logic can be evaluated during one clock cycle using the domino technique [4]. This is done in the usual way, by replacing the logical stage in Fig. 5 with several cascaded logical stages separated by inverters [4], [15]. The single-phase clock thus imposes no limitations on the flexibility of the precharged/domino technique.

Comparing Fig. 2(a) and Fig. 5(a), one notices that the last stage of the positive ETDFF is identical to the last stage (the latching stage) of an N-block and, therefore, they have the same output timing. Thus a positive ETDFF

can always drive a P-block and a negative ETDFF can always drive an N-block.

IV. SINGLE-PHASE SUBSYSTEM CIRCUITS

In this section we will show that all types of circuits used in a CMOS design fit well into the single-phase clocking scheme. To this end, we employ the general concept of the finite state machine used in Section II. It is also important to include array logics, such as memory and PLA circuits, in the same regular clocking scheme.

A. Single-Phase Finite State Machine

When ETDFF's are employed as storage elements, it is relatively simple to fulfill the requirements of the state machine. The structure of the circuit becomes as shown in Fig. 6(a). In this structure, the output signal of the machine is stable after the delays of the flip-flop and the combinational circuit following the leading edge of the clock signal. This output signal can drive another state machine with a similar structure. The combinational logic is normally simple static logic in this case.

We may also form a finite state machine from precharged logic. The feedback loop must then include two precharged stages, one N-block and one P-block. Such a machine is demonstrated in Fig. 6(b).

Furthermore, we may also mix static and precharged logic. For example, static logic may be inserted between the blocks in Fig. 6(b) or a technique as depicted in Fig. 7 may be employed. The only important point to consider is that the timing of the previous latching circuit must fit the timing of the following latching or precharged circuit. In Table I a set of rules covering this problem is given.

When latches are used as storage elements, the block diagram of the state machine may take the configuration shown in Fig. 8. In the combinational logic circuit both

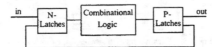

Fig. 8. Block diagram of the system with latches.

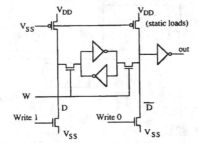

Fig. 9. Structure of a six-transistor RAM cell.

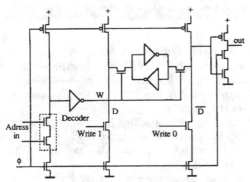

Fig. 10. Single-phase dynamic RAM circuit.

precharge p-units and static circuits can be included. This output signal can drive any system with a similar structure or systems with ETDFF's as storage element.

B. Single-Phase Memory and PLA Structures

The circuit of a normal CMOS RAM cell [2] is shown in Fig. 9. Note that this circuit is not a true CMOS circuit, since the circuit includes a static load (pseudo-nMOS). The memory cell is accessed by making the word select line W high. Often, the address decoder, generating the signals on W, is also designed in pseudo-nMOS in order to save transistors.

Now it will be shown how this memory cell and the address decoder can be designed in a two-stage domino mode with true CMOS (i.e., no static loads) and with the same standard single-phase clock as used above. This is demonstrated in Fig. 10. As shown, the address decoder is the first stage in a domino circuit and the memory cell and the bus circuitry are operated as a second stage. Both stages are then included in an N-block. As in any N-block (see Fig. 5(a)) the read data are latched when the clock signal goes low.

This proposed scheme again gives a block with the same timing behavior as an N-block (see Fig. 5(a)). Both address-in and write 1 and write 0 signals are then considered

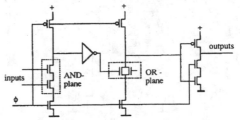

Fig. 11. Single-phase PLA circuit.

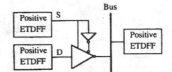

Fig. 12. Circuit for bus structures.

as input signals. The memory or the register stack thus fits exactly into the same timing scheme as any other block.

In addition, a ROM is easily designed in the same scheme. For a ROM we need only one data bus D and the memory cell includes only one transistor, connected between ground and D with its gate connected to W [2]. Data are stored by omitting some of the transistors in the array (see also Fig. 11).

The structure of a PLA also fits into the above scheme. Again, the AND plane can be considered as the first stage of a domino logic and the OR plane as the second stage (see Fig. 11). If some of the outputs are to be fed back to the inputs of the PLA to form a state machine structure, we may use the principle of Fig. 6(b). Note that the evaluation transistor (the clocked n-transistor) can be omitted in the second stage as the signal from the first stage in the domino chain always is zero during precharge. Again, the whole PLA has the same timing as an ordinary N-block.

It is also quite possible to use other variations of circuits to implement RAM's, ROM's, or PLA's. In all cases we may, for example, replace the two-stage domino principle with an N-block followed by a P-block (or vice versa). The latter principle will require more transistors but will give rise to higher speed due to less logical depth.

C. Bus Structures

The problem of selectively delivering data from flip-flops or latches to a bus, or receiving data from a bus, can give rise to timing problems [1, ch. 5]. In our structured single-phase clocking scheme we would prefer to keep the clock signal intact and not gate it or in other ways introduce an extra delay to it. The problem can be solved as shown in Fig. 12, where we access the bus with a tristate inverter, controlled by a signal with the same timing protocol as the data. The bus data will then also follow the same timing protocol and the whole system fits into the scheme already described.

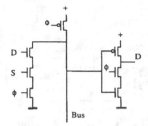

Fig. 13. Precharged bus system.

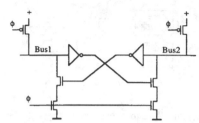

Fig. 14. Bidirectional repeater on a precharged bus.

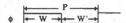

Fig. 15. Parameters of a single-phase clock signal.

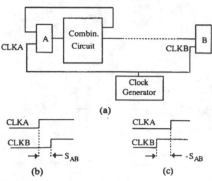

Fig. 16. Demonstration of clock skew in a single-phase system: (a) system diagram, (b) negative skew, and (c) positive skew.

A very useful bus structure is the precharged bus [1]. It is not only used internally on a chip but it may also be used externally, between chips. Again, we find that the single-phase clocking scheme can be applied. The precharged bus may use the same structure and timing as the internal node in a precharged circuit (Fig. 5(a)). This is demonstrated in Fig. 13, where we show a precharged bus with one input circuit and one output latch.

The input circuit is equivalent to the logic part of a precharged gate, where data and select are treated equal (and follow the same timing protocol). This circuit is thus equivalent to an open tristate when select is low. The input circuit may also include logic. If more than one input is active, the bus will perform a wired-OR function and still give valid data. The bus data can be captured either by an output latch as shown in Fig. 13 or by using it directly as input data to a second domino stage. The role of the output latch is to convert the data timing protocol from a precharged form to a normal form.

By using a precharged bus, it is also possible to introduce a bidirectional repeater circuit (which is not possible in a normal bus, see Fig. 14) [8]. This circuit is useful in long buses with large capacitive loads. It can also be used to minimize the effect of the RC delay of the line [9]. During precharge, both bus parts are precharged high. During evaluation, both bus parts will stay high unless one or both are made low, then both will become low. The bidirectional repeater is thus transparent both for data and for the wired-OR function between many input datas.

V. TIMING OF THE SINGLE-PHASE CLOCKING SCHEME

In this section timing considerations of the single-phase clocking scheme are covered. For this purpose we use the circuits of the finite state machine. After a brief discussion about the clock signal, clock skew, and logic delays, we will consider both ETDFF's and latches as storage elements and derive the necessary constraints between the clock signal and different delays to guarantee proper operation of the machine.

The clock signal is specified by its width and period (see Fig. 15). Transition points of the clock signal, i.e., where the signal goes high or low, are also important and may be called reference edges.

In a synchronous system it is ideal if these reference edges occur at the same moment of time at all synchronizing points of the system, that is, where storage elements are placed. In reality, however, clock signals which propagate along routes with variable delays or signals may reach the same point from different paths. This variation of the delays results in clock skew. In a circuit two types of skew are encountered. They may be called self-skew and relative skew [10]. To explain the difference between these two, consider Fig. 16(a). The self-skew of $CLKA$ is the delay variation of the path from the clock generator to register A. Delay variations are caused by imprecision in processing, variation of temperature and power supply, variability among individual storage elements in the register, slopes in the clock pulse edges, etc. The effect of self-skew on timing is insignificant.

The relative skew between the clock signals $CLKA$ and $CLKB$ is the difference between the delays of the paths from the clock generator to registers A and B. $CLKA$ and $CLKB$ timing may then take the form of Fig. 16(b) where we define the skew between clock A and clock B, S_{AB}, as the time lag of $CLKB$'s clock edge after $CLKA$'s corresponding clock edge. Note that S_{AB} may take both positive and negative values. We will further use S_{ABM} for the maximum value of skew and S_{ABm} for its minimum value. Relative skew is evident in the same clock period and is present in nonlocal communications or when different flip-flops in a register receive clock signals from different paths. In estimating clock skew one may use worst-case

358

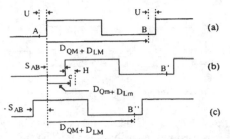

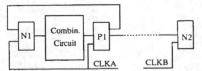

Fig. 18. Block diagram of single-phase finite state machine with latches as storage elements.

Fig. 17. Timing diagram for finite state machine with edge-triggered flip-flop.

analysis or statistical analysis [11]. In the worst-case analysis the skew is the difference between the worst delay (slowest) and the best delay (fastest) of the paths. This is assumed in Fig. 16. Statistical analysis usually results in lower skew but at the cost of a nonzero failure rate. In the following we assume that the skew associated with the leading edge and the trailing edge of the clock signal are equal. This assumption is reasonable since clock signals pass through several inverters before they reach the registers.

The combinational logic circuit included in the state machine structure can also be specified by two delay parameters. The maximum and minimum propagation delays through all circuit paths connecting any input to any output will be called D_{LM} and D_{Lm}. Flip-flops and latches are also characterized by maximum and minimum delays. We now consider the timing of single-phase clocking with edge-triggered flip-flops, latches, and precharged circuits.

A. Edge-Triggered Flip-Flop Timing

The aim of timing calculation is to specify a minimum period for the clock signal such that the state machine circuit operates correctly. For this purpose consider Fig. 16(a) and assume that registers A and B are edge-triggered flip-flops. Now consider a data stored at the input of the flip-flop A. For deterministic behavior of the state machine, this data when triggered in must have enough time to trip round the feedback loop and reach the input of the same flip-flop before the setup time of that device for the next clock cycle. This condition is graphically depicted in Fig. 17(a). An input data at point A in the current clock cycle must reach point B for the next cycle. In this case:

$$P \geqslant D_{QM} + D_{LM} + U \qquad (1)$$

where D_{QM} is the maximum delay of the flip-flop. Furthermore, the data must not trip around the machine too fast, or to be more exact, it must not arrive at the flip-flop input before the end of the hold time for the present clock edge:

$$D_{Qm} + D_{Lm} \geqslant H. \qquad (2)$$

As mentioned above, H is normally very close to zero, hence this condition is always fulfilled.

Now consider a situation where one of the outputs of the machine is feeding a nonlocal flip-flop, say, flip-flop B

in Fig. 16(a). In this case the relative position of the clock signals at the site of flip-flop A and B may be as depicted in Fig. 16(b). Following the same principle stated above and referring to Fig. 17(b):

$$P \geqslant D_{QM} + D_{LM} + U - S_{ABm}. \qquad (3)$$

The other situation must be considered. The minimum delays of the flip-flop and the combinational circuit may be such that the output signal reaches flip-flop B before the hold time of this device (see point C in Fig. 17(b)). Therefore, to avoid violation of the hold time of flip-flop B, we must have

$$D_{Lm} + D_{Qm} \geqslant H + S_{ABM} \qquad (4)$$

where D_{Qm} is the minimum delay of the flip-flop. In other words, one may say that the maximum allowable clock skew is

$$S_{ABM} \leqslant D_{Lm} + D_{Qm} - H. \qquad (5)$$

In the case when the skew is negative, Fig. 17(c), the clock period will increase and is obtained from the relation (3) with the sign of the skew changed. On the other hand, according to (5), the maximum positive skew is limited to the minimum delay of the flip-flop and combinational logic. This class of circuits is thus quite sensitive to positive clock skew [12], [13]. In a general system, (5) and (3) must be used for the maximum allowable skew and minimum clock period, respectively.

It should be mentioned that when a positive ETDFF drives a negative ETDFF, the minimum clock-pulse width, rather than the clock period, must fulfill the above expressions for the clock period. In this case the system will be much less sensitive to positive clock skew.

B. Data Latch Timing

The argument followed here is of the same form as used above. That is, the clock signal width and period are so determined that if data propagates along different paths and round the feedback loop, the setup and hold times of the latches must not be violated. We assume that the input data to a latch do not wait for the enabling clock edge. Rather, latches are open before input data become valid [10]. With this measure the system becomes faster and timing analysis becomes more clear. Fig. 18 shows a block diagram of the single-phase system considered here.

Now consider a data at the input of the latch $N1$. The latest arrival time of this data is the point A in Fig. 19. This data should have enough time to propagate through

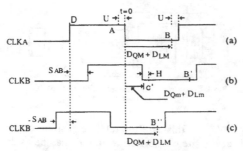

Fig. 19. Timing diagram for the state machine of Fig. 18.

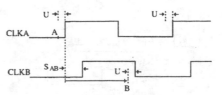

Fig. 20. Timing diagram for precharged circuits.

the latch and the combinational circuit and reach the input of the $P1$ latch before the setup time of this latch (point B in the same figure). For the sake of clarity, we assume that the maximum and minimum delays and setup and hold times of the $N1$ latches are equal to that of the $P1$ latches. Therefore, for the above condition to hold,

$$W' \geqslant D_{QM} + D_{LM} + U \tag{6}$$

where W' is the low part of the clock signal. If another combinational circuit were included in the feedback loop between the $P1$ and the $N1$ latch, a similar relation would apply for the width of the clock signal. In the case of Fig. 18:

$$W \geqslant D_{QM} + U \tag{7}$$

where D_{QM} is the maximum delay of the $P1$ latch. Now consider a situation when one of the outputs of the $P1$ latches is sent to a remote latch $N2$ instead of to $N1$. Relative position of the $CLKA$ and $CLKB$ may then be as in Fig. 19. In this case:

$$W' \geqslant D_{QM} + D_{LM} + U - S_{ABm}. \tag{8}$$

Now we consider the effects of minimum delays of the combinational logic circuit and latches. The minimum delays of the latch and the combinational circuit must be such that the output of the $N1$ latch does not arrive to the $N2$-latch input before the end of the hold time of this latch for the present clock period (point c' in Fig. 19(b)). Therefore, to avoid violation of the hold time of the $N2$ latch, we must have

$$D_{Lm} + D_{Qm} \geqslant H + S_{ABM} \tag{9}$$

where D_{Qm} is the minimum delay of the latch. In other words, one may say that the maximum allowable clock skew is

$$S_{ABM} \leqslant D_{Lm} + D_{Qm} - H. \tag{10}$$

In the case when the skew is negative, Fig. 19(c), the W' will increase and is obtained from relation (8) with the sign of the skew changed. Again, in a general system, (10) and (8) and (7) must be used for the maximum allowable skew and minimum clock period, respectively.

Traditionally, in single-phase clocking using latches, only one latch is used in the feedback loop of the state machine circuit [6]. This is equivalent to taking out latch $P1$ from Fig. 18. In this case the minimum delay of the combinational circuit and latches must be so long that a data arriving at the $N1$-latch input at point D in Fig. 19(a) cannot race through the circuit and reach the $N2$-latch input before point c'; that is:

$$D_{Lm} + D_{Qm} \geqslant W + S_{ABM}. \tag{11}$$

Fulfilling this condition makes the design very expensive and difficult. This constraint on the minimum delay of the combinational circuit and latches is a major drawback of traditional single-phase clocking with latches. This forced the designers to appeal to two-phase clocking (four phases in CMOS) in order to create two reference edges within the same clock period. Then two latches of the same kind but clocked by different phases are used in the feedback loop of the state machine.

In the single-phase scheme considered here, however, instead of using two phases we use two different kinds of latches: one that is transparent during the high part of the clock signal and the other that is transparent during the low part of the clock signal. In fact, the $P1$ latch in Fig. 18 is a barrier that stops the propagation of the data up to point $t = 0$ in Fig. 19(a). In this respect this single-phase scheme is equivalent to a two-phase scheme with no overlap or gap between the phases. Here it is only necessary to guarantee that a data released at point $t = 0$ by the $P1$ latch will not reach the $N2$-latch input before the hold time of this latch (see relation (9)). In Section VI we propose a technique that relaxes this constraint even further.

C. Precharged Circuit Timing

Here again we use the same argument as above to determine the minimum clock width and period. Assume that an N-block clocked by $CLKA$ is communicating with a P-block clocked by $CLKB$ (see Fig. 20). The latest arrival time of a data to the input of the N-block is denoted by point A in this figure. This data must have enough time to be evaluated by the N-block and reach the input of the p-block before the setup time of this block (point B in the figure). Therefore

$$W \geqslant D_{LM} + U - S_{ABm}. \tag{12}$$

W is also constrained by the precharge time T_p, so that $W \geqslant T_p$. Following the same discussion that led to (5), the

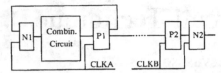

Fig. 21. Circuit technique to cancel the effects of clock skew. An extra latch is inserted on the site of the destination latch.

maximum allowable skew is

$$S_{ABM} \leqslant D_{Lm} - H. \tag{13}$$

Again, for this kind of circuit, $H \approx 0$. If there is a skew, (12) shows that a negative skew will increase the clock period. On the other hand, according to (13) the maximum positive skew is limited to the minimum delay of the block. This class of circuits is thus sensitive to positive clock skew unless some special technique is used (see [4] and [12]).

VI. Signal Resynchronization

It is usually expensive and difficult to satisfy the constraints between the clock skew and the minimum delay of the combinational logic and latches (relation (10)). To reduce the clock skew, great efforts must be exerted, for example, to hand-tune the delays in clock distribution paths, and to control other related factors. On the other hand, to increase the minimum delay of the combinational logic circuit, it is usually necessary to add extra delays (e.g., inverters) at the latch outputs as well as to increase the clock period. In the following, a method is proposed that when used can handle up to half a clock period skew with no constraint on the minimum combinational logic delay.

If we compare Fig. 19(a) and (b), it is evident that the problem is raised because when the data are transmitted by the $P1$ latch at $t = 0$, $CLKB$ is still at high state. But, the data should reach the $N2$ latch first after the $CLKB$ has gone to low state to avoid setup violation of the $N2$ latch. One, for example, can use a negative edge-triggered flip-flop instead of an $N2$ latch that receives data from the remote block. In this way, however, the setup time of the edge-triggered flip-flop may be violated. To solve this problem one can insert an extra P-latch before the $N2$ latch (see Fig. 21). By these methods a clock skew as long as half a clock period can be handled safely without putting any constraint on the minimum delay of the circuit elements (see also [14]). This same technique can also be used in association with edge-tgriggered flip-flops to relax relation (5).

When the clock skew is negative, the above racethrough problem is not present but the clock period will be longer, in accordance with relation (10).

VII. Conclusions

In this paper a unified single-phase clocking scheme was proposed. It was shown that this scheme is general, safe, and simple. The generality makes it possible to use the scheme consistently throughout a digital VLSI system. It provides a structure that can contain all components of a digital VLSI system including static, dynamic, and precharge logic as well as memories and PLA's. This also makes the CMOS circuits and systems well suited for design compilation. Timing expressions for circuits with different storage elements were developed. It was shown that a negative clock skew will just increase the clock period (this is also equivalent to signal delay) and that positive skew may give rise to failure. The problems associated with traditional one-phase clocking using latches are avoided. Techniques were proposed that relax the constraints between the clock skew and the minimum delays of the combinational logic circuits and storage elements. Using the scheme and techniques proposed in this paper should provide a simple and safe clocking for high-speed and compact VLSI digital systems. It also makes the visualization of the system simple and resembles the traditional digital systems where only edge-triggered devices were used.

References

[1] C. Mead and L. Conway. *Introduction to VLSI Systems.* Reading, MA: Addison-Wesley, 1980.

[2] N. Weste and K. Eshraghian, *Principles of CMOS VLSI Design.* Reading, MA: Addison-Wesley, 1985.

[3] Y. Ji-ren, I. Karlsson, and C. Svensson, "A true single phase clock dynamic CMOS circuit technique," *IEEE J. Solid-State Circuits,* vol. SC-22. pp. 899–901, 1987.

[4] J. Yuan and C. Svensson, "High-speed CMOS circuit technique," *IEEE J. Solid-State Circuits,* vol. 24, pp. 62–71, 1989.

[5] D. L. Dietmeyer, *Logic Design of Digital Systems.* Boston: Allyn and Bacon, 1971.

[6] S. H. Unger and C. Tan, "Clocking schemes for high-speed digital systems," *IEEE Trans. Comput..* vol. C-35, pp. 880–895, 1986.

[7] M. Penney and L. Lau, *MOS Integrated Circuits.* Princeton, NJ: Van Nostrand, 1972.

[8] K. Chen and C. Svensson, private communication, 1989.

[9] M. Afghahi and C. Svensson, "A scalable VLSI synchronous system," in *Proc. IEEE Symp. Circuits Syst.* (Espoo, Finland), 1988, pp. 471–474.

[10] G. G. Langdon, *Computer Design.* San Jose, CA: Computeach, 1982.

[11] M. Afghahi and C. Svensson, "Calculation of clock path delay and skew in VLSI systems," in *Proc. European Conf. Circuit Theory Design* (Brighton, England), 1989, pp. 265–269.

[12] I. Karlsson, "True single phase clock dynamic CMOS circuit technique," in *Proc. IEEE Symp. Circuits Syst.* (Espoo, Finland), 1988, pp. 475–478.

[13] M. Hatamian and C. L. Cash, "Parallel bit-level VLSI design for high-speed signal processing," *Proc. IEEE,* vol. 75, pp. 1192–1202, 1987.

[14] C. Svensson, "Signal resynchronisation in VLSI systems," *Integration. The VLSI J.,* vol. 4, pp. 75–80, 1986.

[15] R. H. Krambeck, C. M. Lee, and H. S. Law, "High-speed compact circuits with CMOS." *IEEE J. Solid-State Circuits,* vol. SC-17, pp. 614–619, 1982.

High-Speed CMOS Circuit Technique

JIREN YUAN AND CHRISTER SVENSSON

Abstract — We have demonstrated that clock frequencies in excess of 200 MHz are feasible in a 3-μm CMOS process. This is obtained by means of clocking strategy, device sizing, and logic style selection. We use a precharge technique with a true single-phase clock, which remarkably increases the clock frequency and reduces the skew problems. Device sizing with the help of an optimizing program improves circuit speed by a factor of 1.5–1.8. We minimize the logic depth to one instead of two or more and use pipeline structures wherever possible. The presentation includes experimental demonstrations of several circuits which work at clock frequencies of 200–230 MHz. SPICE simulation shows that some circuits possibly work up to 400–500 MHz.

I. INTRODUCTION

MANY factors control the possible speed of CMOS integrated circuits. There are, for example, device dimensions, logic circuit style, clocking strategy, architecture, clock distribution, etc. To pursue high speed and integration density the dimensions of MOS transistors are scaled down continuously. The delay in a CMOS circuit will be inversely proportional to the scaling factor α if all dimensions are reduced without changing physics [1]. However, there are physical, geometrical [2], [3], and also cost limits on scaling down transistors. Therefore, we should tap the potential of the most popular technique. In fact, we have been investigating the possibilities of increasing speed by combining different circuit techniques in an available and relatively low-cost process, for example, the 3-μm CMOS process. In the present work we will limit ourselves to a discussion of a high-clock-frequency synchronous CMOS circuit technique in a given process (i.e., with a given smallest dimension device). We will assume that the circuit technique rather than the architecture limits the clock frequency. Our results are therefore applicable mainly to simple, pipelineable architectures. In this article, we will present our results by means of both analysis and experiments. Section II describes a new clocking strategy with its accompanying circuit technique and its importance for high clock frequency. We further investigate the robustness and flexibility of this technique and propose some further improvements. Section III describes the effect of logical optimization of the circuits and Section IV describes the effect of device sizing. In Section V

Manuscript received May 25, 1988; revised August 10, 1988.
The authors are with the Department of Physics and Measurement Technology, LSI Design Center, Linköping University, 581 83 Linköping, Sweden.
IEEE Log Number 8824917.

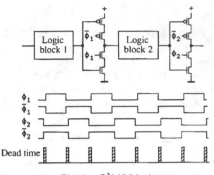

Fig. 1. C²MOS logic.

we present some experiments, illustrating the previous results. Finally we present our conclusions in Section VI.

II. CLOCKING STRATEGY AND CIRCUIT EXAMPLES

A. True Single-Phase-Clock Circuit Techniques

In conventional CMOS circuits both static and dynamic CMOS logic is used. For the purpose of system timing a clocking strategy is always involved except for a self-timed system [4]. The most popular clocking strategy is clocked CMOS logic (C²MOS) [5], [6] which uses a nonoverlapping pseudo two-phase clock as shown in Fig. 1. Four clock signals have to be distributed in such a system and between two pairs of clock signals there should be no overlap. Clock skews in the system will cause serious problems and result in difficulties in increasing circuit speed [7].

The NORA dynamic CMOS technique [8] uses a true two-phase-clock signal ϕ and $\bar{\phi}$ instead of a four-phase-clock signal and can avoid race problems caused by clock skews with some constraints on logic composition. In a NORA pipelined system, ϕ-C²MOS latches and $\bar{\phi}$-C²MOS latches are alternatively used. The most important constraint is that between two C²MOS latches there must be an even number of inversion blocks and if there are static blocks between a precharge block and a C²MOS latch they must also be of an even number. We choose two typical NORA constructions which are called ϕ section and $\bar{\phi}$ section and use an N-precharge block in the ϕ section and a P-precharge block in the $\bar{\phi}$ section for our discussion. These are shown in Fig. 2. Since there is no dead time and no skew problem, it is expected that the NORA dynamic

Reprinted from *IEEE Journal of Solid-State Circuits*, Vol. 24, No. 1, pp. 62-70, February 1989.

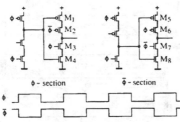

φ - section $\bar{\phi}$ - section

Fig. 2. NORA dynamic CMOS technique.

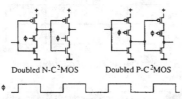

Doubled N-C²MOS   Doubled P-C²MOS

Fig. 3. True single-phase-clock latch stages.

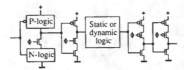

Fig. 4. Logic arrangements using true single-phase-clock latch stages.

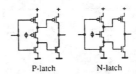

P-latch   N-latch

Fig. 5. Split-output latch stages.

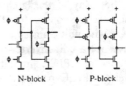

N-block   P-block

Fig. 6. TSPC-1 circuit.

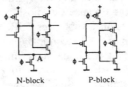

N-block   P-block

Fig. 7. TSPC-2 circuit.

CMOS technique can reach higher clock rates than the C²MOS technique.

A further development in clock strategy should use even less clock signals. The true single-phase-clock dynamic CMOS circuit technique [9] uses only one clock signal which is never inverted. Therefore, no clock skew exists except for clock delay problems and even a higher clock frequency can be reached. As will be explained below, the true single-phase-clock CMOS technique fits not only dynamic but also static CMOS circuits and in most cases can replace the NORA CMOS technique.

Let us discuss only the latch stages first. In Fig. 2, we have seen two C²MOS latch stages controlled by two clock signals φ and $\bar{\phi}$. The necessity of $\bar{\phi}$ lies in controlling transistors M_2 and M_7. However, this can be done by clock signal φ at inverters connected to the two stages as shown in Fig. 3. We call the two different units N-C²MOS stage and P-C²MOS stage, respectively, and if we use doubled N-C²MOS or P-C²MOS in series they become true single-phase-clock latch stages, i.e., N-latch and P-latch. In this system, an N-section (N-latch plus logic blocks) and a P-section (P-latch plus logic blocks) are used alternatively using the same clock signal. Both static and dynamic blocks are accepted and an N- or P-precharge block is used in the N- or P-section, respectively. As long as the clock delay is less than the gate delay the system is reliable. Instead of distributing the $\bar{\phi}$ clock signal there are two transistors more in each latch stage, but the most important thing is that there will be no even inversion constraint either between two latches or between the latch and the dynamic block. Apparently, this is better than the nonoverlapping pseudo two-phase-clock strategy and also, from the point of view of logic constraints, can compete with the NORA two-phase-clock strategy. The logic function blocks can be included in the N-C²MOS or P-C²MOS latch stages or placed between them as shown in Fig. 4, depending on the logic type and the requirement of inversion.

Furthermore, the true single-phase-clock latch stages shown in Fig. 3 can evolve into a simpler version, called split-output latch, as shown in Fig. 5 where only one of the transistors is controlled by the clock. It implies half the clock load. A possible drawback of the split-output technique is that all node voltages do not have a full voltage swing, as some single transistors are used for the transmission of both high and low signals.

In the case of using precharge dynamic logic, let us go back to the NORA circuits in Fig. 2. The P-transistor M_2 in the φ section and the N-transistor M_7 in the $\bar{\phi}$ section are unnecessary because precharge signals will play the same role as $\bar{\phi}$ in both sections, so they can be omitted. In Fig. 6, this evolved into the true single-phase-clock dynamic CMOS technique as described in [9]; we call it true single-phase-clock 1 (TSPC-1). We introduce a further modified circuit as shown in Fig. 7. We call it TSPC-2, which is better than TSPC-1 in performance, as described below. In the following description, we choose the φ section of NORA and N-blocks of TSPC-1 and TSPC-2 dynamic circuits as examples.

Compared with the nonoverlapping pseudo-two-phase clock (NPTC) and the NORA two-phase clock, besides the compact simple clock distribution of TSPC which will naturally lead to a higher clocking speed, we can summarize other features of them as follows.

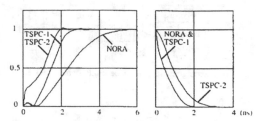

Fig. 8. Step responses of NORA, TSPC-1, and TSPC-2.

B. Step Response

Because the transistor number is reduced from four in a C^2MOS stage to three in either an N-C^2MOS or a P-C^2MOS stage, the delay of the latch stage is reduced. Fig. 8 shows the step responses for the three circuits simulated by SPICE with unit transistor sizes. For the sake of realism, we have put a unit inverter load into all three circuits. In this paper, all SPICE simulations are done by using typical level 2 SPICE parameters for a standard 3-μm double-metal CMOS process [10] and using a power supply voltage of 5 V.

First, we find that omitting the P-transistor in the latch stage is significant because the most critical slope in the circuit is the rise slope. This gives TSPC-1 speed improvement of a factor of 1.8. Second, by omitting the P-transistor the output of the latch stage may continue to rise during the initial part of the succeeding precharge phase, thus allowing a shorter evaluation time. Third, we should explain why TSPC-2 has even better performance. In precharge logic circuits, at the start of evaluation the latch stage tends to output the precharge state first, i.e., to make the output low because of the high precharge node, as can be seen in Fig. 8 at the beginning of the rise slope. This is a common problem for precharge circuits. It is found that the circuit TSPC-2 can solve this problem partly. In Fig. 7, if the logic part is conducting, node A will also be charged to high during the precharge phase, which prevents the output from going low. This makes the rise time shorter. On the other hand, the fall time will increase but this only makes both slopes more balanced in time. Another advantage is that when the output should be kept high, the circuit has much smaller dips in the output as shown in Fig. 9, which is obtained from SPICE simulations. For the P-block of TSPC-2 circuit the top P-transistor should be widened by a factor of 1.5–2 in order to obtain all the advantages mentioned.

C. Sensitivity to the Clock Slope

No maximum slope limit exists in NPTC circuits as long as the clocks are nonoverlapping. This is not true in NORA and TSPC circuits. SPICE simulation shows that NORA and TSPC-1 can accept slopes up to 20 ns for rise and fall edges without disorder. This figure agrees well with experimental results on TSPC-1 in [11]. This is about 20 times larger than the normal gate delay. Because we are interested in the possibility of working at high frequencies,

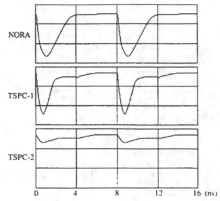

Fig. 9. Output dips of different circuits.

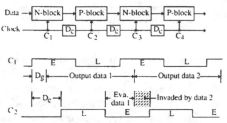

Fig. 10. Problem caused by clock delay. D_g = gate delay; D_c = clock delay; E = evaluation; L = latch.

a slope of 20 ns is quite acceptable. For TSPC-2, the acceptable clock slope reaches 2 ms, which is of the same order of magnitude as the clock period to be accepted by dynamic CMOS circuits. It means that TSPC-2 is more reliable in latching a signal.

D. Clock-Skew Problems

In general terms, no problem caused by clock skew exists in NORA and TSPC circuits so they are better than the NPTC strategy. However, clock skews will compress synchronous margins in NORA circuits, and for both NORA and TSPC circuits the clock delay could be a problem, if it is larger than the gate delay. The problem is caused by data transparency from block 1 to block 2 when both are in the evaluation phase as shown in Fig. 10.

Locally, this is not a problem because the clock delay is usually less than the gate delay. In a pure pipeline structure the condition between the clock delay and the data delay is still satisfactory. In the case of a large system, we propose two ways to solve the problem.

1. Reverse Clock Distribution: It is found that if the clock is distributed in the opposite direction to the data stream the system will be safe. In this case, the evaluation phase of the next block will be completely included in the data stable zone of the last block because the reverse clock distribution creates a safety margin as shown in Fig. 11. This is true for the data stream both from P-block to N-block and from N-block to P-block.

2. D-Latch Type Structure: This can be used in the case of data feedback, e.g., the communication between two

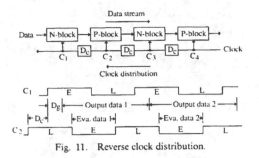

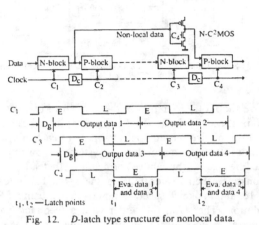

Fig. 11. Reverse clock distribution.

Fig. 12. *D*-latch type structure for nonlocal data.

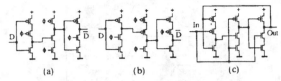

Fig. 13. Circuits constructed by P, N-blocks and P, N-C^{2}MOS stages: (a) positive transition latch; (b) negative transition latch; and (c) divide-by-two circuit.

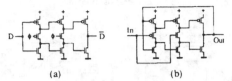

Fig. 14. Circuits constructed by split-output latch stages.

blocks in a large system, where the reverse clock distribution is no solution as there are both forward and backward data streams. Therefore, we propose using a *D*-latch structure instead of the normal alternating P-block and N-block structure for nonlocal data as shown in Fig. 12 (see also [12]). First, if we look at Fig. 10, the problem is caused by the overlap between different evaluation phases of two communicating blocks. If the data for the next block are only latched by the start transition of the evaluation phase of this block at points t_1 and t_2 in Fig. 12, i.e., the communication between two blocks exists only during each start transition, the system will be much more reliable. This can be done simply by placing a P-C^{2}MOS stage before an N-block or an N-C^{2}MOS stage before a P-block for each "nonlocal" data. The *D*-latch structure allows communication between two blocks which have a "clock skew" caused by clock delay, up to almost half a clock cycle. This has been proved by SPICE simulation.

E. Circuit Examples

Because both static and dynamic circuits can be used, including the domino technique, the TSPC strategy has a logic flexibility as high as the NORA technique has. In most cases, NORA circuits can be replaced by TSPC circuits with little modification. One exception is that an N-precharge stage cannot be directly connected to a P-precharge stage without latch stages in the TSPC circuit, which is possible in the NORA technique. However, no even inversion requirement exists in TSPC circuits as in NORA circuits. Because of the compact clock distribution

and the same circuit complexity, TSPC is preferred. We present several circuits below as examples of the TSPC strategy.

If we cascade the P-type and the N-type latches shown in Figs. 3, 5, 6, or 7, they become full dynamic transition *D* latches, each of which includes 12 transistors. However, the *D* latch shown in Fig. 13(a), which consists of nine transistors, is more effective. This is constructed by a P-C^{2}MOS stage, an N-precharge stage and an N-C^{2}MOS stage, and the input data will be latched by the positive transition of the clock signal. If we want a noninversion output an extra inverter can be placed at the output, which gives the circuit driving ability. When a negative transition latch is needed the circuit can be changed to Fig. 13(b). As expected, if the inversion output is connected to the input of the circuit, a divide-by-two circuit is formed as shown in Fig. 13(c).

If we use the split-output latch stages shown in Fig. 5 we can construct a *D* latch and a divide-by-two circuit in an even more efficient way. The resulting circuits have almost the same speed but only half the clock load and are shown in Fig. 14. The latch stages can also be used for building quite effective shift-register chains with both less transistors and less clock load than other techniques, and without output glitches.

Note that in the *D* latches of Figs. 13(a) and 14(a) the P-latches are replaced by P-C^{2}MOS stages with only three transistors. When we need fast P-passing stages in a pipelined structure the three-transistor latch is quite effective. However, it requires an input transition from low to high with a delay more than the evaluation delay of the next N-block, otherwise the active transistor in the N-block may be cut off too early. This is shown in Fig. 15 where the delay of the input signal is changed from 0.8 to 0.5 ns and the output swing is reduced to half the V_{dd}. Nevertheless, as long as these *D* latches are cascaded the delay of the last latch is satisfactory for the requirement of the next latch and the same for the divide-by-two circuits. Simulation shows that a register chain starting with an N-block (the N-block reduces the time constraints) can work at a clock rate of more than 350 MHz using a unit inverter load without critical input time requirement.

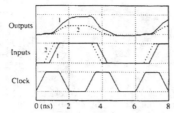

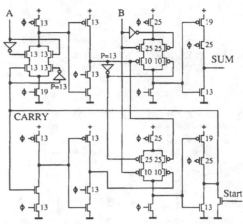

Fig. 15. Different results with the positions of the input transitions for the nine-transistor latch.

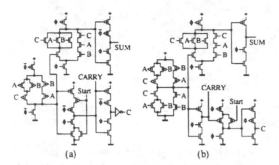

(a)   (b)

Fig. 16. Replaceability between NORA and TSPC: (a) NORA serial full adder, and (b) TSPC serial full adder.

Fig. 17. Serial full adder with logic separated in different half clock cycles (TSPC adder 1).

Another circuit is a serial full adder, which is an example showing the logic replaceability between NORA and TSPC. Fig. 16(a) is the adder using the NORA technique presented in [8]. We can use N-C^{2}MOS instead of C^{2}MOS after an N-precharge stage and a negative transition latch instead of the two C^{2}MOS stages in the NORA circuit for the delay of the carry signal. The P-precharge stage is replaced by a static block as the $\bar{\phi}$ clock is lacking in TSPC, and the static block will evaluate inputs earlier in the previous phase. The resulting circuit is shown in Fig. 16(b), which has the same transistor number but higher speed than NORA and only a single clock. A corresponding circuit with an N-precharge stage instead of the static block, needing more transistors as it will use inverted signals, is presented in [9].

III. LOGIC STYLE

The maximum clock frequency in a clocked system is limited by the delay from one latching instant to the next. This delay depends on the complexity and function of each logic block between these two instants. One way to minimize this delay is to minimize the complexity of each logic block, for example, by decomposing complex blocks into pipelined parts. Such decomposition can also be done between half clock cycles when using the circuit style described above. In such a way it may be possible to reduce the logic depth to one in each block. Although this will cause an initial delay, it is acceptable in a pipelined structure. Note that in CMOS technology inverter delay is very small so that extra inverters can be accepted in the blocks.

The CMOS circuit technique is particularly sensitive to the number of transistors in series [6]. It is quite obvious that we should try to minimize the number of transistors in series in the logic blocks. This normally means that we also need to minimize the number of inputs to each block. We will not attempt to develop a systematic optimization procedure using the above principles. Instead we will demonstrate them by optimizing the serial full adder of Fig. 16 and make a speed comparison by using SPICE simulation. The serial full adder with a feedback path is considered a good example for studying logic style. For possible systematic approaches, see [13].

In Fig. 16(a) and (b), the Boolean functions of CARRY and SUM are

$$CARRY = AB + C(A + B)$$
$$SUM = \overline{CARRY}(A + B + C) + ABC$$
$$= \overline{(AB + C(A + B))}(A + B + C) + ABC.$$

This means that the evaluation of CARRY must be done before the evaluation of SUM and both have to be finished in half a clock cycle. We can, of course, divide the SUM evaluation into two parts and put them into different half clock cycles. Even if we do so, too many steps are still involved in the SUM evaluation and at least three transistors are in series. Let us instead change the above Boolean function to

$$E = A\bar{C} + \bar{A}C$$
$$SUM = E\bar{B} + \bar{E}B$$
$$CARRY = EB + \bar{E}C.$$

Apparently, these Boolean functions are much easier to realize and the intermediate result E can be evaluated during the previous half clock cycle. There will be only two transistors in series in the logic part. The circuit according to the above Boolean functions and using TSPC strategy has been presented in [9]. For convenience, we give it in Fig. 17. The small figures nearby transistors in Fig. 17 as well as in Figs. 18 and 19 are the scaled sizes in micrometers which will be discussed in the next section. The transistors without figures have the minimum width, 7 μm.

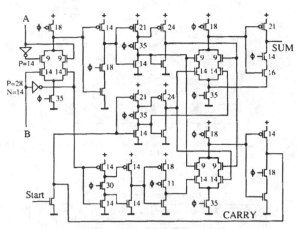

Fig. 18. Serial full adder with all logic in N-blocks (TSPC adder 2).

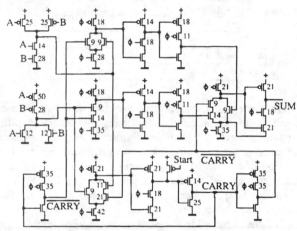

Fig. 19. Fast serial full adder (TSPC adder 3).

TABLE I
SPEED COMPARISON OF SERIAL FULL ADDERS USING DIFFERENT CIRCUIT TECHNIQUES

Performance Technique	Worst delay (ns)	No. of transistor	No. of input load	No. of clock load	Note
Static CMOS	6.2	38	8	6	[14]
NORA	6.0	32	4	10	Fig.16a
Modified NORA	3.8	35	3	16	[14]
DCVS NORA	3.6	38	6	10	[14]
TSPC Adder 1	3.3	42	4	12	Fig.17
TSPC Adder 2	1.8	50	3	13	Fig.18
TSPC Adder 3	1.5	54	4	16	Fig.19

delay will be around 1.2 ns in our simulation using the TSPC-2 circuit with a unit inverter load. A fast serial adder close to the speed limitation is presented in Fig. 19. First, we can see the two-input static AND and OR gates as extensions of the driving stages. The maximum delay variation of these two gates with different input combinations is less than 1.1 ns and this can be reduced by scaling. The delay variation must be less than half the clock period to guarantee that the inputs only change during the precharge phase. The two input gates can also be precharge circuits with more transistors but less input time constraints. Compared with normal, the inputs should precede the evaluation phase with an average delay time caused by the two gates. Second, the inverse carry signal offers both the input of next bit and the sum logic of present bit. Finally, in the critical path, the carry feedback path, two P-latches (three transistors each) are used in parallel for increasing driving capabilities. The worst delay of this adder is about 1.5 ns.

We summarize the above discussion in Table I, where we have also introduced the results from [14] but with our parameters. In [14], normal full adders, not serial full adders, are discussed. In Table I, we have converted them into serial full adders and resimulated. The worst delay is defined as the delay time at the 50-percent level in the critical path. The input load is calculated as the largest number of transistors connected to an input. The clock load is calculated as the number of transistors to which the clock driver is connected since the silicon area is not only related to the number of transistors of the circuit itself but also to the clock driver. The output loads of all these adders are the same, a unit inverter. Note that the domino technique, mentioned in [14], is not included here since this kind of technique means larger logic depth and, therefore, lower clock rate.

Compared with the adder in Fig. 16(a), the adder in Fig. 17 has a speed increase of a factor of 1.86. The critical delay of the adder in Fig. 17 is the carry feedback path, which is about 3.3 ns since a P-block is used. In Fig. 17, the initial delay is half the clock cycle. If an initial delay of one clock cycle is accepted, we can place all of the main logic part into N-blocks and reach higher speed. Fig. 18 shows such a serial full adder combining the principles described above. Note that the P-latches offer complementary signals in an efficient way and that only the sum output N-block is connected in type TSPC-2, while the other two are connected in type TSPC-1 for the delay requirements of the P-latches.

The adder in Fig. 18 has a critical delay of 1.8 ns and it can still be improved by further reducing the logic depth. Since a two-input OR gate or AND gate has the minimum logic depth, it plus a latch stage will determine the ultimate speed of a pipelined logic circuit, which can be seen as a combination of these two gates. Of course, they should be arranged in N-blocks and leave P-blocks as passing stages. In such an arrangement, the AND gate with two N-transistors in series will be critical. If we do so, the ultimate

IV. DEVICE SIZING

The speed of CMOS circuits depends in a complex way on the sizes of all devices used, as the size of each device controls both its current capability and its capacitance [6]. Speed optimization by device sizing has therefore been discussed in several papers [15]–[19]. It is assumed that each device uses its smallest gate length (given by the process used), whereas its gate width is optimized for

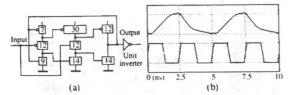

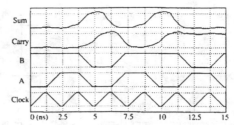

Fig. 20. Scaled eight-transistor divide-by-two circuit: (a) scaled circuit, and (b) input and output at 400 MHz.

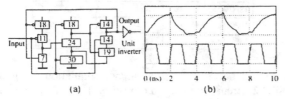

Fig. 21. Scaled nine-transistor divide-by-two circuit: (a) scaled circuit, and (b) input and output at 500 MHz.

Fig. 22. Inputs and outputs of the scaled fast adder (Fig. 19) at a clock rate of 400 MHz.

TABLE II
COMPARISON OF SCALING

Performance / Circuit	Maximum clock rate or worst delay		Sum of widths		No. of input load		No. of clock load		Note
	Unscaled	Scaled	U.	S.	U.	S.	U.	S.	
Divider 1	250 MHz	400 MHz	56	110	2	3.4			Fig.20
Divider 2	250 MHz	400 MHz	63	112	4	9.6			Fig.13c
Divider 3	330 MHz	500 MHz	63	155	4	10.4			Fig.21
TSPC Adder 1	3.3 ns	2.5 ns	294	576	4	7.9	12	28.2	Fig.17
TSPC Adder 2	1.8 ns	1.3 ns	350	820	3	6.0	13	46	Fig.18
TSPC Adder 3	1.5 ns	1.1 ns	378	996	4	14.4	16	54.6	Fig.19

speed. Often, the speed versus device size function does not show an optimum but is monotonic. In such cases cost (as used area) also must be taken into account. For simple cases it is possible to obtain analytical results. It is thus well known that an inverter chain driving a large load can be optimized by device sizing [15], so that each inverter has devices which are about three times larger than the previous inverter. It is also known that an optimal transistor chain connected to ground or supply should be tapered [17]. For more complex circuits it is not possible to obtain analytical results. The effect of the width of a certain transistor depends on the position of the transistor in the network and on the sizes of all other transistors in the network. Its width may affect different delays both through its effect on driving force and self-loading and through its effect on loading of the previous stages. Thus, in some cases we may have critical loops (as in the case of the carry in our serial adder), which means that the loop delay must be analyzed rather than a simple logic delay. In these cases improvements in speed can be obtained by optimization "by hand," by trying different device sizes in a circuit or timing simulator (like SPICE or TMODS [21] [20]. A scaled version of the adder in Fig. 17 is obtained in this way, which is described in [20] and indicated in Fig. 17.

Recently, several computer tools have been developed for automatic optimization through device sizing [16], [18]–[20]. We have used SLOP [20] for this purpose. SLOP is based on a switch-level simulator, TMODS [21], and uses normal switch-level simulation for delay calculation. It can therefore easily handle any kind of CMOS circuit, including circuits with critical loops. The delay calculation algorithm in TMODS is based on the "Elmore delay" without side branches [22] and is similar to the algorithm in CRYSTAL [23].

With the help of the tool SLOP and the confirmation by SPICE simulation, after device sizing, the speed of the circuits described in earlier sections has been increased. Figs. 20 and 21 show the scaled versions of the divided-by-two circuits in Figs. 14(b) and 13(c) and the output waveforms with input signals with frequencies of 400 and 500

MHz, respectively, which are simulated with unit inverter loads. Note that all numbers in boxes are transistor widths in micrometers and that the divider in Fig. 21 has been changed to a type TSPC-2 connection. Since the unscaled version of Fig. 14(b) has an input capacitance equal to a unit inverter and an accepted clock rate of 250 MHz, it then can be put after these scaled version dividers and form ripple counters with the same working frequencies.

For the three adders described in the last section, the scaled sizes are already indicated in the corresponding figures. The worst delays have been reduced to 2.5, 1.3, and 1.1 ns, respectively. Fig. 22 shows the inputs and outputs of the scaled fast adder, TSPC adder 3, at a clock rate of 400 MHz. Note that the sum results from the inputs of the last clock cycle while the carry results from the inputs of the present clock cycle. In this simulation we have used input signals with large slopes (1 ns) to demonstrate the robustness of the circuit technique.

Table II is a comparison of these scaling results. As mentioned in the previous section, the ultimate delay for the pipelined circuit with a minimum logic depth is about 1.2 ns for an unscaled circuit. For a scaled circuit, generally, an improvement of a factor of 1.5 is expected so the ultimate delay will be around 0.8 ns. In principle, this is the maximum speed which can be reached in the 3-μm CMOS process with a 5-V power supply.

V. EXPERIMENTAL RESULTS

In order to verify the above results we designed and fabricated several test circuits. At the moment, three of them are available for testing. The main techniques, i.e., the true single-phase-clock technique and the effective D-latch structure, have been proven experimentally. The

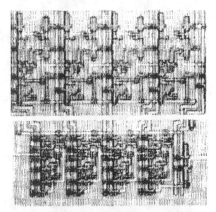

Fig. 23. Layout photograph of two ripple counters.

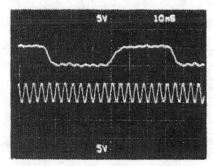

Fig. 24. Photograph of input and output waveforms.

first circuit is a 4-bit register which consists of four D latches (Fig. 13(a) plus an inverter) and the input data are taken from a divider. The chip area is 0.07×0.4 mm and the measured maximum clock rate is more than 230 MHz (limited by the instrumentation). The second circuit is a 4-bit ripple counter which is constructed by successively connecting four divide-by-two circuits as already shown in Fig. 13(c). The input frequency will be divided by 16 so we can measure the output at a lower frequency because it is found that the ordinary output stage and bonding pad are not fit for so high frequencies. Fig. 23 shows a photograph of the layouts of two such ripple counters: one is the unscaled version, chip area 0.28×0.14 mm, and the other is the scaled version, chip area 0.36×0.18 mm.

Since they are directly cascaded, the load for each circuit is more than a unit inverter and the resulting maximum working frequency is then 166 and 250 MHz for these two versions according to SPICE simulations. The measured results are 160 MHz for the unscaled version and more than 230 MHz (limited by the instrumentation) for the scaled version, which are the average results of ten samples from each. Fig. 24, a photograph taken from a sampling oscilloscope, shows the scaled divider successfully working at an input frequency of 230 MHz.

The third circuit is the serial full adder shown in Fig. 17. The chip area of the unscaled one is 0.22×0.16 mm and of the scaled one is 0.22×0.24 mm. The measured maximum

clock rates are 140 and 225 MHz, respectively, which again are the averages from ten samples each.

VI. CONCLUSIONS

1) The true single-phase-clocking strategy has the advantages of simple and compact clock distribution, high speed, and logic design flexibility. In general terms, no clock-skew problem exists in this clocking strategy. The "skew" caused by clock delay between different logic blocks in a large system can be minimized by reverse clock distribution and the D-latch structure.

2) Since the maximum clock frequency in a clocked system is limited by the delay from one latching instant to the next, logic depth minimization in the system will lead to high speed with only a limited increase of the number of latch stages and initial delay.

3) Device sizing will give a significant increase of speed at the cost of increased area. The experimental results from two examples show that the practical ratios between speed and area increases are 1.43/1.65 for a 4-bit ripple counter and 1.60/1.50 for a serial full adder.

4) Both analysis and experiment have demonstrated that clock rates in excess of 200 MHz are feasible in a 3-μm CMOS process by the combination of the above techniques. Some of the circuits possibly reach even the 400–500-MHz range as shown by SPICE simulations.

REFERENCES

[1] C. Svensson, "VLSI physics," *Integration*, vol. 1, pp. 3–19, 1983.
[2] R. W. Keyes, "Physical limits in digital electronics," *Proc. IEEE*, vol. 63, p. 740, 1975.
[3] H. Masuda, M. Nakai, and M. Kubo, "Characteristics and limitation of scaled-down MOSFET's due to two-dimensional field effect," *IEEE Trans. Electron Devices*, vol. ED-26, pp. 980–986, 1979.
[4] C. Seitz, "System timing," in *Introduction to VLSI Systems*, C. Mead and L. Conway, Eds. Reading, MA: Addison-Wesley, 1980, ch. 7.
[5] Y. Suzuki, K. Odagawa, and T. Abe, "Clocked CMOS calculator circuitry," *IEEE J. Solid-State Circuits*, vol. SC-8, pp. 462–469, 1973.
[6] N. Weste and K. Eshraghian, *Principles of CMOS VLSI Design*. Reading, MA: Addison-Wesley, 1985, ch. 5.
[7] M. Shoji, "Electrical design of BELLMAC-32A microprocessor," in *Proc. IEEE Int. Conf. Circuits Comput.*, 1982, pp. 112–115.
[8] N. Goncalves and H. J. De Man, "NORA: A racefree dynamic CMOS technique for pipelined logic structures," *IEEE J. Solid-State Circuits*, vol. SC-18, pp. 261–266, 1983.
[9] Y. Ji-ren, I. Karlsson, and C. Svensson, "A true single phase clock dynamic CMOS circuit technique," *IEEE J. Solid-State Circuits*, vol. SC-22, pp. 899–901, 1987.
[10] "Layout design rules for 3.0 μm P-well CMOS," VTI Technology Inc., San Jose, CA.
[11] I. Karlsson, "True single phase clock dynamic CMOS circuit technique," in *Proc. 1988 IEEE Int. Symp. Circuits Syst.*, vol. 1, pp. 475–478.
[12] C. Svensson, "Signal resynchronization in VLSI systems," *Integration*, vol. 4, pp. 75–80, 1986.
[13] G. De Micheli, "Performance-oriented synthesis of large-scale domino CMOS circuits," *IEEE Trans. Computer-Aided Des.*, vol. CAD-6, pp. 751–765, 1987.
[14] K. M. Chu and D. L. Pulfrey, "A comparison of CMOS circuit techniques: Differential cascode voltage switch logic versus conventional logic," *IEEE J. Solid-State Circuits*, vol. SC-22, pp. 528–532, 1987.
[15] E. T. Lewis, "Optimization of device area and overall delay for CMOS VLSI designs," *Proc. IEEE*, vol. 72, pp. 670–689, 1984.
[16] M. D. Matson and L. A. Glasser, "Macromodeling and optimization of digital MOS VLSI circuits," *IEEE Trans. Computer-Aided Des.*, vol. CAD-5, pp. 659–678, 1986.

[17] M. Shoji, "FET scaling in domino CMOS gates," *IEEE J. Solid-State Circuits*, vol. SC-20, pp. 1067–1071, 1985.

[18] K. S. Hedlund, "Aesop: A tool for automated transistor sizing," in *Proc. 24th ACM/IEEE Design Automation Conf.*, 1987, paper 7.1, pp. 114–120.

[19] M. A. Cirit, "Transistor sizing in CMOS circuits," in *Proc. 24th ACM/IEEE Design Automation Conf.*, 1987, paper 7.2, pp. 121–124.

[20] J. Yuan and C. Svensson, "CMOS circuit speed optimization based on switch level simulation," in *Proc. 1988 IEEE Int. Symp. Circuits Syst.*, vol. 3, pp. 2109–2112.

[21] R. Sundblad and C. Svensson, "Fully dynamic switch level simulation of CMOS circuits," *IEEE Trans. Computer-Aided Des.*, vol. CAD-6, pp. 282–289, 1987.

[22] J. Rubinstein, P. Penfield, Jr., and M. A. Horowitz, "Signal delay in RC tree networks," *IEEE Trans. Computer-Aided Des.*, vol. CAD-2, pp. 202–210, 1983.

[23] J. K. Ousterhout, "A switch-level timing verifier for digital MOS VLSI," *IEEE Trans. Computer-Aided Des.*, vol. CAD-4, pp. 336–349, 1985.

Flow-Through Latch and Edge-Triggered Flip-flop Hybrid Elements

HAMID PARTOVI, ROBERT BURD, UDIN SALIM, FREDERICK WEBER,
LUIGI DIGREGORIO, AND DONALD DRAPER

NEXGEN, INC., MILPITAS, CA

This paper describes a hybrid latch-flipflop (HLFF) timing methodology aimed at a substantial reduction in latch latency and clock load. A common principle is employed to derive consistent latching structures for static logic, dynamic domino and self-resetting logic [1].

HLFF is similar to standard flip-flops in that it samples the data on one edge of the clock and thus eliminates a retardation of data flow on the opposite edge. It is similar to latches because it can provide a soft clock edge which allows for slack passing and minimizes the effects of clock skew on cycle time. At an operating frequency of 500MHz, and presenting half the capacitive load to the clock tree, its latency is about two-thirds the aggregate delays of a transparent low (TLL) and a transparent high (THL) latch. As a result, an improvement of at least 10% in cycle time, in addition to a reduction of about 30% in overall clock load, is achieved.

Flip-flops are commonly designed by cascading TLL and THL latches. To avoid an internal race, a delay element must be inserted between the master and the slave elements. HLFF, on the contrary, operates on a different principle. It is a latch with a brief transparency period. The duration of this period is determined by an integrated one-shot derived from the clock edge.

Figure 1 shows a variation of HLFF that can be modified to a storage element for dynamic circuits. Referring to Figure 1, prior to the rising edge of the clock, N1 and N4 are off while N3 and N6 (both gated by CKDB) in addition to P1 are on. As a result, node X is precharged to VDD and node Q (decoupled from X) holds the previous data. At the rising edge of the clock, N1 and N4 turn on while N3 and N6 stay on for a period determined by the inverter delay chain (I1-I3). It is in this period that the circuit is transparent and data at D can be sampled into the latch. Once CKDB transitions low, node X is decoupled from D and is either held at or begins to precharge to VDD by P3. At the falling edge of the clock, P1 fully precharges or holds X at VDD as long as the clock remains low. HLFF waveforms for data transitions of 1→0 and 0→1 are illustrated in Figure 2. The results were obtained in a 2.5V, 0.35μm technology at 2.3V, 85°C with typical devices.

The transparency period of HLFF also determines its hold time. While it is desirable to minimize hold time, the transparency period should be long enough to allow data to propagate to Q. As is seen from Figure 2, CKDB transitions low about 240ps after the rising edge of the clock, hence a hold time of 240ps. Further, both nodes X and Q evaluate 100ps prior to the falling edge of CKDB. This not only provides sufficient safety margin, but indicates a negative setup-time of the same magnitude. The flip-flop latency of 340ps is 100ps longer than its hold time; however, when variations in output load and clock skew were considered, it was required that two flip-flops be separated by at least three logic gates to eliminate race-through.

The negative setup-time of HLFF illustrates an attractive latch attribute known as the soft-clock edge. It allows a critical path to borrow time from the next stage. The soft edge may alternatively be thought of as the ability to overcome a loss of useful cycle time due to clock skew; with a negative setup-time of 100ps, this design tolerates a clock skew of up to 100ps with minimal impact in cycle time.

Level-sensitive latch pairs used in two high-performance processors (Figure 3) are used to assess the performance of HLFF [2, 3]. All were designed and optimized in the technology described earlier. Simulations were performed with equal output loads under two conditions: 1) equal aggregate clock load for latches and HLFF; 2) HLFF having half the clock load. Table 1 summarizes the aggregate latencies of HLFF and the latch pairs. It also includes the percentage gain in clock frequency for a 500MHz processor using latch pair type 2 when replaced by the other latch elements in the table. As can be seen from the table, HLFF designed at half the clock load (HLFF 1/2) can increase the operating frequency by 10% solely due to lower latency. The overall improvement is, however, larger due to three factors: 1) the reduction in clock load invariably reduces clock skew; 2) a percentage of clock skew is absorbed by the soft edge; and, 3) the retardation in data flow, as is the case with latch pairs, is eliminated at the falling edge of the clock.

In developing latch-flipflop libraries, it is desirable to design elements which incorporate logic functions. In addition, conditionally enabled flip-flops are useful as they provide the ability to deactivate functional blocks which are not used. Figure 4 is the circuit diagram for a conditional flip-flop implemented in HLFF.

Dynamic domino circuits are commonly used when it is not possible to meet timing requirements with static logic. A TLL must precede a dynamic block which is evaluated when clock is high. As an example, Figure 5 comprises a dynamic XOR function fed by a type 2 TLL. As is seen, the pulldown network must be conditioned to the clock because one of the latch outputs will be high during precharge.

Figure 6 illustrates the HLFF-based dual-rail storage element for dynamic domino circuits (DHLFF). It replaces the TLL preceding the dynamic blocks. Referring to the figure, both DHLFF outputs are pre-discharged to ground when the clock is low. At the rising edge of the clock and depending on the state of data at D, either QFH or QFL will be asserted. The outputs are held statically as long as the clock remains high. Unlike the TLL implementation, the XOR pulldown network of Figure 5 need not be conditioned to the clock as DHLFF outputs are both discharged when clock is low.

To evaluate DHLFF, its clock-to-output delay was compared with the setup requirement of TLL. The timing waveforms are included in Figure 7. A setup-time of 400ps was required for TLL such that its deselected output had a clock crossover below a threshold. The latency of DHLFF was only 280ps. Similar to its static counterpart, DHLFF latency is about two-thirds that of TLL.

A single-ended example for self-resetting circuits completes the suite of latching elements for different logic styles (Figure 8). The circuit samples the data on the rising edge of the clock and is reset through PR1 after a predetermined delay period which can be made as long as desired while keeping the hold time consistent with the other HLFF structures.

References:

[1] Chappell, T. I., et al., "A 2ns Cycle, 3.8ns Access 512kb CMOS ECL SRAM with a Fully Pipelined Architecture," J. Solid-State Circuits, Nov., 1991.

[2] Dobberpuhl, D. W., et al., "A 200MHz 64b Dual-Issue CMOS Microprocessor," J. Solid-State Circuits, Nov., 1992.

Reprinted from *Proceedings of 1996 International Solid-State Circuits Conference*, pp. 138-139, February 1996.

[3] Bowhill, W. J., et al., "A 300MHz 64b Quad-Issue CMOS RISC Microprocessor," ISSCC Digest of Technical Papers, pp. 182-183, Feb., 1995.

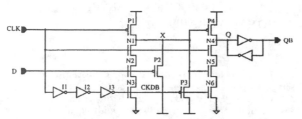

Figure 1: Basic HLFF circuit.

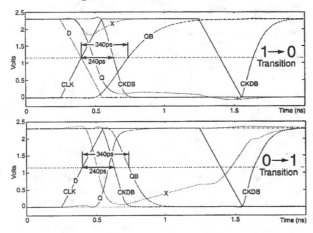

Figure 2: Waveforms for HLFF (Tsu = 0ns).

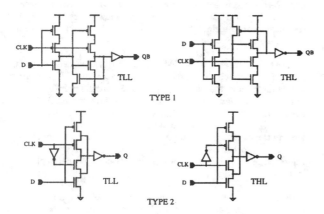

TYPE 1

TYPE 2

Figure 3: Level-sensitive latch pairs.

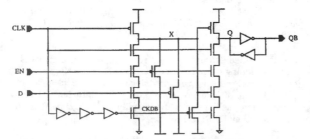

Figure 4: HLFF incorporating ENABLE function.

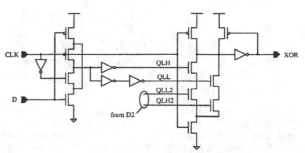

Figure 5: Dynamic XOR circuit driven by TLL type 2.

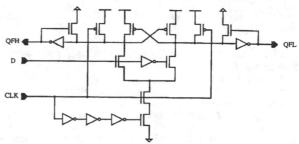

Figure 6: Dual-rail dynamic HLFF.

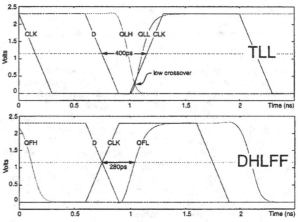

Figure 7: Waveforms of TLL and DHLFF.

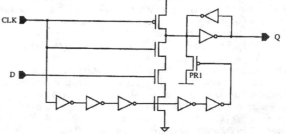

Figure 8: Single-ended dynamic HLFF for self-resetting logic.

	HLFF	HLFF 1/2	Latch pair type 2	Latch pair type 1
Latency(ps)	270	340	530	750
Gain(%)	14	10	---	-10

Table 1: Latch latency and gain in operating frequency.

A 200-MHz 64-b Dual-Issue CMOS Microprocessor

Daniel W. Dobberpuhl, *Member, IEEE*, Richard T. Witek, Randy Allmon, Robert Anglin, David Bertucci, Sharon Britton, Linda Chao, *Member, IEEE*, Robert A. Conrad, Daniel E. Dever, Bruce Gieseke, Soha M. N. Hassoun, Gregory W. Hoeppner, *Member, IEEE*, Kathryn Kuchler, *Member, IEEE*, Maureen Ladd, *Member, IEEE*, Burton M. Leary, Liam Madden, Edward J. McLellan, Derrick R. Meyer, James Montanaro, Donald A. Priore, Vidya Rajagopalan, Sridhar Samudrala, and Sribalan Santhanam

Abstract—A 400-MIPS/200-MFLOPS (peak) custom 64-b VLSI CPU chip is described. The chip is fabricated in a 0.75-μm CMOS technology utilizing three levels of metalization and optimized for 3.3-V operation. The die size is 16.8 mm $\times$ 13.9 mm and contains 1.68M transistors. The chip includes separate 8-kilobyte instruction and data caches and a fully pipelined floating-point unit (FPU) that can handle both IEEE and VAX standard floating-point data types. It is designed to execute two instructions per cycle among scoreboarded integer, floating-point, address, and branch execution units. Power dissipation is 30 W at 200-MHz operation.

I. INTRODUCTION

A RISC-style microprocessor has been designed and tested that operates up to 200 MHz. The chip implements a new 64-b architecture, designed to provide a huge linear address space and to be devoid of bottlenecks that would impede highly concurrent implementations. Fully pipelined and capable of issuing two instructions per clock cycle, this implementation can execute up to 400 million operations per second. The chip includes an 8-kilobyte I-cache, 8 kilobyte D-cache and two associated translation look-aside buffers, a four-entry 32-byte/entry write buffer, a pipelined 64-b integer execution unit with a 32-entry register file, and a pipelined floating-point unit (FPU) with an additional 32 registers. The pin interface includes integral support for an external secondary cache. The package is a 431-pin PGA with 140 pins dedicated to V_{DD}/V_{SS}. The chip is fabricated in 0.75-μm n-well CMOS with three layers of metalization. The die measures 16.8 mm $\times$ 13.9 mm and contains 1.68 million transistors. Power dissipation is 30 W from a 3.3-V supply at 200 MHz.

II. CMOS PROCESS TECHNOLOGY

The chip is fabricated in a 0.75-μm 3.3-V n-well CMOS process optimized for high-performance microprocessor design. Process characteristics are shown in Table I. The thin gate oxide and short transistor lengths result in the fast transistors required to operate at 200 MHz. There are no explicit bipolar devices in the process as the incremental process complexity and cost were deemed too large in comparison to the benefits provided—principally more area-efficient large drivers such as clock and I/O.

The metal structure is designed to support the high operating frequency of the chip. Metal 3 is very thick and has a relatively large pitch. It is important at these speeds to have a low-resistance metal layer available for power and clock distribution. It is also used for a small set of special signal wires such as the data buses to the pins and the control wires for the two shifters. Metal 1 and metal 2 are maintained at close to their maximum thickness by planarization and by filling metal 1 and metal 2 contacts with tungsten plugs. This removes a potential weak spot in the electromigration characteristics of the process and allows more freedom in the design without compromising reliability.

III. ALPHA ARCHITECTURE

The computer architecture implemented is a 64-b load/store RISC architecture with 168 instructions, all 32 b wide [1]. Supported data types include 8-, 16-, 32-, and 64-b integers and 32- and 64-b floats of both DEC and IEEE formats. The two register files, integer and floating point, each contains 32 entries of 64 b with one entry in each being a hardwired zero. The program counter and virtual address are 64 b. Implementations can subset the virtual address size but are required to check the full 64-b address for sign extension. This insures that when later implementations choose to support a larger virtual address, programs will still run and not find addresses that have dirty bits in the previously "unused" bits.

Manuscript received April 13, 1992; revised June 23, 1992.

D. W. Dobberpuhl, R. Allmon, R. Anglin, S. Britton, L. Chao, R. A. Conrad, D. E. Dever, B. Gieseke, K. Kuchler, M. Ladd, B. M. Leary, L. Madden, E. J. McLellan, D. R. Meyer, J. Montanaro, D. A. Priore, V. Rajagopalan, S. Samudrala, and S. Santhanam are with the Semiconductor Engineering Group, Digital Equipment Corporation, Hudson, MA 01749.

R. T. Witek was with the Semiconductor Engineering Group, Digital Equipment Corporation, Hudson, MA 01749. He is now with Apple Computer, Inc., Austin, TX.

D. Bertucci was with the Semiconductor Engineering Group, Digital Equipment Corporation, Hudson, MA 01749. He is now with Sun Microsystems Inc., Mountain View, CA.

S. M. N. Hassoun was with the Semiconductor Engineering Group, Digital Equipment Corporation, Hudson, MA 01749. She is now with the Computer Systems Engineering Department, University of Washington, Seattle, WA 98195.

G. W. Hoeppner was with the Semiconductor Engineering Group, Digital Equipment Corporation, Hudson, MA 01749. He is now with the Advanced Workstation Division, IBM Corporation, Austin, TX.

IEEE Log Number 9202810.

Reprinted from *IEEE Journal of Solid-State Circuits*, Vol. 27, No. 11, pp. 1555–1564, November 1992.

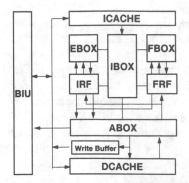

Fig. 1. CPU chip block diagram.

The architecture is designed to support high-speed multi-issue implementations. To this end the architecture does not include condition codes, instructions with fixed source or destination registers, or byte writes of any kind (byte operations are supported by extract and merge instructions within the CPU itself). Also, there are no first-generation artifacts that are optimized around today's technology, which would represent a long-term liability to the architecture.

IV. CHIP MICROARCHITECTURE

The block diagram (Fig. 1) shows the major functional blocks and their interconnecting buses, most of which are 64 b wide. The chip implements four functional units: the integer unit (IRF + EBOX), the floating-point unit (FRF + FBOX), the load/store unit (ABOX), and the branch unit (distributed). The bus interface unit (BIU), described in the next section, handles all communication between the chip and external components. The microphotograph (Fig. 2) shows the boundaries of the major functional units. The dual-issue rules are a direct consequence of the register-file ports, the functional units, and the I-cache interface. The integer register file (IRF) has two read ports and one write port dedicated to the integer unit, and two read and one write port shared between the branch unit and the load store unit. The floating-point register file (FRF) has two read ports and one write port dedicated to the floating unit, and one read and one write port shared between the branch unit and the load store unit. This leads to dual-issue rules that are quite general:

- any load/store in parallel with any operate,
- an integer op in parallel with a floating op,
- a floating op and a floating branch,
- an integer op and an integer branch,

except that integer store and floating operate and floating store and integer operate are disallowed as pairs.

As shown in Fig. 3(a), the integer pipeline is seven stages deep, where each stage is a 5-ns clock cycle. The first four stages are associated with instruction fetching, decoding, and scoreboard checking of operands. Pipeline stages 0 through 3 can be stalled. Beyond 3, however, all pipeline stages advance every cycle. Most ALU operations complete in cycle 4 allowing single-cycle latency, with the shifter being the exception. Primary cache ac-

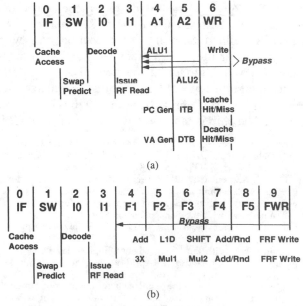

Fig. 2. Chip microphotograph.

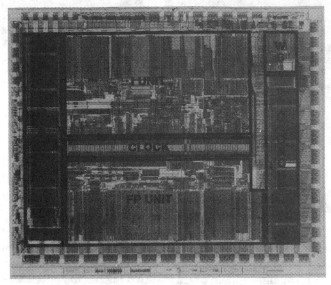

(a)

(b)

Fig. 3. (a) Integer unit pipeline timing. (b) Floating-point unit pipeline timing.

cesses complete in cycle 6, so cache latency is three cycles. The chip will do hits under misses to the primary DCACHE.

374

The ISTREAM is based on autonomous prefetching in cycles 0 and 1 with the final resolution of ICACHE hit not occurring until cycle 5. The prefetcher includes a branch history table and a subroutine return stack. The architecture provides a convention for compilers to predict branch decisions and destination addresses including those for register indirect jumps. The penalty for branch mispredict is four cycles.

The floating-point unit is a fully pipelined 64-b floating-point processor that supports both VAX standard and IEEE standard data types and rounding modes. It can generate a 64-b result every cycle for all operations except division. As shown in Fig. 3(b), the floating-point pipeline is identical and mostly shared with the integer pipeline in stages 0 through 3, however, the execution phase is three cycles longer. All operations, 32 and 64 b, (except division) have the same timing. Division is handled by a nonpipelined, single bit per cycle, dedicated division unit.

In cycle 4, the register file data are formatted to fraction, exponent, and sign. In the first-stage adder, exponent difference is calculated and a $3 \times$ multiplicand is generated for multiplies. In addition, a predictive leading 1 or 0 detector using the input operands is initiated for use in result normalization. In cycles 5 and 6, for add/subtract, alignment or normalization shift and sticky-bit calculation are performed. For both single- and double-precision multiplication, the multiply is done in a radix-8 pipelined array multiplier. In cycles 7 and 8, the final addition and rounding are performed in parallel and the final result is selected and driven back to the register file in cycle 9. With an allowed bypass of the register write data, floating-point latency is six cycles.

The CPU contains all the hardware necessary to support a demand paged virtual memory system. It includes two translation look-aside buffers to cache virtual to physical address translations. The instruction translation buffer contains 12 entries, eight that map 8-kilobyte pages and four that map 4-megabyte pages. The data translation buffer contains 32 entries that can map 8-kilobyte, 64-kilobyte, 512-kilobyte or 4-megabyte pages.

The CPU supports performance measurement with two counters that accumulate system events on the chip such as dual-issue cycles and cache misses or external events through two dedicated pins that are sampled at the selected system clock speed.

V. External Interface

The external interface (Fig. 4) is designed to directly support an off-chip backup cache that can range in size from 128 kilobytes to 16 megabytes and can be constructed from standard SRAM's. For most operations, the CPU chip accesses the cache directly in a combinatorial loop by presenting an address and waiting N CPU cycles for control, tag, and data to appear, where N is a mode-programmable number between 3 and 16 set at power-up time. For writes, both the total number of cycles and the

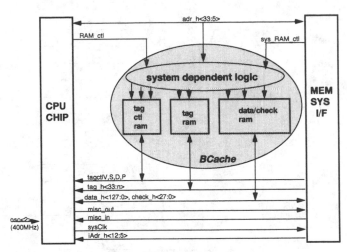

Fig. 4. CPU external interface.

duration and position of the write signal are programmable in units of CPU cycles. This allows the system designer to select the size and access time of the SRAM's to match the desired price/performance point.

The interface is designed to allow all cache policy decisions to be controlled by logic external to the CPU chip. There are three control bits associated with each back-up cache (BCACHE) line: valid, shared, and dirty. The chip completes a BCACHE read as long as valid is true. A write is processed by the CPU only if valid is true and shared is false. When a write is performed the dirty bit is set to true. In all other cases, the chip defers to an external state machine to complete the transaction. This state machine operates synchronously with the SYS_CLK output of the chip, which is a mode-controlled submultiple of the CPU clock rate ranging from divide by 2 to divide by 8. It is also possible to operate without a back-up cache.

As shown in the diagram, the external cache is connected between the CPU chip and the system memory interface. The combinatorial cache access begins with the desired address delivered on the *adr_h* lines and results in ctl, tag, data, and check bits appearing at the chip receivers within the prescribed access time. In 128-b mode, BCACHE accesses require two external data cycles to transfer the 32-byte cache line across the 16-byte pin bus. In 64-b mode, it is four cycles. This yields a maximum backup cache read bandwidth of 1.2 gigabyte/s and a write bandwidth of 711 megabyte/s. Internal cache lines can be invalidated at the rate of one line/cycle using the dedicated invalidate address pins, *iAdr_h* $\langle 12:5 \rangle$.

In the event external intervention is required, a request code is presented by the CPU chip to the external state machine in the time domain of the SYS_CLK as described previously. Fig. 5 shows the read miss timing where each cycle is a SYS_CLK cycle. The external transaction starts with the address, the quadword within block and instruction/data indication supplied on the *cWMask_h* pins, and READ_BLOCK function supplied on the *cReq_h* pins. The external logic returns the first 16 bytes of data on the *data_h* and ecc or parity on the *check_h* pins. The CPU latches the data based on receiving acknowledgment on

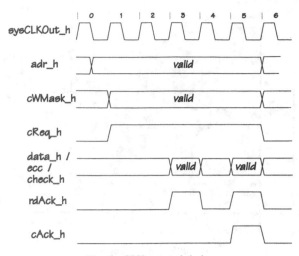

Fig. 5. CPU external timing.

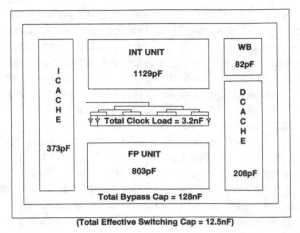

Fig. 6. Clock load distribution.

rdAck_H. The diagram shows a stall cycle (cycle 4) between the request and the return data; this depends on the external logic and could range from zero to many cycles. The second 16 bytes of data are returned in the same way with *rdAck_h* signaling the return of the data and *cAck_h* signaling the completion of the transaction. *cReq_h* returns to idle and a new transaction can start at this time.

The chip implements a novel set of optional features supporting chip and module test. When the chip is reset, the first attempted action is to read from a serial read-only memory (SROM) into the I-cache via a private three-wire port. The CPU is then enabled and the PC is forced to 0. Thus, with only three functional components (CPU chip, SROM, and clock input) a system is able to begin executing instructions. This initial set of instructions is used to write the bus control registers inside the CPU chip to set the cache timing and to test the chip and module from the CPU out. After the SROM loads the I-cache, the pins used for the SROM interface are enabled as serial-in and out ports. These ports can be used to load more data or to return status of testing and setup.

VI. CIRCUIT IMPLEMENTATION

Many novel circuit structures and detailed analysis techniques were developed to support the clock rate in conjunction with the complexity demanded by the concurrence and wide data paths. The clocking method is single wire level sensitive. The bus interface unit operates from a buffered version of the main clock. Signals that cross this interface are deskewed to eliminate races. This clocking method eliminates dead time between phases and requires only a single clock signal to be routed throughout the chip.

One difficulty inherent in this clocking method is the substantial load on the clock node, 3.25 nF in our design. This load and the requirement for a fast clock edge led us to take particular care with clock routing and to do extensive analysis on the resulting grid. Fig. 6 shows the distribution of clock load among the major functional units. The clock drives into a grid of vertical metal 3 and hori-

zontal metal 2. Most of the loading occurs in the integer and floating-point units that are fed from the most robust metal 3 lines. To ensure the integrity of the clock grid across the chip, the grid was extracted from the layout and the resulting network, which contained 63 000 *RC* elements, was simulated using a circuit simulation program based on the AWEsim simulator from Carnegie-Mellon University. Fig. 7 shows a three-dimensional representation of the output of this simulation and shows the clock delay from the driver to each of the 63 000 transistor gates connected to the clock grid.

The 200-MHz clock signal is fed to the driver through a binary fanning tree with five levels of buffering. There is a horizontal shorting bar at the input to the clock driver to help smooth out possible asymmetry in the incoming wavefront. The driver itself consists of 145 separate elements each of which contains four levels of prescaling into a final output stage that drives the clock grid.

The clock driver and predriver represent about 40% of the total effective switching capacitance determined by power measurement to be 12.5 nF (worst case including output pins). To manage the problem of *di/dt* on the chip power pins, explicit decoupling capacitance is provided on-chip. This consists of thin oxide capacitance that is distributed around the chip, primarily under the data buses. In addition, there are horizontal metal 2 power and clock shorting straps adjacent to the clock generator and the thin oxide decoupling cap under these lines supplies charge to the clock driver. *di/dt* for the driver alone is about 2×10^{11} A/s. The total decoupling capacitance as extracted from the layout measures 128 nF. Thus the ratio of decoupling capacitance to switching cap is about 10:1. With this capacitance ratio, the decoupling cap could supply all the charge associated with a complete CPU cycle with only a 10% reduction in the on-chip supply voltage.

A. Latches

As previously described, the chip employs a single-phase approach with nearly all latches in the core of the chip receiving the clock node, CLK, directly. A repre-

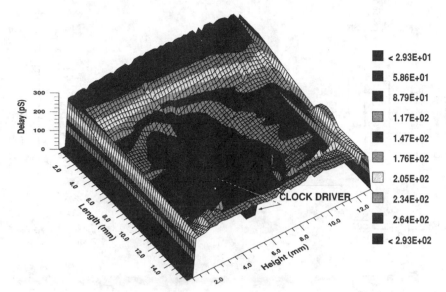

Fig. 7. CPU clock skew.

sentative example is illustrated in Fig. 8. Notice that $L1$ and $L2$ are transparent latches separated by random logic and are not simultaneously active: $L1$ is active when CLK is high and $L2$ is active when CLK is low. The minimum number of delays between latches is zero and the maximum number of delays is constrained only by the cycle time and the details of any relevant critical paths. The bus interface unit, many data-path structures, and some critical paths deviate from this approach and use buffered versions and/or conditionally buffered versions of CLK. The resulting clock skew is managed or eliminated with special latch structures.

The latches used in the chip can be classified into two categories: custom and standard. The custom latches were used to meet the unique needs of data-path structures and the special constraints of critical paths. The standard latches were used in the design of noncritical control and in some data-path applications. These latches were designed prior to the start of implementation and were included in the library of usable elements for logic synthesis. All synthesized logic used only this set of latches.

The standard latches are extensions of previously published work [2] and examples are shown in Figs. 9–11. To understood the operation of these latches refer to Fig. 9(a). When CLK is high, $P1$, $N1$, and $N3$ function as an inverter complementing IN1 to produce X. $P2$, $N2$, and $N4$ function as a second inverter and complement X to produce OUT. Therefore, the structure passes IN1 to OUT. Then CLK is ''low,'' $N3$ and $N4$ are cut off. If IN1, X, and OUT are initially ''high,'' ''low,'' and ''high,'' respectively, a transition of IN1 FALLING pulls X ''high'' through $P1$ causing $P2$ to cut off, which tristates OUT ''high.'' If IN1, X, and OUT are initially ''low,'' ''high,'' and ''low,'' respectively, a transition of IN1 RISING causing $P1$ to cut off, which tristates X ''high'' leaving OUT tristated ''low.'' In both cases, additional transitions of IN1 leave X tristated or driven ''high'' with OUT tristated to its initial value. Therefore,

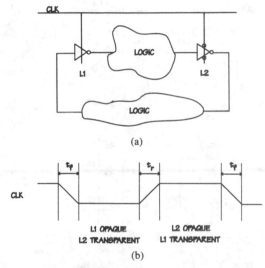

Fig. 8. (a) Latching schema. (b) Latch timing.

the structure implements a latch that is transparent when CLK is ''high'' and opaque when CLK is ''low.'' Fig. 9(c) shows the dual of the latch just discussed; this structure implements a latch that is transparent when CLK is ''low'' and opaque when CLK is ''high.'' Fig. 9(b) and (d) depicts latches with an output buffer used to protect the sometimes dynamic node OUT and to drive large loads.

The design of the standard latches stressed three primary goals: flexibility, immunity to noise, and immunity to race-through. To achieve the desired flexibility, a variety of latches like those in Figs. 9–11 in a variety of sizes were characterized for the implementors. Thus, the designer could select a latch with an optional output buffer and an embedded logic function that was sized appropriately to drive various loads. Furthermore, it was decided to allow zero delay between latches, completely freeing the designer from race-through considerations when designing static logic with these latches.

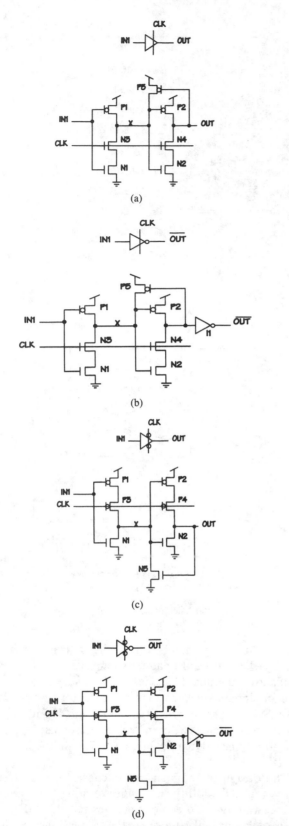

Fig. 9. (a) Noninverting active-high latch. (b) Inverting active-high latch. (c) Noninverting active-low latch. (d) Inverting active-low latch.

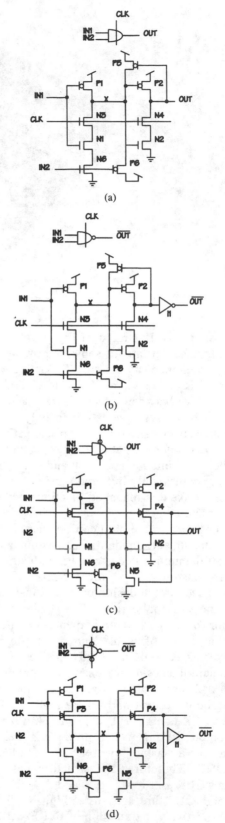

Fig. 10. (a) Two input AND active-high latch. (b) Two-input NAND active-high latch. (c) Two-input AND active-low latch. (d) Two-input NAND active-low latch.

In the circuit methodology adopted for the implementation, only one node, X (Fig. 9(a)), poses inordinate noise margin risk. As noted above, X may be tristated ''high'' with OUT tristated ''low'' when the latch is opaque. This maps into a dynamic node driving into a dynamic gate that is very sensitive to noise that reduces the voltage on X, causing leakage through $P2$, thereby destroying OUT.

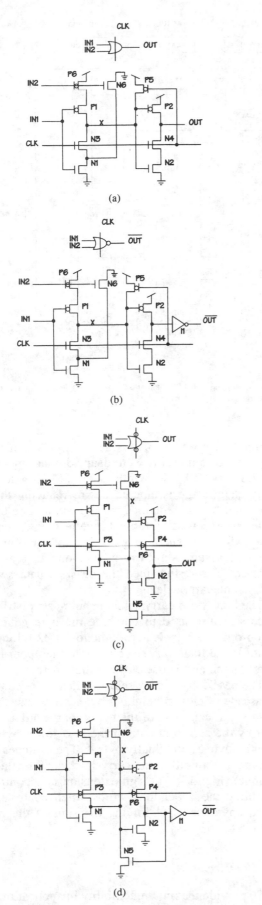

Fig. 11. (a) Two-input OR active-high latch. (b) Two-input NOR active-high latch. (c) Two-input OR active-low latch. (d) Two-input NOR active-low latch.

This problem was addressed by the addition of $P5$. This weak feedback device is sized to source enough current to counter reasonable noise and hold $P2$ in cutoff. $N5$ plays an analogous role in Fig. 9(c).

Race-through was the major functional concern with the latch design. It is aggravated by clock skew, the variety of available latches and the zero delay goal between latches. The clock skew concern was actually the easiest to address. If data propagate in a direction that opposes the propagation of the clock wavefront, clock skew is functionally harmless and tends only to reduce the effective cycle time locally. Minimizing this effect is of concern when designing the clock generator. If data propagate in a direction similar to the propagation of the clock wavefront, clock skew is a functional concern. This was addressed by radially distributing the clock from the center of the chip. Since the clock wavefront moves out radially from the clock driver toward the periphery of the die, it is not possible for the data to overtake the clock if the clock network is properly designed.

To verify the remaining race-through concerns, a mix and match approach was taken. All reasonable combinations of latches were cascaded together and simulated. The simulations were stressed by eliminating all interconnect and diffusion capacitance and by pushing each device into a corner of the process that emphasized race-through. Then many simulations with varying CLK rise and fall times, temperatures, and power supply voltages were performed. The results showed no appreciable evidence of race-through for CLK rise and fall times at or below 0.8 ns. With 1.0-ns rise and fall times, the latches showed signs of failure. To guarantee functionality, CLK was specified and designed to have an edge rate of less than 0.5 ns. This was not a serious constraint since other circuits in the chip required similar edge rates of the clock.

A last design issue worth noting is the feedback devices, $N5$ and $P5$, in Figs. 10(c), 10(d), 11(a), and 11(b). Notice that these devices have their gates tied to CLK instead of OUT like the other latches. This difference is required to account for an effect not present in the other latches. In these latches a stack of devices is connected to node X without passing through the clocked transistors $P3$ or $N3$. Referring to Fig. 11(a), assume CLK is "low," X is "high," and OUT is "low." If multiple random transitions are allowed by IN1 with IN2 "high," then coupling through $P1$ can drive X down by more than a threshold even with weak feedback, thereby destroying OUT. To counter this phenomenon, $P5$ cannot be a weak feedback device and therefore cannot be tied to OUT if the latch is to function properly when CLK is "high." Note that taller stacks aggravate this problem because the devices become larger and there are more devices to participate in coupling. For this reason stacks in these latches were limited to three high. Also, note that clocking $P5$ introduces another race-through path since X will unconditionally go "high" with CLK falling and OUT must be able to retain a stored ONE. So there is a two-sided constraint: $P5$ must be large enough to counter coupling and

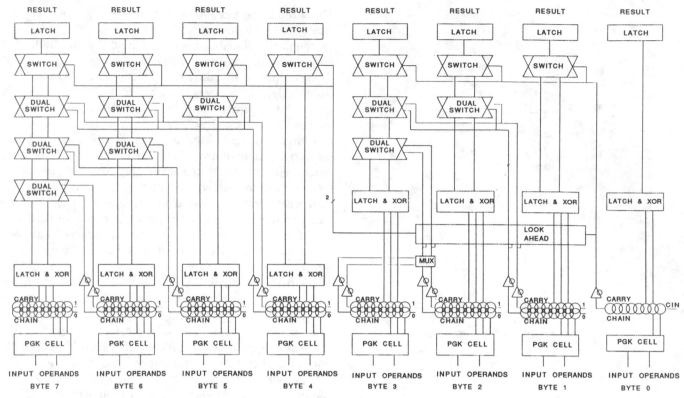

Fig. 12. 64-b adder block diagram.

small enough not to cause race-through. These trade-offs were analyzed by simulation in manner similar to the one outlined above.

B. 64-b Adder

A difficult circuit problem was the 64-b adder portion of the integer and floating-point ALU's. Unlike a previous high-speed design [3], we set a goal to achieve single-cycle latency in this unit. Fig. 12 has an organizational diagram of its structure. Every path through the adder includes two latches, allowing fully pipelined operation. The result latches are shown explicitly in the diagram, however, the input latches are somewhat implicit, taking advantage of the predischarge characteristics of the carry chains. The complete adder is a combination of three methods for producing a binary add: a byte-long carry chain, a long-word (32 b) carry select, and local logarithmic carry select [4]. The carry select is built as a set of nMOS switches that direct the data from byte carry chains. The 32-b long-word lookahead is implemented as a distributed differential circuit controlling the final stage of the upper longword switches. The carry chains are organized in groups of 8 b.

Carry chain width was chosen to implement a byte compare function specified by the architecture. The carry chain implemented with nMOS transistors is shown in Fig. 13(a). Operation begins with the chain predischarged to V_{SS}, with the controlling signal an OR of CLK and the kill function. Evaluation begins along the chain length without the delay associated with the $V_{gs}-V_t$ threshold found in a chain precharged to V_{DD}. An alternative to a predis-

charged state was to precharge to $V_{DD}-V_t$, but the resulting low noise margins were deemed unacceptable. From the LSB to MSB, the width of the nMOS gates for each carry chain stage is tapered down, reducing the loading presented by the remainder of the chain. The local carry nodes are received by ratioed inverters Each set of propagate, kill, and generate signals controls two carry chains, one that assumes a carry-in and one that assumes no carry-in. The results feed the bit-wise data switches as well as the carry selects.

The long-word carry select is built as a distributed cascode structure used to combine the byte generate, kill, and propagate signals across the lower 32-b long word. It controls the final data selection into the upper long-word output latch and is out of the critical path.

The nMOS byte carry select switches are controlled by a cascade of closest neighbor byte carry-outs. Data in the most significant byte of the upper long word are switched first by the carry-out data of the next lower byte, byte 6, then by byte 5, and finally byte 4. The switches direct the sum data from either the carry-in channel or the no-carry channel (Fig. 13(b)). Sign extension is accomplished by disabling the upper long-word switch controls on longword operations and forcing the sign of the result into both data channels.

C. I/O Circuitry

To provide maximum flexibility in applications, the external interface allows for several different modes of operation all using common on-chip circuitry. This includes

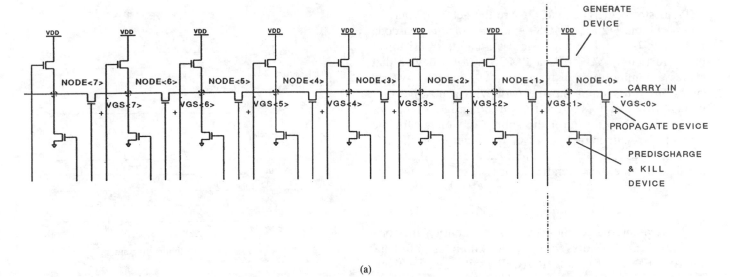

(a)

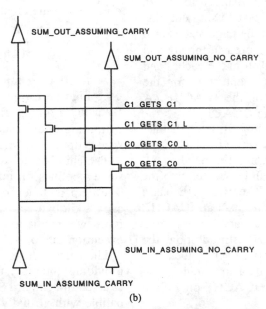

(b)

Fig. 13. (a) Adder carry chain. (b) Adder carry-select switches.

choice of logic family (CMOS/TTL or ECL) as well as bus width (64/128 b), external cache size and access time, and BIU clock rate. These parameters are set into mode registers during chip power-up. The logic family choice provided an interesting circuit challenge. The input receivers are differential amplifiers that utilize an external reference level which is set to the switching midpoint of the external logic family. To maintain signal integrity of this reference voltage, it is resistively isolated and RC filtered at each receiver.

The output driver presented a more difficult problem due to the 3.3-V V_{DD} chip power supply. To provide a good interface to ECL, it is important that the output driver pull to the V_{DD} rail (for ECL operation $V_{DD} = 0$ V, $V_{SS} = -3.3$ V). This precludes using NMOS pull-ups. PMOS pull-ups have the problem of well-junction forward bias and PMOS turn-on when bidirectional outputs are connected to 5-V logic in CMOS/TTL mode. The solution, as shown in Fig. 14, is a unique floating-well

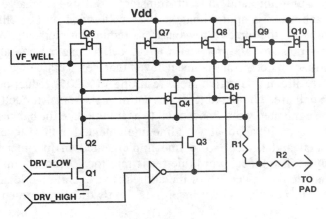

Fig. 14. Floating-well driver.

driver circuit that avoids the cost of series PMOS pull-ups in the final stage [5], while providing direct interface to 5-V CMOS/TTL as well as ECL.

381

Transistors $Q1$, $Q2$, and $Q6$ are the actual output devices. $Q1$ and $Q2$ are NMOS devices arranged in cascode fashion to limit the voltages across a single transistor to no more than 4 V. $Q6$ is a PMOS pull-up device that shares a common n-well with $Q7$–$Q10$, which have responsibility for supplying the well with a positive bias voltage of either V_{DD} or the I/O pin potential, whichever is higher. $Q3$–$Q5$ control the source of voltage for the gate of $Q6$—either the output of the inverter or the I/O pad if it moves above V_{DD}. $R1$ and $R2$ provide 50-Ω series termination in either operating mode.

D. Caches

The two internal caches are almost identical in construction. Each stores up to 8 kilobytes of data (DCACHE) or instruction (ICACHE) with a cache block size of 32 bytes. The caches are direct mapped to realize a single cycle access, and can be accessed using untranslated bits of the virtual address since the page size is also 8 kilobytes. For a read, the address stored in the tag and a 64-b quadword of data are accessed from the caches and sent to either the memory management unit for the DCACHE or the instruction unit for the ICACHE. A write-through protocol is used for the DCACHE.

The DCACHE incorporates a pending fill latch that accumulates fill data for a cache block while the DCACHE services other load/store requests. Once the pending fill latch is full, an entire cache block can be written into the cache on the next available cycle. The ICACHE has a similar facility called the stream buffer. On an ICACHE miss, the IBOX fetches the required cache block from memory and loads it into the ICACHE. In addition, the IBOX will prefetch the next sequential cache block and place it in the stream buffer. The data are held in the stream buffer and are written into the ICACHE only if the data are requested by the IBOX.

Each cache is organized into four banks to reduce power consumption and current transients during precharge. Each array is approximately 1024 cells wide by 66 cells tall with the top two rows used as redundant elements. A six-transistor 98-μm^2 static RAM cell is used. The cell utilizes a local interconnect layer that connects between polysilicon and active area, resulting in a 20% reduction in cell area compared to a conventional six-transistor cell. A segmented word line is used to accommodate the banked design, with a global word line implemented in third-level metal and a local word line implemented in first-metal layer. The global word line feeds into local decoders that decode the lower 2 b of the address to generate the local word lines. As shown in Fig. 15, the word lines are en-

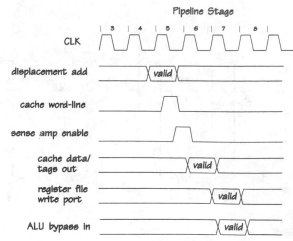

Fig. 15. DCACHE timing diagram.

abled while the clock is high and the sense amplifiers are fired on the falling edge of the clock.

VII. SUMMARY

A single-chip microprocessor that implements a new 64-b high-performance architecture has been described. By using a highly optimized design style in conjunction with a high-performance 0.75-μm technology, operating speeds up to 200 MHz have been achieved.

The chip is superscalar degree 2 and has seven and ten stage pipelines for integer and floating-point instructions. The chip includes primary instruction and data caches, each 8 kilobytes in size. In each 5-ns cycle, the chip can issue two instructions to two of four units yielding a peak execution rate of 400 MIPS and 200 MFLOPS.

The chip is designed with a flexible external interface providing integral support for a secondary cache constructed of standard SRAM's. The interface is fully compatible with virtually any multiprocessor write cache coherence scheme, and can accommodate a wide range of timing parameters. It can interface directly to standard TTL and CMOS as well as ECL technology.

REFERENCES

[1] *Alpha Architecture Handbook*, EC-H1689-10, Digital Equipment Corp., 1992.
[2] J. Yuan and C. Svensson, "High-speed CMOS circuit techniques," *IEEE J. Solid-State Circuits*, vol. SC-24, no. 1, pp. 62–70, Feb. 1989.
[3] R. Conrad et al., "A 50 MIPS (peak) 32/64-b microprocessor," in *ISSCC Dig. Tech. Papers*, Feb. 1989, pp. 76–77.
[4] J. Sklansky, "Conditional-sum addition logic," *IRE Trans. Electron. Comput.*, vol. EC-9, pp. 226–231, 1960.
[5] H. Lee et al., "An experimental 1 Mb CMOS SRAM with configurable organization and operation," in *ISSCC Dig. Tech. Papers*, Feb. 1988, pp. 180–181.

A 300–MHz 64–b Quad-Issue CMOS RISC Microprocessor

Bradley J. Benschneider, Andrew J. Black, *Member, IEEE*, William J. Bowhill, *Member, IEEE*, Sharon M. Britton, Daniel E. Dever, Dale R. Donchin, *Member, IEEE*, Robert J. Dupcak, Richard M. Fromm, Mary K. Gowan, Paul E. Gronowski, Michael Kantrowitz, *Member, IEEE*, Marc E. Lamere, Shekhar Mehta, Jeanne E. Meyer, Robert O. Mueller, Andy Olesin, Ronald P. Preston, *Member, IEEE*, Donald A. Priore, Sribalan Santhanam, Michael J. Smith, and Gilbert M. Wolrich

Abstract—This 300 MHz quad-issue custom VLSI implementation of the Alpha architecture delivers 1200 MIPS (peak), 600 MFLOPS (peak), 341 SPECint92, and 512 SPECfp92. The 16.5 mm × 18.1 mm die contains 9.3 M transistors and dissipates 50 W at 300 MHz. It is fabricated in a 3.3 V, four-layer metal, 0.5 μm, CMOS process. The upper metal layers (metal-3 and metal-4) are primarily used for power, ground, and clock distribution. The chip supports 3.3 V/5.0 V interfaces and is packaged in a 499-pin ceramic IPGA. It contains an 8-kbyte instruction cache; an 8-kbyte, dual-ported, data cache; and a 96-kbyte, unified, second-level, 3-way set associative, fully pipelined, write-back cache. This paper describes the circuit and implementation techniques that were used to attain the 300 MHz operating frequency.

I. INTRODUCTION

A SECOND-GENERATION Alpha RISC microprocessor has been designed that operates at an internal clock frequency of 300 MHz. The 16.5 mm × 18.1 mm die contains 9.3 million transistors and delivers a peak performance of 1.2 billion instructions per second (BIPS) and 600 million floating point operations per second (MFLOPS). This chip has attained measured performance of 341 SPECint92 and 512 SPECfp92. The chip is implemented in a 3.3 V, 4-layer metal, 0.5 μm, CMOS process and is housed in a 499-pin interstitial pin grid array (IPGA) package. Power dissipation is 50 W from a 3.3 V supply at 300 MHz. Fig. 1 shows a photomicrograph of the chip with an overlay showing all major sections.

The high performance of this second-generation implementation results from many factors, including:

- 0.5 μm CMOS process technology;
- 300 MHz internal clock frequency;
- grid based power and clock distribution;
- fast and versatile latching scheme;
- innovative circuit techniques;
- advanced design and verification tools.

In addition, several architectural improvements over the first Alpha implementation [1] are included in this design. The key architectural performance features are four-way superscalar instruction issue; a high-throughput, nonblocking memory sub-

Manuscript received May 4, 1995; revised August 24, 1995.
The authors are with Digital Semiconductor, Hudson, MA 01749 USA.
IEEE Log Number 9415232.

system with low latency primary caches; a large second-level on-chip write-back cache; and reduced operational latencies in all of the functional units.

II. ARCHITECTURE

As shown in Fig. 2, the chip is functionally partitioned into the following major sections: the instruction unit (I-Box), the integer execution unit (E-Box), the floating point unit (F-Box), the memory management unit (M-Box), and the cache control and bus interface unit (C-Box). The chip features two levels of on-chip cache. The first level consists of an 8-kbyte instruction cache (I-Cache) and an 8-kbyte data cache (D-Cache). The second level is a 96-kbyte unified instruction and data cache.

The I-Box contains the 8-kbyte, direct-mapped I-Cache, an instruction prefetcher and associated refill buffer, branch prediction logic, and a 48-entry, fully associative instruction translation buffer. The I-Box can issue up to two integer and two floating point instructions per cycle. Instructions are issued in-order but may complete out-of-order.

The E-Box contains two execution pipelines and a register file for integer operands. Both E-Box pipelines execute load, arithmetic, and logical instructions. In addition, one of the pipelines executes shift and store instructions, while the other pipeline completes jumps and branches. Multiply instructions are executed in a separate unit attached to one of the pipelines. Both pipelines implement full register bypassing, allowing the results from all function units to be available for immediate use. All integer instructions except multiply complete in one cycle.

The F-Box contains a register file for floating point operands and two execution pipelines. One pipeline executes multiply instructions while the other executes all remaining instructions. Divide instructions are executed in a separate unit attached to one of the pipelines. All floating point instructions except divide execute in four cycles, a two-cycle reduction from the previous implementation.

The M-Box contains the 8-kbyte, direct-mapped D-Cache, a fully-associative, 64-entry, data translation buffer (DTB), a miss address file for queuing and merging misses from the first-level caches, and a write buffer. The M-Box processes load, store, and memory barrier instructions.

Reprinted from *IEEE Journal of Solid-State Circuits*, Vol. 30, No. 11, pp. 1203–1211, November 1995.

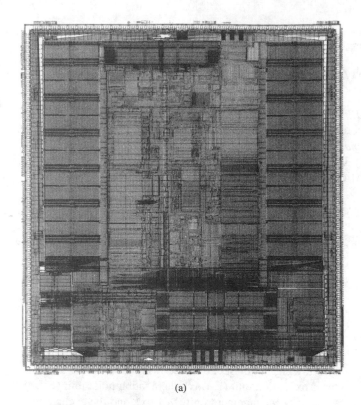

(a)

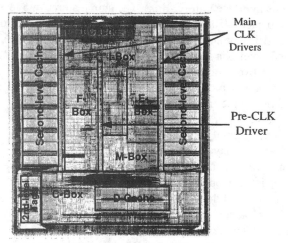

Main CLK Drivers

Pre-CLK Driver

Die Size : 16.5 mm x 18.1 mm = 298mm²

(b)

Fig. 1. Photomicrograph with overlay.

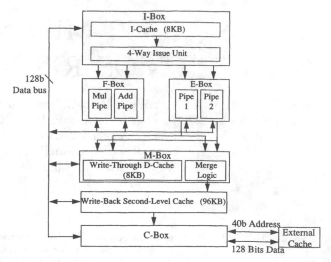

Fig. 2. Block diagram.

TABLE I
TECHNOLOGY CHARACTERISTICS.

Feature size	0.5 μm
Channel length (L_{eff})	0.365 μm
Gate oxide	9.0 nanometers (nm)
Vtn/Vtp	0.5/-0.5 V
Power supply	3.3 V (supports 3.3/5.0 V interface)
Substrate	P-epi with n-well
Salicide	Cobalt di-silicide in diffusions and poly
RAM cell	6T, TiN buried contact
Metal-1	0.84 μm thick AlCu, 1.50 μm pitch (contacted)
Metal-2	0.84 μm thick AlCu, 1.75 μm pitch (contacted)
Metal-3	1.53 μm thick AlCu, 5.00 μm pitch (contacted)
Metal-4	1.53 μm thick AlCu, 6.00 μm pitch (contacted)

III. GLOBAL IMPLEMENTATION

The high internal clock frequency, power requirements, and transistor count of the chip required substantial global planning prior to circuit implementation. Technology development, floorplanning, power and clock distribution, and latching strategies were areas of particular importance.

A. Process Technology

The chip is fabricated in a 0.5 μm, 3.3 V, n-well, CMOS process. The major process characteristics are shown in Table I. Technology development occurred in parallel with chip development. Close cooperation between the two development teams allowed both the technology and the chip to be optimized for performance. Metal-4 was added to the technology for power and clock distribution.

B. Floorplanning

The process of floorplanning was initiated during the microarchitectural definition of the chip and was key to achieving the performance goals. Several critical speed paths were iden-

The C-Box controls the on-chip, second-level cache and an optional, off-chip, third-level cache and implements a flexible, user-configurable interface to the system. The second-level cache is a fully pipelined, 96-kbyte, three-way set associative, write-back cache. It reads or writes 16 bytes per cycle, providing a peak bandwidth of 4.8 Gb/s at 300 MHz. A 128-b data bus is shared between the optional, off-chip, third-level backup cache and the memory system.

This microprocessor also contains a number of on-chip testability features, including built-in self-test and self-repair of the I-Cache, linear feedback shift registers placed throughout the chip to improve fault coverage, a parallel debug port for monitoring internal chip nodes in real time during chip and system debug, and an IEEE 1149.1 test access port.

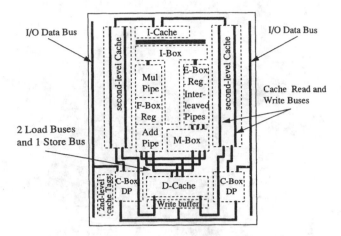

Fig. 3. Floorplan.

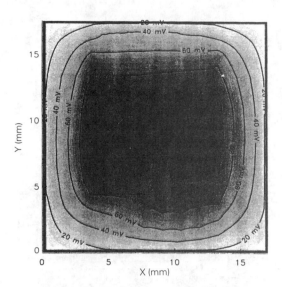

Fig. 4. VDD voltage drop contour plot.

tified and the final floorplan was formulated to optimize these paths. Global interconnect channels were defined early in the project to give priority routing to critical signals. With reference to Fig. 1, metal-4 and metal-2 are routed vertically and metal-3 and metal-1 are routed horizontally. The final floorplan of the chip is shown in Fig. 3.

The second-level cache, the C-Box datapath, and the I/O pins were split into two halves along the left and right edges of the chip. This was done to optimize the routing of data into and out of the chip. The I-Cache was positioned at the top to feed the I-Box decoders and to receive fill data from the pins on the same buses as the second-level cache, minimizing both fill latency and interconnect area. The D-Cache was placed at the bottom of the chip, beneath the E-Box, to minimize the interconnect length of data signals to the E-Box register file. The adder used to calculate the address displacement in the E-Box, and the DTB in the M-Box were placed next to each other. This design allowed the result of the address addition to flow directly into the DTB with little interconnect delay, facilitating a two-cycle D-Cache hit latency.

Different performance requirements led to different floorplan solutions in the F-Box and E-Box. The critical nature of the E-Box data bypass loop required the two integer pipes to be interleaved. This optimized the bypass bus at the expense of widening the pitch of the datapath. The F-Box, having fewer bypass paths, was organized with the register file in the center of the two pipes, effectively sharing vertical metal slots and hence minimizing the area of the datapath.

C. Power Distribution

A high quality power distribution network was essential to meet the power dissipation and performance goals of this chip. Metal-3 and metal-4 are used extensively to distribute power and ground across the chip. These metal layers are approximately twice as thick as lower level metals, thereby offering substantially lower resistance, and are used to form dense grids. Alternating power and ground lines comprise the majority of the grids with a single clock line interspersed every few pairs. The typical drawn line width for V_{DD} and V_{SS} is 12 μm. Power and ground lines in local cells are connected to

metal-2 lines that are long enough to span two or more of the metal-3 grid lines. This allowed for a simple and automated procedure to connect local logic to the grid. In addition, 160 nF of on-chip distributed decoupling capacitance was added to minimize switching noise. The clock drivers are closely surrounded by 35 nF of the decoupling capacitance.

The power grid underwent significant electrical verification to ensure its compliance with strict long-term reliability requirements and to guarantee minimal voltage drops in the grid. The analysis was based on full chip capacitance and resistance nodal extracts and used nodal switching data from logic simulations to estimate how much current was conducted through the supply networks. This analysis showed a maximum voltage drop of only 110 mV per rail, demonstrating that the metal-3 and metal-4 grid could effectively distribute power to all devices. Fig. 4 is a contour plot that shows the simulated average voltage drop in the V_{DD} network across the chip. The V_{SS} network exhibited similar characteristics.

Power is supplied to the package through 205 of the 499 pins. The connection between the die and the package is achieved using 282 bond wires. The package outer lead bond pads are arranged in two tiers around the die cavity. The die bond pads are evenly spaced around the chip using an alternating pattern of power and signal (i.e., V_{SS} signal V_{DD} signal V_{SS} etc. . . .). This configuration allowed all supply pads to be bonded to the lower tier of landing pads in the package with short, low inductance bond wires. To help dissipate the heat generated on chip, a custom ceramic IPGA package was used. It incorporates an intrusive copper tungsten slug that provides a low thermal impedance contact between the die and a removable heat sink. The slug results in a θ_{JC} for the package of only 0.45°C/W. Although the chip dissipates 50 W, it is effectively cooled by conventional thermal management techniques.

D. Clock Distribution

Another critical aspect of the global chip implementation was the buffering and distribution of the two-phase, single wire

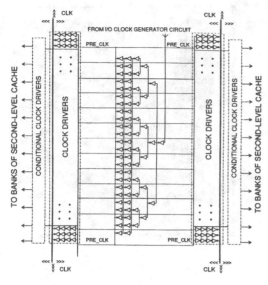

Fig. 5. Clock generation network.

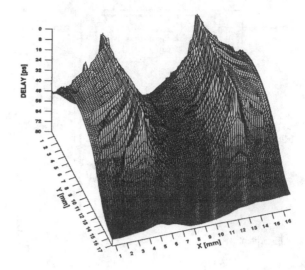

Fig. 6. Clock *RC* contour plot.

clock. The chip receives a 600 MHz differential ECL oscillator signal that is level shifted and divided by two to produce a 300 MHz, 50% duty cycle clock. As shown in Fig. 5, this clock is routed to the center of the chip where it is buffered up and fanned out through a balanced tree of inverters to generate the PRE_CLK signal that drives the main clock drivers. The main clock drivers provide four additional levels of buffering to generate the single wire clock (CLK) that is distributed over the entire chip using a grid of metal-3 and metal-4. The final CLK driver inverter has a transistor width of 58 cm and drives a load of 3.75 nF.

In addition to the single wire clock, two sets of conditional clocks are generated for the second-level cache. These clocks conserve power by conditioning the bank enable signals with preliminary address decode information when accessing the cache. This is described in more detail in Section IV-D.

A low-skew clock signal was critical to meeting the cycle time goal of the chip. The CLK network was designed to minimize the following three components of clock skew: driver and RC delay variations in the PRE_CLK driver network; variations in the transistor characteristics of the main CLK drivers; and RC delay through the global metal grid. Techniques used to reduce the individual clock skew components included: connecting the common nodes of the PRE_CLK and CLK drivers with low resistance upper level metal; tuning the PRE_CLK driver by adjusting the widths of metal wires to balance the RC delays; working with the process development group to design a modular clock driver block to minimize polysilicon processing variations; and specifying design rules that limited the RC delay of the global metal grid.

The CLK driver network was subjected to extensive simulation to evaluate the impact of RC delay on the chip's various circuits. The simulation used layout extracted resistance and capacitance data and contained one million resistors and two million capacitors. Fig. 6 is a three-dimensional inverted contour plot that shows the results of the RC delay simulations. The two peaks seen on the contour plot represent the locations of the main clock drivers. As shown on the plot, the maximum

RC delay is 80 ps. The PRE_CLK network and the distributed CLK drivers were also analyzed with SPICE to assess the impact of the power supply noise and cross-chip device characteristic variations on CLK skew.

E. Latch Design

This chip primarily uses dynamic, level-sensitive, pass-transistor latches. This type of latch was chosen to minimize the propagation delay of data through the latch. Since the chip uses a single wire, two-phase clocking scheme, two types of latches are utilized in the chip: an A-phase latch, which is open when CLK is a logic-1; and a B-phase latch, which is open when CLK is a logic-0. Fig. 7(a) and (b) shows examples of the standard A-phase and B-phase latches. These latches were carefully designed and laid out to minimize capacitive coupling effects onto the dynamic node (ZZ_XD in the figures). Several basic latches were designed prior to implementation and were maintained in a library for use by all designers to minimize the number of latch variants.

The overhead associated with latching data in critical paths was further reduced by building simple logic functions into the input and output gates of the latch. An example of a standard A-phase latch with a NAND function at the front end is shown in Fig. 7(c). Fig. 7(d) shows a sample configuration where logic has been built into both stages of the latch, effectively reducing the latching delay to that of a single pass transistor. When this configuration was used, steps were taken to prevent the output node from coupling back onto the ZZ_XD nodes (through the output transistors) while the ZZ_XD nodes are in the dynamic state. The inputs to the final logic gate are required to come from the same latch type (A-phase or B-phase), and the output nodes of the pass transistors must only drive the final 2-input logic gate using minimal routing. These requirements, together with the latch setup time, ensure that the output gate will switch while the ZZ_XD node is statically driven.

While these level-sensitive latches are fast, they are susceptible to data race-through. This problem was managed by a

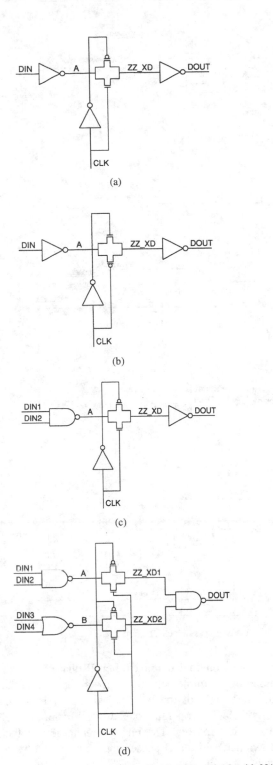

Fig. 7. (a) A-phase latch. (b) B-phase latch. (c) A-Phase latch with NAND function. (d) A-Phase latch with embedded logic.

of delays between latches. A suite of postlayout checks were performed to ensure that the internal dynamic latch nodes were not adversely affected by noise sources.

IV. IMPLEMENTATION EXAMPLES

The operating frequency of this chip lead to some very challenging implementation problems that were encountered during the design of the chip. This section of the paper describes four example circuits.

A. I-Box Issue Stage Domino Logic Circuit

The issue stage of the I-Box coordinates the release of instructions into the E-Box, F-Box, and M-Box pipelines. Issuing four instructions per cycle in a machine with deep pipelines and a complex memory system presented several design challenges. The four result and eight source registers of the issuing instructions must be compared against the 37 possible outstanding instructions (seven integer instructions, nine floating point instructions, and 21 load instructions that missed) within the machine. Concurrent with the register checks, 44 possible data bypass calculations must also be performed to ensure the most up-to-date data is forwarded to the issuing instruction.

Domino logic was used to implement the register scoreboard and bypass structures in order to meet the chip performance and area requirements. During instruction issue, each source and result register address is decoded and loaded into a 31-b wide pipeline that mimics the execution pipeline. Checks are performed for stalls and bypasses by selecting the appropriate bits from each stage of the pipeline and comparing them to the decoded register addresses of the new instructions. Integer and floating point instructions are handled in separate pipelines.

Decoding the register addresses into a 31-b wide vector, prior to entering the pipeline, allows the result register addresses to be logically ORed together to create a dirty register vector. This allows the stall calculations to be performed using only 38 comparators and the bypass calculations using only 44 comparators.

The comparators were implemented in three dynamic domino stages, as shown in Fig. 8. The first stage is a two-input multiplexer that selects the source/result decode field for the new instruction or the source/result decode address of the previous cycle based on whether or not a stall had been detected. The dirty bit vector is created in a similar logical OR structure. The second stage detects if there is a register conflict. The register conflict wire is discharged when there is a matching source/result decode and dirty vector bit. The transmission gate qualifies the detected register conflict with an instruction valid signal. The third stage is used to further qualify the detected conflict with instruction type decode information and to start combining the 38 conflict outputs down to a single stall wire. In the case of bypasses, the third stage is used to ensure that only the most up-to-date data is bypassed by priority-encoding the individual bypass indications.

The dynamic circuit implementation of the issue logic required careful analysis of several key circuit issues, such

combination of techniques, including controlling the skew on CLK, precisely sizing the local clock buffer inside the latch, placing strict rules on the use of the locally buffered clock, and requiring a minimum number of gate delays between latches. A custom verification tool was developed to check that the latching rules were followed throughout the design. This CAD tool performed a static analysis that identified places where buffered clocks were used outside of the standard latch structure and where there was an insufficient number

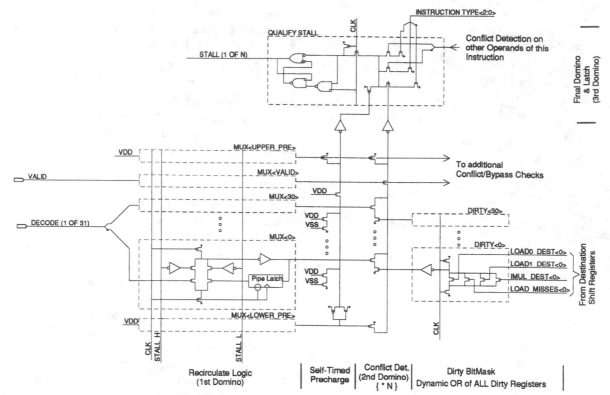

Fig. 8. I-Box domino logic circuit diagram.

as power dissipation, noise margins, and coupling. Power dissipation resulting from overlap between the precharge and the evaluate phases of the domino stages was minimized by self-timing the precharge enable signals for the second and third domino stages. Noise margins were maintained by locally buffering the input signals of dynamic gates and ensuring a common ground. To reduce the lateral capacitive coupling between the many dynamic signals, the space between the signals was maximized and wires were arranged to take advantage of signals with mutually exclusive switching characteristics.

B. E-Box Bypass Circuit

The E-Box contains two 64-b integer pipelines. Each pipeline produces a result from an operation requiring two input operands. A total of four distinct input operands must be provided to the two execution pipelines. The E-Box performance is significantly increased by implementing a bypass scheme that allows the result from any function unit in either pipeline to be bypassed to each of these four input operands. Through this scheme, every result is immediately available to the next instruction in each pipeline.

Fig. 9 shows the block diagram for one of the four E-Box operand bypasses. The operand bus can be sourced from the register file or from one of seven other bypasses. A bypass select signal from the I-Box controls that bypass input to use and a B-phase latch stores the chosen data. The function unit then performs its operation, latches the result with an A-phase latch, and drives it onto the result bus completing the loop.

The circuit implementation of one of the E-Box operand bypass buses is shown in Fig. 10. The A-phase latch is

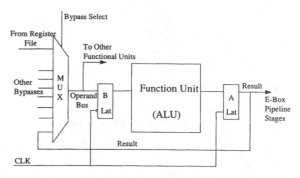

Fig. 9. E-box bypass block diagram.

immediately obvious but no explicit B-phase latch is used. The operand bus is implemented as a dynamic differential bus that spans the height of the E-Box datapath. It is driven by one of eight clocked operand bus drivers forming a distributed dynamic multiplexer. The bus is precharged in the A-phase and evaluated in the B-phase when one of the bypass enable signals is asserted.

The function units use a static receiver followed by a dynamic gate to receive the operand data. The dynamic gate evaluates in the A-phase and precharges during the B-phase. A race exists between the precharge of the operand bus and the evaluation of the operand bus in the function unit, since they both occur in the same phase. The race could have been avoided by placing a B-phase latch between the bypass bus and function unit receivers. However, the latch was eliminated to save time in the bypass critical speed path. The race was managed by including a bus precharge delay that provides additional hold time for the data so that all receivers have

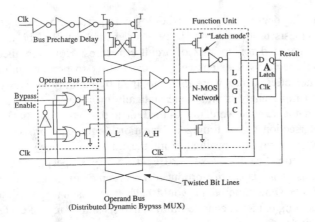

Fig. 10. E-box bypass circuit diagram.

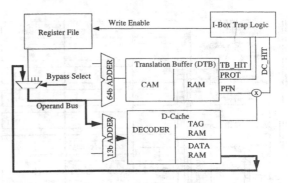

Fig. 11. D-cache access critical path.

enough time to capture the operand data before the bus is precharged.

A further concern of the bypass scheme is the possibility that crosstalk could corrupt data on a bus due to interconnect coupling capacitance. This is a consequence of the E-Box datapath being densely populated with eight dynamic differential buses per bit. Three design techniques were used to manage this problem. First, cross-coupled P-devices were connected across each differential bus pair to retain the precharged "1" value of the dynamic bus during the evaluate phase. Second, the beta-ratios of the bypass receiver gates were skewed to maximize the noise margins. Lastly, the operand buses were arranged in a manner similar to memory arrays with twisted bit lines, thus reducing worst-case capacitive crosstalk by approximately one half.

C. D-Cache Access Critical Paths

This microprocessor has a two-cycle latency for load instructions that hit in the D-Cache. A dependent instruction can issue two cycles after the load instruction has issued. This is a one cycle improvement over the previous implementation. This gives rise to two major critical speed paths. Fig. 11 shows a logical representation of these two paths. The D-Cache is accessed with a 13-b index and returns the data and its associated address tag. Simultaneously, the DTB translates the full address, which is then compared with the tag. Signals that indicate whether the load hit in the D-Cache are sent to the I-Box, which drives write enable signals to the E-Box register file. Data returned to the E-Box may be speculatively used by a consuming instruction, but it is only written into the register file if the load hits in the D-Cache.

One critical path delivers the lower 13 b of the address that is used as the index to the D-Cache. The index is calculated by a dedicated 13-b adder that drives the large capacitive load of the D-Cache decoders. The 13-b adder is placed outside of the E-Box datapath, between the register file and the D-Cache. The index output from the adder is routed to the D-Cache in metal-3 and metal-4.

The other critical path determines whether a reference hits in the D-Cache. In the first of the two cycles needed for this determination, a 64-b adder in the E-Box calculates the full address that is then compared against all entries in the CAM array of the DTB. This array is carefully positioned below the E-Box adders to avoid unnecessary RC delays. The adder, which is also used for basic integer arithmetic instructions, normally drives its output to the E-Box's result bus. However, in this case, the output is driven directly from the adder into the CAM array with a large dynamic driver. To save power, this dynamic driver is conditioned to evaluate only during address calculations.

In the first phase of the next cycle, an address tag is read from the D-Cache and delivered to a comparator prior to the arrival of the translated address. This address is read from the DTB and driven directly into the comparator by the sense-amps at the output of the DTB. The remainder of the cycle is used to perform the address comparison, to check the success of the address translation, and to return data back to the E-Box.

D. Cache Design

Special design considerations were given to the three on-chip caches. Redundancy was included in all the arrays to improve yield and the caches were designed to minimize power consumption. In addition, the general cache design was shared among the three on-chip caches to reduce the amount of design and verification effort.

The I-Cache and D-Cache both include two sets of fuse-programmable redundant rows to improve yield. These fuses are constructed of a titanium nitride layer that can be electrically blown using a laser. The I-Cache features built-in self-test logic to identify bad rows and built-in self-repair logic to automatically map the redundant rows over failing rows during wafer probe [2], [3]. This approach allows a more extensive test of the chip at wafer probe. After wafer probe the redundant rows are permanently mapped over the failing rows by blowing the fuses with the laser.

The large 96-kbyte, 3-way set associative, second-level cache is 128 b wide and is partitioned into two 64-b wide data arrays each containing twelve 4-kbyte banks (4 per set). The two arrays are placed on the left and right sides of the chip. Read and write buses pass over the two arrays in metal-4, connecting the data banks together and allowing access to the primary caches at the top and bottom of the chip. Fig. 12 is a block diagram showing the arrangement of the twelve 4-kbyte banks on the right side of the chip. There are also three separate tag arrays (1 per set) that are placed in the bottom left corner of the chip (see Fig. 1). Each bank of the tag and

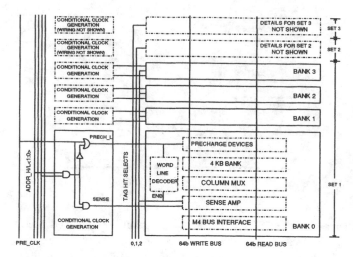

Fig. 12. Block diagram of right-half of the second-level cache.

data arrays implements row redundancy. The data arrays also implement column redundancy.

The second-level cache operates in a three-stage pipeline, two stages for tag lookup and modification, and one for data access. Pipelining the tag access ahead of the data access limits the number of data banks that are concurrently accessed. Partial address decoding during the tag lookup enables two banks per set (six total) by using the conditional clocks discussed earlier. Banks that have not been selected are held in the precharge state. The "hit" signals from the three tag arrays gate the word lines and sense amplifiers of the banks. Therefore, of the six banks enabled, only the two banks for the set that hit are activated and discharged. This results in an estimated power savings of 10 W.

The 8-kbyte D-cache supports two loads per cycle, requiring a dual-read-ported design. The D-cache was implemented as two single-ported caches containing identical data instead of one dual-ported cache. The major consideration that led to this decision was the ability to share the single-ported design with the I-Cache at the cost of a small increase in area. Sharing the design also reduced the overall analysis and verification required. In addition, the aspect ratio of this D-Cache configuration was ideal for the floorplan.

V. CAD Tools and Verification

A. Design Methods and Tools

An extensive suite of proprietary in-house CAD tools contributed significantly to the successful design of this chip. These tools were particularly effective in supporting design entry and rigorously checking the large amount of full-custom circuitry that was employed.

Tools that aided schematic generation included a schematic editor, a logic synthesis tool, and a device sizing tool. Postschematic tools included a latching methodology checker, a circuit verifier that highlighted electrical design methodology violations (e.g. dynamic node noise susceptibility), and a timing verifier that identified and analyzed potential critical paths. The use of the design tools varied across the chip based on the degree of customized logic required. For example, synthesis tools were not heavily used in the F-Box because of the need for optimized circuit structures. However, these tools were used extensively in the C-Box to produce initial schematics, which were then modified by hand as necessary.

Timing analysis was done statically on a per-section basis and the results were maintained in a database that tracked intersection signal timing information. Significant productivity improvement resulted from the timing analyzer's ability to automatically invoke SPICE[1] to verify suspected critical paths. This capability produced an order of magnitude more SPICE simulations than would have been manually possible, resulting in more accurate timing analysis and increased design confidence.

As previously discussed, minimizing clock skew across the chip was critical in meeting the operating frequency target. This could not have been achieved without robust simulation tools for accurately analyzing and reporting the skew during the clock design. Extracted RC time constants were analyzed by an asymptotical waveform evaluation tool. Clock skew results were displayed as a function of X, Y, and time using three-dimensional visualization tools that generated an animated display of the clock as it propagated from the drivers to the receivers across the chip. Fig. 6 shows a representation of the results of this analysis.

Detailed physical verification and reliability checks included analysis of electromigration, hot carriers, latch-up, power supply noise, coupling, and the detection of structures susceptible to charge damage from plasma etch processing during manufacturing. The CAD programs used to automate these tasks were frequently updated to provide additional features and enhanced circuit analysis. It is noteworthy that no electrical sensitivities were found in the chip during prototype debug.

B. Functional Verification

The complexity of the chip necessitated extensive functional verification prior to mask generation. A two-state RTL behavioral model was the primary verification vehicle. This model provided a balance of accurate design representation and simulation speed. The RTL model of the CPU was augmented with an abstract behavioral model of the remaining system components, including an off-chip cache, main memory, and I/O devices. This allowed normal system traffic to be applied to the simulation model. Once the circuit design was complete, a two-state gate-level model was extracted from the transistor netlist. This was used to verify that the schematics matched the RTL model. In addition, a three-state switch-level model was used to check for proper reset initialization.

The simulation models were verified with a variety of stimuli. The most effective techniques used targeted-random exercisers. These exercisers consisted of developing an outline of a test and allowing the specifics to be generated randomly every time the test was run. The general outline could be

---

[1] SPICE is a general-purpose circuit simulator program developed by Lawrence Nagel and Ellis Cohen of the Department of Electrical Engineering and Computer Sciences, University of California at Berkeley.

executed repeatedly, generating a different test stimuli each time. These outlines could be combined with each other to create complicated random test generators called exercisers. In addition to the random exercisers, hand-crafted focused tests were also created to verify specific areas of the design.

Coverage analysis techniques were used to guide the verification process. This process consisted of analyzing each piece of logic and determining which sequences needed to be stimulated. For instance, in a state machine, it is important to exercise all transitions between states. For sections of the design where specific coverage goals were not achieved, either additional focused tests were created or random exercisers were tuned to fill in the coverage holes.

By the time the design was released to manufacturing, over 14 billion cycles of random stimuli were run on the RTL model, and over 500 million cycles were run on the gate-level models. In addition, over 400 focused tests were developed and executed on the RTL and gate-level models.

Many of the focused tests were used to generate manufacturing test patterns. Fault simulation was performed using these test patterns. The results indicated that tests that do an excellent job covering design faults achieve about 85% coverage of gate-level stuck-at faults. The fault simulation data is being used to direct test enhancements, leading to a steady increase in fault coverage.

VI. CONCLUSION

Employing a custom circuit design style, coupled with an optimized high-performance 0.5 μm technology, this second-generation, high-end microprocessor was designed to operate at a target clock frequency of 300 MHz. This was accomplished by employing a dense, low resistance power grid, maintaining minimal clock skew, using fast latches, and employing high-speed circuit techniques. The chip encompasses an area of 299 mm^2 and contains 9.3 million transistors, including a 96-kbyte second-level cache.

Within four weeks of silicon fabrication, the OpenVMS[2]operating system was booted, quickly followed by Digital UNIX[3] and Windows NT[4] operating systems. Currently, all three operating systems are running successfully on a number of different system platforms operating at 300 MHz. This microprocessor has been incorporated into a number of available system products since April 1995.

Fig. 13 shows a shmoo plot of operating frequency versus supply voltage for this microprocessor functioning at a case temperature of 85°C. The plot shows the pass/fail boundary for various speeds and voltages for a chip with typical target process parameters. The plot demonstrates that the chip functions at frequencies greater than 300 MHz under normal operating conditions. Additionally, the results demonstrate that full speed operation is attainable even at high temperatures and voltages less than the normal 3.3 V supply.

---

[2]OpenVMS is a trademark of Digital Equipment Corporation.

[3]UNIX is a registered trademark in the United States and other countries licensed exclusively through the X/Open Company Limited. Digital UNIX for Alpha is a UNIX 93 branded product.

[4]Windows NT is a trademark of Microsoft Corporation.

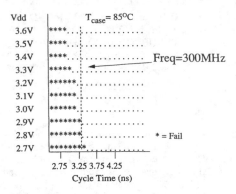

Fig. 13. Shmoo plot.

At 300 MHz, the four-way superscalar microprocessor is capable of achieving a peak execution rate of 1.2 BIPS and 600 MFLOPS. It has also attained measured performance of 341.4 SPECint92 and 512.9 SPECfp92 [4].

ACKNOWLEDGMENT

The authors would like to acknowledge the contributions of many people who helped make this chip possible. These include W. Herrick and P. Rubinfeld for project management support; F. Fox for technical consultation; A. Cave and R. Cvijetic for invaluable CAD assistance; and L. Bair, N. Arora, L. Gruber, and B. Zetterlund for device and technology modeling. Designers include R. Allmon, R. Badeau, P. Bannon, S. Bell, T. Benninghoff, R. Blake-Campos, D. Brasili, K. Broch, T. Broch, R. Castelino, M. Charnoky, E. Cooper, J. Edmondson, H. Fair, T. Fischer, A. Jain, J. Keller, J. Kowaleski, P. Kroesen, T. Mast, S. Mehta, A. Murphy, J. Mylius, T. Pham, V. Rajagopalan, T. Shedd, C. Somanathan, S. Strickland, S. Thierauf, and J. White.

REFERENCES

[1] D. Dobberpuhl et al., "A 200-MHz 64-b dual-issue CMOS microprocessor," IEEE J. Solid State Circuits, vol. 27, pp. 1555–1564, Nov. 1992.
[2] D. K. Bhavsar and J. H. Edmondson, "Testability strategy of the Alpha AXP 21164 microprocessor," in Int. Test Conf., Oct. 1994, pp. 50–59.
[3] D. K. Bhavsar and R. M. Fromm, "Testability features and testability access of the Alpha 21164 microprocessor," in Custom Integrated Circuits Conf., May 1995.
[4] SPEC Newsletter: vol. 7, no. 2, June 1995.

A 600MHz Superscalar RISC Microprocessor with Out-of-Order Execution

BRUCE A. GIESEKE, RANDY L. ALLMON, DANIEL W. BAILEY, BRADLEY J. BENSCHNEIDER,
SHARON M. BRITTON, JOHN D. CLOUSER, HARRY R. FAIR III, JAMES A. FARRELL, MICHAEL K. GOWAN,
CHRISTOPHER L. HOUGHTON, JAMES B. KELLER, THOMAS H. LEE[1], DANIEL L. LEIBHOLZ,
SUSAN C. LOWELL, MARK D. MATSON, RICHARD J. MATTHEW, VICTOR PENG, MICHAEL D. QUINN,
DONALD A. PRIORE, MICHAEL J. SMITH, AND KATHRYN E. WILCOX

DIGITAL SEMICONDUCTOR, DIGITAL EQUIPMENT CORPORATION, HUDSON, MA
[1]STANFORD UNIVERSITY, PALO ALTO, CA

A six-issue, four-fetch, out-of-order execution, 600MHz Alpha microprocessor achieves an estimated 40SpecInt95, 60SpecFP95 and 1800MB/s on McCalpin Stream. The 16.7x18.8mm² die contains 15.2M transistors and dissipates an estimated 72W. It is in 2.0V, 6-metal, 0.35µm CMOS with CMP planarization (Table 1) [1]. The chip is in a 587-pin ceramic IPGA with 198 pins for VDD/VSS that includes a CuW heat slug for low thermal resistance between die and detachable heat sink. An on-chip PLL performs frequency multiplication of a differential PECL reference and synchronizes I/O by phase-aligning a CPU clock to the reference. Figure 1 is a detailed floorplan of the chip. Figure 2 depicts a block/pipeline diagram of major sections and functions.

The instruction fetcher (Figure 3) reads four instructions per cycle plus a next-address pointer from a 64kB, 2-way pseudo-set associative, virtual instruction cache. The next-address pointer predicts the address of the subsequent four instructions and indexes the cache in the next cycle. In parallel, a branch predictor resolves the prediction. It contains three tables: a PC-indexed prediction table, a path-indexed prediction table, and a path-indexed table that dynamically chooses one of the former two predictions, based on the success of previous predictions.

Fetched instructions are dispatched to integer/memory (INT/MEM) and floating point (FP) pipelines, issued and executed out of order and retired in order. During dispatch, register specifiers are renamed to eliminate false dependencies by two twelve-port register mappers that dynamically map the architectural registers into a pool of physical registers (80 integer and 72 FP). Resulting map state is retained in an array until the instruction retires. Pre-retire map state is used to generate a list of remaining free physical registers. Buffered map state is restored when the CPU is redirected following a branch mispredict or exception.

Mapped instructions enter a 20-entry INT/MEM or a 15-entry FP issue queue. The INT/MEM queue arbiter identifies the 4 oldest data-ready instructions. They issue to the integer execution unit (EBOX) and are removed from the INT/MEM queue. Similarly, the FP queue issues the 2 oldest data-ready instructions to the FP execution unit (FBOX) and removes them from the FP queue.

The EBOX (Figure 4) is divided into two clusters, CL0 and CL1; each cluster contains 2 independent execution pipelines surrounding an 80-entry register file. Coherency between the two register file copies is maintained by broadcasting results across intercluster buses. Each of the four pipelines executes and bypasses arithmetic and logical operations in one cycle. Bypassed results between clusters take an additional cycle. The upper pipelines handle branches and shifts; CL0 contains a pipelined multimedia engine (3-cycle latency) and CL1 contains a pipelined multiplier (7-cycle latency). The lower pipelines handle displacement address calculations for memory operations. The FBOX contains 2 independent execution pipelines surrounding a 72-entry register file. One datapath implements a 4-cycle pipelined multiply and the second datapath, a 4-cycle pipelined add, a 12/15-cycle non-pipelined divide and a 18/33-cycle non-pipelined square root ("/" denotes single/double precision).

Memory references access a 64kB, 2-way set associative dual-ported virtually-indexed physically-tagged write-back data cache. The cache supports either two 8B reads to arbitrary addresses per CPU cycle by phase pipelining the word and bit-lines or one (up to 16B) combined victim extract and write to a single address. Byte/word stores use one cache cycle. Load latency is 3 cycles to both EBOX clusters and 4 cycles to FBOX. The memory reorder unit (MBOX) ensures correct behavior of out-of-order load and store instructions. It contains two 32-entry reorder units for loads and stores, an 8-entry miss address file, an 8-entry victim address file and two 128-entry translation look-aside buffers. Memory references missing the internal caches are looked up in an off-chip, second-level, direct-mapped cache (1 to 16MB) through a 128b data-port. The interface supports clock-forwarded, late-write, dual-data, synchronous SRAMs and peak transfer rates up to 4GB/s. Addresses, commands, and data are transferred to and from memory controllers and data switches via a separate address and 64b data-port at a peak rate of 2GB/s.

Tradeoffs between micro-architecture and circuits optimize performance. For example, the two-cluster EBOX (Figure 4) incurs a penalty of 1% in architectural efficiency (due to an additional cycle of latency for intercluster transfers) versus a proposed unified, four-pipeline cluster. However, the unified cluster requires 4 additional differential operand buses and 4 additional register file read ports. The resulting increases of 22% in total area, 47% in datapath width and 75% in operand bus length make the unified cluster unworkable at the target cycle time.

Figure 5 illustrates the circuit design of each EBOX cluster. Each functional unit has a differential sense amplifier stage, a dynamic computation stage and a differential driver stage. The multimedia engine and the multiplier share the shifter receivers and drivers to reduce operand bus length at the cost of an additional cycle of latency (Figure 4). Functional unit outputs, register file contents, load data, and inter-cluster results may all drive the twisted, small signal (200mV), 3.3mm operand buses directly. Up to six results per cycle may be written to the register file through dedicated single-ended buses (Figures 4, 5).

Reference planes provide a low-impedance dc path for supply current to the core of the chip and a controlled-impedance environment for signal and clock routing [2, 3]. The addition of two dedicated metal layers (Table 1) to the process dramatically simplifies analysis of capacitive and inductive interconnect parasitics that might otherwise lead to unpredictable signal timing, overshoot, undershoot or crosstalk. Integrated power supply decoupling capacitance of 250nF is included beneath routing channels and near large clock drivers to minimize power supply noise components above the chip operating frequency.

The highly-parallel conditionally-clocked design exposes indirect control of ΔIdd through program and data dependencies. Maximum ΔIdd is estimated at ~25A over several cycles. A 1µF wirebond-attached chip capacitor (WACC) augments on-chip decoupling to manage power supply noise components near the chip/package supply-loop resonant frequencies (Figure 6). The WACC is a 2cm², accumulation-mode pMOS device in 1µm 2-metal MOS. Effective series resistance (ESR) is targeted with on-WACC series resistors and low effective series inductance (ESL) is achieved with ~160 VDD/VSS bondwire pairs between WACC and microprocessor. With the WACC on top of the microprocessor, standard wirebond and PGA technologies are used without impacting package thermal or I/O performance.

Reprinted from *Digest of Technical Papers, 1997 International Solid-State Circuits Conference,* pp. 176-177, 451, February 1997.

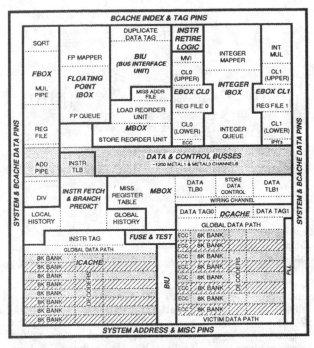

Figure 1: Microprocessor chip floorplan.

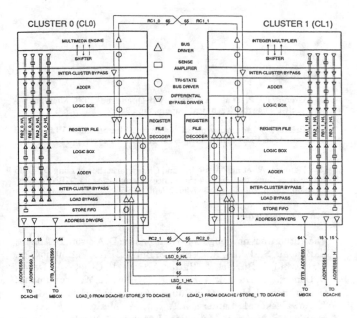

Figure 4: Integer execution unit organization.

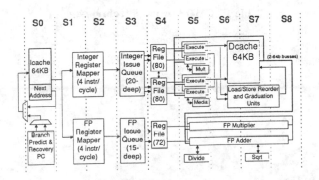

Figure 2: Microprocessor block/pipeline diagram.

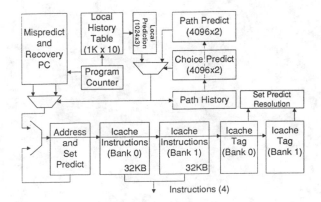

Figure 3: Instruction fetch and branch prediction.

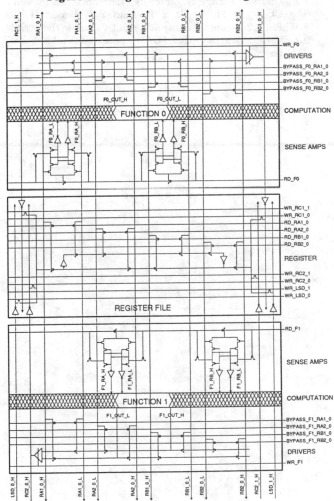

Figure 5: Integer execution cluster design.

References:

[1] Gronowski, P., et al., "A 433MHz 64b Quad-Issue RISC Microprocessor," ISSCC Digest of Technical Papers, pp. 222-223, Feb.,1996.

[2] Priore, D., "Inductance on Silicon for Sub-Micron CMOS VLSI," 1993 Symposium on VLSI Circuits Digest of Technical Papers, pp. 17-18, May, 1993.

[3] Deutsch, A., et al., "Modeling and Characterization of Long On-Chip Interconnections for High-Performance Microprocessors," IBM J. Res. Develop., no. 5, vol. 39, pp. 547-567, Sept., 1995.

Acknowledgments:

The authors acknowledge contributions by: D. Akeson, D. Ament, R. Badeau, S. Bakke, A. Barber, S. Bell, L. Biro, V. Bolkhovsky, W. Bowhill, E. Burdick, S. Butler, A. Cave, R. Cvijetic, D. Dever, D. Donchin, R. Dupcak, T. Fischer, K. Gillespie, W. Grundmann, H. Harkness, J. Heath, C. Holub, A. Hughes, D. Jackson, J. Keefe, J. Kowaleski, M. Lamere, J.T. Lin, W. McGee, S. Meier, A. Mendoza, S. Miller, B. Molhollen, J. Mylius, D. Noorlag, L. Olsen, J. Pickholtz, D. Ramey, N. Raughley, N. Rethman, M. Saldana, J. Siegel, C. Stark, M. Tareila, S. Thierauf, Y. Watanabe, G. Watt, J. White, T. Zou, and the many other contributors in the Architecture, Verification, Layout and CAD teams, and thank the CSEM PLL design team.

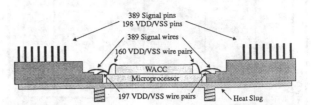

Figure 6: Microprocessor, WACC and IPGA assembly.

Feature size	0.35µm
Channel length	0.25µm
Gate oxide	6.0nm
VTXn / VTXp	0.35V / -0.35V
Power supply	2.0V
Substrate	P-EPI with N-Well
Salicide	Cobalt-disilicide
Metal 1	5.7kÅ AlCu, 1.225µm contacted pitch
Metal 2	5.7kÅ AlCu, 1.225µm contacted pitch
Reference plane 1	14.8kÅ AlCu, VSS plane
Metal 3	14.8kÅ AlCu, 2.80µm contacted pitch
Metal 4	14.8kÅ AlCu, 2.80µm contacted pitch
Reference plane 2	14.8kÅ AlCu, VDD plane

Table 1: CMOS process technology.

High-Performance Microprocessor Design

Paul E. Gronowski, *Member, IEEE*, William J. Bowhill, *Member, IEEE*, Ronald P. Preston, *Member, IEEE*,
Michael K. Gowan, and Randy L. Allmon, *Associate Member, IEEE*

Abstract— Three generations of Alpha microprocessors have been designed using a proven custom design methodology. The performance of these microprocessors was optimized by focusing on high-frequency design. The Alpha instruction set architecture facilitates high clock speed, and the chip organization for each generation was carefully chosen to meet critical paths. Digital has developed six generations of CMOS technology optimized for high-frequency design. Complex circuit styles were used extensively to meet aggressive cycle time goals. CAD tools were developed internally to support these designs. This paper discusses some of the technologies that have enabled Alpha microprocessors to achieve high performance.

Index Terms— Alpha, CMOS digital integrated circuits, computer architecture, flip-flops, integrated circuit design, logic design, microprocessors.

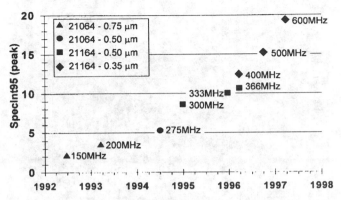

Fig. 1.   Alpha performance versus time.

I. INTRODUCTION

DIGITAL introduced the Alpha 21064 in 1992, the highest performance microprocessor in the industry at that time [1]. Digital has delivered three generations of high-performance Alpha microprocessors through process advancements, architectural improvements, and aggressive circuit design techniques. Fig. 1 shows the integer performance of the 21064 and 21164 as a function of time. Over the last five years, the clock frequency of the Alpha microprocessor has increased from 150 to 600 MHz. The 21264, the third-generation Alpha, has been designed to operate at 600 MHz with improved performance over the 21164 [2]. Table I contains the key features for each of these microprocessors.

The 21064 (Fig. 2) was the first implementation of the Alpha architecture. It was designed to operate at 200 MHz in a 0.75-μm n-well CMOS process, allowing for roughly 16 gate delays per cycle including latching. Power dissipation is 30 W from a 3.3-V power supply at 200 MHz. The die measures 2.3 cm^2, and contains 1.68 million transistors, half of which are dedicated to noncache logic.

The second-generation Alpha microprocessor, the 21164 (Fig. 3), is fabricated in a 0.5-μm n-well CMOS process [3]. It was designed to operate at 300 MHz using a 3.3-V supply, and it dissipates 50 W. The number of gate delays per cycle was reduced from 16 to 14 on this design to provide an additional 10% reduction in cycle time beyond process scaling. The die is roughly 3.0 cm^2 and contains 9.3 million total transistors. The noncache transistor count is tripled from the previous generation design to 2.5 million. Although originally designed

TABLE I
MICROPROCESSOR FEATURES

	21064	21164	21264
Transistor Count (million)	1.68	9.3	15.2
Die Size (mm^2)	16.8x13.9	18.1x16.5	16.7x18.8
Process Technology	0.75 μm	0.50 μm	0.35 μm
Power Supply (Volts)	3.3	3.3	2.2
Power Dissipation (Watts)	30	50	72
Target Design Frequency (MHz)	200	300	600
Typical Gate Delays/Cycle	16	14	12
On-chip Cache	8-KB L1-I	8-KB L1-I	64-KB L1-I
	8-KB L1-D	8-KB L1-D	64-KB L1-D
		96-KB L2	
Instruction Issue/Cycle	2	4	6
Execution Flow	in-order	in-order	out-of-order

to operate at 300 MHz, migration of this design to a 0.35-μm process has increased the operating frequency to 600 MHz.

The 21264 (Fig. 4) is the third-generation Alpha microprocessor. It is designed in a 0.35-μm n-well CMOS process, and is targeted to operate at 600 MHz. The number of gate delays per cycle has been further reduced to 12, again providing an additional 10% reduction in cycle time relative to the previous design. A nominal supply voltage of 2.2 V is used to limit power dissipation to an estimated 72 W, but the design and process can operate reliably up to 2.5 V. The die is 3.1 cm^2, and contains 15.2 million transistors. The noncache transistor count is more than double that of the 21164.

To achieve high performance without impacting time-to-market, a careful balance among microarchitectural features, process complexity, and circuit design style was required on each of these microprocessors. The use of high-performance circuit design techniques required the development of many custom CAD tools, and added to the complexity of the circuit verification task.

Manuscript received September 1997; revised December 17, 1997.

The authors are with Digital Semiconductor, Digital Equipment Corporation, Hudson, MA 01749 USA.

Publisher Item Identifier S 0018-9200(98)02228-8.

Reprinted from *IEEE Journal of Solid-State Circuits*, Vol. 33, No. 5, pp. 676-685, May 1998.

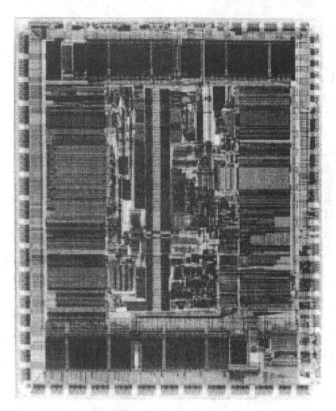

Fig. 2. 21064 die photo.

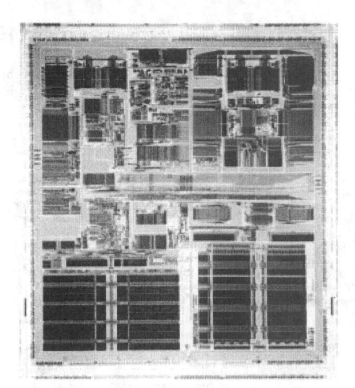

Fig. 4. 21264 die photo.

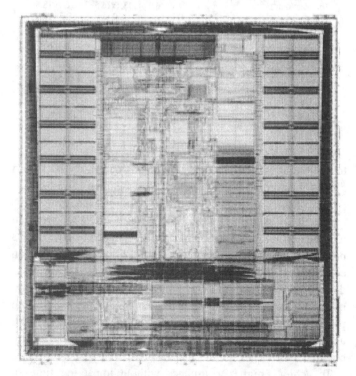

Fig. 3. 21164 die photo.

II. ARCHITECTURE

The Alpha instruction set architecture is a true 64-bit load/store RISC architecture designed with emphasis on high clock speed and multiple instruction issue [4]. Fixed-length instructions, minimal instruction ordering constraints, and 64-bit data manipulation allow for straightforward instruction decode and a clean microarchitectural design. The architecture does not contain condition codes, branch delay slots, adaptations from existing 32-bit architectures, and other bits of architectural history that can add complexity. The chip organization for each generation was carefully chosen to gain the most advantage from microarchitectural features while maintaining the ability to meet critical circuit paths.

The 21064 is a fully pipelined in-order execution machine capable of issuing two instructions per clock cycle. It contains one pipelined integer execution unit and one pipelined floating-point execution unit. Integer instruction latency is one or two cycles, except for multiplies which are not pipelined. Floating-point instruction latency is six cycles for all instructions except for divides. The chip includes an 8-kB instruction (I) cache and an 8-kB data (D) cache. The emphasis of this design was to gain performance through clock rate while keeping the architecture relatively simple. Subsequent designs rely more heavily on aggressive architectural enhancements to further increase performance.

The quad-issue, in order execution implementation of the 21164 was more complex than the 21064, but simpler than an out-of-order execution implementation [5]. It contains two pipelined integer execution units and two pipelined floating-point execution units. The first-level cache was changed to nonblocking. A second-level 96-kB unified I and D cache was added on-chip to improve memory latency without adding excessive complexity. Integer latency was reduced to one cycle for all instructions, and was roughly halved for all MUL instructions. The floating-point unit contains separate add and multiply pipelines, each with a four-cycle latency [6]. Floating-point divide latency is reduced by 50%.

The trend of increased architectural complexity continues with Digital's latest Alpha microprocessor. The 21264 gains

TABLE II
TECHNOLOGY FEATURES

Technology	CMOS4	CMOS5	CMOS6
Flagship CPU	21064	21164	21264
Feature Size (μm)	0.75	0.50	0.35
Channel L_{eff} (μm)	0.50	0.35	0.25
Gate Oxide (nm)	10.5	9.0	6.0
V_{tn}/V_{tp}	0.5/-0.5	0.5/-0.5	0.35/-0.35
Power Supply (V)	3.3	3.3	2.0-2.5
M1 thick/pitch (μm)	.75/2.25	.85/1.5	.62/1.225
M2 thick/pitch (μm)	.75/2.675	.85/1.75	.62/1.225
Reference Plane 1	n/a	n/a	Vss
M3 thick/pitch (μm)	2.0/7.5	1.53/5.0	1.53/2.8
M4 thick/pitch (μm)	n/a	1.53/6.0	1.53/2.8
Reference Plane 2	n/a	n/a	Vdd

significant performance from six-way-issue and out-of-order execution. It contains four integer execution units and two floating-point execution units. The size of the $L1$ instruction and data caches was increased from 8 to 64 kB, eliminating the need for an on-chip $L2$ cache. Integer multiply latency was reduced and full pipelining improved throughout. The floating-point latency remained at four cycles, but the divide latency was reduced by another 50%. In addition, the ISA was extended to include square root and to support multimedia instructions.

Despite the added architectural complexity, clock frequencies have continued to improve due to circuit design enhancements and advances in process technology.

III. TECHNOLOGY

Digital Semiconductor has developed six generations of CMOS process technology, with a new technology for each major microprocessor design. The microprocessor design occurs in parallel with the development of the manufacturing process. Therefore, close cooperation is required between the process development and microprocessor design teams to perform this concurrent design and ensure optimum chip performance. Table II highlights the key features of the three process technologies used to produce these three microprocessors. The processes were optimized for high-frequency microprocessor design. In particular, emphasis is placed on low V_t's and very short L_{eff}'s which increase drive current at the cost of higher leakage.

Close interaction between the circuit design team and the process development team also results in the following benefits.

1) Early process information and timely updates of technology parameters are provided to the design teams, allowing circuit design to start before the process is fully defined.
2) Early design work provides valuable feedback to the process team to ensure that target process performance is met.
3) Major process features such as number of interconnect layers, interconnect pitch, and device characteristics are managed in the context of the overall chip design.
4) The design of critical structures such as RAM arrays and data paths can be optimized through process and circuit design.

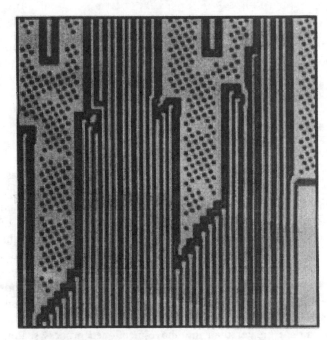

Fig. 5. 21264 metal fill.

5) Scaling issues for future process shrinks may be uncovered.

A. Definition of Design Rules

One of the key areas where close collaboration is required between design and process development teams is the definition of layout design rules. Aggressive design rules can result in increased circuit density, and can potentially improve overall chip performance. However, design rules that are too aggressive will complicate manufacturing, and may impact yield. On the other hand, slack design rules may result in increased die size, resulting in increased distances between critical structures. This increased distance results in higher capacitance, larger RC routing delays, and lower chip performance.

Often, the process team can be more aggressive if limits are placed not only on the minimum widths and spaces of structures, but also on the maximum widths and spaces. The 21264 implements metal fillers to limit the maximum spacing between adjacent lines. The fill metal is automatically placed in the design and tied to V_{DD} or V_{SS}. For large areas of fill metal, stress relief holes are automatically placed in the fill pattern. Metal fill may increase the capacitance of nearby signal lines, but they also result in improved interlayer dielectric uniformity. The improved uniformity allows the process to be targeted more aggressively. Fig. 5 shows filler polygons inserted in the gaps between widely spaced lines.

IV. CIRCUIT DESIGN

Advanced process technologies have allowed the Alpha microprocessor designers to increase performance on each new generation chip through two means, better electrical properties and higher densities. First, scaling the physical dimensions of the transistors and interconnect has reduced the nodal

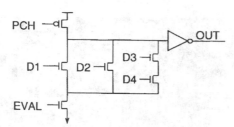

Fig. 6. Dynamic logic example.

capacitance, and has allowed the circuit speed to increase proportionally with the inverse of the technology physical scale factor. Second, the reduced size of the devices and interconnect have enabled designs to double the amount of circuitry that can be implemented in the same die size. The increase in circuit area is inversely proportional to the square of the technology scaling.

A primary objective of each new microprocessor development has been to further increase the performance of each generation of microprocessor by increasing the clock frequency by an amount greater than the above-mentioned process scaling factor. This has been achieved through the use of sophisticated circuit design techniques.

Full-custom circuit design methodologies have been used universally by the microprocessor design teams. All circuits are designed at the transistor level, and are uniquely sized to meet speed and area goals. Custom layout design techniques are used to optimize parasitic capacitance for all circuits. Automatic synthesis approaches for logic and circuit design have been used for fewer than 10% of the circuits.

The use of a full-custom design methodology gives the designer flexibility. The Alpha microprocessors have been implemented with a wide range of circuit styles including conventional complementary CMOS logic, single- and dual-rail dynamic logic, cascode logic pass transistor logic, and ratioed static logic [7].

Dynamic circuits are one of the most commonly used circuit styles, and are present in both data path and random control structures. Dynamic logic has many advantages, but it requires careful analysis to ensure functionality. Fig. 6 shows a simple dynamic domino gate. Dynamic gates allow wide OR structures to be implemented in a single gate which otherwise would require many levels of complementary logic. Dynamic gates are also faster than their complementary gate equivalents for several reasons. First, eliminating the PMOS transistor network reduces both the gate fan-in and fan-out capacitance. Second, the switching point of the dynamic gate is set by the NMOS device threshold voltage. If the timing of the inputs is such that they are not asserted until after the precharge clock has been deasserted, there in no crossover current during the output transition. Finally, removing the PMOS transistor network reduces the layout area, which results in lower interconnect capacitance, further increasing the speed of the circuit. However, dynamic circuits are very sensitive to noise, and require very careful design and extensive verification to ensure functionality. Much of this verification has been automated, and will be discussed in the CAD tool section of this paper.

Another circuit style that is widely used in the micropro-

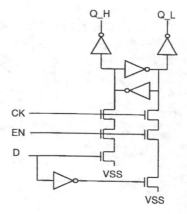

Fig. 7. Cascode logic example.

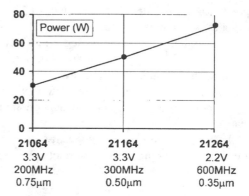

Fig. 8. Power dissipation.

cessors is dual-rail cascode logic [8]. A simple gate is shown in Fig. 7. Like dynamic logic, cascode logic has been used in both data path and random control logic areas. It has many of the same advantages that dynamic logic possesses over complementary CMOS. The fan-in and fan-out capacitance are both lower, thus reducing delay. Large complex functions such as multiplexers and XOR gates can be easily implemented in a single cascode gate with both true and complement outputs. Finally, a latch function can easily be constructed with the addition of one pair of transistors (see Fig. 7).

A. Power Dissipation and Supply Distribution

Power consumption has been an important design constraint for the Alpha microprocessor designers. Fig. 8 shows the average power dissipation, process technology, and nominal supply voltage for the first three Alpha microprocessor designs. The graph clearly illustrates a steady increase in power despite the use of advanced technologies and scaled power supply voltages. This increase in power has resulted from a number of factors. First, more complex architectural features have been included into the design of each generation microprocessor. Second, the use of sophisticated circuit techniques has allowed clock frequencies to increase faster than pure process scaling would have provided. Third, aggressive transistor design (high Id_{sat}) has increased the magnitude of subthreshold leakage significantly. Finally, improvements in compiler and software technologies have increased the switching activity within the microprocessor.

398

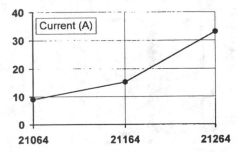

Fig. 9. Power supply current.

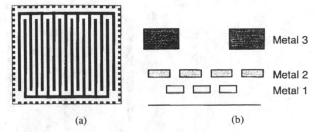

Fig. 10. (a) 21064 metal layers and (b) power distribution.

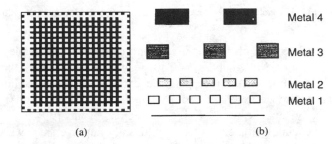

Fig. 11. (a) 21164 metal layers and (b) power distribution.

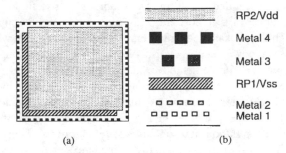

Fig. 12. (a) 21264 metal layers and (b) power distribution.

Power supply scaling has been an important lever to slowing the rise in power consumption. However, if current rather than power is considered, a much more disturbing trend is seen (see Fig. 9). Supply current is roughly doubling with each new generation. The problem of on-chip power distribution is increasing with each generation microprocessor. A design goal has been to limit the combination of dc IR drops and inductive ringing on the chip to 10% of the total supply voltage. As a side effect of scaling V_{dd} to achieve lower power levels, the amount of acceptable power supply noise is also reduced. Therefore, each generation of microprocessor has required additional process options to be added to each technology generation to lower the power supply impedance.

In CMOS logic, circuit speed is directly related to supply voltage. Therefore, in order to achieve high clock frequencies, the power supply networks must be designed to supply the required current with minimal IR drops. The 21064 consumed 30 W, and is fabricated in a 0.75-μm process. Distributing the necessary supply current across the 2.3-cm^2 die, with an acceptable IR drop, would not have been possible using the existing two-level metal process. Therefore a third, low-resistance aluminum interconnect layer ($M3$) was added to the process. Adding layers to a fabrication process increases the cost of the die both through extra masking steps and reduction in yield. Therefore, relatively large minimum geometries were selected for the third metal layer to minimize yield impact. This new layer was used mainly for power, ground, and clock distribution.

V_{dd} and V_{ss} $M3$ lines were alternated to form two interleaved combs as shown in Fig. 10. To further reduce the effective resistance of the power grid, the $M3$ lines were routed with a pitch of 15 μm or about twice the process minimum. The second metal layer ($M2$) was used to strap together the $M3$ lines, forming a grid for power, ground, and clock. One disadvantage of this power-routing scheme is that the V_{dd} and V_{ss} bond pads on the left and right sides of the die are connected to the grid using the higher resistance $M2$.

The 21164 consumed nearly twice the power of the 21064. The $M2/M3$ grid structure used on the 21064 was not adequate to meet the 21164 power requirements. Additionally, reliability concerns in connecting V_{dd} and V_{ss} to $M3$ would have required dedicating all of the bond pads on two sides of the chip to power and ground. The floor plan constraints of this bond pad allocation would have significantly complicated chip layout. Therefore, a fourth layer ($M4$) of aluminum interconnect was added to the new 0.5-μm process. $M4$ was

routed perpendicular to the $M3$ to form a two-dimensional (2-D) grid of V_{dd} and V_{ss} as shown in Fig. 11. A CAD tool was used to automatically contact the intersections between $M3$ and $M4$ with a maximum number of contacts. The 2-D grid allowed for bond pads on all four sides of the die to contribute to supplying the current demand of the 21164. The $M3$ and $M4$ routing layers were also used to rout a limited number of global signals and for clock distribution.

The power dissipation on the 21264 increased to 72 W despite a V_{dd} reduction from 3.3 to 2.2 V, increasing the supply current to 33 A. In addition, conditional clocking exaggerates the cycle-to-cycle current variation causing a maximum delta-I_{dd} of 25 A between adjacent cycles. As a result, the two-dimensional grid used on the 21164 was no longer sufficient.

In order to meet the very large cycle-to-cycle current variations, two thick low-resistance aluminum reference planes were added to the process. Reference plane 1, tied to V_{ss}, was added between $M2$ and $M3$. Reference plane 2 was added above $M4$, and is connected to V_{dd}. Fig. 12 provides a cross section of the metallization and power routing of the 21264. Contacting the reference planes to adjacent layers was automated.

The use of planes has several beneficial effects. Nearly the entire die area is available for power distribution. Second, the

399

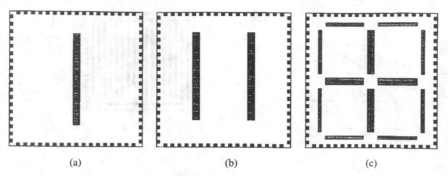

(a)   (b)   (c)

Fig. 13. Final clock driver location: (a) 21064, (b) 21164, and (c) 21264.

lower reference plane inductively and capacitively decouples $M2$ from $M3$ signal lines. This reduces on-chip crosstalk and simplifies CAD tool parasitic extraction. Finally, solid planes provide excellent current return paths which minimize inductive noise caused by signal-switching events [9].

The additional metal layers significantly improve the power distribution on the chips, and help to reduce the voltage loss in the center of the die due to dc IR drop. However, the high clock frequencies of these microprocessors result in large fluctuations in I_{dd} current. This current must be supplied through the package lead and bond wire inductance, and which results in power supply noise on chip. On-chip decoupling capacitance is implemented to help reduce this noise.

The power supply is decoupled using the gate oxide of NMOS transistors to form a capacitor. This type of structure was chosen as it provides the highest efficiency in capacitance per unit area without introducing additional process steps. The capacitor design provides a low-impedance path to the V_{ss} terminal, improving the bandwidth of the capacitor enough to efficiently decouple high-frequency noise on the power grids. A decoupling capacitor standard cell was designed and automatically placed in the chip. Whenever possible, the capacitor was placed in vacant areas of the chip where it did not impact die area, e.g., underneath global metal 1 buses. However, in areas close to the clock generators, it was required to dedicate die area to decoupling capacitance to supply the large clock switching currents. In total, 15–20% of the die area is used for decoupling.

The 21264's conditionally clocking scheme and the super-scalar architecture of the microprocessor exaggerated the data and program dependencies that result in large variations in supply current. For the 21264, It was not possible to integrate enough decoupling capacitor on chip to manage this noise. Therefore, an additional source of decoupling was added to the chip and package network. A 1-μF 2-cm^2 wirebond attached chip capacitor (or WACC), implemented as a p-type accumulation mode MOS capacitor, was bonded on top of the microprocessor die [3]. This silicon capacitor helped control power supply noise.

B. Clock Distribution

The high frequencies of the Alpha microprocessors have required the generation and distribution of a very high-quality clock signal and the use of fast (low-latency) latches. The primary objective of the clock system is to not limit the per-

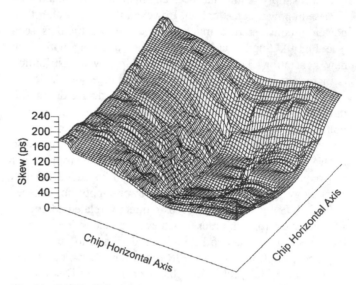

Fig. 14. 21064 clock skew.

formance of the microprocessor. Uncertainties in clock edges resulting from power supply noise, process variation, and interconnect RC delay lower the maximum clock frequency of the microprocessor. In addition, slow clock edges introduce uncertainties in latch timing which further limit performance and can lead to functional failures due to latch race-through.

The 21064 uses a two-phase single wire clocking scheme. The driver is located in the center of the die as shown in Fig. 13(a). The final clock load was 3.5 nF, and it required a final driver with a gate length of 35 cm. To handle the large di/dt transient currents in the power grid when the clock driver switched, on-chip decoupling structures (NMOS transistor with the gate tied to V_{dd} and the source and drain tied to ground) were placed around the clock driver. Roughly 10% of the chip area was allocated to decoupling capacitance. Fig. 14 shows the results of the 21064 clock skew analysis.

Fig. 15 shows the approximate power breakdown of the first two microprocessors. Since the main clock drivers consumed 40% of the chip power, thermal management was a major concern. Fig. 16 shows that the temperature of the main clock driver is elevated about 30°C relative to the rest of the die. The elevated temperature in the clock driver area reduces the performance of the clock drivers and other local logic, directly impacting performance.

The primary goals of the 21164 clock design were to reduce the clock skew by 30% and to reduce the thermal gradients.

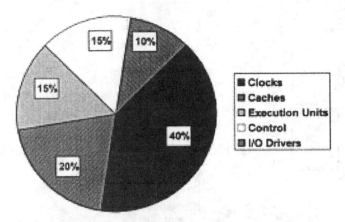

Fig. 15.   Approximate power breakdown of the 21064 and the 21164.

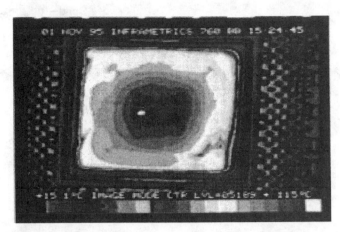

Fig. 18.   21164 thermal image.

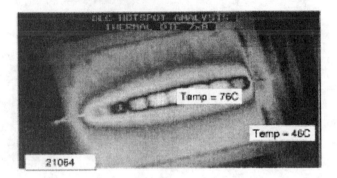

Fig. 16.   21064 thermal image.

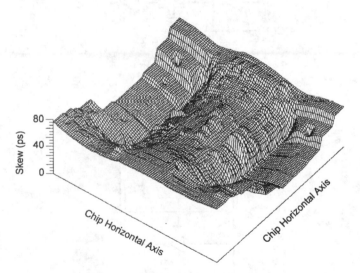

Fig. 17.   21164 clock skew.

Fig. 13(b) illustrates the location of the main clock drivers on the 21164. The main clock driver is split into two banks and is placed midway between the center of the die and the edges. A predriver is located in the center of the die to distribute the clock to the two main drivers. The clock skew was reduced by a factor of 2 using this approach (Fig. 17). In addition, by distributing the main clock driver over a larger area, the localized heating seen on the 21064 was reduced. The thermal image of the 21164 in Fig. 18 shows reduced temperature gradient.

As more aggressive circuit techniques and complex microarchitectural features were implemented in the 21264, power consumption became a major concern in designing the clocking system. A single wire global clock (GLK) is routed over the entire chip as a global timing reference. The GCLK drivers are distributed in a window pane pattern as shown in Fig. 13(c) to reduce clock grid *RC* delay and distribute clock power.

GCLK is the root of a hierarchy of thousands of buffered and conditioned local clocks used across the chip as shown in Fig. 19. There are several advantages to this clocking scheme. First, conditioning the local clocks saves power. Second, circuit designers can take advantage of multiple clocks. For example, a phase path can be extended by initiating it with GCLK and terminating it with a delayed section clock. This approach significantly complicates race and timing verification, which will be discussed later in this paper. Finally, using local buffering significantly lowers the GCLK load, which reduces GCLK skew. Extensive electrical analysis was performed on the clock distribution network. The GCLK skew is less than 75 ps as shown in Fig. 20.

C. Latch Design

Latch design is another important element of the microprocessor circuit design strategy [10]. In order to ensure proper operation across all operating conditions, clock and latch circuits cannot be designed independently. Each generation of the Alpha microprocessor has utilized latches with improved characteristics combined with the improvements to the clock distribution networks previously described.

The high clock frequencies and small number of gate delays available per cycle on Alpha implementations has made low-latency latch design essential. In addition to high speed, other primary goals in latch design are minimal area and clock loading, low power dissipation, and low setup and hold times. The capability of embedding a logic function directly in the latch also helps reduce the number of gate delays per cycle. The 21064 was Digital's first microprocessor to use a two-phase, single wire-clocking scheme. This was a radical departure from the four-phase scheme that allowed for race-free design in previous versions of Digital's microprocessors. This major change in design methodologies has required designers to develop new strategies to manage noise and race-through issues.

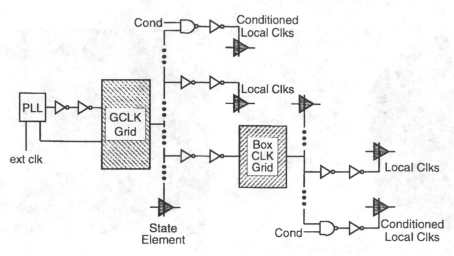

Fig. 19. 21264 clock hierarchy.

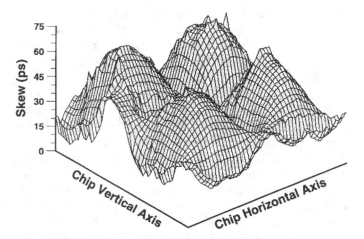

Fig. 20. 21264 GCLK skew.

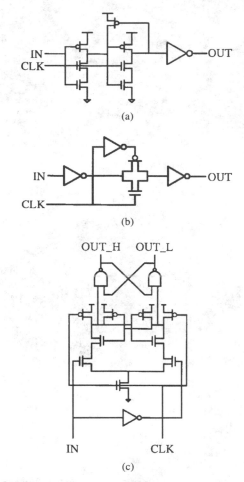

Fig. 21. Latches.

To reduce the chance of data race-through on the 21064, a variation of the true single-phase clocked (TSPC) level-sensitive latch developed by Svensson and Yuan was used [11], [12]. An example of this latch is shown in Fig. 21(a). These latches use the unbuffered main clock directly, significantly increasing race immunity. As long as the main clock edge rate is kept reasonably fast compared to the latch delay, there is little chance for data race-through. An additional benefit of this latch is that the first stage can incorporate a simple logic function. The combined delay of the high- and low-phase latches consumed about 25% of the cycle time.

One goal of the 21164 design was to increase the clock frequency by more than could be provided by process scaling. To accomplish this goal, the number of typical gate delays per cycle was reduced from 16 to 14. To offset the reduction in available gates per cycle, a lower latency latch was employed. Fig. 21(b) shows the basic dynamic CMOS transmission gate latch used on the 21164 [13]. This latch requires true and complementary clock signals, one of which is generated locally. The clock buffer for each latch type was custom designed so that its delay and edge rate characteristics could be tightly controlled. This additional buffer delays the clocking of the latch by one gate delay after the global clock transitions.

However, since the preceding latch opens with the global clock transition, the possibility of latch race-through is significantly increased. In order to minimize the possibility of data race-through with the use of these latches, at least one minimum logic delay element was required between all latches. This constraint was easily verified with a simple CAD tool.

Consistent with the previous latch family, simple logic elements were built into the input stage, thus lowering the overhead associated with the use two of these latches to

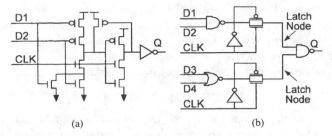

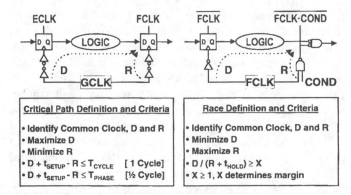

(a)   (b)

Fig. 22. Embedding logic into latches: (a) 21064 Function Latch: one level of logic; (b) 21164 Function Latch: two levels of logic.

Fig. 23. Critical path and race analysis.

15% of the cycle time. In some instances, both inverters were replaced with logic gates, further reducing the latching overhead. Examples of embedded logic in the two latch families are shown in Fig. 22. It can be seen that the cost of latching in the 21164 could be reduced to a minimum of one pass transistor.

Again, a primary goal of the 21264 design was to further increase the clock frequency by more than the process scale factor. Additionally, the 21264 utilized many conditional clocks to reduce power, thus requiring a static latch design. A family of edge-triggered flip-flops, based on the dynamic flip-flip show in Fig. 21(c), was developed to simplify the timing and race issues that were exacerbated by the addition of conditional clocking. Despite reducing the cycle time, the latching overhead was kept constant by using a flip-flop-based design that required only one latching element per clock cycle. The change from level-sensitive to edge-triggered design techniques and the use of multiple clock buffers introduced a number of new timing issues to the design.

The capability of buffering and conditioning the main clock is possible as long as each circuit satisfies both its critical path and race requirements. The left example in Fig. 23 illustrates the use of two buffered section clocks, ECLK and FCLK, in a one-cycle path. The circuit on the right of the figure combines a buffered local clock and a conditioned local clock to define a one-phase path. For both examples, critical path and race analysis start with the identification of the common clock initiating both the receive and drive paths, denoted R and D, respectively. Every critical path or race is defined by a single common clock and a pair of receive and drive paths. In the examples, GCLK and FCLK are the respective common clocks. Critical path analysis verifies that the difference in delay between the drive path D plus the receiver setup time and the receive path R does not exceed the phase or cycle time of the common clock. For worst case analysis, effects that minimize R and maximize D are considered. The converse is true for races; effects that maximize R and minimize D are considered. When the ratio of delay of the drive path D to the receive path R, including hold time, is equal to 1, the circuit is on the verge of failing. Ratios (denoted by X) exceeding 1 imply a margin that may be used to account for effects not included in the analysis. Given the clocking hierarchy and timing methodology used on the 21264, the ability to control and predict, in relative terms, the minimum and maximum path delays was essential in accurately predicting

cycle time and ensuring functionality. Factors that could not be managed by the design were accounted for by additional margin.

V. CAD TOOLS AND VERIFICATION

Custom circuit techniques allow designers to build very high-speed circuits. However, the use of these custom circuits requires design expertise and detailed postlayout electrical verification. Commercially available EDA design systems and point CAD tools have very limited support for these techniques. Therefore, Digital developed an extensive suite of in-house CAD tools to facilitate the design of custom VLSI microprocessors. Internally developed CAD tools are used extensively in all phases of microprocessor design, from initial performance evaluation, through circuit implementation and final design verification. The internally developed CAD suite includes tools for schematic and layout entry, two-state and three-state logic simulation, RTL versus schematic equivalence checking, static timing analysis and race verification, parameter and netlist extraction, and electrical analysis and verification.

A. Electrical Verification

Electrical verification covers all circuit issues that are not related to logic functionality such as timing behavior, electrical hazards, and reliability. Electrical hazards result from noise sources interfering with the logical functions of the chip, and include charge sharing, interconnect capacitive and inductive coupling, power supply IR noise, and noise-induced minority-carrier charge injection [14]. Reliability checks include metal and via electromigration, transistor hot-carrier damage, ESD, and latch-up failure.

The primary goal of the electrical verification tools is to verify that all circuits conform to the project design methodology. The design methodology defines an acceptable set of circuit styles and sizing rules that, when followed, ensures functionality with minimal analysis. The methodology also forces a consistent design style to be used project-wide, which has the added benefit of simplifying CAD design. However, occasionally, there is a need to design circuits outside the methodology to meet performance or area goals. In these instances, additional manual verification is required to ensure functionality.

Detailed analysis requires complex models with many process and circuit variables. Many checks are complex and hard to define as procedures that can be completely automated. Exact chip-wide analysis is impractical; instead, the tools perform design filtering. The tools filter out all circuits that can easily be validated while identifying the small number of circuits that may have problems and require additional analysis. This approach focuses design attention on potential problem areas, and therefore helps improve overall design efficiency. The CAD tools perform over 100 unique electrical checks. Some of the major areas of focus are circuit topology violations, dynamic node checks, including charge sharing, IR noise, injection, and leaker usage, interconnect coupling, noise margin checks, writeability checks, latch checks, beta ratio checks, gate fan-in and fan-out restrictions, transistor size and stack height limitations, max/min edge rates and delays, and power consumption. Some of the checks are applied to all circuit styles, while other checks are required for specific circuit types. The CAD tools require a large amount of design information to perform these checks, including electrical parameter extraction from layout, device electrical characteristics, relative circuit locations, timing information, and transistor connectivity.

B. Functional and Logical Verification

The functional complexity and time-to-market pressures of microprocessor design necessitated the development of an extensive functional verification strategy covering all phases of the design process from functional definition through manufacturing tests.

During the microarchitectural design phase of chip development, a two-state RTL behavioral model is the primary verification vehicle. This model provides a balance between design detail and simulation speed. The model is combined with abstract behavioral models of the other system components to verify correct operation of the processor in the system environment. Once logic design is complete, a two-state gate-level simulation model is extracted from the circuit schematics. This model is used to ensure that the schematics match the RTL model. Finally, to verify correct initialization of the circuits at power-up a three-state switch-level model is extracted from the circuit schematics.

A wide variety of simulation stimuli is used to verify the design, including hand-coded test patterns and randomly generated test patterns. Coverage analysis guides the verification process. Many of the manufacturing test patterns are derived from the simulation stimuli. Fault simulation is used to direct test enhancement.

VI. FUTURE CHALLENGES

As clock frequencies continue to increase with each new technology, many significant challenges are on the horizon. Power consumption has increased, despite voltage scaling, to the point of being a first-order concern. Designers must identify innovative solutions to lower power without significantly impacting performance. In addition, the distribution of a chipwide timing reference with extremely fast edge rates poses a significant challenge. The size and architectural complexity of the next generation microprocessor will require improved design and verification methodologies. Finally, designer productivity must increase significantly, requiring innovative CAD tools to meet time-to-market constraints.

VII. CONCLUSION

This paper has reviewed Digital's approach to high-performance microprocessor design. Three generations of Alpha microprocessors have been designed and optimized for performance by focusing on high-speed design using fully custom circuit design techniques, incorporating state-of-the-art architectural features, and utilizing high-performance CMOS technology for fabrication. An extensive suite of in-house CAD tools is used to analyze complex electrical and logical behavior to ensure functionality. All three microprocessors booted multiple operating systems using first-pass silicon, validating the design and verification methodologies.

ACKNOWLEDGMENT

The microprocessors described in this paper have resulted from the work of a tremendous number of people. The authors would like to recognize the technical contributions of the 21064 design team, including D. Dobberpuhl, R. Witek, J. Montanaro, L. Madden, and S. Samudrala; the 21164 design team, including J. Edmondson and P. Rubinfeld; and the 21264 design team, including B. Gieseke, J. Keller, and D. Priore. Finally, they would also like to acknowledge the significant contributions made to these microprocessors by the verification, CAD, process development, and manufacturing teams.

REFERENCES

[1] D. Dobberpuhl et al., "A 200 MHz 64 b dual-issue CMOS microprocessor," IEEE J. Solid-State Circuits, vol. 27, pp. 106–107, Nov. 1992.
[2] W. Bowhill et al., "A 300 MHz 64 b quad-issue CMOS microprocessor," in ISSCC Dig. Tech. Papers, Feb. 1995, pp. 182–183.
[3] B. Gieseke et al., "A 600 MHz superscalar RISC microprocessor with out-of-order execution," in ISSCC Dig. Tech. Papers, Feb. 1997, pp. 176–177.
[4] R. Sites and R. Witek, Alpha AXP Architecture Reference Manual, 2nd ed. Boston, MA: Digital, 1995.
[5] J. Edmondson et al., "Superscalar instruction execution in the 21164 Alpha microprocessor," IEEE Micro, vol. 15, pp. 33–43, Apr. 1995.
[6] J. Kowaleski et al., "A dual-execution pipelined floating-point CMOS processor," in ISSCC Dig. Tech. Papers, Feb. 1996, pp. 358–359.
[7] B. Benschneider et al., "A 300-MHz 64-b quad-issue CMOS RISC microprocessor," IEEE J. Solid-State Circuits, vol. 30, pp. 1203–1214, Nov. 1995.
[8] L. Heller and W. Griffin, "Cascode voltage switch logic: A differential CMOS logic family," in ISSCC Dig. Tech. Papers, Feb. 1984, pp. 16–17.
[9] D. Priore, "Inductance on silicon for sub-micron CMOS VLSI," in 1993 Symp. VLSI Circuits, Dig. Tech. Papers, May 1993, pp. 17–18.
[10] P. Gronowski and W. Bowhill, "Dynamic logic and latches: Practical implementation methods and circuit examples used on the ALPHA 21164," in 1996 Symp. VLSI Circuits—Proc. VLSI Circuits Workshop.
[11] D. Dobberpuhl et al., "A 200-MHz 64-bit dual-issue CMOS microprocessor," Digital Tech. J., vol. 4, no. 4, pp. 35–50, 1992.
[12] P. Larsson and C. Svensson, "Impact of clock slope on true single phase clocked (TSPC) CMOS circuits," IEEE J. Solid-State Circuits, vol. 29, pp. 723–726, June 1994.
[13] W. Bowhill et al., "Circuit implementation of a 300-MHz 64-bit second-generation CMOS alpha CPU," Digital Tech. J., vol. 7, no. 1, pp. 100–118, 1995.
[14] P. Gronowski et al., "A 433 MHz 64 b quad-issue CMOS RISC microprocessor," in ISSCC Dig. Tech. Papers, Feb. 1996, pp. 222–223.

Chapter 5

High-Performance
Arithmetic Units

HIGH-PERFORMANCE ARITHMETIC UNITS

Computer arithmetic has an intrinsic relationship to technology and the implementation of algorithms in a digital computer, owing to the fact that the value of a particular algorithm is directly related to the actual speed with which this computation is performed. There are other measures of the value of the algorithm: the area of the VLSI chip taken for its implementation, regularity of the implementation, and wireability (the ability to connect the pieces with relatively short and regular interconnections), as well as power consumption, which has recently became a very important feature. A direct and strong relationship exists between the technology in which digital logic is implemented and the way the computation is organized. This relationship is one of the guiding principles in developing computer arithmetic.

Number Representation

Information is represented in a digital computer by a string of bits, that is, zeroes and ones. This relationship is defined by the rule that associates one numerical value designated as X (in the text we will use capital X for the numerical value) with the corresponding bit string designated as x.

$$x = \{x_{n-1}, x_{n-2}, \ldots \ldots, x_0\}$$

where : $x_i \in 0, 1$

In this case the word is n bits long.

When for every value X one and only one corresponding bit string x exists, we define the number system as *nonredundant*. If, however, we can have more than one bit string x that represents the same value X, the number system is said to be *redundant*.

The digit set x_i can be both redundant and nonredundant. If the number of different values x_i can assume is $n \leq r$, then we have a nonredundant digit set; otherwise, if $n > r$, we have a redundant digit set. Use of the redundant digit set has its advantages in efficient implementation of algorithms (multiplication and division in particular).

We also define *explicit value x_e* and *implicit value X_i* of a number represented by a bit-string x. The implicit value is the only value of interest to the user, whereas the explicit value provides the most direct interpretation of the bit-string x. Mapping of the explicit value to the implicit value is obtained by an arithmetic function that defines the number representation used. The arithmetic designer's task is to devise algorithms that result in the correct implicit value of the result for operations on the operand digits representing the explicit values. In other words, the arithmetic algorithm needs to satisfy the *closure* property. The relationship between implicit value and explicit value is illustrated by Table 5.1.

Representation of Signed Integers. The two most common representations of signed integers are Sign and Magnitude (SM) and True and Complement (TC). Although the SM representation might be easier to understand and convert to and from, it has implementation problems. Therefore, we will find that TC representation is more commonly used.

Sign and Magnitude Representation (SM). In SM representation, the signed integer X_i is represented by sign bit x_s and magnitude x_m (x_s, x_m). Usually, 0 represents a positive sign (+), and 1 represents a negative sign (−). The magnitude of the number x_m can be represented in any way chosen for the representation of positive integers. The disadvantage of SM representation is that two representations of zero exist, positive and negative zero: $x_s = 0, x_m = 0$, and $x_s = 1, x_m = 0$.

True and Complement Representation (TC). TC representation does not use a separate bit to represent the sign. Mapping between the explicit and implicit value is defined as

$$X_i = \begin{cases} x_e & x_e \leq C/2 \\ x_e - C & x_e > C/2 \end{cases}$$

For an illustration of TC mapping, see Table 5.2.

In this representation positive integers are represented in the *true form,* whereas negative are represented in the *complement form.*

TABLE 5.1   The Relationship Between the Implicit Value and Explicit Value for $x = 11011$, $r = 2$

Implied Attributes: Radix Point, Negative Number Representation, Others	Expression for Implicit value X_i as a function of explicit value x_e	Numerical implicit value X_i (in decimal)
Integer, Magnitude	$X_i = x_e$	27
Integer, Two's Complement	$X_i = -2^5 + x_e$	−5
Integer, One's Complement	$X_i = -(2^5 - 1) + x_e$	−4
Fraction, Magnitude	$X_i = 2^{-5} x_e$	27/32
Fraction, Two's Complement	$X_i = 2^{-4}(-2^{-5} + x_e)$	−5/16
Fraction, One's Complement	$X_i = 2^{-4}(-2^{-5} + 1 + x_e)$	−4/16

*A. Avizienis, "Digital Computer Arithmetic: A Unified Algorithmic Specification," Symp. Computers and Automata, Polytechnic Institute of Brooklyn, April 13–15, 1971.

TABLE 5.3   Mapping of the Explicit Value x_e into RC and DRC Number Representations

x_e	X_i(RC)	X_i(DRC)
0	0	0
1	1	1
2	2	2
—	—	—
—	—	—
$1/2r^n - 1$	$1/2r^n - 1$	$1/2r^n - 1$
$1/2r^n$	$-1/2r^n$	$-(1/2r^n - 1)$
—	—	—
—	—	—
—	—	—
$r^n - 2$	−2	−1
$r^n - 1$	−1	0

With respect to how the complementation constant C is chosen, we can further distinguish two representations within the TC system:

- If the complementation constant is chosen to be equal to the range of possible values taken by x_e, $C = r^n$ in a conventional number system where $0 \le x_e \le r^n - 1$, then we have defined the *Range Complement* (RC) system.
- If, on the other hand, the complementation constant is chosen to be $C = r^n - 1$, we have defined the *Diminished Radix Complement* (DRC), which is also known as the *Digit Complement* (DC) number system. The RC and DRC number representation systems are shown in Table 5.3.

As can be seen in Table 5.3, the RC system provides for one unique representation of zero because the complementation constant $C = r^n$ falls outside the range. There are two representations of zero in the DRC system, $x_e = 0$ and $r^n - 1$. The RC representation is not symmetrical, and it is not a closed system under the change of sign operation. The range for RC is $[-1/2r^n, 1/2r^n - 1]$. The DRC is symmetrical and has the range of $[-(1/2r^n - 1), 1/2r^n - 1]$.

For the radix $r = 2$ RC and DRC number representations are commonly known as the *Two's Complement* and *One's Complement* number representation systems. These two representations are illustrated Table 5.4 for the range of values $-(4 \le X_i \le 3)$.

Implementation of Elementary Arithmetic Operations

The algorithms for the arithmetic operation are dependent on the number representation system used. Therefore, their implementation should be examined for each number representation system separately, given that the complexity of the algorithm, as well as its hardware implementation, is dependent on it.

Implementation of Fast Addition in VLSI. The first significant speed improvement in the implementation of a parallel adder was the Carry-Lookahead-Adder (CLA) developed by Weinberger and Smith in 1963. Even today the CLA adder is one of the fastest schemes used for the addition of two numbers, given that the delay incurred to add two numbers is logarithmically dependent on the size of the operands (delay = log[N]). The CLA concept is illustrated in Fig. 5.1 *a, b*. For each bit position of the adder, a pair of signals (p_i, g_i) is generated in parallel. Local carries can be generated using (p_i, g_i), as seen in the equations. Those signals are designated as: p_i—*carry-propagate* and g_i—*carry-generate,* because they take part in the propagation and generation of carry signal C_{out}. However, each bit position requires an incoming carry signal C_{in} in order to generate the outgoing carry C_o. This makes the addition slow because the carry signal has to ripple from stage to stage as shown in Fig. 5.1*a*.

The adder can be divided into groups, and *carry-generate* and *carry-propagate* signals can be calculated for the entire group

TABLE 5.2   True and Complement Mapping

x_e	X_i
0	0
1	1
2	2
—	—
—	—
$C/2 - 1$	$C/2 - 1$
$C/2 + 1$	$-(C/2 - 1)$
—	—
—	—
$C - 2$	−2
$C - 1$	−1
C	0

TABLE 5.4   Two's Complement and One's Complement Representations

	Two's Complement	$C = 8$	One's Complement	$C = 7$
X_i	x_e	X_i (2's complement)	x_e	X_i (1's complement)
3	3	011	3	011
2	2	010	2	010
1	1	001	1	001
0	0	000	0	000
−0	0	000	7	111
−1	7	111	6	110
−2	6	110	5	101
−3	5	101	4	100
−4	4	100	3	—

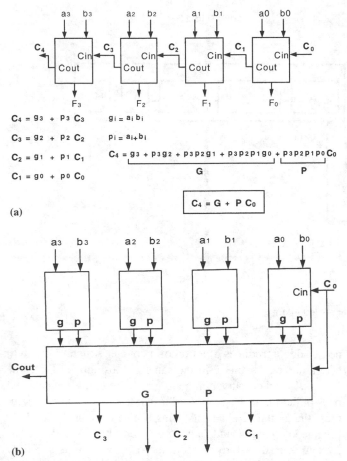

$C_4 = g_3 + p_3 C_3$ $g_i = a_i b_i$

$C_3 = g_2 + p_2 C_2$ $p_i = a_i + b_i$

$C_2 = g_1 + p_1 C_1$

$C_4 = g_3 + p_3 g_2 + p_3 p_2 g_1 + p_3 p_2 p_1 g_0 + p_3 p_2 p_1 p_0 C_0$

$C_1 = g_0 + p_0 C_0$

$$C_4 = G + P C_0$$

(a)

(b)

Fig. 5.1. Lookahead Adder structure. **(a)** Generation of carry generate and propagate signals, **(b)** Generation of group signals G, P and intermediate carries.

(G, P). This will take an additional time equivalent to and AND-OR delay of the logic. However, now we can calculate each group's carry signals in an additional AND-OR delay. For the generation of the carry signal from the adder, only the incoming carry signal into the group is now required. Therefore, the rippling of the carry is limited to the groups. In the next step we may calculate *generate* and *propagate* signals for the group of groups (G^*, P^*) and continue in that fashion until we have only one group left, generating the C_{out} signal from the adder. This process will terminate in a log number of steps, given that we are generating a tree structure for the generation of carries. The computation of carries within groups is done individually as illustrated in Fig. 5.1a, and this process requires only the incoming carry into the group.

The logarithmic dependence on the delay (delay = log[N]) is valid only under the assumption that the gate delay is constant without depending on the fan-out and fan-in of the gate. In practice, this is not true, and even when the bipolar technology (which does not exhibit strong dependence on the fan-out) is used to implement the CLA structure, the further expansion of the carry-block is not possible given the practical limitations of the fan-in of the gate.

In CMOS technology, this situation is much different given that the CMOS gate has a strong dependency not only on fan-in

but on fan-out as well. This limitation takes away many of the advantages gained by using the CLA scheme. However, by clever optimization of the critical path and appropriate use of dynamic logic, the CLA scheme can still be advantageous, especially for adders of a larger size as demonstrated in the paper by Naini et al.

Conditional-Sum Addition. Another fast scheme for the addition of two numbers, which predates CLA, is the Conditional-Sum Addition method proposed by Sklansky in 1960. The essence of CSA is the realization that we can add two numbers without waiting for the carry signal to be available. Simply put, the numbers are added in two instances: one assuming $C_{in} = 0$ and the other assuming $C_{in} = 1$. The results: Sum^0, Sum^1 and $Carry^0$, and $Carry^1$ are presented at the input of a multiplexer. The final values are selected at the time C_{in} arrives at the "select" input of a multiplexer. As in CLA, the input bits are divided into groups, which are added "conditionally." It is apparent that starting from the Least Significant Bit (LSB) position, the hardware complexity starts to grow rapidly. Therefore, in practice, the full-blown implementation of the CSA is seldom seen.

The idea of adding the Most Significant Bit (MSB) portion conditionally and selecting the results once the carry-in signal is computed in the LSB portion is attractive, however. Such a scheme (which is a subset of CSA) is known as Carry-Select Adder. A 26-b Carry-Select Adder consisting of two 13-bit portions is shown in Fig. 5.2.

As can be seen in Fig. 5.2, the block sizes in the particular 13-bit adders used consist of rather unusual numbers: 1–3–5–3–1. Each block is an optimized Carry-Skip Adder (CSA) known as a Variable Block Adder (VBA) as presented in the 1985 paper by Oklobdzija and Barnes. The sizes of the blocks are determined using linear programming techniques, with objective function being the carry signal delay. In this way, not only was the speed of the CSA reduced but also the delay dependency was changed from that of linear in CSA to a square root (delay = Sqrt(N)).

Further optimization of the critical path in a VLSI adder can be obtained if the multiple levels of *"skip"* paths are introduced and optimized. Such optimization does not yield a close form solution, however; in some instances, the speed of such an optimized VBA surpasses that of the CLA (with the groups of regular sizes).

The delay of the critical path in the VBA adder is calculated taking fan-in and fan-out (as well as wire delays) into account, making it more realistic for a VLSI implementation. A particular implementation of 16-bit and 32-bit VBA adders is presented in the paper by Oklobdzija and Barnes. The comparison with CLA and adders based on the Recurrence Solving Scheme (HPA) shows that in spite of its hardware simplicity the VBA adder outperforms both: CLA and HPA (for the 32-bit size). This demonstrates that in VLSI the complexity of the algorithm may quickly pass the point of diminishing returns.

The observation that the multilevel VBA scheme outperformed CLA led to an investigation of the CLA and the way the group sizes have traditionally been determined. The paper by Lee and Oklobdzija shows that the CLA (as traditionally known) is a suboptimal structure. This explains why CLA has been outperformed by a multilevel VBA, which is indeed an

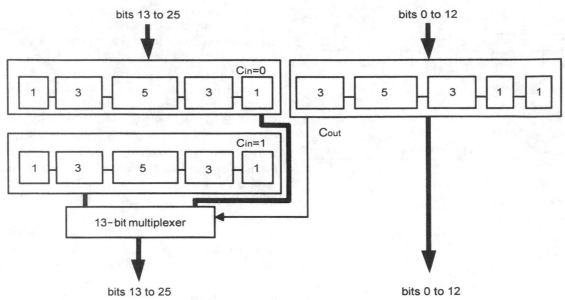

Fig. 5.2. 26-bit Carry-Select Adder.

optimized subset of CLA. The paper by Lee and Oklobdzija also shows how to determine an optimal group size for a speed-optimized CLA. Such optimized CLA outperforms the traditional CLA in terms of speed.

Detection of Leading Zero. In a floating-point arithmetic, normalization is required. In order to "normalize" the number (which consists of shifting the fraction to the left until a nonzero bit is reached), it is necessary to know in advance the number of zeroes before a nonzero bit is reached. Such an operation is known as the *counting of leading-zeroes* and is performed by a circuit known as the Leading Zero Detector (LZD). A LZD circuit consists of a block with N inputs and $\log[N]$ outputs, of which each output is dependent on all of the inputs. When the number of inputs is large (usual case), the design and optimization of such a circuit is not a straightforward problem. Oklobdzija showed how such a design can be treated in a modular and hierarchical way, which greatly simplifies the design of an LZD. In order to determine the structure of the LZD, an algorithm for computing an LZD is devised first. Implementation of such an algorithm from a number of optimized modular blocks not only yields the minimal number of transistors used, but also results in the fastest LZD circuit. Arriving at the design solution in an algorithmic way is important because such an algorithm can be easily integrated into the logic synthesis tools. This helps the logic synthesis tools to produce better and more efficient hardware.

Multiplication. The speed of the multiply operation is of great importance in Digital Signal Processors (DSPs), as well as in general-purpose processors. Therefore, research in building a fast parallel multiplier has been ongoing since C. S. Wallace published the first paper on this subject in 1964. In his historic paper, Wallace outlined a new approach to summing the partial product bits in parallel using a tree of Carry-Save Adders, which later became known as the Wallace Tree shown in Fig. 5.3.

A paper by Dadda suggests a speed improvement of the process of adding partial product bits in parallel. In his 1965 pa-

per, Dadda introduces a notion of a counter structure that will take a number of bits p in the same bit position (of the same *"weight"*) and output a number q which represents the count of ones in the input. Dadda describes a number of ways to compress the partial product bits using such a counter, which later became known as Dadda's counter. Stenzel and Kubitz undertook an extensive study of the use of Dadda's counters in their 1977 study in which they also demonstrate a parallel multiplier built using ROM to implement counters used for partial product summation [Ling, Naffziger].

Amazingly, the quest for an even faster parallel multiplier continued after almost 30 years. But the search for the fastest "counter" did not succeed in yielding a faster partial product summation than that which used the Full-Adder (FA) cell, or "3:2 counter." Therefore, use of the Wallace Tree became almost prevalent in implementing parallel multipliers.

In 1981 Weinberger disclosed a structure that he called the "4-2 carry-save module." This structure contained a combination of FA cells in an intricate interconnection structure, which was yielding faster partial product compression than the use of 3:2 counters. This scheme was further investigated by Santoro, and it became a popular feature in several VLSI implementations. The 4:2 structure compresses five partial product bits into three. However, it is connected in such a way that four of the inputs are coming from the same bit position of the weight j, while one bit is stemming from the neighboring position $j-1$ (known as carry-in) shown in Fig. 5.4. The output of such a 4-2 module consists of one bit in the position j and two bits in the position $j+1$. This structure does not represent a counter (though it became erroneously known as the "4-2 counter") but a "compressor," which would compress four partial-product bits into two (while using one bit laterally connected between adjacent 4-2 compressors).

The efficiency of such a structure is obviously higher in that it reduces the number of partial product bits by one-half. The speed of such a 4-2 compressor has been determined by the

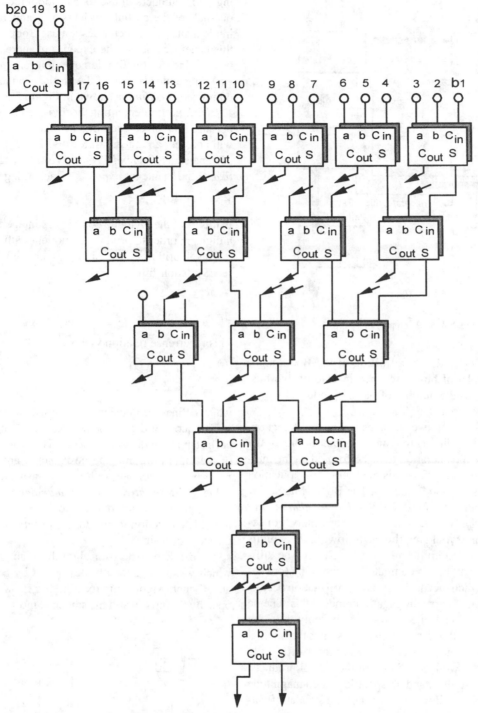

Fig. 5.3. The Wallace Tree.

speed of 3 XOR gates in series (in the redesigned version of 4-2 compressor), making such a scheme more efficient that the one using 3:2 counters in a regular Wallace Tree. The other equally important feature of the 4-2 compressor is that the interconnections between such a cells follow a more regular pattern than in case of the Wallace Tree.

The Booth Recoding Algorithm. Various modifications of the multiplication algorithm exist, but one of the most famous is the Booth Recoding Algorithm described by Booth in 1951.

This algorithm allows for the reduction of the number of partial products, thus speeding up the multiplication process. Generally, the Booth algorithm is a case of using the redundant number system with the radix higher than 2.

Booth's algorithm is widely used in the implementation of hardware or software multipliers because its application makes it possible to reduce the number of partial products. It can be used for both sign-magnitude numbers as well as 2's complement number, with no need for a correction term or a correction step.

409

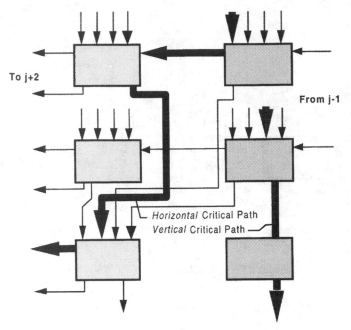

Fig. 5.4. 4-2 compressors.

MacSorley has proposed a modification of the Booth algorithm whereby a triplet of bits is scanned instead of two bits. This technique reduces the number of partial products by one-half regardless of the inputs (see Table 5.5).

Recoding is performed in two steps: *encoding* and *selection*. The purpose of the encoding is to scan the triplet of bits of the multiplier and define the operation to be performed on the multiplicand, as shown in Table 5.5. This method is an application of a sign-digit representation in radix 4. The Booth-MacSorley algorithm, usually called the *Modified Booth algorithm* or simply the *Booth algorithm,* can be generalized to any radix. For example, a 3-bit recoding would require the following set of digits to be multiplied by the multiplicand : $0, \pm1, \pm2, \pm3$. The difficulty lies in the fact that $\pm3Y$ is computed by summing (or subtracting) Y to $\pm2Y$, which means that a carry propagation occurs. The delay caused by the carry propagation renders this scheme slower than a conventional one. Consequently, only the 2-bit Booth recoding is used. Booth recoding necessitates the internal use of the 2's complement representation in order to efficiently perform subtraction as well as addition of the partial products. However, the floating-point standard specifies a sign magnitude representation, which is followed by most of the standard float-

ing-point numbers in use today. Booth recoding generates only one-half of the partial products as compared to the multiplier implementation, which does not use Booth recoding. However, this benefit comes at the expense of increased hardware complexity. Indeed, this implementation requires hardware for the *encoding* and *selection* of the partial products $(0, \pm Y, \pm 2Y)$. An optimized *encoding* is shown in Figure 5.5.

Division. Implementing the division process is more complex because it involves guessing the digits of the quotient. Here, we will consider an algorithm for division of two positive integers designated as *dividend Y*, and *divisor X*, and resulting in a *quotient Q* and an integer *remainder Z* according to the relation given:

$$Y = XQ + Z$$

In this case, the dividend contains $2n$ integers and the divisor has n digits in order to produce a quotient with n digits.

The algorithm for division is given with the following recurrence relationship:

$$z^{(0)} = Y$$
$$z^{(j+1)} = rz^{(j)} - Xr^n Q_{n-1-j} \quad \text{for } j = 0, \ldots, n-1$$

This recurrence relation yields

$$z^{(n)} = r^n(Y - XQ)$$
$$Y = XQ + z^{(n)}r^{-n}$$

which defines the division process with the remainder $Z = z^{(n)}r^{-n}$.

The quotient digit is selected by ensuring that $0 \le Z < X$ at each step in the division process. This selection is a crucial part of the algorithm, and the best known are *restoring* and *nonrestoring* division algorithms. In the restoring algorithm the value of the *tentative partial remainder $z^{(j)}$* is restored after the wrong guess is made of the quotient digit q_j. In the nonrestoring, this correction is not done in a separate step, but rather in the step following.

The best known division algorithm is so-called SRT algorithm, which was independently developed by Sweeney, Robertson, and Tocher. Algorithms for a higher radix were further developed by Robertson and his students, most notably Ercegovac.

TABLE 5.5 Modified Booth Recording

$x_{i+2}x_{i+1}x_i$	Add to Partial product
000	$+0Y$
001	$+1Y$
010	$+1Y$
011	$+2Y$
100	$-2Y$
101	$-1Y$
110	$-1Y$
111	$-0Y$

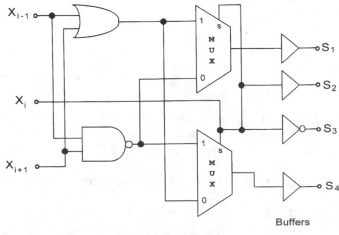

Fig. 5.5. Booth Encoder.

Some Optimal Schemes for ALU Implementation in VLSI Technology

VOJIN G. OKLOBDZIJA AND EARL R. BARNES
IBM T.J. WATSON RESEARCH CENTER
P.O. BOX 218, YORKTOWN HEIGHTS, NY 10598

ABSTRACT

An efficient scheme for carry propagation in an ALU implemented in n-MOS technology is presented. An algorithm that determines the optimum division of the carry chain of a parallel adder for various data path sizes is developed. This yields an implementation of a fast ALU which due to its regular structure occupies a modest amount of silicon. The speed of the implementation described is comparable to the carry look-ahead scheme. Our method is based on the optimization of the carry path implemented in n-MOS technology but the results can be applied to other technologies.

1. INTRODUCTION

An efficient implementation of an ALU in VLSI technology depends on many parameters. We consider an *efficient* implementation to be one which is fast, of a small and regular area, and low power. In many VLSI designs achieving the ultimate speed is not always important goal especially if this is achieved by consuming excessive area and power. Therefore Carry-Lookahead (CLA) scheme is not very attractive for VLSI implementations where area, regularity of structure and power are important. Also the determining factor, in case of an ALU, is wheather it is part of a critical path or not.

In this paper we considered Carry-Skip (CSA) scheme [1],[2],[3],[4] because it met our objective of reasonable performance achieved with a relatively small and regular area. This adder is in essence a Carry Lookahead for which the carry-generate portion which consumes a large amount of logic, has been eliminated. As in a Carry Lookahead adder the bits to be added are divided into groups. A circuit is provided for detecting when a carry signal entering a group will ripple through the group.

When this condition is detected, the carry is allowed to skip over the group. Carry-Skip Adder (CSA) does not require excessive amount of logic (area) and the "skip" portion of the logic can be added to the existing carry chain of Ripple-Carry Adder (RCA), therefore not disturbing the inherently regular bit-slice implementation of an RCA (Fig.1.).

The power requirements of a CSA are considerably lower than that of a CLA-ALU. In this paper we show how the carry chain in a CSA can be optimized to yield better speed which in the case of a multi-level optimized CSA-ALU can even outperform the speed of a CLA divided into the groups of constant size.

Lehman and Burla [3] studied a design of a CSA and suggested varying the size of the groups. By varying the sizes of the groups one can influence the maximum delay a carry signal can experience in propagating through the adder. Lehman and Burla [3] posed the problem of determining the optimal group sizes for minimizing the maximum delay [9]. They gave a heuristic method for obtaining economical group sizes. However, they did not solve the problem of determining optimal group sizes. For example, for a 48 bit adder they gave the group sizes 4 5 6 7 8 8 7 6 5 4 yielding the maximum delay of 14 [3]. We show that, under the same assumptions an optimal subdivision results in a delay of 12.

They also discussed the problem of achieving even faster addition by allowing carry signals to skip over blocks of groups. The problem of optimizing the carry chain is now complicated by having to choose both the optimal number of blocks and the optimal sizes of groups within blocks. Some rules for choosing economical block and group sizes are given [3]. However, the problem of determining the optimal sizes remains unsolved.

In this paper we consider the problem of designing a carry-skip adder in FET technology and give some optimal solutions. Actually, our solutions are more general in that we generally assume

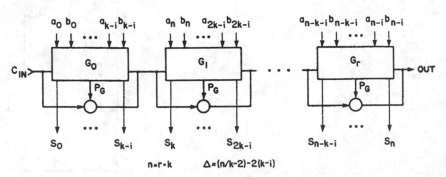

Fig.1. Carry-Skip Adder

Reprinted from *Proceedings of 7th Symposium on Computer Arithmetic*, pp. 2-8, June 1985.

411

that the time required for a carry signal to skip over a group of bits is longer than the time required for the carry to ripple through a single bit. This assumption is relevant for adders designed in n-MOS technology. Lehman and Burla assumed their adder to be designed using discrete components where these two times are equal. Our analysis will include their problem as a special case.

2. DIVIDING THE ADDER INTO GROUPS

Let n denote the number of bits in a carry skip adder and let m denote the number of groups into which the bits are divided. Let $x_1, \ldots, x_m$ denote the sizes of the groups beginning with the most significant bit. Let T denote the time required for a carry signal to skip over a group of bits. To be precise we should write $T = T(x)$ to indicate that T depends on the size x of the group over which the carry is skipped. However, T changes very slowly with x over the range of group sizes that concern us. So we assume that T is constant.

For a given n, the following three-step procedure gives an optimal way of dividing an n bit adder into groups of bits.

Procedure 2:

2(i)   Let m be the smallest positive integer such that

$$n \leq m + \frac{1}{2}mT + \frac{1}{4}m^2T + (1 - (-1)^m)\frac{T}{8}. \quad (1)$$

2(ii)   Let

$$y_i = \min\{1 + iT, \ 1 + (m + 1 - i)T\}, \quad i = 1, \ldots, m$$

and construct a histogram whose i-th column has height y_i. For example, for $T = 3$ and $n = 48$, we have $m = 7$ and the following histogram.

2(iii)   It is easily verified that the area of the histogram in (ii) is

$$m + \frac{1}{2}mT + \frac{1}{4}m^2T + (1 - (-1)^m)\frac{T}{8} \geq n$$

so these are at least n unit squares in the histogram. Starting with the first row, shade in n of the squares, row by row. Let x_i denote the number of shaded squares in column i of the histogram, $i = 1, \ldots, m$.

Then $x_1, \ldots, x_m$ is an optimal division of the adder. The maximum delay of a carry signal is mT.

Example 1. For a 48 bit adder we have, from Figure 2. $x_1 = x_7 = 4$, $x_2 = x_6 = 7$, $x_3 = 8$ and $x_4 = x_5 = 9$.

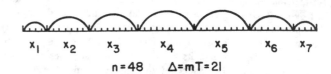

Fig.3.   Carry chain of a 48-bit adder: n=48, mT=21

The maximum delay is experienced by a signal generated in the second bit position and terminating in the 47th bit position. The delay is $mT = 21$.

Example 2. Consider a 54 bit adder. From 2(i) we see that again $m = 7$. If we shade 54 squares in Figure 2, we see that

$$x_1 = x_7 = 4, \ x_2 = x_6 = 7, \ x_3 = x_5 = 10 \text{ and } x_4 = 12$$

yields an optimal division of the adder. Again the maximum delay is $mT = 21$.

Example 3. Consider a 64 bit adder. From 2(i) we compute $m = 8$. The corresponding histogram is shown in Figure 4. The optimal group sizes are:

$$x_1 = x_8 = 4, \ x_2 = x_7 = 7, \ x_3 = x_6 = 10, \ x_4 = x_5 = 11.$$

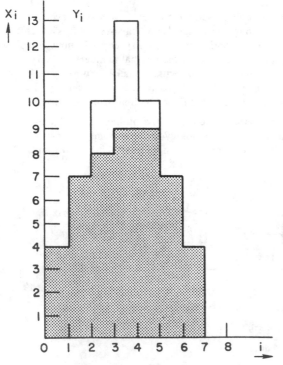

Fig.2.   Histogram of segments y_i, x_i

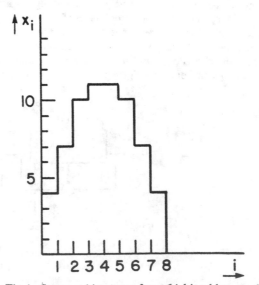

Fig.4.   Segment histogram for a 64-bit adder: m=8

412

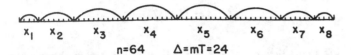

Fig.5. Carry chain of a 64-bit adder: m=8, mT=24

The delay of the longest signal is $mT = 24$.

Proof of Optimality We are going to prove optimality of the division of the carry chain described in 2(i)-(iii). First we need a lemma.

Lemma 1. *When the bits of a carry skip adder are grouped according to the scheme (i)-(iii), the maximum propagation time of a carry signal is mT.*

Proof. The carry generated at the 2nd bit position and terminating at the (n-1) clearly has propagation time mT. We must show that any other signal has propagation time smaller than or equal to mT. Consider a signal originating in the ith group and terminating in the jth, $i < j$. Denote its propagation time by P. Clearly

$$P \leq (x_i - 1) + (j - i - 1)T + (x_j - 1).$$

By construction, $x_i \leq \min\{1 + iT, 1 + (m + 1 - i)T\}$ for each i so

$$P \leq \min\{1 + iT, 1 + (m + 1 - i)T\}$$
$$+ \min\{1 + jT, 1 + (m + 1 - j)T\} + (j - i - 1)T - 2.$$

There are three cases to consider.

1. First assume

$$\min\{1 + iT, 1 + (m + 1 - i)T\} = 1 + iT$$

and

$$\min\{1 + jT, 1 + (m + 1 - j)T\} = 1 + jT.$$

In this case $1 + jT \leq 1 + (m + 1 - j)T \Rightarrow 2jT - T \leq mT$. It follows that

$$P \leq 1 + iT + 1 + jT + (j - i - 1)T - 2$$
$$= 2jT - T \leq mT.$$

2. Now assume that

$$\min\{1 + iT, 1 + (m + 1 - i)T\} = 1 + iT$$

and

$$\min\{1 + jT, 1 + (m + 1 - j)T\} = 1 + (m + 1 - j)T.$$

In this case we have

$$P \leq 1 + iT + 1 + (m + 1 - j)T + (j - i - 1)T - 2 = mT.$$

3. Finally, assume that

$$\min\{1 + iT, 1 + (m + 1 - i)T\} = 1 + (m + 1 - i)T$$

and

$$\min\{1 + jT, 1 + (m + 1 - j)T\} = 1 + (m + 1 - j)T.$$

It follows that

$$P \leq 1 + (m + 1 - i)T + 1 + (m + 1 - j)T + (j - i - 1)T - 2$$
$$= 2mT - (2iT - T) \leq 2mT - mT = mT.$$

This completes the proof of the lemma.

Lemma 2. *Let Δ denote the maximum delay of a carry signal in a n bit carry skip adder with group sizes chosen optimally. Then*

$$(m - 1)T \leq \Delta \leq mT.$$

Proof. Since we have exhibited a division of the carry chain into groups in such a way that the maximum delay of a carry signal is mT we clearly have $\Delta \leq mT$.

Let $x_1, x_2, \ldots, x_r$ denote the optimal group sizes corresponding to Δ. For the moment assume that $r = 2k$ is even. By considering carries originating in group i and terminating in group $r - i + 1$, $i = 1, \ldots, k$, we deduce the following system of inequalities.

$$\begin{aligned}
(x_1 - 1) + (r - 2)T + (x_r - 1) &\leq \Delta \\
(x_2 - 1) + (r - 4)T + (x_{r-1} - 1) &\leq \Delta \\
&\vdots \\
(x_k - 1) + (r - 2k)T + (x_{k+1} - 1) &\leq \Delta \\
rT &\leq \Delta
\end{aligned}$$

If we add these inequalities and use the fact that $\sum_{i=1}^{r} x_i = n$, we obtain the inequality

$$n - 2k + (k + 1)rT - k(k + 1)T \leq (k + 1)\Delta$$

which simplifies to

$$\frac{n - 2k}{k + 1} + kT \leq \Delta.$$

The left hand side of this inequality assumes its minimum value at

$$k + 1 = \frac{\sqrt{n + 2}}{T}.$$

It follows that

$$\Delta \geq -(T + 2) + \sqrt{4nT + 8T}. \quad (2)$$

Assume now that $r = 2k + 1$ is odd. We then have the system of inequalities

$$\begin{aligned}
(x_1 - 1) + (r - 2)T + (x_r - 1) &\leq \Delta \\
(x_2 - 1) + (r - 4)T + (x_{r-1} - 1) &\leq \Delta \\
&\vdots \\
(x_k - 1) + (r - 2k)T + (x_{k+1} - 1) &\leq \Delta \\
(x_{k+1} - 1) + kT &\leq \Delta,
\end{aligned}$$

which implies that

$$\frac{n - 2k - 1}{k + 1} + kT \leq \Delta.$$

By minimizing the left hand side of this inequality with respect to k we find that

$$\Delta \geq -(T + 2) + \sqrt{4nT + 4T}. \quad (3)$$

By comparing this with (2) we see that (3) holds in all cases.

We will now produce an upper bound on mT. Since m is the smallest positive integer satisfying (1) we have

$$n > (m - 1) + \frac{1}{2}(m - 1)T + \frac{1}{4}(m - 1)^2 T + (1 - (-1)^{m-1})\frac{T}{8}.$$

413

Since each size of this inequality is an integer, we can increase the right hand side by 1 to obtain

$$n \geq m - \frac{1}{4}T + \frac{1}{4}m^2 T + (1 - (-1)^{m-1})\frac{T}{8}.$$

Solving this inequality for mT gives

$$mT \leq -2 + \sqrt{4nT + 4 + T^2 - (1 - (-1)^{m-1})\frac{T^2}{2}}. \quad (4)$$

Combining this with (2) and (3) gives

$$mT - \Delta \leq \begin{cases} T + \dfrac{T^2 - 8T + 4 - (1-(-1)^{m-1})\dfrac{T^2}{2}}{\sqrt{4nT + 8T} + \sqrt{4nT + 4 + T^2 - (1-(-1)^{m-1})\dfrac{T^2}{2}}}, & r\text{-even} \\[4ex] T + \dfrac{T^2 - 4T + 4 - (1-(-1)^{m-1})\dfrac{T^2}{2}}{\sqrt{4nT + 4T} + \sqrt{4nT + 4 + T^2 - (1-(-1)^{m-1})\dfrac{T^2}{2}}}, & r\text{-odd} \end{cases} \quad (5)$$

For n sufficiently large we have $mT - \Delta < T + 1$ and since $mT - \Delta$ is an integer, $mT - \Delta \leq T$. This completes the proof of the lemma.

Theorem 1 *The scheme 2(i)-2(iii) given above for dividing the bits of a carry skip adder into groups is optimal for $2 \leq T \leq 7$.*

Proof. Assume the scheme is not optimal and let Δ be the maximum delay corresponding to an optimal division of the bits into groups. Assume there are r groups in the optimal division. Since a carry in signal to the least significant bit group can skip over each group we have $rT \leq \Delta \leq mT$ so $r \leq m$. If $r = m$ then $\Delta = mT$ and the theorem holds by Lemma 1. If $r = m - 1$ m and r have different parities and it follows from (5) that $mT - \Delta < T$ for $2 \leq T \leq 7$ so that $\Delta > (m-1)T = rT$. This means that a signal which skips over each of the r groups has delay less than the maximum Δ. Similarly, if $r < m - 1$, $\Delta \geq (m-1)T > rT$ so that a signal which skips over each group has delay $< \Delta$. It follows that a signal with delay Δ must start in a group i, ripple to the end of this group, then skip over $s < r$ groups and either terminate, or ripple through the first few bits of a group $j > i$. Let x_i and x_j denote the lengths of the ith and jth groups respectively. Assume that i is chosen as small as possible and j as large as possible. A signal originating in group i, rippling to the end of this group and then skipping over the next s group has delay

$$\Delta \leq (x_i - 1) + sT \leq (x_i - 1) + (r-1)T$$
$$\leq (x_i - 1) + (m-2)T.$$

Since $\Delta \geq (m-1)T$ this implies that $x_i \geq T + 1$. Divide group i into two groups such that the group containing the most significant bits has size T. Since the i-th group is the first group in which a signal having maximum delay can originate, this subdivision does not increase the delay of any carry signal of maximum delay. However, it increases the number of groups by 1.

Suppose now that a carry signal originates in a group i, ripples to its end, skips over $s \leq r - 2$ groups and finally ripples through the first few bits of a group j and terminates. We then have

$$\Delta \leq (x_i - 1) + sT + (x_j - 1)$$
$$\leq x_i + x_j - 2 + (m-3)T$$

So that either $x_i \geq T + 1$ or $x_j \geq T + 1$. This means that we can subdivide one of the groups i,j without increasing Δ. Continuing in this way, we can always increase the number r of group in an optimal division of a carry chain by 1 without increasing Δ if $r < m$. This means that we can arrive at an optimal division of the carry chain into m groups. We must then have $\Delta \geq mT$ which, together with Lemma 2, implies $\Delta = mT$. This completes the proof of the theorem.

3. DIVIDING GROUPS INTO BLOCKS

It is clear that the maximum delay of a carry signal in a carry skip adder can be further reduced if signals are allowed to skip over blocks of groups. We define a block to be an additional path allowing carry signal to skip directly over groups. In this section we will describe an efficient scheme for dividing the carry chain into blocks of groups. We assume that the time required for a carry signal to skip over a block of groups is T_b. Actually, the time required for a carry to skip over a block T_b is slightly longer than the time T_g required to skip over a group. But for the sake of simplifying the analysis we will assume these two times to be equal i.e. $T_b = T_g$. However, our technique extends to the case where $T_g \neq T_b$.

Let M denote the number of blocks into which the groups of bits are divided. Let Δ denote the maximum delay a carry signal can have in an adder divided into M blocks. Clearly, $\Delta \geq MT$. We will show how to choose the blocks such that $\Delta = MT$. We will also show how to choose M for an adder of length n.

Our blocks are chosen in such a way that the maximum delay of a signal originating and terminating in block i and $M + 1 - i$ is iT.

Consider a signal originating in the first of these blocks and terminating in the second. Such a signal will skip over $M - 2i$ blocks and will accordingly have delay $\leq (iT) + (M - 2i)T + iT = MT$ as desired. It follows from our work in Section 2 that in order for a signal originating and terminating in block i to have delay less or equal iT we must choose the length of the ith and $(M + 1 - i)th$ blocks to be less or equal the number of unit squares in a histogram with base of width i. Thus the maximum length of the ith and $(M + 1 - i)th$ blocks must be

$$i + \frac{1}{2}iT + \frac{1}{4}i^2 T + (1 - (-1)^i)\frac{T}{8}, \quad i \leq \lceil \frac{M}{2} \rceil.$$

(Here we use the symbol $\lceil I \rceil$ to denote the smallest integer $\geq I$.) It follows that the maximum length of an adder divided into M blocks must be

414

$$2 \sum_{i=1}^{\lceil \frac{M-1}{2} \rceil} \left\{ i + \frac{1}{2}iT + \frac{1}{4}i^2T + (1-(-1)^i)\frac{T}{8} \right\} \qquad (3.1)$$

$$+ \frac{(1-(-1)^M)}{2} \left\{ \lceil \frac{M}{2} \rceil + \frac{1}{2}\lceil \frac{M}{2} \rceil T + \frac{1}{4}\lceil \frac{M}{2} \rceil^2 T \right.$$

$$\left. + (1-(-1)^{\lceil M/2 \rceil})\frac{T}{8} \right\}.$$

Thus for a given adder length n, we choose M to be the smallest positive integer such that the expression (3.1) exceeds or equals n. M is then the number of blocks into which our adder must be divided. The formal statement of our algorithm is as follows.

3(i) Choose M to be the smallest positive integer such that

$$n \le 2 \sum_{i=1}^{\lceil \frac{M-1}{2} \rceil} \left\{ i + \frac{1}{2}iT + \frac{1}{4}i^2T + (1-(-1)^i)\frac{T}{8} \right\}$$

$$+ \frac{(1-(-1)^M)}{2} \left\{ \lceil \frac{M}{2} \rceil + \frac{1}{2}\lceil \frac{M}{2} \rceil T + \frac{1}{4}\lceil \frac{M}{2} \rceil^2 T \right.$$

$$\left. + (1-(-1)^{\lceil M/2 \rceil})\frac{T}{8} \right\}.$$

3(ii) Form M blocks labeled $1, 2, \ldots, M$, with blocks i and $M+1-i$ each containing

$$i + \frac{1}{2}iT + \frac{1}{4}i^2T + (1-(-1)^i)\frac{T}{8}$$

bits, $i \le \lceil \frac{M}{2} \rceil$.

This construction is analogous to the construction of the histogram in 2(ii). If necessary, delete bits from the largest blocks in this chain until a total of exactly n bits remain in the M blocks.

3(iii) Treat each of the final blocks in 3(ii) as a complete carry chain and divide it into groups optimally using the algorithms 2(i)-2(iii).

Example 3.1. Consider a 32-bit adder. For $i = 1, 2, 3, \ldots$, and $T = 3$ the numbers

$$i + \frac{1}{2}iT + \frac{1}{4}i^2T + (1-(-1)^i)\frac{T}{8}$$

take on the values 4, 8, 15, 22, 32,... respectively. Since

$$32 \le 2\{4+8\} + 15$$

we must have $M = 5$ blocks in step 3(ii). These blocks have sizes 4, 8, 15, 8, 4 respectively. If we delete 7 units from the middle block we obtain block sizes 4, 8, 8, 8, 4 which add up to 32. Di-

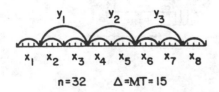

Fig.6. Carry chain of an 32-bit adder: MT=15

viding each block into groups by the procedure 2(i)-2(iii) we obtain the following chain where each group has size 4. The maximum delay of a carry signal is $MT = 15$.

Example 3.2. Consider a 48 bit adder. Again we assume $T = 3$. Since

$$48 \le 2\{4+8+15\}$$

we must take $M = 6$ corresponding to block sizes 4, 8, 15, 15, 8, 4. The total number of units is 54. So we reduce the size of the two middle blocks by 3 each. This gives block sizes 4, 8, 12, 12, 8, 4 adding up to 48. If we divide each block into groups by the procedure 2(i)-2(iii), each group has size 4. The maximum delay of a carry signal is given by $MT = 18$.

Example 3.3. Consider a 64 bit adder and assume $T = 3$. Since

$$64 \le 2\{4+8+15\} + 22$$

we take $M = 7$ and start with blocks of sizes 4, 8, 15, 22, 15, 8, 4 respectively. The lengths of these blocks total 76. So we reduce the middle block by 12. The new block sizes are 4, 8, 15, 10, 15, 8, 4. The optimal division of these blocks into groups is given in Figure 7. We could have just

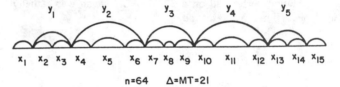

Fig.7. Carry chain of a 64-bit adder: MT = 21

as well reduced the sizes of the three middle blocks obtaining final blocks of sizes 4, 8, 13, 14, 13, 8, 4. The maximum delay of a carry signal would still be $MT = 21$.

4. COMPARISON

In the literature comparison of schemes for ALU implementation is done mainly on the basis of the number of gates, propagation delay per gate and power consumed per gate [5]. However, if a VLSI implementation of a high-speed ALU is considered, these measures are not easily applied. For example, the propagation delay in terms of number of gates is not an adequate measure unless care is taken to implement the function exactly as specified by its logic (gate) representation. This is often not the case since the function is frequently merged into a group of transistors or implemented by using pass-transistors, precharge or other techniques applied by the circuit designer in order to minimize delay and power.

In general, if the function is implemented in two levels of logic, the delay is not necessarily smaller than the implementation of the same function in three or more logic levels. This is due to the fact that in n-MOS technology the delay is heavily dependent on several factors:

1. *Gate type* : NOR gates are faster than NAND gates.

2. *Fan-in* : for a NAND gate, the delay is directly proportional to the number of inputs, since inside the n-input NAND gate the signal has to propagate through n-transistors. In case of

a NOR gate, delay is not strongly affected by the number of inputs, and therefore the use of NOR gate is preferable.

3. *Fan-out* : the speed of a gate will be different if the fan-out is larger than in the case of small fan-outs.

4. *Wiring* : speed is also dependent on the length of the wires i.e. "wiring capacitance" and the resistance of the long wires.

4.1 Comparison with Carry-Lookahead Scheme

For the purpose of our analysis we consider as a rough measure of delay the number of FET transistors through which the signal needs to propagate in order to reach the destination point. This seems to be satisfactory for measuring delay through a VLSI FET network. In addition we modify the delay equations for the points known to have either a substantial fan-out or considerable capacitive loading due to the long wires carrying signal to the distant points within the VLSI macro block. For example, we assume that the time required for a carry signal to skip over a block of groups is three times the time to ripple through one bit position of the carry-chain. These assumptions were confirmed by simulation performed on a 32-bit ALU which was actually implemented in n-MOS technology.

The Case T = 1.

In order to compare the speed of our carry skip adder with that of the Carry Lookahead Adder (CLA) we take $T = 1$. This modification is made in order to compare our results with the estimates for the CLA adder which are based on the gate delay calculations in which case it is assumed that each gate level introduces a delay $t_G = 1$ regardless of the gate fan-in or fan-out [5]. This assumption works favorably for CLA when calculating its speed. In practice we expect CLA to show worse performance than estimated while the calculation for our CSA schemes should be more accurate.

For our comparison we consider full-CLA with the groups of size $G_s = 4$. CLA adder of size n=32 bits is shown in the Fig.8. with the delays indicated at each signal input and output to or from the block. Delay calculation for CLA of the sizes n = 16, 32, 48, 64 shows delays for the critical path of the carry signal to be $\Delta = 6$, 8, 10, 10 respectively [5].

First we consider our case where the adder is simply divided into groups of bits. The procedure in Section 2 shows that the maximum delay of a carry signal in an n bit adder is m, where m is the smallest integer satisfying

$$n \leq m + \frac{1}{2}m + \frac{1}{4}m^2 + (1 - (-1)^m)\frac{1}{8}.$$

The values of a delay m for our method compared with the CLA delays for several values of n are shown in the Table 1.

n	$\Delta(CSA)$	$\Delta(CLA)$
16	6	6
32	9	8
48	12	10
64	14	10

Table 1. Delay comparison with the CLA for the method of Section 2.

Consider now the case where the adder is divided into blocks of groups according to the procedure of Section 3. This algorithm shows that the maximum delay of a carry signal in an n bit carry skip adder is M, the smallest positive integer satisfying

$$n \leq 2 \sum_{i=1}^{\lceil \frac{M-1}{2} \rceil} \{i + \frac{1}{2}i + \frac{1}{4}i^2 + (1 - (-1)^i)\frac{1}{8}\}$$
$$+ \frac{(1 - (-1)^M)}{2}\{\lceil \frac{M}{2} \rceil + \frac{1}{2}\lceil \frac{M}{2} \rceil + \frac{1}{4}\lceil \frac{M}{2} \rceil^2$$
$$+ (1 - (-1)^{\lceil M/2 \rceil})\frac{1}{8}\}.$$

Recall that $\lceil r \rceil$ is the smallest integer $\geq$ r for any real number r.

Table 2. shows the values of delay M for CSA of Section 3 compared to that of CLA adder corresponding to several values of n.

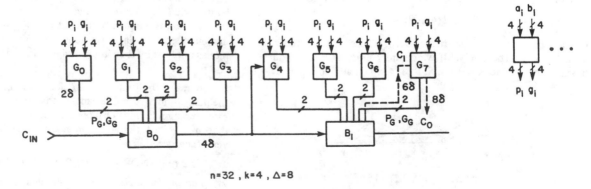

n=32 , k=4 , Δ=8

Fig.8. Carry Signal Delay of a CLA Adder of the size n=32 bits

416

n	$\Delta(CSA)$	$\Delta(CLA)$
16	5	6
32	7	8
48	9	10
64	10	10

Table 2. Delay comparison between CSA of Section 3. and CLA

In the first case (Table 1.) we notice that CLA is faster than our method of Section 1. resulting in equal delay for n=16 and ending up with 40% advantage for n=64. However, in the second case (Table 2.) our method of Section 3. shows advantage in speed over CLA of approximately 20% for n=16 and no advantage for n=64.

Our scheme exhibits regularity in fan-in and fan-out throughout the entire carry chain and therefore yields simpler and regular implementation. The amount of logic required to implement our scheme should therefore be smaller then that of CLA. Therefore its implementation in VLSI technology should yield even better results.

5. CONCLUSION

Our goal in implementing this scheme was to achieve a regular structure without using an excessive chip area. Comparison of various ALU implementation schemes for area and power as a parameter show that after some point small improvement in speed is achieved by a large investment in area or power [6],[7]. In our approach we argue that this incremental improvement in speed is diminished due to the overhead in the wiring capacitance and device size. Therefore our approach is to maximize the speed not by increasing the power or adding a substantial amount of logic but rather by optimizing on the path of the signal with the critical timing (carry signal) by designing the ALU around the carry path.

The method described in Sections 2. and 3. is especially applicable for the floating-point fraction ALU which is usually of large size n.

REFERENCES

[1] T.Kilburn, D.B.G.Edwards and D.Aspinall, *Parallel Addition in Digital Computers: A New Fast "Carry" Circuit*, Proc. IEE, Vol.106, Pt.B. P.464, September 1959.

[2] M.Lehman, *A Comparative Study of Propagation Speed-up Circuits in Binary Arithmetic Units*, Information Processing, 1962, Elsevier-North Holand, Amsterdam, 1963, p.671.

[3] M.Lehman and N.Burla, *Skip Techniques for High-Speed Carry-Propagation in Binary Arithmetic Units*, IRE Transactions on Electronic Computers, December 1961, p.691.

[4] L.P.Morgan and D.B.Jarvis, *Transistor Logic Using Current Switching and Routing Techniques and its Application to a Fast "Carry" Propagation Adder*, Proc. IEE, pt.B, Vol.106, p.467, September 1969.

[5] Kai Hwang , *Computer Arithmetic: Principles Architecture and Design*, John Wiley and Sons, 1979.

[6] S.Ong and D.E.Atkins, *A Comparison of ALU Structures for VLSI Technology*, Proc. of the 6th Symposium on Computer Arithmetic, June 20-22, 1983, Aarhus University, Aarhus, Denmark.

[7] Robert K. Montoye, P.W.Cook, *Automatically Generated Area, Power, and Delay Optimized ALUs*, IEEE ISSCC Digest of Technical Papers, February 23-24, 1983, New York.

[8] A.Bilgory and D.D.Gajski, *Automatic Generation of Cells for Recurrence Structures*, Proc. of 18th Design Automation Conference, Nashville, Tennessee 1981.

[9] V.F.Demyanov, V.N.Malozemov, *Vvedenie v Minimaks*, Izdatelstvo Nauka , Fiziko-Matematicheskoi Literaturi, Moskva 1972.

A 4.5 NS 96B CMOS Adder Design

AJAY NAINI, DAVID BEARDEN AND WILLIAM ANDERSON

MICROPROCESSOR AND MEMORY TECHNOLOGIES GROUP, MOTOROLA INC.

6501 WILLIAM CANON DRIVE WEST, AUSTIN, TEXAS 78735-8598

ABSTRACT

A new approach to the design of high-speed adders using the carry look-ahead principle is presented. These techniques discuss a new organization for the carry-chain, with minimal area impact, which minimizes the latency while maintaining modularity. Application of these techniques to a 96-bit adder, implemented in a CMOS process with 1.0 µm design rules, shows a critical path delay of 4.5 ns.

INTRODUCTION

Carry look-ahead (CLA) addition has become one of the more popular techniques for implementing fast adders because of the resulting speed, area, and modularity[1]. However, as the word size of the adders increase, the conventional carry-chain delay easily limits cycle time. In an attempt to decrease carry-chain delay, new circuit techniques have been applied to CLA addition[2]. This paper discusses a new organization for the carry-chain to minimize latency. As an example, a 96-bit adder design will be discussed.

The recursive method of CLA addition is well known. In general, fan-in limits carry look-ahead to groups of four bits. Because of this, multi-level look-ahead structures are used for larger words. Fig. 1a shows the structure for a 64-bit conventional CLA carry-chain in a group size of four, and the CLA equations are shown in Fig. 1b. Each carry-block generates three local carry-out terms from the P,G terms and the carry-in. Fig. 1c shows a circuit implementation of a 4-bit group PG-block and carry block. There are three levels of delay to the most significant carry-out term. The PG-block generates

the 4-bit group P,G terms.

The critical path for the adder which is highlighted in Fig. 1a is:

$$A,B - G_0 - G_{3:0} - G_{15:0} - C_{16} - C_{32} - C_{48} - C_{52} - C_{56} - C_{60} - C_{61} - C_{62} - C_{63} - S_{63}$$

MODIFIED CLA STRUCTURE

Fig. 2a shows the modified CLA structure. The PG-block here generates intermediate group P terms ($P_{2:0}$, $P_{1:0}$) and group G terms ($G_{2:0}$, $G_{1:0}$), in addition to the 4-bit group P,G terms. A Manchester chain structure is used in the generation of the group G terms to reduce the transistor count. The carry-block generates three local carry-out terms (C_1, C_2, C_3). The associated circuit structure for the PG-block and carry block is shown in Fig. 2b. The generation of the intermediate group P,G terms means the delay to any carry-out term is only one level.

The critical path for the modified CLA structure which is highlighted in Fig. 2a is:

$$A,B - G_0 - G_{3:0} - G_{15:0} - G_{47:0} - C_{48} - C_{60} - C_{63} - S_{63}$$

Table 1 details the timing associated with each of the above critical paths. These timing numbers are based upon a 1 µm design rules CMOS process ($L_{eff} = 0.8$ µm), 5.0 V, 25° C. As can be seen, the 64-bit modified CLA addition is faster than the conventional 64-bit addition. The reason for the faster speed is the availability of the intermediate group P,G terms at every PG-block, which enabled the most significant carry-out term delay of every carry-block to be a one-level delay. The PG-block of the modified CLA design has 41 transistors as opposed to 19 transistors for the conventional CLA

Reprinted from *IEEE 1992 Custom Integrated Circuits Conference*, pp. 25.5.1-25.5.4, 1992.

design. However, this will minimally impact area because these transistors arise in the intermediate bits, which will not be the limiting factor in determining the bit cell size.

CARRY INPUT REQUIREMENTS

In the critical path for the conventional 64-bit CLA adder, the carry-in, C_0, is needed in the generation of C_{16} (Fig. 1a). In the modified 64-bit CLA adder, the carry-in is needed in the generation of C_{48} (Fig. 2a). This means the carry-in to the respective adders should be valid by 1.7 ns and 2.9 ns (Refer to Table 1 for timing numbers). This tolerance of a late carry-in of the modified adder may be applied in the construction of larger adders.

A 96-BIT ADDER USING MODIFIED CLA

The 96-bit adder shown in Fig. 3 consists of a high-order 64-bit modified adder and a low-order 32-bit conventional adder. The low-order 32-bit adder has the carry-out, C_{32}, valid by 2.35 ns (refer to Table 1 for timing numbers). Since this carry-out delay falls within the carry-in requirement (2.9 ns) of the 64-bit adder, the 32-bit adder timing does not affect the critical path. Therefore, the low-order 32-bit addition is effectively transparent to the operation of the higher-order 64-bit addition. This phenomenon enables the 96-bit addition to be performed in 4.5 ns, which is the same time as the modified 64-bit addition, and in a lesser time than the 64-bit conventional CLA addition.

REFERENCES

[1] S. Waser and M. Flynn, *Introduction to Arithmetic for Digital Systems Designers,* Holt, Rinehart, and Winstion, New York, 1982, pp. 83-88.

[2] I. Hwang and A. Fisher, "A 3.1 ns 32b CMOS Adder in Multiple Output Domino Logic," *ISSCC Digest of Technical Papers,* Feb. 1988, pp. 140-141.

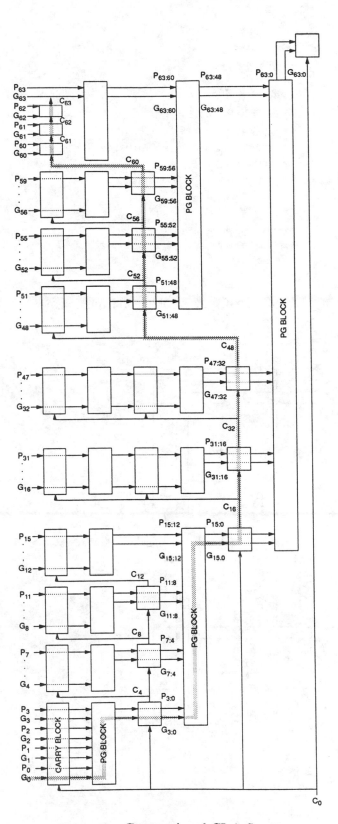

Figure 1a. Conventional CLA Structure

$P = A \oplus B$ (propogate)

$G = A \cdot B$ (generate)

C_0 (carry-in)

$C_1 = G_0 + C_0 \cdot P_0$

$C_2 = G_1 + C_1 \cdot P_1 = G_1 + G_0 \cdot P_1 + C_0 \cdot P_0 \cdot P_1$

$C_3 = G_2 + C_2 \cdot P_2 = G_2 + G_1 \cdot P_2 + G_0 \cdot P_1 \cdot P_2 + C_0 \cdot P_0 \cdot P_1 \cdot P_2$

$C_4 = G_3 + C_3 \cdot P_3 = G_3 + G_2 \cdot P_3 + G_1 \cdot P_2 \cdot P_3 + G_0 \cdot P_1 \cdot P_2 \cdot P_3 + C_0 \cdot P_0 \cdot P_1 \cdot P_2 \cdot P_3$

$\quad = G_{3:0} + C_0 \cdot P_{3:0}$

$C_8 = G_{7:4} + C_4 \cdot P_{7:4} = G_{7:4} + G_{3:0} \cdot P_{7:4} + C_0 \cdot P_{3:0} \cdot P_{7:4}$

$\ldots$

$\ldots$

$C_{16} = G_{15:0} + C_0 \cdot P_{15:0}$

$C_{64} = G_{63:0} + C_0 \cdot P_{63:0}$

$S_n = P_n \oplus C_n$

Figure 1b. Carry-Lookahead Equations

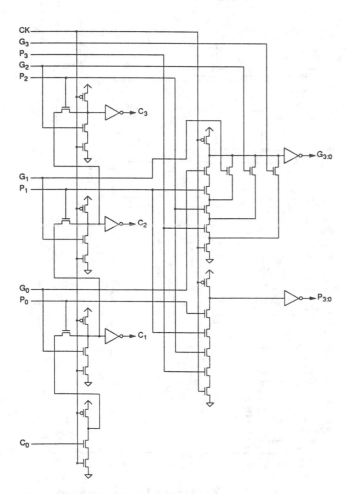

Figure 1c. Circuit Implementation of Conventional
4b Group PG and Carry Blocks

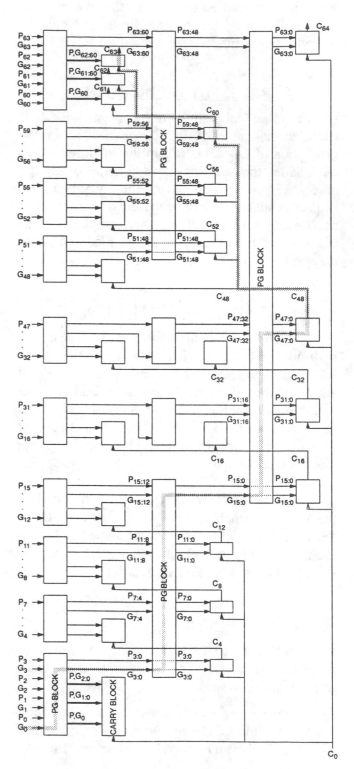

Figure 2a. Modified CLA Structure

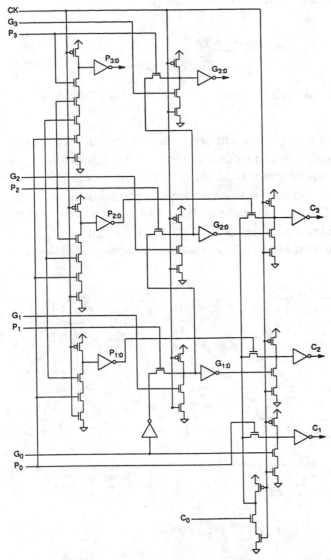

Figure 2b. Circuit Implementation of Modified 4b
Group PG and Carry Blocks

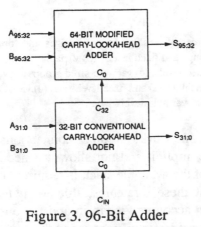

Figure 3. 96-Bit Adder

Table 1. Timing Characteristics for 1.0 µm CMOS process with VDD = 5V and 25° C

CONVENTIONAL CLA ADDER (FIGURE 1A)													
CRITICAL PATH	$A,B - C_0$	$C_0 - G_{3:0}$	$G_{3:0} - G_{15:0}$	$G_{15:0} - C_{16}$	$C_{16} - C_{32}$	$C_{32} - C_{48}$	$C_{48} - C_{52}$	$C_{52} - C_{56}$	$C_{56} - C_{60}$	$C_{60} - C_{61}$	$C_{61} - C_{62}$	$C_{62} - C_{63}$	$C_{63} - S_{63}$
DELAY (NS)	0.4	0.65	0.65	0.3	0.35	0.35	0.35	0.35	0.35	0.35	0.35	0.35	0.6

TOTAL = 5.4NS

MODIFIED CLA ADDER (FIGURE 2A)								
CRITICAL PATH	$A,B - C_0$	$C_0 - G_{3:0}$	$G_{3:0} - G_{15:0}$	$G_{15:0} - G_{47:0}$	$G_{47:0} - C_{48}$	$C_{48} - C_{60}$	$C_{60} - C_{63}$	$C_{63} - S_{63}$
DELAY (NS)	0.4	0.9	0.9	0.7	0.3	0.35	0.35	0.6

TOTAL = 4.5NS

Improved CLA Scheme with Optimized Delay

BRIAN D. LEE AND VOJIN G. OKLOBDZIJA*

Department of Electrical Engineering and Computer Science, University of California, Berkeley

Received January 31, 1991; Revised April 25, 1991.

Abstract. The delay characteristics of carry-lookahead (CLA) adders are examined with respect to a delay model that accounts for fan-in and fan-out dependencies. Though CLA structures are considered among the fastest topologies for performing addition, they have also been characterized as providing marginal speed improvement for the amount of hardware invested. This analysis shows that this inefficiency can be explained by the suboptimal nature of common CLA implementations. Simulation results show that the CLA structures in wide use can be improved by varying the block sizes and the number of levels within each adder. Examples of optimal CLA structures are given and heuristic methods for finding these structures are presented.

1. Introduction

Analysis of carry-lookahead adders is important in the design of high performance machines. In these designs, processor speed is a primary concern and carry-lookahead structures are often used because their delay times exhibit log dependence on the size of the adder and they are considered among the fastest circuit topologies for performing addition. However, adder comparisons [1], [2] have ranked CLA structures low on effective hardware utilization and this apparent inefficiency raises concerns over the optimality of current CLA implementations. Simulation results from this research show that the commonly used CLA structures can be improved by varying block sizes and levels within the adder.

Typical CLA implementations are made of lookahead units of relatively fixed sizes. This artificial constraint produces slack in the circuit and results in poor hardware utilization. The strategy of varying group sizes to reduce slack and improve performance is a promising idea and has been used successfully on carry-skip adders [3], [4]. A natural extension of this method is to also vary the number of lookahead levels [5], [6]. The example structures in figure 1 illustrate what it means to vary group sizes and lookahead levels. Each box corresponds to a BCLA/CLA unit and the enclosed number gives the unit size in bits. The dotted lines in figure 1(d) represent a connection to a primary input

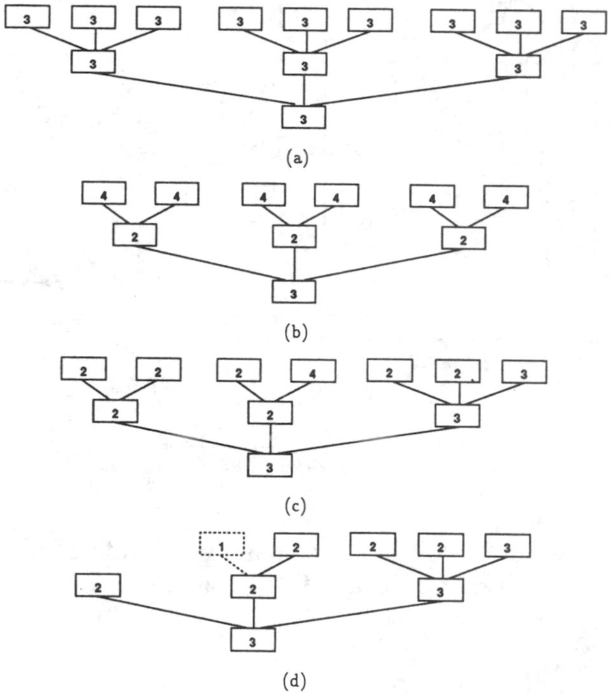

Fig. 1. Degrees of freedom in group sizes and lookahead levels: (a) fixed groups and fixed levels, (b) variable-sized groups between levels and fixed levels, (c) variable-sized groups anywhere (between levels and within levels) and fixed levels, and (d) variable-sized groups anywhere and variable levels.

of the CLA network. All other connections to primary inputs are implicit and not shown. Since an accurate measure of the available slack is required to effectively implement these strategies, this work uses a delay model that accounts for fan-in and fan-out dependencies. The parameter values used in the model are based on industrial data.

*Currently with IBM Thomas J. Watson Research Center, Yorktown Heights, NY 10598.

A simulation program has been written to compare different CLA structures. Preliminary data on varying block sizes was obtained through exhaustive search. Based on this data and an analysis of delay and slack in the CLA scheme, heuristics were chosen to find structures with completely variable block sizes and levels. The structures found by these heuristics are faster than more constrained topologies.

2. Carry Lookahead Structure

The simulation results in this paper are based on carry-lookahead adders that implement full lookahead. A description of the basic organization of carry-lookahead adders can be found in references such as [7]. Each adder consists of three main components—the propagate and generate generation circuitry, the carry-lookahead network, and the sum generation circuitry. This work concerns varying the sizes of circuit blocks and the number of levels in the carry-lookahead network to optimize adder delay.

Given an n-bit adder with inputs, A and B, the logic equations for producing the initial propagate and generate signals and the final sum signals are

$$P_i = A_i \oplus B_i$$

$$G_i = A_i B_i$$

and

$$S_i = P_i \oplus C_{i-1}$$

for $0 \leq i \leq n - 1$. The simulations assume that P_i and S_i are produced by monolithic XOR gates instead

of two levels of NAND gates and the sum XOR gate is assumed to have a fan-out of one.

The carry-lookahead network in a full carry-lookahead adder consists of a tree of block carry-lookahead (BCLA) units rooted at a single carry-lookahead (CLA) unit. Two different implementations of BCLA/CLA units are analyzed. Their performance differences are discussed later.

The first implementation generates carry signals in two levels of logic. Four-bit versions of these circuits are shown in figures 2 and 3. A k-bit carry-lookahead unit of this type generates

$$C_j = G_j + G_{j-1}P_j + G_{j-2}P_{j-1}P_j$$
$$+ \ldots + G_0 P_1 \ldots P_j + C_{-1}P_0 \ldots P_j$$

where $0 \leq j \leq k - 1$ and a k-bit block carry-lookahead unit of this type generates

$$P^* = P_0 P_1 \ldots P_{k-1}$$

$$G^* = G_{k-1} + G_{k-2}P_{k-1} + \ldots + G_0 P_1 \ldots P_{k-1}$$

$$C_j = G_j + G_{j-1}P_j + G_{j-2}P_{j-1}P_j$$
$$+ \ldots + G_0 P_1 \ldots P_j + C_{-1}P_0 \ldots P_j$$

where $0 \leq j \leq k - 2$.

The second implementation generates carry signals in three levels of logic. Four-bit versions of these circuits are shown in figures 4 and 5. A k-bit carry-lookahead unit of this type generates

$$C_j = G_j^* + C_{-1}P_j^*$$

where $0 \leq j \leq k - 1$ and a k-bit block carry-lookahead unit of this type generates

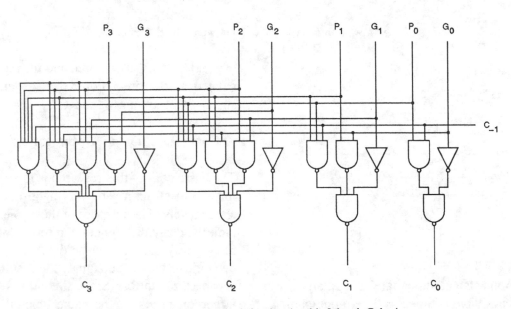

Fig. 2. A 4-bit carry-lookahead unit with 2-level C_i logic.

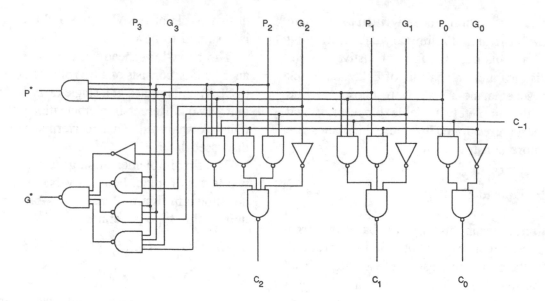

Fig. 3. A 4-bit block carry-lookahead with 2-level C_i logic.

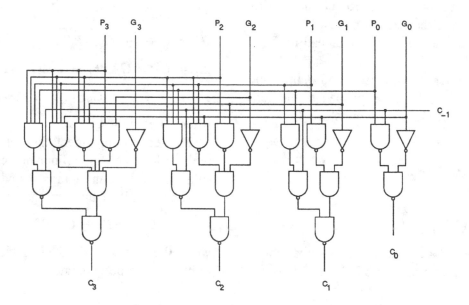

Fig. 4. A 4-bit carry-lookahead unit with 3-level C_i logic.

$$P^* = P^*_{k-1}$$

$$G^* = G^*_{k-1}$$

$$C_j = G^*_j + C_{-1}P^*_j$$

where $0 \leq j \leq k - 2$. For both circuit blocks,

$$G^*_j = G_j + G_{j-1}P_j + G_{j-2}P_{j-1}P_j$$

$$+ \ldots + G_0P_1 \ldots P_j$$

and

$$P^*_j = P_0P_1 \ldots P_j.$$

The simulations also assume that C_{-1} and all A_i, B_i are latched and available at time $t = 0$. Fan-out loading of A_i, B_i and C_{-1} is ignored and the adder delay is calculated as the time required to generate the slowest signal from among S_i and C_{n-1}.

3. Carry Lookahead Optimization

The basic goal of this research is to show how to optimize CLA structures by varying group sizes and lookahead levels. The purpose of these operations is to exploit the delay differences that represent under-utilized time. Early signals can be delayed by modifying the network to make the signals the result of more lookahead computation. Since this allows the addition of larger operands in the same amount of time, slack

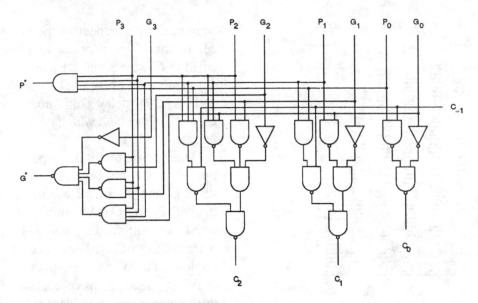

Fig. 5. A 4-bit block carry-lookahead unit with 3-level C_i logic.

reduction corresponds to adder structure optimization. The optimization requires a delay model that accurately measures slack in the circuit, an examination of the delay characteristics of the circuit blocks, and an analysis of critical delay paths to identify slack in the circuit. These requirements lead to a heuristic method for determining optimal CLA structures.

3.1. Delay Model

Logic gate delays are modeled as

$$\text{delay} = f(\text{fi, fo})$$

$$= A + B \cdot \text{fi} + (D + E \cdot \text{fi}) \cdot \text{fo}.$$

Simpler delay models that use unit gate delays are inadequate because they do not reveal all the slack in the circuit.

The simulations use the following delay functions:

$$t_{NAND} = 0.1058 + 0.1175\text{fi} + (0.0825 + 0.0148\text{fi})\text{fo}$$

$$t_{AND} = 0.2825 + 0.1675\text{fi} + (0.0911 + 0.0037\text{fi})\text{fo}$$

$$t_{INV} = 0.265 + 0.1016\text{fo}$$

$$t_{XOR} = 0.945 + 0.05645\text{fo}.$$

The constants in these functions are based on LSI Logic Corporation's 1.5 Micron Compacted Array™ Technology [8]. To limit complexity, the models assume that that all logic gates are single, possibly large, gate structures rather than multiple levels of smaller gates.

3.2. BCLA/CLA Delay Characteristics

An understanding of BCLA/CLA delay characteristics is important for finding opportunities to reduce slack in CLA structures. In particular, the fan-in and fan-out properties of the circuits must be analyzed.

The input and output loading on the propagate and generate signals of a BCLA can be derived by induction on the circuits of figures 2, 3, 4, and 5. By inspection, the following results are obtained. Given a k-bit BCLA unit connected to the jth input of an m-bit BCLA unit

- G^* of the k-bit BCLA unit has fan-out $(m - j)$.
- G^* of the k-bit BCLA unit has worst case fan-in k. Specifically, in the two-level implementation of G^*, the first level NAND gate i associated with input G_i has fan-in $k - i$.
- P^* of the k-bit BCLA unit has fan-in k.
- P^* of the k-bit BCLA unit has fan-out $(m - j)(j + 1)$

Examples of these relationships are shown graphically in figures 6 and 7. The analysis for a BCLA unit feeding into a CLA unit is similar and gives the same results.

The loading on carry signals may be derived by a similar analysis of the circuit diagrams. Each carry signal, C_i, of a BCLA/CLA unit is the C_{-1} of BCLA units on previous levels. An example of this is shown in figure 8. In particular, it connects to its $(i + 1)th$ fan-in unit, the zeroth fan-in unit of its $(i + 1)th$ fan-in unit, the zeroth fan-in unit of the zeroth fan-in unit of its $(i + 1)th$ fan-in unit, etc. Each carry signal also connects to an XOR gate in the sum generation circuitry. A BCLA unit of size k contributes $k - 1$ to the fan-out loading of its input C_{-1} signal. In the two-level implementation

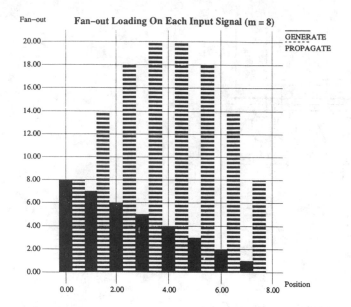

Fan-out Loading On Each Input Signal (m = 8)

Fig. 6. Fan-out loading on input signals of an 8-bit BCLA/CLA unit.

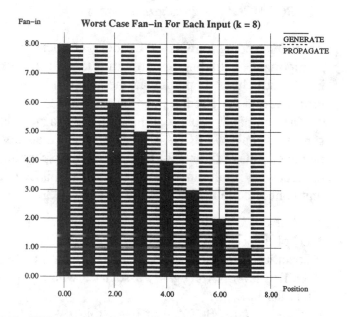

Worst Case Fan-in For Each Input (k = 8)

Fig. 7. Worst case fan-in gates for the inputs of an 8-bit BCLA unit.

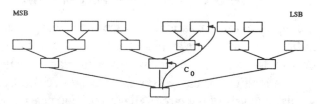

Fig. 8. Fan-out path within the CLA network of C_0 of the final lookahead unit.

of C_i, the worst case first-level NAND gate has fan-in $i + 2$. In the three-level implementation of C_i, C_{-1} always connects to an input NAND gate with a fan-in of 2.

These fan-in and fan-out properties have an important effect on the critical delay analysis in BCLA/CLA units. Unlike typical analyses, the P^* delay cannot be neglected with respect to the G^* delay. Even though computing P^* requires one less level of logic than calculating G^*, the potentially high group propagate fan-out loading may place generation of P^* on the critical path. The P^* delay path should be compared to the G_0-to-G^* delay path which contains the worst case fan-in gate.

Another important delay path is the carry propagation (assimilation) path. This is the path from C_{-1} to some C_i and is critical when C_{-1} arrives much later than all P_i and G_i. Assuming a k-bit BCLA unit, the worst case path for the two-level implementation of C_i is from C_{-1} to C_{k-2}. For the corresponding three-level implementation of C_i, the paths are equal for all i because C_{-1} feeds gates of constant fan-in. When C_{-1}-to-C_i delays are a significant fraction of the total adder delay, the three-level implementation should produce faster adders than the two-level implementation because of its superior fan-in properties. This condition should hold for larger adders.

An issue related to BCLA delay is the relationship between group size and number of lookahead levels. Clearly, larger BCLA units have longer delays than smaller units. Also, adding more levels of BCLA units tends to increase delay because of the extra logic levels. However, at some size, implementing a single large BCLA unit as multiple levels of BCLA units is advantageous. Unfortunately, determining this breakpoint is difficult because the delay of a BCLA unit depends on the block sizes on the next level of lookahead and typically, those sizes are determined by the number and sizes of blocks on all previous levels.

3.3. Critical Path Analysis

Identifying critical paths in CLA structures is important because the remaining noncritical delay paths represent opportunities for slack reduction. Standard CLA analyses assume that the critical path in the adder is always as shown in figure 9. Unfortunately, the validity of this assumption is not guaranteed when variable group sizes and lookahead levels are allowed. However, the actual critical path will have an analogous form. The first part of the critical path is the delay to the generation of a carry signal in the CLA unit and the last part of the critical path is the propagation of this carry signal back through some subtree of the BCLA network. Furthermore, each subtree of the network has an analogous critical path. Opportunities to reduce slack can be found in each portion.

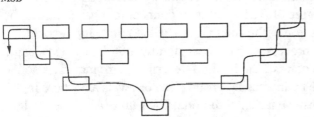

Fig. 9. Critical path assumed by typical analyses.

The delay to the CLA unit depends only on G^* and P^* delays. Most of this delay is expected to be from G_0-to-G^* delays because this path has both high fan-out loading and worst case fan-in gates. On a given level, the critical path might depend on the generation of P^*, but on the next level, the critical path will most likely be the group generate computation because G^* is a function of all the input propagate signals except P_0. The adder delay after the CLA unit depends only on carry propagation delays. All P_i and G_i signals have already settled and the critical path follows the worst C_{-1}-to-C_i path in a BCLA unit on each level.

Assuming that the generation of G^* is the critical path through a BCLA unit, then the group feeding G_i can be larger than the group feeding G_j for $i > j$, since G_j feeds a higher fan-in gate. This will reduce the slack between the different G_i-to-G^* delays. This argues for fewer levels of smaller lookahead units in the least significant bits than in the most significant bits of every subtree of lookahead units.

The slack in the second portion of a critical path arises from different C_i delay times in a lookahead unit. The subtrees fed by C_j should be faster than those fed by C_i for $j > i$. This indicates that fewer levels of smaller lookahead units should handle the most significant bits of each subtree than should handle the least significant bits.

The criteria for each portion of the critical path contradict each other. This indicates that the best opportunities to increase group sizes or add levels of lookahead may occur in the middle bits of each subtree rather than at the ends. Unfortunately, this criteria cannot be used to make specific decisions except for trivial cases. The problem is that the fan-in and fan-out dependencies of the delay model interfere with local optimizations. The whole adder must be analyzed because decisions on each level depend on the sizes of the next level which depend on the structure of the previous levels. A dynamic programming method is used to avoid the complexity of analyzing size combinations for entire adder structures.

3.4. Heuristics

The basic algorithm for finding an optimal CLA structure is based on dynamic programming. The generic pseudo-code for this approach is shown in figure 10. The problem of finding the optimal n-bit adder structure is reduced to a series of subproblems. Each subproblem requires finding an optimal $(k + b)$-bit adder structure given a k-bit adder structure. Note that the number of subproblems is bounded by $n - k_0$. The main components of the algorithm are the transformation step and the choice of initial adder structure.

The transformation step increases adder size by adding levels and/or increasing group sizes. Greedy heuristics are used to determine the location and magnitude of these increments. When increasing the size of a group, the extra bits are added to the most significant end of the group and the inputs to these new positions are connected directly to the initial propagate and generate generation circuitry. The solution of the transformation problem is simplified by formulating the decision to increase the number of lookahead levels as a decision to increase group sizes. This reduction is obtained by viewing all the inputs into the carry-lookahead network as groups of size one. Thus, increasing the size of a 1-bit *input* group often corresponds to increasing the number of lookahead unit levels in the adder. Increasing the size of such an input group does not always increase the number of levels because for any given network, multiple input groups may exist which will increase the number of lookahead levels and only the first one of these groups which is enlarged will actually increase the number of levels.

The choice for the initial adder structure is constrained to single CLA units. Ideally, the initial structure should be optimal and should admit a transformation path to a final optimal structure. The results show that the simple constraint of starting with a single CLA unit is effective.

Many different heuristics can be used to perform the transformation step and to choose the initial CLA unit size. Two different sets of heuristics were used to implement two different versions of the basic algorithm.

```
let i = 0
let initial structure = k₀-bit adder

while ( kᵢ ≤ n )
    transform the current kᵢ-bit adder into a (kᵢ + bᵢ)-bit adder, bᵢ ≥ 1
    let kᵢ₊₁ = kᵢ + bᵢ
    let i = i + 1
end
```

Fig. 10. Basic algorithm for finding the structure of an n-bit adder.

The first version runs the basic algorithm for different starting CLA unit sizes and chooses the best structure from among all the runs. The current range of starting sizes is from 2 to 16. The transformation step constrains $b_i = 1$, $\forall i$. Each block in the network except for the CLA unit is considered for receiving this extra bit. For each block, the delay of the structure that results from increasing the block size by one is calculated. The transformation is performed by enlarging the block that results in the adder structure with the shortest delay.

The second version always starts with a 2-bit CLA unit and then uses a more complex transformation step. The basic goal of the transformation heuristic is to achieve the maximum increase in adder size per unit delay increase. Increasing by one the size of some group in an adder of size k and delay d results in an adder of size $k + 1$ and delay $d + \delta$. Depending on the location of the enlarged block, the size of other groups may also be increased without further increase in adder delay. In general, it is possible to increase the size of other groups to produce an adder of size $k + b$ and delay $d + \delta$ where $b \geq 1$. Given the selection of the initial enlarged group, the location of the remaining $b - 1$ bits can be found by adding as many extra bits as possible to each group of the network such that the adder delay remains $d + \delta$. The value of $b - 1$ is maximized by processing groups by level, starting with the final CLA unit and working back to the initial propagate and generate circuitry. Within levels, groups are processed from least significant to most significant. The transformation heuristic at step i calculates for each group j in the network

$$\text{benefit}_j = \frac{\min(b_j, n - k_i)}{\delta_j}$$

where

δ_j = increase in delay caused by increasing the size of group j by 1

b_j = maximum increase in adder size possible given the selection of group j for the initial 1 bit increment and a delay increase limit of δ_j

k_i = size of adder before transformation step i

and

n = desired final adder size.

The group j with the largest benefit value is chosen to transform the k_i-bit adder into a $(k_i + \min(b_j, n - k_i))$-bit adder.

A similar method has been implemented by [5]. That method also uses dynamic programming and the recursive step increases adder size by determining an optimal subtree of BCLA units to connect to a particular input position of the final CLA (BCLA) unit. Input positions are processed from least significant to most significant. In contrast, the method described here increases adder size by increasing group sizes or levels anywhere in the transitive fan-in of the final CLA unit. Also, the method of [5] requires delay models that are linear monotone nondecreasing non-negative functions of fan-in and fan-out and only optimizes the carry-lookahead network. This work allows delay models that are any arbitrary function of fan-in and fan-out and optimizes the whole adder. Simulating the whole adder exploits the delay differences between the initial propagate and generate signals and properly accounts for loading by the sum generation circuitry. Some comparisons with the results of the method in [5] are given in the next section.

4. Results

Figures 11 and 12 show results for the two different implementations of BCLA/CLA units. Each graph has three curves corresponding to requiring fixed-sized groups and fixed number of levels (**Fixed**), allowing different sized groups only on different levels and requiring fixed levels (**Inter-level**), and allowing variable sizes and levels anywhere (**Variable**). The delay values are for the best structures of each category. The **Fixed** and **Inter-level** curves were obtained by simulation of all possible combinations. If an integral number of groups of the chosen size for a level handled more

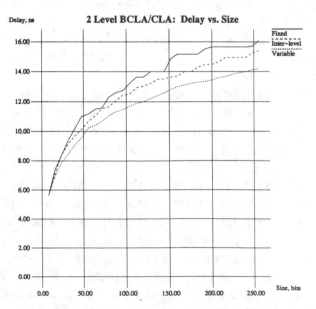

Fig. 11. Delay versus adder size in adders using a 2-level BCLA/CLA carry implementation.

428

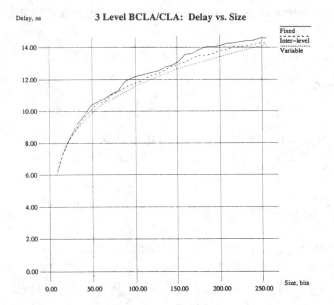

Fig. 12. Delay versus adder size in adders using a 3-level BCLA/CLA carry implementations.

signals on that level than was necessary, then the size of the group corresponding to the most significant bits of the adder was decreased to eliminate the excess capability. The **Variable** curve for the two-level BCLA/CLA carry implementation (figure 11) was obtained using the first set of heuristics. This version of the algorithm worked well for this lookahead implementation but not as well for the three-level implementation. The **Variable** curve for the three-level BCLA/CLA carry implementation (figure 12) was obtained using the second

set of heuristics since it produced better results than the first set.

Figure 13 gives the optimal 32-bit adder structure found by each algorithm version. The 1's represent input from the initial bit positions.

The results correspond well to theory. Adders of fixed-size groups are slower than adders allowing variable inter-level groups. Adders with variable groups and levels are faster than both other types. The optimal **Variable** structures tend to have more levels and larger group sizes in the middle of the adder than on the ends. Results of comparing the different lookahead implementations are mixed. The **Fixed** and **Inter-level** three-level implementations performed much better than the two-level implementations. However, the heuristics improved the performance of the two-level implementations more than they improved the three-level implementations. The delay differences between the **Variable** adders of the two implementations is smaller, though the three-level implementation is still faster for larger adders.

Table 1 compares the delays of adders generated from the carry-lookahead network configurations of [5] with the delay adder structures found by the first version of the basic algorithm. The comparisons are based on the two-level implementations of the BCLA/CLA units. The delay models are those used in [5] plus an equivalent XOR delay model. The delay functions used are:

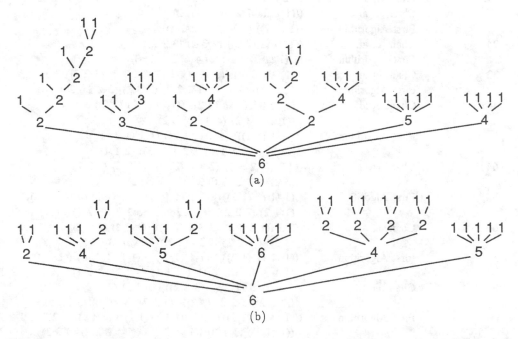

Fig. 13. 32-bit adders found by (a) the first set of heuristics using 2-level BCLA/CLA units (delay = 8.5 ns) and (b) the second set of heuristics using 3-level BCLA/CLA units (delay = 8.9 ns).

$$t_{NAND} = 5fi + 20fo$$

$$t_{AND} = 17 + 20fi + 5fo$$

$$t_{INV} = 12 + 5fo$$

$$t_{XOR} = 5fo.$$

The adder structures found by the first version of the basic algorithm are faster in all cases.

Table 1. Adder *sum* delays for configurations from [1] and for adders found by the first version of the basic algorithm.

Adder Size, bits	Delay	
	Chan, et al.	First Algorithm
8	407	394
16	524	514
24	627	584
32	652	637
48	736	721
64	821	779
66	824	779
84	877	834

Table 2 compares the adder structures corresponding to the delays in table 1. The structures are represented in parentheses notation as in [5]. For example, the adder structures shown in figures 13(a) and 13(b) are represented as

$$((1 \ (1 \ (1 \ (1 \ 2)))) \ (1 \ 1 \ 3) \ (1 \ 4) \ ((1 \ 2) \ 4) \ 5 \ 4)$$

and

$$(2 \ (1 \ 1 \ 1 \ 2) \ (1 \ 1 \ 1 \ 2) \ 6 \ (2 \ 2 \ 2 \ 2) \ 5),$$

respectively. The numbers in the expression represent the block sizes at the highest level.

5. Conclusion

Varying group sizes and lookahead levels improves the performance of commonly used CLA implementations. Unfortunately, finding these improved structures is difficult because of delay fan-in and fan-out dependencies. In general, the whole adder structure must be known before a decision to increase group sizes or the number of lookahead levels can be made. Fortunately, simple heuristics can deal effectively with this problem. Simulation results show that heuristic methods can find CLA adder structures with variable group sizes and levels that are faster than more constrained carry-lookahead adders.

Table 2. Adder structures corresponding to the delay values in table 1.

Adder size, bits	Source	Adder Structure
8	Chan, et al.	(1 1 2 2 2)
	First Algorithm	(1 2 3 2)
16	Chan, et al.	(((1 2) (1 1 2) 4 (2 1)) 2)
	First Algorithm	((1 (1 2)) (1 1 2) 4 (2 1 1))
24	Chan, et al.	((1 3) (2 (2 2)) (2 3 2) (3 2 2))
	First Algorithm	(1 (1 (1 2)) (1 1 2) 4 4 (2 1 1) 3)
32	Chan, et al.	(((1 (1 (1 2))) ((1 2) ((1 2) 3)) (2 3 2) (3 2 2)) 2 2)
	First Algorithm	(((1 (1 2)) (1 1 3) (1 1 2 2) (2 3 2) (3 2 2)) 2 1)
48	Chan, et al.	(((1 2) ((1 (1 2) (2 2)) ((1 2) (1 2) (3 2)) ((1 2) 3 3 2) (4 3 3)) 3 2)
	First Algorithm	(((1 (1 (1 2))) (1 1 (1 1 2)) (1 1 (1 2) 4) ((1 2) 3 (2 1 1)) (4 (2 1 1) 3)) (2 1 1) 3)
64	Chan, et al.	((2 3 ((1 2 3 3) (2 3 4 (2 2)) ((1 2 2) (2 2 2) (3 2)) ((2 2 2) (3 2) 3)) (3 2)) 2)
	First Algorithm	(((1 (1 (1 (1 2)))) (1 1 (1 1 3)) (1 1 (1 2) 4) ((1 (1 2)) (1 1 2) (2 2 2)) ((1 1 2 1 1) (3 2 2) (3 2 1))) (2 2 1) (2 1 1))
66	Chan, et al.	((2 3 ((1 2 3 3) (2 3 4 (2 2)) ((1 2 2) (2 2 2) (3 2)) ((3 2 2) (3 2) (2 2))) (3 2)) 2)
	First Algorithm	(((1 (1 (1 (1 2)))) (1 1 (1 1 3)) (1 1 (1 2) 4) ((1 (1 2)) (1 1 3) (2 3 2)) ((1 1 2 1 1) (3 2 2) (3 2 1))) (2 2 1) (2 1 1))
84	Chan, et al.	((3 4 ((2 3 (1 2 2) (3 2 2)) ((1 2 2) (2 3 2) (3 2 2)) ((2 3 2) (3 2 2) (3 2)) ((3 2 2) (3 2) 3)) (3 2)) 2)
	First Algorithm	(((1 (1 (1 (1 2)))) (1 1 (1 1 (1 1 2))) (1 1 (1 (1 2)) (1 1 2 2)) ((1 (1 2)) (1 1 3) ((1 2) 3 3)) ((1 1 2 2.2) (3 3 3) ((2 1 1) 3 2))) (3 3 2) (3 2 1))

430

Work is in progress to re-run the simulations in this paper for ECL delay data. This is particularly important in the domain of high performance machines where bipolar is the dominant technology. Also, more heuristics and implementations will be examined. For example, a hybrid of the two sets of heuristics presented here will be tried and an adder structure that uses the three-level BCLA carry implementation with the two-level CLA carry implementation will be tested.

Acknowledgments

The authors would like to thank Professor Pak Chan for sharing his ideas and insight on carry-lookahead adders. His many helpful discussions and suggestions are greatly appreciated.

References

1. R. Sherburne, Jr., "Processor design tradeoffs in VLSI," Technical Report UCB/CSD 84/173, University of California, Berkeley, 1984.
2. J. Sklansky, "An evaluation of several two-summand binary adders," *IRE Trans.*, EC-9(2), 1960, pp. 213–226.
3. V.G. Oklobdzija and E.R. Barnes, "On implementing addition in VLSI technology," *Journal of Parallel and Distributed Computing*, 5, 1988, pp. 716–728.
4. S. Turrini, "Optimal group distribution in carry-skip adders," In *Proceedings of the 9th Symposium on Computer Arithmetic*, 1989, pp. 1–18.
5. P.K. Chan, M.D.F. Schlag, C.D. Thomborson, and V.G. Oklobdzija, "Delay optimization of carry-skip adders and block carry-lookahead adders," in *Proceedings of the 10th IEEE Symposium on Computer Arithmetic*, June 1991, pp. 159–164.
6. B.D. Lee and V.G. Oklobdzija, "Optimization and speed improvement analysis of carry-lookahead adder structure," In *Proceedings of the Asilomar Conference on Signals, Systems, and Computers*, 1990, pp. 918–922.
7. K. Hwang, *Computer Arithmetic: Principles, Architecture, and Design*, New York: Wiley & Sons, 1979.
8. LSI Logic Corporation, *Databook: 1.5 Micron Compacted Array™ Technology*, 1987.

High-Speed Binary Adder

HUEY LING

Based on the bit pair (a_i, b_i) truth table, the carry propagate p_i and carry generate g_i have dominated the carry-look-ahead formation process for more than two decades. This paper presents a new scheme in which the new carry propagation is examined by including the neighboring pairs $(a_i, b_i; a_{i+1}, b_{i+1})$. This scheme not only reduces the component count in design, but also requires fewer logic levels in adder implementation. In addition, this new algorithm offers an astonishingly uniform loading in fan-in and fan-out nesting.

Introduction

The traditional recursive formula for carry propagation has dominated the carry handling process in the computer industry for more than two decades. Today, adder designs based on a similar technique include Amdahl V6, IBM 168, and IBM 3033.

The recursive formulation of carry is based on the bit pair (a_i, b_i) truth table. By examining the local bit pair, carry propagate p_i and carry generate g_i are formed. The high-order carries are generated by nesting the p_i and g_i together. By considering the adjacent bit pairs $(a_i, b_i; a_{i+1}, b_{i+1})$, a new recursive formula is obtained for new carry propagation. The comparison between this new scheme and the existing scheme will be discussed in the following sections. The detailed implementation, circuits, and logic level count are also included. Surprisingly, this method offers an astonishingly uniform loading in fan-in/fan-out nesting.

The formation of new carry and sum

This paper introduces a new approach to represent the new carry formation and propagation based on the concept of the complementing signal which was introduced in 1965 [1]. To examine the impact of this complementing signal in performing binary addition and complementing signal look-ahead, one should evaluate the formation of H_i and H_{i+1} as a function of neighboring bit pairs $(i, i + 1)$. Let us consider adding two binary numbers A and B together, where

$$A = a_0 2^n + a_1 2^{n-1} + a_2 2^{n-2} + \cdots + a_i 2^{n-i} + \cdots + a_n 2^0 ;$$

$$B = b_0 2^n + b_1 2^{n-1} + b_2 2^{n-2} + \cdots + b_i 2^{n-i} + \cdots + b_n 2^0 .$$

The relation among the new carry (H_i, H_{i+1}) and the neighboring bit pairs $(a_i, b_i; a_{i+1}, b_{i+1})$ can be expressed as in Table 1 [1]; all of these are generated by a_i, b_i or transmitted through the low-order bits, $i + 1, i + 2, \cdots$, with the transmitting-enable switch ON. This signal or new carry can only be terminated when the inhibitor is ON $(a_{i+1} + b_{i+1} = 0)$. H_i plays both regular carry and complementing signal roles in performing binary addition.

By grouping all the H_i, we obtain

$$H_i = f(1, 2, 3, 5, 6, 7, 9, 10, 11, 12, 13, 14, 15)$$
$$= a_i b_i + H_{i+1}(\bar{a}_{i+1} b_{i+1} + a_{i+1}\bar{b}_{i+1} + a_{i+1} b_{i+1})$$
$$= a_i b_i + H_{i+1}(a_{i+1} + b_{i+1}) = k_i + H_{i+1}T_{i+1} , . \qquad (1)$$

where k_i is the new complementing signal, H_{i+1} is the previous complementary signal, and T_{i+1} is the previous carry enable switch or the previous stage propagate.

Equation (1) shows that new carry H_i can be formed locally by k_i or produced remotely; H_{i+1} can be produced with the remote stage carry inhibitor not ON $(a_{i+1} + b_{i+1}$

Reprinted with permission from *IBM Journal of Research Development*, H. Ling, "High-Speed Binary Adder," Vol. 25, No. 3, p 156-166, May 1981. © 1981 International Business Machines.

= 1). The formation of sum S_i can be expressed by a similar process. The truth table for S_i is shown in Table 2.

By grouping all the S_i, we obtain

$$S_i = f(1, 2, 3, 4, 5, 6, 8, 9, 10, 13, 14, 15) \ ;$$

$$1, 2, 3 \rightarrow \bar{a}_i \bar{b}_i H_{i+1}(\bar{a}_{i+1} b_{i+1} + a_{i+1} \bar{b}_{i+1} + a_{i+1} b_{i+1})$$

$$= \bar{a}_i \bar{b}_i H_{i+1}(a_{i+1} + b_{i+1})$$

$$= [a_i b_i + H_{i+1}(a_{i+1} + b_{i+1})]\bar{a}_i \bar{b}_i$$

$$= H_i \bar{T}_i \ ;$$

$$5, 6, 9, 10 \rightarrow (a_i \veebar b_i)(a_{i+1} \veebar b_{i+1})\bar{H}_{i+1}$$

$$= (a_i \veebar b_i)(a_{i+1} \veebar b_{i+1})\bar{H}_{i+1} \ ;$$

$$4, 8 \rightarrow (a_i \veebar b_i)(\bar{a}_{i+1} \bar{b}_{i+1}) \ ;$$

$$4, 5, 6, 8, 9, 10 \rightarrow (a_i \veebar b_i)[\bar{H}_{i+1}(a_{i+1} \veebar b_{i+1}) + \bar{a}_{i+1} \bar{b}_{i+1}]$$

$$= (a_i + b_i)(\bar{a}_i + \bar{b}_i)[\bar{H}_{i+1}(a_{i+1} \veebar b_{i+1})$$

$$+ \bar{a}_{i+1} \bar{b}_{i+1} \bar{H}_{i+1} + \bar{a}_{i+1} \bar{b}_{i+1}]$$

$$= (a_i + b_i)(\bar{a}_i + \bar{b}_i)(\bar{H}_{i+1} + \bar{a}_{i+1} \bar{b}_{i+1}) \ ;$$

$$H_i = a_i b_i + H_{i+1}(a_{i+1} + b_{i+1}) \ ;$$

$$\bar{H}_i = (\bar{a}_i + \bar{b}_i)(\bar{H}_{i+1} + \bar{a}_{i+1} \bar{b}_{i+1}) \ ;$$

$$4, 5, 6, 8, 9, 10 \rightarrow (a_i + b_i)\bar{H}_i = T_i \bar{H}_i \ ;$$

$$13, 14, 15 \rightarrow a_i b_i H_{i+1}(\bar{a}_{i+1} b_{i+1} + a_{i+1} \bar{b}_{i+1} + a_{i+1} b_{i+1})$$

$$= a_i b_i H_{i+1}(a_{i+1} + b_{i+1})$$

$$= k_i H_{i+1} T_{i+1} \ ;$$

$$S_i = f(1, 2, 3, 4, 5, 6, 8, 9, 10, 13, 14, 15)$$

$$= H_i \bar{T}_i + T_i \bar{H}_i + k_i H_{i+1} T_{i+1}$$

$$= (H_i \veebar T_i) + k_i H_{i+1} T_{i+1} \ . \tag{2}$$

We have obtained a set of recursive formulae for both new carry H_i and sum S_i. They are different from the conventional process. Before discovering the difference, let us examine the carry-look-ahead process.

New carry-look-ahead

For ease of discussion, let us consider $i = 31$. We have

$$H_{31} = k_{31} + H_{32} T_{32} \ . \tag{3a}$$

By substituting $i = 30, 29,$ and 28 in (3a), we obtain

$$H_{28} = k_{28} + T_{29} k_{29} + T_{29} T_{30} k_{30} + T_{29} T_{30} T_{31} k_{31}$$

$$+ T_{29} T_{30} T_{31} T_{32} k_{32} \ . \tag{3b}$$

By following a similar process, we obtain

$$H_{24} = k_{24} + T_{25} k_{25} + T_{25} T_{26} k_{26} + T_{25} T_{26} T_{27} k_{27}$$

$$+ T_{25} T_{26} T_{27} T_{28} H_{28} \ ;$$

Table 1 The relation of new carry H_i with H_{i+1} and its neighboring bit pairs $(a_i, b_i; a_{i+1}, b_{i+1})$.

i	$H_i = 1$ in relation with H_{i+1}	a_i	b_i	a_{i+1}	b_{i+1}
0	$H_i = 0$	0	0	0	0
1	$H_{i+1} = 1$	0	0	0	1
2	$H_{i+1} = 1$	0	0	1	0
3	$H_{i+1} = 1$	0	0	1	1
4	$H_i = 0$	0	1	0	0
5	$H_{i+1} = 1$	0	1	0	1
6	$H_{i+1} = 1$	0	1	1	0
7	$H_{i+1} = 1$	0	1	1	1
8	$H_i = 0$	1	0	0	0
9	$H_{i+1} = 1$	1	0	0	1
10	$H_{i+1} = 1$	1	0	1	0
11	$H_{i+1} = 1$	1	0	1	1
12	$H_{i+1} = X$	1	1	0	0
13	$H_{i+1} = X$	1	1	0	1
14	$H_{i+1} = X$	1	1	1	0
15	$H_{i+1} = X$	1	1	1	1

Table 2 Sum S_i formation.

i	S_i	a_i	b_i	a_{i+1}	b_{i+1}
0	0	0	0	0	0
1	$H_{i+1} = 1$	0	0	0	1
2	$H_{i+1} = 1$	0	0	1	0
3	$H_{i+1} = 1$	0	0	1	1
4	1	0	1	0	0
5	$H_{i+1} = 0$	0	1	0	1
6	$H_{i+1} = 0$	0	1	1	0
7	0	0	1	1	1
8	1	1	0	0	0
9	$H_{i+1} = 0$	1	0	0	1
10	$H_{i+1} = 0$	1	0	1	0
11	0	1	0	1	1
12	0	1	1	0	0
13	$H_{i+1} = 1$	1	1	0	1
14	$H_{i+1} = 1$	1	1	1	0
15	$H_{i+1} = 1$	1	1	1	1

$$H_{20} = k_{20} + T_{21}k_{21} + T_{21}T_{22}k_{22} + T_{21}T_{22}T_{23}k_{23}$$
$$+ T_{21}T_{22}T_{23}T_{24}H_{24} ; \quad (4)$$

$$H_{16} = k_{16} + T_{17}k_{17} + T_{17}T_{18}k_{18} + T_{17}T_{18}T_{19}k_{19}$$
$$+ T_{17}T_{18}T_{19}T_{20}H_{20}. \quad (5)$$

By substituting (3b) for (5), we obtain

$$H_{16} = H_{16}^* + I_{16}^*H_{20}, \quad (6)$$

where

$$H_{16}^* = k_{16} + T_{17}k_{17} + T_{17}T_{18}k_{18} + T_{17}T_{18}T_{19}k_{19} ; \quad (7)$$

$$I_{16}^* = T_{17}T_{18}T_{19}T_{20} . \quad (8)$$

By substituting (3a) and (3b) for (5), we obtain

$$H_{16} = H_{16}^* + I_{16}^*H_{20}^* + I_{16}^*I_{20}^*H_{24}^* + I_{16}^*I_{20}^*I_{24}^*H_{28} . \quad (9)$$

The asterisk of H_{16}^* represents the fact that H_{16}^* can be implemented with one level of logic. Based on current switching technology, both fan-in and fan-out are equal to four with eight-emitter dotting; H_{16} can be implemented with two levels of logic.

Comparison with the existing scheme

Based on the local bit pair (a_i, b_i), carry C_i and sum S_i can be written in the form

$$C_i = g_i + C_{i+1}p_i, \qquad g_i = a_ib_i ; \quad (10)$$

$$S_i = a_i \lor b_i \lor C_{i+1}, \qquad p_i = a_i + b_i .$$

For $i = 16$, we have

$$C_{16} = g_{16} + C_{17}p_{16} .$$

By substituting $i = 17, 18, \cdots, 19$, C_{16} can be rewritten as

$$C_{16} = g_{16} + p_{16}(g_{17} + p_{17}g_{18} + p_{17}p_{18}g_{19} + p_{17}p_{18}p_{19}C_{20})$$
$$= g_{16} + p_{16}g_{17} + p_{16}p_{17}g_{18} + p_{16}p_{17}p_{18}g_{19}$$
$$+ p_{16}p_{17}p_{18}p_{19}C_{20}$$
$$= G_{16p} + P_{16p}C_{20} , \quad (11)$$

where G_{16p} and P_{16p} are the grouping of the following terms:

$$G_{16p} = g_{16} + p_{16}g_{17} + p_{16}p_{17}g_{18} + p_{16}p_{17}p_{18}g_{19} ; \quad (12)$$

$$P_{16p} = p_{16}p_{17}p_{18}p_{19} . \quad (13)$$

Similarly, C_{16} can be written in terms of C_{28}:

$$C_{16} = G_{16p} + P_{16p}G_{20p} + P_{16p}P_{20p}G_{24p}$$
$$+ P_{16p}P_{20p}\overline{P}_{24p}G_{28p} .$$

Equations (6), (7), and (8) are similar to (11), (12), and (13); however, H_{16}^* can be implemented with one level of logic, whereas G_{16p} cannot. By expanding (7) and (12) we obtain

$$H_{16}^* = a_{16}b_{16} + (a_{17} + b_{17})a_{17}b_{17}$$
$$+ (a_{17} + b_{17})(a_{18} + b_{18})a_{18}b_{18}$$
$$+ (a_{17} + b_{17})(a_{18} + b_{18})(a_{19} + b_{19})a_{19}b_{19}$$
$$= a_{16}b_{16} + a_{17}b_{17} + a_{17}a_{18}b_{18} + b_{17}a_{18}b_{18}$$
$$+ a_{17}a_{18}a_{19}b_{19} + a_{17}b_{18}a_{19}b_{19}$$
$$+ b_{17}a_{18}a_{19}b_{19} + b_{17}b_{18}a_{19}b_{19} ; \quad (14)$$

$$G_{16p} = a_{16}b_{16} + (a_{16} + b_{16})a_{17}b_{17}$$
$$+ (a_{16} + b_{16})(a_{17} + b_{17})a_{18}b_{18}$$
$$+ (a_{16} + b_{16})(a_{17} + b_{17})(a_{18} + b_{18})a_{19}b_{19}$$
$$= a_{16}b_{16} + a_{16}a_{17}b_{17} + b_{16}a_{17}b_{17}$$
$$+ a_{16}a_{17}a_{18}b_{18} + a_{16}b_{17}a_{18}b_{18}$$
$$+ b_{16}a_{17}a_{18}b_{18} + b_{16}b_{17}a_{18}b_{18} + a_{16}a_{17}a_{18}a_{19}b_{19}$$
$$+ a_{16}a_{17}b_{18}a_{19}b_{19} + a_{16}b_{17}a_{18}a_{19}b_{19} + a_{16}b_{17}b_{18}a_{19}b_{19}$$
$$+ b_{16}a_{17}a_{18}a_{19}b_{19} + b_{16}a_{17}b_{18}a_{19}b_{19}$$
$$+ b_{16}b_{17}a_{18}a_{19}b_{19} + b_{16}b_{17}b_{18}a_{19}b_{19} . \quad (15)$$

Equation (14) contains eight terms, whereas (15) contains fifteen. With current available technology, the former can be implemented with one level of logic (this is shown in detail in the next section); the latter can only be implemented with two levels of logic.

Let us further examine the ith-digit carry formation. For (1), the carry is generated by local complementing signal k_i, and the remote carry H_{i+1} is controlled by remote bit pair $(a_{i+1} + b_{i+1})$; whereas for (10), the carry is generated by local carry g_i, and the remote carry C_{i+1} is controlled by local bit pair $(a_i + b_i)$. From the carry-look-ahead point of view, (1) offers faster resolution, whereas the latter is one stage slower. That is why (14) contains only eight terms, and (15), fifteen.

To illustrate the step-by-step operation, two examples are given.

Example 1   Assume the contents of A and B registers to be as shown and find their sum;

A register   0000000001101010111110110 0011001

B register   0000000001101101110101010 10110111

The k_i and T_i can be implemented with one level of logic:

k_i 00000000011010001101000100010001

T_i 00000000011011111111111101011111

434

The complementary signals can be implemented by grouping k_i and T_i together. This process requires one level of logic:

H_i 00000000111111111111111100111111

The sum digit S_i is implemented in parallel with H_i; the result of H_i will force S_i to select one value between $H_i = 0$ and $H_i = 1$:

S_i 00000000110110001101000001110000

This example demonstrates that it is possible to implement a 32-bit adder with three levels of logic with the hardware constraints indicated in the previous section. The detailed implementation of S_i is discussed in the next section.

Example 2   Assuming that the contents of index, base, and displacement registers are as shown, compute the virtual address. (To test the generality of this scheme, odd contents are purposely chosen; in the normal mode of operation, an EXCPN will occur.)

Index register   00000000000011101011101110101101 1

Base register   00000000000001011100100111011101

Displacement register   101111010111

To implement the carry-save adder (CSA) requires one level of logic:

s_i 00000000000011000111101010101010001

c_i 00000000000010100001011110111110

Implementation of k_i and T_i requires one additional level of logic:

k_i 00000000000010000001010100010000

T_i 00000000000110101111011111111111

Implementation of the complementary signal requires one level of logic to group k_i and T_i together:

H_i 00000000000111001111111111110000

The address digit S_i is implemented in parallel with H_i; the result of H_i will force S_i to select one value between $H_i = 0$ and 1:

S_i 00000000000100011000011010000 1111

This example demonstrates that the AGEN adder can be implemented with four rather than six levels of logic, as is the case in current machine organization. The detailed

implementation of S_i (the address) is discussed in the next section. The logic implementation of every fourth bit ($i =$ 31, 27, 23, 19, 16, 15, 11, 7, 3, 0) is shown in the Appendix of this paper.

Implementation

The detailed implementation can be divided into two categories: binary addition and subtraction, and address generation.

• Addition and subtraction

Equation (3b) is a general representation of the new carry-look-ahead process. For ADDITION, $k_{32} = 0$; therefore, the fifth term in (3b) is dropped and H_{28} can be written as

$$H_{28} = k_{28} + T_{29}k_{29} + T_{29}T_{30}k_{30} + T_{29}T_{30}T_{31}k_{31} \,. \qquad (4a)$$

For SUBTRACTION, there is a HOT ONE carry input from bit 31; thus H_{28} can be written as

$$H_{28} = k_{28} + T_{29}k_{29} + T_{29}T_{30}k_{30} + T_{29}T_{30}T_{31}k_{31}$$
$$+ T_{29}T_{30}T_{31} \,. \qquad (4b)$$

Equation (2) shows that S_i is a function of H_i and H_{i+1}. For ease of implementation, this equation is rewritten in the form

$$S_i = (H_i \lor T_i) + k_i H_{i+1} T_{i+1}$$
$$= [(k_i + H_{i+1}T_{i+1}) \lor T_i] + k_i H_{i+1} T_{i+1}$$
$$= H_{i+1}(\overline{T}_i T_{i+1} + k_i T_{i+1})$$
$$+ \overline{H}_{i+1}\overline{k}_i T_i + k_i \overline{T}_i + \overline{k}_i T_i \overline{T}_{i+1} \,. \qquad (16)$$

Equation (16) demonstrates that S_i can be written in the conditional form

$$S_i(H_{i+1} = 0) = k_i \lor T_i \,;$$
$$S_i(H_{i+1} = 1) = \overline{T}_i T_{i+1} + k_i T_{i+1} + k_i \overline{T}_i + \overline{k}_i T_i \overline{T}_{i+1} \,.$$

The general expression of SUM S_i can be written as

$$S_i = H_{i+1}(a_{i+1} + b_{i+1})(\overline{a_i \lor b_i}) + \overline{H}_{i+1}(a_i \lor b_i)$$
$$+ (\overline{a_{i+1} + b_{i+1}})(a_i \lor b_i) \,.$$

For $i = 31$, we have

$$S_{31} = H_{32}(a_{32} + b_{32})(\overline{a_{31} \lor b_{31}}) + \overline{H}_{32}(a_{31} \lor b_{31})$$
$$+ (\overline{a_{32} + b_{32}})(a_{31} \lor b_{31}) \,.$$

For $i = 0$, we obtain

$$S_0 = H_1(a_1 + b_1)(\overline{a_0 \lor b_0}) + \overline{H}_1(a_0 \lor b_0)$$
$$+ (\overline{a_1 + b_1})(a_0 \lor b_0) \,; \qquad (17)$$
$$H_1 = k_1 + T_2k_2 + T_2T_3k_3 + T_2T_3T_4H_4$$
$$= H_1^* + I_1^*H_4 \,; \qquad (18)$$

$$H_4 = k_4 + T_5k_5 + T_5T_6k_6 + T_5T_6T_7k_7 + T_5T_6T_7T_8H_8$$
$$= H_4^* + I_4^*H_8 ; \tag{19}$$

$$H_8 = H_8^* + I_8^*H_{12} ; \tag{20}$$

$$H_{12} = H_{12}^* + I_{12}^*H_{16} . \tag{21}$$

By substituting (21) for (20), we have

$$H_8 = H_8^* + I_8^*H_{12}^* + I_8^*I_{12}^*H_{16} . \tag{22}$$

Similarly, we obtain H_4 and H_1:

$$H_4 = H_4^* + I_4^*H_8^* + I_4^*I_8^*H_{12}^* + I_4^*I_8^*I_{12}^*H_{16}; \tag{23}$$

$$H_1 = H_1^* + I_1^*H_4^* + I_1^*I_4^*H_8^* + I_1^*I_4^*I_8^*H_{12}^*$$
$$+ I_1^*I_4^*I_8^*I_{12}^*H_{16}. \tag{24}$$

By substituting Eqs. (22), (23), and (24) for Eqs. (18), (19), (20), and (21), Eq. (17) can be written as

$$S_0 = (H_1^* + I_1^*H_4^* + I_1^*I_4^*H_8^* + I_1^*I_4^*I_8^*H_{12}^*$$
$$+ I_1^*I_4^*I_8^*I_{12}^*H_{16})(a_1 + b_1)\overline{(a_0 \, \forall \, b_0)}$$
$$+ \overline{(H_1^* + I_1^*H_4^* + I_1^*I_4^*H_8^* + I_1^*I_4^*I_8^*H_{12}^*}$$
$$+ \overline{I_1^*I_4^*I_8^*I_{12}^*H_{16})}(a_0 \, \forall \, b_0) + \overline{(a_1 + b_1)}(a_0 \, \forall \, b_0) ;$$
$$= (H_1^* + I_1^*H_4^* + I_1^*I_4^*H_8^* + I_1^*I_4^*I_8^*H_{12}^*)(a_1 + b_1)$$
$$\times \overline{(a_0 \, \forall \, b_0)} + I_1^*I_4^*I_8^*I_{12}^*H_{16}(a_1 + b_1)\overline{(a_0 \, \forall \, b_0)}$$
$$+ \overline{(H_1^* + I_1^*H_4^* + I_1^*I_4^*H_8^* + I_1^*I_4^*I_8^*H_{12}^*)}$$
$$\times (\overline{I_1^*I_4^*I_8^*I_{12}^*} + \bar{H}_{16})(a_0 \, \forall \, b_0)$$
$$+ \overline{(a_1 + b_1)}(a_0 \, \forall \, b_0) . \tag{25}$$

By using the Sklansky conditional-sum method [2], (25) can be written as

$$S_0 = (H_1^* + I_1^*H_4^* + I_1^*I_4^*H_8^* + I_1^*I_4^*I_8^*H_{12}^*)$$
$$\times (a_1 + b_1)\overline{(a_0 \, \forall \, b_0)} + \overline{(a_1 + b_1)}(a_0 \, \forall \, b_0)$$
$$+ \overline{(H_1^* + I_1^*H_4^* + I_1^*I_4^*H_8^* + I_1^*I_4^*I_8^*H_{12}^*)}$$
$$\times (a_0 \, \forall \, b_0) \quad [H_{16} = 0]$$
$$+ I_1^*I_4^*I_8^*I_{12}^*(a_1 + b_1)\overline{(a_0 \, \forall \, b_0)} \quad [H_{16} = 1]$$
$$+ \overline{(H_1^* + I_1^*H_4^* + I_1^*I_3^*H_8^* + I_1^*I_4^*I_8^*H_{12}^*)}$$
$$\times (\overline{I_1^*I_4^*I_8^*I_{12}^*})(a_0 \, \forall \, b_0) \quad [H_{16} = 1].$$

The hardware implementation of S_0 is included in the Appendix.

Equation (9) has indicated that H_{16} can be implemented with two levels of logic. Let us examine the individual terms of S_0. It is clearly pointed out that they also require only two levels of logic to be implemented. That is to say, when H_{16} is ready, S_0 can be obtained by using one additional level of logic. We have proved, by using current switching logic, that one can implement a 32-bit adder by consuming only three levels of logic.

- *Address generation*

In the address generation process, we are dealing with positive numbers only. Therefore, $k_{32} = 0$. The output of the (3, 2) carry-save adder provides the s_i and c_{i+1} corresponding to the a_i and b_i bit pairs. In addition, X_i and B_i both are 32 bits in length. However, D_i has only a 12-bit width. For $i = 0$–18, the output of the carry-save adder has a special pattern; s_i and c_{i+1} will not have the form

$$111—101—1111—11$$

$$101—111—1111—11 .$$

In general, the output of CSA will appear as

$$1111—01—001—10110$$

$$0001—01—001—10010 .$$

Therefore, for $i = 19$–31, S_i appears as usual:

$$S_i = (H_i \, \forall \, T_i) + k_iH_{i+1}T_{i+1} .$$

For $i = 0$–18, S_i appears as

$$S_i = (H_i \, \forall \, T_i) .$$

The detailed implementations for i from 0, 3, 7, 11, 15, 16, 19, 23, 27, and 31 are shown in the Appendix.

Summary

It is intended in this paper to speed up the carry propagation for examining two bit pairs. The formulation of H_{16}^* contains eight terms as compared to that of the regular carry-look-ahead process, where $G_{16\nu}$ contains fifteen terms. It is possible to implement H_{16}^* with one level of logic, whereas it is not possible with $G_{16\nu}$. The formulation of sum S_i in this new process will contain slightly more terms; however, they are not in the critical path.

References
1. H. Ling, "High Speed Binary Parallel Adder," *IEEE Trans. Electron. Computers* EC-15, 799–802 (1966).
2. J. Sklansky, "Conditional-Sum Addition Logic," *IEEE Trans. Electron. Computers* EC-9, 226–231 (1960).

$i = 0$

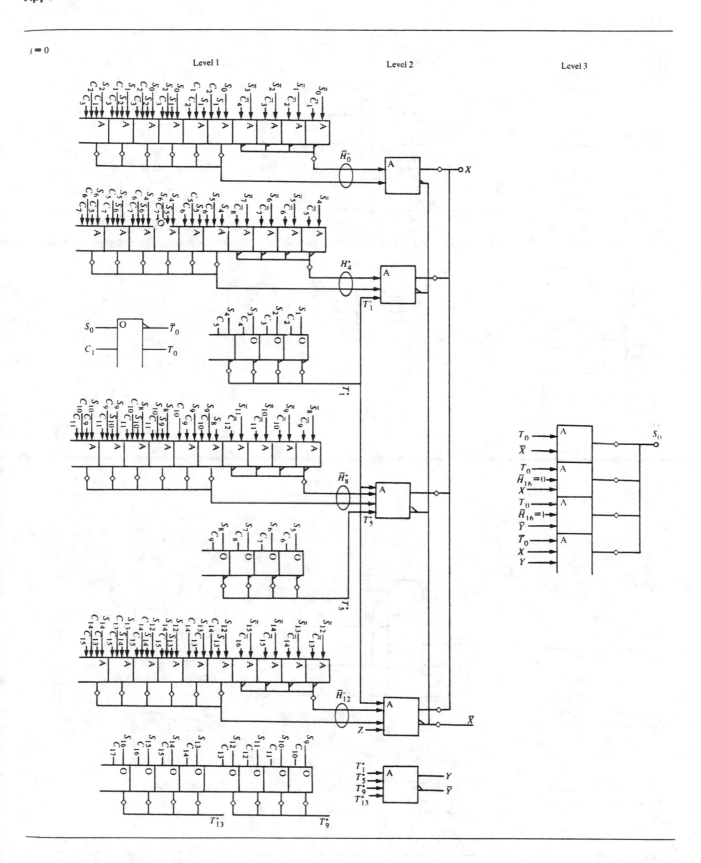

$i = 3$

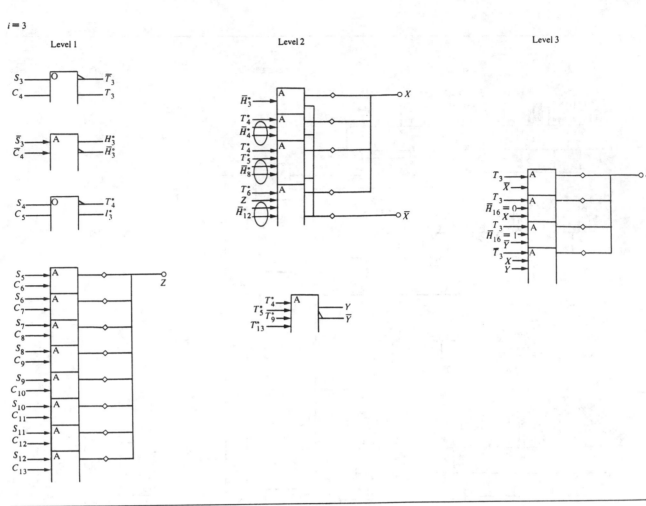

$i = 7$

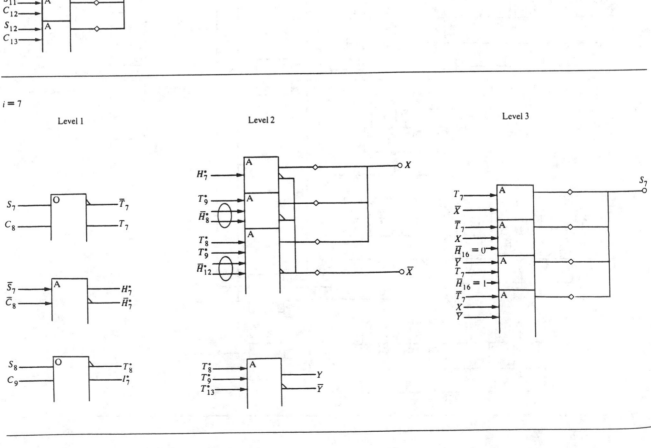

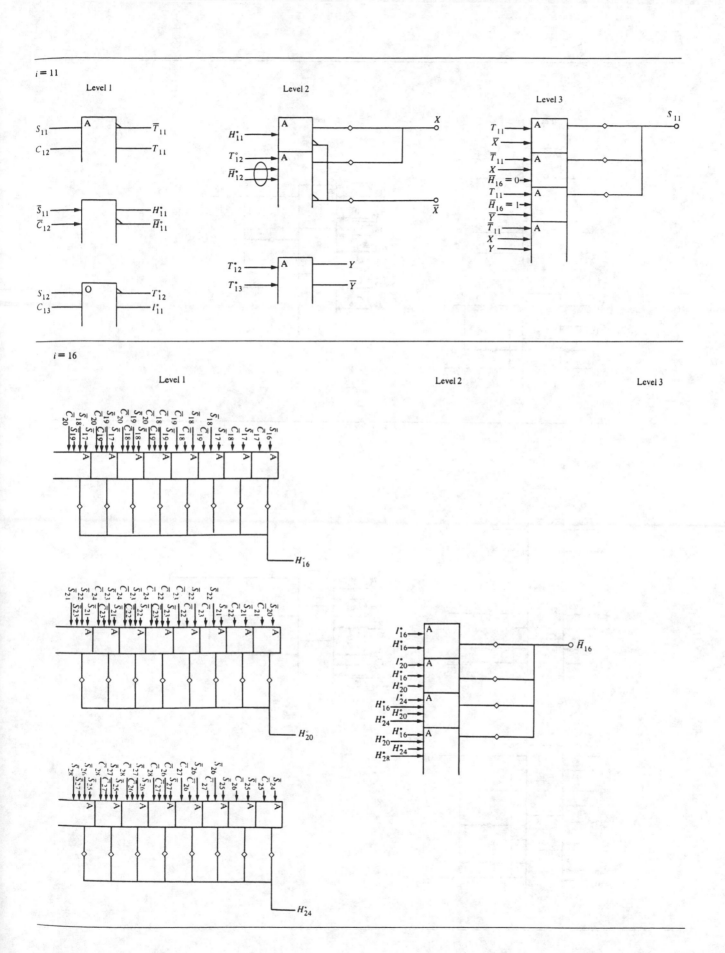

$i = 15$

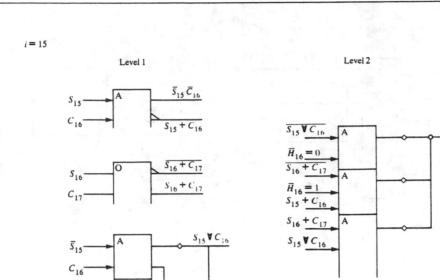

$i = 27$

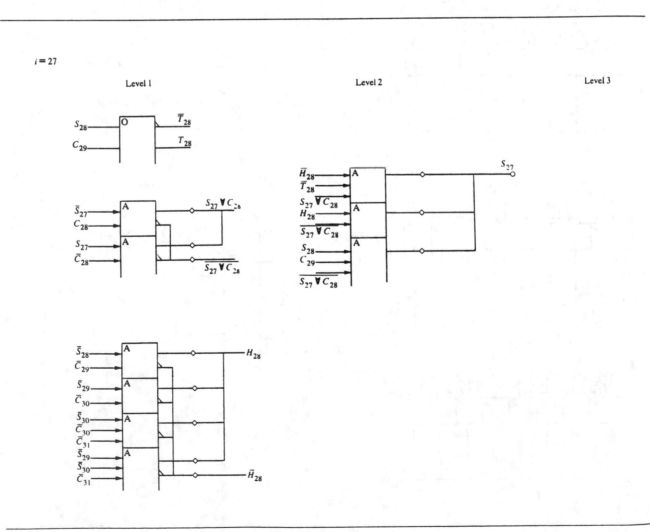

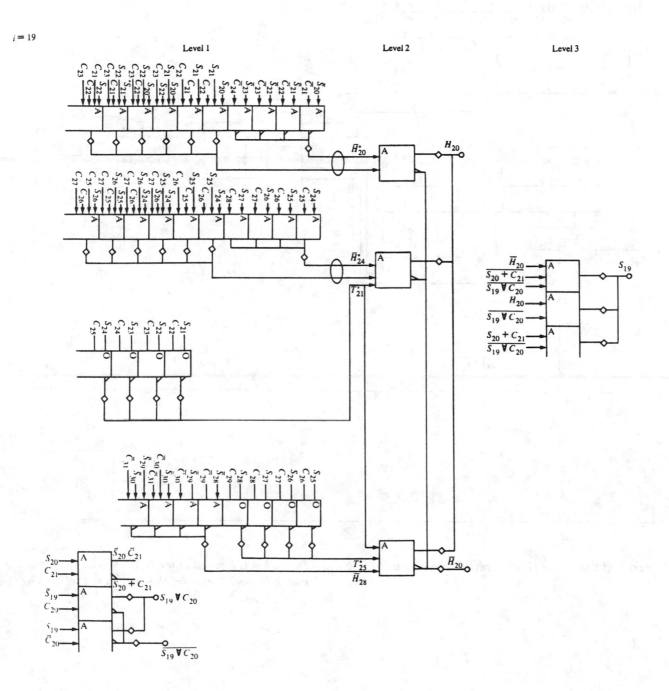

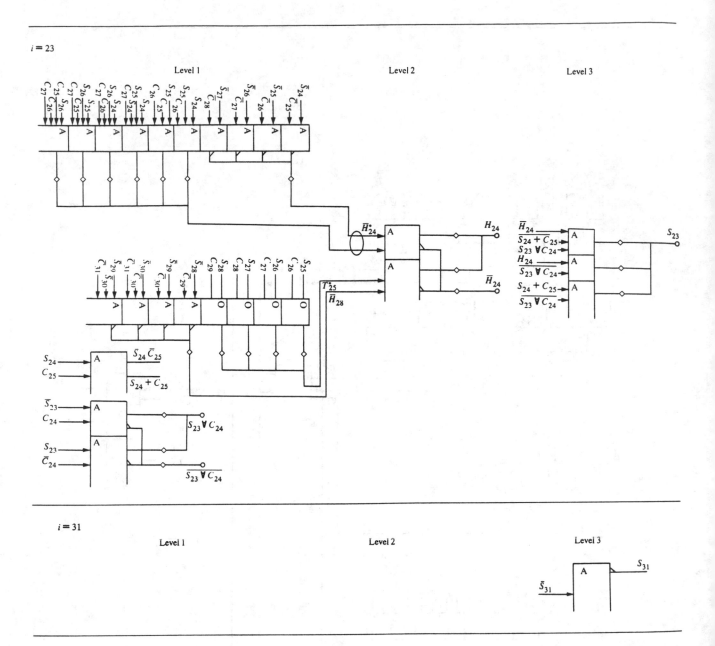

Received September 11, 1980; revised November 26, 1980

The author is located at the IBM Thomas J. Watson Research Center, Yorktown Heights, New York 10598.

A Sub-Nanosecond 0.5μm 64b Adder Design

SAMUEL NAFFZIGER

HEWLETT-PACKARD CO., FORT COLLINS, CO

A sub-nanosecond 64b adder in 0.5μm CMOS forms the basis for the integer and floating point execution units. Integrating dual-rail dynamic CMOS and use of Ling's equations, the adder is composed of 7k FETs in 0.246mm² and performs a full 64b add, operands to result in <1ns (7 fanout of 4 inverter delays) under nominal conditions.

Addition time is important to CPU design. Add latency is in the critical path of such important areas as memory address calculation, ALU evaluation and floating point computation. Current state of the art addition hardware [1-3] is not adequate for 64b computing, and this prompted a new design that forms the core of the execution hardware. It incorporates architecture and circuit design techniques that enable it to achieve low latency for adders of width from 13b to 112b. These techniques for the 64b adder design are described here.

Conceptually, an adder is:

$$Sum = A \oplus B \oplus Carry\_in$$
$$Carry = (A * B) + (A + B) * Carry\_in$$

Rippling the carry one bit at a time however is clearly unacceptable for a 64b adder so a fusion of carry-look ahead and carry-select techniques is implemented. Carry-look-ahead techniques involve parallel calculation of groups of carries (in this case, 4 at a time) in a modular fashion that reduces the carry calculation time to $Log_r[n] + 2$ gate delays where r is the group size and n is the width of the adder. Most current adder designs are based on this scheme [1, 2]. The equations for a group of 4 carry look-ahead are:

$$C4 = G3 + G2*P3 + G1*P2*P3 + G0*P1*P2*P3$$

Where the G and P terms are the generate and propagate terms derived from the operands (G=A*B, P=A+B). These groups of carries can then be combined at another level to produce a longer carry look-ahead using the same equations. Hence, for a traditional group of 2 carry look-ahead adder delay:

1 (generate GPK terms) + $log_2[64]$ + 1 (Final XOR) = 8 gate delays. If the technology of implementation allows the greater fanout and fanin of group of 4 carry look-ahead, the gate delays are:

$$1 + log_4[64] + 1 = 5 \text{ gate delays.}$$

A sum select method is often used that goes ahead and calculates two sums based on Carry_in = 0 and 1 for a group using duplicate carry chains. When the actual carry into that group becomes available via look ahead, the correct sum is selected.

Ling's equations can be used to reduce delay further. Ling developed a series of equations specifically to make use of the dot or capability of ECL logic [4]. The result is the definition of a psuedo-carry calculated quickly with dot-or circuits. The relevant equations are as follows for a normal 4b carry:

$$C4 = G3 + G2*P3 + G1*P2*P3 + G0*P1*P2*P3$$

For a Ling "psuedo carry" [5]

$$H4 = G3 + G2*P2 + G1*P1*P2 + G0*P2*P1*P0$$

The propagate terms are simply shifted by one so C4 = H4*P3. When we define P as the OR of the operands, the G2*P2 term is redundant since G2 implies P2. This redundancy can be used to simplify all of the terms for H4 yielding an equation based on the actual operands, not P and G terms and can be calculated in one gate delay with fanin 4 logic:

$$H4 = A3*B3 + A2*B2 + A2*A1*B1 + A1*B2*B1 + A2*A1*A0*B0$$
$$+ A2*B1*A0*B0 + B2*A1*A0*B0 + B2*B1*A0*B0$$

This is simpler (8 terms, fanin of 4 vs.15 terms, fanin of 5) than the expansion of the C4 equation in terms of the operands. With careful use of X terms in the carry propagation path, the conversion of H4 into C4 can be accomplished with no penalty.

To perform a CLA, 4b propagate terms must be calculated in at least the same time as the H4 terms that can then be combined for higher level look-aheads. These 4b propagate terms (I4 = P3*P2*P1*P0) can be calculated using a wired-or to generate P_n. This wired-or capability can be duplicated in dynamic CMOS using multiple NFET pulldown legs on a precharged node. The resulting circuits that generate H4 and I4 directly off the input operands are shown in Figure 1. The H4 and I4 terms are combined in a distributed Manchester gate to produce the block of 16 carry signals:

$$C16 = (H0+I0)*I1*I2*I3 + H1*I2*I3 + H2*I3 + H3$$

Where the Hx Ix terms are for the x'th group of 4 in the 16b quadrant. These 16b carries are combined in 1 more gate to produce the final carry select signals to the upper quadrants of the 64b adder. In parallel with this long carry generation, a carry ripple is performed in each of the 16b quadrants that generates both values of the carry to be selected. To hide the psuedo-carry to real-carry conversion, the carry chain is shifted over one bit and a special C3 (or carry out of 3b) term is used from the H4 gate in combination with G0 (Figure 2). In this way the I4 terms can be used directly in the carry-ripple chain as well. Finally, the carry select and sum generate are performed in a single gate (Figure 3). The critical path in the 64b adder (Figure 4) involves 4 total gate delays: (H4/I4 generation) + (C16 gen) + (Long carry gen) + (Sum select). The fast fanout of 1 carry ripple that occurrs in parallel with the look-ahead, produces the local carries just ahead of the long carry select signal at each sum gate.

The high gain and fanin capability of dynamic CMOS along with the flexibility of dual rail are key enablers to the design. The adder requires dual monotonic (DCVS) inputs and produces a dual monotonic sum. The number of transistors for a 64b sum is 6924, that when layed out in 0.6μm geometry occupies 96x2560μm or 0.246mm² in 3 layers of metal. This adder occupies several critical paths in a microprocessor whose frequency of operation confirms simulated time at nominal process and voltage conditions of 0.93ns data in to sum out (Figure 5) [6].

References:

[1] Inoue, A., et al., "A 0.4mm 1.4ns 32b Dynamic Adder using Non-precharge Multiplexers and Reduced Precharge Voltage Technique," Symp. VLSI Circuits Digest of Tech. Papers, pp. 9-10, June, 1995.

[2] Dobberpuhl, et al., "A 200 MHz 64b dual-issue CMOS microprocessor," IEEE J. of Solid-State Circuits, Vol. 27, No. 11, Nov., 1992.

[3] Suzuki, M., et al., "A 1.4ns 32b CMOS ALU in Double Pass-Transistor Logic," IEEE J. of Solid-State Circuits, Vol. 28, No. 11, Nov., 1993.

[4] Ling, H., "High Speed Binary Adder," IBM J. Reasearch. Dev., Vol. 25, No. 3, p.156, May, 1981.

[5] Flynn, M., "Topics in Arithmetic For Digital Systems Designers," (Preliminary Second Edition) pp. 104-105, 1995.

[6] Heikes, C., G. Colon-Bonet, "Dual floating-Point Coprocessor with an FMAC Architecture," ISSCC Digest of Technical Papers, pp. 354-356, Feb., 1996.

Reprinted from *Digest of Technical Papers, 1996 IEEE International Solid-State Circuits Conference*, pp. 362-363, February 1996.

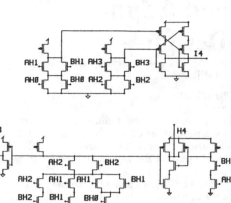

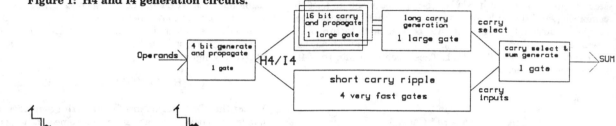

Figure 1: H4 and I4 generation circuits.

Operands

H4/I4

Figure 4: 64b add critical path.

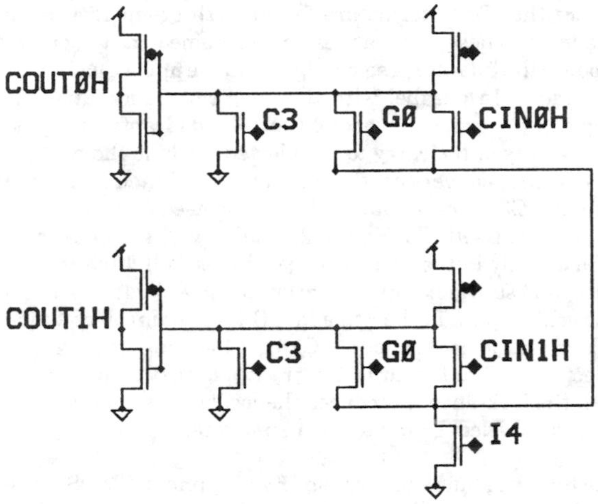

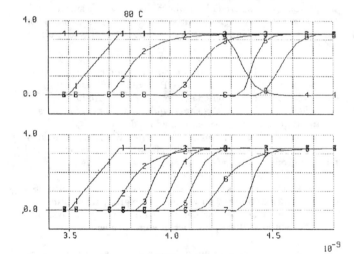

Figure 2: Carry ripple circuit.

Figure 5: Critical path SPICE waveforms:

Top key: (1) operands, (2) I4 , (3) C16, (5) long carry select, (6) sum.

Bottom key: (1) operands, (2) I4, (6) short carry ripple, (7) long carry select.

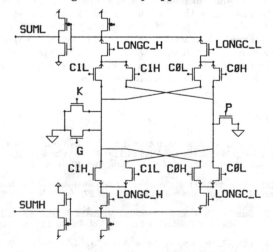

Figure 3: Sum select gate.

444

A Parallel Algorithm for the Efficient Solution of a General Class of Recurrence Equations

PETER M. KOGGE AND HAROLD S. STONE

Abstract—An mth-order recurrence problem is defined as the computation of the series $x_1, x_2, \cdots, x_N$, where $x_i = f_i(x_{i-1}, \cdots, x_{i-m})$ for some function f_i. This paper uses a technique called recursive doubling in an algorithm for solving a large class of recurrence problems on parallel computers such as the Illiac IV.

Recursive doubling involves the splitting of the computation of a function into two equally complex subfunctions whose evaluation can be performed simultaneously in two separate processors. Successive splitting of each of these subfunctions spreads the computation over more processors.

This algorithm can be applied to any recurrence equation of the form $x_i = f(b_i, g(a_i, x_{i-1}))$ where f and g are functions that satisfy certain distributive and associative-like properties. Although this recurrence is first order, all linear mth-order recurrence equations can be cast into this form. Suitable applications include linear recurrence equations, polynomial evaluation, several nonlinear problems, the determination of the maximum or minimum of N numbers, and the solution of tridiagonal linear equations. The resulting algorithm computes the entire series $x_1, \cdots, x_N$ in time proportional to $\lceil \log_2 N \rceil$ on a computer with N-fold parallelism. On a serial computer, computation time is proportional to N.

Index Terms—Parallel algorithms, parallel computation, recurrence problems, recursive doubling.

INTRODUCTION

A. Definition of Problem

IT FREQUENTLY occurs in applied mathematics that the solution to some problem is a sequence $x_1, x_2, \cdots, x_N$, where each x_i is a function of the previous m x's, namely $x_{i-1}, \cdots, x_{i-m}$. A common example of such a problem is a time-varying linear system, where the state of the system at time i is x_i, and can be computed from the equations

Manuscript received October 7, 1972; revised March 21, 1973. This work was supported in part by NSF Grant GJ 1180 and in part by an IBM Corporation fellowship.

P. M. Kogge was with the Department of Electrical Engineering, Digital Systems Laboratory, Stanford University, Stanford, Calif. He is now with the Systems Architecture Department, IBM Corporation, Owego, N.Y. 13827.

H. S. Stone is with the Department of Electrical Engineering and the Department of Computer Science, Digital Systems Laboratory, Stanford University, Stanford, Calif.

Reprinted from *IEEE Transactions on Computers*, Vol. SC-22, No. 8, pp. 786–793, August 1973.

$$x_1 = B_1$$
$$x_2 = A_2 x_1 + B_2$$
$$x_3 = A_3 x_2 + B_3$$
$$\cdot$$
$$\cdot$$
$$x_i = A_i x_{i-1} + B_i$$
$$\cdot$$
$$\cdot$$
$$x_N = A_N x_{N-1} + B_N \tag{1}$$

where A_i and B_i represent the internal dynamics of the system. A_i and B_i can be real or complex numbers, constant or time-varying matrices, etc., depending on the problem.

The equation used to compute x_i is called a recurrence equation and, together with some initial values for some of the x_i, represents a complete problem description. Formally, a recurrence problem consists of a set of recurrence equations:

$$x_i = f_i(x_{i-1}, \cdots, x_{i-m}), \quad i = m+1, \cdots, N \tag{2}$$

and some boundary values, which may consist of the following.

1) $x_1, \cdots, x_m$. This is an initial value problem.
2) $x_{N-m+1}, \cdots, x_N$. This is a final value problem.
3) A mixture of m initial and final values.

This paper discusses an algorithm for solving a particular class of initial value recurrence problems on parallel computing systems such as the Illiac IV. This class of problems includes the computation of the sequence $x_1, \cdots, x_N$ when the expression for x_i is a linear recurrence equation of the form of (1), the calculation of the maximum or minimum of N numbers, the evaluation of Nth-degree polynomials, and several nonlinear problems. Such problems as these can be solved in a very straightforward manner on serial processors in time proportional to N. Some have also been solved on parallel computers with special-purpose algorithms tailored to those problems, e.g., polynomial evaluation (Munro and Paterson [3]). With a computer having N-fold parallelism, the algorithm in this paper solves all these problems and others in time proportional to $\lceil \log_2 N \rceil$.[1]

B. Computer Model

The algorithm to be described in this paper is designed for a computer of the Illiac IV class. The major assumptions about the computer's architecture are as follows.

Assumption 1: There are p identical processors, each able to execute the usual arithmetic and logical operations, and each with its own memory.

Assumption 2: Each processor can communicate with every other processor. The exact method of data exchange between processors can affect the algorithm's computational complexity and will be discussed in a future report.

---

[1] $\lceil x \rceil$ is the ceiling function and represents the smallest integer not smaller than x.

Assumption 3: Each processor has a distinct index by which it is referenced.

Assumption 4: All processors obtain their instructions simultaneously from a single instruction stream. Thus all processors execute the same instruction, but they operate on data stored in their own memories.

Assumption 5: Any processor may be "blocked" or "masked" from performing some instruction. This mask may be set by an explicit instruction directed to that processor via its index, or by the result of some test instruction such as "set mask if accumulator = 0."

Assumption 6: Elementary arithmetic operations have two operands.

It is assumed throughout this paper that the number of processors p is greater than N, the maximum number of elements to be computed. In reality when p is less than N, this algorithm can be used $\lceil N/p \rceil$ times to calculate p elements of the series at a time.

II. General First-Order Recurrence Equation

A. Example

In this section we develop a parallel solution to a simple first-order recurrence problem. The solution is a special case of the general algorithm, but its development is not obscured by the notation needed to describe the general algorithm.

Given $x_1 = b_1$, find $x_2, \cdots, x_N$, where

$$x_i = a_i x_{i-1} + b_i. \tag{3}$$

Before solving this problem let us give the following definition.

Definition: The function $\hat{Q}(m, n)$, $n \leq m$, is defined to be

$$\hat{Q}(m, n) = \sum_{j=n}^{m} \left(\prod_{r=j+1}^{m} a_r \right) b_j$$

where the vacuous product $(\prod_{r=m+1}^{m} a_r)$ is given the value 1. Stone [4] first used this notation in the derivation of this algorithm. The basic algorithm involves a concept called *recursive doubling*, which consists of breaking the calculation of one term into two equally complex subterms.

Now we can write the solution to (3) as follows:

$$x_1 = b_1 \qquad = \hat{Q}(1, 1)$$
$$x_2 = a_2 x_1 + b_2 \qquad = a_2 b_1 + b_2 = \hat{Q}(2, 1)$$
$$x_3 = a_3 x_2 + b_3 \qquad = a_3 a_2 b_1 + a_3 b_2 + b_3 = \hat{Q}(3, 1)$$
$$\cdot$$
$$\cdot$$
$$\cdot$$
$$x_i = a_i x_{i-1} + b_i \qquad = \hat{Q}(i, 1)$$
$$\cdot$$
$$\cdot$$
$$x_N = a_N x_{N-1} + b_N = \hat{Q}(N, 1). \tag{4}$$

We can also write this solution as

$$\hat{Q}(1, 1) = x_1 = b_1$$

$$\hat{Q}(2,1) = x_2 = a_2 x_1 + b_2 = a_2 \hat{Q}(1,1) + \hat{Q}(2,2)$$

.
.

$$\begin{aligned}
\hat{Q}(4,1) = x_4 &= a_4 x_3 + b_4 \\
&= a_4 a_3 x_2 + a_4 b_3 + b_4 \\
&= a_4 a_3 (a_2 b_1 + b_2) + (a_4 b_3 + b_4) \\
&= a_4 a_3 \hat{Q}(2,1) + \hat{Q}(4,3)
\end{aligned}$$

.
.

. .

In general

$$\hat{Q}(2i,1) = x_{2i} = \left(\prod_{r=i+1}^{2i} a_r \right) \hat{Q}(i,1) + \hat{Q}(2i, i+1). \quad (5)$$

Equation (5) gives us our recursive doubling. Both $\hat{Q}(i,1)$ and $\hat{Q}(2i, i+1)$ are identical in structure since they both require the same number and sequence of multiplications and additions. Also, each of these terms involves i a's and i b's, exactly one-half the number of a's and b's used in $\hat{Q}(2i,1)$. Thus if at the kth step we want to compute x_{2i}, then at the $k-1$st step we should have one processor compute $\hat{Q}(i,1)$ and another compute $\hat{Q}(2i, i+1)$. We then continue this splitting operation recursively. The resulting computation graph for the case $N = 8$ is given in Fig. 1.

Note that when we compute $\hat{Q}(2i,1)$ from the two equally complex subterms $\hat{Q}(i,1)$ and $\hat{Q}(2i, i+1)$, we also need the additional product $(\prod_{r=i+1}^{2i} a_r)$. This is not a serious hindrance since we can compute the products using the scheme shown in Fig. 2. We see that in all cases the correction products needed at one level of the tree in Fig. 1 are always available just after the previous level in Fig. 2. Figs. 1 and 2 show the computation of $\hat{Q}(8,1)$. However, it is straightforward to extend the computation to eight processors, and compute $\hat{Q}(i,1)$ for $1 \le i \le 8$ in parallel. The algorithm solves (3) in a time proportional to $\lceil \log_2 N \rceil$.

An example of the complete solution of (3) for the case $N = 8$ is given in detail in Table I.

B. A General Class of First-Order Recurrence Equations

In this section we define a general class of first-order recurrence equations for which we develop a parallel algorithm. The limitation to first-order equations is not as restrictive as it might first appear, since it is often the case that we can very easily reformulate a more general mth-order problem as a first-order problem. Section III-B describes such a reformulation.

The general parallel algorithm developed in Section II-C solves all recurrence equations that can be placed in the following form:

$$\begin{aligned}
x_1 &= b_1 \\
x_i &= f_i(x_{i-1}) = f(b_i, g(a_i, x_{i-1})), \quad 2 \le i \le N \quad (6)
\end{aligned}$$

where b_i and a_i are arbitrary constants and f and g are index-independent functions that satisfy the following restrictions.

Restriction 1: f is associative. $f(x, f(y, z)) = f(f(x, y), z)$.

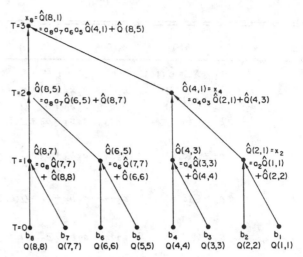

Fig. 1. Parallel computation of x_8 in the sequence $x_i = a_i x_{i-1} + b_i$.

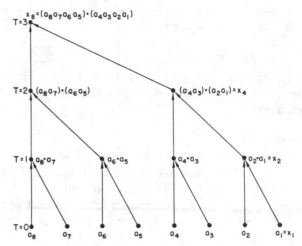

Fig. 2. Parallel computation of $x_8 = \prod_{j=1}^{8} a_j$.

Restriction 2: g distributes over f. $g(x, f(y, z)) = f(g(x, y), g(x, z))$.

Restriction 3: g is semiassociative, that is, there exists some function h such that $g(x, g(y, z)) = g(h(x, y), z)$.

The previous restrictions on f and g are the only ones necessary to prove the correctness of the general parallel algorithm. However, these restrictions may also limit the domains from which a_i and b_i and the variables x_i can be chosen. For most normal arithmetic operators like $+$ or $\cdot$ there is no problem, but more exotic operations such as floor, ceiling, modulo division, etc., may constrain the permissible domains and should be checked carefully.

The semiassociative property of g forces h to behave as if it were associative. In particular, we have

$$\begin{aligned}
g(h(h(a, b), c), d) &= g(h(a, b), g(c, d)) \\
&= g(a, g(b, g(c, d))) \\
&= g(a, g(h(b, c), d)) \\
&= g(h(a, h(b, c)), d).
\end{aligned}$$

Hence, iterated compositions of h when used as the first argument of the function g can be evaluated as if h were as-

TABLE I

Processor	T=0 A(i)	B(i)	T=1 A(i)	B(i)	T=2 A(i)	B(i)	T=3 A(i)	B(i)
1	**	$b_1 = X_1 = \hat{Q}(1,1)$	**	$X_1 = \hat{Q}(1,1)$	**	$X_1 = \hat{Q}(1,1)$	**	X_1
2	a_2	$b_2 = \hat{Q}(2,2)$	a_2	$a_2 x_1 + b_2 = x_2 = \hat{Q}(2,1)$	a_2	$X_2 = \hat{Q}(2,1)$	a_2	X_2
3	a_3	$b_3 = \hat{Q}(3,3)$	$a_3 a_2$	$a_3 b_2 + b_3 = \hat{Q}(3,2)$	$a_3 a_2$	$(a_3 a_2)X_1 + (a_3 b_2 + b_3) = X_3 = \hat{Q}(3,1)$	$a_3 a_2$	X_3
4	a_4	$b_4 = \hat{Q}(4,4)$	$a_4 a_3$	$a_4 b_3 + b_4 = \hat{Q}(4,3)$	$a_4 a_3 a_2$	$(a_4 a_3)X_2 + (a_4 b_3 + b_4) = X_4 = \hat{Q}(4,1)$	$a_4 a_3 a_2$	X_4
						$a_5 a_4(a_3 b_2 + b_3) + a_5 b_4 + b_5 = \hat{Q}(5,2)$		
5	a_5	$b_5 = \hat{Q}(5,5)$	$a_5 a_4$	$a_5 b_4 + b_5 = \hat{Q}(5,4)$	$a_5 a_4 a_3 a_2$	$= \sum_{w=2}^{5}\left(\prod_{m=w+1}^{5} a_m\right) b_w$	$\prod_{m=2}^{5} a_i^*$	X_5
6	a_6	$b_6 = \hat{Q}(6,6)$	$a_6 a_5$	$a_6 b_5 + b_6 = \hat{Q}(6,5)$	$a_6 a_5 a_4 a_3$	$\sum_{w=3}^{6}\left(\prod_{m=w+1}^{6} a_m\right) b_w = \hat{Q}(6,3)$	$\prod_{m=2}^{6} a_i^*$	X_6
7	a_7	$b_7 = \hat{Q}(7,7)$	$a_7 a_6$	$a_7 b_6 + b_7 = \hat{Q}(7,6)$	$a_7 a_6 a_5 a_4$	$\sum_{w=4}^{7}\left(\prod_{m=w+1}^{7} a_m\right) b_w = \hat{Q}(7,4)$	$\prod_{m=2}^{7} a_i^*$	X_7
8	a_8	$b_8 = \hat{Q}(8,8)$	$a_8 a_7$	$a_8 b_7 + b_8 = \hat{Q}(8,7)$	$a_8 a_7 a_6 a_5$	$\sum_{w=5}^{8}\left(\prod_{m=w+1}^{8} a_m\right) b_w = \hat{Q}(8,5)$	$\prod_{m=2}^{8} a_i^*$	X_8

* Not really needed to compute $X_1, \cdots, X_8$.
** Arbitrary.

sociative without altering the output value of g. In all interesting practical problems discovered thus far, the function h is associative.

C. Parallel Algorithm

The principle of recursive doubling can be applied in a natural way to any recurrence equation that satisfies the restrictions of Section II-B. In fact, the resulting general algorithm bears a very strong resemblance to the example of Section II-A. Before giving the algorithm, however, we first give two definitions.

Definition: For any function q of two arguments define the generalized composition of q as $q_{j=n}^{(m)}(a_j)$, where

$$q_{j=n}^{(n)}(a_j) = a_n, \qquad \text{for } n \geq 1$$
$$q_{j=n}^{(m)}(a_j) = q(a_m, q_{j=n}^{(m-1)}(a_j)), \qquad \text{for } m > n \geq 1.$$
$$= q(a_m, q(a_{m-1}, \cdots, q(a_{n+2}, q(a_{n+1}, a_n)) \cdots).$$

If we let $q(a, b) = a + b$ (scalar addition), then

$$q_{j=n}^{(m)}(a_j) = (a_m + (a_{m-1} + \cdots + (a_{n+2} + (a_{n+1} + a_n)) \cdots)$$
$$= \sum_{j=n}^{m} a_j.$$

Likewise, if $q(a, b) = a \cdot b$ (scalar multiplication), then

$$q_{j=n}^{(m)}(a_j) = \prod_{j=n}^{m} a_j.$$

Definition: Define $Q(m, n)$ as

$$Q(m, n) = f_{j=n}^{(m)}(g[h_{r=j+1}^{(m)}(a_r), b_j]), \qquad m \geq n \geq 1$$

where we define

$$g(h_{r=m+1}^{(m)}(a_r), b_j) = b_j.$$

If we consider the case where f is scalar addition, and g and h

are scalar multiplication, then the $Q(m, n)$ defined previously is exactly the same as the $\hat{Q}(m, n)$ defined for the example in Section II-A.

The similarities between Q and $\hat{Q}$ carry even further. The function $Q(i, 1)$ is the solution of the general recurrence equation (6), that is,

$$x_i = Q(i, 1), \qquad \forall \, 1 \leq i \leq N. \tag{7}$$

Also, as in the example, we can derive a formula computing $Q(2i, 1)$ strictly in terms of two equally complex subterms, namely,

$$Q(2i, 1) = f(Q(2i, i+1), g(h_{j=i+1}^{(2i)}(a_j), Q(i, 1))). \tag{8}$$

Both (7) and a more general version of (8) are proved in the Appendix.

Equation (8) is a perfect candidate for recursive doubling. $Q(2i, i+1)$ and $Q(i, 1)$ are identical in terms of the number of unique a's and b's referenced and require the same sequence of f, g, and h function calls to evaluate them. As with the second example, the only hindrance in implementing (8) directly as a recursive doubling algorithm is the correction term, the h composition. However, since h can be treated as an associative function, we can use a scheme similar to Fig. 2 to compute these correction terms exactly as they are needed.

Fig. 3 is a computation graph using (8) and the h composition algorithm to compute x_8. Despite its increased complexity, the general structure of this graph is identical to Figs. 1 and 2 and can be extended to solve for all elements of the sequence $x_1, \cdots, x_N$ in parallel.

We can now state the complete algorithm for solving our general recurrence equations. The detailed proof of the correctness of this algorithm is given in the Appendix.

Algorithm A—General Algorithm: This algorithm solves for $x_1, x_2, \cdots, x_N$ where $x_i = f(b_i, g(a_i, x_{i-1}))$ and f and g satisfy the restrictions of Section II-B.

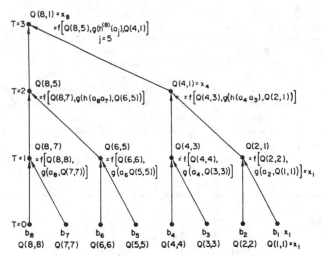

Fig. 3. Parallel computation of x_8 from the general recurrence equation.

The algorithm requires two vectors A and B of N elements. The ith component of each vector, namely $A(i)$ and $B(i)$, is stored in the memory of processor (i). The actual data structure required to represent $A(i)$ and $B(i)$ depends on the definition of the domain of the entities a_i and b_i in the basic equation (8) and may be scalars, matrices, lists, etc., depending on the problem.

Let $A^{(k)}(i)$ and $B^{(k)}(i)$ represent respectively, the contents of $A(i)$ and $B(i)$ after the kth step of the following algorithm.

Initialization Step (k = 0):

$B^{(0)}(i) = b_i$ for $1 \leq i \leq N$.
$A^{(0)}(i) = a_i$ for $1 < i \leq N$.

$A(1)$ is never referenced and may be initialized arbitrarily.

Recursion Steps: For $k = 1, 2, \cdots \lceil \log_2 N \rceil$ do each of the following assignment statements:

$$B^{(k)}(i) = f(B^{(k-1)}(i), g(A^{(k-1)}(i), B^{(k-1)}(i - 2^{k-1}))),$$
$$\text{for } 2^{k-1} < i \leq N. \quad (9)$$

$$A^{(k)}(i) = h(A^{(k-1)}(i), A^{(k-1)}(i - 2^{k-1})),$$
$$\text{for } 2^{k-1} + 1 < i \leq N. \quad (10)$$

Each statement is assumed to be evaluated simultaneously by all processors whose indices lie in the specified interval.

After the $\lceil \log_2 N \text{th} \rceil$ step, $B(i)$ contains x_i for $1 \leq i \leq N$.

End of Algorithm A.

Several things should be noted about any implementation of Algorithm A. First, when the ith processor executes (9) and (10) in that order, it must have the old values of $B(i - 2^{k-1})$ and $A(i - 2^{k-1})$, which can only be obtained from processor $(i - 2^{k-1})$. Thus at the beginning of the kth recursion step, all processors must shift their values of A and B to the processors with index 2^{k-1} greater than their own. Exactly how this data routing is performed depends on the processor interconnection pattern available in a given computer system.

Another problem with implementing Algorithm A lies in limiting the processors that execute (9) and (10) to just those with the proper indices. The masking feature (Section I-B) is the most direct way. This, however, requires executing explicit mask instructions during each recurrence step. If the

number of available processors is greater than about $3N/2$, another method can avoid these extra instructions. The N processors with the highest indices are allocated to the solving of $x_1, \cdots, x_N$, and the next $N/2$ processors are initialized so that when one of the top N processors references their data, the values returned cause no change in the higher processor's values for $A(i)$ and $B(i)$. These bottom $N/2$ processors are completely masked off initially so that these initial values never change. These initial values are

$$A(i) = I, \quad \text{for } -N/2 \leq i \leq 1$$
$$B(i) = Z, \quad \text{for } -N/2 \leq i \leq 0$$

where for all a and b, $h(a, I) = a$ and $f(b, g(a, Z)) = b$. For the example of Section II-A, I is simply 1, and Z is 0.

III. Applications

A. Various First-Order Problems

As has been mentioned before, Algorithm A is applicable to a rather wide class of problems. Table II gives a collection of such problems that satisfy the functional constraints stated in earlier sections.

An interesting case occurs when we constrain all the a_i of Example 1 in Table II to be the same number z as indicated in Example 5 in Table II. We then get the recursion

$$x_i = zx_{i-1} + b_i$$

which, if we solve for x_N, yields

$$x_N = b_1 z^{N-1} + b_2 z^{N-2} + \cdots + b_{N-1}z + b_N.$$

But this is simply the evaluation of the polynomial $b_1 x^{N-1} + \cdots + b_N$ at $x = z$. In fact, Algorithm A in this case is simply the parallel evaluation of polynomials (Munro and Paterson [3]).

B. Extension to mth-Order Equations

The algorithm given in the previous sections is applicable to a class of first-order recurrence equations. However, a little manipulation of the description of a problem can often convert an mth-order recurrence equation into a first-order equation with a slightly more complicated data structure. The clue to how this is done can be found in the third example in Table II, a matrix or "state variable" problem.

As an example, consider the problem

$$x_i = a_{i,1} x_{i-1} + \cdots + a_{i,m} x_{i-m} + b_i. \quad (11)$$

We wish to reformulate it in a form amenable to Algorithm A.

The first step is to see that we can collapse the m x's that are needed in (11) into a single new "variable" by using state variable notation as follows.

Let

$$Z_i = \begin{bmatrix} x_i \\ \cdot \\ \cdot \\ \cdot \\ x_{i-m+1} \end{bmatrix}. \quad (12)$$

Now we can rewrite (11) as

449

TABLE II
APPLICATIONS OF ALGORITHM A

Example	Domain of b	Domain of a	Domain of x	$f(a,b)$	$g(a,b)$	$h(a,b)$	Comments
1)		real numbers		$a+b$	$a \cdot b$	$a \cdot b$	$x_{i+1} = b_{i+1} + a_{i+1} x_i$
2)		real numbers		$a \cdot b$	$b \uparrow a$	$a \cdot b$	$x_{i+1} = b_{i+1} \cdot (x_i \uparrow a_{i+1})$, "$\uparrow$" is exponentiation
3)	$m \times 1$ matrix	$m \times m$ matrix	$m \times 1$ matrix	vector addition	mult. of matrix by vector	matrix mult.	$x_{i+1} = B_{i+1} + A_{i+1} x_i$ where A is $m \times m$ and x, B are $m \times 1$
4)		real numbers		b	$\min(a,b)$	$\min(a,b)$	x_i is the smallest of $a_1, \cdots, a_i$
5)		real numbers		b	$\max(a,b)$	$\max(a,b)$	x_i is the largest of $a_1, \cdots, a_i$
6)	real number	any real number z	real number	$a+b$	$a \cdot b$	$a \cdot b$	$x_i = x_{i-1} \cdot z + b_i$, polynomial evaluation $x_N = P(z) = b_1 z^{N-1} + b_2 z^{N-2} + \cdots + b_{N-1} z + b_N$

$$Z_i = \begin{bmatrix} a_{i,1} \ldots a_{i,m} \\ 1\,0\ldots 0 \\ 0\,1\,0\ldots 0 \\ 0\ldots 0\,1\,0 \end{bmatrix} \begin{bmatrix} x_{i-1} \\ . \\ . \\ . \\ x_{i-m} \end{bmatrix} + \begin{bmatrix} b_i \\ 0 \\ . \\ . \\ . \\ 0 \end{bmatrix} \qquad (13)$$

$$= \hat{A}_i Z_{i-1} + \hat{B}_i \qquad (14)$$

where $\hat{A}_i$ and $\hat{B}_i$ are the $m \times m$ matrix and $m \times 1$ vector respectively. The first row of $\hat{A}_i$ represents the original (11) and the remaining rows simply select the proper x_j to make Z_i be consistent.

Equation (14), however, is in exactly the right format for Example 3 of Table II to be applied. The variables in the recursion are m-element vectors, the $\hat{A}_i$ are $m \times m$ matrices, and the B_i are m-element vectors. The function f is vector addition, g is multiplication of a matrix by a vector, and h is matrix multiplication. Thus if we rewrite (11) into (14) we can apply Algorithm A to get a parallel solution to the original problem (11).

This particular formulation, however, is not very efficient in its use of the parallel processors. At the end of the calculation we have N m-element vectors $Z_1, \cdots, Z_N$. Only one mth of each Z_i, namely its first component x_i, represents new calculations not available from previous Z's. Most of the matrix calculations done in the recurrence steps are redundant.

We can increase the amount of parallelism in the problem by propagating (14) forward m steps before using Algorithm A. This results in a new formulation of the problem, which yields $Z_{(k+1)m} = (x_{(k+1)m}, \cdots, x_{km+1})'$ directly from $Z_{km} = (x_{km}, \cdots, x_{(k-1)m+1})'$.

It is easy to show by induction that Z_{km+m} can be computed as follows:

$$Z_{km+m} = \left(\prod_{j=km+1}^{km+m} \hat{A}_j \right) Z_{km} + \sum_{r=km+1}^{km+m} \left(\prod_{j=r+1}^{km+m} \hat{A}_j \right) \hat{B}_r,$$

$$k = 1, \cdots, N/m - 1. \qquad (15)$$

This equation can be restated in a form directly usable by Algorithm A as follows.

Let

$$X_{k+1} = Z_{(k+1)m} = \begin{bmatrix} x_{(k+1)m} \\ . \\ . \\ . \\ x_{km+1} \end{bmatrix}$$

$$A_{k+1} = \left(\prod_{j=km+1}^{(k+1)m} \hat{A}_j \right) \qquad B_{k+1} = \sum_{r=km+1}^{(k+1)m} \left(\prod_{j=r+1}^{(k+1)m} \hat{A}_j \right) \hat{B}_r. \qquad (16)$$

Now (15) becomes

$$X_{k+1} = A_{k+1} X_k + B_{k+1}, \qquad k = 1, \cdots, N/m \qquad (17)$$

which again is our familiar first-order linear matrix recurrence equation.

Now to compute all N elements of (11), we need only compute N/m elements of the series $X_1, \cdots, X_{N/m}$ using (17). Using Algorithm A we can compute these N/m elements with $\lceil \log_2 N/m \rceil$ applications of the recurrence step, plus some initial time to compute the initial A's and B's given by (16). Further, since there are only N/m elements to compute, Algorithm A also calls for only N/m processors.

The important aspect of this reformulation is not that the number of steps has been reduced, but that the number of processors has dropped. Equation (17) takes $\lceil \log_2 m \rceil$ fewer recurrence iterations to evaluate than does (14), but about $\lceil \log_2 m \rceil$ additional iterations are required to set up (17) from (14) with N processors. Thus we have not reduced the time to solve the problem, but we have reduced redundant computations to the point where we need only N/m processors after the initial setup.

IV. SUMMARY AND CONCLUSION

Various researchers have developed parallel algorithms for specific problems, such as polynomial evaluation (Munro and Paterson [3]), and the solution of tridiagonal systems of equations (Buneman [1], Buzbee et al. [2], and Stone [4]). As with Algorithm A, these algorithms typically require execution times proportional to $\lceil \log_2 N \rceil$. None of them, however, is applicable to any wider class of problems than the particular ones they were designed to solve. Algorithm A, on the other hand, solves any problem for which the solution can be stated in terms of a recurrence equation satisfying a few simple restrictions. It is worthwhile mentioning that the running time for Algorithm A can vary widely from problem to problem even though the time is always proportional to $\lceil \log_2 N \rceil$. The constant of proportionality depends on the time it takes to evaluate f, g, and h. These functions can be as simple as a magnitude comparison or floating-point addition and can be as complex as a matrix multiplication, with very large differences in their respective constants of proportionality.

The power of Algorithm A comes from the generalization of the technique of recursive doubling. This technique seems to hold an important key to understanding exactly how parallelism can be extracted from what appear to be highly serial problems. The major results of this paper indicate that the class of serially stated problems that are amenable to parallel solutions is a large one, and includes some problems that have been thought to be poorly suited to parallel processors.

APPENDIX
VALIDITY OF ALGORITHM A

This Appendix contains some basic theorems that establish the validity of Algorithm A. We assume we are solving equations of the form of (6), where the functions f, g, and h all satisfy the restrictions of Section II-B. We also assume that the concept of generalized composition and the definition of the function $Q(m, n)$ carry over from Section II-C.

Theorem 1: For any i, k such that $1 \leqslant k < i \leqslant N$ then for any j such that $1 \leqslant j \leqslant k$

$$Q(i, i - k) = f(Q(i, i - j + 1), g(h^{(i)}_{r=i-j+1}(a_r), Q(i - j, i - k))).$$

Proof: Assume $1 \leqslant j \leqslant k$. Then

$f(Q(i, i - j + 1), g(h^{(i)}_{r=i-j+1}(a_r), Q(i - j, i - k)))$

$= f(Q(i, i - j + 1), g(h^{(i)}_{r=i-j+1}(a_r), f^{(i-j)}_{m=i-k}(g(h^{(i-j)}_{r=m+1}(a_r), b_m))))$

[by definition of $Q(i - j, i - k)$]

$= f(Q(i, i - j + 1), f^{(i-j)}_{m=i-k}(g(h^{(i)}_{r=i-j+1}(a_r), g(h^{(i-j)}_{r=m+1}(a_r), b_m))))$

[g distributes over f]

$= f(Q(i, i - j + 1), f^{(i-j)}_{m=i-k}(g(h^{(i)}_{r=m+1}(a_r), b_m)))$

[g is semiassociative]

$= f(f^{(i)}_{m=i-j+1}(g(h^{(i)}_{r=m+1}(a_r), b_m)), f^{(i-j)}_{m=i-k}(g(h^{(i)}_{r=m+1}(a_r), b_m)))$

[definition of $Q(i, i - j + 1)$]

$= f^{(i)}_{m=i-k}(g(h^{(i)}_{r=m+1}(a_r), b_m))$ [associativity of f]

$= Q(i, i - k).$

End of proof.

Theorem 2: For $1 \leqslant i \leqslant N$, $x_i = Q(i, 1)$.

Proof: By induction on i.

Basis Step: $i = 1$.

$$Q(1, 1) = b_1 = x_1 \quad \text{[by definition]}.$$

Induction Step: Assume $Q(j, 1) = x_j$ for $j < i$. Then

$x_i = f(b_i, g(a_i, x_{i-1}))$ [recurrence equation (6)]

$= f(Q(i, i), g(h^{(i)}_{r=i}(a_r), Q(i - 1, 1)))$

[inductive hypothesis, definition of Q and h composition]

$= Q(i, 1)$ [by Theorem 1].

End of proof.

We can now state a theorem that demonstrates the validity of Algorithm A.

Theorem 3: For all $1 \leqslant i \leqslant N$, $0 \leqslant k \leqslant \lceil \log_2 N \rceil$,

a) $A^{(k)}(i) = \begin{cases} h^{(i)}_{r=2}(a_r), & 1 < i \leqslant 2^k + 1 \\ h^{(i)}_{r=i-2^k+1}(a_r), & 2^k + 1 < i \leqslant N \end{cases}$

b) $B^{(k)}(i) = \begin{cases} Q(i, 1), & 1 \leqslant i \leqslant 2^k \\ Q(i, i - 2^k + 1), & 2^k < i \leqslant N. \end{cases}$

Proof: Directly by induction and Theorems 1 and 2. The proof of Theorem 3 is direct but tedious; we omit it here.

Using part b) of Theorem 3 we get the immediate result.

Corollary: After the $\lceil \log_2 N \rceil$th iteration of Algorithm A, $B(i)$ contains x_i for $1 \leqslant i \leqslant N$.

Thus we have shown that not only does Algorithm A compute the solution $x_1, \cdots, x_N$ to (6), but also that it terminates in exactly $\lceil \log_2 N \rceil$ iterations.

ACKNOWLEDGMENT

Recursive doubling solutions to the first-order problem of Section II-A were discovered independently by H. R. Downs of Systems Control, Inc., and H. Lomax of NASA Ames Research Center. Recursive doubling solutions to second-order linear recurrences have been known to J. J. Sylvester as early as 1853. The authors wish to thank D. Knuth for pointing out Sylvester's work and for several stimulating suggestions while this research was in progress and the referees for pointing out that the h function need not be associative.

After this paper had been reviewed and accepted for publication, the authors encountered the report by Trout [5], which has several similar results. His work was done independently of the work reported here and carries the research beyond the limits of this paper.

REFERENCES

[1] O. Buneman, "A compact non-iterative Poisson solver," Stanford Univ. Inst. Plasma Res., Stanford, Calif., Rep. 294, 1969.
[2] B. L. Buzbee, G. H. Golub, and C. W. Nelson, "On direct methods for solving Poisson's equations," *SIAM J. Numer. Anal.*, vol. 7, pp. 627-656, Dec. 1970.
[3] I. Munro and M. Paterson, "Optimal algorithms for parallel polynomial evaluation," in *Conf. Rec., 1971 12th Annu. Symp. Switching and Automata Theory*, IEEE Publ. 71 C 45-C, pp. 132-139.
[4] H. S. Stone, "An efficient parallel algorithm for the solution of a tridiagonal linear system of equations," *J. Ass. Comput. Mach.*, vol. 20, pp. 27-38, Jan. 1973.
[5] H. R. G. Trout, "Parallel techniques," Dep. Comput. Sci., Univ. Illinois, Urbana, Rep. UIUCDCS-R-72-549, Oct. 1972.

An Algorithmic and Novel Design of a Leading Zero Detector Circuit: Comparison with Logic Synthesis

VOJIN G. OKLOBDZIJA

Abstract— A novel way of implementing the Leading Zero Detector (LZD) circuit is presented. The implementation is based on an algorithmic approach resulting in a modular and scalable circuit for any number of bits. We designed a 32 and 64 bit leading zero detector circuit in CMOS and ECL technology. The CMOS version was designed using both: logic synthesis and an algorithmic approach. The algorithmic implementation is compared with the results obtained using modern logic synthesis tools in the same 0.6 μm CMOS technology. The implementation based on an algorithmic approach showed an advantage compared to the results produced by the logic synthesis. ECL implementation of the 64 bit LZD circuit was simulated to perform in under 200 pS for nominal speed.

I. INTRODUCTION

In any floating-point processor *normalization* is a required operation. It consists of an appropriate left shift until the first nonzero digit is in the left-most position. The amount of shift is determined by counting the number of zero digits from the left-most position until the first nonzero digit is reached. The exponents are appropriately decremented for the shift amount. The normalization is usually performed before storing the numbers in the register file (memory), commonly referred to as *post-normalization*, and referred to as *pre-normalization* before the operation is performed. In both cases, the special circuit implemented (in hardware) to detect the number of leading zeros is referred to as a *Leading Zero Detector* (LZD).

Applying a straight forward combinatorial approach in designing the LZD circuit is a rather complicated process because each bit of the result is dependent on all of the input bits, which in the case of 64 bit word, consists of 64 inputs. For example a 64 bit LZD circuit would consist of six outputs, each dependent on 64 inputs. It is obvious that such large fan-in dependencies are a problem and that the resulting circuit is likely to be complicated and slow. To design such a circuit using computer aided Boolean minimization techniques or the Karnaugh map method is cumbersome and slow and the resulting design does not exhibit any structure. An immediate solution would be to resort to the use of the logic synthesis tools and "let the tool do the job". This is perhaps the most commonly practiced approach, for any of the complex and complicated circuit of which LZD is a very good example.

Characteristic of the LZD circuit is its concise functional description. It is also very easy to describe the circuit using any of the common hardware description languages. Such a circuit naturally leads itself to the use of logic synthesis by describing the expected functionality in VHDL and simply letting the computer to do the rest.

On the other hand, we can use some intelligence in designing the circuit by attempting first to identify some common modules and impose hierarchy on the design. The LZD circuit in particular, is quite suitable for exploring possibilities of hierarchical and structural design. This yields to substantial improvement of the circuit regularity and speed, compared to straight-forward minimization. The resulting circuit performs well with low and regular fan-in and fan-out,

Manuscript received January 6, 1993; revised July 21, 1993. This work was supported by a minigrant from the Office of Research, University of California, Davis.

The author is with the Department of Electrical and Computer Engineering, University of California, Davis, CA 95616.

IEEE Log Number 9214288.

TABLE I
TWO BIT TRUTH TABLE FOR LZD

Pattern	Position	Valid
1X	0	yes
01	1	yes
00	X	no

leading to better wireability and better layout. However, LS tools have not yet reached a level of sophistication which can deal with hierarchical structures and create or impose hierarchy in the design. Their approach is to rather expand the logic in one level and optimize it via elaborate and laborious logic minimization, using hours of CPU time yet not being able to identify common modules and hidden structure. Therefore, the design presented in this paper results not only in an efficient and fast LZD, but also provides an efficiency measure of the logic synthesis tools and their limitations.

II. DESIGN USING AN ALGORITHMIC APPROACH

In our approach we use the inherent hierarchy associated with the leading zero detection process and map it into a hierarchical and modular design. In order to understand this design process, let us begin by first examining only the two bit case, as shown in the Table I. The pattern on the left designate the possible two bit combination. If the left-most bit is "1" we assign "0" to the *Position* and "1" to the *Valid* bit indicating that there is *zero* distance from the left-most bit to the first *nonzero* bit. If both bits are "0", we would set *Valid* bit to 0 indicating that this is not a valid position. Not only is this because we have only one bit to indicate position, but the next two-bit group might have more zeroes to follow and therefore position information is not complete. It is straightforward to construct the logic for the two bits representing the valid bit (V) and the position bit (P) as shown in Table I.

We can easily extend the two bit case to the four bit case. Let us designate the position bits (4 bits total) as $P0$ for the left-most two bits and $P1$ for the right-most two bits as shown in Fig. 1(b). Also, we will designate $V0$ and $V1$ as the valid bits for the first two and second two bits respectively starting from the left to right. The leading zero position can be represented as a function of those four bits as shown in the fifth column of Fig. 1(a) (minus sign represents complementation). Also, it should be noted that we are using "Big Endian" notation, i.e., we start indexing from left to right (this notation is used by IBM).

The 4 bit circuit has a depth of 2 logic levels; in the second level the valid bit is formed as a logical OR of the valid bits from the previous level. In other words, if there is a "valid" string of bits within the group in the previous level, then this group has a valid position bit. If all of the groups, however, do not show a "valid" output, this simply means that there are a string of zeros and that the first nonzero bit can be expected only within one of the groups to the "right". The left valid bit V_1 is inverted and concatenated with P_0 if $V_0 = 1$, or with P_1 if $V_0 = 0$ and $V_1 = 1$. This is achieved

Reprinted from *IEEE Transactions on VLSI Systems*, Vol. 2, No. 1, pp. 124–128, March 1994.

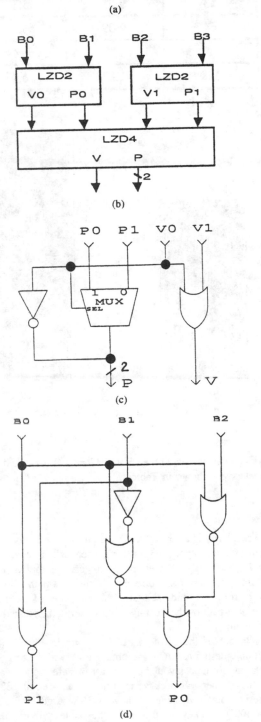

pattern	position	position (binary)	valid	position
1011	0	00	yes	(-V0)P0
0100	1	01	yes	(-V0)P0
0011	2	10	yes	(-V0)P1
0001	3	11	yes	(-V0)P1
0000	X	XX	no	XX

(a)

(b)

(c)

(d)

Fig. 1. Design of a 4 bit LZD. (a) Truth table. (b) Using two 2 bit LZD's. (c) The logic sturcture of the 4 bit LZD block. (d) One level implementation of the 4 bit LZD.

by simply multiplexing $P0$ and $P1$ to the output of the multiplexer. The logic structure of the 4 bit LZD group (LZD4) is shown in Fig.

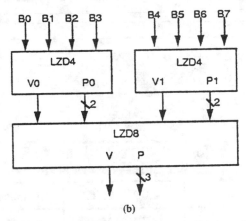

8-bit LZD				"left" nibble		"right" nibble	
bit pattern		position	valid	P0	V0	P1	V1
1XXX	XXXX	000	yes	00	yes		
01XX	XXXX	001	yes	01	yes		
001X	XXXX	010	yes	10	yes		
0001	XXXX	011	yes	11	yes		
0000	1XXX	100	yes		no	00	yes
0000	01XX	101	yes		no	01	yes
0000	001X	110	yes		no	10	yes
0000	0001	111	yes		no	11	yes
0000	0000	XXX	no		no		no

(a)

(b)

Fig. 2. Design of an 8 bit LZD. (a) Truth table. (b) Using two 4 bit LZD's.

1(c). This implementation is particularly fast because the propagation delay of the multiplexer is smaller than the propagation delay of a gate. This is the case in several CMOS and ASIC libraries such as the ASIC library from LSI logic corporation [3].

A 4 bit LZD can also be implemented in one level directly from Fig. 1(a), avoiding the need to go into the above exercise. A one level implementation of a 4 bit LZD is shown in Fig. 1(d). It is not necessary that we limit our design to 2 bits per level. Naturally, the implementation depends on the technology used. The entire concept can be grouped in 4 bit groups. In a technology that tolerates high fan-in and fan-out we can compress even more bits into one level or one logic tree (such as in the case of ECL technology).

Now we can take two groups of 4 bits and form a LZD for an 8 bit word by simply following the same concept that we did in the example of 4 bits. The truth table is given in Fig. 2(a) and the design in Fig. 2(b).

From the above discussion we can deduce the hierarchical structure for the LZD and arrive at the following algorithm for generating the number of leading zeros:

Algorithm for generating LZ count:
(1) Form the pair of bits B_i, B_{i+1} for $i = 0$ to
    N-2 with bit 0 being the leftmost one
(2) Determine P and V bits for each pair
(3) for the next level determine the P_g and
    V_g bits as function of two pairs of inputs
    P and V in this level in the following way:

$$V_g = V_l + V_r \text{ where "+" is logical}$$
OR operation of the left and right inputs

if $V_l = 1$ then $P_g = 0, P_l$ where "," designates concatenation
    else if $V_r = 1$ then $P_g = 1, P_r$
    othervise $V_g = 0$

Repeat step (3) $\log(N) - 2$ times

453

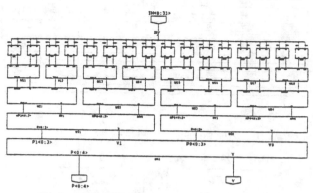

Fig. 3. 32 bit LZD circuit composed of 2 bit groups.

TABLE II
SIMULATION CONDITIONS

T_{ox}=150A, V_T=0.6V, L_{eff}=0.6u $R_{metal\ 1}$=120 mohm/sq		
NC	4.0 V, 125 C	Nominal
WC	2.8 V, 125 C	Worse Case

It can easily be concluded that the logical depth of this circuit is $\log_2(N)$ stages where the path through each stage is of the complexity of the multiplexer or one level of equivalent logic. The multiplexer is actually implemented using a pass-transistor structure and therefore is even faster than a regular CMOS gate. This is the reason for the extraordinary speed of this scheme for the LZD implemented in CMOS. In addition, we might want to save one level and implement this scheme in $\log(N) - 1$ levels instead of $\log(N)$ levels. This is achieved by starting with the groups of 4 bits and proceeding in the way described by the algorithm. Using CMOS technology, this is usually the best that can be achieved in terms of logic levels, because any further compression of the number of levels would pass the point of diminishing returns by increasing the fan-in and fan-out of the circuits, thus negatively affecting the speed. However, the same concept can be applied to groups of 4 bits instead of 2 bits, which is more appropriate for ECL and BiCMOS technologies. In that case there is an additional speed advantage, because the speed of this LZD implementation is proportional to $\log_4(N)$ stages and therefore faster. The structure of the 32 bit LZD composed of the 2 bit groups is shown in Fig. 3.

III. IMPLEMENTATION AND LOGIC SYNTHESIS EXPERIMENT

We implemented six LZD prototypes of various sizes. They were laid out and simulated for worst case conditions. In addition, we repeated those designs using logic synthesis tools, produced layouts and simulated the results. Our second goal was to gain performance by resizing the transistors as we went down the path. Larger transistors result in better driving capability and their size will be increased where space is available. Given that our structure is tree like, as the signal moves from the first stage to the second and third, the available space increases. By filling this space with transistors of larger size, the resulting LZD layout takes more of a rectangular shape. Such an approach is used successfully in an adder based on recurrence solving [1]. Therefore our objective was to have some performance gain by applying this findings. This provided enough data for performance

TABLE III
PERFORMANCE OF THE NEW LZD CIRCUIT
UNDER NOMINAL AND WORSE-CASE CONDITION

Bits	WC [nS]	NC [nS]
25	7.69	4.49
32	7.7	4.52
53	9.08	5.35
64	9.09	5.37
112	10.7	6.41
128	10.7	6.43

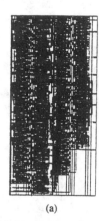

(a)   (b)

Fig. 4. Layout of the 32 bit LZD. (a) for the algorithmic design and (b) LS result.

analysis and evaluation of the LS tool. The technology and simulation parameters are shown in Table II.

A. The Layout

The regularity of the novel LZD design can also be used to produce a more efficient layout. By creating each cell as a basic building block for each stage, the entire circuit can be routed primarily in metal lines flowing in the direction of the data-path. This results in better performance and facilitates the inclusion of the LZD in the regular data-path of the floating-point or any other unit that needs a LZD circuit.

The layout of the 32 bit LZD is shown in Fig. 4, (a) for the algorithmic design and, (b) one obtained as a result of LS. Observe that the algorithmic layout has 4 rows of cells which were placed using the Timberwolf package resulting in an area of $163 \times 340\ \mu$. The results obtained using LS have a 35% larger area resulting in $186 \times 403\ \mu$m. They are plotted to the same scale and placed next to each other for easy comparison. In terms of speed, the algorithmic LZD introduces a delay of $T_a = 4.5$ nS, while the LZD resulting from LS has a delay of $T_{ls} = 5.8$ ns (both for the typical case) which is 29% slower. Therefore we can say that the algorithmically designed LZD is roughly one–third smaller and faster than the equivalent LZD resulting from logic synthesis.

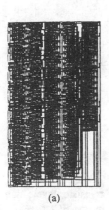 

(a)   (b)

Fig. 5. "Rectangular" layout of the LZD circuit produced using (a) the algorithmic approach and (b) LS

TABLE IV
PERFORMANCE COMPARISON OF REGULAR VERSUS RECTANGULAR 32 BIT LZD LAYOUT WITH (a) 1 pF LOAD. (b) NO LOAD CONDITION.

Speed NC (WC) for 1.0pF load [nS]	
Regular Layout (163X340μ)	Rectangular Layout (206X363μ)
7.95	6.9
(12.8)	(11.4)

(a)

Speed NC (WC) for 0pF load [nS]	
Regular Layout (163X340μ)	Rectangular Layout (206X363μ)
4.5	4.93
(7.7)	(9.5)

(b)

TABLE V
COMPARISON OF THE ALGORITHMIC LZD AND LOGIC SYNTHESIS RESULTS

Comparison of the Algorithmic LZD circuit with the LZD obtained via Logic Synthesis										
Algorithmic LZD (regular layout)					LZD obtained with Logic Synthesis					
	no load		1.0 pF load		regular	no load		rectang.	1.0 pF load	
Area[u]	WC [nS]	NC [nS]	WC [nS]	NC [nS]	Area[u]	WC [nS]	NC [nS]	Area[u]	WC [nS]	NC [nS]
163×340 86mils	7.7	4.5	12.8	7.95	186×403 116mils	12	5.8	181×473 133mils	13.6	7.7

We have also tried to explore the regular and hierarchical structure of this design by applying the approach described by Vuillemin and Guibas [1] to the layout of this circuit. The idea in [1] is to lay the tree-like structures like this one on a rectangular pattern in such a way that as the signal progresses down the levels, the size of the cells is made larger, increasing their driving capability so that the signal can drive more inputs in a shorter time. We implemented this idea and laid out the 32 bit LZD circuit such that the width of the circuit was kept constant. The layout of the "rectangular" LZD structures obtained using the algorithmic approach and LS are shown in Fig. 5. They are plotted on the same scale (as in Fig. 4) to point to the difference in size between the algorithmic LZD (a) and LS (b). The "rectangular" layout of the algorithmic LZD is 206×363 μm [Fig. 5(a)] while the LS resulted in 181×473 μm. The algorithmic approach has resulted in a 14.5% smaller layout compared to the LS result in this case.

B. Performance

The performance of the novel LZD was simulated under nominal and worst case conditions, denoted NC and WC, respectively. The NC and WC conditions are characterized in Table II together with the parameters typical for this CMOS process. The table shows the speed of the LZD for different sizes starting from $N = 25$ to $N = 128$ bits. This is also shown in Table I(a) for the unbuffered LZD (without the output buffers) for nominal and worst case conditions.

In terms of performance, the "rectangular" layout introduced a delay of 6.9 nS for the algorithmic LZD and 7.7 nS for the one resulting from LS (for the nominal case). The performance advantage of the algorithmic LZD was 12% over LS for the NC. For worst-case conditions, this advantage was 19%. However, when we compared the "rectangular" layout versus the regular layout in terms of performance in both cases, the results showed that overall, the "rectangular" approach did not improve the performance in every case.

The performance of the "rectangular" layout was better than the regular layout in case of 1 pF output load. The difference was 15% for the nominal case and 12% for the worst case which favors the "rectangular" layout. The performance difference is shown in Table IV(a). In the case of no load (or light load) on the output, the "rectangular" layout was worse: 10% (nominal case) and 23% (worse case). This is explained by the fact that this circuit maintains regular fan-in and fan-out, and neither one of them increases when reaching the levels closer toward the output. The performance was not affected

because the input capacitance grew proportionally, as did the driving capacity of the gate. The increase in delay caused by the capacitance increase, more than offset the improved driving capability, resulting in a slight loss. Therefore, the idea [1] which worked well for a CLA-like adder structure did not work in our case. This is shown in Table IV(b). We observe that under the no-load conditions, the LZD circuit obtained via regular layout performs better. However with 1.0 pF load, the rectangular LZD outperforms the regular one in both NC and WC. This is attributed to the stronger driving capabilities of the final stages. However, we feel that just adding stronger buffers at the output nodes would still make the regular case perform better.

In Table V we compare the results for a 32 bit LZD using our approach with the one obtained using logic synthesis without load and with 1.0 pF loading capacitance on the outputs. The algorithmic LZD outperforms the one obtained via LS under all conditions. The only exception has been the marginal difference of 0.25 nS in favor of the implementation obtained by synthesis which is attributed to the rectangular layout rather than the way LZD has been implemented. With a 1.0 pF load and use of rectangular layout, algorithmic LZD is 12% faster under nominal conditions (6.9 versus 7.7 nS) and 15% faster under worst conditions (11.4 versus 13.6 nS). In any one case, we have demonstrated that our algorithmic LZD circuit outperforms LS, and that the improvement in performance ranges from 12% (NC rectangular layout, 1 pF load) to 56% (WC regular layout, no load). The improvement in the area is from 14.5% (rectangular layout) up to 35% (regular layout). This clearly demonstrates the superiority of the algorithmic approach.

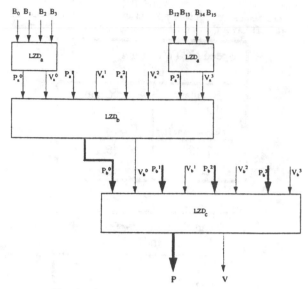

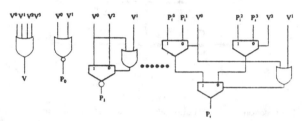

Fig. 6. Structure of 64 bit ECL LZD circuit.

Fig. 7. ECL structure of the second level LZD circuit.

IV. HIGH-PERFORMANCE ECL IMPLEMENTATION

In a Floating-Point processor a *critical path* consists of: *leading zero detector, shift* and *rounding* operation, where each of the operations contributes approximately equally to the delay. The technology of implementation may generally differ from low power CMOS to ECL technology which is very relevant even today due to the renewed attention given to it. Implementation of LZD by the algorithm described is challenging because the rules that are applied for ECL technology are very different from those applied to CMOS. Also ECL has a tendency to combine as much logic as possible into one *level* or *logic tree*, generally allowing for larger fan-in. Finding a suitable ECL structure for any algorithmic and technology independent design is not easy. Therefore we had to solve two problems:

1) compress the design in as few levels as possible
2) identify a common and characteristic structure that also implements itself well in ECL

The first problem has been to define a 4-way LZD structure which leads to 3 levels (or 3 ECL trees) for a 64 bit LZD circuit. The first level is trivial to design and it is very much similar to the 4 bit group design shown in Fig. 1(d) except by using wired-OR (in ECL), we were able to implement everything in on the level of gates. The structure of 64 LZD implemented in ECL technology is shown in Fig. 6.

The second level will indicate the LZD position based on the input from the 4 LZD groups from the previous level. The structure of this group is not as regular as in the CMOS case. However, it was possible to identify a common multiplexer based structure that can be applied in general for any position bit ($i = 2, 3 \ldots k$). This structure is shown in Fig. 7.

Fortunately our algorithmic approach favors multiplexer structure which suits an ECL circuit very well. The bits P_0 and P_1 (as

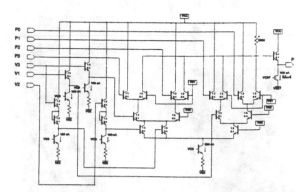

Fig. 8. ECL circuit implementation of the second level (64 bit LZD).

well as V) are implemented separately. The ECL tree used for this computation is shown in Fig. 8. It is three transistor levels high, which is about as much as we can implement in one ECL tree. Therefore the decision to calculate P and V based on the input of the previous four groups, rather than two or eight, seems to be optimal. The depth of the n-bit LZD circuit implemented in this way is $\log_4(n)$ levels. In our case for 64 bit LZD, we have a depth of 3 ECL *trees*. Using the advanced Motorola BiCMOS process, this circuit produces the result in $T_{cr} = 200$ pS (nominal time) for a 64 bit LZD circuit. It should be noted that the first level, which is one gate deep, could be integrated into the second level yielding an implementation in only 2 ECL trees.

V. CONCLUSION

In this paper we have described an algorithmic approach to designing a leading zero detector. This circuit has been implemented in 0.6 μ CMOS technology and is compared to the results obtained using logic synthesis under various conditions and for different layout approaches. The algorithmic approach outperformed LS consistently, with improvements in speed ranging from 12%–56% and improvements in layout area ranging from 14.5%–35%. We have clearly demonstrated the superiority of the algorithmic approach on this circuit. The results generally indicate that careful analysis of the problem and clever management of the hierarchy pays big dividends in the performance of critical circuits, especially data-paths. Although very useful, LS tools are still not capable of managing hierarchies and making intelligent choices when it comes to design, and therefore they should be treated accordingly. The resulting LZD circuit has remarkable performance, which is important since it is often a part of the critical path in the floating-point unit.

ACKNOWLEDGMENT

I gratefully acknowledge V. Chang for running the simulation and Timberwolf program. I thank the referees and R. Maeder for careful reading and useful remarks. The idea was conceived of in June 1987 while the author was on the TF-1 project at IBM T. J. Watson Research Center in New York. I am grateful to M. Denneau of IBM for his ideas and discussion during the project. The author is thankful for the generous support from Sun Microsystems Laboratories, G. Taylor, D. Ditzel, and P. Hansen in particular.

REFERENCES

[1] J. Vuillemin and L. Guibas, "On fast binary addition in MOS technology," in *Proc.*, ICCC'82, New York, Sept. 28, 1982.
[2] V. G. Oklobdzija, "An implementation algorithm and design of a novel leading zero detector circuit," presented at 26th Asilomar Conf. on Signals, Systems and Computers, Oct. 26–28, 1992.
[3] LSI Logic, *1.0 micron cell-based product databook.* 1991.

SPIM: A Pipelined 64 × 64-bit Iterative Multiplier

MARK R. SANTORO, STUDENT MEMBER, IEEE, AND MARK A. HOROWITZ, MEMBER, IEEE

Abstract — A 64 × 64-bit iterating multiplier, the Stanford Pipelined Iterative Multiplier (SPIM), is presented. The pipelined array consists of a small tree of 4:2 adders. The 4:2 tree is better suited than a Wallace tree for a VLSI implementation because it is a more regular structure. A 4:2 carry-save accumulator at the bottom of the array is used to iteratively accumulate partial products, allowing a partial array to be used, which reduces area. SPIM was fabricated in a 1.6-μm CMOS process. It has a core size of 3.8 × 6.5 mm and contains 41 000 transistors. The on-chip clock generator runs at an internal clock frequency of 85 MHz. The latency for a 64 × 64-bit fractional multiply is under 120 ns, with a pipeline rate of one multiply every 47 ns.

I. INTRODUCTION

THE DEMAND for high-performance floating-point co-processors has created a need for high-speed, small-area multipliers. Applications such as DSP, graphics, and on-chip multipliers for processors require fast area efficient multipliers. Conventional array multipliers achieve high performance but require large amounts of silicon, while shift and add multipliers require less hardware but have low performance. Tree structures achieve even higher performance than conventional arrays but require still more area.

The goal of this project was to develop a multiplier architecture which was faster and more area efficient than a conventional array. As a test vehicle for the new architecture, a structure capable of performing the mantissa portion of a double extended precision (80 bit) floating-point multiply was chosen. The multiplier core should be small enough such that an entire floating-point co-processor, including a floating-point multiplier, divider, ALU, and register file, could be fabricated on a single chip. A core size of less than 25 mm^2 was determined to be acceptable. This paper presents a 64 × 64-bit pipelined array iteratively accumulating multiplier, the Stanford Pipelined Iterative Multiplier (SPIM), which can provide over twice the performance of a comparable conventional full array at one-fourth of the silicon area.

Manuscript received July 1, 1988; revised September 25, 1988 and November 21, 1988. The development of SPIM was supported in part by the Defense Advanced Project Research Agency (DARPA) under Contracts MDA903-83-C-0335 and N00014-87-K-0828.

The authors are with the Center for Integrated Systems, Stanford University, Stanford, CA 94305.

IEEE Log Number 8826243.

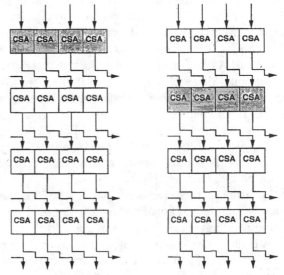

Fig. 1. Conventional array multiplier. Shaded areas represent intermediate partial product flowing down array.

II. ARCHITECTURAL OVERVIEW

Conventional array multipliers consist of rows of carry-save adders (CSA) where each row of CSA's sums up one additional partial product (see Fig. 1).[1] Since intermediate partial products are kept in carry-save form there is no carry propagate, so the delay is only dependent upon the depth of the array and is independent of the partial-product width. Although arrays are fast, they require a large amount of hardware which is used inefficiently. As the sum is propagated down through the array, each row of CSA's is used only once. Most of the hardware is doing no useful work at any given time. Pipelining can be used to increase hardware utilization by overlapping several calculations. Pipelining greatly increases throughput, but the added latches increase both the required hardware and the latency.

Since full arrays tend to be quite large when multiplying double or extended precision numbers, chip designers have used partial arrays and iterated using the system clock. This structure has the benefit of reducing the hardware by increasing utilization. At the limit, an iterative structure

[1] Carry-save adders are also often referred to as full adders or 3:2 adders.

Reprinted from *IEEE Journal of Solid-State Circuits*, Vol 24, No. 2, pp. 487-493, April 1989.

457

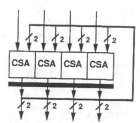

Fig. 2. Minimal iterative structure using a single row of CSA's Black bars represent latches.

would have one row of CSA's and a latch. Fig. 2 shows a minimal iterative structure. Clearly, this structure requires the least amount of hardware and has the highest utilization since each CSA is used every cycle. An important observation is that iterative structures can be made fast if the latch delays are small, and the clock is matched to the combinational delay of the CSA's. If both of these conditions are met the iterative structure approaches the same throughput and latency as the full array. This structure does, however, require very fast clocks. For a 2-μm process clocks may be in the 100-MHz range. A few companies use iterative structures in their new high-performance floating-point processors [5].

In an attempt to increase performance of the minimal iterative structure additional rows of CSA's could be added, resulting in a bigger array. For example, addition of a row of CSA cells to the minimal structure would yield a partial array with two rows of CSA's. This structure provides two advantages over the single row of CSA cells: it reduces the required clock frequency, and requires only half as many latch delays.[2] One should note, however, that although we doubled the number of CSA's, the latency was only reduced by halving the number of latch delays. The number of CSA delays remains the same. Increasing the depth of the partial array by simply adding additional rows of CSA's in a conventional structure yields only a slight performance increase. This small reduction in latency is the result of reducing the number of latches.

To increase the performance of this iterative structure we must make the CSA cells fast and, more importantly, decrease the number of series adds required to generate the product. Two well-known methods for the latter are Booth encoding and tree structures [2], [9]. Modified Booth encoding, which halves the number of series adds required, is used on most modern floating-point chips, including SPIM [7], [8]. Tree structures reduce partial products much faster than conventional methods, requiring only order $\log N$ CSA delays to reduce N partial products (see Fig. 3). Though trees are faster than conventional arrays, like conventional arrays they still require one row of CSA cells for each partial product to be retired. Unfortunately, tree structures are notoriously hard to lay out, and require large wiring channels. The additional wiring makes full trees even larger than full arrays. This has caused designers to look at permutations of the basic tree structure [1], [11].

[2]In fact one rarely finds a multiplier array that consists of only a single row of CSA's. The latch overhead in this structure is extremely high.

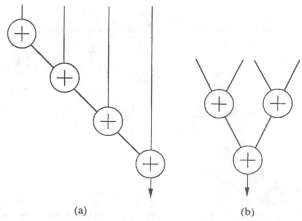

(a)   (b)

Fig. 3. (a) A conventional structure has depth proportional to N, while (b) a tree structure has depth proportional to $\log N$

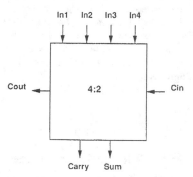

Fig 4. Block diagram of a 4:2 adder.

Unbalanced or modified trees make a compromise between conventional full arrays and full tree structures. They reduce the routing required of full trees but still require one row of CSA's for each partial product. Ideally one would want the speed benefits of the tree in a smaller and more regular structure. Since high performance was a prerequisite for SPIM, a tree structure was used. This left two problems. The first was the irregularity of commonly used tree structures. The second was the large size of the trees.

Wallace [9], Dadda [4], and most other multiplier trees use a CSA as the basic building block. The CSA takes three inputs of the same weight and produces two outputs. This 3:2 nature makes it impossible to build a completely regular tree structure using the CSA as the basic building block. A binary tree has a symmetric and regular structure. In fact, any basic building block which reduces products by a factor of 2 will yield a more regular tree than a 3:2 tree. Since a more regular tree structure was needed, the solution was to introduce a new building block: the 4:2 adder, which reduces four partial products of the same weight to 2 bits. Fig. 4 is a block diagram of the 4:2 adder. The truth table for the 4:2 adder is shown in Table I. Notice that the 4:2 adder actually has five inputs and three outputs. It is different from a 5:3 counter which takes in five inputs of the same weight and produces three outputs of different weights. The sum output of the 4:2 has weight 1 while the carry and C_{out} both have the same weight of 2. In addition, the 4:2 is not a simple counter as

TABLE I

TRUTH TABLE FOR THE 4:2 ADDER
n is number of inputs (from $In1$, $In2$, $In3$, $In4$) which $= 1$, C_{in} is the input carry from the C_{out} of the adjacent bit slice, C_{out} and carry both have weight 2, and sum has weight 1.

n	Cin	Cout	Carry	Sum
0	0	0	0	0
1	0	0	0	1
2	0	*	*	0
3	0	1	0	1
4	0	1	1	0
0	1	0	0	1
1	1	0	1	0
2	1	*	*	1
3	1	1	1	0
4	1	1	1	1

*Either C_{out} or Carry may be ONE for two or three inputs equal to 1 but NOT both.

C_{out} may NOT be a function of the C_{in} from the adjacent block or a ripple carry may occur.

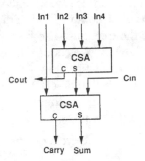

Fig. 5. A 4:2 adder implemented with two CSA's.

the C_{out} output must NOT be a function of the C_{in} input or a ripple carry could occur. As for the name, 4:2 refers to the number of inputs from one level of a tree and the number of outputs produced at the next lower level. That is, for every four inputs taken in at one level, two outputs are produced at the next lower level. This is analogous to the binary tree in which for every two inputs one output is produced at the next lower level. The 4:2 adder can be implemented directly from the truth table, or with two CSA cells as in Fig. 5.[3]

A 4:2 tree will reduce partial products at a rate of $\log_2 (N/2)$ whereas a Wallace tree requires $\log_{1.5} (N/2)$, where N is the number of inputs to be reduced. Though the 4:2 tree might appear faster than the Wallace tree, the basic 4:2 cell is more complex so the speed is comparable. The 4:2 structure does, however, yield a tree which is much more regular. In addition the 4:2 adder has the advantage that two CSA's are in each pipe in place of one. This reduces both the required clock frequency and the latch overhead.

---

[3]SPIM implemented the 4:2 adder with two CSA cells because it permits a straightforward comparison with other architectures on the basis of CSA delays. By knowing the size and speed of the CSA cells in any technology, a designer can predict the size and speed advantages of this method over that currently used

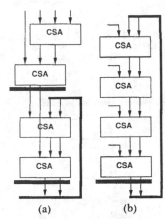

(a) (b)

Fig. 6. With the same four CSA cells a four-input partial tree structure with a (a) carry-save accumulator will attain almost twice the throughput of a (b) partial piped array. In (a) the carry-save accumulator is placed under the 4:2 adder.

To overcome the size problem SPIM uses a partial 4:2 tree, and then iteratively accumulates partial products in a carry-save accumulator to complete the computation. The carry-save accumulator is simply a 4:2 adder with two of the inputs used to accumulate the previous outputs. The carry-save accumulator is much faster than a carry-propagate accumulator and requires only one additional pipe stage.

Fig. 6 compares a single 4:2 adder with carry-save accumulator to a conventional partial piped array.[4] Both structures reduce four partial products per cycle. Notice, however, that the tree structure is clocked at almost twice the frequency of the partial piped array. It has only two CSA cells per pipe stage, whereas the partial piped array has four. Consequently, the partial array would require 32 CSA delays to reduce 32 partial products whereas the tree structure would need only 18 CSA delays. Using the 4:2 adder with carry-save accumulator is almost twice as fast as the partial piped array, while using roughly the same amount of hardware.

The 4:2 adder structure can be used to construct larger trees, further increasing performance. In Fig. 7 we use the same 4:2 adder structure to form an eight-input tree. This allows us to reduce eight partial products per cycle. Notice that we still pipeline the tree after every two carry-save adds (each 4:2 adder). In contrast, if we clocked the tree every four carry-save adds it would double the cycle time and only decrease the required number of cycles by one. The overall effect would be a much slower multiply.

Fig. 8 shows the size and speed advantages of different sized 4:2 trees with carry-save accumulators versus conventional partial arrays. This plot is a price/performance plot where the price is size and the performance is speed (latency $= 1$/speed). The plot assumes we are doing a 64×64-bit multiply. Booth encoding is used, thus we must retire 32 partial products. Size has been normalized such

---

[4]In Figs. 6, 7, and 9 the detailed routing has not been shown. Providing the exact detailed routing, as was done in Fig. 5, would provide more information; however, it would significantly complicate the figures and would tend to obscure their purpose, which is to show the data flow in terms of pipe stages and CSA delays.

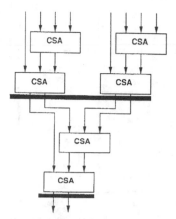

Fig. 7. An eight-input tree constructed from 4·2 adders can reduce eight partial products per cycle.

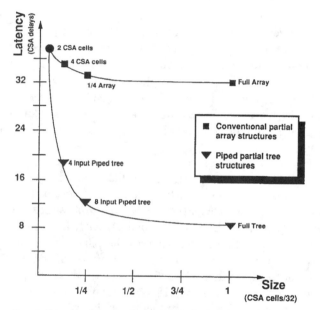

Fig. 8 Architectural comparison of piped partial tree structure with carry-save accumulator versus conventional partial array

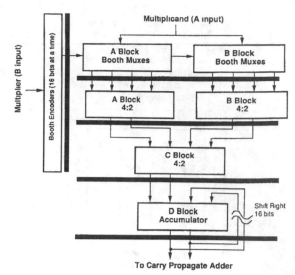

Fig. 9. SPIM data path.

$\log_2(K/2)+(N/K)$ where N is the operand size and K is the partial tree size. If Booth encoding is used N would be one-half the operand size since Booth encoding has already provided a factor of 2 compression. Start-up times and pipe stages before the tree must also be taken into account when determining latency. We choose the eight-input piped tree with Booth encoding for SPIM, as we feel this provides the best area speed trade-off for our purpose. The number of cycles required to reduce 64 bits using Booth encoding and an 8-bit tree is

$$\log_2(8/2)+(32/8)+\text{one cycle overhead} = 7 \text{ cycles.}[6]$$

III. SPIM Implementation

Fig. 9 is a block diagram of the SPIM data path. The Booth encoders, which encode 16 bits per cycle, are to the left of the data path. The Booth-encoded bits drive the Booth select MUX's in the A and B block. The A and B block Booth select MUX outputs drive an eight-input tree structure constructed of 4:2 adders which are found in the A, B, and C blocks. Each pipe stage uses one 4:2 adder which consists of two CSA's. The D block is a carry-save accumulator. It also contains a 16-bit hard-wired right shift to align the partial sum from the previous cycle to the current partial sum to be accumulated.

Fig. 10 is a die photograph of SPIM. The A block inputs are preshifted allowing the A block to be placed on top of the B block. Using 4:2 adders in a partial tree allows the array to be efficiently routed, and laid out as a bit slice, thus making the SPIM array a very regular structure. Interestingly, the CSA cells occupy only 27 percent of the core area. The Booth select MUX's used in the A and B blocks make these blocks three times as large as the C block. Each Booth MUX with its corresponding latch is larger than a single CSA. Also, due to the routing required for the 16-bit shift, the D block is twice as large as the C block. The array area can be split into four main components: routing, CSA cells, MUX's, and latches. The routing

that 32 rows of CSA cells (a full array) has a size of one unit.[5] In the upper left corner is the structure using only two rows of CSA cells. In this case the tree and conventional structures are one and the same and can be seen as a partial array two rows deep, or as a two-input partial tree. We can see that adding hardware to form larger partial arrays provides very little performance improvement. A full array is only 15 percent faster than the iterative structure using two rows of CSA's. Adding hardware in a tree-type structure, however, dramatically improves performance. For example, using a four-input tree, which uses four rows of CSA's, is almost twice as fast as the two-input tree. Using an eight-input tree is almost three times as fast as a two-input tree and only one-fourth the size of the full array.

The latency of the multiplier is determined by the depth of the partial 4:2 tree and the fraction of the partial products compressed each cycle. The latency is equal to

---

[5]Latency is in terms of CSA delays. We have assumed a latch is equivalent to one-third of a CSA delay in an attempt to take the latch delays into account. Size is the number of CSA cells used. It does not include the latch or wiring area.

[6]The one-cycle overhead is used for the Booth select MUX's.

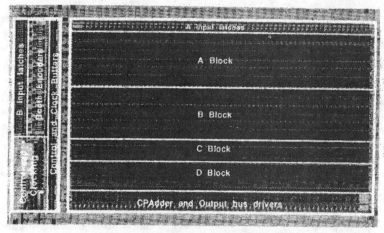

Fig. 10. Microphotograph of SPIM.

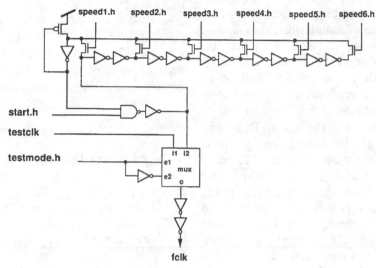

Fig. 11. SPIM clock generator circuit.

required 20 percent of the area, while the other 75 percent was equally split between the CSA cells, MUXs, and latches.

The critical path in the SPIM data path is through the D block. The D block contains the slowest path because of the added routing at the output, and the additional control MUX at its input. The input MUX is needed to reset the carry-save accumulator. It selects ZERO to reset, or the previous shifted output when accumulating. The final critical path through the D block includes two CSA cells, a master–slave latch, a control MUX, and the drive across 16 bits (128 μm) of routing.

IV. CLOCKING

The architecture of SPIM yields a very fast multiply; however, the speed at which the structure runs demands careful attention to clocking issues. Only two CSA's (one 4:2 adder) are found in each pipe stage, yielding clock rates on the order of 100 MHz. The typical system clock is not fast enough to be useful for this type of structure. To produce a clock of the desired frequency, SPIM uses a controllable on-chip clock generator. The clock is generated by a stoppable ring oscillator. The clock is started

when a multiply is initiated, and stopped when the array portion of the multiply has been completed. The use of a stoppable clock provides two benefits. It prevents synchronization errors from occurring and it saves power as the entire array is powered down upon completing a multiply. The actual clock generator used on SPIM is shown in Fig. 11. It has a digitally selectable feedback path which provides a programmable delay element for test purposes. This allows the clock frequency to be tuned to the critical path delay. In addition, the clock generator has the ability to use an external test clock in place of the fast internally generated clock.

When a multiply signal has been received, a small delay occurs while starting up the clocks. This delay comes from two sources. The first source is the logic which decodes the run signal and starts up the ring oscillator. The second source is from the long control and clock lines running across the array. They have large capacitive loads and require large buffer chains to drive them. The simulated delay of the buffer chain and associated logic is 6 ns, almost half a clock cycle. Since the inputs are latched before the multiply is started, SPIM does the first Booth encode before the array clocks become active (cycle 0). Thus, the start-up time is not wasted. After the clocks have

TABLE II
SPIM PIPE TIMING
Numbers indicate which partial products are being reduced. 0 is the least significant bit.

Cycle / Action	0	1	2	3	4	5	6	7
Booth Encode	startup 0-15	16-31	32-47	48-63				
A and B block Booth Muxs		0-15	16-31	32-47	48-63			
A Block CSA's			0-7	16-23	32-39	48-55		
B Block CSA's			8-15	24-31	40-47	56-63		
C Block				0-15	16-31	32-47	48-63	
D Block					clear 0-15	16-31	32-47	48-63

been started SPIM requires seven clock cycles (cycles 1–7) to complete the array portion of a multiply.

The detailed timing is shown in Table II. In the time before the clocks are started (cycle 0) the first 16 bits are Booth encoded. During cycle 1, the first 16 Booth-coded partial products from cycle 0 are latched at the input of the array. The next four cycles are needed to enter all 32 Booth-coded partial products into the array. Two additional cycles are needed to get the output through the C and D blocks. If a subsequent multiply were to follow it would have been started on cycle 4, giving a pipelined rate of four cycles per multiply. When the array portion of the multiply is complete the carry-save result is latched, and the run signal is turned OFF. Since the final partial sum from the D block is latched into the carry-propagate adder only every fourth cycle, several cycles are available to stop the clock without corrupting the result.

The clock generator is located in the lower left-hand side of the die (see Fig. 10). The clock signal runs up a set of matched buffers, along the side of the array, which are carefully tuned to minimize skew across the array. Wider than minimum metal lines are used on the master clock line to reduce the resistance of the clock line relative to the resistance of the driver. The clock and control lines driven from the matched buffers then run across the entire width of the array in metal.

V. TEST RESULTS

To accurately measure the internal clock frequency, the clock was made available at an output allowing an oscilloscope to be attached. SPIM was then placed in continuous (loop) mode where the clock is kept running and multiplies are piped through at a rate of one multiply every four cycles. Since the clock is continuously running its frequency can be accurately determined.

Three components determine the actual performance of SPIM: 1) the start-up time, when the clocks are started and the first Booth encode takes place (cycle 0); 2) the array time, which includes the time through the partial array plus the accumulation cycles (cycles 1–7); and 3) the carry-propagate addition (cpadd) time, when the final carry-propagate addition converts the carry-save form of the result from the accumulator to a simple binary representation. Due to limitations in our testing equipment, only the array time could be accurately measured. Since the array time requires seven cycles, and the array clock frequency was 85 MHz, the array time is simply $7 \cdot (1/85$ MHz$) = 82.4$ ns. The start-up and cpadd times, based upon simulations, were 6 and 30 ns, respectively. In flowthrough mode the total latency is simply the sum of the start-up time (6 ns), the array time (82.4 ns), and the cpadd time (30 ns), for a total of 118.4 ns. Thus SPIM has a total latency under 120 ns. SPIM has a throughput of one multiply every four cycles or $4 \cdot (1/85$ MHz$) = 47$ ns, for a maximum pipelined rate in excess of 20-million 80-bit floating-point multiplies per second.

The performance range of the parts tested was from 85.4 to 88.6 MHz at a room temperature of 24.5°C and a supply voltage of 4.9 V. One of the parts was tested over a temperature range of 5–100°C. At 5°C it ran at 93.3 MHz with speeds of 88.6 and 74.5 MHz at 25 and 100°C. The average power consumed at 85 MHz was 72 mA while an average of only 10 mA was consumed in standby mode.

VI. FUTURE IMPROVEMENTS

The Booth select MUX's with their corresponding latches account for 38 percent of the array area. This was larger than expected. Though Booth encoding reduces the number of partial products by a factor of 2, the same result could be achieved by adding one more level of 4:2 adders to the tree. Since much of the routing already exists for the Booth MUX's, adding another level to the tree requires replacing each two Booth select MUX's with a 4:2 adder and four AND gates (see Fig. 12). Since the CSA cells are slightly larger than the Booth select MUX's the array size will grow slightly (by about 7 percent). However, if we take the whole picture into account, the core will remain about the same size, as we would no longer need the Booth encoders. Replacing the Booth encoders and Booth select MUX's with an additional level to the tree would also reduce the latency by one cycle from seven cycles to six. This occurs because the cycle required to Booth encode is now no longer needed. There are other advantages in addition to the increase in speed. Perhaps the greatest gain is the reduction in complexity. Both the Booth encoders and Booth select MUX's are now unnecessary, thus the number of cells has been reduced. In addition, Booth encoding generates negative partial products. An increase in complexity results in the need to handle the negative partial products correctly. Replacing the Booth encoders with an additional level of 4:2 adders would remove the negative partial products. Our observation is that an increase in speed and reduction in complexity can be obtained with little or no increase in area.[7]

[7]Replacing the Booth encoders and select MUX's with an additional level of 4:2 compressors is a viable alternative on more conventional, i.e., nonpiped and noniterative, trees as well. The nonpipelined speed gain depends upon the relative speed of the Booth encode plus Booth select MUX versus the delay through one 4:2 compressor and a NAND gate.

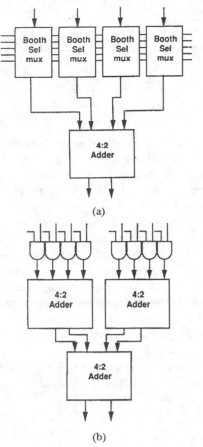

(a)

(b)

Fig. 12. Booth encoding versus additional tree level. (a) The Booth encoders and Booth select MUX's can be replaced with (b) an additional level of 4:2 adders and AND gates.

SPIM uses full static master–slave latches for testing purposes. These latches are quite large, accounting for 27 percent of the array size. In addition, they are slow, requiring 25 percent of the cycle time. Since the SPIM architecture has been proven, these latches are not required on future versions. One obvious choice is simply to replace the full static master–slave version with dynamic latches. Another option is to split the master–slave latches into two separate half latches and incorporate them into the CSA cells. This would reduce area and increase speed. A still more efficient structure is the use of single-phase dynamic latches. The balanced pipe nature of the multiplier makes the use of single-phase latches possible. Since only half as many latches are required in the pipe, single-phase dynamic latches would reduce the cycle time and decrease latch area.

Research on piped 4:2 trees and accumulators has continued. A test circuit consisting of a new clock generator and an improved 4:2 adder has been fabricated in a 0.8-μm CMOS technology. Preliminary test results have demonstrated performance in the range of 400 MHz.

VII. Conclusion

SPIM was fabricated in a 1.6-μm CMOS process through the DARPA MOSIS fabrication service. It ran at an internal clock speed of 85 MHz at room temperature. The latency for a 64×64-bit fractional multiply is under 120 ns. In piped mode SPIM can initiate a multiply every four cycles (47 ns), for a throughput in excess of 20-million multiplies per second. SPIM required an average of 72 mA at 85 MHz, and only 10 mA in standby mode. SPIM contains 41 000 transistors with a core size of 3.8×6.5 mm, and an array size of 2.9×5.3 mm.

The 4:2 adder yields a tree structure which is as efficient and far more regular than a Wallace-type tree and is therefore better suited for a VLSI implementation. By using a partial 4:2 tree with a carry-save accumulator a multiplier can be built which is both faster and smaller than a comparable conventional array. Future designs implemented in a 0.8-μm CMOS technology should be capable of clock speeds approaching 400 MHz.

Acknowledgment

Fabrication support through MOSIS is gratefully acknowledged.

References

[1] S. F. Anderson et al., "The IBM system/360 model 91: Floating-point execution unit," IBM J., vol. 11, no. 1, pp. 34–53, Jan. 1967.
[2] A. D. Booth, "A signed binary multiplication technique," Quart. J. Mech. Appl. Math., vol. 4, Part 2, 1951.
[3] J. F. Cavanagh, Digital Computer Arithmetic Design and Implementation. New York: McGraw-Hill, 1984.
[4] L. Dadda, "Some schemes for parallel multipliers," Alta Freq., vol. 34, no. 5, pp. 349–356, Mar. 1965.
[5] B. Elkind, J. Lessert, J. Peterson, and G. Taylor, "A sub 10 ns bipolar 64 bit integer/floating point processor implemented on two circuits," in Proc. IEEE Bipolar Circuits and Technology Meeting, Sept. 1987, pp. 101–104.
[6] K. Hwang, Computer Arithmetic: Principles, Architecture, and Design. New York: Wiley, 1979.
[7] P. Y. Lu et al., "A 30-MFLOP 32b CMOS floating-point processor," in ISSCC Dig. Tech. Papers, vol. XXXI, Feb. 1988, pp. 28–29.
[8] W. McAllister and D. Zuras, "An nMOS 64b floating point chip set," in ISSCC Dig. Tech. Papers, Feb. 1986, pp. 34–35.
[9] C. S. Wallace, "A suggestion for fast multipliers," IEEE Trans. Electron. Computers, vol. EC-13, pp. 14–17, Feb. 1964.
[10] S. Waser and M. J. Flynn, Introduction to Arithmetic for Digital Systems Designers. New York: CBS Publishing, 1982.
[11] D. Zuras and W. McAllister, "Balanced delay trees and combinatorial division in VLSI," IEEE J. Solid-State Circuits, vol. SC-21, no. 5, pp. 814–819, Oct. 1986.

A 54 × 54-b Regularly Structured Tree Multiplier

Gensuke Goto, Tomio Sato, Masao Nakajima, and Takao Sukemura

Abstract—A 54 × 54-b parallel multiplier is implemented in 0.8-μm CMOS using the new, regularly structured tree (RST) design approach. The circuit is basically a Wallace tree, but the tree and the set of partial-product-bit generators are combined into a recurring block which generates seven partial-product bits and compresses them to a pair of bits for the sum and carry signals. This block is used repeatedly to construct an RST block in which even wiring among blocks included in wire shifters is designed as recurring units. By using recurring wire shifters, we can expand the level of repeated blocks to cover the entire adder tree, which simplifies the complicated Wallace tree wiring scheme. In addition to design time savings, layout density is increased by 70% to 6400 transistors/mm², and the multiplication time is decreased by 30% to 13 ns.

I. INTRODUCTION

RAPID progress in VLSI technology has enabled the speed and performance of computers to be increased by a factor of 10 every 5 years. In addition to improvements in device technology, these advances are enhanced by improvements in processor architecture. Third-generation 32-b microprocessors integrate a floating-point processing unit on the same die as the integer unit [1], [2]. For double-precision multiplication, however, the multiplier unit is not fully implemented in hardware but is operated repeatedly for one operation. This scheme results in multiplication which is time consuming as compared to other arithmetic operations.

Further improvement in arithmetic operations will require a fresh approach. A microprocessor based on the very long instruction word (VLIW) concept [3] is a candidate for the upcoming fourth- or fifth-generation architectures. With VLIW, several arithmetic units are implemented on a single chip for parallel operation. Small high-performance units are expected to be in great demand for enhanced-performance processing systems. An improved multiplier unit is essential because the multiplier is the limiting factor in both the performance and die size of most chips today.

In the conventional Wallace tree multiplier construction [4], multi-input partial-product bits, at the same bit position, are consecutively compressed to a final sum and carry signal pair by using a series of single-bit full adders. This reduction process differs at each bit position because of the variety in the number of the partial products to be compressed. A very complicated design procedure is thus required. Using 4–2 compressors rather than full adders does simplify the design, but 100 000 manually routed interconnections are still required to design a 27 × 53-b multiplier tree [5].

An array-type multiplier consists of an array of units with full adders (3–2 compressors) and partial-product-bit generators, where each of the units processes single-bit data consecutively [6]. The entire array can be regularly laid out with only a few unit circuits, making the design very easy. Although most automatically generated multipliers are of this type [7], they are too slow for application to high-speed systems. An array and tree hybrid design may result in a multiplier of intermediate speed and complexity [8], but the difficulties in design would not be reduced much since a holding procedure is necessary to minimize the waste area.

II. REGULARLY STRUCTURED TREE (RST) CONCEPTS

We propose a simple design method for a high-speed 54 × 54-b multiplier [9]. This multiplier is used for the mantissa multiplication of two double-precision numbers as outlined in the IEEE standard [10], where the mantissa of a double-precision number is represented by 52 b, and there are a hidden bit and a sign bit which are used for two's complement operations in Booth's algorithm [11]. A maximum number of 28 b may be added at the same bit position. Of these, 27 come from the partial-product-bit generators and one is from the sign bit of second-order Booth encoding.

To simplify the design process of the multiplier adder tree, we divide the tree into subcircuit modules that are reused in the construction of the complete tree. This approach has already been tried. For instance, Hokenek *et al.* [12] adopted a 7/3 counter (7 input bits reduced to a 3-b binary sum), reducing the connections among counters to half that of a tree of full adders. Even in this approach, however, time-consuming wiring design must be done on individual modules for optimization. The same is true for methods using 4–2 compressors [13], [14].

Therefore, a simplified process must contain a wiring scheme that is repeated among modules. With this in mind, we divide the adder tree and partial-product-bit generators into two equivalent blocks, 14D, to extract the maximum identity as shown in Fig. 1. The blocks are halved again to yield the 7D subblocks which have 7 input data bits. The subblocks consist of seven partial-product-bit generators and a 7–2 compressor. As they are, the sub-

Manuscript received November 26, 1991; revised May 11, 1992.

G. Goto and T. Sato are with Fujitsu Laboratories Ltd., 10-1 Morinosato-Wakamiya, Atsugi 243-01, Japan.

M. Nakajima and T. Sukemura are with Fujitsu Limited, 1015 Kamikodanaka, Nakahara-ku, Kawasaki 211, Japan.

IEEE Log Number 9202179.

Reprinted from *IEEE Journal of Solid-State Circuits*, Vol. 27, No. 9, pp. 1229–1235, September 1992.

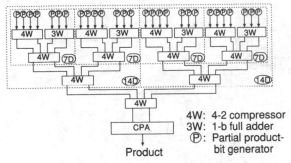

Fig. 1. Division scheme of partial-product-bit generators and adders in the 54 × 54-b tree multiplier.

4W: 4-2 compressor
3W: 1-b full adder
Ⓟ: Partial product-
bit generator

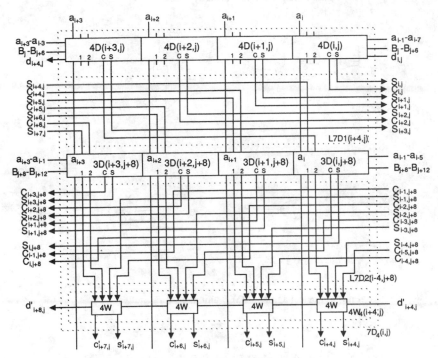

Fig. 2. Construction of the 7D₄ (i, j) block.

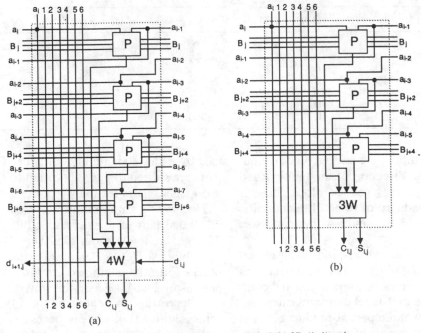

Fig. 3. Construction of (a) 4D (i, j) unit and (b) 3D (i, j) unit.

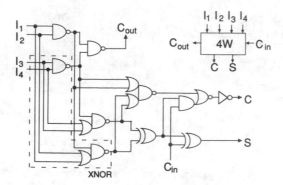

Fig. 4. Logic diagram of the 4W unit.

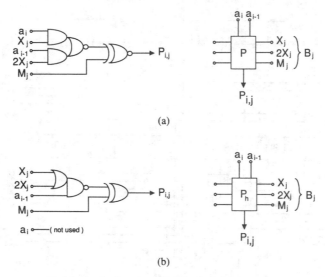

(a)

(b)

Fig. 5. Logic diagram of (a) P unit and (b) P_h unit.

TABLE I
MODIFICATIONS TO $7D_4$ BLOCK

Block	4D Unit				3D Unit			
	$(i+3, j)$	$(i+2, j)$	$(i+1, j)$	(i, j)	$(i+3, j+8)$	$(i+2, j+8)$	$(i+1, j+8)$	$(i, j+8)$
$7D_4$	4D	4D	4D	4D	3D	3D	3D	3D
$7D_{A4}$	$4D_F$	$4D_G$	4D	4D	$3D_I$	$3D_H$	3D	3D
$7D_{B4}$	$4D_F$	$4D_G$	4D	4D	$3D_F$	$3D_G$	$3D_B$	$3D_A$
$7D_{C4}$	4D	4D	4D	4D	$3D_B$	$3D_A$	$3D_B$	$3D_A$
$7D_{D4}$	4D	4D	$4D_E$	$4D_E$	3D	3D	3D	3D
$7D_{E4}$	4D	4D	$4D_E$	$4D_E$	$3D_B$	$3D_A$	$3D_B$	$3D_B$
$7D_{F4}$	$4D_D$	$4D_C$	$4D_B$	$4D_A$	$3D_A$	$3D_E$	$3D_C$	$3D_D$

blocks are not repeatable, although a packed block of four contiguous subblocks in the direction of the multiplicand bit can be used repeatedly. This packed block, $7D_4$, is shown in Fig. 2. $7D_4$ is made up of three different units: a 4D (Fig. 3(a)), a 3D (Fig. 3(b)), and a 4–2 compressor, 4W (Fig. 4).

The 4D unit consists of four partial-product-bit generators (P in Fig. 5(a)) and a 4W unit. This unit generates four partial-product bits at the same bit position and compresses them into a pair of first-level sum and carry signals. The 3D unit generates three partial-product bits and compresses them into another pair of sum and carry sig-

nals by using a single-bit full adder (3W). Output signal pairs are shifted to the right or left to generate the second-level signal pairs for the 4W units.

Two wiring pattern units are used: one for shifting the 4D signal pair to the right four bits and one to shift the 3D signal to the left four bits. The $7D_4$ block and its modified blocks, $7D_{A4}$–$7D_{F4}$, are listed in Table I. These blocks form the combined array of an adder tree and the partial-product-bit generator units.

To construct the entire 54×54-b multiplier circuit, some additional blocks are necessary as shown in Fig. 6, which outlines the design of our regularly structured tree

466

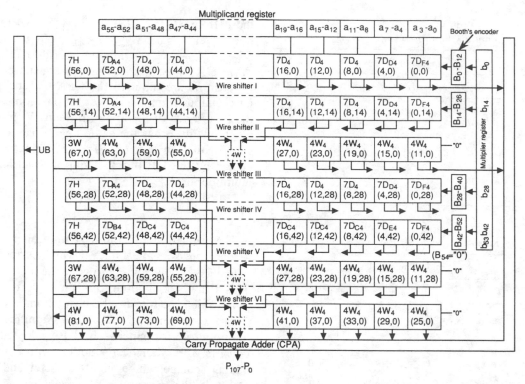

Fig. 6. Organization of regularly structured tree (RST) multiplier.

(RST) multiplier. This approach reduces multiplier design time since most blocks are repeatable in highly regular patterns which include the wiring among blocks.

III. REPEATABLE BLOCK AND UNIT DESIGN

Fig. 4 shows the 4–2 compressor (4W). Our full CMOS circuit uses 60 transistors. Although the gate count is the same as in the single-bit full adders in a conventional four-input Wallace tree, the speed is increased by 33%. This is because conventional adders have four EXCLUSIVE-OR (XOR) gate delays for the sum signal of four binary numbers, while 4W gives the sum in three XOR-gate delays. This may be explained by considering the outlined section in Fig. 4 as an XNOR gate whose speed is equivalent to that of an XOR gate. Simulations of the 4W with a fan-out of 4 give a delay time of 1.0 ns, in contrast to the conventional circuit delay of 1.6 ns. This verifies the above assumption.

Use of n-channel pass transistors in XOR circuits [14] may decrease the gate count to 58, but would necessitate keeping the input node wiring as short as possible. This is because the pass transistor's large resistance and the wiring capacitance distort input signals and delay the output. In most practical situations, these effects cannot be reduced without adopting buffering gates.

Fig. 5(a) shows the partial-product-bit generator (P unit) logic which is based on the second-order Booth algorithm. X_j, $2X_j$, and M_j indicate Booth encoder output for a term $b_j + b_{j-1} - 2b_{j+1}$. They represent $\times 1$, $\times 2$, and the negation of the partial product P_j for the jth bit of the multiplier. a_i and a_{i-1} are the multiplicand bits. Fig.

5(b) shows the sign bit generator (P_h unit) for the partial product. The P_h unit corrects the sign bit without expanding it to the most significant (107th) bit [15].

The construction of the $7D_4$ block must be modified for $7D_{A4}$, $7D_{B4}$, $7D_{C4}$, $7D_{D4}$, $7D_{E4}$, and $7D_{F4}$ as listed in Table I, to complete an RST block in Fig. 6. $7D_4$ is modified for $7D_{A4}$ by replacing $4D$ $(i + 2, j)$ and $4D$ $(i + 3, j)$ with $4D_G$ $(i + 2, j)$ and $4D_F$ $(i + 3, j)$. Likewise, $3D$ $(i + 2, j + 8)$ and $3D$ $(i + 3, j + 8)$ must also be replaced with $3D_H$ $(i + 2, j + 8)$ and $3D_I$ $(i + 3, j + 8)$. Modifications to other blocks may be determined similarly using Table I. Table II lists the modifications of the 4D unit for input signal variation to the 4W unit contained in the 4D unit. Table III similarly lists modifications for 3D input variation. These unit modifications are needed to fit input signals to units where there is no corresponding input on the lower and upper part of the RST—where the corrected sign bit is generated and where $+1$ is added if the Booth encoder outputs a negative partial product. The modifying process is not as cumbersome as it first appears, and the regular structure of the RST multiplier is retained.

Once $7D_4$ and the modified blocks are obtained, we can proceed with the wire shifter design as shown in Fig. 6. To complete the RST structure, we designed four kinds of wire shifters. The first kind of shifter (shifters I and IV) is for connecting second-level paired sum and carry signals output from the first (third) $7D_4$ row to the first (second) $4W_4$ row as shown in Fig. 6. This kind of shifter moves the sum signals to the right 7 b and the carry signals by 6 b, and connects the signal lines to the through-wire pairs numbered 3 and 4 in Fig. 3 in the 4D and 3D units which belong to the second (fourth) $7D_4$ row.

TABLE II
MODIFICATIONS TO 4D UNIT

Unit	Input to 4W			
	(i, j)	$(i-2, j+2)$	$(i-4, j+4)$	$(i-6, j+6)$
4D	P	P	P	P
$4D_A$	P	M_j	0	0
$4D_B$	P	0	0	0
$4D_C$	P	P	M_{j+2}	0
$4D_D$	P	P	0	0
$4D_E$	P	P	P	0
$4D_F$	0	P	P	P
$4D_G$	P_h	P	P	P

TABLE III
MODIFICATIONS TO 3D UNIT

Unit	Input to 4W		
	(i, j)	$(i-2, j+2)$	$(i-4, j+4)$
3D	P	P	P
$3D_A$	P	P	0
$3D_B$	P	P	1
$3D_C$	P	0	0
$3D_D$	P	M_j	0
$3D_E$	P	P	M_{j+2}
$3D_F$	0	P	1
$3D_G$	P_h	P	0
$3D_H$	P_h	P	P
$3D_I$	0	P	P

The second kind of shifter (shifters II and V) shifts the second-level sum signal from the second (fourth) $7D_4$ row to the left 7 b and the carry signal to the left by 8 b. Thus two sets of paired signals are added to compress the third-level paired sum and carry signals output from the first (second) $4W_4$ row.

The third type of shifter (shifter III) shifts the third-level sum signal from the first $4W_4$ row to the right 14 b and the carry signal by 13 b, and then connects to the through-wire pairs 5 and 6 in the 4D and 3D units of the third and fourth $7D_4$ row.

The fourth kind of shifter (shifter VI) shifts similar signals from the second $4W_4$ row to the left 14 and 15 b. The shifted third-level paired signals are compressed in the third $4W_4$ row to yield the fourth-level paired signals for each bit, and the output signals are input to the carry-propagate adder (CPA) for final product generation.

Wire shifters I, II, IV, and V are constructed of repeatable patterns which deal with two adjacent $7D_4$ outputs. Similarly, wire shifters III and VI are constructed of repeatable patterns over four $7D_4$ blocks. Since all the shifters form repeatable patterns, there is a significant saving in wiring design effort.

The 7H block on the left edge of the RST structure in Fig. 6 generates the second-level sum and carry signals $S'_{i+4,j}$ to $S'_{i+10,j}$, $C'_{i+4,j}$ to $C'_{i+9,j}$, and the first-level signals $S_{i,j}$ to $S_{i+3,j}$ and $C_{i,j}$ to $C_{i+2,j}$ by a combination of P, P_h, 4W, 3W, and 2W (half adder) units.

The UB block in Fig. 6 is constructed of 37 2W's and two 3W's. The UB receives these second-level signals and

those from 3W(67, 28) and 4W(81, 0) to generate the third-level signals $S''_{68,0}$ to $S''_{80,0}$, $C''_{68,0}$ to $C''_{80,0}$, $S''_{97,0}$ to $S''_{107,0}$, $C''_{97,0}$ to $C''_{107,0}$, and the fourth-level signals $S'''_{82,0}$ to $S'''_{96,0}$, $C'''_{82,0}$ to $C'''_{96,0}$. All output signals beyond the 81st bit position are input to the CPA to generate final products. Output signals at lower bit positions are returned to the RST. Each of the upper thirteen 2W blocks in the UB block of Fig. 6 has the same layout pattern. The same is true for the lower twenty-four 2W blocks. The area occupied by the UB block is only a fraction of the entire multiplier area.

IV. EVALUATION

The critical path of the RST multiplier consists of four 4–2 compressors at each level which correspond to 12 XOR gate delays, plus a delay to generate a partial-product bit. This delay is equivalent to results reported by Mori et al. [14], which seem to be the best among reported circuits.

The speed of the 108-b CPA significantly influences multiplier performance. The CPA adopted in our design consists of a Manchester adder with a new bypass and adding scheme which yields a faster and smaller adder [16] than the conventional carry select adder. Assuming a 125% delay of 0.8-μm full-CMOS technology, the estimated delay is 8.5 ns.

The 54 × 54-b RST multiplier was fabricated on a test chip using triple-level-metal, single-polysilicon, and 0.8-μm CMOS. Both the n- and p-channel transistors have 0.8-μm gates, and the metal pitch is 2.5 μm for the first level, 3.1 μm for the second, and 5.0 μm for the third. Fig. 7 is a photomicrograph of the test chip which includes the multiplier and several sets of registers for monitoring delay time.

The wiring patterns in the repeatable blocks such as $7D_4$, $4W_4$, and all the wire shifters are implemented using the first- and the second-level metal lines. The third-level metal lines are used for multiplicand bit routing over the repeatable blocks, which run vertically on the die to minimize wiring capacitance. They are also used to supplement the power supply lines. Without the third-level wires, the die area would be 10% larger.

The rows at the top and the bottom are data registers. The RST multiplier is located in the narrow central strip. Data flow from the registers on the top and go down to the CSA (adjacent to the output registers) on the bottom edge. The multiplier measures 3.36 × 3.85 mm. There are 82 500 transistors.

Fig. 8 shows the shmoo plot at room temperature. About 20 000 random patterns and 12 specific patterns were used for this measurement. For specific patterns, the critical patterns for the RST part, the CPA, and the combined ones are included. The most critical pattern and its related output node were detected by scanning patterns. We measured the multiplier response with respect to this critical pattern and output node, using an electron beam tester (Fig. 9). The critical path delay was 13.0 ns for room temperature with a 5-V supply.

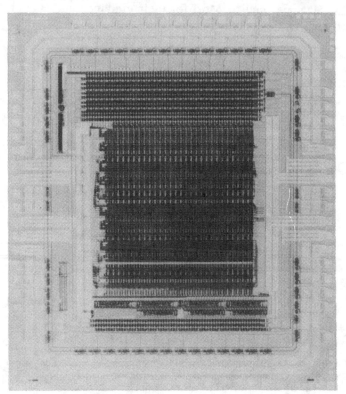

Fig. 7. Photograph of the RST multiplier chip.

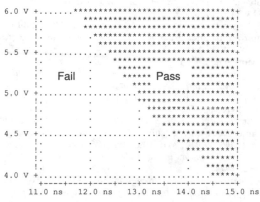

Fig. 8. Shmoo plot.

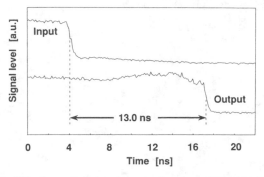

Fig. 9. Waveform of electron-beam testing on an RST multiplier.

Table IV summarizes the multiplier's features. We will now present a brief comparison of the RST multiplier with others. The 32-b multiplier in [13] uses almost the same technology as ours, but the delay time of our multiplier is

TABLE IV
RST MULTIPLIER FEATURES

Multiplier and multiplicand	54 b (including sign)
Product	107 b (including sign)
Multiplication time	13.0 ns
Power at 40 MHz	875 mW
Block size	3.36 × 3.85 mm
Transistors	82 500
Process technology	0.8-μm CMOS triple metal

30% less and the density 70% higher in a comparison on a 54-b basis. The 54-b multiplier in [14] employed 0.5-μm process technology, so its density should be far higher than the density we obtained using 0.8-μm technology. The densities, however, are almost the same. It has a smaller delay because it uses advanced 0.5-μm transistors and pseudo-CMOS logic to enhance the CPA speed. There is, however, a sacrifice in standby power consumption. Since our multiplier uses fully static CMOS gates, it does not exhibit this problem.

An analysis using shrunk 0.5-μm CMOS devices with 3.3-V power supplies produced a 54 × 54-b RST multiplier with an active area of 8.1 mm^2 and a 10-ns delay. The area is 36% smaller and the speed is comparable to the multiplier in [14], which is the smallest and fastest multiplier for double-precision numbers reported to date. The speed may be further improved by optimizing the circuits using 0.5-μm technology.

We define the quality factor Q for multiplier circuit evaluation considering the performance/cost ratio. The performance is inversely related to delay time T which is normalized by $L \cdot \log (N/2)$, where L is the gate length of the FET used, and N is the bit length to be processed. The log term is derived when using the second-order Booth algorithm and Wallace tree in the array [8] and a carry lookahead adder for the CPA [17]. The cost is proportional to the occupied area S which is normalized by $L \cdot N^2$. Thus, the Q_{54} quality factor (normalized to $N = 54$) is defined by

$$Q_{54} = (L^2 \cdot N^2 \cdot \log [N/2]) /$$
$$(T \cdot S \cdot 4.17 \times 10^3) \quad (\text{s}^{-1}).$$

The factors are plotted in Fig. 10 for various multipliers reported in the literature. The factor increases a little with N as a result of circuit improvement of larger multipliers. The factors of most multipliers reported to date are below 2.0. In contrast, the RST factor is 3.8—well above ratings in other designs. The RST multiplier has an excellent performance/cost ratio.

The regular pattern of the RST multiplier allows us to implement faster and denser designs than achieved by other approaches. The design time was reduced to less than one-fourth (from three months to three weeks) that required for conventional Wallace tree design because little effort was required to wire the 3W, 4W, P, and P_h circuit units, and because optimization and verification of

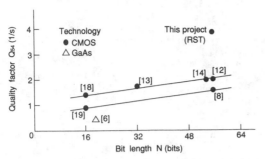

Fig. 10. Quality factors of various multipliers. The numbers in the brackets are references.

the tree circuits was much easier due to the recurring structure in the RST block.

V. CONCLUSION

Because of the highly regular structure of the RST multiplier, a tree multiplier with a logical hierarchy can now be laid out easily with tightly coupled recurring blocks. As a result, the tree multiplier allows a quick design which achieves the fastest and densest architecture to date. It is flexible in the sense that the repeatable $7D_4$ block can be replaced by another block to generate and compress signals of different numbers from the one presented in this paper.

The common conception that the tree multiplier is fast but large and difficult to design is no longer tenable as it is now one of the most promising approaches for implementing high-speed arithmetic units. The architecture is especially suitable for future single-chip supercomputers with strong parallel processing capability.

As this approach is systematic, a module generator based on the RST multiplier will produce multiplier designs quickly, with quality compatible to designs optimized manually.

ACKNOWLEDGMENT

The authors thank H. Okada and M. Sakate for their CPA design contributions, and K. Fujita and M. Kimura for their useful suggestions. We also thank H. Ishikawa, S. Hazama, and S. Mori for their guidance and support throughout this study.

REFERENCES

[1] R. W. Edenfield et al. "The 68040 processor, part 1," IEEE Micro, pp. 66–78, Feb. 1990.
[2] J. H. Crawford, "The i486 CPU: Executing instructions in one clock cycle," IEEE Micro, pp. 27–36, Feb. 1990.
[3] J. Labrousse and G. A. Slavenburg, "A 500 MHz microprocessor with a very long instruction word architecture," in ISSCC Dig. Tech. Papers, Feb. 1990, pp. 44–45.
[4] C. S. Wallace, "A suggestion for fast multipliers," IEEE Trans. Electron. Comput., vol. EC-13, pp. 14–17, Feb. 1964.
[5] L. Korn and S.-W. Fu, "A 1,000,000 transistor microprocessor," in ISSCC Dig. Tech. Papers, Feb. 1989, pp. 54–55.
[6] H. P. Singh et al., "A 6.5-ns GaAs 20 × 20-b parallel multiplier with 67-ps gate delay," IEEE J. Solid-State Circuits, vol. 25, pp. 1226–1231, Oct. 1990.
[7] C. Asato et al. "A data-path multiplier with automatic insertion of pipeline stages," IEEE J. Solid-State Circuits, vol. 25, pp. 383–387, Apr. 1990.
[8] C. C. Stearns and P. H. Ang, "Yet another multiplier architecture," in Proc. IEEE Custom Integrated Circuits Conf. (Boston), May 1990, pp. 24.6.1–24.6.4.
[9] T. Sato et al., "A regularly structured 54-bit modified Wallace tree multiplier," in Proc. IFIP Int. Conf. VLSI (VLSI91) (Edinburgh), Aug. 1991, pp. 1.1.1–1.1.9.
[10] ANSI/IEEE Standard 754—1985 for Binary Floating—Point Arithmetic. Los Alamitos, CA: IEEE Computer Soc., 1985.
[11] A. D. Booth, "A signed binary multiplication technique," Quart. J. Mech. Appl. Math., vol. 4, pp. 236–240, 1951.
[12] E. Hokenek et al., "Second-generation RISC floating point with multiply–add fused," IEEE J. Solid-State Circuits, vol. 25, pp. 1207–1213, Oct. 1990.
[13] M. Nagamatsu et al., "A 15-ns 32 × 32-b CMOS multiplier with an improved parallel structure," IEEE J. Solid-State Circuits, vol. 25, pp. 494–497, Apr. 1990.
[14] J. Mori et al., "A 10-ns 54 × 54-b parallel structured full array multiplier with 0.5-μm CMOS technology," IEEE J. Solid-State Circuits, vol. 26, pp. 600–606, Apr. 1991.
[15] The TTL Databook for Design Engineers, 2nd ed., Texas Instruments Inc., 1976, pp. 7–385.
[16] T. Sato et al., "An 8.5-ns 112-bit transmission gate adder with a conflict-free bypass circuit," in Symp. VLSI Circuits, Dig. Tech. Papers (Oiso, Japan), May 1991, pp. 105–106.
[17] R. Turn, Computers in the 1980s. New York: Columbia University Press, 1974.
[18] R. Sharma et al., "A 6.75 ns single metal CMOS 16 × 16 multiplier IC," in Symp. VLSI Circuits, Dig. Tech. Papers (Tokyo), Aug. 1988, pp. 91–92.
[19] K. Yano et al., "A 3.8-ns CMOS 16 × 16-b multiplier using complementary pass-transistor logic," IEEE J. Solid-State Circuits, vol. 25, pp. 388–395, Apr. 1990.

A 4.4 ns CMOS 54 × 54-b Multiplier Using Pass-Transistor Multiplexer

Norio Ohkubo, Makoto Suzuki, *Member, IEEE,* Toshinobu Shinbo, Toshiaki Yamanaka, *Member, IEEE,*
Akihiro Shimizu, Katsuro Sasaki, *Member, IEEE,* and Yoshinobu Nakagome, *Member, IEEE*

Abstract—A 54 × 54-b multiplier using pass-transistor multiplexers has been fabricated by 0.25 μm CMOS technology. To enhance the speed performance, a new 4-2 compressor and a carry lookahead adder (CLA), both featuring pass-transistor multiplexers, have been developed. The new circuits have a speed advantage over conventional CMOS circuits because the number of critical-path gate stages is minimized due to the high logic functionality of pass-transistor multiplexers. The active size of the 54 × 54-b multiplier is 3.77 × 3.41 mm. The multiplication time is 4.4 ns at a 2.5-V power supply.

I. INTRODUCTION

ENHANCING the performance of floating-point operation is indispensable for current high-performance microprocessors. In particular, high-speed multiplication is becoming increasingly important in RISC's, DSP's, graphics accelerators, and so on, because of increasing demand for multimedia applications. Recent high-end microprocessors call for an operating frequency of 200 MHz or over. Furthermore, a multiplier will be required for single-clock-cycle operation. However, no CMOS 54 × 54-b multiplier with a delay time less than 5 ns has yet been reported [1], [2].

This paper describes a 54 × 54-b multiplier macro developed formantissa multiplication of 2 double-precision numbers, as outlined in the IEEE standard [3]. The target multiplication time is less than 5 ns. To reduce the multiplication time, a new 4-2 compressor and a carry lookahead adder (CLA), both featuring pass-transistor multiplexers, have been developed. The new circuits provide a speed advantage over conventional CMOS circuits because the number of critical-path gate stages is minimized due to the high logic functionality of pass-transistor multiplexers. In addition, power reduction is important in attaining such high performance. For this purpose, we employed 0.25 μm CMOS technology and reduced the supply voltage to 2.5 V.

The architecture of the 54 × 54-b multiplier is described in Section II. In Section III, the circuit design based on Booth's algorithm, as well as the design of the pass-transistor

Manuscript received August 1, 1994; revised 10/21/94.
N. Ohkubo, M. Suzuki, T. Yamanaka, and Y. Nakagome are with the Central Research Laboratory, Hitachi, Ltd., Kokubunji, Tokyo 185, Japan.
T. Shinbo and A. Shimizu are with the Hitachi VLSI Engineering Corporation, Kodaira, Tokyo 187, Japan.
K. Sasaki is with the R&D Division, Hitachi America, Ltd., Brisbane, CA 94005 USA.
IEEE Log Number 9408745.

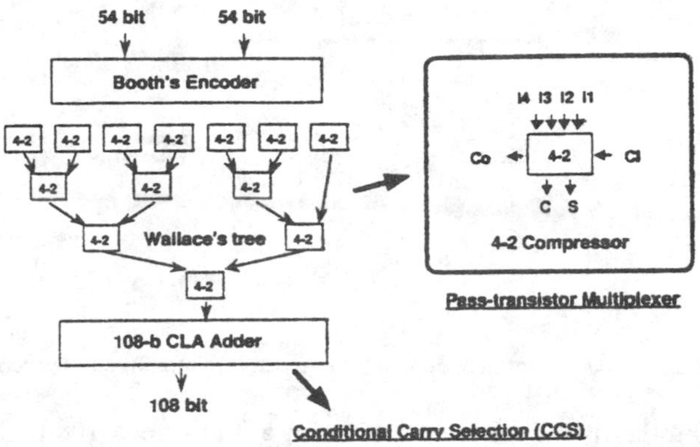

Fig. 1. Block diagram of the 54 × 54-b multiplier usingpass-transistor multiplexers.

multiplexer, 4-2 compressor, and carry lookahead adder are discussed. Section IV describes the fabrication of a test chip. Some experimental results are shown in Section V, and the conclusions are summarized in Section VI.

II. ARCHITECTURE

The block diagram of the 54 × 54-b multiplier is shown in Fig. 1. It employs Booth's algorithm [4], Wallace's tree [5], and a conditional carry-selection (CCS) adder [6]. The number of partial products is halved by Booth's algorithm. The partial products are summed by Wallace's tree, without carry propagation. The summed results are then added by the CCS adder with high-speed carry propagation.

Reducing the delay of Wallace's tree is important in reducing multiplication time, so we used a 4-2 compressor, which has 5 inputs and 3 outputs. The carry-out (Co) is connected to the next higher bit 4-2 compressor's carry-in (Ci), as shown in Fig. 1. Without propagating the carry to a higher bit, the 4-2 compressor can add four partial products (I1-I4) because the carry-out (Co) does not depend on the carry-in (Ci). By using this 4-2 compressor, only four addition stages are needed for Wallace's tree, as shown in Fig. 1. It is known that the 4-2 compressor has a speed advantage over full-adder-based designs, because of the reduced number of addition stages [1], [2]. For further improvement, we have developed a new 4-2 compressor that reduces the critical path gate stages by exploiting the high logic functionality of the pass-transistor multiplexer.

Reprinted from *IEEE Journal of Solid-State Circuits,* Vol. 30, No. 3, pp. 251-256, March 1995.

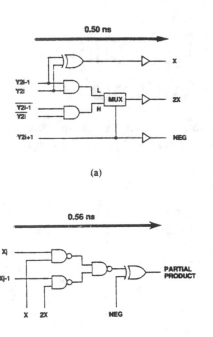

(a)

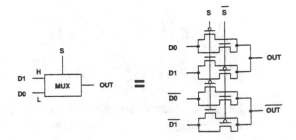

Fig. 3. Pass-transistor multiplexer circuit.

(b)

Fig. 2. Booth's algorithm. (a) Booth's encoder. (b) Partial-product generator.

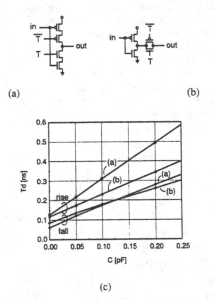

(a)   (b)

(c)

Fig. 4. Comparison of CMOS and pass-transistor logic circuits. (a) CMOS tristate inverter. (b) Pass-transistor tristate inverter. (c) Comparison of delay time.

Furthermore, we have developed a high-speed 108-b CLA adder, which is another important component of the high-speed multiplier. We have already reported a 32-b ALU with a 4-b lookahead carry scheme called conditional carry-selection [6]. To apply this scheme to the final adder of the multiplier, the 4-b CLA was modified into an 8-b CLA.

III. CIRCUIT DESIGN

A. Booth's Algorithm

The purpose of using Booth's algorithm in this design is not to reduce delay time, but to reduce the chip area. If the full-adder-based design is applied in Wallace's tree, the delay time can be shortened by using Booth's algorithm because it reduces the number of addition stages. However, because one addition stage of the 4-2 compressor halves the number of partial products, the extra delay time without Booth's algorithm is only that of one 4-2 compressor. Since the delay time in generating the partial products without Booth's encoder is shorter than that with Booth's encoder, the multiplication time, either with or without Booth's algorithm, are almost the same.

Booth's encoder is shown in Fig. 2(a). The multiplicands, $Y2_{i-1}$, $Y2_i$, and $Y2_{i+1}$, are encoded by this circuit. Encoding the data halves the number of partial products. The simulated propagation delay time is 0.50 ns. The partial-product generator is shown in Fig. 2(b). A multiplier, either X_j or X_{j-1}, is selected depending on whether encoded data, X or $2X$ is high and inverted by encoded data, NEG. The simulated propagation delay time is 0.56 ns.

B. Pass-Transistor Multiplexer

The pass-transistor multiplexer used in the 4-2 compressor and 108-b CLA adder is shown in Fig. 3. When the control signal S is low, data $D0$ is selected, and when the control

signal S is high, data $D1$ is selected. The output is used as the control signal input for the next-stage multiplexer. Thus, the multiplexer has both positive and negative output. It can reduce the propagation delay by eliminating an inverter.

Several pass-transistor logic circuits have been proposed to improve the performance of CMOS circuits. The NMOS pass-transistor logic circuits [7] is one example. It has been shown to result in high speed due to its low input capacitance and high logic functionality. However, particularly in reduced supply voltage designs, it is important to take into account the problems of noise margins and speed degradation. These are caused by mismatches between the input signal levels and the logic threshold voltage of the CMOS gates, which fluctuates with process variations. To avoid these problems, the multiplexer in this design consists of both NMOS and PMOS pass transistors.

The delay time of the pass-transistor multiplexer is shorter than that of a CMOS gate, because of the pass-transistor-based design where both the NMOS and PMOS are turned on. The CMOS tristate inverter and pass-transistor tristate inverter are shown in Fig. 4(a) and (b), which are used in CMOS and pass-transistor multiplexers, respectively. The number of transistors in both circuits is the same, and both have

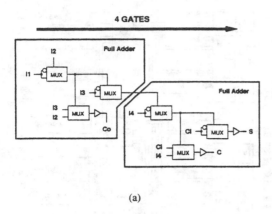

(a)

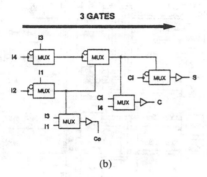

(b)

Fig. 5. 4-2 compressor circuits using pass-transistor multiplexers. (a) Full-adder-based construction. (b) Proposed construction.

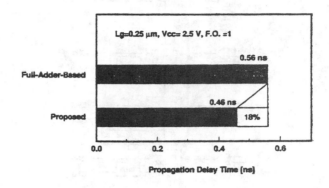

Fig. 6. Simulated comparison of 4-2 compressor circuits.

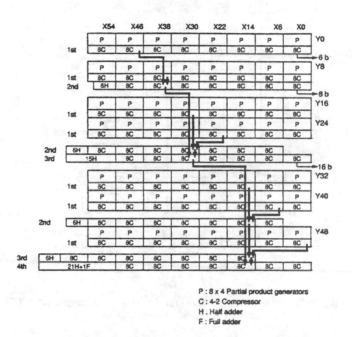

P : 8 x 4 Partial product generators
C : 4-2 Compressor
H : Half adder
F : Full adder

Fig. 7. Construction of Wallace's tree.

equal input capacitance. A simulated comparison is shown in Fig. 4(c), showing the dependence of the delay time from "in" to "out" on the output load capacitance. The low driving-source impedance attained by using the pass-transistor makes the delay time of the pass-transistor shorter than that of a CMOS gate.

C. 4-2 Compressor Circuit

The 4-2 compressor circuits using pass-transistor multiplexers are shown in Fig. 5. The signal lines in the figure represent positive and negative signals. The inputs of the multiplexers are either two different signals, or one signal and logical invert. All outputs (S, C, and Co) have buffers to enhance driving ability. The 4-2 compressor circuits add four partial products (I1–I4) and generate a sum signal (S) and two carry signals (C and Co).

Since the pass-transistor multiplexer circuit shown in Fig. 3 has high logic functionality, a full-adder circuit is constructed from three pass-transistor multiplexers. The 4-2 compressor is constructed from 2 full adders, such that there are 4 critical-path gate stages, as shown in Fig. 5(a). This circuit is faster than conventional CMOS circuits due to the use of pass-transistor multiplexers.

For further speed improvement, we developed a new 4-2 compressor. Though the number of multiplexers is the same, the number of critical-path gate stages in this circuit is reduced to 3 by exploiting parallelism, as shown in Fig. 5(b). In this new configuration, the carry-out (Co) does not depend on the carry-in (Ci), so, the advantage of the 4-2 compressor is maintained with this new configuration. The simulated delay

comparison for these 4-2 compressor circuits is shown in Fig. 6 The proposed circuit reduces the propagation delay time by 18% from that of a full-adder-based circuit.

Wallace's tree, shown in Fig. 7, is constructed from partial-product generators, 4-2 compressors, full adders, and half adders. Using the 4-2 compressor simplifies the construction of Wallace's tree. The Wallace's tree consist of only five kinds of blocks and four kinds of wire shifters. The five kinds of blocks include an 8×4 partial-product generator (P), eight 4-2 compressors ($8C$), six half adders ($6H$), 15 half adders ($15H$), and 21 half adders and a full adder ($21H + 1F$). The four kinds of wire shifters include an 8-b right shifter, an 8-b left shifter, a 16-b right shifter, and a 16-b left shifter. The sum and carry signals of 4-2 compressors are shifted by the wire shifters. In addition, the carry signals are shifted to the left one more bit.

D. Conditional Carry-Selection Circuit

A number of fast-adder architectures have been proposed [8]–[14], some of which use pass-transistor logic circuits for carry propagation [13], [14]. These circuits gain their

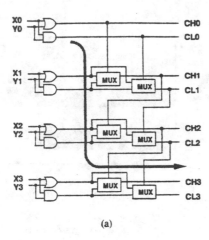

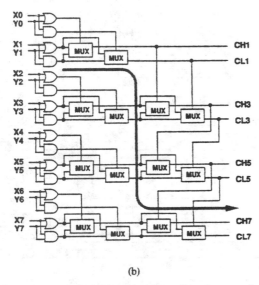

(b)

Fig. 8. Carry lookahead circuits. (a) 4-b conditional carry-select (CCS) circuit. (b) 8-b modified conditional carry-select circuit.

speed advantage over CMOS circuits due to their high logic functionality. However, a problem with this architecture is the serial connection of the pass transistors in the carry propagation path.

We have already reported a new look ahead carry scheme called conditional carry-selection (CCS) [6]. The 4-b CLA is constructed so as to have three critical-path gate stages, as shown in Fig. 8(a). In the CCS architecture, conditional carry signals for each bit, CL_j (assuming an incoming group-carry of "0") or CH_j (assuming an incoming group-carry of "1"), are selected by the multiplexers depending on the conditional carry signals of the previous bit, CL_{j-1} or CH_{j-1}, as expressed by

$$CL_j = G_j + P_j \cdot C_{j-1} = G_j + P_j \cdot 0$$
$$= G_j = X_j \cdot Y_j \tag{1}$$
$$CH_j = G_j + P_j \cdot C_{j-1} = G_j + P_j \cdot 1$$
$$= G_j + P_j = X_j + Y_j \tag{2}$$

where G_j is carry generation and P_j is carry propagation. The CCS adder is faster than the conventional pass-transistor-based design because the pass-transistors are not serially connected

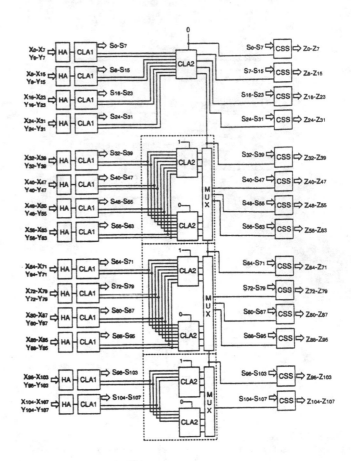

Fig. 9. Construction of 108-b adder.

in the carry propagation path. This is because the multiplexer's output is directly connected to the next multiplexer's control signal input.

To apply the CCS scheme to the final 108-b adder, the 4-b CLA is modified into an 8-b CLA, as shown in Fig. 8(b). In the modified CCS architecture, conditional carry signals for each bit, CL_j or CH_j, are selected by the multiplexers depending on the conditional carry signals of the two previous bits, CL_{j-2} or CH_{j-2}, as expressed by

$$CL_j = G_j + P_j \cdot C_{j-1} = G_j + P_j \cdot (G_{j-1} + P_{j-1} \cdot 0)$$
$$= G_j = X_j \cdot Y_j, \quad \text{if } G_{j-1} = 0 \tag{3}$$
$$= G_j + P_j = X_j + Y_j, \quad \text{if } G_{j-1} = 1 \tag{4}$$
$$CH_j = G_j + P_j \cdot C_{j-1} = G_j + P_j \cdot (G_{j-1} + P_{j-1} \cdot 1)$$
$$= G_j = X_j \cdot Y_j, \quad \text{if } G_{j-1} + P_{j-1} = 0 \tag{5}$$
$$= G_j + P_j = X_j + Y_j, \quad \text{if } G_{j-1} + P_{j-1} = 0. \tag{6}$$

In this configuration, there are several series-connected pass-transistors, as shown in Fig. 8(b). However, the critical path of addition in the multiplier has no series-connected pass-transistors. This critical path is created because the $X0$ in the critical path of multiplier arrives later than the other bits. The new 8-b CLA has only four critical-path gate stages because it makes use of parallelism. It reduces the 108-b addition time to 1.52 ns from 1.82 ns.

Fig. 9 shows a construction of the 108-b adder based on carry-selected architecture [10]. DPL gates [6] are used for the

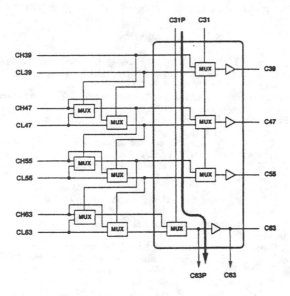

Fig. 10. Block carry lookahead circuit CLA2.

TABLE I
PROCESS TECHNOLOGY

Technology	0.25-μm CMOS Triple metal
Gate length	0.25 μm
Gate oxide	6.5 nm
1st Metal W/S	0.5 μm / 0.4 μm
2nd Metal W/S	0.5 μm / 0.4 μm
3rd Metal W/S	0.7 μm / 0.6 μm

TABLE II
CHARACTERISTICS OF THE 54 × 54-BIT MULTIPLIER

Organization	54 x 54-b
Delay	4.4 ns
Active Area	3.77 X 3.41 mm
Transistors	100,200
Supply Voltage	2.5 V

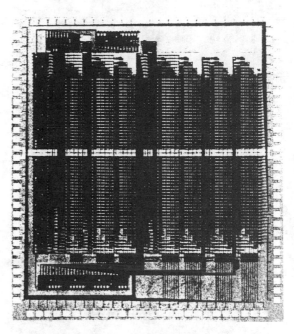

Fig. 11. Micrograph of the 54 × 54-b multiplier.

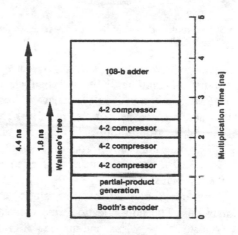

Fig. 12. Multiplication time of the 54 × 54-b multiplier.

half adders (HA) circuits. The conditional-sum selection (CSS) circuit consists of a multiplexer which selects the conditional sums, $S_j(0)$ or $S_j(1)$, according to the incoming block carry signal. The CCS architecture is applied not only to the 8-b carry lookahead circuit CLA1, but also to the block carry look-ahead circuit CLA2, as shown in Fig. 10. The block carry signals are generated by CLA2 assuming the carry of the lower block carry signals to be 0 or 1, and are then selected by the multiplexer according to the incoming true carry. To shorten the carry propagation delay, the multiplexer of the critical carry propagation path is separated from the other multiplexers, as shown in Fig. 10. This architecture enhances parallelism and results in fast operation, because the carry signals of the upper 32 b are calculated in parallel with those of the lower 32 b, and the carry signals of the upper 32 b are generated after the delay time of a single multiplexer.

IV. FABRICATION

A 54 × 54-b multiplier test chip was fabricated by triple-metal 0.25-μm CMOS technology. The major process parameters are summarized in Table I. The first metal is tungsten, and the second and third metals are aluminum. It operates on a supply voltage of 2.5 V. Fig. 11 shows a micrograph of the chip. The top and right-hand side are the input sections, Booth's encoder is also positioned on the right-hand side. The final adder and the output are at the bottom. 100 200 transistors are integrated in an active area of 3.77 × 3.41 mm. The area is mostly occupied by the 4-2 compressors, the partial-product generators, and wire shifters.

V. EVALUATION

The simulated multiplication time of the 54 × 54-b multiplier is shown in Fig. 12. It is only 4.4 ns with a 2.5-V power supply and typical process at room temperature. It

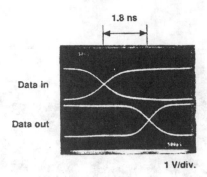

1.8 ns

Data in

Data out

1 V/div.

Fig. 13. Measured waveforms of Wallace's tree.

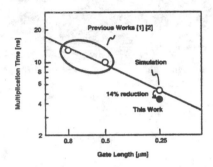

Fig. 14. 54 × 54-b multiplication time versus device dimensions.

shows excellent characteristics at such a low voltage because of the pass-transistor-multiplexer-based design where both the NMOS and PMOS are tuned on. The new 4-2 compressor reduces the Wallace's tree delay time to 1.8 ns. The characteristics of this 54 × 54-b multiplier test chip are summarized in Table II. The measured waveforms of Wallace's tree are shown in Fig. 13. The measured delay was almost the same as that of the simulated delay.

Fig. 14 shows the multiplication time plotted against the device dimensions. The multiplication time without the new circuits is estimated to be 5.1 ns, and the area of the multiplier does not increase significantly with the new circuits. Therefore, this multiplier achieves 14% improvement in multiplication time with the use of the new circuits having pass-transistor multiplexers.

VI. Conclusion

A new 4-2 compressor and CLA using pass-transistor multiplexers have been developed to shorten multiplication time. The multiplication time of the 54 × 54-b multiplier is reduced by 14% due to the reduction in the number of critical-path gate stages resulting from the use of pass-transistor multiplexers. A 4.4-ns multiplication time was achieved with 0.25-μm CMOS technology.

Acknowledgment

The authors wish to thank Dr. E. Takeda, Dr. K. Shimohigashi, Dr. T. Nishimukai, and Dr. T. Nagano for their useful discussions, and K. Ueda and K. Takasugi for their assistance and support. The authors are also greatly indebted to T. Nishida, N. Hashimoto, N. Ohki, and H. Ishida for their assistance with the process technology and device fabrication.

References

[1] G. Goto et al., "A 54- × 54-bit regularly structured tree multiplier," IEEE J. Solid-State Circuits, vol. 27, pp. 1229–1236, Sept. 1992.
[2] J. Mori et al., "A 10-ns 54- × 54-bit parallel structured full array multiplier with 0.5-μm CMOS technology," IEEE J. Solid-State Circuits, vol. 26, pp. 600–606, Apr. 1991.
[3] Binary Floating-Point Arithmetic, ANSI/IEEE Standard 754-1985, IEEE Computer Society, Los Alamitos, CA, 1985.
[4] A. D. Booth, "A signed binary multiplication technique," Quart. J. Mech. Appl. Math., vol. 4, pp. 236–240, 1951.
[5] C. S. Wallace, "A suggestion for a fast multiplier," IEEE Trans. Electron. Comput., vol. EC-13, pp. 14–17, Feb. 1964.
[6] M. Suzuki et al., "A 1.5-ns 32-b CMOS ALU in double pass-transistor logic," IEEE J. Solid-State Circuits, vol. 28, pp. 1145–1151, Nov. 1993.
[7] K. Yano et al., "A 3.8-ns CMOS 16 × 16-bit multiplier using complementary pass-transistor logic," IEEE J. Solid-State Circuits, vol. 25, pp. 388–395, Apr. 1990.
[8] J. Sklansky, "An evaluation of several two-summand binary adders," IRE Trans. Electron. Comput., vol. EC-9, pp. 213–226, June 1960.
[9] ____, "Conditional-sum addition logic," IRE Trans. Electron. Comput., vol. EC-9, pp. 226–231, June 1960.
[10] O. J. Bedrij, "Carry-select adder," IRE Trans. Electron. Comput., vol. EC-11, pp. 340–346, June 1962.
[11] C. L. Chen, "2.5-V bipolar/CMOS circuits for 0.25-μm BiCMOS technology," IEEE J. Solid-State Circuits, vol. 27, no. 4, pp. 485–491, Apr. 1992.
[12] K. Yano et al., "3.3-V BiCMOS circuit techniques for 250-MHz RISC arithmetic modules," IEEE J. Solid-State Circuits, vol. 27, pp. 373–381, Mar. 1992.
[13] H. Hara et al., "0.5-μm 3.3-V BiCMOS standard cells with 32-kilobyte cache and ten-port register file," IEEE J. Solid-State Circuits, vol. 27, pp. 1579–1584, Nov. 1992.
[14] A. Rothermel et al., "Realization of transmission-gate conditional-sum (TGCS) adders with low latency time," IEEE J. Solid-State Circuits, vol. 24, pp. 558–561, June 1989.

A Method for Speed Optimized Partial Product Reduction and Generation of Fast Parallel Multipliers Using an Algorithmic Approach

VOJIN G. OKLOBDZIJA, FELLOW, IEEE, DAVID VILLEGER, AND SIMON S. LIU

Abstract—This paper presents a method and an algorithm for generation of a parallel multiplier, which is optimized for speed. This method is applicable to any multiplier size and adaptable to any technology for which speed parameters are known. Most importantly, it is easy to incorporate this method in silicon compilation or logic synthesis tools. The parallel multiplier produced by the proposed method outperforms other schemes used for comparison in our experiment. It uses the minimal number of cells in the partial product reduction tree. These findings are tested on design examples simulated in 1μ CMOS ASIC technology.

Index Terms—Parallel multiplier, partial product reduction, Wallace tree, Dadda's counter, VLSI arithmetic, Booth encoding, 3:2 counter, 4:2 adder, array multiplier.

---

1 INTRODUCTION

THE increased level of integration brought by modern VLSI and ULSI technology has rendered possible the integration of many components that were considered complex and were implemented only partially or not at all in the past. The multiplication operation is certainly present in many parts of a digital system or digital computer, most notably in signal processing, graphics and scientific computation. Therefore, it became quite common to see a multiplier implemented in full in many parts where it was not found before. Examples of such are floating-point processors and recently graphics processor, various kinds of digital signal processors used for user interfaces, communication or code compression. Parallel multipliers have even migrated into the fixed-point processor of digital computers for the purpose of speeding up and facilitating address calculation needed for fast and efficient indexing through arrays of data. The speed of the parallel multiplier has always been a critical issue and, therefore, the subject of many research projects and papers [1], [2].

Several popular and well-known schemes, with the objective of improving the speed of the parallel multiplier, have been developed in the past. The first departure from the iterative array structure [3] has been described in a paper by Wallace [7]. Wallace has introduced a notion of a carry-save tree constructed from one-bit Full Adders as a way of reducing the number of partial product bits in a fast and efficient way. The notion of counters and a generaliza-tion of the Wallace scheme have been described in the pa-per by Dadda [8] who also proposed a method that mini-mized the number of counters in a compression tree. A good survey of several possible schemes based on Dadda's method can be found in the paper by Stenzel [9].

In 1981, Weinberger [10] introduced a 4:2 compressor as a way of reducing the bits in the parallel multiplier array. His compressor was used by Santoro [12], Nagamatsu et al. [13], and Mori et al. [14]. The introduction of the 4:2 com-pressor (as an alternative to counters) was a departure from the traditional path which resulted in an improvement over the traditionally used Wallace and Dadda's scheme.

The use of larger compressors and families of compres-sors was explored by Song and DeMichelli [15]. They have also developed a 9:2 compressor and made a comparison with respect to their implementation and layout.

Use of higher order compressors yielded mixed results in some cases, however, it showed a general trend toward building compressors of larger sizes as a way of making incremental improvements in multiplier speed. In our re-search we took the approach of generating the compressors of maximal possible size, (i.e., the size of the multiplier) which yielded better results than the use of any previous compressor or counter family. Therefore, we gradually abandoned the notion of levels and undertook a design of an optimized one-level compressor which evolved into an optimization process involving the entire array.

Our method is based on the fact that not all inputs and outputs from a device used as a compressor (or counter) contribute equally to the delay. Therefore, we sort them in a way which favors the use of fast inputs and outputs in the paths that are critical to the speed while we assign slow inputs to the signal paths which belong to the domain where an increase in the delay is tolerable. In the creation process we examine the entire multiplier array and all the signals that are entering the compressor which lead to a

- V.G. Oklobdzija is with Integration, Berkeley, California, and with the Department of Electrical and Computer Engineering, University of California, Davis, CA 95616. E-mail: vojin@ece.ucdavis.edu.
- D. Villeger is with Ecole Superieure d'Ingenieurs en Electrotechnique et Electronique, 93162 Noisy le Grand Cedex, France
- S.S. Liu is with Advanced Micro Devices, Sunnyvale, CA 94088-3453.

Manuscript received Oct. 21, 1993; revised May 19, 1994.
For information on obtaining reprints of this article, please send e-mail to: transactions@computer.org, and reference IEEECS Log Number C96005.

Reprinted from *IEEE Transactions on Computers*, Vol. 45, No. 3, pp. 294-305, March 1996.

global rather than local optimization. This is characterized by the use of a particular compressor optimized for a minimal delay with respect to its inputs and outputs only. Our method also leads to a general solution for any chosen device (compressor or counter). By being used interactively in conjunction with improvement of a basic compressor device, this leads to an array which is optimized for speed down to the device level.

Although Booth recoding [4], [5] is widely used in parallel multipliers, it does not change the structure of the reduction tree. Moreover, its efficiency has been denied by several authors which was consistent with our findings [6], [11, [15]. As it has been shown in [6], one row of 4:2 compressors has the same effect as Booth encoding, which is achieved in less time. Given that the method presented here achieves this compression in even less time than the one using 4:2 compressors, the use of Booth encoding in parallel multipliers ceases to be advantageous for the range of technologies discussed in this paper.

We very recently became aware of an effort to build multipliers automatically via a program which attempts to compensate for the different speed of different paths in the multiplier by by-passing some stages [16]. While the approach taken is correct, which basically abandons the idea of counters or compressors, it does not take into account that different inputs (not only outputs) have different contribution to speed, as does our method. Unlike our method, half adders are used extensively in [16] which leads to worse results in terms of speed and cell number. Moreover, that method is less general and does not allow extension to device optimization. Similar approaches have also been applied in the process of "timing optimization" used in "logic synthesis," most notably in [28] and [29].

In the last section of this paper, we consider the final carry-propagate adder used to sum the partial products which have been reduced to two rows. The Final Adder delay is an addendum to the multiplier delay and, therefore, is critical. The approaches taken in designing the Final Adder were mainly concentrated on the raw speed of the adder and did not consider the specifics related to the uneven signal arrival profile of its inputs. Our selection of the Final Adder is based on these specifics. The structure of the Final Adder is *tuned* into the signal arrival profile so that the delay of the adder is minimized under these conditions which are specific to its application. We conclude that *tuning* of the Final Adder is more important than the its raw speed and that any application of a complex and hardware consuming scheme which is not optimized with respect to the signal arrival, would only constitute a waste of resources.

Finally, in the examples of various multipliers designed in 1.0μ CMOS ASIC technology, we show the advantages of our method and how it compares to the others.

2 COMPARISON OF DIFFERENT PARTIAL PRODUCT REDUCTION SCHEMES

2.1 Wallace and Dadda Schemes

A major departure from iterative array realization of parallel multipliers has been introduced in papers by Wallace [7] and Dadda [8]. Common to both is the use of Carry-Save form in a compression tree in order to reduce the number of partial product bits to two rows. The advantage of trees is that their speed increases with the *log* of the operand length, while this increase is linear in the case of iterative arrays. Dadda introduced the notion of *counters* and a methodology of designing trees that are optimized in terms of cell number. Thus, Dadda has treated the bit reduction process as a number of *levels* (steps), the application of which results in the reduction of the number of rows of partial products. At every level the partial product bits are passed through devices designated as *counters*, the purpose of which has been the reduction of the number of rows after a pass through a level. A (p, q) counter is defined as a combinatorial network that determines the number of ones (active signals) among its p inputs producing a result on its q bit output in the form of a binary number (count). Essential to the counter is a process of summation of the input bits. The number of output bits must be sufficient to represent all the possible sums of n bits: $2^q - 1 \geq p$. A Full Adder can be treated as a $(3, 2)$ counter, leading to a representation of the Wallace tree as a special case of Dadda. It would be ideal if it were possible to develop a higher order counter such as $(7, 3)$ or $(15, 4)$ with a speed that surpasses the speed of the same counters built by using Full Adders. There were several studies undertaken in the past, most notably by Stenzel [9], however, the use of a Full Adder as a $(3, 2)$ counter is still the most prevailing.

The process applied by Wallace and Dadda can be summarized as follows: after generating the partial products, a set of counters reduces the partial product matrix but does not propagate the carries. The resulting matrix is composed of the sums and the carries of the counters. Another set of counters then reduces this new matrix and so on, until a two-row matrix is generated. Those two rows are summed up with a Final Adder to which we will refer to as a Carry Propagate Adder (CPA). This method takes advantage of the carry save form to avoid the carry propagation until the Final Adder. In this scheme the number of levels is crucial and will determine the speed of the multiplier.

2.2 Use of 4:2 and Higher Order Compressor

A major departure from the Wallace-Dadda scheme has been the introduction of a 4:2 compressor by Weinberger of IBM [10]. His idea was made visible by Santoro who used it in his PhD thesis [11] and by Mori et al. [14] who later implemented it. The 4:2 compressor, as shown in Fig. 3, made a major difference in the way cells are interconnected by introducing a *horizontal* path and, therefore, a limited propagation of the carry signal in the multiplier. Indeed 4:2 structure is not a counter, since two output bits cannot represent five possible sums of 4 bits. Thus, a carry out is necessary and subsequently a carry in. However, since the carry out is not dependent on the carry in, only a limited carry propagation occurs. There were several benefits from using a 4:2 compressor, such as simpler and more regular wiring of the multiplier tree. More notable is the introduction of the notion of a *horizontal* and a *vertical* signal path. Naturally application of 4:2 compressors led to faster realization of a multiplier. The existence of a *horizontal* path has led to the idea of moving the *critical path* of the multiplier tree away from the center toward the most significant end where the depth of the column of partial product bits is smaller than in the middle [18]. The exploration of this idea has brought some mixed results, mainly because a major redesign of the compressors was necessary. However, it has shed enough light on the real problem of the reduction of the partial product bits.

Naturally, researchers tried to explore the idea of 4:2 compressors further and research in this direction, most notably by Song and DeMichelli [15] led to the introduction of the 6:2 and 9:2 family. The term family is being used because the new compressors introduced were simply built on 4:2 compressors and (3, 2) counters. For example, the 9:2 compressor consists of one 6:2 compressor and three (3, 2) counters, where in turn, the 6:2 compressor contains one 4:2 compressor and two (3, 2) counters. Nevertheless, a multiplier built using 9:2 compressors showed a speed advantage over one built using 4:2 compressors.

The speed comparison for three different multipliers using, respectively, (3, 2) counters, 4:2 and 9:2 compressors for the multiplier sizes ranging from 0 to 100 bits in equivalent XOR delays in the *critical* path is shown in Fig. 1. We redesigned the 4:2 and 9:2 compressors at the gate level in order to obtain the best possible performance. The difference in speed favors 9:2 over 4:2 and (3, 2). These results are in agreement with findings in [11], [13], and [15]. We use a normalized XOR delay as a measure of speed for a particular implementation of an algorithm or a method. There are several reasons for doing this:

1) the critical path in the multiplier consists of a path through a series of XOR gates independent the algorithm is used.
2) the speed comparison is made independent on the technology, which is characterized by its ability to realize a fast XOR function.

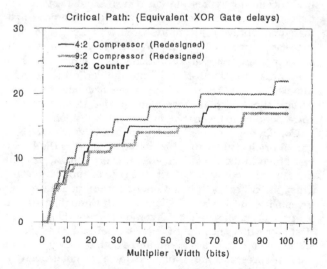

Fig. 1. Speed comparison for three different compressors (counters) used for partial product reduction: 9:2 is the best, followed by 4:2, and (3, 2).

In our early work, we followed this idea and came to the conclusion that the most efficient way to build a compressor is to start with the size of the multiplier (N), in other words to use the largest applicable size [18], [20]. The multipliers built in such a way resulted in better speed than those for which compressors of a smaller size were used [17], [18]. Though we could not claim that compressors built in such a way were optimal and would lead to the fastest realization, we identified the path to the solution. This path led us to the point where we were able to realize that the real problem is not in application or existence of a different counter or a compressor, but in the way the compressor tree has been interconnected. As we will see in the next section, any compressor can be designed with Full Adders with the speed of one designed at the gate level. Simply speaking, inside of each applied compressor, there is a Full Adder used as a building block. If that is the case, then a question arises to what is the difference? The answer is that the difference in using 4:2 and higher order compressors is not in the structure of the compressor but in the way they were interconnected.

2.3 Unequal Delays in a Full Adder: Existence of Fast Inputs and Fast Outputs

It is known that the delay from an input to an output in a Full Adder is not the same. This delay is even dependent on a particular transition (0-to-1, 1-to-0). It is also possible to come up with different realizations of a Full Adder where a specific signal path is favored with respect to the others and is designed in such a way that a signal propagation of this path takes a minimal amount of time. This is sometimes done even at the expense of other possible signal paths. For example, a ripple carry adder is designed so that the carry-in to carry-out delay is minimized. In that respect let us analyze a particular implementation of a (3, 2) counter as shown in Fig. 2. This particular case is taken from LSI 100K 1μ CMOS-ASIC cell [30]. It is used only for the illustration of the algorithm. In the case of a parallel multiplier, our design objective would be to minimize the delay from the *Input s* to the *Sum*, of the Full Adder which has direct effect on the *critical path*.

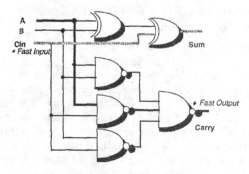

Fig.2. Signal delays in a Full Adder ((3, 2) counter).

In this example, the delay from input A or B to the Sum is equal to two equivalent XOR delays. The delay for the path from Cin to the output Sum is equal to one XOR

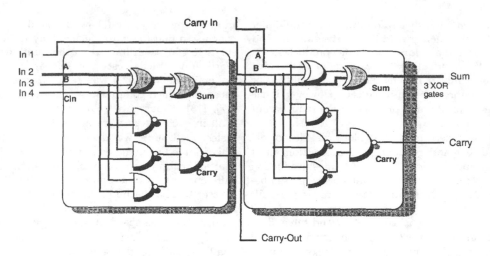

Fig. 3. Modified 4:2 compressor obtained by optimally interconnecting two Full Adders with fast input and output.

equivalent delay. We define Cin as a *fast input*. For this case, the propagation delay from A or B to the output Sum is twice as long as the propagation delay from input Cin to the Sum output. Considering the delay at the output Sum, in this particular technology delay from input A or B to Sum output is equivalent to two XOR delays. However, delay from inputs (A, B, or Cin) to the output Carry is equivalent to one XOR delay. We define Carry as a *fast output*. The value of those delays varies with technology and particular circuit implementation. In general we can use any (noninteger) values in our algorithm.

If we were to construct a 4:2 compressor by simply stacking two Full Adders together, as done by Santoro [11] the *critical path* of such a counter would be equal to four equivalent XOR delays. The researchers from Toshiba have simply redesigned the entire 4:2 compressor and treated it as a single cell. Their design resulted in three equivalent XOR gate delays and, therefore, they claimed 25% speed improvement over a conventional Full Adder Wallace tree realization.

In the next section, we will show how with proper interconnection the same speed can be achieved by using two regular Full Adders.

2.4 Improved 4:2 Compressor with Optimized Interconnections

The advantage of proper interconnection of *fast inputs* and *fast outputs* can be illustrated in the example of a 4:2 compressor. Fig. 3. shows an optimized 4:2 cell which is a result of applying our delay model and properly interconnecting *fast inputs* and *fast outputs* with the objective of minimizing the *critical path* of the 4:2 compressor. In our case, the delay through a 4:2 compressor level is equivalent to three XOR gate delays regardless of the path. If used to reduce the partial product bits in an 24 × 24-bit multiplier, the application of this modified 4:2 compressor would result in a delay of 12 equivalent XOR gates versus 16 if a regular 4:2 compressor, as used by Santoro, were applied (this numbers become 11 and 14, respectively, if a level of Full Adders is used every time the column size is three).

The use of the 4:2 compressor permits the reduction of the vertical critical path while the path involving the carry propagation, that we call *horizontal* path, is not changed. However, the *horizontal* propagation is fast and limited to 1 bit per level.

3 THREE DIMENSIONAL REDUCTION METHOD

3.1 A New Approach

Instead of developing an efficient compressor structure and then using it in the process of partial product reduction, we took a global approach. In our method, we treat the entire multiplier array as a whole. The compressor consists of a vertical slice where the partial product bit array is represented in space and time (reduction steps). The vertical cross section in our representation (Fig. 4.), represents the partial product compressor of the multiplier array. The vertical slice in this representation is further interleaved with the compressors that are used in the reduction process and, therefore, represents a compressor structure corresponding to the appropriate bit position. We will refer to it as a Vertical Compressor Slice or VCS. It should be noted that there are a number of input signals into the vertical slice and a number of output signals originating from the particular vertical slice which are then being passed to the next VCS corresponding to the first higher order bit position.

Considering just one VCS we can see how the matrix of partial products is reduced by a tree of counters (Full Adders shown in the Fig. 4). However, every Full Adder produces a carry out which affects the slice of superior weight. Thus, the critical path is not only a *vertical* path through a given slice, but is also a *horizontal* path through the slices. As previously shown, the 4:2 compressor shortens the vertical path while including the horizontal path. The goal of the following scheme is to minimize both paths by building vertical slices that are optimized for a minimal delay. A method of designing VCS by considering both *vertical* and *horizontal* critical paths will be discussed.

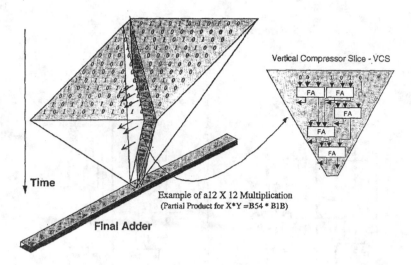

Time

Final Adder

Vertical Compressor Slice - VCS

Example of a 12 X 12 Multiplication
(Partial Product for X*Y =B54 * B1B)

Fig. 4. A three-dimensional view of partial product reduction.

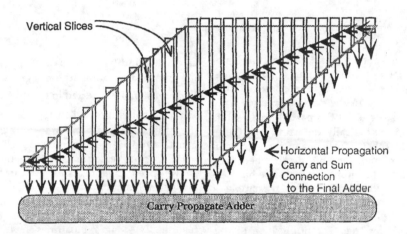

Vertical Slices

Horizontal Propagation

Carry and Sum
Connection
to the Final Adder

Carry Propagate Adder

Fig. 5. The partial products matrix is divided into vertical compression slices.

The reduction of partial product bits in this new scheme is performed by creating a bit slice compressor VCS whose size is equal to the size of a given vertical cross-section of the partial product matrix, and then by assembling those compressors into an integral structure, as illustrated in Fig. 5. Since VCS are optimized by taking the neighboring VCS and their signals into account, as a contiguous process, the optimization is truly a three dimensional optimization process. We will refer to it as a Three Dimensional Minimization (TDM).

This approach is very different from Dadda's since all the partial products are compressed into a single step. This means that no intermediate partial products are considered in our case. However, TDM still produces a carry save number to be translated to a conventional form with a fast carry propagate adder (CPA). Indeed, each VCS reduces the partial products to a Sum and a Carry. Since it does not use the carry save form for the intermediate partial products, this scheme involves a carry propagation through the VCS. Therefore, it is necessary to design the

VCS such that this propagation does not introduce a long delay. The main concern becomes the minimization of the *vertical* and *horizontal* critical paths rather than the number of Full Adder levels.

In addition, this scheme simplifies the design of the multiplier and its description in a hardware description language (VHDL). It also leads to an automatic generation of the partial product array by producing a net list of its signals and components. However, its efficiency will depend on the features of the VCS. The next section introduces a method of designing the VCS and subsequently the whole tree with Full Adders and half adders. This method, which automatically generates a net-list of the partial product array has been implemented in C language.

3.2 Method

The basic idea of this method is to make proper connections globally so that the delay throughout each path is approximately the same. The long delay path originating from the previous compressor should be connected to the short

Example of Delay Optimization

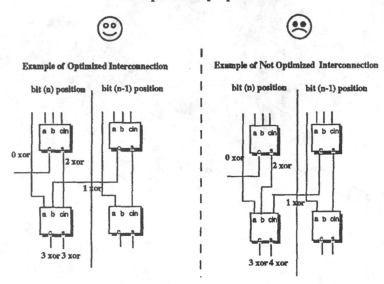

Fig. 6. Example of delay minimization by proper ordering of the input signals.

delay path of the next one, and so on. In general, this is not always possible since each output function has its unique characteristics and requires specific logic cells in its path. It is feasible to apply this idea to the partial product array using Full Adders (FA) since all of the partial product bits in the same bit position are logically the same and, therefore, interchangeable. In other words, all of the signals in any bit position can be interchanged no matter where they are originally coming from (as inputs in the same bit position, or as carries from a lower bit position). An example of how speed improvement can be achieved by application of this principle is shown in Fig. 6. The picture represents a small section of a multiplier tree. The application of this same principle resulted in the optimized 4:2 compressor shown in Fig. 3.

The presented method first creates a data structure consisting of $2N - 1$ lists L_i containing names and delays of partial products. Each VCS is represented by a list of pairs $<d_j, n_j>_i$ containing delay and name information of a node. Initially, L_i represents the inputs (names and delays) of the corresponding VCS. Consequently, its length is the number of partial products from the corresponding slice and the delays are the delays produced by the partial product generator. Initially, we assume that all of the partial product bits are generated at the same time and we chose this to be a reference point by initializing all d_j to 0. If the partial products do not arrive at the same time, d_j will be assigned corresponding delay values. Some partial products do not need to be summed in the tree, and they are directly connected to the CPA.

After sorting the elements of L_i in ascending order by the values of the delays contained in the records of the list, a FA is connected to the first three nodes of L_i. The third node, that is the slowest one, is connected to the fast input (Carry-In) of the FA. The delays of the Sum and Carry are

calculated and two new pairs $<d_j, n_j>_i$ are created containing information about those signals. The pair concerning the Sum signal is inserted in L_i while the one concerning the Carry signal is inserted in L_{i+1}. The size of L_i and L_{i+1} are then adjusted. This same procedure can be applied for any general type of (p, q) compressor cell used. The use of such a (p, q) compressor is advantageous only if such a cell shows speed advantages over FA in the particular technology of implementation.

The process stops when the size of L_i reaches three. The last three signals are then connected to a FA whose signals feed the CPA. Fig. 7 illustrates the effect of this method on different arrangements of signals that have different delays.

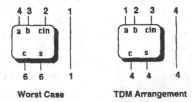

Worst Case   **TDM Arrangement**

Fig. 7. Delay improvement with a different signal arrangement.

The delays of the Sum and Carry signal of the FA are calculated with the following equations:

$$\text{Delay(S)}$$
$$= \text{MAX} \{\text{Delay(A)} + D_{A-S}, \text{Delay(B)} + D_{B-S}, \text{Delay(C}_{in}) + D_{Cin-S}\}$$

$$\text{Delay(C)}$$
$$= \text{MAX} \{\text{Delay(A)} + D_{A-C}, \text{Delay(B)} + D_{B-C}, \text{Delay(C}_{in}) + D_{Cin-C}\}$$

where Delay(X) represents the delay attached to the signal X and the constants D_{u-v} represent the delays of a signal crossing a FA from u to v. It should be noted that delays D_{u-v} can have any value. Those values are determined by the technology of implementation and circuit techniques used.

It is not always possible to use only FA in our design. In some cases, the use of a HA is necessary. The following demonstration shows that this use depends on the number of inputs of a VCS.

Let I_{CN} be the number of signals resulting from the application of CN counters. I_0 is the number of inputs of a VCS. Consequently, I_0 is the number of partial products at this particular bit position plus the number of carry signals originating from the previous VCS. It is not difficult to show that $I_{CN} = I_0 - CNR$ where R is defined as the difference between the number of inputs of the compressor and the number of Sum signals. In our case using FAs, R = 2 and $I_{CN} = 1$ (representing the Sum signal from the VCS). Then, $CN = (I_0 - 1)/2$. Since CN must be an integer, this expression is true if I_0 is odd. If I_0 is even, the number of Full Adders used to produce two signals is $CN = (I_0 - 2)/2$ which is an integer where the remaining two signals are reduced using a HA to produce a Sum and a Carry. Therefore, the number of cells can be expressed as:

$$CN_i = \lfloor I_i/2 \rfloor$$

In other words, a HA will be used if the number of inputs of a VCS is even. This HA is positioned near the partial product generator because carries that are generated at this position propagate through several slices. By taking advantage of the small carry delay generated by a HA, this positioning results in the gain of one XOR delay in the critical path of the multiplier.

The method, which generates the parallel multiplier bit compression array structure, is presented below. This method can be used as a basis for a program which generates a logic file containing the interconnection list as was done in our case. Such a program can be easily integrated into a silicon compiler or logic synthesis tool used for automatic generation of fast and efficient multiplier structure of any size.

3.3 Algorithm for Automatic Generation of Partial Product Array

```
Initialize:
    Form 2N - 1 lists L_i (i = 0, 2N - 2) each
        consisting of p_i elements where:
    p_i = I + 1 for i ≤ N - 1 and p_i = 2N - 1 - i
        for i ≥ N

    An element of a list L_i (j = 0, ..., p_{i-1}) is
        a pair: <d_j, n_j>_i where:

    nj : is a unique node identifying name
    d_j : is a delay associated with that node
        representing a delay of a signal arriving
        to the node n_j with respect to some
        reference point.

    For i = 0, 1 and 2N - 2: connect nodes from
```

```
    the corresponding lists L_i directly to
    the CPA;
For I = 2 to I = 2N - 3 {Partial Product
Array Generation}
    Begin For
        if length of L_i is even Then
            Begin If
                sort the elements of L_i in
                    ascending order by the values of
                    delay d_j;
                connect an HA to the first 2
                    elements of L_i starting with the
                    slowest input;
                    Ds = max {d_A+d_{A-s}, d_B+d_{B-s}}
                    Dc = max {d_A+d_{A-c}, d_B+d_{B-c}}
                remove 2 elements from L_i;
                insert the pair <Ds, NetName> into L_i;
                insert the pair <Dc, NetName> into L_{i+1};
                decrement the length of L_i;
                increment the length of L_{i+1};
            End If;

        while length of L_i > 3

            Begin While
                sort the elements of L_i in ascending
                    order by the values of delay d_j;
                connect an FA to the first 3 elements
                    of L_i starting with the slowest
                    input of the FA:
                    Ds = max {dc_A+dc_{A-s}, dc_B+dc_{B-s},
                        dc_{Ci}+dc_{Ci-s}};
                    Dc = max {dc_A+dc_{A-c}, dc_B+dc_{B-c},
                        dc_{Ci}+dc_{Ci-c}};
                remove 3 elements from L_i;
                insert the pair <Ds, NetName> into L_i;
                insert the pair <Dc, NetName> into L_{i+1};
                subtract 2 from the length of L_i;
                increment the length of L_{i+1};
            End While;
        sort the elements of L_i;
        connect an FA to the last 3 nodes of L_i;
        connect the S and C to the bit i and I + 1
            of the CPA;
    End For;
End Method;
```

The delay constants are technology dependent and are defined by the user.

3.4 Discussion of the Algorithm

The presented algorithm takes into account any delay value Ds and Dc as determined for the particular counter structure and technology of implementation. An average short wire delay and the effect of *loading* is included into Ds and Dc and calculated as a part of

$$Ds = \max \{dc_A + dc_{A-s}, dc_B + dc_{B-s}, dc_{Ci} + dc_{Ci-s}\}$$

and

$$Dc = \max \{dc_A + dc_{A-c}, dc_B + dc_{B-c}, dc_{Ci} + dc_{Ci-c}\}$$

expressions. It would be easy to generalize the same algorithm for use of higher order compressors (p, q). In such a case, we calculate the delay D_i ($i = 1, \ldots q$) as:

$$D_i = \max \{d_j + d_{j-i} \text{ for } j = 1 \text{ to } p\}$$

Once the delays were calculated and list L_i has been sorted, we proceed by eliminating p elements from the list L_i and adding q delays to it. The objective of the algorithm is to minimize the delay of the signals that go to the next VCS. The critical path in the multiplier is usually found to be either in the largest VCS (in the middle of the multiplier tree), or as a path that starts at the least significant VCS and is being passed to the next until it ends up in the middle VCS. Therefore, our objective is to minimize:

1) the direct vertical path in a VCS (signals that are passed vertically to the FA),
2) the delay and the number of signals that are passed from VCS_i to VCS_{i+1}.

The problem of finding the multiplier structure with the minimal delay using arbitrary (p, q) counter is an NP hard problem, as is for finding a minimum delay wire layout. In the particular case presented in this paper, the problem is currently not known to be NP hard or polynomially solvable. Resolving this question is an interesting problem, and indeed we were able to prove that our algorithm produces an optimal and the best achievable structure as far as the speed of the multiplier is concerned [34].

The algorithm can easily be implemented to run in $O(n^2 \log n)$ time using priority queues, and in $O(n^2)$ time if the delays are small integers. We have not found a counter example which would yield better results for the cases of N < 64.

3.5 Example

In our case, the delays are :

$$FA_{A \to S} = FA_{B \to S} = 2 \ XOR \ delays$$

$$FA_{Cin \to S} = FA_{A \to C} = FA_{B \to C} = FA_{Cin \to C} = 1 \ XOR \ delay.$$

In the case of a HA, the delays become :

$$HA_{A \to S} = HA_{B \to S} = 1 \ XOR$$
$$delay \ while \ HA_{A \to C} = HA_{B \to C} = 0.5 \ XOR \ delay.$$

This is because our examples utilize LSI 100K: 1μ CMOS ASIC technology [30].

It is not difficult to generalize this method for the use of a general p:q compressor rather than a FA. This assumes that such a compressor can be built more efficiently than the one built from the FAs and that delay dependencies are known for the given technology of implementation.

An example representing the ninth VCS of a 12×12 multiplier is shown in Fig. 8. The numbers represent the delays that are associated to the nodes. The zero delays are the partial products and the nonzero delays at the top are carry signals originating from the previous slice. Since the number of inputs is even, the first cell used by the program is a HA. The reduction is achieved in five XOR delays.

In our example, we choose to normalize all the signal delays with respect to XOR gate delay and express delays as fractions of it. When different technology is used or the circuits are designed differently than in our case, the ratio of these delays can be quite different. For example, in [31]

the sum delay is 1.5 times longer than the delay of the carry bit. However, even in such a case, re-connecting *fast-inputs* and *fast-outputs* according to the method presented here may result in the speed improvement of up to 18%.

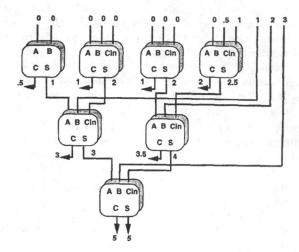

Fig. 8. The ninth VCS of a multiplier.

Although no Full Adder arrangement was assumed, the method produces a tree structure. This result was predictable since the tree structure is theoretically the fastest structure. As illustrated in the Fig. 8, the signals that have the longest delays will actually skip the next level of Full Adders. When it becomes impossible to skip more levels, they are connected to C_{in}, and when all C_{in} are occupied, they are connected to A or B. Since the delay of the carries are also calculated and used, the scheme minimizes both vertical and horizontal critical paths.

Table 1 compares the critical path in the partial product array produced by TDM and of the other schemes. The length of critical paths is estimated in the number of XOR levels. The delay of the *critical path* is dependent not only on the technology of implementation but also on the circuit family (dynamic or static) as well as layout and wiring delays. Therefore, it is difficult to give exact comparison, and the ones given in Table 1 are only reasonable estimates of the relative differences in delays with respect to other schemes [7], [11], [16]. We have decided to normalize all the delays to that of the XOR gate and use the XOR gate delay to represent respective delays in the multiplier tree structure. This decision is justified by the observation that the *critical path* in the multiplier tree indeed consists of a path through a series of XOR gates and that the ultimate speed of the multiplier (Final Adder included) indeed depends on how fast XOR gate can be implemented in a particular technology. A good example supporting this decision can be found in the pass-transistor multiplier implementation by Okhubo et al. [33].

This however, does not affect our method which is based on the *real delays* as calculated for the particular *counter* used in the multiplier under consideration.

TABLE 1
COMPARISON BETWEEN TDM
AND OTHER REPRESENTATIVE SCHEMES,
IN XOR LEVELS USED IN THE PARTIAL PRODUCT ARRAY

Multiplier Word-length	Wallace Tree [7]	4:2 Tree [11]*	Fadavi-Ardekani [16]	TDM
3	2	2	2	2
4	4	3	3	3
6	6	6 (5)	5	5
8	8	6	7	5
9	8	8	7	6
11	10	9 (8)	8	7
12	10	9 (8)	8	7
16	12	9	10	8
19	12	12 (11)	11	9
24	14	12 (11)	12	10
32	16	12	13	11
42	16	15 (14)	14	12
53	18	15	15	13
64	20	15	16	14
95	20	18 (17)	17	15

* Number in parenthesis represent delays when a Full Adder is used (instead of 4:2 compressor) every time the column size is found to be three.

Our algorithm and method can be extended further by going one level deeper into the hierarchy of design. We can treat the multiplier tree as a collection of *interconnected gates* or even *transistors* (as done in [33]) with designated *fast* and *slow* inputs and outputs. We leave this problem for the future extensions of this work suggesting this as a direction for possible further speed optimizations. However, assuming that we have optimized the signal paths using all the available circuit techniques in a particular counter, our method should yield the same results. Wiring delays were not given full consideration in our method.

The average wire delay is rather included into the gate delay as a function of fan-out. If the method presented here is to be fully integrated into the silicon compiler system, wire delay can be included and delays recalculated in an iterative process as each level of the bit reduction matrix is produced. In such a case it might be necessary to re-run the algorithm several time for possible corrections of the interconnection pattern and for fine-tuning of the delay list.

In the Wallace tree implementation, the number of XOR levels is simply the number of Full Adder levels multiplied by two. The improvement using TDM is up to 30%.

3.6 Hardware Complexity

In the following section, we show that the number of cells used is the same as the number produced by Dadda optimization.

The number of inputs to VCS_i is the sum of the number of partial products, called P_i, and the number of carries out from VCS_{i-1} which is actually $C_{i-1} - 1$, since every cell of VCS_{i-1} produces a carry out except the one that is connected to the CPA. We have already shown that: $CN_i = \lfloor I_i/2 \rfloor$

THEOREM 1. *For $i \in [3, N-1]$ the number of cells in VCS_i, $CN_i = CN_{i-1} + 1$.*

PROOF.

The proof is by mathematical induction.

1) Initial step: It is true for I = 3. In this case, $CN_3 = 2$ and $CN_2 = 1$. Therefore, the theorem is true for the initial step

2) Induction hypothesis: $CN_i = CN_{i-1} + 1$.

3) Using the relation $CN_{i+1} = \lfloor P_{i+1} + CN_i - 1/2 \rfloor$ and, since the induction hypothesis is $CN_i = CN_{i-1} + 1$ and $P_{i+1} = P_i + 1$ for i Œ [3, N − 1], it follows that:

$$CN_{i+1} = \lfloor (P_i + CN_{i-1} - 1)/2 + 1 \rfloor = \lfloor (P_i + CN_{i-1} - 1)/2 \rfloor + 1.$$

Hence, $CN_{i+1} = CN_i + 1$ for i Œ [3, N − 1].

THEOREM 2. *For $i \in [N, 2N-2]$ the number of cells in VCS_i, $CN_i = CN_{i-1} - 1$.*

PROOF.

1) Initial step: $CN_{N-1} = N - 2$ (application of the Theorem 1). For i = N,
$$CN_N = \lfloor (P_N + CN_{N-1} - 1)/2 \rfloor = \lfloor (2N - 5)/2 \rfloor = N - 2.$$
Then, for i = N + 1
$$CN_{N+1} = \lfloor (P_{N+1} + CN_N - 1)/2 \rfloor = \lfloor (2N - 6)/2 \rfloor = N - 3.$$
Therefore, the theorem is verified for the initial step

2) Induction hypothesis: $CN_i = CN_{i-1} - 1$.

3) Using the relation $CN_{i+1} = \lfloor (P_{i+1} + CN_i - 1)/2 \rfloor$ and, since the induction hypothesis is $CN_i = CN_{i-1} - 1$ and $P_{i+1} = P_i - 1$ if i ∈ [N, 2N − 2], we find:

$$CN_{i+1} = \lfloor (P_i + CN_{i-1} - 1)/2 - 1 \rfloor = \lfloor (P_i + CN_{i-1} - 1)/2 \rfloor - 1.$$

Hence, $CN_{i+1} = CN_i - 1$ for i ∈ [N, 2N − 2].

THEOREM 3. *The total number of cells in the partial product array is $(N-1)(N-2)$*

PROOF.

VCS_0 and VCS_1 have no cells. The numbers of cells for VCS_2 to VCS_{N-1} represent an arithmetic progression $1 + 2 + 3 + ... + N - 2$. That is:

$$\sum_{i=1}^{N-2} i = \frac{(N-1)(N-2)}{2}.$$

The number of cells in VCS_N to VCS_{2N-2} is similarly shown to be $(N - 1)(N - 2)/2$ by using arithmetic progression $N - 2 + N - 1 + ... + 1$.

Therefore, the total number of cells is:

$$\sum_{i=0}^{2N-2} i = \sum_{i=0}^{N-1} i + \sum_{i=N}^{2N-2} i$$
$$= \frac{(N-1)(N-2)}{2} + \frac{(N-1)(N-2)}{2} = (N-1)(N-2).$$

The number of FA and HA produced by our program is exactly $(N - 1)(N - 2)$. Although Dadda did not give any formula [8], the results of his optimization are 110 cells for 12-bit multiplication and 506 cells for 24-bit multiplication. Those numbers correspond to our formula $(N - 1)(N - 2)$. Therefore, we claim that our method is also optimized in terms of number of cells. Fadavi-Ardekani (F-A) scheme produces 121 and 582 cells, respectively, for 12- and 24-bit multiplier sizes.

3.7 Comparison and Experimental Results

We have designed 4:2, 9:2, F-A, and TDM partial product arrays in 1μ CMOS-ASIC technology for a 24-bit multiplier. The results that were obtained by simulation using timing simulator from LSI Logic and LSI 100K [30] timing information under nominal conditions [T = 25°C, V_{cc} = 5V]. The results for the critical path delay in the partial product array are summarized in the Table 2. LSI Logic simulator takes into account loading due to different fan-outs and includes an estimated wire load. This estimation is based on the fan-out of the particular output node. Our past experience with the simulator based on comparisons with actual fabrications has been very good. However, our results are limited to simulation and they do not represent actual measurements. It is also appropriate to mention that our results do not reflect the complexity of wiring. The wiring required for a multiplier produced using the TDM method is not substantially different from other schemes such as Wallace or Dadda, knowing that the number of counters used by our method is comparable to Dadda's [8] and that the main difference is in the way inputs and outputs of the counters are connected locally, not in the additional connections. However, using an improved Dadda's scheme a multiplier can be designed with only local interconnections as done in [32]. TDM scheme uses more complex wiring than [32] and an *array multiplier* such as [31]. The use of (4, 2) compressors results in less complex wiring and layout.

TABLE 2
CRITICAL PATH DELAY [CMOS: LEFF = 1μ, T = 25°C, V_{CC} = 5V]

N = 24 bits	4:2 Design	9:2 Design	Fadavi-Ardekani	TDM Design
Delay [nS]	14.0	13.0	11.7	10.5

Fig. 9 depicts a comparison in terms of XOR levels between a regular Wallace/Dadda scheme, an optimized 4:2, the Fadavi-Ardekani scheme and the TDM scheme. The delay for the partial product array includes partial product generator delay consisting of a simple row of AND gates. In the next section, we consider the delay component of an optimal CPA.

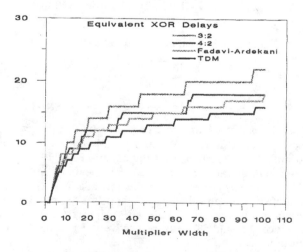

Fig. 9. Comparison of (3, 2), 4:2, F-A, and TDM schemes.

For this particular 24-bit CMOS implementation, the TDM Scheme is 33% faster than 4:2 and 25% and faster than 9:2. The improvement over F-A method is 11% in terms of delay and 15% in terms of the number of cells used.

4 SPEED IMPROVEMENT IN THE FINAL ADDER

Finally, multiplier speed can be further improved via optimization of the CPA to the nonuniform signal arrival profile of its inputs. It is well-known that the signals applied to the inputs of the CPA arrive first at the ends of the CPA and the last ones are the signals fed to the bits in the middle of the CPA [26]. The shapes of the signal arrival profile originating from the CPA and the multiplier tree of an 13 × 13-bit ASIC multiplier [19] are shown in Fig. 10.

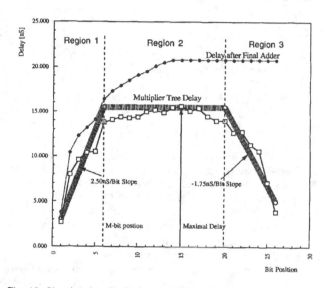

Fig. 10. Signal Arrival Profile and selection of the adder types in the three regions of the multiplier.

4.1 The Choice of Final Adder

All the known schemes for fast addition are developed under the assumption of an uniform signal arrival profile. The first problem is in selecting one of the CPA schemes that are most adequate to be used in the multiplier. It is obvious that we should use the fastest scheme since final addition time is a significant addendum to the critical path of the multiplier.

There are three regions to be considered with respect to the worst case signal arrival profile from the multiplier tree as shown in Fig. 10. Region 1 has a positive slope with respect to bit position. To use an adder which adds faster than this slope would not make much sense and would be a waste of hardware resources. The type of adder used in Region 1 is determined by this slope. This slope is determined by the fact that the arrival of bits is incrementally delayed by a path traversing a FA used in bit compression. Therefore, using any of the more powerful and, therefore, more complex, schemes for this part of the multiplier is not justified. If ripple-carry adder speed can not match this slope, good choices are simple and VLSI efficient schemes such as Variable Block Adder (VBA) [23]. A simple analysis shows that CLA has worse performance in this region [27].

From the point of the maximal delay (M) which is usually in the middle bit position (or skewed a few bits toward the most significant side) the addition has to be performed in the fastest possible way. Therefore, in Regions 2 and 3, addition has to be very fast because the addition time Δ_{ADD_2} is a direct addendum to the multiplier delay. Analysis shows that the best choices for the adder in this region are: Conditional Sum Adder (CSA) and the Carry Select Adder (CSLA) [21], [27]. Choice of CSLA scheme results in less complex implementation simply because CSLA is a subset of CSA. The difference in speed between CSA and CSLA diminishes as the input arrival profile is changed from uniform to nonuniform [21]. Though the CSLA adder is still slower than the CSA adder, this difference is offset by the relative simplicity of its implementation which is preferred by most designers. Smaller size and simpler layout of CSLA would further affect the relative speed difference, reducing the advantage of CSA. Given that fact, Carry Select Addition (CSLA) is the best choice for Region 2 [22].

The adder constructed for this part (Regions 2 and 3) can be further subdivided into two. Region 2 requires the fastest addition available, the choice of which is Carry-Lookahead (CLA) adder, or optimized derivatives of CLA [24] combined into a CSLA, all of which depend on the particular circuit and technology used.

In the region of the negative slope (Region 3), where the most significant bits arrive first, the most suitable choice is again CSLA. However, CSLA adds time for selection multiplexers, ΔMUX, which is not negligible given that the adder sections are already constructed to be in the same speed range. It is also obvious that due to the steep negative slope in this region, there is substantial time left to add the bits in the most significant position before a selection process occurs. Simple analysis shows that for this part, VBA is a much better candidate than CLA or for that matter any other scheme [27].

Determination of bit positions separating the Regions 1, 2, and 3 is based on intersecting the bit arrival time from the adders used in the corresponding regions, so that the selection in CSLA is done at the appropriate time. This is a relatively complex iterative process which is illustrated in Fig. 11. As can be inferred from Fig. 11, this is an iterative technique which does not require a large number of iterations, given that the bit positions S_1 and S_2 are integers. The total delay of the multiplier is:

$$\Delta_{MULT} = \Delta_{TREE} + \Delta_{ADD_2} + \Delta_{MUX}$$

The resulting CPA structure is shown in Fig. 12a. The signal arrival profiles originating from the tree of an 13×13-bit multiplier for two different input patterns applied are shown in Fig. 12b. It is interesting to observe that while the pattern B results in a faster signal arrival from the multiplier array than pattern A, the product originating from the CPA for the pattern B has a longer delay than the pattern A. This example illustrates the point that it is more important to *tune* the CPA into the

signal arrival profile than to apply the fastest available addition scheme. Further analysis of CPA optimization can be found in [27].

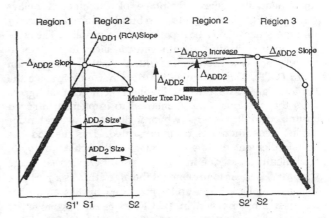

Fig. 11. Determination of bit positions S_1 and S_2 determining the size of the adders used: Left—Structure of the Final Adder; Right—Signal Arrival Profile.

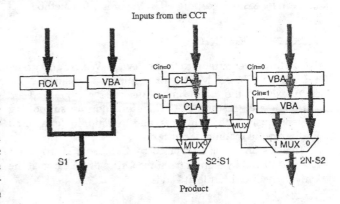

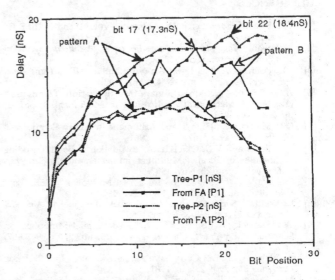

Fig. 12. Structure of the Final Adder (top); Signal Arrival Profile from the column compression tree and the Final Adder (bottom).

487

5 Conclusion

The presented algorithm and method for parallel multiplier implementation takes advantage of the uneven delays through a Full Adder in order to build a global compressor that minimizes the critical path of the multiplier. The compression tree is divided into vertical slices that are optimized globally to produce individual Vertical Compression Slices (VCS). This minimization does not only involve the vertical signal path, but also involves the horizontal signals from the previous VCS. A method to implement a speed optimized multiplier tree which includes an algorithm for net-list generation has been presented. This method has been implemented using C language, although it would not be difficult to describe in a hardware description language such as VHDL, or to implement the algorithm as a part of a silicon compiler. A multiplier produced by this algorithm has been implemented in 1μ CMOS technology together with several competing schemes. The results obtained using our algorithm are better not only in terms of speed but surprisingly also in terms of the number of cells used. The use of a compressor optimized at the transistor level instead of a Full Adder could result in further improvements of speed and can be easily incorporated in the presented method.

ACKNOWLEDGMENTS

The authors are grateful to the reviewers. We are also grateful to Prof. Charles Martel for helping us understand the complexity of the algorithm. We express our gratitude to LSI Logic and Cascade Design Automation for supporting us generously with their tools, and to Antonio de la Serna for his careful reading of our text.

This research was supported by California Micro Grant No. 93-118 and by a grant from the Office of Research at the University of California at Davis.

REFERENCES

[1] E.E. Swartzlander, *Computer Arithmetic*, vols. 1 and 2, IEEE CS Press, 1990.

[2] K. Hwang, *Computer Arithmetic: Principles, Architecture, and Design*. John Wiley and Sons, 1979.

[3] S.D. Pezaris, "A 40ns 17-bit Array Multiplier," *IEEE Trans. Computers*, vol. 20, no. 4, pp. 442-447, Apr. 1971.

[4] A.D. Booth, "A Signed Binary Multiplication Technique," *Quarterly J. Mechanical Applications in Math.*, vol. 4, part 2, pp. 236-240, 1951.

[5] O.L. MacSorley, "High Speed Arithmetic in Binary Computers," *IRE Proc.*, vol. 49, pp. 67-91, Jan. 1961.

[6] D. Villeger and V.G. Oklobdzija, "Evaluation of Booth Encoding Techniques for Parallel Multiplier Implementation," *Electronics Letters*, vol. 29, no. 23, Nov. 1993.

[7] C.S. Wallace, "A Suggestion for a Fast Multiplier," *IEEE Trans. Computers*, vol. 13, no. 2, pp. 14-17, Feb. 1964.

[8] L. Dadda, "Some Schemes for Parallel Multipliers," *Alta Frequenza*, vol. 34, pp. 349-356, Mar. 1965.

[9] W.J. Stenzel, "A Compact High Speed Parallel Multiplication Scheme," *IEEE Trans. Computers*, vol. 26, no. 2, pp. 948-957, Feb. 1977.

[10] A. Weinberger, "4:2 Carry-Save Adder Module," *IBM Technical Disclosure Bull.*, vol. 23, Jan. 1981.

[11] M.R. Santoro, "Design and Clocking of VLSI Multipliers," PhD dissertation, Technical Report No. CSL-TR-89-397, Oct. 1989.

[12] M.R. Santoro, "A Pipelined 64×64b Iterative Array Multiplier," *Digest of Technical Papers, Int'l Solid-State Circuits Conf.*, Feb. 1988.

[13] M. Nagamatsu et al., "A 15nS 32×32-bit CMOS Multiplier with an Improved Parallel Structure," *Digest of Technical Papers, IEEE Custom Integrated Circuits Conf.*, 1989.

[14] J. Mori et al., "A 10nS 54×54-b Parallel Structured Full Array Multiplier with 0.5-u CMOS Technology," *IEEE J. Solid State Circuits*, vol. 26, no. 4, Apr. 1991.

[15] P. Song and G. De Michelli, "Circuit and Architecture Trade-Offs for High Speed Multiplication," *IEEE J. Solid State Circuits*, vol. 26, no. 9, Sept. 1991.

[16] J. Fadavi-Ardekani, "M x N Booth Encoded Multiplier Generator Using Optimized Wallace Trees," *IEEE Trans. VLSI Systems*, vol. 1, no.2, June 1993.

[17] G. Bewick, "High Speed Multiplication," *Proc. Electronic Research Laboratory Seminar*, Stanford Univ., Mar. 12, 1993 (also private communications).

[18] V.G. Oklobdzija and D. Villeger, "Multiplier Design Utilizing Improved Column Compression Tree and Optimized Final Adder in CMOS Technology," *Proc. 10th Anniversary Symp. VLSI Circuits*, Taipei, Taiwan, May 1993.

[19] T. Soulas, D. Villeger, and V.G. Oklobdzija, "An ASIC Multiplier for Complex Numbers," *Proc. EURO-ASIC 93, The European Event in ASIC Design*, Paris, France, Feb. 22-25, 1993.

[20] D. Villeger, "Fast Parallel Multipliers," *Final Report, Ecole Superieure d'Ingenieurs en Electrotechnique et Electronique*, Noisy-le-Grand, France, May 11, 1993.

[21] A.K.W. Yeung and R.K. Yu, "A Self-Timed Multiplier with Optimized Final Adder," *Final Report for CS 292I* (Prof. Oklobdzija), Univ. of California at Berkeley, Fall 1989.

[22] O.J. Bedrij, "Carry-Select Adder," *IRE Trans. Electronic Computers*, June 1962.

[23] V.G. Oklobdzija and E.R. Barnes, "On Implementing Addition in VLSI Technology," *IEEE J. Parallel Processing and Distributed Computing*, no. 5, 1988.

[24] B.D. Lee and V.G. Oklobdzija, "Delay Optimization of Carry-Lookahead Adder Structure," *J. VLSI Signal Processing*, vol. 3, no. 4, Nov. 1991.

[25] M. Suzuki et al., "A 1.5nS 32b CMOS ALU in Double Pass-Transistor Logic," *Digest of Technical Papers, 1993 IEEE Solid-State Circuits Conf.*, San Francisco, Feb. 24-26, 1993.

[26] K. Fai-Pang et al., "Generation of High-Speed CMOS Multiplier-Accumulators," *Proc. ICCD-88, Int'l Conf. Computer Design*, Rye, N.Y., Oct. 1988.

[27] V.G. Oklobdzija, "Design and Analysis of fast Carry-Propagate Adder under Non-Equal Input Signal Arrival Profile," *Proc. 28th Asilomar Conf. Signals, Systems, and Computers*, Oct. 30-Nov. 2, 1994.

[28] K.J. Singh et al., "Timing Optimization of Combinational Logic," *Proc. ICCAD 88, Int'l Conf. CAD*, Nov. 1988.

[29] H.J. Touati et al., "Delay Optimization of Combinational Logic Circuits by Clustering," *Proc. ICCAD 91, Int'l Conf. CAD*, Oct. 1991.

[30] *1.0-Micron Array-Based Products Databook*, LSI Logic Corp., Sept. 1991.

[31] J.Y. Lee et al., "A High-Speed High-Density Silicon 8X8-bit Parallel Multiplier," *IEEE J. Solid State Circuits*, vol. 22, no. 1, Feb. 1987.

[32] Z. Wang, G.A. Julien, and W.C. Miller, "Column Compression Multipliers for Signal Processing Applications," *VLSI Signal Processing*. New York: IEEE Press, 1992.

[33] N. Okhubo et al., "A 4.4nS CMOS 54×54-b Multiplier Using Pass-Transistor Multiplexer," *Proc. Custom Integrated Circuits Conf.*, San Diego, Calif., May 1-4, 1994.

[34] C. Martel, V.G. Oklobdzija, R. Ravi, and P. Stelling, "Design Strategies for Optimal Multiplier Circuits," *Proc. 12th IEEE Symp. Computer Arithmetic*, Bath, England, July 19-21, 1995.

A 4.1-ns Compact 54 × 54-b Multiplier Utilizing Sign-Select Booth Encoders

Gensuke Goto, Atsuki Inoue, Ryoichi Ohe, Shoichiro Kashiwakura, Shin Mitarai, Takayuki Tsuru, and Tetsuo Izawa

Abstract—A 54 × 54-b multiplier with only 60 K transistors has been fabricated by 0.25-μm CMOS technology. To reduce the total transistor count, we have developed two new approaches: sign-select Booth encoding and 48-transistor 4-2 compressor circuits both implemented with pass transistor logic. The sign-select Booth algorithm simplifies the Booth selector circuit and enables us to reduce the transistor count by 45% as compared with that of the conventional one. The new compressor reduces the count by 20% without speed degradation. By using these new circuits, the total transistor count of the multiplier is reduced by 24%. The active size of the 54 × 54-b multiplier is 1.04 × 1.27 mm and the multiplication time is 4.1 ns at a 2.5-V power supply.

Index Terms— Algorithm, CMOS digital integrated circuits, computer graphics, encoding, floating point numbers, IEEE standard, mulitplication, multiplying circuits.

I. INTRODUCTION

BECAUSE of the rapid progress in very large scale integration (VLSI) design technologies, consecutive improvements in operational speed and design integration have been made. As a result of these improvements, interactive real-time three-dimensional (3-D) graphics applications have become available even on personal computers. In the geometric conversion processes of 3-D graphics applications, huge amounts of floating point operations are required and parallel data processing is inevitable. The amount of these arithmetic operations in real-time 3-D graphics applications depends on the quality of the 3-D objects and the frame rate. For example, in the case of a screen with a 1600 × 1280 resolution and at 60 frames/s, if 3-D objects are formed with small polygons (25 pixels/polygon) and 80% of the entire screen is constantly being processed with 3-D objects, approximately 4 Mp/s (million polygons per second) are processed. Our evaluations so far indicate that, to achieve 1 Mp/s of 3-D geometric conversion performance, 250 to 300 MFLOPS of floating point arithmetic performance are required.

Consequently, to achieve the above mentioned 4 Mp/s operation, more than 1 GFLOPS of floating point computation power is needed. In addition to 3-D graphics applications, a drastic improvement in the floating-point performance of general-purpose microprocessors and DSP's has been hoped for to meet the user demand for multimedia data processing. In such circumstances as above, a cost-effective, high-speed float-ing point computational core takes on a greater importance. In particular, the multiplier is one of the largest components of the datapath area, and a reduction of this multiplier core area is effective for this purpose.

This paper describes the method to reduce the transistor count and area of the multiplier core with the use of the new Booth encoding algorithm [1] and a new 4-2 compressor circuit [2], [3] for a carry-save adder (CSA) tree [4]. We applied this algorithm to a 54 × 54-b multiplier macro for mantissa multiplication of double-precision floating-point numbers, as outlined in the IEEE standard [5]. In our earlier design of a compact 54 × 54-b multiplier [6], nearly 90% of the transistor count was occupied with Booth selector circuits and the carry-save adders. We adopted Booth encoding using both negative and positive selection signals as encoded outputs. Thus, we call this method sign-select Booth encoding. With this encoder, the Booth selector circuit can be simplified and we succeeded in significantly reducing the transistor count in a multiplier. We have also developed new circuit implementation technology of 4-2 compressors that also enable us to reduce the transistor count in a carry-save adder tree.

The architecture of the 54 × 54-b multiplier and the signed Booth encoding algorithm is discussed in Section II. Section III describes the carry-save adder tree and circuit design based on a new 4-2 compressor. The fabrication of a test chip and some experimental results are described in Section IV. In Section V, a comparison with the other multiplier designs is made. Tne conclusions are summarized in Section VI.

II. SIGN-SELECT BOOTH ENCODING ALGORITHM

The block diagram of a 54 × 54-b tree-type multiplier is shown in Fig. 1. It employs the modified second-order Booth algorithm and a carry-save adder. The division scheme of Booth selectors (partial-product-bit generators) and adders in the multiplier is shown in Fig. 2. The carry-save adders are realized with the combination of 4-2 compressors 4 W and one-bit full adders 3 W. This is because 28 partial-product bits at maximum need to be compressed. Partitioning them into 7-b units composed of a 4- and a 3-W unit is useful for simplifying the complicated layout design. In the earlier 54 × 54-b multiplier design [6], carry-save adders occupy about 55% and the Booth selectors about 35% of the transistor count as shown in Fig. 1. This is because we need a large number of components for these two modules. In that design, 1527 Booth selectors and 625 4-2 compressors are used. However, the Booth encoder circuits consume 1.2%, as only 27 Booth encoder modules are necessary in that multiplier.

Manuscript received April 17, 1997; revised June 28, 1997.

G. Goto, A. Inoue, R. Ohe, and S. Kashiwakura are with Fujitsu Laboratories Ltd., System LSI Development Laboratories, Nakahara-ku, Kawasaki 211-88, Japan.

S. Mitarai, T. Tsuru, and T. Izawa are with Fujitsu Limited, System LSI Development Laboratories, Nakahara-ku, Kawasaki 211-88, Japan.

Publisher Item Identifier S 0018-9200(97)08032-3.

Reprinted from *IEEE Journal of Solid-State Circuits*, Vol. 32, No. 11, pp. 1676-1681, November 1997.

489

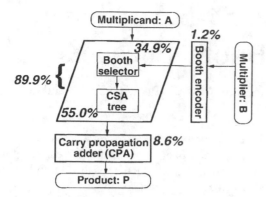

Fig. 1. Block diagram of multiplier. The italic numbers shows the percentage of number of transistors used in each part with respect to the total transistor count in our earlier multiplier design [6].

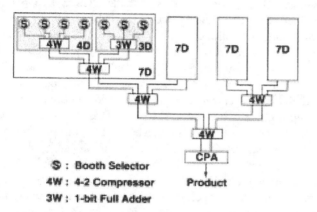

S : Booth Selector
4W : 4-2 Compressor
3W : 1-bit Full Adder

Fig. 2. Division scheme of Booth selectors and carry save adder tree in 54 × 54-b multiplier.

Thus, minimizing the volume of Booth selectors and carry-save adders is the key to establishing a compact multiplier.

The use of the Booth algorithm reduces the numbers of partial products and the carry-save adders significantly. In the second-order Booth encoding algorithm, we can reduce them by half. Consider the multiplication of two n-bit numbers in 2's complement, A and B, where we assume that n is an even number. The relative bit positions of the multiplicand and the multiplier are represented by a_i and b_j, respectively. The truth table of the usual second-order Booth encoder is summarized in Table I with that of the sign-select encoder. In the usual Booth encoder, three signals, X_j, $2X_j$, and M_j, are generated from the three adjacent multiplier bits, b_{j-1}, b_j, and b_{j+1} for selecting a partial product, that is, one of 0, $+A$, $-A$, $+2A$, and $-2A$. Here A is a multiplicand value of n bit width. The X_j and $2X_j$ signals show whether or not the partial product is doubled and M_j means the negative partial product. Thus, the multiplier value, either A or $2A$, is selected depending on whether the encoded data, X_j or $2X_j$, is high and the selected signal is inverted by encoded data, if M_j is the logical one. In this algorithm, a logical equation for the output signal, $P_{i,j}$, of the Booth selector at the ith multiplicand bit, a_i, and the jth multiplier bit, b_j, is given by

$$P_{i,j} = (a_i \cdot X_j + a_{i-1} \cdot 2X_j) \oplus M_j$$
$$(i = 0, 1, 2, \cdots, n-1 \quad j = 0, 2, 4, \cdots, n-4, n-2). \quad (1)$$

TABLE I
TRUTH TABLE FOR SECOND-ORDER MODIFIED BOOTH ENCODING

Inputs				Usual			Sign select			
b_{j+1}	b_j	b_{j-1}	Func.	X_j	$2X_j$	M_j	X_j	$2X_j$	PL_j	M_j
0	0	0	0	0	0	0	0	1	0	0
0	0	1	+A	1	0	0	1	0	1	0
0	1	0	+A	1	0	0	1	0	1	0
0	1	1	+2A	0	1	0	0	1	1	0
1	0	0	-2A	0	1	1	0	1	0	1
1	0	1	-A	1	0	1	1	0	0	1
1	1	0	-A	1	0	1	1	0	0	1
1	1	1	0	0	0	1	0	1	0	0

$P = A \times B$

$A = -a_{n-1}2^{n-1} + \sum_{i=0}^{n-2} a_i 2^i$

$B = -b_{n-1}2^{n-1} + \sum_{j=0}^{n-2} b_j 2^j$

$(j = 0, 2, 4, \cdots, n-4, n-2)$

PL_j : Positive partial product
M_j : Negative partial product
X_j : Not doubled
$2X_j$: Doubled

Thus, the equation includes an exclusive-or (XOR) circuit to determine whether the partial product is inverted or not. A Booth encoder and a Booth selector require 32 and 18 transistors, respectively, for static CMOS logic implementation.

We developed a new Booth encoding algorithm with four output signals. We added an extra control signal PL_j to represent the selection of positive partial product. We introduced a sign-select Booth encoder with four output signals as shown in Fig. 3(a). In the usual encoder, only M_j, which represents the negative partial product, is generated. Conversely, in our sign-select encoder, signals for both the negative and positive partial products are generated. PL_j and M_j become active when the partial product is positive and negative, respectively. The X_j and $2X_j$ signals show whether or not the partial product is doubled. Using these encoded signals, we simplified the partial product generation logic. For example, when $+2A$ is necessary for the partial product, the PL_j and $2X_j$ signals are active, and the logical product of PL_j and $2X_j$ chooses $+2A$ as the partial product. When $-A$ is necessary, the logical product of M_j and X_j chooses the correct partial product, and so on. The logical equation for $P_{i,j}$ in this case is given by

$$P_{i,j} =$$
$$(a_i \cdot PL_j + \overline{a_i} \cdot M_j) \cdot X_j + (a_{i-1} \cdot PL_j + \overline{a_{i-1}} \cdot M_j) \cdot 2X_j$$
$$(i = 0, 1, 2, \cdots, n-1 \quad j = 0, 2, 4, \cdots, n-4, n-2). \quad (2)$$

From this equation, you will find that when both PL_j and M_j are inactive, "0" multiple as $P_{i,j}$ can be derived.

Thus, in our new Booth selector, exclusive OR logic is no longer necessary. Also, by utilizing pass transistor selectors, the Booth selector circuit is formed with fewer transistors. The circuit diagram is shown in Fig. 3(b) and (c) for both our new Booth selector and the conventional one. The selector (SEL) in Fig. 3(b) is implemented using pass transistor and consists of four transistors. The transistor count of the Booth selectors can be reduced from 18 transistors/b to 10 transistors/b when we combine two Booth selectors at consecutive multiplicand bit positions into a single module and sharing two inverters

TABLE II
SIMULATED CIRCUIT PERFORMANCE

Circuit	Transistor count in 4D	Delay time (ns)			
		Bit buffer +Booth encoder	Booth selector	4-stage 4-2 compressors	Total delay time
(1) OPPG+OD4W	132	0.34	0.40	1.99	2.73
(2) OPPG+O6T4W	116	0.34	0.41	3.05	3.80
(3) OPPG+DPL4W	124	0.31	0.46	2.05	2.82
(4) NPPG+MY4W	88	0.32	0.39	2.10	2.81

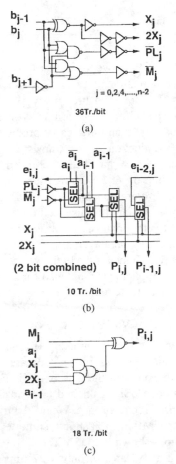

Fig. 3.   Circuit diagrams for (a) Booth encoder, (b) new Booth selector used in this work, and (c) the conventional Booth selector.

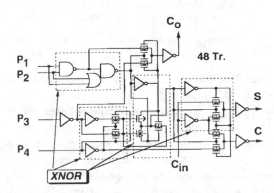

Fig. 4.   Circuit diagram for 48-transistor 4-2 compressor.

$\times$ 54-b multiplier compared to only 1.2% for Booth encoders. The reduction in the number of transistors in Booth selectors thus has a greater impact on the total transistor count. The sign-select Booth encoding method enables us to reduce the transistor count by about 44% in our Booth selectors.

III. CARRY-SAVE ADDER TREE

The carry-save adder tree is the largest among the multiplier components. Therefore, we have to reduce the transistor count for the 4-2 compressors which are the constituents along with a one-bit full adder. The reduction is not easy, however, because a 4-2 compressor with fewer transistors usually needs a larger delay than that with more transistors. Careful optimization technology is required for the reduction of the transistor count for a 4-2 compressor without speed degradation.

There are four XOR circuits in a 4-2 compressor, and they are the major components in it. To implement an XOR circuit with fewer transistors and to share some transistors with an intermediate carry-generating circuit is key to overcoming these difficulties. There is an XOR circuit that is composed of six transistors [7]. A series-connected six-transistor XOR circuit is too slow to be applicable to a high-speed multiplier. However, we found that it is not too slow when sandwiched with other ten-transistor XOR circuits. We then adopted it in the middle part of the three-stage XOR chain, as shown in Fig. 4. In one of the first stages, we adopted an XOR circuit composed of a NAND gate and an OR–AND–inverter gate to share the NAND gate with an intermediate carry (C_o) generating circuit. By this sharing, an extra inverter is saved in the carry generator circuit. The other two XOR circuits are of a ten-transistor type composed of pass transistors and inverters. With this scheme

through which we provide the selector with PL_j and M_j signals. By selecting these two signals by the multiplicand bit signal, a_i, sharing the inverters becomes possible with the selectors at the same jth multiplier bit position. In Fig. 3(b), the inverters are shared with two Booth selectors at consecutive multiplicand bit positions. Sharing the two inverters with more than two Booth selectors at consecutive multiplicand bit positions is logically possible, but at the cost of slower speed due to the larger RC time constant. The $\overline{a_i}$ and a_i signals used in the Booth selectors are buffered globally and the buffer circuits do not affect the total transistor count significantly. Although four more transistors are needed for each bit in the Booth encoder, eight transistors are eliminated from each bit in the Booth selector. As described before, the transistor count of Booth selectors occupies 35% of the space in a 54

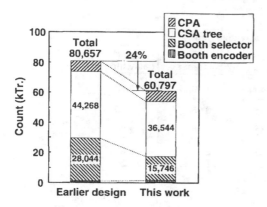

Fig. 5. Total transistor count compared with earlier design multiplier.

TABLE III
PROCESS TECHNOLOGY

Technology	0.25 μm CMOS 3-metal layer Co-Salicide
Gate Length	0.25 μm
Gate Oxide	5.5 nm
1st Metal (Line/Space)	0.44 μm / 0.46 μm
2nd Metal (Line/Space)	0.44 μm / 0.46 μm
3rd Metal (Line/Space)	0.44 μm / 0.46 μm
Planarization	CMP
Supply voltage	2.5 V

we reduced the transistor count of the 4-2 compressor from 60 to 48: a 20% reduction.

Table II shows the result of analysis using an HSPICE circuit simulator, and it compares our new circuits with the existing ones with respect to delay time of the 54 × 54-b multiplier except the final carry-propagate adder. The analysis is based on one process condition in 0.25-μm CMOS technology and all the n-channel or p-channel transistors are assumed to have the same width except for bit buffers. The wire lengths between the same kinds of blocks are assumed to be equal to one another among the models, and are in the same order as those of the test chip described in Section IV. OPPG and NPPG denote the Booth selectors that are implemented according to logical equations (1) and (2), respectively. OD4W, O6T4W, DPL4W, and MY4W are 4-2 compressors that are composed of the earlier design [6], all six-transistor XOR's [7], pass-transistor multiplexer logic [8], and the new 48-transistor logic, respectively. A 4D block consists of four Booth selectors and a 4-2 compressor 4W and generates and compresses four encoded bits at the same logical bit position. It is one of a basic block which is repeatedly arranged in a matrix to constitute a core part of the 54 × 54-b multiplier. In the table, the circuit (1), the combination of OPPG and OD4W is fastest, but the transistor count is also the highest. The difference in delay from circuits (3) and (4), however, is within 5% and it may be irrelevant as the accuracy of modeling for critical delay time is scattered among the circuits. Then we found that the three circuits (1), (3), and (4) have almost the same speed, and that circuit (2) is far slower than the other circuits, although the transistor count in a 4D block is lower than in circuits (1) and (3). Note that circuit (4), a combination of the new circuits, a Booth selector, and a 4-2 compressor, can constitute a 4D unit with fewer transistors than the alternative circuits and without any speed degradation as compared with the other competitors.

The total transistor count for our new 54 × 54-b multiplier is shown in Fig. 5, as compared with that of the earlier design of the regularly structured tree (RST) multiplier. By adopting the RST, we simplified the complicated carry-save adder tree wiring scheme and achieved a very compact design in less design time. We also obtained an area-efficient block because of the repetitive structure of the wiring, and thus the dedicated wiring area was minimized. For the final carry-propagate adder (CPA), we used the pass-transistor adder based on the combination of carry-skip and modified carry-select schemes that are used previously [9]. This adder can give sufficient speed for a performance-oriented multiplier with fewer transistors than other carry-propagate adders.

About a 24% reduction in the transistor count compared with the earlier design has been achieved for the whole multiplier. A total of 60 797 transistors in a 54 × 54-b multiplier is the minimum count reported to date [6], [8], [10]–[13]. When we apply this method to a 26 × 26-b multiplier used for the mantissa multiplication of two single-precision numbers, we again obtain a 23% reduction in the transistor count, which means the logic we used is effective not only for larger multipliers [14], but also for smaller multipliers to reduce the transistor count.

IV. FABRICATION

We fabricated this multiplier using 0.25-μm CMOS technology. The major process parameters are summarized in Table III. The thickness of the gate oxide is 5.5 nm and salicidation process of $CoSi_2$ is utilized to form the source/drain area to reduce parasitic resistance. We used a triple-metal layer process with the CMP planarization technique. A photomicrograph of the chip is shown in Fig. 6. The chip integrates 60 797 transistors in an active area of only 1.27 mm^2 except for the testing circuits. Trends regarding the chip area of a 54 × 54-b multiplier are summarized in Fig. 7. We can see that this is the minimum reported area for a 54 × 54-b multiplier, and we estimated a 21% reduction in the area as compared with the earlier RST multiplier made using the same fabrication technology. When we compare the area with the previously reported multiplier [10] with equivalent technology, it is ten times larger than ours. This is caused not only by larger transistor count but also by larger transistor size. Transistor count of the previously reported multiplier is 100 200 which is 1.6 times larger than this work. The area of one transistor can be calculated as 128 μm^2/Tr. for the previously reported and 21 μm^2/Tr. for this work. Then the transistor size is six times larger than that of this work. The designed chip area reduction is directly reflected by the reduction in transistor count. Using the RST structure with the triple-metal process, we can minimize the wiring overhead in the Booth selector and CSA tree area. The measured wiring overhead by power

492

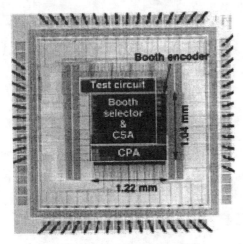

Fig. 6. A photomicrograph of the 54 × 54-b multiplier.

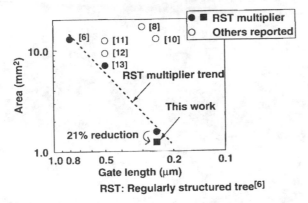

Fig. 7. Chip area trend for the 54 × 54-b multiplier.

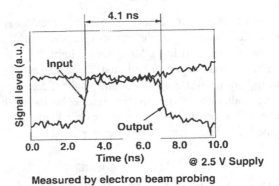

Fig. 8. Signal waveform for critical path.

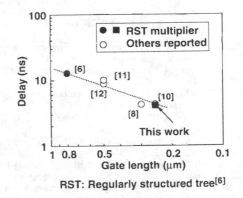

Fig. 9. Delay time versus gate length for 54 × 54-b multiplier.

lines and other wirings in this area is only 6.7%, thus the cell area dominates the entire array area in our RST layout.

V. EVALUATION

Fig. 8 shows the measured waveform along the critical path using electron beam probing. A critical path delay of 4.1 ns was obtained at room temperature with a 2.5-V supply. This is a little faster than those of the multipliers fabricated with the equivalent technology [8], as summarized in Fig. 9. The measured delay time is 15% smaller than the simulated value in Section III, probably because of the parameter variation. By this measurement, we confirmed that our success in reducing the logic size does not diminish the speed performance. Since the simulated delay time of this circuit was almost the same as with others, the speed advantage is thought to be the smaller area of the present design compared to others as a result of reducing the transistor count, and also to the exclusion of wiring with the adoption of the RST structure. The previous design [10] necessitates a ten-times larger area than ours under a similar process condition, and this may require nearly a three-times larger wiring capacitance than that assumed in the model in Table II. The simulated delay time for all the circuits in the table is increased by 15% if the wiring capacitance between all the circuit units is doubled.

Taking these results into consideration, the data (OPPG + DPL4W) in Table II matches the previously reported data [8]. Since the earlier design of the RST multiplier in itself has an advantage over the other multipliers in the performance/cost ratio [6], the new design of the RST multiplier makes possible a ratio in excess of 2.5.

The evidence that the 4-2 compressor can be constructed of only 48 transistors means that it has a speed advantage without a handicap in the transistor count against a series-connected pair of one-bit full adders, each consisting of 24 transistors. Furthermore, the evidence that the Booth selector can be constructed of ten transistors per bit indicates that the necessary number of transistors for generating two partial product bits in the second-order Booth encoding (20 transistors) is smaller than when directly generating four partial product bits using AND logic with the multiplicand bit and the multiplier bit at each bit position (24 transistors). This situation forces us to utilize the Booth algorithm even for a smaller multiplier such as 8 × 8-b, since this algorithm is no longer a consumer of transistors but an accelerator without hardware overhead. The measured power dissipation of this multiplier was as small as 2.23 mW/MHz at 2.5-V supply.

VI. CONCLUSION

In summary, we have developed sign-select Booth encoders and simplified the logic of Booth selectors that occupy a large area of the multipliers, in order to reduce the total transistor count. We also proposed a 48-transistor 4-2 compressor to save transistors without compromising in speed in the carry-save adder tree. We applied these ideas to a 54 × 54-b multiplier and fabricated a chip with 0.25-μm CMOS technology. A 54 × 54-b multiplier with 4.1-ns latency at 2.5-V supply and a 1.27 mm² active area was implemented. The total transistor count was 24% less than the previously designed multiplier, and this made our device the world's smallest, fastest multiplier using CMOS technology.

The approaches described in this paper are applicable to multipliers in 3-D graphics engines with floating-point arithmetic units and those in the integer units of high-speed general-purpose microprocessors, many kinds of DSP's, and embedded controllers.

ACKNOWLEDGMENT

The authors thank K. Fujita, K. Kubota, M. Kanazawa, S. Sugatani, and S. Hijiya for their continuous encouragement.

REFERENCES

[1] A. D. Booth, "A signed binary multiplication technique," *Quart. J. Mech. Appl. Math.,* vol. 4, pp. 236–240, 1951.

[2] A. Weinberger, "4-2 carry-save adder module," *IBM Tech. Discl. Bulletin,* vol. 23, Jan. 1981.

[3] V. G. Oklobdzija and D. Villeger, "Multiplier design utilizing improved column compression tree and optimized final adder in CMOS technology," in *Int. Symp. VLSI Technology, Systems and Applications,* Taipei, Taiwan, May 5–8, 1993, pp. 209–212.

[4] C. S. Wallace, "A suggestion for fast multipliers," *IEEE Trans. Electron. Comput.,* vol. EC-13, pp. 14–17, Feb. 1964.

[5] ANSI/IEEE Standard 754-1985 for Binary Floating-Point Arithmetic, IEEE Computer Soc., Los Alamitos, CA, 1985.

[6] G. Goto *et al.,* "A 54 × 54-b regularly structured tree multiplier," *IEEE J. Solid-State Circuits,* vol. 27, pp. 1229–1236, Sept. 1992.

[7] Japanese Unexamined Patent Application no. 59-211138, Nov. 1984.

[8] K. Ohkubo *et al.,* "A 4.4 ns CMOS 54 × 54-b multiplier using pass-transistor multiplexer," *IEEE J. Solid-State Circuits,* pp. 251–257, vol. 30, Mar. 1995.

[9] T. Sato *et al.,* "An 8.5-ns 112-b transmission gate adder with a conflict-free bypass circuit," *IEEE J. Solid-State Circuits,* vol. 27, pp. 657–659, Apr. 1992.

[10] M. Hanawa *et al.,* "A 4.3 ns 0.3 μm CMOS 54 × 54-b multiplier using precharged pass transistor logic," in *ISSCC Dig. Tech. Papers,* Feb. 1996, pp. 364–365.

[11] J. Mori *et al.,* "A 10 ns 54 × 54-b parallel structured full array multiplier with 0.5-μm CMOS technology," *IEEE J. Solid-State Circuits,* pp. 600–606, Apr. 1991.

[12] H. Makino *et al.,* "A design of high-speed 4-2 compressor for fast multiplier," *IEICE Trans. Electron.,* pp. 538–547, Apr. 1996.

[13] H. Iino *et al.,* "A 289 MFLOPS single-chip supercomputer," in *ISSCC Dig. Tech. Papers,* Feb. 1992, pp. 112–113.

[14] M. Santoro and M. Horowitz, "SPIM: A pipelined 64 × 64 bit iterative multiplier," *IEEE J. Solid-State Circuits,* vol. 24, pp. 487–493, Apr. 1989.

Division and Square Root: Choosing the Right Implementation

PETER SODERQUIST AND MIRIAM LEESER

Peter Soderquist

Cornell University

Miriam Leeser

Northeastern University

Although frequently neglected by microprocessor designers, division and square root are critical operations. Here, we explore the four major design issues in implementing these operations: algorithm, issue rate, connectivity, and replication.

Floating-point support has become a mandatory feature of new microprocessors due to the prevalence of business, technical, and recreational applications that use these operations. Spreadsheets, CAD tools, and games, for instance, typically feature floating-point-intensive code. Over the past few years, the leading architectures have incorporated several generations of floating-point units (FPUs). However, while addition and multiplication implementations have become increasingly efficient, support for division and square root has remained uneven. The design community has reached no consensus on the type of algorithm to use for these two functions, and quality and performance of the implementations vary widely. This situation originates in skepticism about the importance of division and square root and an insufficient understanding of the design alternatives.

Division and square root have long been considered minor, bothersome members of the floating-point family. Microprocessor designers frequently perceive them as infrequent, low-priority operations, barely worth the trouble of implementing; they allocate design effort and chip resources accordingly. The survey of microprocessor FPU performance in Table 1 shows some of the uneven results of this philosophy. While multiplication requires from two to five machine cycles, division latencies range from nine to sixty. The variation is even greater for square root, which several of the designs listed do not support in hardware. This data only hints, however, at the significant variation in algorithms and topologies among the different implementations.

The error in the Intel Pentium FPU, the accompanying publicity, and the $475-million write-off illustrate some of the hazards of an incorrect division implementation.[1] But correctness is not enough; poor performance causes enough problems by itself.

Though divide and square root are relatively infrequent operations in most applications, they are indispensable, particularly in many scientific programs. Compiler optimizations tend to increase the frequency of these operations by virtue of eliminating accompnaying nonarithmetic instructions.[2] This means that poor implementations disproportionately penalize code that uses them at all.[3] Furthermore, as the latency gap grows between addition/multiplication on the one hand and division/square root on the other, the latter operations increasingly become performance bottlenecks.[4]

Programmers have attempted to get around this problem by rewriting algorithms to avoid divide/square root operations, but the resulting code generally suffers from poor numerical properties, such as instability or overflow.[4-6] In short, division and square root are natural components of many algorithms, and they are best served by implementations with good performance.

Quantifying what constitutes good performance is challenging. One rule of thumb, for example, states that the latency of division should be three times that of multiplication; this figure is based on division frequencies in a selection of typical scientific applications.[7] Even if we accept this doctrine at face value, implementing division—and square root—involves much more than relative latencies. We must also consider area, throughput, complexity, and the interaction with other operations. This article explores the various trade-offs involved and illuminates the consequences of different design choices, thus enabling designers to make informed decisions.

Add-multiply configurations

Add-multiply circuits are the foundation of any FPU. Addition (which subsumes subtraction) and multiplication are not only the

Reprinted from *IEEE Micro*, Vol. 17, No. 4, pp. 56-66, July/August 1997.

most frequently occurring arithmetic operations, but together they can support all other operations required by the IEEE 754 floating-point standard.[8] We can regard all other functions, including division and square root, as additions to or enhancements of the add-multiply hardware. For these reasons, the implementation of addition and multiplication largely determines the overall performance of an FPU.

Add-multiply configurations in the latest generation of microprocessors fall into two basic categories. Figure 1 illustrates the first variety, which we have termed the independent configuration. It features add and multiply units that operate independently, reading operands from the register file and returning correctly rounded results. Although the illustration shows sharing of inputs and outputs between functional units, some implementations give each unit individual access to the register file.

The table accompanying Figure 1a shows the typically high performance figures for this type of configuration. Frequently, the adder is actually a floating-point ALU, performing comparisons, complementation, and format conversions as well as addition and subtraction. Processors that feature independent add-multiply configurations include the DEC 21164, Hewlett-Packard PA7200, Intel Pentium Pro, and Sun UltraSparc.

Figure 1b shows the second type of add-multiply structure, known as the multiply-accumulate configuration, along with typical performance figures. This configuration combines both functions into a single atomic operation on three operands, with the multiplication of two operands followed by addition of the product and the third operand. Typically, this unit maintains the product at a very high precision and adds without intermediate rounding. The motivation for the multiply-accumulate configuration is that a large proportion of scientific computations, such as series and matrix operations, involve many instances of multiplication followed immediately by addition. The IBM RS/6000 series, the Mips R8000, the HP PA8000, and the Hal Sparc64 use add-multiply structures of this type.

Design issues

Given the add-multiply infrastructure, there are many ways to incorporate division and square root functionality into an FPU. The following is a brief introduction to the major design decisions, which we will explore in greater detail in subsequent sections. The primary issues fall into four categories:

- *Algorithm.* The choice of a divide/square root algorithm is by far the most important design decision. It involves a series of smaller decisions, and once made, largely determines the subsequent range of choices and the area

Table 1. Arithmetic performance of recent microprocessor FPUs for double-precision operands.

Design	Cycle time (ns)	Latency/throughput (cycles/cycles)			
		$a \pm b$	$a \times b$	$a \div b$	$\sqrt{a}$
DEC 21164 Alpha AXP	2.0	4/1	4/1	22-60/22-60*	†
Hal Sparc64	6.49	4/1	4/1	8-9/7-8	†
HP PA7200	7.14	2/1	2/1	15/15	15/15
HP PA 8000	5	3/1	3/1	31/31	31/31
IBM RS/6000 Power2	13.99	2/1	2/1	16-19/15-18*	25/24*
Intel Pentium	5.0	3/1	3/2	39/39	70/70
Intel Pentium Pro	7.52	3/1	5/2	30*/30*	53*/53*
Mips R8000	13.33	4/1	4/1	20/17	23/20
Mips R10000	3.64	2/1	2/1	18/18	32/32
PowerPC 604	5.56	3/1	3/1	31/31	†
PowerPC 620	7.5	3/1	3/1	18/18	22/22
Sun SuperSparc	16.67	3/1	3/1	9/7	12/10
Sun UltraSparc	4	3/1	3/1	22/22	22/22

* Inferred from available information

† Not supported

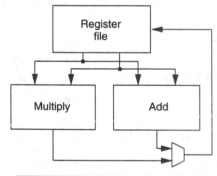

(a)

Operation	Latency	Throughput
Addition	2	1
Multiplication	2	1

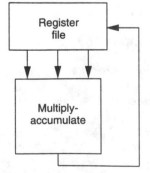

(b)

Operation	Latency	Throughput
Multiply-accumulate	2	1

Figure 1. Typical independent add-multiply configuration (a) and typical multiply-accumulate configuration (b).

and performance of the implementation in general.

- *Issue rate.* Although some chips execute floating-point instructions strictly in series, many of the most recent microprocessors can issue up to two floating-point instructions per cycle. Therefore, the next issue worthy of attention is the impact of floating-point issue rate on the efficiency of divide/square root implementations.
- *Connectivity.* Closely related to issue rate is the manner in which the divide/square root hardware interfaces with the rest of the FPU, either with independent access to the register file, or coupled to or integrated with one of the other functional units.
- *Replication.* Finally, it is becoming more common for microprocessors to duplicate floating-point functional units, including the divide/square root hardware. The performance and area effects of such replication deserve closer consideration.

These issues are not independent but interact with each other in complex and nonobvious ways. The following sections explain the connections between the issues, while trying to isolate and quantify their effects on the design trade-offs.

Simulation

To provide a uniform, quantitative foundation for the analysis of the design trade-offs, we implemented an FPU-level simulator that executes a benchmark based on Givens rotations. (For a description of Givens rotations, please see the supplement to this article on the World Wide Web at http://www.computer.org/pubs/micro/micro.htm.) Given an FPU configuration, including its structure and performance on individual operations, the simulator calculates the number of cycles required to transform an arbitrary matrix into triangular form using Givens rotations. The simulations use a standard set of test matrices ranging in size from 10×10 elements to 200×100 elements.[4]

We chose the Givens rotation algorithm because it is rich in divide and square root operations, and because it is a real program with important numerical applications. The primary task of many scientific programs is the solution of partial differential equations. Givens rotations are a central component of several frequently used solution techniques. They also feature prominently in signal processing algorithms, and the rotation and projection algorithms common in graphics and solid modeling use similar combinations of operations.

The Givens rotation algorithm is not typical in its use of divide/square root operations. In fact, there are probably few other real applications whose performance is so dependent on the efficiency of division and square root. Nevertheless, what it lacks in representativeness, it makes up for in importance. Finally, as a kind of divide/square root torture test, Givens rotations clearly illustrate the effects of different implementation choices.

Divide/square root algorithms

There are many possible algorithms for computing division and square root, but only a subset are currently practical for microprocessor implementation. These fall into the two categories of multiplicative and subtractive algorithms.[9]

Multiplicative methods use hardware integrated with the floating-point multiplier and have low to moderate latencies, while subtractive methods generally employ separate circuitry and have low to high latencies. We consider only the case that implements division and square root in hardware using the same algorithm. We cover the arguments for this decision in more detail elsewhere;[4] in summary, division and square root tend to have very similar implementations and allow for extensive hardware sharing. Furthermore, this is the most common case in actual practice.

Multiplicative techniques. Multiplicative algorithms compute an estimate of the quotient or square root using iterative refinement of an initial guess. They reduce divide and square root operations into a series of multiplications, subtractions, and bit shifts. In addition, multiplicative algorithms typically converge at a quadratic rate, which means the number of accurate digits in the estimate doubles at each iteration. The two techniques used in recent microprocessors are the Newton-Raphson method and Goldschmidt's algorithm. With many features in common, they differ primarily in operation ordering, which affects their mapping onto pipelined hardware.

Newton-Raphson method. An algorithm with a long history, the Newton-Raphson method[10,11] has been implemented in many machines, including the IBM RS/6000 series. To find a/b, set the "seed value" x_0 to an estimate of $1/b$ and iterate over

$$x_{i+1} = x_i \times (2 - b \times x_i)$$

until x_{i+1} is within the desired precision of $1/b$. Then $a \times x_{i+1} \approx a/b$.

For square root computation, set the seed to approximate $1/\sqrt{a}$ and refine it with the iteration

$$x_{i+1} = 1/2 \times x_i \times (3 - a \times x_i^2).$$

The product $a \times x_{i+1}$ yields an approximation of $\sqrt{a}$.

Goldschmidt's algorithm. Recent implementations of this algorithm[11] have included the Sun SuperSparc and arithmetic coprocessors from Texas Instruments. To compute $a/b = x_0/y_0$, iteratively calculate

$$x_{i+1} = x_i \times r_i \text{ and } y_{i+1} = y_i \times r_i.$$

Here, $r_i = 2 - y_i$; then $y_i \to 1$ and $x_i \to a/b$. Both x_0 and y_0 should be prescaled by a seed value approximating $1/b$ to ensure rapid convergence. To compute $\sqrt{a}$, set $x_0 = y_0 = a$, prescale by an estimate of $1/\sqrt{a}$, and iterate over

$$x_{i+1} = x_i \times r_i^2 \text{ and } y_{i+1} = y_i \times r_i,$$

where $r_i = (3 - x_i)/2$, such that

$$x_{i+1}/y_{i+1}^2 = x_i/y_i^2 = 1/a.$$

Then $x_i \to 1$, and hence $y_i \to \sqrt{a}$.

Implementations. Multiplicative divide/square root implementations generally take the form of modifications to the floating-point multiplier. This is because divide and square root functionality alone do not justify the area required for a second multiplier. The designers of the IBM RS/6000 FPU chose the Newton-Raphson method as being best suited to the multiply-accumulate structure. In independent add-multiply configurations, Goldschmidt's algorithm is a natural choice for pipelined multipliers. Since the numerator and denominator operations are independent, clever scheduling allows the multiplier to run at maximum efficiency. The Newton-Raphson method suffers from dependencies between successive operations, which hobble pipelining.

The block diagram in Figure 2 shows one possible implementation, namely an independent floating-point multiplier modified for Goldschmidt's algorithm.

Although the details will vary from case to case, most multiplicative implementations feature one or more of the following hardware enhancements:

- *Extra routing and storage.* These are required to render divide and square root as atomic operations independent of the register file.
- *Seed value lookup tables.* An algorithm's execution time relates directly to the accuracy of the initial guess. An accurate double-precision value can be produced in three iterations with an 8-bit seed, while a 16-bit seed can bring the number of iterations down to two. Unfortunately, the lookup table for the latter case must store 512 times as many bits.[4]
- *Constant subtraction/shifting logic.* For independent multipliers, supporting the algorithm's nonmultiply operations directly improves performance and maintains independence from the floating-point adder. Multiply-accumulate structures have these capabilities built in.
- *Last-digit rounding support.* Values derived from reciprocal estimates, as in the Newton-Raphson method and Goldschmidt's algorithm, have an inherent error that can lead to inaccuracies in the last result digit. The several strategies for solving this nontrivial problem carry associated area and performance trade-offs.[4] Multiply-accumulate units can avoid extra hardware at the expense of extra operations.

Multiplicative divide/square root implementations can incur a penalty beyond the area needed for extra logic, storage, and routing. Since some of the required modifications lie on the multiplier's critical path, they can negatively affect

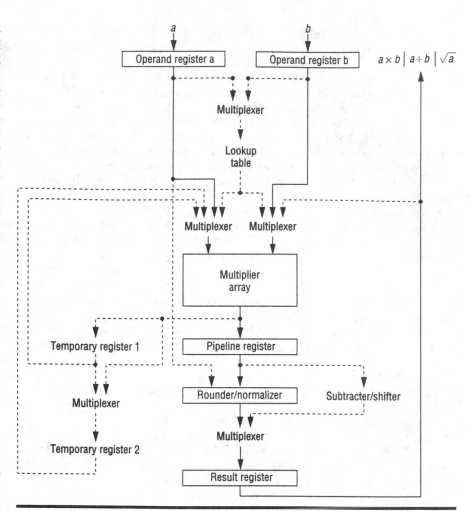

Figure 2. A floating-point multiplier enhanced for multiplicative divide/square root computation. Shading indicates new components, and dashed lines indicate new routing required for divide-square root.

the latency of multiplication itself.

Subtractive techniques. The subtractive divide/square root algorithms employed in current microprocessors can generally be classified as SRT methods. These are named for D. Sweeny, J.E. Robertson, and K.D. Tocher, who independently developed very similar algorithms for division.[10] Subtractive division and square root compute the quotient q or square root s directly, one digit at a time. The conventional procedure for long division by hand is an algorithm of this type. In practice, subtractive algorithms typically use redundant representations of values and treat groups of consecutive bits as higher-radix digits to enhance performance.[12] For example, a radix-4 divider scans successive pairs of bits as single digits—interpreting two radix-2 values as a single radix-4 one.

Algorithm. For division, let $q[j]$ be the partial quotient at step j (where $q[n] = q$), and let $w[j]$ be the residual or partial remainder. The goal of the algorithm is to find the sequence of quotient digits that minimizes the residual at each step. To compute $q = x \div d$ for n-digit, radix-r values, set $w[0] = x$ and evaluate the recurrence

$$w[j + 1] = rw[j] - dq_{j+1},$$

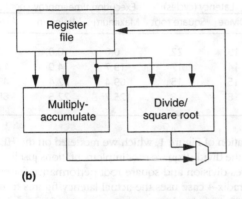

(a)

(b)

Figure 3. FPU topologies with subtractive divide/square root functional units and independent add-multiply (a) or multiply-accumulate configurations (b).

where q_{j+1} is the $(j + 1)$th quotient digit.

For square root computation, let $s[j]$ denote the jth partial root. To find $s = s[n] = \sqrt{x}$, set $w[0] = x$ and evaluate

$$w[j + 1] = rw[j] - 2s[j]s_{j+1} - s^2_{j+1}r^{-(j+1)}.$$

Define $f[j] = 2s[j] - s_{j+1}r^{-(j+1)}$; then

$$w[j + 1] = rw[j] - f[j]s_{j+1}$$

has the same form as the division recurrence. In practice, $f[j]$ is simple to generate, which facilitates combined division and square root implementations.

Implementations. Subtractive divide/square root implementations generally rely on dedicated hardware that executes in parallel with the remainder of the FPU, as shown in Figure 3. A minority of designs, such as the IBM/Motorola PowerPC line and the Mips R4000 series, use enhancements of the addition hardware. However, a separate functional unit offers superior performance, since the multiplication and addition hardware can continue to operate on independent instructions while the other unit computes the quotient or root. If multiplication hardware, for example, supports divide and square root, then these operations—which typically have long latencies—tie up the unit, holding up subsequent multiply instructions until the quotient or root is available.

Figure 4 shows a basic radix-4 divide/square root unit, including most of the features critical to its high performance. The residual is stored in redundant form as vectors of sum and carry bits; this enables the use of a low-latency carry-save adder to calculate the subtraction in the recurrence. The result-digit selection table returns the next quotient/square root digit on the basis of the residual and divisor/partial root values; the redundant result-digit set allows truncated input values, keeping the table small and fast. Factor generation logic keeps all possible multiples of d or $f[j]$ and every result digit (qj or $sj \in \{-2, -1, 0, 1, 2\}$) available at all times for immediate multiplexer selection. Finally, the on-the-fly conversion logic[12] operates concurrently with quotient computation and off of the critical path. This maintains updated values of $s[j]$ and $f[j]$ and incrementally converts the partial result from redundant into conventional representation.

While radix-2 divide/square root units operate on one bit at a time, radix-4 implementations retire two bits of the result at every iteration. Higher-radix

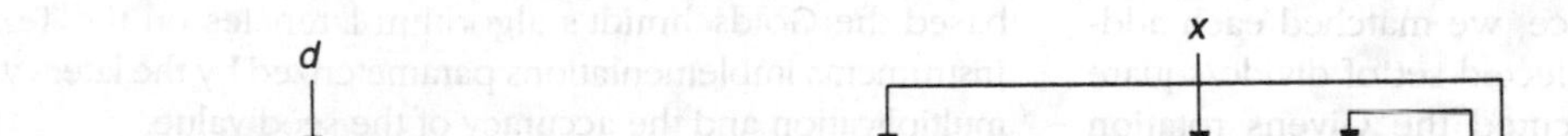

Figure 4. Radix-4 SRT divide/square root unit.

units process even larger groups of bits at once. Unfortunately, for radices greater than 8, the latency and area requirements of result digit selection and factor generation become prohibitive. To circumvent these effects, we can combine lower-radix stages into a single higher-radix unit. We can obtain radix-16 division, for example, by overlapping two radix-4 dividers, with only a modest increase in area and iteration delay over radix-4 alone.[12]

Certain methods achieve even higher radix operations; the majority of these are restricted to theoretical treatment or experimental implementations. Ercegovac, Lang, and Montuschi summarize and compare several of these approaches.[13]

One way to improve the performance of any subtractive implementation is to boost the number of iterations per machine cycle, if possible. In the HP PA7200, for example, the divide/square root unit's low iteration delay means that it can cycle at twice the system clock rate. Another technique that offers extremely low latencies is self-timing, whereby a functional unit operates asynchronously with respect to the rest of the FPU, completing each iteration at the highest rate its internal logic permits.[14] The Hal Sparc64 implements a version of self-timed division.

Performance impact. To determine how the algorithms compare in terms of performance, we matched each add-multiply configuration with a selected set of divide/square root implementations and executed the Givens rotation benchmark on the standard matrix data set. For each configuration (independent and multiply-accumulate) we assumed a floating-point issue rate of one instruction per cycle. We closely modeled the FPU structures and performance figures for individual operations on real examples. We compared the following divide/square root implementations:

- *8-bit seed multiplicative.* This is a baseline multiplicative implementation with an 8-bit seed lookup table, as used in many actual chips. The independent configuration uses Goldschmidt's algorithm, while the multiply-accumulate structure implements the Newton-Raphson method.
- *16-bit seed multiplicative.* As before, we matched this algorithm to the add-multiply configuration; the larger lookup table reduces the latency of divide/square root computation.
- *Radix-4 SRT.* The subtractive equivalent of the 8-bit seed case, this is a basic, practical implementation using a separate functional unit.
- *Radix-16 SRT.* Overlapping radix-4 units provide lower-latency operations.

The first set of experiments matches the independent con-

Table 2. Divide/square root performance of independent implementations.

Algorithm	Latency (cycles)		Execution time improvement (%)		
	Divide	Square root	Maximum	Minimum	Average
8-bit seed Goldschmidt	9	13	0.0	0.0	0.0
16-bit seed Goldschmidt	7	10	9.9	1.6	5.0
Radix-4 SRT	15	15	20.9	7.2	15.2
Radix-16 SRT	8	8	46.0	7.2	23.4

Table 3. Divide/square root performance of multiply-accumulate implementations.

Algorithm	Latency (cycles)		Execution time improvement (%)		
	Divide	Square root	Maximum	Minimum	Average
8-bit seed Newton-Raphson	19	22	0.0	0.0	0.0
16-bit seed Newton-Raphson	14	17	19.7	4.9	11.6
Radix-4 SRT	15	15	69.4	22.4	48.3
Radix-16 SRT	8	8	125.7	22.5	68.8

figuration of Figure 1, which we modeled on the HP PA7200, with the divide/square root implementations just listed. Table 2 gives division and square root performance for each case. The radix-4 case uses the actual latency figures from the HP PA7200. This implementation computes four quotient digits per machine cycle by clocking the divide/square root unit at double speed. For radix-16 operations, the figures assume that the configuration can accommodate the slightly longer cycle time of the overlapped implementation without difficulty. We based the Goldschmidt's algorithm latencies on the Texas Instruments implementations parameterized by the latency of multiplication and the accuracy of the seed value.

Table 2 also summarizes the relative performance of the different implementations on the Givens rotation benchmark, normalized to the 8-bit seed Goldschmidt case. The subtractive implementations clearly dominate the multiplicative ones, even in cases where the latter have superior latencies. It is the ability to execute in parallel with the rest of the FPU that gives the radix-4 and radix-16 units their decisive performance advantage.

The second set of experiments pairs the multiply-accumulate structure shown in Figure 2 with the standard set of divide/square root implementations. This configuration and the first divide/square root implementation in Table 3 are based on the IBM RS/6000 series FPU, which uses specially adapted versions of the Newton-Raphson method. The 16-bit seed version uses these same algorithms but assumes a reduced number of iterations. As for the subtractive implementations, we simply carried them over from the HP PA7200 case. Not only does this permit a more uniform comparison, but it seems like a feasible implementation, since the RS/6000 actually has a longer cycle time than the PA7200 while using a comparable fabrication technology.

The Givens rotation results, also shown in Table 3, display the same pattern as for the independent add-multiply case, albeit with even greater contrast. The Newton-Raphson

Table 4. Area comparison of two divide/square root implementations.

Device	Algorithm	Chip area (mm²)	Transistor count	Divider/square root area (mm²)
Weitek 3364	Radix-4 SRT	95.2	165,000	4.8
TI 8847	8-bit seed Goldschmidt	100.8	180,000	6.3

Table 5. Relative cost of different divide/square root implementations.

Algorithm	Area factor
8-bit seed multiplicative	1.0
16-bit seed multiplicative	22
Radix-4 SRT	1.5
Radix-16 SRT	2.2

operations' longer latencies mean that the subtractive implementations perform even better by comparison. Note also the significant improvement that the radix-16 implementation affords over the radix-4 case.

Area considerations. There are vast differences in the implementation styles of multiplicative and subtractive methods. The former are primarily enhancements of existing circuitry, while the latter inhabit completely new hardware. Furthermore, the core operations are entirely different. In the interest of an objective evaluation, it is essential to compare the circuit area required by these very distinct approaches.

Estimating area in a way that yields a fair comparison between chips is difficult because of basic differences in the implementation technologies. Nevertheless, it is possible to give some basis for evaluating the different implementations. Table 4 compares the size of the hardware dedicated to division in the Weitek 3364 and Texas Instruments 8847 arithmetic coprocessors; we based the figures on measurements of microphotographs.[11]

The chips have similar die sizes and device densities, and we can assume the feature sizes are comparable as well. Although the multiplication algorithms are different, both have two-pass arrays that take up approximately 22% of the chip area. In short, apart from their divide/square root implementations, these two chips have a lot in common. However, the area devoted to division hardware is little more than 6% of the chip size in either case. Also, the relative area requirements differ by 1.5%. These figures represent only two particular designs whose implementation technology is by now somewhat out of date. However, they suggest that 8-bit seed Goldschmidt and radix-4 SRT implementations are both economical, and that the area differences between them can be kept small.

An alternative approach uses standard-cell technology to estimate the areas of different implementations, including the more sophisticated options.[4] Table 5 shows the results of this study, with the values normalized to the size of the 8-bit seed Goldschmidt variant. According to these results, a radix-16 SRT unit need only be 45% larger than a radix-4

design in the same technology. A 16-bit seed Goldschmidt implementation, by contrast, could be over 20 times larger than the 8-bit seed implementation, and 10 times larger than the radix-16 SRT unit. This calls into question the practicality of such a solution, especially since it leads to a relatively small improvement in latency as evidenced by Tables 2 and 3.

Summary. Multiplicative implementations can provide low latency and lower cost than subtractive ones. However, their overall performance on our benchmark is inferior, mainly because the multiplication hardware is responsible for additional nonpipelined, multicycle operations. With the multiply-accumulate configuration, the same pipeline must accommodate addition, multiplication, division, and square root. Although 8-bit implementations tend to require less area than radix-4 hardware, 16-bit lookup tables dwarf even radix-16 implementations. Furthermore, the incredibly expensive upgrade from an 8-bit seed table to a 16-bit one offers only a modest reduction in divide/square root latency, and a downright meager improvement in benchmark execution time.

Subtractive divide/root implementations provide superior benchmark performance at a reasonable cost, for both independent and multiply-accumulate configurations. This is primarily due to the fact that subtractive hardware operates independently and does not tie up other FPU resources. The parallel operation also means that improvements in divide/square root latency have a more decisive impact than in multiplicative implementations. Upgrading from radix-4 to radix-16 can improve performance by as much as 21% for the independent case or 56% for the multiply-accumulate structure, at a cost of 33% more area.

Subtractive implementations also fit in nicely with the current trend toward decoupled superscalar processors with high instruction issue rates and growing transistor budgets. We conclude that implementations of subtractive divide/square root algorithms are the most sensible choice, and focus on them for the remainder of the discussion.

Issue rate

The simulations in the previous section assumed an issue rate of one floating-point instruction per machine cycle. This situation holds in such processors as the PowerPC series, where the FPU is a single pipeline, and the Intel Pentium Pro, where the arithmetic functional units share a single set of inputs and outputs. However, many recent machines, including the Mips 10000 and Sun UltraSparc, can generate up to two floating-point instructions at once.

Dual-issue operation only makes sense when there are at least two independent functional units. Furthermore, if one of the two functional units is a dedicated divide/square root unit, dual issue is not worthwhile because these instructions are not frequent enough to keep a separate unit busy. By this reasoning, dual floating-point instruction issue is appropriate for an FPU with an independent adder, multiplier, and divide/square root unit, but not one with a multiply-accumulate structure and separate divide/square root unit.

Performance impact. To explore the interaction of higher issue rates and divide/square root performance, we ran a series of experiments using the Givens rotation benchmark. For the reasons given earlier, we restricted our focus to independent add-multiply configurations with independent, subtractive divide/square root implementations. The variables are the instruction issue rate (single or dual) and the algorithm of the divide/square root implementation (radix-4 or radix-16). Performance figures for individual operations are the same as from previous experiments.

From the results in Table 6, it is apparent that increasing the floating-point instruction issue rate leads to a genuine performance boost. This effect tends to overshadow the effects of divide/square root latency. The dual-issue radix-4 case outperforms the single-issue radix-16 one, even though the latter is faster on individual operations. Also, the difference between the dual-issue radix-4 and radix-16 cases is much less significant than the contrast between cases with the same algorithm but differing issue rates. This is another instance, not unlike the contest between multiplicative and subtractive algorithms, where parallelism wins over latency. By keeping all of the units busier, especially the adder and multiplier, dual-instruction issue softens the impact of division latency.

We also performed a set of simulations to determine how dual-instruction issue affects the performance balance between multiplicative and subtractive methods. As shown in Table 7, multiple instruction issue strengthens the performance advantage of subtractive implementations. This is what one would expect, given the increased parallelism of operations afforded by a separate functional unit for divide/square root operations. These data support the claim that subtractive methods are better suited to superscalar implementations.

Area considerations. The choice of how many floating-point instructions to dispatch per cycle is a much larger issue than divide/square root implementation alone, and probably not entirely up to the floating-point designer. Nevertheless, if the option is available, there are several issues to consider. Obviously, there must be at least as many functional units as the maximum number of instructions issued per cycle. Also, allocating one instruction per cycle to divide/square root operations alone is a waste of resources. In addition, there is a cost to boosting the floating-point issue rate that we cannot readily quantify. Routing must be added to deliver operands and return results. The register file needs extra ports, which add to its area and access time. Finally, there are the necessary changes to the dispatch logic, which affect the processor as a whole.

Table 6. Effect of floating-point instruction issue rate on benchmark performance; execution time improvement for an independent configuration.

Algorithm	Issue rate	Execution time improvement (%)		
		Maximum	Minimum	Average
Radix-4 SRT	Single	0.0	0.0	0.0
Radix-16 SRT	Single	26.9	0.0	7.0
Radix-4 SRT	Dual	50.0	5.5	35.4
Radix-16 SRT	Dual	52.1	22.3	44.4

Table 7. Effect of algorithm choice on benchmark performance for dual-issue configurations.

Algorithm	Execution time improvement (%)		
	Maximum	Minimum	Average
8-bit seed Goldschmidt	0.0	0.0	0.0
16-bit seed Goldschmidt	11.4	2.3	6.2
Radix-4 SRT	37.2	4.5	20.1
Radix-16 SRT	46.7	14.1	28.9

Connectivity

In an FPU with a dual-issue, independent add-multiply configuration, there are several ways to connect subtractive divide/square root hardware. We can either package divide/square root circuitry as an independent functional unit as in the Sun UltraSparc, share ports with the adder as in the DEC 21164, or share ports with the multiplier as in the Mips R10000. Figure 5 (next page) illustrates the possible configurations. Each of these implementations computes division and square root concurrently with other FPU operations. However, certain configurations lead to contention for shared input and/or output ports.

Multiplicative divide/square root hardware is always integrated with the multiplier, so the connectivity issue is moot for those types of implementations. Sharing subtractive divide/square root unit ports with a multiply-accumulate unit is equivalent to single-issue FPU, a case we have already covered.

Performance impact. At first glance, sharing functional unit ports between divide/square root and other operations seems sensible, since addition and multiplication occur much more frequently. It is important, however, to minimize collisions between divide/square root and the host operations. The simulations we performed explore the impact of different interconnection schemes on benchmark performance.

In every case, we assumed an independent add-multiply configuration and dual floating-point instruction issue. The variables are the algorithm (radix-4 or radix-16) and the connectivity (independent, adder-coupled, or multiplier-coupled). In cases where operations share functional units, we pushed divide and square root operations back one cycle in the event of a port conflict.

The simulation results, given in Table 8, are somewhat

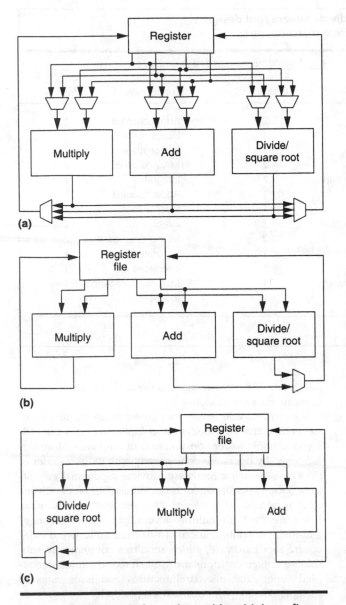

(a)

(b)

(c)

Figure 5. Dual-issue, independent add-multiply configurations with independent (a), adder-coupled (b), and multiplier-coupled (c) divide/square root units.

counterintuitive. First, for both algorithms, the independent and adder-coupled cases show exactly the same performance. This is a feature of the Givens rotations algorithm, which has approximately twice as many multiplies as adds, and, therefore, allocates many free issue slots to the adder. The adder-coupled case can take advantage of these openings, yielding the same performance as the independent case. The multiply-coupled configuration is hampered by frequent collisions between multiply and divide/square root operations, but the effects are relatively mild, especially for the radix-4 cases. With the shorter latency of radix-16 operations, the collisions have comparatively greater impact on performance, but still less than 7% on average.

Although the Givens rotation benchmark favors the adder-coupled design, we must not take its near 2:1 ratio of multiply to add operations as typical. In a survey of floating-point operations in the SPEC92fp benchmark suite, the multiply:add ratio is closer to 2:3.[2] This supports the intuitive suggestion that addition rather than multiplication dominates many applications, and that multiplier-coupled divide/square root is therefore the most appropriate choice. In every other design decision we have covered so far, improving the performance of the benchmark is not in conflict with enhancing the speed of all programs; this case is an exception.

Area considerations. The primary cost of a separate, independent divide/square root unit is the extra routing required, including the extra buses and the multiplexing/selection logic. Compare the FPU routing complexity of the independently connected divide/square root unit in Figure 5a with the adder- and multiplier-coupled cases in Figures 5b and 5c. When the divide/square root unit is coupled to one of the other functional units, we can readily split the two sets of ports on the register file between the two clusters of components. Connecting two sets of ports to three independent units is considerably more complicated. Similarly, independent divide/square root units complicate the instruction dispatch logic by increasing the number of possible paths and destinations for instructions.

Replication

Several microprocessor manufacturers have recently attempted to boost numerical performance by replicating floating-point functional units. The IBM RS/6000 Power2 and Mips R8000 feature dual multiply-accumulate units, and the Mips R10000 couples its independent multiplier with two other units for divide and square root. The HP PA8000 FPU consists of two multiply-accumulate units and two divide/square root units. Increasingly compact technology and higher maximum issue rates facilitate this trend.

There are also many different ways to replicate floating-point functionality and connect it with the rest of the system. For processors with FPUs organized as single, atomic pipelines or blocks, such as the PowerPC and IBM RS/6000 series, the most natural alternative is to replicate the entire

Table 8. Effect of divide/square root connectivity on benchmark performance; execution time improvement for an independent configuration.

Algorithm	Connectivity	Execution time improvement (%)		
		Maximum	Minimum	Average
Radix-4 SRT	Multiplier-coupled	0.0	0.0	0.0
Radix-4 SRT	Adder-coupled	4.4	1.4	2.8
Radix-4 SRT	Independent	4.4	1.4	2.8
Radix-16 SRT	Multiplier-coupled	12.1	0.1	3.8
Radix-16 SRT	Adder-coupled	26.3	2.3	10.6
Radix-16 SRT	Independent	26.3	2.3	10.6

Table 9. Summary of divide/square root design decisions for different architectures.

Design	Add-multiply configuration	Division/square root design decisions			
		Algorithm	FP issue rate (instr./cycle)	Connectivity	Units (no.)
DEC 21164 Alpha AXP	Independent	SRT	2	Adder-coupled	1
Hal Sparc64	MAC*	SRT	2	Independent	1
HP PA7200	Independent	SRT	2	Independent	1
HP PA8000	MAC	SRT	2	MAC-coupled	2
IBM RS/6000 Power2	MAC	Newton-Raphson	2	Integrated	2
Intel Pentium	Independent	SRT	1	Adder-coupled	1
Intel Pentium Pro	Independent	SRT	1	Independent	1
Mips R8000	MAC	Multiplicative	2	Integrated	2
Mips R10000	Independent	SRT	2	Multiplier-coupled	2
PowerPC 604	MAC	SRT	1	Integrated	1
PowerPC 620	MAC	SRT	1	Integrated	1
Sun SuperSparc	Independent	Goldschmidt	1	Multiplier-integrated	1
Sun UltraSparc	Independent	SRT	2	Independent	1

* Multiply-accumulate

unit. Decoupled superscalar designs like the Intel Pentium Pro and Sun SuperSparc are more at liberty to add individual units, such as an extra divide/square root unit.

Performance impact. There are many possible ways to analyze how replicating floating-point hardware affects divide/square root performance. We considered one example particularly worthy of examination—namely, the duplication of divide/square root hardware in a fully independent FPU. However, from a preliminary investigation, we concluded that the addition would have no effect on Givens rotation benchmark performance, unless we heavily modified the method of scheduling operations. Even with a much more aggressive schedule, we expect the effects would be insignificant.

Addressing the performance impact of replicating floating-point functionality in general is beyond the scope of this investigation. However, since the Givens rotation benchmark represents an unusually high level of divide/square root use, replicating the hardware for these functions alone seems to offer little potential benefit.

Area considerations. It is a more efficient use of area to improve a slow divide/square root unit than to replicate it. For example, in a dual-issue, independent FPU, upgrading from radix-4 to radix-16 gives a maximum improvement of 22.5% and an average improvement of 7.5% on the Givens rotation benchmark. Duplication of the radix-4 unit, on the other hand, gives no appreciable performance benefit. The upgrade to radix-16 costs only 45% of the radix-4 area, as opposed to 100% for replication, not including all of the extra routing and any changes to the register file or instruction issue logic.

FLOATING-POINT DIVISION AND SQUARE ROOT

are important operations warranting fast, efficient imple-

mentations. Table 9 summarizes the design decisions made in recent microprocessors.

Subtractive divide/square root implementations offer latencies competitive with the fastest multiplicative ones, but with a considerably lower consumption of chip area. More significantly, by operating concurrently with the remainder of the FPU, subtractive hardware provides a potentially significant boost in performance, as our simulations on the Givens rotation benchmark evidence. This performance advantage holds true for both multiply-accumulate and independent add-multiply configurations. Unlike multiplicative divide/square root hardware, which serializes computation, subtractive implementations are well suited to exploit decoupled superscalar microarchitectures and issue rates of multiple floating-point instructions per cycle.

For subtractive implementations, increasing the instruction issue rate from one to two instructions per cycle can also dramatically improve performance—by over 35% on average for the Givens rotation benchmark. As long as the implementation does not squander the extra instruction on divide/square root operations alone, this seems like a worthwhile improvement, taking into account the cost of routing and upgrading the register file and instruction issue logic.

The cost and complexity of adding a fully independent divide/square root unit to a dual-issue, independent add-multiply configuration are considerable. Therefore, coupling this functionality to one of the existing structures by sharing connections to the register file seems like a reasonable economizing measure. The worst-case performance impact seems modest compared with the possible explosion in routing costs and the additional scheduling complications incurred. The balance of the evidence suggests that the multiplier is the best candidate for sharing ports with the divide/square root unit.

The merits of replicating floating-point functional units,

however, remain to be seen. Certainly, even with highly parallel algorithms, one can expect significantly less than twice the performance from a doubling of hardware. In particular, duplicating divide/square root hardware alone appears to hold few advantages, even for the Givens rotation benchmark with its heavy use of these operations.

The growing popularity of 3D graphics and multimedia applications, and the allocation of the computational burden to the CPU (as for example, in Intel's MMX technology), mean ever-increasing demands for microprocessor arithmetic performance. The optimal integration of multimedia functionality with existing floating-point infrastructure is an area ripe for future research. ⊓⊔

Acknowledgments

The National Science Foundation partly supported this research under contract CCR-9257280. Peter Soderquist's support included a fellowship from the National Science Foundation. Miriam Leeser's support included an NSF Young Investigator award.

References

1. "The Pentium Papers," Mathworks, Inc., Natick, Mass., Nov. 1994, http://www.mathworks.com/README.html.
2. S.F. Oberman and M.J. Flynn, "Design Issues in Floating-Point Division," Tech. Report CSL-TR-94-647, Stanford University, Departments of Electrical Engineering and Computer Science, Stanford, Calif., Dec. 1994.
3. S.E. McQuillan, J.V. McCanny, and R. Hamill, "New Algorithms and VLSI Architectures for SRT Division and Square Root," *Proc. 11th IEEE Symp. Computer Arithmetic*, IEEE, Piscataway, N.J., 1993, pp. 80-86.
4. P. Soderquist and M. Leeser, "Area and Performance Tradeoffs in Floating-Point Division and Square Root Implementations," *ACM Computing Surveys*, Vol. 28, No. 3, Sept. 1996, pp. 518-564.
5. S.E. McQuillan and J.V. McCanny, "A VLSI Architecture for Multiplication, Division, and Square Root," *Proc. 1991 Int'l Conf. Acoustics, Speech and Signal Processing*, IEEE, 1991, pp. 1205-1208.
6. W. Kahan, *Using MathCAD 3.1 on a Mac*, unpublished article, Aug. 1994; available upon request from author.
7. B.K. Bose et al., "Fast Multiply and Divide for a VLSI Floating-Point Unit," *Proc. Eighth IEEE Symp. Computer Arithmetic*, IEEE, 1987, pp. 87-94.
8. *ANSI/IEEE Std. 754–1985, Binary Floating-Point Arithmetic*, IEEE, New York, 1985.
9. P. Soderquist and M. Leeser, "An Area/Performance Comparison of Subtractive and Multiplicative Divide/Square Root Implementations," *Proc. 12th IEEE Symp. Computer Arithmetic*, IEEE, 1995, pp. 132-139.
10. N.R. Scott, *Computer Number Systems and Arithmetic*, Prentice Hall, Englewood Cliffs, N.J., 1985.
11. D. Goldberg "Appendix A, Computer Arithmetic," in J.L. Hennessy and D.A. Patterson, *Computer Architecture: A Quantitative Approach*, Morgan Kaufmann Publishers, San Francisco, 1990.
12. M.D. Ercegovac and T. Lang, *Division and Square Root: Digit Recurrence Algorithms and Implementations,* Kluwer Academic Publishers, Norwell, Mass., 1994.
13. M.D. Ercegovac, T. Lang, and P. Montuschi, "Very High Radix Division with Selection by Rounding and Prescaling," *Proc. 11th IEEE Symp. Computer Arithmetic*, IEEE, 1993, pp. 112-119.
14. T.E. Williams, "A Zero-Overhead Self-Timed 160-ns 54-b CMOS Divider," *IEEE J. Solid-State Circuits*, Vol. 26, No. 11, Nov. 1991, pp. 1651-1661.

167 MHz Radix-8 Divide and Square Root Using Overlapped Radix-2 Stages

J. Arjun Prabhu and Gregory B. Zyner

SPARC Technology Business, Sun Microsystems, Inc.
Mountain View, California

Abstract - UltraSPARC's IEEE-754 compliant floating point divide and square root implementation is presented. Three overlapping stages of SRT radix-2 quotient selection logic enable an effective radix-8 calculation at 167 MHz while only a single radix-2 quotient selection logic delay is seen in the critical path. Speculative partial remainder and quotient calculation in the main datapath also improves cycle time. The quotient selection logic is slightly modified to prevent the formation of a negative partial remainder for exact results. This saves latency and hardware as the partial remainder no longer needs to be restored before calculating the sticky bit for rounding.

I. INTRODUCTION

The SRT algorithm provides a means of performing non-restoring division [1], [2]. More bits of quotient are developed per iteration with higher radices, but at a cost of greater complexity. A simple SRT radix-2 floating point implementation (Fig. 1) requires that the divisor and dividend both be positive and normalized, $1/2 \le D, Dividend < 1$. The initial shifted partial remainder, $2PR[0]$, is the dividend. Future partial remainders are developed according to the following equation,

$$PR_{i+1} = 2PR_i - q_{i+1}D. \qquad (1)$$

where q is the quotient digit {-1, 0, or +1} which is solely determined by the value of the previous partial remainder and is independent of the divisor, an attractive feature for square root. Discussion of the quotient digit selection function will be deferred until the next section. The partial remainder is often kept in redundant form so that carry-save adders can be used instead of slower and larger carry-propagate adders. The partial remainder is converted to non-redundant form after the desired precision is reached. The quotient digits can also be kept in redundant form and converted to non-redundant form at the end, or the quotient and quotient minus one (Q and $Q-1$) can be generated on the fly according to rules developed by Ercegovac [3].

The SRT algorithm can also be used for square root

allowing utilization of existing division hardware. The simplified square root equation looks surprisingly similar to that of division [4]:

$$PR_{i+1} = 2PR_i - q_{i+1}(2Q_i + q_{i+1}2^{-(i+1)}). \qquad (2)$$

The terms in parentheses are the effective divisor. For square root, the so-called divisor is a function of the previous quotient bits (root bits to be more precise) [4], hence on-the-fly quotient generation is required.

Quotient selection logic (qslc) for a radix-2 SRT implementation will be discussed in Sections II and III with emphasis on how it can be modified to better constrain the partial remainder in the case of exact results. Preventing the partial remainder from unnecessarily going negative for exact results leads to a one cycle savings in generating the sticky bit.

It will be demonstrated in Section IV that overlapping radix-2 quotient selection logic stages can provide an effective timing solution for generating multiple bits of quotient per cycle. The timing benefits of speculative datapath calculations of the partial remainder, quotient, and next divisor occurring in conjunction with quotient digit selection will be shown. A comparison of overlapped radix-2 versus radix-4 implementations will also be discussed.

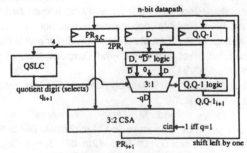

Fig. 1. Simple SRT Radix-2 Divide, Square Root Implementation

II. QUOTIENT SELECTION LOGIC

The logic which generates quotient selection digits is the central element of an SRT division implementation.

Reprinted from *Proceedings of the 12th Symposium on Computer Arithmetic*, pp. 155-162, July 1995.

Early research indicated that only the upper three bits of the redundant partial remainder are necessary inputs for a radix-2 quotient digit selection function [5], [6]. However, more recent studies have shown that four bits are required to correctly generate quotient digit selection digits and keep the partial remainder within prescribed bounds [7], [8], [9], [10]. The selection rules can be expressed as:

$$q_{i+1} = \begin{cases} 1, & \text{if } 0 \le 2PR_i \le 3/2 \\ 0, & \text{if } 2PR_i = -1/2 \\ -1, & \text{if } -5/2 \le 2PR_i \le -1 \end{cases} \quad (3)$$

The quotient selection logic is designed to guess correctly or overestimate the true quotient result. e.g. predicting 1 instead of 0, or 0 instead of -1. The SRT algorithm corrects itself later if the wrong quotient digit has been chosen.

TABLE I
TRUTH TABLE FOR RADIX-2 QUOTIENT SELECTION LOGIC

$2PR_{i,estimated}$	quotient digit$_{i+1}$	comments
100.0	don't care	2PR never < -5/2
100.1	don't care	2PR never < -5/2
101.0	-1*	2PR never < -5/2, but 2PR could be 101.1 when 2PR$_{est}$ is 101.0
101.1	-1	
110.0	-1	
110.1	-1	
111.0	-1	2PR could = 111.1
111.1	0	2PR could = 000.0
000.0	+1	
000.1	+1	
001.0	+1	
001.1	+1	
010.0	don't care	2PR never > 3/2
010.1	don't care	2PR never > 3/2
011.0	don't care	2PR never > 3/2
011.1	don't care	2PR never > 3/2

The truth table for SRT radix-2 quotient selection logic has several don't care inputs because the partial remainder is constrained $-5/2 \le 2PR_i \le 3/2$. The estimated partial remainder is always less than or equal to the true partial remainder because the lower bits are ignored. There is a single case, $2PR_{est} = 101.0$, where the estimated partial remainder can appear to be out of bounds. By construction, the real partial remainder is within the negative bound, so -1 is the appropriate quotient digit to select. There are two other cases where the quotient digit selected based on the estimated partial remainder differs from what would be chosen based on the real partial remainder. However, in both of these instances of "incorrect" quotient digit selection, the quotient is not underestimated and the partial remainder is kept within prescribed bounds, so the final result will still be generated correctly.

For increased testability, most designs today are scannable. All registers are stitched together in a chain to allow a special test mode known as scan. During scan, external values can be optionally be shifted into these registers, the system can be clocked for one or more cycles, and the new register values can be shifted out and observed. These capabilities aid in functional and timing debug.

While loading the scan chain, the partial remainder flip flops can take on any value. Logic simplification for don't care cases should ensure that a unique quotient digit is always selected (i.e. the quotient digits selects are 1-hot) for all input combinations to prevent contention of multiplexer selects. The simplified truth table follows.

TABLE II
SIMPLIFIED QSLC TRUTH TABLE

$2PR_{i,estimated}$	quotient digit$_{i+1}$
0xx.x	+1
111.1	0
1xx.x	-1

III. STICKY BIT CALCULATION

Floating point operations generate a sticky bit along with the result in order to indicate whether the result is inexact. The sticky bit is also used with the guard and round bits for rounding according to IEEE Standard 754 [11]. For divide and square root operations, the sticky bit is determined by checking if the final partial remainder is non-zero. The final partial remainder is defined as the partial remainder after the desired number of quotients bits have been calculated. Since the partial remainder is in redundant form, a carry-propagate addition is performed prior to zero-detection (See Fig. 3a).

A. Exact Results

At first glance, the above solution seems perfectly reasonable for all final partial remainder possibilities, positive or negative. However, in the rare case where the result is exact, the final partial remainder will be equal to the negative divisor. For example, consider a number divided by itself (Fig.2). Since the dividend is positive and normalized, the quotient digit from the first iteration is one. For the next iteration, the partial remainder is zero which causes the second quotient digit to be one. For all subsequent iterations, the partial remainder will equal the negative divisor and quotient digits of minus one will be selected. After the last iteration, performing a sign detect on the final partial remainder indicates that $Q-1$ should be chosen which is in fact the correct result. However, this same final partial remainder is non-zero which erroneously suggests an inexact result.

507

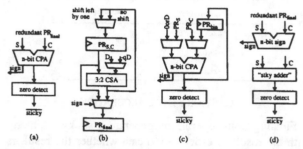

PR[0] = init dividend/2	= D/2	q[1]	=+1	Q=1	Q-1=0
PR[1] = 2(D/2) - (1)D	= 0	q[2]	=+1	Q=11	Q-1=10
PR[2] = 2(0) - (1)D	=- D	q[3]	=- 1	Q=101	Q-1=100
PR[3] = 2(-D) - (-1)D	=- D	q[4]	=- 1	Q=1001	Q-1=1000
PR[n] = 2(-D) - (-1)D	=- D	q[n+1]	=- 1	Q=100..001	Q-1=100..000

Fig. 2. Divide iterations for a number divided by itself.

This problem extends to any division operation for which the result should be exact because the quotient selection logic is defined to guess positive for a zero partial remainder and correct for it later. The simplest solution is to restore negative final partial remainders by adding the divisor before performing zero-detection. Given the area expense of an additional carry-propagate adder, the solution should try to take advantage of existing hardware. Two ways to achieve this are shown in Figs. 3b and 3c.

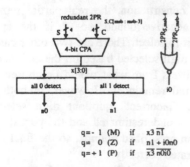

Fig. 3. Zero-detection and sign-detection on the final partial remainder.

Option 1 (Fig. 3b) takes advantage of existing csa hardware for restoration while option 2 (Fig. 3c) re-uses the carry-propagate adder. Both alternatives add extra multiplexer hardware and require the sticky bit calculation to take an additional cycle when the preliminary final partial remainder is negative. The first option especially impacts cycle time for basic iterations since the multiplexer is on the partial remainder formation critical path. Variable latency instructions in a pipelined superscalar processor make instruction scheduling and bypass control logic much more complex and is generally undesirable. Thus the net effect of restoring negative partial remainders is to add another cycle of latency for all divide and square root operations.

B. Improved Quotient Selection Algorithm

Enhancing the quotient digit selection function to prevent formation of a negative partial remainder for exact results is an ideal solution because it saves hardware and improves latency. This can be achieved by choosing a quotient digit of zero instead of one when the partial remainder is zero. This suggests choosing $q=0$ for $2PR_{est}$=000.0. Since the quotient digit selection function works on an estimated partial remainder, caution is required. An estimated partial remainder (shifted) could appear to be less than 1/2 when, in reality, adding the

lower bits causes a 1 to propagate into the lowermost bit of the upper four bit partial remainder.

If the full partial remainder is 1/2 or above, $q=1$ should always be chosen since the divisor is constrained $1/2 \le D < 1$. The true quotient bit could be one. It will be corrected later if the divisor was greater than the partial remainder. There is no way to correct for underestimation if $q=0$ is selected when $q=1$ was the correct quotient digit. The result will be irreversibly incorrect and the next partial remainder will be out of bounds.

Performing binary addition on the full partial remainder eliminates the estimation problem, but defeats the timing benefits of SRT division. $q=0$ could be chosen only when the full partial remainder is zero, but such detection would also be detrimental to timing.

A simple alternative is to detect a possible carry propagation into the least significant bit of the partial remainder. This can be done by looking at the fifth most significant bits of the redundant partial remainder, $PR_{S,msb-4}$ and $PR_{C,msb-4}$. If they are both zero, then propagation is not possible, and $q=0$ should be chosen; otherwise $q=1$ should be chosen. Even though lower bits of the partial remainder could be non-zero, the partial remainder is still within prescribed bounds and the correct result will be generated. As far as timing is concerned, the carry-propagate addition is still performed on four bits.

TABLE III
REFINED QSLC TRUTH TABLE

2PR$_{i,estimated}$	quotient digit$_{i+1}$
000.0 (2PR$_{S,C,msb-4}$ both 0)	0
0xx.x	+1
111.1	0
1xx.x	-1

Fig. 4 shows a logic implementation of this enhanced quotient digit selection function. In practice, the four bit adder and subsequent logic are merged into five stages of logic for more efficient timing and area utilization.

redundant 2PR S,C[msb : msb-3]

4-bit CPA

x[3:0]

all 0 detect   all 1 detect

$q=-1$ (M) if x3 $\overline{n1}$
$q=0$ (Z) if n1 + i0n0
$q=+1$ (P) if $\overline{x3}$ $\overline{n0i0}$

Fig. 4. Simple implementation of modified QSLC.

508

The number of additional gates needed to implement the new radix-2 quotient digit selection logic is relatively small. From Spice analysis, the impact on the qslc critical timing path was under five percent. There is an implementation dependent timing trade-off between slower quotient selection logic and eliminating the partial remainder restoration cycle at then end. Notice that if slowing down qslc doesn't limit the processor cycle time, there will always be a performance gain from saving one cycle of latency.

C. Parallel Sign Calculation and Zero-detection

It is possible to save hardware while also performing sign detection and sticky bit calculation in parallel as shown in Fig. 3d. Instead of using a full 59 bit adder, a 59 bit sign detect adder can be used, slightly improving timing, but mainly saving area. Zero-detection can be done without an explicit addition to convert the redundant partial remainder into binary.

$$t_i = \left(s_i \oplus c_i\right) \oplus \left(s_{i-1} + c_{i-1}\right) \quad (4)$$

where s_i and c_i are the sum and carry values of the final partial remainder. The sticky bit is computed by:

$$sticky = t_0 + t_1 + \dots + t_n \quad (5)$$

This method generates inputs to the zero-detector with a 3-input xor delay instead of the delay of a 59 bit carry-propagate adder, a significant net savings.

IV. OVERLAPPING RADIX-2 STAGES

The biggest performance gain is achieved by maximizing the number of iterations per cycle. As the number of result bits formed per cycle increases, the relative importance of saving one cycle of latency, as described in Section III, grows. A straightforward way of generating n result bits per cycle is to serialize the basic SRT radix-2 implementation. This solution is not attractive since the critical path includes n quotient selection logic delays and n carry save adder delays.

A. Optimal Timing via QSLC Overlapping

Overlapping quotient selection logic for the first and second iterations [12] as shown in Fig. 5 yields better timing results since only one qslc is in the critical path. Overlapping is achieved by performing $+D$ and $-D$ operations on the upper bits of the partial remainder while the first quotient selection bit is being determined. In this way, the second quotient digit selection can start before the first is finished.

Maximal overlapping can be extended to three bits per cycle to have an effective radix-8 implementation.

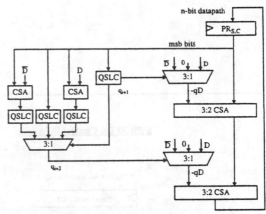

Fig. 5. Two overlapped SRT radix-2 QSLC stages.

Analyzing the possible partial remainders needed for the third quotient digit selection, as depicted in Fig. 6, shows that only seven csa's and qslc's are needed rather than nine. Since quotient selection logic is area intensive, a 22% reduction is quite beneficial.

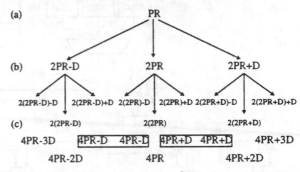

Fig. 6. Possible partial remainder values after the first and second iterations. (a) initial PR_i, (b) three possible PR_{i+1}, (c) seven possible PR_{i+2}.

Further timing improvement is also possible. By speculatively calculating the next partial remainder, quotient, and divisors for each possible quotient selection digit, the delay of a datapath carry-save adder can be masked by the longer qslc operation occurring in parallel. Fig. 7 shows the overall *UltraSPARC* floating point divide, square root implementation. The critical path is reduced to 1 qslc, 2 csa's, and 3 muxes.

There is a timing, area trade-off. Overlapping improves timing at a cost of additional speculative hardware. The focus of this study is to optimize timing to the utmost with area minimization as a secondary goal.

B. Extension

In theory, n qslc stages can be overlapped. Assuming speculative datapath partial remainder calculations each iteration, the critical path for n bits per cycle is 1 qslc, (n-1) csa's, and n muxes. The incremental timing cost is one csa and one mux. There are 2^n-1 partial remainder

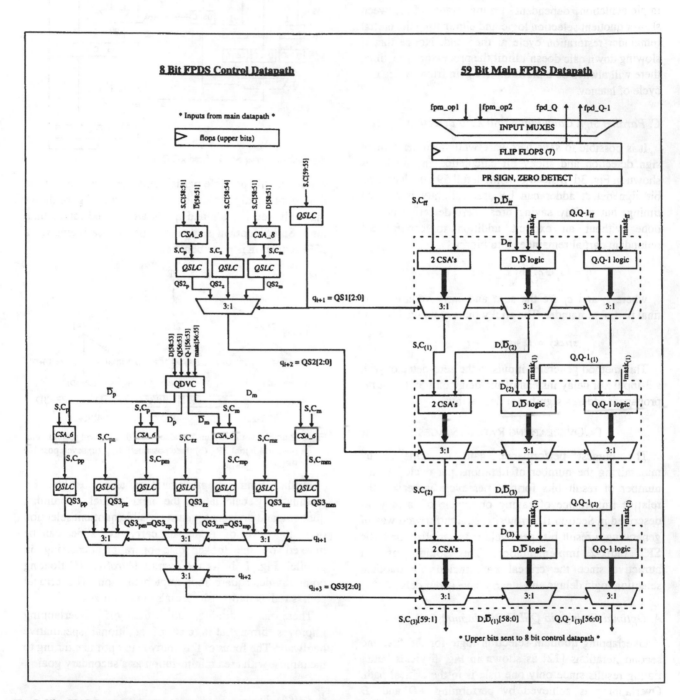

Fig. 7. *Ultra*SPARC radix-8 floating point divide, square root implementation with three overlapped radix-2 stages and speculative datapath calculation.

possibilities for the nth qslc stage. They are in the range:

$$2^{n-1}PR_i +[-(2^{n-1}-1)D,..,0,..,(2^{n-1}-1)D]. \qquad (6)$$

Therefore, the incremental qslc cost for the nth overlapped stage is 2^n-1 rather than 3^{n-1} as suggested by Taylor [12]. In practice, overlapping two to four stages makes the most sense. As n gets higher, the number of qslc's grows exponentially making the area cost prohibitive. In addition, greater fanout leads to increased wire and gate loads which significantly lessen the timing benefits. Table IV summarizes timing and qslc cost considerations for overlapped radix-2 stages.

TABLE IV
PERFORMANCE, COST TABLE FOR MAXIMUM OVERLAPPING

stages	critical path	tot qslc's	delta critical path	delta qslc's
1	qs + mux	1	-	-
2	qs + pr + 2mux	4	pr + mux	3
3	qs + 2pr + 3mux	11	pr + mux	7
4	qs + 3pr + 4mux	26	pr + mux	15
n	qs + (n-1)pr +(n)mux	$\sum_{i=1}^{n} 2^i - 1$	pr + mux	2^n - 1

There is a limit on how much overlapping is necessary to achieve the optimal radix-2 implementation critical path as illustrated in a four bit per cycle example. Suppose the quotient selection logic delay is three times a carry-save adder delay. Then the quotient digits from the first stage qslc will be ready at the same time as the third stage partial remainder bits are entering the fourth stage qslc's. The first level of three-to-one muxes following the fourth stage of qslc's can be moved before the quotient selection logic as shown in Fig. 8. Thus, the fourth stage number of qslc's is reduced from fifteen to seven while achieving the same timing as with maximum overlapping. In general, the optimal degree of overlapping will depend on the relative csa and qslc delays, and need only be sufficient to mask the delay of previous quotient selection logic stages.

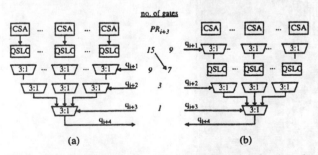

Fig. 8. Optimal radix-2 timing with (a) maximum overlapping, (b) reduced overlapping.

C. Overlapped Radix-2 versus Radix-4

Overlapping radix-2 stages yields better timing results than overlapping radix-4 stages. With radix-4, the upper eight bits of the redundant partial remainder and upper four bits of the divisor are sent to the quotient selection logic. The logic cannot be merged as with radix-2, so there is an explicit eight bit binary addition followed by a ten input, forty-four product term PLA [12].

Comparison of the critical path gate delays (gd) for two bits per cycle and four bits per cycle confirms better timing from a radix-2 based solution. The critical paths for two bits per cycle are as follows.

$$radix\text{-}2: \quad qslc + csa + 2\ mux3. \qquad (8.8gd) \qquad (7)$$
$$\phantom{radix\text{-}2:}\quad 5 \qquad 1.8 \qquad (2 \times 1)$$

$$radix\text{-}4: \quad add8 + pla + mux5. \qquad (7gd+pla) \qquad (8)$$
$$\phantom{radix\text{-}4:}\quad 6 \qquad 1$$

Two overlapped radix-4 stages are implemented in the same way as depicted for radix-2 in Fig. 5. The critical paths for four bits per cycle are shown below.

$$radix\text{-}2: \quad qslc + 3\ csa + 4\ mux3. \qquad (14.4gd) \qquad (9)$$
$$\phantom{radix\text{-}2:}\quad 5 \qquad (3 \times 1.8) \qquad (4 \times 1)$$

$$radix\text{-}4: \quad csa + add8 + pla + 2\ mux5. \qquad (10)$$
$$\phantom{radix\text{-}4:}\quad 1.8 \qquad 6 \qquad (2 \times 1) = (9.8gd+pla)$$

A 10-input, 44-product term pla takes more than five gate delays, so for both two bits per cycle and four bits per cycle, overlapping radix-2 stages produces better timing. This analysis implies that a combined radix-4, radix-2 approach to yield an effective radix-8 solution is not faster than simply overlapping three radix-2 stages.

V. PROCESSOR IMPLEMENTATION

UltraSPARC performs non-blocking divide and square root operations. Generating three result bits per cycle at 167 Mhz, the latency is twelve cycles for single precision (SP) and twenty-two cycles for double precision (DP). The divide and square root unit contains seventy-thousand transistors implemented in 0.5um CMOS technology (Fig. 9).

Instructions are both issued and completed through the pipelined floating point multiplier which formats operands, calculates the exponent, and performs rounding. The floating point multiplier and divider share the same register file read ports for operands, so divide and multiply instructions are never issued at the same time. Therefore the multiplier is always available for operand formatting and exponent calculation when a divide instruction is issued. Multiply instructions can be issued during subsequent cycles while the divider is iterating. Four cycles prior to division completing, a

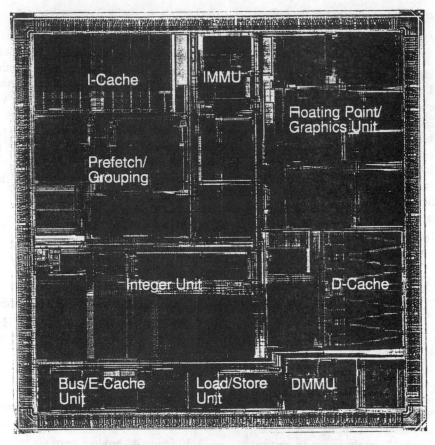

Fig. 9. *Ultra*SPARC (above), *Ultra*SPARC floating point, graphics unit (below).

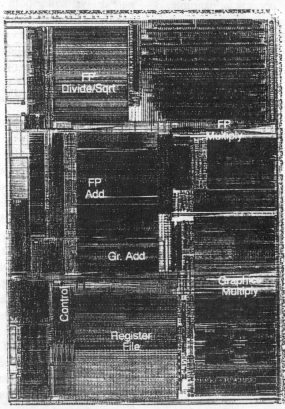

multiply slot is reserved so that the divider can re-use the final stage multiplier rounding hardware. Since *UltraSPARC* performs in-order execution, there will be a one cycle delay only in the unlikely event that the superscalar processor has been able to issue independent instructions for 8 cycles (SP) or 18 cycles (DP) and an independent multiply instruction is ready to be issued during this exact cycle. Thus making use of the multiplier yields significant hardware savings with near zero performance impact.

The modified quotient selection logic algorithm slowed the internal floating point divider critical path by approximately 100ps, or by less than two percent. The limiting timing path on the processor was slower, so there was no negative impact from the improved qslc. The full benefit of eliminating the cycle for restoring the partial remainder prior to sticky detection was realized.

Division and square root operations were thoroughly tested with 100% toggle coverage and 100% finite state machine arc coverage in a stand-alone test (SAT) environment, at the cpu level, and in silicon. Over 860,000 directed vectors and 140,000 random vectors were applied at the SAT level. Included were pseudo-exhaustive double precision tests for division in which all combinations of the upper eight bits of the dividend and divisor were sequenced (64K vectors). The same was done for square root in which all combinations of the upper thirteen bits of square root operands for odd and even exponents were sequenced (16K vectors).

A radix-2 solution was preferable over radix-4 for a number of reasons including timing, as described earlier, and greater flexibility. From the outset, it was known that it might be necessary to scale back the number of bits per cycle due to timing or area considerations. That raised the specter of having to reduce to two bits per cycle in a radix-4 implementation. Going with radix-2 provided a better safety option since three bits per cycle yields 6% better overall floating point performance (SPECfp92) than two bits per cycle [13].

Implementing square root did not come for free. While it is true that square root completely re-uses existing divide hardware, additional logic was required to generate the so-called divisors used in square root. In addition, the quotient had to be developed on the fly, as opposed to using a shift register and assimilating the redundant bits during the final cycle, since it was needed for the divisor calculation. Additional datapath feedthrough tracks were also necessary. An estimated 15% of the area went to support square root. With careful design, the logical critical path for square root was made the same as for divide. The additional area required to support square root, though, did impact timing since wire delays were greater.

VI. CONCLUSION

High-speed floating point division and square root is achieved by overlapping radix-2 quotient selection logic stages and speculatively calculating the partial remainder, quotient, and next divisor. An enhanced quotient digit selection function prevents the working partial remainder from becoming negative if the result is exact. This translates into a one cycle savings since negative partial remainders no longer need to be restored before calculating the sticky bit.

ACKNOWLEDGMENT

The authors would like to thank Marc Tremblay and Guy Steele for reviewing the preliminary draft, and Nasima Parveen and Richard Landes for their contributions to verification and physical design.

REFERENCES

[1] J. E. Robertson, "A new class of digital division methods," *IEEE Trans. Comput.*, vol. C-7, pp. 218-222, Sept. 1958.

[2] K. D. Tocher, "Techniques of multiplication and division for automatic binary computers," *Quart. J. Mech. Appl. Math.*, vol. 11, pt. 3, pp. 364-384, 1958.

[3] M. D. Ercegovac and T. Lang, "On-the-fly rounding," *IEEE Trans. Comput.*, vol. 41, no. 12, pp. 1497-1503, Dec. 1992.

[4] M. D. Ercegovac and T. Lang, "Radix-4 square root without initial PLA," *IEEE Trans. Comput.*, vol. 39, no. 8, pp. 1016-1024, Aug. 1990.

[5] S. Majerski, "Square root algorithms for high-speed digital circuits," *Proc. Sixth IEEE Symp. Comput. Arithmetic.* pp. 99-102, 1983.

[6] D. Zuras and W. McAllister, "Balanced delay trees and combinatorial division in VLSI," *IEEE J. Solid-State Circuits*, vol. SC-21, no. 5, pp. 814-819, Oct. 1986.

[7] M. D. Ercegovac and T. Lang, *Division and Square Root: Digit-recurrence Algorithms and Implementations*. Kluwer Academic Publishers, 1994, ch 3.

[8] S. Majerski, "Square-rooting algorithms for high-speed digital circuits," *IEEE Trans. Comput.*, vol. C-34, no. 8, pp. 724-733, Aug. 1985.

[9] P. Montuschi and L. Ciminiera, "Simple radix 2 division and square root with skipping of some addition Steps," *Proc. Tenth IEEE Symp. Comput. Arithmetic.* pp. 202-209, 1991.

[10] V. Peng, S. Samudrala, and M. Gavrielov, "On the implementation of shifters, multipliers, and dividers in floating point units," *Proc. Eighth IEEE Symp. Comput. Arithmetic.* pp. 95-101, 1987.

[11] "IEEE standard for binary floating-point arithmetic," ANSI/IEEE Standard 754-1985, New York, The Institute of Electrical and Electronic Engineers, Inc., 1985.

[12] G. S. Taylor, "Radix 16 SRT dividers with overlapped quotient selection stages," *Proc. Seventh IEEE Symp. Comput. Arithmetic.* pp. 95-101, 1985.

[13] M. Tremblay, "A fast and flexible performance simulator for micro-architecture trade-off analysis on UltraSPARC," Submitted to the 1995 *Design Automation Conference*.

Radix-4 Square Root Without Initial PLA

MILOŠ D. ERCEGOVAC, MEMBER, IEEE, AND TOMAS LANG

Abstract—A systematic derivation of a radix-4 square-root algorithm using redundant residual and result is presented. Unlike other similar schemes it does not use a table lookup or PLA for the initial step, resulting in a simpler implementation without any time penalty. The scheme can be integrated with division and also incorporates an on-the-fly conversion and rounding of the result, thus eliminating a carry-propagate step to obtain the final result. The result-digit selection uses 3 bits of the result and 7 bits of the estimate of the residual.

Index Terms—Digital arithmetic, digit-recurrence, on-the-fly conversion, radix-4 square root, redundant representation.

I. INTRODUCTION

SEVERAL implementations of radix-4 square root have been presented in the literature [9], [6], [5], [10], [8]. In all these cases, a table lookup or a special PLA is included for the determination of the first few bits of the result, whereas another PLA implements the result-digit selection for the remaining radix-4 digits. In this paper, we present a systematic derivation of the algorithm and show that this initial PLA is not necessary; this results in a simpler implementation without any penalty in execution time. As in the other designs, the implementation can be combined with division: the result-digit selection and the recurrence form are identical in all steps.

To obtain a fast implementation, as done in [6], [5], [10], and [8], redundant addition is used and the result-digit selection depends on low-precision estimates of the residual and of the partial result. This requires that the result digit be from a redundant digit-set. As the other referenced implementations, we use the set $\{-2, -1, 0, 1, 2\}$ to simplify the multiple formation required by the recurrence.

Two alternative redundant adders are possible: carry-save adder or signed-digit adder, each having advantages and drawbacks. The advantage of using a carry-save adder is that the adder slice is just a full adder, whereas the signed-digit adder is more complex and slower [7]. On the other hand, since the result is produced in a signed-digit representation and the partial result is one input to the adder, if carry-save addition is used this partial result has to be converted to conventional representation, whereas it can be used directly for signed-digit addition. We choose the carry-save alternative because

Manuscript received October 26, 1989; revised March 6, 1990. This work was supported in part by NSF Grant MIP-8813340, Composite Operations Using On-Line Arithmetic for Application-Specific Parallel Architecture: Algorithms, Design, and Experimental Studies.

The authors are with the Department of Computer Science, School of Engineering and Applied Science, University of California, Los Angeles, CA 90024.

IEEE Log Number 9036137.

the result has to be converted anyhow to conventional representation and to do this we use an on-the-fly converter, which after each iteration produces the partial result in conventional form. Moreover, during this conversion on-the-fly rounding is performed [3].

We describe a sequential implementation, in which the hardware of one iteration step is reused for each of the digits of the result. Of course, it is possible to have a combinational implementation, in which the iteration step is replicated. The selection between these alternatives is influenced by cost, speed, and throughput considerations.

The operation is defined by

$$s = x^{1/2} - \epsilon$$

where x is the operand, s is the result, and ϵ is the error. The algorithm is presented for *fractional* operand x and result s. For floating-point representation and normalized operand, it is necessary to scale the operand to have an even exponent (to allow the computation of the exponent). Consequently,

$$1/4 \leq x < 1 \quad 1/2 \leq s < 1.$$

If s has m fractional radix-4 digits then for a correct result the error is bounded so that

$$|\epsilon| < 4^{-m}.$$

In Section II, we describe the algorithm and in Sections III and IV we discuss its implementation.

II. RECURRENCE AND SQUARE ROOT STEP

The algorithm is based on a continued-sum recurrence. We now develop the digit-recurrence and show the implementation of the corresponding iteration step.

A. Recurrence and Bound

Each iteration of the recurrence produces one digit of the result, most significant digit first. Let us call $S[j]$ the value of the partial result after j iterations, that is

$$S[j] = \sum_{i=0}^{j} s_i 4^{-i} \quad s_i \in \{-2, -1, 0, 1, 2\} \quad (1)$$

(the digit s_0 is needed to represent a result value $s \geq 2/3$, since the representation of s is in signed-digit form and the maximum value of s_i is 2).

The final result is then

$$s = S[m] = \sum_{i=0}^{m} s_i r^{-i}$$

Reprinted from *IEEE Transactions on Computers*, Vol. 39, No. 8, pp. 1016-1024, August 1990.

and the result has to be correct for m-digit precision, that is,

$$|x^{1/2} - s| < 4^{-m}. \tag{2}$$

We define an error function ϵ so that its value after j steps is

$$\epsilon[j] = x^{1/2} - S[j]. \tag{3}$$

To have a correct final result this error has to be bounded. Since

$$s = S[j] + \sum_{i=j+1}^{m} s_i 4^{-i}$$

we get from (2) and (3)

$$\min\left(\sum_{i=j+1}^{m} s_i 4^{-i}\right) - 4^{-m} < \epsilon[j] < \max\left(\sum_{i=j+1}^{m} s_i 4^{-i}\right) + 4^{-m}.$$

Since the minimum (maximum) digit value is -2 (2), we get

$$-\frac{2}{3} 4^{-j} \leq \epsilon[j] \leq \frac{2}{3} 4^{-j}. \tag{4}$$

Introducing (3) in (4) and transforming to eliminate the square root operation (add $S[j]$ and obtain the square), we get

$$\frac{4}{9} 4^{-2j} - \frac{4}{3} 4^{-j} S[j] + S[j]^2 \leq x \leq \frac{4}{9} 4^{-2j}$$
$$+ \frac{4}{3} 4^{-j} S[j] + S[j]^2.$$

Subtracting $S[j]^2$ we obtain

$$\frac{4}{9} 4^{-2j} - \frac{4}{3} 4^{-j} S[j] \leq x - S[j]^2 \leq \frac{4}{9} 4^{-2j} + \frac{4}{3} 4^{-j} S[j]. \tag{5}$$

That is, we have to compute $S[j]$ such that $x - S[j]^2$ is bounded according to (5). We now define a residual (or partial remainder) w so that

$$w[j] = 4^j(x - S[j]^2). \tag{6}$$

From (5) the bound on the residual is

$$-\frac{4}{3} S[j] + \frac{4}{9} 4^{-j} \leq w[j] \leq \frac{4}{3} S[j] + \frac{4}{9} 4^{-j}. \tag{7}$$

Since from (1) $S[-1] = 0$, we get from (6) the initial condition

$$w[-1] = 4^{-1} x. \tag{8}$$

From (6) and (1) the basic recurrence of the square root algorithm is

$$w[j+1] = 4w[j] - 2S[j]s_{j+1} - s_{j+1}^2 4^{-(j+1)}. \tag{9}$$

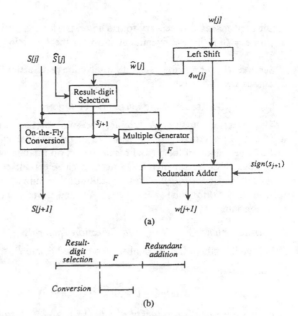

Fig. 1. (a) Square root step. (b) Timing.

B. Implementation of Square Root Step

The square root algorithm consists in performing m iterations of the recurrence (9). Moreover, each iteration consists of four subcomputations [Fig. 1(a)].

1) One digit arithmetic left-shift of $w[j]$ to produce $4w[j]$.

2) Determination of the result digit s_{j+1} using a result-digit selection function $Select$. The value of the digit s_{j+1} is selected so that the application of the recurrence produces a $w[j+1]$ that satisfies the bound (7). The function has as arguments $\hat{w}[j]$ (an estimate of $4w[j]$) and $\hat{S}[j]$ (an estimate of $S[j]$) and produces s_{j+1}. That is,

$$s_{j+1} = Select(\hat{w}[j], \hat{S}[j]).$$

3) Formation of

$$F = -2S[j]s_{j+1} - s_{j+1}^2 4^{-(j+1)}. \tag{10}$$

4) Addition of F and $4w[j]$ to produce $w[j+1]$.

The four subcomputations are executed in sequence as indicated in the timing diagram of Fig. 1(b). Note that no time has been allocated for the arithmetic shift since it is performed just by suitable wiring. Moreover, the relative magnitudes of the delay of each of the components depend on the specific implementation.

To have a fast recurrence step we use a carry-save adder (a signed-digit adder could also be used) and a result-digit selection that depends on low-precision estimates of the residual and of the partial result. To achieve this, it is necessary to have a redundant representation of the result digit. As indicated before, we use the symmetric digit-set

$$s_i \in \{-2, -1, 0, 1, 2\} \tag{11}$$

because it allows a simpler implementation of the adder input F (no multiple of three is required). Moreover, the signed

515

result-digit makes it necessary to use an on-the-fly conversion to produce $S[j]$ in a conventional form for the formation of F.

Section III presents the result-digit selection and Section IV discusses the formation of F.

III. RESULT-DIGIT SELECTION FUNCTION

The selection function determines the value of the result digit s_{j+1} as a function of an estimate of the residual $w[j]$ and of an estimate of the partial result $S[j]$. Two fundamental conditions must be satisfied by a result-digit selection: *containment* and *continuity*. These conditions determine a *selection interval* for each value of s_{j+1}. We now develop these selection intervals.

A. Containment Condition and Selection Intervals

Let the bounds of the residual $w[j]$ be called $\underline{B}$ and $\bar{B}$, that is,

$$\underline{B}[j] \leq w[j] \leq \bar{B}[j]. \tag{12}$$

Define the *selection interval* of $4w[j]$ for $s_{j+1} = k$ to be $[L_k, U_k]$. That is, L_k (U_k) is the smallest (largest) value of $4w[j]$ for which it is *possible* to choose $s_{j+1} = k$ and keep $w[j + 1]$ bounded. Therefore, from the recurrence

$$L_k[j] \leq 4w[j] \leq U_k[j] \rightarrow \underline{B}[j + 1] \leq 4w[j]$$
$$-2S[j]k - k^2 4^{-(j+1)} \leq \bar{B}[j + 1]. \tag{13}$$

Consequently,

$$U_k[j] = \bar{B}[j + 1] + 2S[j]k + k^2 4^{-(j+1)}$$
$$L_k[j] = \underline{B}[j + 1] + 2S[j]k + k^2 4^{-(j+1)}. \tag{14}$$

Since $\bar{B}[j]$ and $\underline{B}[j]$ are the upper bound of the interval for $k = 2$ and the lower bound for $k = -2$, respectively, we get

$$U_2[j] = 4\bar{B}[j] \quad L_{-2}[j] = 4\underline{B}[j].$$

Introducing these values in (14) we get

$$\bar{B}[j + 1] + 2S[j] \times 2 + 2^2 4^{-(j+1)} = 4\bar{B}[j]$$

$$\underline{B}[j + 1] - 2S[j] \times 2 + 2^2 4^{-(j+1)} = 4\underline{B}[j]. \tag{15}$$

This results in

$$\bar{B}[j] = \frac{4}{3} S[j] + \frac{4}{9} 4^{-j}$$

$$\underline{B}[j] = -\frac{4}{3} S[j] + \frac{4}{9} 4^{-j}. \tag{16}$$

To show that (16) is correct it is sufficient to replace in (15). Note that these bounds are identical to those obtained in (7). If this were not the case, we would use the tighter bounds.

From (14) and (16), the selection intervals are given by the

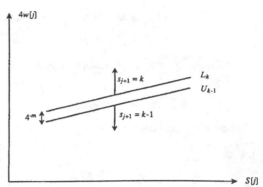

Fig. 2. Continuity.

expressions

$$U_k[j] = \frac{4}{3} S[j + 1] + \frac{4}{9} 4^{-(j+1)} + 2S[j]k + k^2 4^{-(j+1)}$$

$$L_k[j] = -\frac{4}{3} S[j + 1] + \frac{4}{9} 4^{-(j+1)} + 2S[j]k + k^2 4^{-(j+1)}.$$

Since $S[j + 1] = S[j] + k \times 4^{-(j+1)}$ we get

$$U_k[j] = 2S[j] \left(k + \frac{2}{3} \right) + \left(k + \frac{2}{3} \right)^2 4^{-(j+1)} \tag{17a}$$

$$L_k[j] = 2S[j] \left(k - \frac{2}{3} \right) + \left(k - \frac{2}{3} \right)^2 4^{-(j+1)}. \tag{17b}$$

B. Continuity Condition and Overlap Between Selection Intervals

A second requirement for the selection interval is the *continuity condition*. It states that for any value of $4w[j]$ between $4\underline{B}[j]$ and $4\bar{B}[j]$ it must be possible to select *some* value for the result digit. This can be expressed as

$$U_{k-1} \geq L_k - 4^{-m}$$

where, as shown in Fig. 2, the term 4^{-m} appears because of the granularity of the representation of the residual with m radix-4 digits.

Moreover, to use estimates of $4w[j]$ and $S[j]$ for the result-digit selection, it is necessary to have an overlap between adjacent selection intervals. For the square root operation with digit-set $\{-2, -1, 0, 1, 2\}$ from (17) the overlap is

$$U_{k-1} - L_k = \frac{1}{3} (2S[j] + (2k - 1)4^{-(j+1)}). \tag{18}$$

Note that the bounds, selection intervals, and the overlap depend on j.

C. Staircase Result-Digit Selection for Residual in Redundant Form

As indicated before, to have a fast recurrence step, a redundant adder is used. We now determine a staircase result-digit selection using an estimate of the partial result and an estimate of the shifted residual obtained by truncating the redundant form.

516

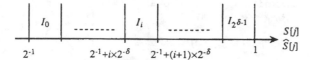

2⁻¹   2⁻¹+i×2⁻ᵟ   2⁻¹+(i+1)×2⁻ᵟ   1

Fig. 3.   Generic intervals.

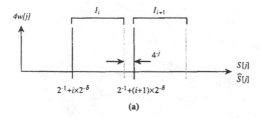

(a)

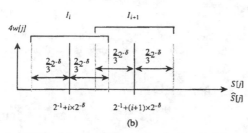

(b)

Fig. 4.   Selection intervals. (a) Truncation of conventional form. (b) Truncation of signed-digit form.

As illustrated in Fig. 3, the values of the estimate of $S[j]$, called $\hat{S}[j]$, divide the range of $S[j]$ into intervals I_i (one interval per value of the estimate) so that

If $\hat{S}[j] = 2^{-1} + i \times 2^{-\delta}$ then $S[j] \in I_i$ for $0 \leq i \leq 2^{\delta-1}$
(19)

where δ is the number of fractional bits of the estimate $\hat{S}[j]$. Note that the value of $\hat{S}[j]$ for $i = 0$ is 2^{-1}, since the result is normalized, and that $\hat{S}[j] = 1$ for $i = 2^{\delta-1}$.

Fixed and variable estimate of S[j]: Since the result is produced one digit per iteration in signed-digit form, the following alternatives can be used to form the estimate of $S[j]$ [1] (this is in contrast to division, where the estimate is always obtained by truncating the conventional representation of the divisor):

a) Use the truncated conventional representation of $S[j]$. As discussed in Section IV, this conventional representation has to be formed on-the-fly anyhow since we use a carry-save adder for the recurrence. In this case, if $\hat{S}[j]$ is obtained by truncating $S[j]$ to δ fractional bits, then as shown in Fig. 4(a), I_i (the ith interval) is defined by

$$2^{-1} + i \times 2^{-\delta} \leq S[j] < 2^{-1} + (i + 1) \times 2^{-\delta}$$

$$0 \leq i \leq 2^{\delta-1}.$$

However, since $S[j]$ has only j fractional radix-4 digits, the upper bound of the interval is restricted so that I_i becomes

$$I_i = [2^{-1} + i \times 2^{-\delta}, 2^{-1} + (i + 1) \times 2^{-\delta} - 4^{-j}). (20)$$

b) Use directly the truncated signed-digit representation or

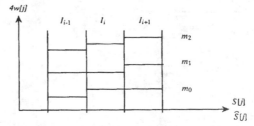

Fig. 5.   Staircase result-digit selection.

(equivalently) the fixed value of $S[\lceil \delta/2 \rceil]$ (that is, the value of $S[j]$ immediately after at least δ fractional bits are produced). In this case [Fig. 4(b)], the ith interval is defined by

$$I_i = \left(2^{-1} + i \times 2^{-\delta} - \frac{2}{3} \times 2^{-\delta}, 2^{-1} + i \times 2^{-\delta} + \frac{2}{3} \times 2^{-\delta}\right).$$
(21)

The result-digit selection depends (slightly) on which of the two methods is used. However, since the difference is not significant, we will use method a) in our implementation.

Staircase Result-Digit Selection: Using the estimate of the result $\hat{S}[j]$, the result-digit selection is described by the set of *selection constants*

$$\{m_k(i) | 2^{-1} + i \times 2^{-\delta} \in \text{set of values of } \hat{S}[j],$$

$$k \in \{-2, -1, 0, 1, 2\}\}. (22)$$

That is, there is one selection constant per value of $\hat{S}[j]$ and per value of the result digit. In terms of these selection constants, the *staircase result-digit selection* is defined by

$$s_{j+1} = k \text{ if } \hat{S}[j] = 2^{-1} + i \times 2^{-\delta} \text{ and}$$

$$m_k(i) \leq \hat{w}[j] < m_{k+1}(i) (23)$$

where $\hat{w}[j]$ is an estimate of the shifted residual $4w[j]$ obtained by truncating the redundant form to t fractional bits. This is illustrated in Fig. 5.

For the case of the residual represented in carry-save form, the error introduced by using the estimate (with respect to the truncated residual in conventional two's complement representation) is

$$0 \leq \text{error} < 2^{-t}$$

as shown in Fig. 6(a). Consequently, the result-digit selection has to satisfy the relations [see Fig. 6(b)] [4]

$$m_k(i) \geq \max(L_k(I_i))$$

$$m_k(i) + 2^{-t} \leq \min(U_{k-1}(I_i)). (24)$$

Note that the relations depend on j, the iteration number; consequently, the selection constants can, in general, be different for different j.

From these expressions, the minimum overlap required for a feasible result-digit selection is

$$\min(U_{k-1}(I_i)) - \max(L_k(I_i)) \geq 2^{-t}. (25)$$

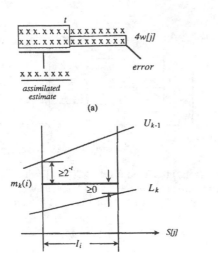

(a)

(b)

Fig. 6. (a) Error in the estimate—carry-save form. (b) Conditions on selection constants.

For method a) of formation of the estimate $\hat{S}[j]$, from (17) and (20) we get for $\min(U_{k-1}(I_i))$ and $\max(L_k(I_i))$

For $k > 0$

$$\min(U_{k-1}(I_i)) = 2(2^{-1} + i \times 2^{-\delta})\left(k - \frac{1}{3}\right)$$
$$+ \left(k - \frac{1}{3}\right)^2 4^{-(j+1)}$$

$$\max(L_k(I_i)) = 2(2^{-1} + (i+1) \times 2^{-\delta})\left(k - \frac{2}{3}\right)$$
$$- \left(k - \frac{2}{3}\right)\left(\frac{26}{3} - k\right)4^{-(j+1)}. \quad (26a)$$

For $k \leq 0$

$$\min(U_{k-1}(I_i)) = 2(2^{-1} + (i+1) \times 2^{-\delta})\left(k - \frac{1}{3}\right)$$
$$+ \left(\frac{1}{3} - k\right)\left(\frac{25}{3} - k\right)4^{-(j+1)}$$

$$\max(L_k(I_i)) = 2(2^{-1} + i \times 2^{-\delta})\left(k - \frac{2}{3}\right)$$
$$+ \left(k - \frac{2}{3}\right)^2 4^{-(j+1)}. \quad (26b)$$

These expressions are used to determine the result-digit selection. However, since they depend on j, a different selection function might result for different j. If we want to have a single selection function, we need to develop expressions that are independent of j. For $\min(U_{k-1}(I_i))$, the term depending on j is always positive and approaching zero for large j; therefore, this term can be neglected. On the other hand, for $\max(L_k(I_i))$, the term depending on j is negative for $k > 0$ and can be neglected, but is positive for $k \leq 0$ so it cannot be neglected and we have to use its maximum value (which occurs for $j = 0$). Consequently, the corresponding expressions

independent of j are as follows:

For $k > 0$

$$\min(U_{k-1}(I_i)) = 2(2^{-1} + i \times 2^{-\delta})\left(k - \frac{1}{3}\right)$$

$$\max(L_k(I_i)) = 2(2^{-1} + (i+1) \times 2^{-\delta})\left(k - \frac{2}{3}\right). \quad (27a)$$

For $k \leq 0$

$$\min(U_{k-1}(I_i)) = 2(2^{-1} + (i+1) \times 2^{-\delta})\left(k - \frac{1}{3}\right)$$

$$\max(L_k(I_i)) = 2(2^{-1} + i \times 2^{-\delta})\left(k - \frac{2}{3}\right) + \left(k - \frac{2}{3}\right)^2 4^{-1}. \quad (27b)$$

To determine whether a single selection function is possible we apply (25) to (27). The worst case is $i = 0$ and $k = -1$, resulting in

$$\min(U_{-2}(I_0)) - \max(L_{-1}(I_0))$$
$$= -\frac{8}{3}(2^{-1} + 2^{-\delta}) + \frac{10}{3}(2^{-1}) - \frac{25}{9}(4^{-1})$$
$$= -\frac{13}{36} - \frac{8}{3}2^{-\delta} \geq 2^{-t}. \quad (28)$$

Since there is no pair of values of t and δ that satisfy this inequality, no single selection function exists for all j. A possible alternative is to find a value J so that a single selection can be used for $j \geq J$ and then consider separately the cases for $j < J$.

For the case $j \geq J$, the same considerations given before produce

For $k > 0$

$$\min(U_{k-1}(I_i)) = 2(2^{-1} + i \times 2^{-\delta})\left(k - \frac{1}{3}\right)$$

$$\max(L_k(I_i)) = 2(2^{-1} + (i+1) \times 2^{-\delta})\left(k - \frac{2}{3}\right). \quad (29a)$$

For $k \leq 0$

$$\min(U_{k-1}(I_i)) = 2(2^{-1} + (i+1) \times 2^{-\delta})\left(k - \frac{1}{3}\right)$$

$$\max(L_k(I_i)) = 2(2^{-1} + i \times 2^{-\delta})\left(k - \frac{2}{3}\right)$$
$$+ \left(k - \frac{2}{3}\right)^2 4^{-(J+1)}. \quad (29b)$$

Introducing these expressions in (25) (for the worst case $i = 0$, $k = -1$) we get [similar to (28)]

$$\frac{1}{3} - \frac{25}{36}(4^{-J}) - \frac{8}{3}(2^{-\delta}) \geq 2^{-t}. \quad (30)$$

TABLE I
SELECTION INTERVALS FOR $j \geq 3$ ($t = 4$)

i	0	1	2	3	4	5	6	7
$\hat{S}[j]$	8/16	9/16	10/16	11/16	12/16	13/16	14/16	15/16, 16/16
$ML_2(i), m\hat{U}_1(i)$	3/2, 77/48	5/3, 29/16	11/6, 97/48	2, 107/48	13/6, 39/16	7/3, 127/48	15/6, 137/48	8/3, 49/16
$ML_1(i), m\hat{U}_0(i)$	3/8, 29/48	5/12, 11/16	11/24, 37/48	1/2, 41/48	13/24, 15/16	7/12, 49/48	5/8, 53/48	2/3, 19/16
$ML^*_0(i), m\hat{U}_{-1}(i)$	-2/3, -7/16	-3/4, -23/48	-5/6, -25/48	-11/12, -9/16	-1, -29/48	-13/12, -31/48	-7/6, -11/16	-5/4, -35/48
$ML^*_{-1}(i), m\hat{U}_{-2}(i)$	-5/3, -25/16	-15/8, -83/48	-25/12, -91/48	-55/24, -33/16	-15/6, -107/48	-65/24, -115/48	-35/12, -41/16	-75/24, -133/48

A solution is $\delta = 4$, $J = 3$, $t = 3$. However, when applying the conditions (24), one of the selection constants is $-29/16$ [4]; consequently, it is necessary to use $t = 4$. Table I shows the corresponding limits of the intervals. The following notation is used

$$m\hat{U}_{k-1}(i) = \min(U_{k-1}(I_i)) - \frac{1}{16}$$

$$ML_k(i) = \max(L_k(I_i)).$$

As shown in the expressions (29), for $k \leq 0$ the expressions for $ML_k(i)$ contain the term $(k - 2/3)^2 4^{-4}$. This term has the following values:

k	0	-1
$\dfrac{(k - 2/3)^2}{256}$	1/576	1/90

Since these values are small (compared to $2^{-t} = 1/16$), it is simpler to present in the table

$$ML^*_k(i) = \max(L_k(I_i)) - \left(k - \frac{2}{3}\right)^2 4^{-4}$$

instead of $ML_k(i)$, with the restriction that the selection constant $m_k(i)$ cannot be equal to $ML^*_k(i)$.

Now we have to consider the cases $j < 3$. One possible approach is to have an initial PLA to determine from a truncated x the value of $S[3]$. Another possibility is to analyze the cases $j < 3$ and then find a combined result-digit selection (which might depend on the value of j). As indicated in the Introduction, unlike other reported implementations, we follow this second approach because it potentially results in a simpler implementation.

For all cases, we use $\delta = 4$ and $t = 4$ to match with the case $j \geq 3$.

As indicated in expression (1), since the maximum value of the radix-4 digit is 2, it is necessary to have

$$s_0 = 1$$

to be able to represent values of $s \geq 2/3$. Consequently, $S[0] = 1$. Moreover, the values of s_1 are restricted to the set $s_1 \in \{0, -1, -2\}$ (to have $s < 1$).

Therefore, for $j = 0$ and $S[0] = 1$ we obtain

$$\hat{U}_{-1} = -2 \times 1 \times (1/3) + (1/9)(1/4) - 1/16 = -(101/144)$$

$$L_0 = -2 \times 1 \times (2/3) + (4/9)(1/4) = -(44/36)$$

TABLE II
SELECTION INTERVALS FOR $j = 1$

i	0	4	8
$\hat{S}[1] = S[1]$	1/2	3/4	1
$L_2(i), \hat{U}_1(i)(\times 144)$	208, 256	304, 376	400, 496
$L_1(i), \hat{U}_0(i)(\times 144)$	49, 91	73, 139	97, 187
$L_0(i), \hat{U}_{-1}(i)(\times 144)$	-	-140, -80	-188, -104
$L_{-1}(i), \hat{U}_{-2}(i)(\times 144)$	-	-335, -281	-455, -377

$$\hat{U}_{-2} = -2 \times 1 \times (4/3) + (16/9)(1/4) - 1/16$$

$$= -(329)/(144)$$

$$L_{-1} = -2 \times 1 \times (5/3) + (25/9)(1/4) = -(95)/(36).$$

For $j = 1$ we can have $s_2 \in \{-2, -1, 0, 1, 2\}$. Moreover, since $s_1 \in \{-2, -1, 0\}$ and $S[0] = 1$ the only possible values of $S[1]$ are 1/2, 3/4, and 1. Since $\delta = 4$ (because of the case $j \geq 3$), the estimate $\hat{S}[1]$ coincides with $S[1]$ and we need consider only the values of U_{k-1} and L_k for 1/2, 3/4, and 1 and not consider min or max values in intervals. The corresponding values of L_k and U_{k-1} are given in Table II. Note that it is not possible to select $s_2 < 0$ when $S[1] = 1/2$, because this would make $S[2] < 1/2$.

For $j = 2$, since we use $\delta = 4$ and the granularity of $S[2]$ is 1/16 again $\hat{S}[2] = S[2]$, so we use exact values of $S[2]$ instead of intervals for the computation of L_k and U_{k-1}. These values are shown in Table III.

We now need to obtain a single result-digit selection for $j = 0$, $j = 1$, $j = 2$, and $j \geq 3$. To do this we combine all previous tables into Table IV.

From Table IV we can see that for all entries except for $m_{-1}(8)$ it is possible to have a single selection function for all values of j; that is, for each pair (k, i) it is possible to find a selection constant that is inside the allowed intervals for all j. The single nonconforming case is $k = -1$, $i = 8$, and $j = 0$, for which the selection interval is $[-1520/576, -1216/576]$ which does not overlap with the corresponding intervals for $j \neq 0$. We have chosen to solve this problem by changing the estimate for $j = 0$ ($\hat{S}[0]$) from its normal value 1 to either 12/16 or 13/16; this satisfies the only two cases which are significant for $j = 0$, namely $m_0(8)$ and $m_{-1}(8)$.

In addition, to reduce the number of bits required for the estimate $\hat{S}[j]$, we convert $\hat{S}[j] = 1$ into $\hat{S}[j] = 15/16$ (for $j \neq 0$), that is we fold interval $i = 8$ onto $i = 7$; this can

TABLE III
Selection Intervals for $j = 2$

i $\hat{S}[2] = S[2]$	0 8/16	1 9/16	2 10/16	3 11/16	4 12/16	5 13/16	6 14/16	7 15/16	8 16/16
$L_2, U_1(\times576)$	784, 922	880, 1042	976, 1152	1072, 1262	1168, 1372	1264, 1482	1360, 1592	1456, 1702	1552, 1812
$L_1, U_0(\times576)$	193, 325	217, 373	241, 421	265, 469	289, 517	333, 565	357, 613	381, 661	405, 709
$L_0, \hat{U}_{-1}(\times576)$	-380, -254	-428, -278	-476, -302	-524, -326	-572, -350	-620, -374	-668, -398	-716, -422	-754, -444
$L_{-1}, \hat{U}_{-2}(\times576)$	-935, -815	-1045, -911	-1155, -1007	-1265, -1103	-1375, -1199	-1485, -1295	-1595, -1391	-1696, -1487	-1815, -1583

TABLE IV
Selection Intervals and Selection Constants for all Values of j

i $\hat{S}[j]$	0 8/16	1 9/16	2 10/16	3 11/16	4 12/16	5 13/16	6 14/16	7 15/16	8 16/16
$ML_2(i), m\hat{U}_1(i)$ $(\times576)$									
$j \geq 3$	864, 924	960, 1044	1056, 1164	1152, 1284	1248, 1404	1334, 1524	1440, 1644	1536, 1764	1536, 1764
$j = 2$	784, 922	880, 1042	976, 1152	1072, 1262	1168, 1372	1264, 1482	1360, 1592	1456, 1702	1552, 1812
$j = 1$	832, 1024				1216, 1504				1600, 1984
$j = 0$									-
$m_2(i)$	3/2	7/4	2	2	9/4	5/2	5/2	11/4	3
$m_2(i)\times576$	864	1008	1152	1152	1296	1440	1440	1584	1728
$ML_1(i), m\hat{U}_0(i)$ $(\times576)$									
$j \geq 3$	216, 348	240, 396	264, 444	288, 492	312, 540	336, 588	360, 636	384, 684	384, 684
$j = 2$	193, 325	217, 373	241, 421	265, 469	289, 517	333, 565	357, 613	381, 661	405, 709
$j = 1$	196, 364				292, 556				388, 748
$j = 0$									-
$m_1(i)$	1/2	1/2	1/2	1/2	3/4	3/4	1	1	1
$m_1(i)\times576$	288	288	288	288	432	432	576	576	576
$ML^*_0(i), m\hat{U}_{-1}(i)$ $(\times576)$									
$j \geq 3$	-384, -252	-432, -306	-570, -300	-528, -324	-576, -348	-624, -372	-672, -396	-720, -420	-720, -420
$j = 2$	-, -	-428, -278	-476, -302	-524, -326	-572, -350	-620, -374	-668, -398	-716, -422	-754, -444
$j = 1$	-				-560, -320				-752, -416
$j = 0$									-704, -404
$m_0(i)$	-1/2	-5/8	-3/4	-3/4	-3/4	-1	-1	-1	-1
$m_0(i)\times576$	-288	-360	-432	-432	-432	-576	-576	-576	-576
$ML^*_{-1}(i), m\hat{U}_{-2}(i)$ $(\times576)$									
$j \geq 3$	-960, -900	-1080, -996	-1200, -1092	-1320, -1188	-1440, -1284	-1560, -1380	-1680, -1476	-1800, -1596	-1800, -1596
$j = 2$	-, -	-1045, -911	-1155, -1007	-1265, -1103	-1375, -1199	-1485, -1295	-1595, -1391	-1696, -1487	-1815, -1583
$j = 1$	-				-1340, -1124				-1820, -1408
$j = 0$									-1520, -1216
$m_{-1}(i)$	-13/8	-7/4	-2	-17/8	-9/4	-5/2	-11/4	-23/8	*
$m_{-1}(i)\times576$	-936	-1008	-1152	-1224	-1296	-1440	-1584	-1656	

be done because the selection intervals are satisfied. Therefore, the estimate used in the selection $\hat{S} = (\hat{S}_1, \hat{S}_2, \hat{S}_3, \hat{S}_4)$ is obtained as follows:

$$(\hat{S}_1, \hat{S}_2, \hat{S}_3, \hat{S}_4)$$
$$= \begin{cases} (1, 1, 0, -) & \text{if } (j = 0) \\ (1, 1, 1, 1) & \text{if } (A_0 = 1) \text{ and } (j \neq 0) \\ (1, A_2, A_3, A_4) & \text{if } (j \neq 0) \end{cases}$$

where $(A_0, A_1, A_2, A_3, A_4)$ are the most significant bits of A, the conventional representation of $S[j]$ (as discussed in Section IV), and for $j = 0$, $A_0 = 1$ and $A_2 = A_3 = A_4 = 0$.

The resulting selection function is given in Table V and its implementation shown in Fig. 7. From the possible selection constants, those which minimize the number of fractional bits in their representation are chosen. Since the selection constants are of the form $D \times 2^{-3}$ (D integer), 3 fractional bits of the estimate of the shifted residual are used in the result-digit

TABLE V
RESULT-DIGIT SELECTION FOR ALL j

i	0	1	2	3	4	5	6	7
$\hat{S}[j]$	8/16	9/16	10/16	11/16	12/16	13/16	14/16	15/16
$m_2(i)$	3/2	7/4	2	2	9/4	5/2	5/2	11/4
$m_1(i)$	1/2	1/2	1/2	1/2	3/4	3/4	1	1
$m_0(i)$	-1/2	-5/8	-3/4	-3/4	-3/4	-1	-1	-1
$m_{-1}(i)$	-13/8	-7/4	-2	-17/8	-9/4	-5/2	-11/4	-23/8

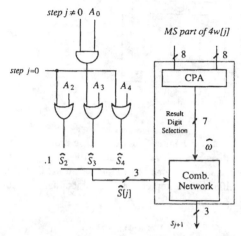

Fig. 7. Result-digit selection implementation.

selection. Moreover, since $t = 4$, 4 fractional bits of the shifted carry-save residual are used, resulting in a total of 8 bits (4 integer bits and 4 fractional bits).

Alternative implementations exist for the combinational network of the result-digit selection. One possibility is to use a ten-input/three-output PLA; on the other hand, some preprocessing can be done to reduce the size of the PLA, as proposed for example in [5].

IV. GENERATION OF ADDER INPUT F

As part of the implementation of the recurrence (9) it is necessary to form the adder input F with value

$$F[j] = -2S[j]s_{j+1} - s_{j+1}^2 4^{-(j+1)}.$$

Since the digit of the result is produced in signed-digit form, the partial result $S[j]$ is also in this form. However, for the case we are considering, which uses a carry-save adder, the input F has to be in two's complement representation. Consequently, $S[j]$ is converted to this form on-the-fly using a variation of the scheme presented in [2]. It requires that two conditional forms $A[j]$ and $B[j]$ are kept, such that

$$A[j] = S[j]$$

$$B[j] = S[j] - 4^{-j}.$$

These forms are updated with each result-digit as follows:

$$A[j+1] = \begin{cases} A[j] + s_{j+1}4^{-(j+1)} & \text{if } s_{j+1} \geq 0 \\ B[j] + (4 - |s_{j+1}|)4^{-(j+1)} & \text{otherwise.} \end{cases}$$

(31a)

$$B[j+1] = \begin{cases} A[j] + (s_{j+1} - 1)4^{-(j+1)} & \text{if } s_{j+1} > 0 \\ B[j] + (3 - |s_{j+1}|)4^{-(j+1)} & \text{otherwise.} \end{cases}$$

(31b)

The implementation of this conversion requires two registers for A and B, appending of one digit, and loading. For controlling this appending and loading, a shift register K is used, containing a moving 1. This implementation is shown in Fig. 8. In terms of these forms, the value of F is given by the following expressions:

For $s_{j+1} > 0$

$$F[j] = -2S[j]s_{j+1} - s_{j+1}^2 4^{-(j+1)}$$
$$= -(2A[j] + s_{j+1}4^{-(j+1)})s_{j+1}. \quad (32a)$$

For $s_{j+1} < 0$

$$F[j] = 2S[j]|s_{j+1}| - s_{j+1}^2 4^{-(j+1)}$$
$$= 2(B[j] + 4^{-j})|s_{j+1}| - s_{j+1}^2 4^{-(j+1)}$$
$$= (2B[j] + (4 - |s_{j+1}|)4^{-(j+1)})|s_{j+1}|. \quad (32b)$$

Note that these expressions are obtained by concatenation and multiplication by a radix-4 digit. The resulting bit-strings are given in Table VI, where $a \cdots aa$ and $b \cdots bb$ are the bit-strings representing $A[j]$ and $B[j]$, respectively (shifted one position). The location of the trailing string is also controlled by the moving 1 of register K.

Fig. 8 also shows a block to perform on-the-fly rounding, as described in [3].

V. OVERALL ALGORITHM, IMPLEMENTATION, AND TIMING

The overall algorithm is as follows (not including rounding):

begin Square root
   Initialization
      $w[0] = x - 1$ *since $w[-1] = S[-1] = 0$ and $s_0 = 1$ (Load x and make 1 the sign position)*
      $A[0] \leftarrow 1.000...000$ *$S[0] = 1$*
      $B[0] \leftarrow 0.000...000$ *$B[0] = A[0] - 1$*
      $K[0] \leftarrow 0.100...000$
   Iterations
      for $j = 0$ **to** m
      begin
         $s_{j+1} = \text{SELECT}(\hat{w}[j], \hat{S}[j])$ *see Table V*
         $F[j] = f(A[j], B[j], s_{j-1})$ *see Table VI*
         $w[j+1] \leftarrow 4w[j] + F[j]$
         $A[j+1] \leftarrow g_a(A[j], B[j], s_{j+1})$
                 see expression (31a)
         $B[j+1] \leftarrow g_b(A[j], B[j], s_{j+1})$
                 see expression (31b)
         $K[j+1] \leftarrow \text{shift-right}(K[j])$
      end for
   Result
      $s = S[m] = A[m]$
end Square root

521

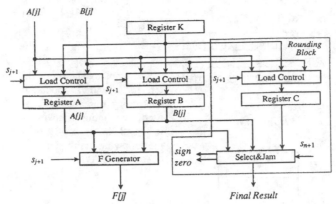

Fig. 8. Network for generating F and rounding.

TABLE VI
GENERATION OF $F[j]$

s_{j+1}	$F[j]$ Value	$F[j]$ Value	Bit-string
0	0	0	$0\cdots00000$
1	$-2S[j]-4^{-(j+1)}$	$-2A[j]-4^{-(j+1)}$	$\bar{a}\cdots\bar{a}\bar{a}111$
2	$-4S[j]-4\times4^{-(j+1)}$	$-4A[j]-4\times4^{-(j+1)}$	$\bar{a}\cdots\bar{a}1100$
-1	$2S[j]-4^{-(j+1)}$	$2B[j]+7\times4^{-(j+1)}$	$b\cdots bb\,111$
-2	$4S[j]-4\times4^{-(j+1)}$	$4B[j]+12\times4^{-(j+1)}$	$b\cdots b\,1100$

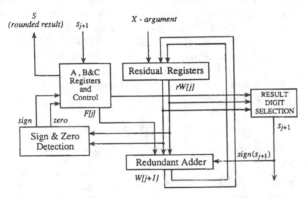

Fig. 9. Block diagram of the square root scheme (mantissa part).

The overall implementation at the block-diagram level is shown in Fig. 9. The cycle time is

$$T_{\text{cycle}} = t_{\text{result\_digit\_select}} \quad \{8\text{-bit CPA} + 10\text{-input comb. net}\}$$

$$+t_{F\_\text{generate}} \quad \{4\text{-to-1 multiplexer}\}$$

$$+t_{\text{CSA}} \quad \{3\text{-to-2 carry-save adder}\}$$

$$+t_{\text{load}} \quad \{\text{register loading}\}.$$

This is comparable to the cycle time of a radix-4 division with carry-save adder.

VI. CONCLUSIONS

We have shown a radix-4 square root algorithm that does not require an initial PLA. More precisely, this PLA is replaced by the four gates of Fig. 7. This is of theoretical interest and also produces a simpler implementation. It should be noted that the number of iterations required in this algorithm is the same as in those that use an initial PLA. That

is, the fact that the initial bits of the result are obtained from the PLA does not reduce the number of iterations, since the residual has to be obtained from the iterations. Consequently, the simplification in implementation is obtained without time penalty.

ACKNOWLEDGMENT

We thank P. Montuschi for helpful suggestions.

REFERENCES

[1] L. Ciminiera and P. Montuschi, "Higher radix square rooting," Intern. Rep., Politecnico di Torino, Dipartimento di Automatica e Informatica, I.R. DAI/ARC 4-87, Dec. 1987.
[2] M. D. Ercegovac and T. Lang, "On-the-fly conversion of redundant into conventional representations," IEEE Trans. Comput., vol. C-36, no. 7, pp. 895–897, July 1987.
[3] ——, "On-the-fly rounding for division and square root," in Proc. 9th Symp. Comput. Arithmetic, 1989, pp. 169–173.
[4] ——, "Division and square root algorithms and implementations," monograph in preparation.
[5] J. Fandrianto, "Algorithm for high speed shared radix-4 division and radix-4 square root," in Proc. 8th Symp. Comput. Arithmetic, 1987, pp. 73–79.
[6] J. B. Gosling and C. M. S. Blakeley, "Arithmetic unit with integral division and square-root," IEE Proc., vol. 134, pt. E, no. 1, pp. 17–23, Jan. 1987.
[7] S. Kuninobu et al., "Design of high speed MOS multiplier and divider using redundant binary representation," in Proc. 8th Int. Symp. Comput. Arithmetic, 1987, pp. 80–86.
[8] P. Montuschi and L. Ciminiera, "On the efficient implementation of higher radix square root algorithms," in Proc. 9th Symp. Comput. Arithmetic, Sept. 1989, pp. 154–161.
[9] M. B. Vineberg, "A radix-4 square-rooting algorithm," Rep. 182, Dep. Comput. Sci., Univ. of Illinois, Urbana-Champaign, June 1965.
[10] J. H. Zurawski and J. B. Gosling, "Design of a high-speed square root multiply and divide unit," IEEE Trans. Comput., vol. C-36, pp. 13–23, Jan. 1987.

167 MHz Radix-4 Floating Point Multiplier

Robert K. Yu and Gregory B. Zyner

SPARC Technology Business, Sun Microsystems Inc.
Sunnyvale, California

Abstract - An IEEE floating point multiplier with partial support for subnormal operands and results is presented. Radix-4 or modified Booth encoding and a binary tree of 4:2 compressors are used to generate the 53x53 double-precision product. Delay matching techniques were used in the binary tree stage and in the final addition stage to reduce cycle time. New techniques in rounding and sticky-bit generation were also used to reduce area and timing. The overall multiplier has a latency of 3 cycles, a throughput of 1 cycle, and a cycle time of 6.0ns. This multiplier has been implemented in a 0.5um static CMOS technology in the *Ultra*SPARC RISC microprocessor.

I. INTRODUCTION

Multiplier units are commonly found in digital signal processors and, more recently, in RISC-based processors[1]. Double-precision floating point operations involve the inherently slow operation of summing 53 partial products together to produce the product. IEEE-compliant multiplication also involves the correct rounding of the product, adjustment to the exponent, and generation of correct exception flags. Multiplier units embedded in modern RISC-based processors must also be pipelined, small, and *fast*. Judicious functional and physical partitioning are needed to meet all these requirements[5][6][14]. In this paper, Section II describes the partial subnormal support. Section III describes details of the implementation, including the 4:2 compressor design, the binary tree composition, final addition, rounding, and sticky-bit generation. Finally, Section IV concludes this paper.

II. SUBNORMAL OPERATIONS

Multiplier units that handle subnormal or denormal operands and results often require the determination of leading zeros, adjustments to mantissas (shifting) and exponent, and rounding. Overall timing and area are affected when subnormal operations are fully supported. However, by introducing a modest amount of hardware to compare only the value of the exponents, we can provide support for a large subset of the subnormal operations. This partial subnormal support does not require detection of leading zeros, adjustments of subnormal mantissas, and does not introduce an extra cycle penalty.

Figure 1 shows the floating point data formats, and Figure 2 shows their value definitions according to IEEE standards. The value of the mantissa is formed by concatenating the implicit bit with the fraction. For normalized values, the implicit bit is one, and for subnormal values, the implicit bit is zero. Note that for subnormal values, the convention is to have the exponent field zeroed, but the *value of the exponent* is taken to be one.

Single Precision:		
s	exp[7:0]	fraction[22:0]

Double Precision:		
s	exp[10:0]	fraction[51:0]

Fig. 1. Floating point formats.

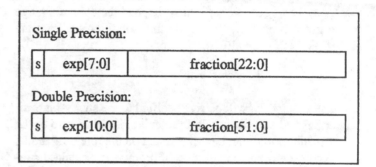

s = sign
e = biased exponent
f = fraction
E = number of bits in exponent (8 for single, 11 for double)
F = number of bits in fraction (23 for single, 52 for double)
B = exponent bias (127 for single, 1023 for double)

Normalized Value ($0 < e < 2^E-1$):
$$(-1)^s \times 2^{e-B} \times 1.f$$

Subnormal Value ($e = 0$):
$$(-1)^s \times 2^{1-B} \times 0.f$$

Zero:
$$(-1)^s \times 0$$

Fig. 2. Floating point format definition.

In multiplication, the resultant exponent e_r is calculated by:

$$e_r = e_1 + e_2 - B - z_1 - z_2 \qquad (1)$$

where e_1 and e_2 are the biased exponents, and z_1 and z_2 are the leading zeros in the mantissas to the multiplicand and mul-

Reprinted from *Proceedings of the 12th Symposium on Computer Arithmetic*, pp. 149-154, July 1995.

tiplier, respectively. We can simplify Equation 1 by noting that if both operands are subnormal, then the value of e_r underflows to a value no greater than $-B$, and cannot be represented in the given precision[1]. We can therefore impose the constraint that only one operand can be subnormal without the loss of generality and rewrite this equation as:

$$e_r = e_1 + e_2 - B - z \qquad (2)$$

where z is the number of leading zeros of the subnormal operand, if any. We further note that if the resulting value e_r is less than one, then the amount by which the resulting mantissa needs to be right-shifted to set $e_r = 1$ is:

$$rshift = 1 - e_r \qquad (3)$$

If the $rshift$ value is greater than the number of bits in the mantissa, then again we underflow beyond the dynamic range of the given precision. We define this condition as "extreme underflow."[2] Specifically, if

$$rshift \geq (F + 3) \qquad (4)$$

then the mantissa will be right-shifted to the sticky-bit position or beyond and the resulting mantissa is either zero or the smallest subnormal number, depending on the rounding mode. The two cases where rounding will produce the smallest subnormal number are 1) rounding to plus infinity and the result is positive and 2) rounding to minus infinity and the result is negative. All other rounding modes including rounding to nearest and rounding to zero produces zero as the result.

Rewriting Equation 4 in terms of e_r, extreme underflow occurs when

$$e_r \leq - (F + 2) \qquad (5)$$

We note from Equation 2 that e_r requires the detection of leading zeros z. If we ignore z altogether, which greatly simplifies the implementation, then Equation 5 becomes a conservative criterion for extreme underflow. That is, using only

$$e_r = e_1 + e_2 - B \qquad (6)$$

to determine extreme underflow ignores those cases where extreme underflow would occur because of leading zeros present in the subnormal mantissa.

Equation 5 and Equation 6 form the basis used to provide partial subnormal support in this design. If Equation 5 is satisfied with the appropriate rounding mode, then the multiplier generates a zero result. The multiplier does not support the case where Equation 5 is not satisfied and a subnormal operand is encountered.

1. Does not apply to single precision multiplication resulting in a double precision result, or "fsmuld".
2. Sometimes loosely referred to as "gross underflow".

III. IMPLEMENTATION

A. Folded 3-Stage Pipeline

The multiplier operates over three stages. In the first stage, the multiplicand and multiplier operands go through the radix-4 encoding and multiply tree[3][4][7]. The intermediate summations of partial products and the result of the tree are in carry-save format. In the second stage, a conditional-sum adder is used to determine the product, converting the result from carry-save to binary form. In the third stage, rounding and flag generation is performed. The multiplier has a single-cycle throughput and a three-cycle latency.

B. Stage 1: Multiply Tree

1) Interleaved Binary Tree with Delay Matching

The first stage of a 53x53 bit multiplication of the mantissas is performed by a radix-4 Booth encoded binary tree of 4:2 compressors, or 5:3 counters. Figure 3 shows a schematic of the binary tree. The encoding scheme produces 27 partial products which are generated at blocks 0,1,3,4,7,8, and 10. Since 4:2 compressors are used, each block generates 4 partial products, except for block 10 which generates 3.

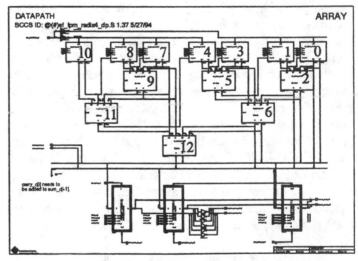

Fig. 3. Schematic of radix-4 binary tree.

Unlike traditional implementations, where inputs flow starting from the top and side of the tree to the bottom of the tree, this implementation has the multiplier and staged results placed on the same side of the tree. That is, pipeline register are embedded in the tree and are routed to the same side as the multiplicand, as shown in Figure 4. The advantage of this approach is to reduce interconnect lengths and to push some of the interconnect delay to the next stage.

The complexity of a binary tree does not lend itself to a straight-forward layout[2]. To minimize the delay through the tree due to interconnects, the placement of the rows of partial product generators and adders are done such that wire lengths are balanced among the rows. Both the vertical distance and the horizontal distance, which comes about because the tree is "left-justified" and is significant in some cases, were taken into account.

Table I shows the vertical row distances and horizontal bit distances between cells on different rows. The critical path through the array involve the rows with large horizontal shifts, namely rows 0, 2, 6, and 12. These rows have been placed close together to reduce this path. Figure 4 shows the placement used[8].

TABLE I
DISTANCE BETWEEN CELLS

Row Transition	Horizontal Distance	Vertical Distance
0 -> 2	9	1
1 -> 2	1	2
3 -> 5	9	1
4 -> 5	1	3
7 -> 9	9	1
8 -> 9	1	2
2 -> 6	17	3
5 -> 6	1	3
9 -> 11	3	3
10 -> 11	0	7
6 -> 12	18	4
11 -> 12	0	4

2) Folded Adder Rows

Another problem presented by the irregular tree structure is the differing number of bits in each row, which varied from 61 to 76 bits. In order to reduce the area of the tree, some of the adders in the larger rows were "folded" to rows with fewer cells. The folding was done such that timing was not affected. More folding could be done but not without impacting the critical path through the tree.

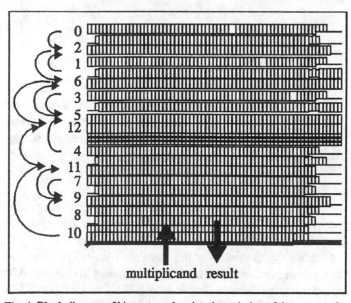

Fig. 4. Block diagram of binary tree showing the ordering of the rows used to balance the interconnect delays.

3) 4:2 Compressor Design

The 4:2 compressor schematic is shown in Figure 5. This adder takes 5 inputs {x3, x2, x1, x0, and cin} and generates 3 outputs {carry, cout, and sum}. All inputs and the sum output have a weight of one, and two outputs carry and cout have a weight of two[10][12]. That is,

$$2^0 \cdot (x3 + x2 + x1 + x0 + cin)$$
$$= 2^1 \cdot (carry + cout) + 2^0 \cdot sum \qquad (7)$$

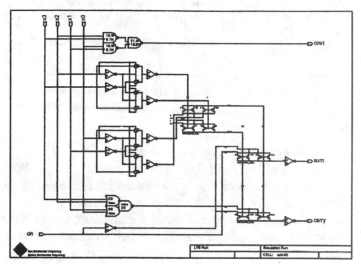

Fig. 5. Circuit schematic of 4:2 compressor.

The 4:2 compressor has been designed according to Table II. Since the cout signal is independent of the cin input and only dependent on the x inputs, a row of such adders hooked up together as shown in Figure 6 will not exhibit any

TABLE II
TRUTH TABLE FOR 4:2 COMPRESSOR

x3	x2	x1	x0	cout	carry	sum
0	0	0	0	0	0	cin
0	0	0	1	0	cin	$\overline{cin}$
0	0	1	0	0	cin	$\overline{cin}$
0	0	1	1	1	0	cin
0	1	0	0	0	cin	$\overline{cin}$
0	1	0	1	0	1	cin
0	1	1	0	0	1	cin
0	1	1	1	1	cin	$\overline{cin}$
1	0	0	0	0	cin	$\overline{cin}$
1	0	0	1	0	1	cin
1	0	1	0	0	1	cin
1	0	1	1	1	cin	$\overline{cin}$
1	1	0	0	1	0	cin
1	1	0	1	1	cin	$\overline{cin}$
1	1	1	0	1	cin	$\overline{cin}$
1	1	1	1	1	1	$\overline{cin}$

rippling of carries from cin to cout. The adder has also been designed such that the delay from xi to sum or carry is approximately the same as the combined delay from xi to cout, and cin to *sum* or *carry* of an adjacent adder.

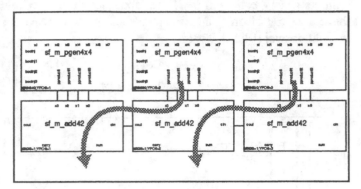

Fig. 6. Interconnect of 4:2 compressors with no horizontal ripple carry.

C. Stage 2: Final Addition

1) Optimized Conditional-Sum Adder

The final add stage of the multiplier makes use of a 52-bit conditional sum adder that is partitioned for minimum delay. Figure 7 shows a block diagram of the recursive structure of the conditional sum adder. As shown in the diagram, an N-bit conditional sum adder is made up of two smaller conditional sum adders, one that is j-bits, and one that is N-j bits wide. Two 2:1 muxes are used to output the upper sum and carry results; these outputs are selected by the carries from the lower j-bit adder. Typically the selects to the muxes are buffered up to handle the capacitive loading due to large fanouts. These smaller adders are, in turn, made up of smaller conditional sum adders[11].

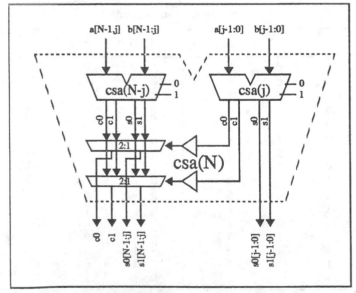

Fig. 7. Recursive structure of conditional sum adder.

The delay through this adder is affected by how the adders are partitioned. We define the delay for an N-bit adder partitioned at position j as $T(N,j)$:

$$T(N, j) = $$ \hfill (8)
$$max[T_{opt}(j) + T_{buf}(N-j) + T_{sel}, T_{opt}(N-j) + T_{mux}]$$

In Equation 8, $T_{opt}(i)$ is the optimal delay for an i-bit adder, T_{buf} is the buffer delay and is a function of the number of bits on the left adder, or N-j, T_{sel} is the select to out delay of the mux, and T_{mux} is the data to out delay of the mux. The minimum delay for an N-bit adder $T_{opt}(N)$ is simply:

$$T_{opt}(N) = min[T(N,j)]$$ \hfill (9)

where j varies from 1 to N-1. The problem of finding $T_{opt}(N)$ is a recursive min-max problem and lends itself well to an efficient dynamic programming solution developed internally for this implementation.

2) Sticky-Bit Generation from Carry-Save Format

The *sticky* bit, which is needed to perform the correct rounding, is generated in the second stage. Typically, the lower 51 bits in carry-save format are summed and OR'ed together. However, in our technique, we are able to generate the *sticky* bit directly from the outputs of the tree in carry-save format without the need for any 51-bit adder to generate the sum beforehand, resulting in significant timing and area savings. We define:

$$p_i = s_i \oplus c_i$$ \hfill (10)

$$h_i = s_i + c_i$$ \hfill (11)

$$t_i = p_i \oplus h_{i-1}$$ \hfill (12)

where s_i and c_i are the sum and carry outputs from the tree. The *sticky* bit is then computed directly by using a ones-detector[9]:

$$sticky = t_0 + t_1 + ... + t_{50}$$ \hfill (13)

D. Stage 3: Rounding

1) Using Conditional Sum Adders

By using a conditional sum adder to generate both the sum and sum+1, we remove the need for an incrementer to perform the rounding operation. Only multiplexing is needed to select the correct result after rounding[13].

2) Overflow After Rounding

In double precision multiplication, after the 106-bit product (in carry-save format) has been generated by the array, the decimal point occurs between bits 104 and 103, and only the upper 53-bits are used for the mantissa result. The lower 53-bits are needed only to perform the correct rounding. After rounding is performed, either bits 105-53 or bits 104-52 are used depending on the value of bit 105 or MSB. If this MSB is set, then the mantissa is taken from bits 105-53, and the value of the exponent is incremented. Otherwise, if the MSB is not set, then bits 104-52 are used, and the exponent is not incremented. Note that the rounding itself may propagate to set the MSB; this is the case of overflow *after* rounding. Figure 8 shows how the mantissa is selected from the array result depending on bit 105 after rounding. In the figure, L, G, R, and S represents the LSB, guard, round, and sticky bits, respectively.

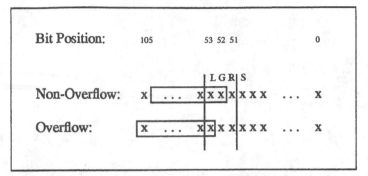

Fig. 8. Mantissa selection for overflow and non-overflow.

Figure 9 shows a block diagram of the rounding datapath and logic. The dotted line represents pipeline registers: the final add operation and rounding are done in separate stages. The lower 50 bits from the array are used to generate two signals: c51 and S. The c51 signal is the carry into bit 51. Bits 53 through 51 along with c51 are added to create the L, G, and R bits, and the rest of the bits 105:54 are added using a conditional sum adder to form two results sum0[105:54] and sum1[105:54].

Because of the c51 signal and a rounding that may occur at bits 53 or 52, there is a possibility of introducing two carries into bit position 54. To ensure that only one carry is propagated into bit 54, a row of half-adders is used at bits 105 through 51.

To correctly handle the case of overflow *after* rounding, our implementation makes use of two adders, ovf and novf, to generate the signals c54_v and c54_n, respectively, which are needed by the final selection logic. The c54_v and c54_n are the carries into bit position 54 assuming an overflow and non-overflow, respectively. The L, G, R, S, and rounding mode bits are used by the round logic to generate two rounding values. One value assumes a mantissa overflow, and the other assumes no mantissa overflow. These rounding bits are added to the L and {L,G} bits to form the lower one and two bits of the resulting mantissa for overflow (manv) and non-overflow (mann), respectively.

3) Final Selection

The final select logic combines the appropriate sum0 and sum1 from the conditional sum adder with either manv or mann to form the final mantissa. Table III shows the truth table to the selection logic. The key to the table is the expression for the *Overflow* signal, shown in Equation 14. The first expression refers to the case where the MSB is set as a result of a carry within the addition of the 51 bits without a carry into bit 54. The second expression refers to the case where the MSB is set due to some carry into bit 54 in the non-overflow case. This carry may be due to rounding itself, or the case of overflow after rounding.

$$Overflow = sum0[105] + (c54\_n \cdot sum1[105]) \qquad (14)$$

TABLE III
SELECTION LOGIC

Overflow	c54_n	c54_v	Select
0	0	x	sum0[104:54], mann[1:0]
0	1	x	sum1[104:54], mann[1:0]
1	x	0	sum0[105:54], manv[0]
1	x	1	sum1[105:54], manv[0]

4) Shared Hardware with Divide and Square Root

In our implementation of the floating point unit, the rounding for multiplication is similar to that needed for division and square root. In the interest of saving area, multiplication, division, and square root all share the same rounding hardware. Only additional muxing between the multiply, divide, or square root results is required before the inputs to the block shown in Figure 9.

One difference, however, between multiplication, division, and square root is the handling of the mantissas overflow. In

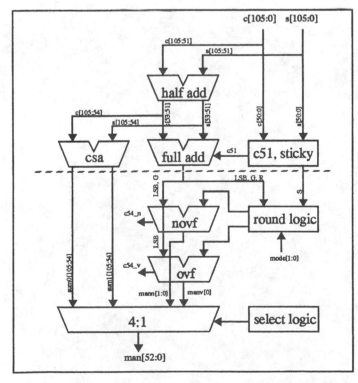

Fig. 9. Rounding.

multiplication, the incremented exponent is used if an overflow occurs. In division and square root, however, the decimal point is taken to be immediately to the right of the MSB. Therefore, if the mantissa's MSB is zero, then the decremented exponent is selected. Table IV shows how the exponent is selected for multiplication, division, and square root.

TABLE IV
EXPONENT SELECTION FOR MULTIPLY AND DIVIDE

Mantissa	Multiply	Divide/Sqrt
Overflow	$e_r + 1$	e_r
Non-overflow	e_r	$e_r - 1$

IV. CONCLUSION

We have presented the design and implementation of a high-speed floating point multiplier. Partial subnormal support has been implemented with minimal addition to hardware and penalty on performance. Delay matching techniques were used in the multiplier tree and in the final addition stages. The rounding hardware is shared with the divide and square root units.

V. ACKNOWLEDGEMENT

The authors would like to thank Marc Tremblay, Arjun Prabhu, and the Program Committee for their valuable suggestions, and Nasima Parveen for her contributions in verification.

REFERENCES

[1] M. Mehta, et al, "High-speed multiplier design using multi-input counter and compressor circuits," *Proceedings 10th Symposium on Computer Arithmetic*, pp. 43-50, 1991.

[2] M. Nagamatsu, et al, "A 15 ns 32x32 bit cmos multiplier with an improved parallel structure," *Custom Integrated Circuits Conference*, pp. 10.3.1-10.3.4, 1989

[3] L. P. Rubinfield, "A proof of the modified Booth's algorithm for multiplication," *IEEE Trans. Comput.*, pp. 1014-1015, Oct. 1975

[4] C. S. Wallace, "A suggestion for parallel multipliers," *IEEE Trans. Electron. Comput.*, vol. EC-13, pp. 14-17, Feb. 1964

[5] L. Dadda, "Some schemes for parallel multipliers," *Alta Frequenza*, vol. 34, pp. 349-356, 1965

[6] L. Dadda, "On parallel digital multipliers," *Alta Frequenza*, vol. 45, pp. 574-580, 1976.

[7] A. D. Booth, "A signed multiplication technique," *Quarterly J. Mechan. Appl. Math*, vol. 4, pt. 2, pp. 236-240, 1951

[8] D. Zuras and W. McAllister, "Balanced delay trees and combinatorial division in VLSI," *IEEE J. Solid-State Circuits*, vol. SC-21, no. 5, pp.814-819, Oct. 1986

[9] Email correspondence with Vojin Oklobdzija, 1993.

[10] D. T. Shen and A. Weinberger, "4-2 carry-save adder implementation using send circuits", *IBM Technical Disclosure Bulletin*, vol. 20, no. 9, Feb 1978.

[11] J. Sklansky, "Conditional sum addition logic", *Trans. IRE*, vol. EC-9, no. 2, pp. 226-230, June 1960.

[12] M. Santoro and M. Horowitz, "A pipelined 64x64b iterative array multiplier", *IEEE Int. Solid-State Circuits conf.*, pp.35-36, Feb. 1988.

[13] M. Santoro, G. Bewick, and M. Horowitz, "Rounding algorithms for IEEE multipliers", *IEEE 9th Symposium on Computer Arithmetic* proceedings, pp. 176-183, Sept. 1989.

[14] V. Peng, S. Sanudrala, M. Gavrielov, "On the implementation of shifters, multipliers, and dividers in VLSI floating point units", *IEEE 8th Symposium on Computer Arithmetic* proceedings, pp. 95-102, May 1987.

Author Index

Subject Index

About the Editor

Vojin G. Oklobdzija obtained his Ph.D. in computer science from the University of California, Los Angeles, in 1982 and his MSc degree in 1978. He obtained his Dipl. Ing. (MCcEE) degree in electronics and telecommunication from the Electrical Engineering Department, University of Belgrade, Yugoslavia, in 1971. He came to the United States as a Fulbright scholar in 1976. From 1982 to 1991 he was a research staff member at IBM T. J. Watson Research Center in New York. Earlier, he was at Xerox Microelectronics (1979–1982), on the faculty of Electrical Engineering at the University of Belgrade (1974–1976), and the Institute of Physics (1971–1974). From 1988 to 1990 he was visiting faculty at the University of California, Berkeley, while on leave from IBM.

Professor Oklobdzija has made contributions to the development of RISC architecture and processors and super-scalar RISC. He is a co-holder of a patent on the IBM RS/6000 (PowerPC) and one of the initiators of the supercomputer project at IBM.

Since 1991, Professor Oklobdzija has held various consulting and academic positions. He is the president of Integration Corporation, which has been engaged in various projects. He was consultant to Sun Microsystems Laboratories in 1992, AT&T Bell Laboratories in 1994, Siemens Corporation in 1996–1997 and others. He is consultant to Hitachi Research Laboratories where he works on low-power electronic design.

Professor Oklobdzija has had academic appointments with the University of Belgrade in Yugoslavia, and the University of California and visiting appointments with the Universidad National de Ingenieria in Peru, San Francisco State University, and State University of New York. In 1991 he spent time in Peru and Bolivia as a Fulbright professor, and in 1995 and 1996 he lectured and helped the universities in South America. During 1996–1998 he taught courses in the Silicon Valley through the University of California, Berkeley Extension, and Hewlett-Packard.

Professor Oklobdzija's interest is in high-performance systems and architectures, VLSI and fast circuits, and efficient implementations of algorithms and computation. His hobbies are electronics projects, amateur-radio and judo.

Professor Oklobdzija holds four U.S. and four European patents in the area of computer design, and he has four others pending. He is a Fellow of IEEE and a member of the American Association for Advancement of Science, and the American Association of University Professors. He serves or has served on the editorial boards of the *Journal of VLSI Signal Processing, IEEE Transaction of VLSI Systems*, and the International Conference on Computer Design. He was a general chair of the 13th Symposium on Computer Arithmetic, and he is a program committee member of the International Solid-State Circuits Conference and the International Symposium on VLSI Technology. He has published over 100 papers in the areas of circuits and technology, computer arithmetic, and computer architecture, and he has given over 100 invited talks and short courses in the United States, Europe, Latin America, Australia, China, and Japan.